Carnivore Behavior, Ecology, and Evolution

Carnivore Behavior, Ecology, and Evolution

John L. Gittleman, EDITOR

*Department of Zoology and Graduate Programs in Ecology
and Ethology, The University of Tennessee, Knoxville*

Comstock Publishing Associates, a division of

Cornell University Press | Ithaca, New York

First published 1989 by Cornell University Press.

Printed in the United States of America

Library of Congress Cataloging-in-Publication Data

Carnivore behavior, ecology, and evolution.

  Bibliography: p.
  Includes index.
  1. Carnivora.   I. Gittleman, John L.
QL737.C2C33   1989   599.74'045   88-47725
ISBN 0-8014-2190-X (alk. paper)
ISBN 0-8014-9525-3 (pbk. : alk. paper)

Contents

Preface

Carnivores have always fascinated us, even though they make up only 10% of all mammalian genera and only about 2% of all mammalian biomass. In Greek mythology most of the gods adorned their robes and helmets with depictions of carnivores, and the great hero Hercules' most famous feat was killing the "invulnerable" lion with his bare hands.

Part of our fascination with carnivores stems from fright and intrigue, and sometimes even hatred because of our direct competition with them. Cases of "man-eating" lions, bears, and wolves, as well as carnivores' reputation as killers of livestock and game, provoke communities and governments to adopt sweeping policies to exterminate them. Even President Theodore Roosevelt, proclaimer of a new wildlife protectionism, described the wolf as "the beast of waste and desolation." The sheer presence and power of carnivores is daunting: they can move quickly yet silently through forests, attaining rapid bursts of speed when necessary; their massive muscles are aligned to deliver powerful attacks, their large canines and strong jaws rip open carcasses, and their scissor-like carnassials slice meat. Partly because of our fear of these attributes, trophy hunting of carnivores has been, and to a certain extent still is, a sign of bravery and skill. Among some Alaskan Inuit, for example, a man is not eligible for marriage until he has killed a succession of animals of increasing size and dangerousness, culminating with the most menacing, the polar bear. Carnivore fur, glands, and musk are still treasured even though alternative synthetic materials have been developed. Despite our close relationship and fascination with carnivores, humans are still relatively ignorant of most species in the order. This book synthesizes some of the recent advances in research in the biology of carnivores.

The mammalian order Carnivora is characterized by great morphological, ecological, and behavioral variation. Body sizes range from the 100-g least weasel to the gigantic polar bear, weighing as much as 800 kg. Reproductive rate may be as low as one offspring every five to seven years, as in some black bears, or as high as three litters a year with eight offspring in each litter, as in some populations of dwarf mongooses. Carnivores inhabit every habitat or vegetational zone, from short grassland (meerkat) to sparse woodland (dwarf mongoose) to desert (fennec fox) to thick forest (banded palm civet) to oceanic waters (sea otter). And the size of the home range may be relatively small (0.55 km^2: coati) or extremely large and nondefensible (1500–2000 km^2: African hunting dog). In terms of behavior, species range from those that live alone with only brief encounters between adults during breeding (ermine) to those that form monogamous pair bonds (golden jackal) to those that live in large extended packs with as many as 80 individuals (spotted hyena).

Although scientists have emphasized widespread interspecific differences among carnivores, considerable variation and flexibility also occur within species. In wolves, for example, adults weigh from 31 to 78 kg, litter size varies from one to 11, home ranges extend between 103 and 12,950 km^2, populations are found in every vegetational zone except tropical forests and arid deserts, and adults live as solitaries or within extended packs comprising up to 21 individuals. Within carnivores, therefore, we can look for examples of evolutionary adaptation at the interspecific level as well as investigate natural selection at the intraspecific level.

From the time of R. I. Pocock's and Dwight Davis's classic studies in comparative anatomy and of G. G. Simpson's pioneering work on the fossil record, the carnivores have formed the centerpiece for many studies in comparative and evolutionary biology. But until recently studies of carnivores which required detailed knowledge of their habits and lifestyles in the wild were blocked by our inability to find and track individuals easily. Because of their elusive, nocturnal, fast, solitary, and often dangerous nature, details on most species remained obscure. Today, as a result of improved research techniques (e.g., radiotelemetry and infrared spotting scopes), conservation and captive management programs, and a surge of interest in the ecological and evolutionary significance of carnivore behavior, many species are better known.

Anyone who has tried to compile comparative data on the behavior, ecology, and morphology of carnivores has quickly learned that no volume that critically summarizes and evaluates recent research in carnivore biology has been available. To meet that need, I have assembled this volume; it presents critical reviews in rapidly developing and expanding areas of carnivore behavior, ecology, and evolution. It also elucidates the features of carnivores which distinguish the group from other mammalian lineages so that nonspecialists will come to know them better.

I and all the contributors feel a deep gratitude to R. F. Ewer for her monumental volume *The Carnivores* (1973, Cornell University Press), which laid the foundation for modern carnivore studies. It is testimony to the longlasting effect of her work that most of the contributors, though using very different methodologies and theoretical predictions, refer to *The Carnivores* for framing their questions.

Following a general introduction to the carnivores by John F. Eisenberg, the volume is organized in three parts: (1) behavior—acoustic and olfactory communication; behavioral development; behavioral ecology of hyaenids and canids; modes of solitary living; and group living; (2) ecology—the feeding ecology of the giant panda and Asiatic black bear; adaptations for aquatic living; ecological constraints on predation by felids; the consequences of small size in mustelids; the rate of basal metabolism and food habits; and reproductive output; (3) evolution—morphological constraints on locomotion; dentition and diet; the physiology of delayed implantation; molecular and morphological approaches to phylogeny; and the fossil record. Each part is preceded by a brief introduction outlining the main themes presented in the chapters and explaining why certain subjects are included. Within each part, chapters proceed from specific areas to more general ones; therefore each part first deals with specific mechanisms that drive or constrain general evolutionary systems. At the end of the volume is an appendix by W. Chris Wozencraft which presents a classification of the Recent Carnivora, including the pinnipeds. This listing, which is derived from Honacki et al. (1982, *Mammal Species of the World*, Lawrence, Kans.: Association of Systematic Collections), is not meant to present a new or final word on classification; indeed, some authors have preferred the nomenclature of other taxonomies and have so stated in their text. Rather, the appendix is intended as a reference guide, modeled after that presented in Ewer's volume, for translating species names into common ones (or vice versa), for showing familial membership of species, and for pointing out general taxonomic changes made in recent years.

Because this volume covers a broad taxonomic group and includes many disparate research topics, several editorial decisions were necessary. The first question addressed was general: what is a carnivore, or which species should be included in a book dealing with the Carnivora? There is historical precedent either for combining the terrestrial carnivores and aquatic pinnipeds in the order Carnivora or for splitting off the terrestrial carnivores to form an independent, ordinal group. Studies of the origin and phylogeny of carnivores and pinnipeds continue to produce lively debate, as chapters in this volume attest. The chapters deal mainly with the terrestrial (fissiped) carnivores or species included in the following taxonomic families: Canidae (wolves, wild dogs, jackals, foxes), Procyonidae (raccoons, coatis, kinkajou), Ursidae (bears), Ailuropodidae (giant panda), Ailuridae (red panda), Mustelidae (weasels, martens, fisher, tayra, ratel, badgers, skunks, otters), Viverridae (civets, genets), Herpestidae (mongooses), Hyaenidae (hyenas, aardwolf), and Felidae (wild cats, ocelots, serval, caracal, lynxes, puma, leopards, jaguar, lion, tiger, cheetah). I have thus excluded the pinnipeds from discussion; the behavioral, ecological, and evolutionary features of adaptations for aquatic living set this group apart from the terrestrial carnivores. A further, more practical reason is that other volumes nicely synthesize recent advances in research on the pinnipeds. By considering only the terrestrial carnivores, this book avoids duplicating other publications.

Even when the taxonomic limitations were set, another potential source of confusion remained. The word "carnivore" in everyday language refers to a dietary proclivity for flesh eating. Thus, describing a species as a carnivore could refer to its taxonomic affiliation or to its dietary preference. Because the general focus of this volume is on species taxonomically included as the terrestrial carnivores, "carnivore" refers to the taxonomic usage unless otherwise stated. For example, where diet is analyzed, some ursids are described as being herbivorous carnivorans or simply herbivores.

My primary criteria in making these decisions was consistency and defensibility of argument throughout—that is, questions regarding taxonomy, priority of theoretical questions, method of citations, and other recurring features across chapters were to be presented in the same fashion and as rigorously as possible in each context so that at least ambiguity would be avoided. Because consistency was a goal, however, the problem of overlap between chapters became an issue. Even though repetition was eliminated whenever possible, some overlap had to remain if each chapter was to stand on its own. Such repetition reduces the burden of cross-referencing between chapters. Whenever chapters deal with similar subjects, one chapter serves as the main source for a given set of data or particular theoretical perspective. When other chapters refer to the same material, a brief summary is given, along with cross-references to the chapter(s) where more detailed treatment occurs.

Another editorial decision concerned the presentation of unorthodox views, unpublished information, or disagreement among review chapters, which are meant to present only facts and references. And indeed, the data compilations, descriptions of field or experimental studies, and theoretical discussions in most chapters are based on previously published, refereed work, but some chapters include original data or theoretical arguments that offer a unique perspective. I decided that these views should be aired also, so that future work in carnivore biology might more rapidly test their validity. Nevertheless, in cases where new information is presented, I have required the authors to refer to background literature to introduce the conceptual problem or new methodology and to elucidate why these new data are important in the broad context here, wherein previous studies already established in a particular area are being reviewed.

It is my hope that this collection of chapters brings the excitement and beauty of carnivores closer to those who have not had the opportunity to study them, especially to students in the behavioral, ecological, and evolutionary sciences who are looking for a diverse and intriguing group to work on. And, for my scientific colleagues, it is my intention that the problems and questions raised by taking stock of what we have learned about carnivores will spur us on to discover what we need to know in order to preserve them.

JOHN L. GITTLEMAN

Knoxville, Tennessee

Acknowledgments

I am extremely grateful to the contributors to this volume for accepting the challenge of synthesizing diverse findings and creating new questions in carnivore biology. With heartfelt thanks, I acknowledge their tolerance, patience, and sustained encouragement throughout the development of this book.

In forming my thoughts on the comparative morphology, behavioral ecology, and evolution of carnivores, I have been most influenced by three teachers: Paul Harvey, who taught me the strengths and weaknesses of the comparative approach; John Eisenberg, who constantly encouraged me to test my conclusions on carnivores with other mammalian taxa; and John Maynard Smith, who gave me a deep appreciation for the value and insight of evolution theory. I owe a special debt to each of these individuals.

Each chapter was reviewed for content, style and accuracy by at least one reader. I am very grateful to the following individuals who kindly gave of their time to examine manuscripts: Gordon Burghardt, Timothy Caro, John Eisenberg, Sam Erlinge, Richard Golightly, Theodore Grand, Paul Harvey, Robert Hunt, Richard Kay, Hans Kruuk, Björn Kurtén, David Macdonald, Larry Martin, Gene Morton, Craig Packer, Katherine Ralls, Mel Sunquist, Richard Tedford, Steven Thompson, David Webb, Christen Wemmer, and Phillip Wright. I thank Robb Reavill and Helene Maddux of Cornell University Press and Barbara Reitt of Reitt Editing Services for enthusiastically supporting this project from beginning to end. And I am indebted to my wife, Karen E. Holt, for discussions on editorial matters and for persistently trying to improve my writing.

During the preparation of the volume I received financial support from the following institutions: the National Zoological Park (Smithsonian Institution); the Friends of the National Zoo (FONZ); the National Institutes of Health, which provided a Training Grant (T32-HD-07303); and the Department of Zoology and Graduate Programs in Ecology and Ethology, University of Tennessee.

I especially thank my colleagues at the University of Tennessee—Dewey

Bunting, Gordon Burghardt, James Drake, Sandy Echternacht, Gary Mc-Cracken, Stuart Pimm, and Susan Riechert—for their constructive criticisms, assistance, and friendship throughout all stages of editing.

Finally, I thank my father, David L. Gittleman, and my mother, Joan Frentz, for guidance, support, and love during some bleak periods in the preparation of this book. I dedicate this volume to their efforts.

J. L. G.

Contributors

MARC BEKOFF, Department of Environmental, Population, and Organismic
Biology, University of Colorado, Boulder, Colorado 80309-0334, USA

RAOUL E. BENVENISTE, Laboratory of Viral Carcinogenesis, National
Cancer Institute, Frederick, Maryland 21701, USA

JOHN F. EISENBERG, The Florida State Museum, University of Florida,
Gainesville, Florida 32611, USA

JAMES A. ESTES, U.S. Fish and Wildlife Service, Institute of Marine Sciences,
University of California, Santa Cruz, California 95064, USA

JOHN L. GITTLEMAN, Department of Zoology and Graduate Programs in
Ecology and Ethology, The University of Tennessee, Knoxville, Tennessee
37916, USA

MARTYN L. GORMAN, Department of Zoology, Culterty Field Station,
University of Aberdeen, Newburgh, Grampian, AB4 0AA, Scotland

HU JINCHU, Department of Zoology, Nanchong Normal College,
Nanchong, Sichuan, China

DIANNE N. JANCZEWSKI, Laboratory of Viral Carcinogenesis, National
Cancer Institute, Frederick, Maryland 21701, USA

KENNETH G. JOHNSON, Department of Forestry, Wildlife, and Fisheries,
University of Tennessee, Knoxville, Tennessee 37901-1071, USA, and
World Wildlife Fund–International, Gland, Switzerland

CAROLYN M. KING, P.O. Box 598, Royal Society of New Zealand,
Wellington, New Zealand

BRIAN K. McNAB, Department of Zoology, University of Florida,
Gainesville, Florida 32611, USA

LARRY D. MARTIN, Museum of Natural History, Department of Systematics
and Ecology, University of Kansas, Lawrence, Kansas 66045, USA

RODNEY A. MEAD, Department of Biological Sciences, University of Idaho,
Moscow, Idaho 83843, USA

M. G. L. MILLS, National Parks Board, Kruger National Park, Private Bag
X402, Skukuza 1350, South Africa

PATRICIA D. MOEHLMAN, Wildlife Conservation International, New York Zoological Society, Bronx Zoo, Bronx, New York 10460, USA

STEPHEN J. O'BRIEN, Laboratory of Viral Carcinogenesis, National Cancer Institute, Frederick, Maryland 21701, USA

OLAV T. OFTEDAL, Department of Zoological Research, National Zoological Park, Smithsonian Institution, Washington, D.C. 20008, USA

GUSTAV PETERS, Zoologisches Forschungsinstitut und Museum Alexander Koenig, Adenauerallee 150-164, D-5300 Bonn 1, West Germany

MIKAEL SANDELL, Department of Wildlife Ecology, Swedish University of Agricultural Sciences, S–901 83 Umea, Sweden

GEORGE B. SCHALLER, Wildlife Conservation International, New York Zoological Society, New York, New York 10460, USA

SHEN HEMING, Tangjiahe Natural Reserve, Qingchuan County, Sichuan, China

FIONA C. SUNQUIST, Florida State Museum and Department of Wildlife and Range Sciences, University of Florida, Gainesville, Florida 32611, USA

MEL E. SUNQUIST, Florida State Museum and Department of Wildlife and Range Sciences, University of Florida, Gainesville, Florida 32611, USA

MARK E. TAYLOR, Department of Zoology, University of Toronto, Toronto, Ontario M5S 1A1, Canada

TENG QITAO, Tangjiahe Natural Reserve, Qingchuan County, Sichuan, China

BEVERLEY J. TROWBRIDGE, Department of Zoology, Culterty Field Station, University of Aberdeen, Newburgh, Grampian AB4 0AA, Scotland

BLAIRE VAN VALKENBURGH, Department of Biology, University of California, Los Angeles, California 90024-1606, USA

WANG XIAOMING, Department of Zoology, Nanchong Normal College, Nanchong, Sichuan, China

ROBERT K. WAYNE, Department of Biology, University of California, Los Angeles, California 90024-1606, USA

W. CHRIS WOZENCRAFT, Division of Mammals, National Museum of Natural History, Smithsonian Institution, Washington, D.C. 20560, USA

Carnivore Behavior, Ecology, and Evolution

An Introduction to
the Carnivora

John F. Eisenberg

The carnivores are a fascinating group. Trends in their evolution and the convergent and parallel developments of life history strategies have intrigued us all (Eisenberg 1986). Carnivora literally means "eaters of flesh." Thus, the ordinal name describes an attribute or aspect of a niche that some but not all members of the order Carnivora occupy. The first flesh-eating mammal group to appear in the fossil record, however, is not at all closely related to the modern-day carnivores. The Deltatheridia appeared in the Paleocene and dominated the carnivore niche for a considerable period of time (Van Valen 1966). At the time of the Upper Eocene the first members of the order Carnivora may be found as fossils (see Martin, this volume). These are generally assigned to the family Miacidae. The miacids persisted until the Oligocene. When they are first recognizable in the fossil record, they show enlarged canine teeth and specialized shearing carnassial teeth. The shearing teeth involved the opposition of the fourth upper premolar with the first lower molar. The miacids did not have an ossified tympanic bulla and the carpal bones remained unfused. In the Late Eocene and Early Oligocene the more advanced carnivores make their appearance, with an ossified bulla and a fusion of the scapholunar in the carpus (Dawson and Krishtalka 1984).

The terrestrial carnivores, or Fissipedia, are often placed either in their own order or as a suborder in opposition to the Pinnipedia, or aquatic carnivores. That the two groups are related is not to be doubted. The pinnipeds are usually divided into three families, the walruses, the true seals, and the eared seals. King (1964) has argued for an independent origin for the eared seals and the true seals. In short, she believed the eared seals to be more closely related to bears and dogs and the earless seals more closely related to the weasel family. This implies that adaptation to an aquatic existence occurred twice. Thus, from her standpoint the Pinnipedia are an artificial assemblage.

Sarich (1969), however, presented biochemical evidence indicating that the pinnipeds are a natural grouping deriving from a common ancestor. I have found no one since Sarich who has disputed this viewpoint. Indeed, there has

1

been growing support for the monophyletic origin of the pinnipeds. Given a common origin for the Pinnipedia and in full recognition that they derive ultimately from carnivore stock, it is convenient to consider the pinnipeds as a separate taxon. But whether this taxon be given an ordinal name or be subsumed as a suborder under the order Carnivora is somewhat arbitrary.

Wozencraft (this volume) reevaluates the evidence for and against a monophyletic origin for the pinnipeds. It is clear from his taxonomic arrangement and some of his comments that he leans toward the diphyletic origin. In any event, more research needs to be done before the question can be resolved. Research on pinnipeds is progressing at a rapid rate. Recent advances in radiotelemetry have allowed pinnipeds to be studied while at sea. A good summary of recent research is included in the volume edited by Gentry and Kooyman (1986).

Because this volume is concerned with the fissiped carnivores, I will confine my discussion to this group. The terrestrial carnivores are classically divided into two groups: the Arctoidea, in which the tympanic bulla is single chambered, and the Canoidea, in which the bulla is incompletely divided by a low septum. There is much dispute concerning the taxonomic validity of this division (Stains 1984). Suffice it to say that there was a rapid early radiation, and the subsequent lineages were well differentiated in the Eocene and Oligocene.

The Arctoidea (Feloidea) include civets, mongooses, hyenas, and cats. The civets (Viverridae) are considered to be the most conservative in terms of carrying forward ancestral characters into the present time. That civets and mongooses are closely allied is not to be disputed, but the mongooses present such a uniform assemblage with some derived characters that I choose to consider them a separate family, Herpestidae. Wozencraft (this volume) has affirmed the validity of separating the mongooses from the civets. The civets then would be united within the family Viverridae.

Modern-day civets are confined to the Old World tropics, with one genus extending into the Mediterranean region of Europe. Within this assemblage a wide range of trophic specialization is shown; some members are frugivorous, others more carnivorous, but generalist omnivores dominate. There is a strong trend within the civets for adaptation to an arboreal way of life. However, some members such as the African civet (*Civettictis civetta*) are terrestrial and digitigrade. Nocturnality and solitary foraging dominates within the group.

The Herpestidae, on the other hand, tend to be diurnal, although there are a few exceptions. Some members are semi-arboreal but many are terrestrial. Some species of mongoose have become quite social and live in cohesive groups that forage and defend exclusive territories.

The island of Madagascar has an interesting array of herpestid and viverrid carnivores, suggesting successive colonization events. The ring-tailed mongoose (*Galidia elegans*) and its allies clearly are herpestids, whereas the Fanaloka civet (*Fossa fossa*) and its allies are clearly viverrids. The enigmatic fossa

(*Cryptoprocta ferox*) shows affinities with the viverrids but has diverged so far from the ancestral stock as to obscure its exact relationships. It is the largest carnivore extant on Madagascar and is a semi-arboreal predator.

The Hyaenidae include the aardwolf (*Proteles cristatus*) and the true hyenas. This Old World radiation is most strongly expressed in Africa, although the striped hyena (*Hyaena hyaena*) ranges through the Middle East on into peninsular India. The aardwolf is specialized as a termite feeder. The other three extant species of hyenas are scavengers or active predators. The spotted hyena (*Crocuta crocuta*), is highly social, living in matriarchal groups that show territorial defense (Kruuk 1972).

The Felidae, or cat family, have members strongly specialized for a carnivorous way of life. Obviously having an origin in forested habitats, most forms still retain the ability to climb well. With the exception of the cheetah (*Acinonyx jubatus*), which is a pursuit hunter over short distances, most felids are specialized for concealment and a rapid rush to overcome their prey. An enduring tendency to live and hunt in a solitary fashion characterizes the group; however, the African lion (*Panthera leo*) is a notable exception since it is a highly social species (Kleiman and Eisenberg 1973).

The rest of the carnivores are grouped into a rather heterogenous assemblage, the Canoidea. Several distinct lines of descent may be noted. If we exclude Australia and Antarctica, the Mustelidae, or weasel family, shows at the present time a worldwide distribution, even South America having been colonized at the end of the Pliocene by the long-tailed weasel (*Mustela frenata*) and allies, as well as by the otters (*Lutra* species and the giant river otter, *Pteronura brasiliensis*). Classically, the family is divided into five subfamilies. The Lutrinae, or otters, are distributed worldwide with the exception of Australia. Specialized for an aquatic life, they primarily exploit fish and shellfish in their diet. The Melinae, or badgers and their kin, are a northern hemisphere group broadly distributed and showing strong adaptations for digging. The Mephitinae, or skunks and their relatives, are North American, but some Old World Mustelinae resemble them. Noted for their bold markings in black and white and their anti-predator defense system involving strong secretions from specialized anal glands, skunks represent a distinctive group of terrestrial omnivores. The Mellivorina, or honey badgers, are a distinct Old World group showing affinities to the Mephitinae as well as the Melinae. Finally, the Mustelinae are the typical northern hemisphere weasels and are among the most highly specialized predators for feeding on rodents and ground-nesting birds. Members of this subfamily have successfully colonized Africa and South America.

The next easily distinguished family is the Canidae, or dog family. It was widely distributed on all continents except Antarctica and Australia, but humans introduced *Canis* species to Australia 10,000 years B.P. The foxes, wolves, and their kin represent an old lineage adapted for the cursorial pursuit of prey. The larger members, such as the gray wolf (*Canis lupus*), are often highly social and exhibit group hunting, which allows them to bring down prey much larger than themselves.

The remaining canoid taxa have presented some puzzles to taxonomists. One can clearly distinguish the family Procyonidae with a New World distribution. They were early entrants into South America before the completion of the Pliocene land bridge. These include the raccoons (*Procyon* spp.), coatis (*Nasua* spp.), the kinkajou (*Potos flavus*), and their allies. Although the procyonids show affinities with the canids, they also share characters with the bear family, Ursidae; and bears in turn show affinities with the canids. Contemporary bears are distributed worldwide, with the exception of Australia and Africa. This group includes the largest extant members of the order Carnivora. Most species of bears are generalized omnivores, but some, such as the sloth bear (*Melursus ursinus*), have specialized for feeding on ants and termites; the polar bear (*Ursus maritimus*) is the top carnivore of the region bounding the Arctic Ocean, where it is a specialist feeder on seals.

Finally, we come to two puzzling genera, the lesser or red panda (*Ailurus fulgens*) and the giant panda (*Ailuropoda melanoleuca*). These are Asian in their present distribution and are presently found on the eastern and southern escarpment of the Tibetan plateau. The red panda has specialized for herbivory and includes a great deal of young bamboo in its diet, although it also preys on small birds and mammals. The giant panda is a specialist bamboo feeder and well known to one and all as the symbol of the World Wildlife Fund. Though bearlike in its anatomy, the giant panda shows differences in its genital structure that lead one to believe it is not closely related to bears, although surely in some way related. The red panda has been variously classified as a procyonid or allied with the giant panda. I am inclined to follow Pocock and place each of these unique genera in its own family, the Ailuridae and the Ailuropodidae (Eisenberg 1981).

The giant panda continues to be controversial with respect to its taxonomic status. That pandas are in some way related to bears is not to be doubted, but it is the degree of relationship that remains in question. O'Brien et al. (1984) have demonstrated by the study of allozymes that the giant panda branched off from the true bears as early as the Miocene. If their interpretation is correct, then, to my mind, to call the panda a bear would require that we call the orangutan a human. It seems to me to serve no useful purpose to lump the giant panda along with the bears in a single family. Of course, many people would agree that the orangutan is not very dissimilar from a human. After all, the origin of the name "orangutan" is from the Malay meaning "man of the forest."

I believe when disputes arise concerning the manner in which species are classified, one could well turn to Simpson (1945:12): "It is often stated that the purpose of classification is or should be to express phylogenetic relationships, but in the first place no one has ever devised a method of classification that could express phylogeny sufficiently or consistently, and in the second place, the system that is actually used in zoology was not devised for that purpose and is notably incapable of serving it."

"This is, as I see it, the primary purpose of classification: simply to provide a convenient, practical means by which zoologists may know what they are talking about and others may find out. It is helpful for this purpose, and it is also a secondary but still essential aim that that classification be consistent with the most important thing that evolutionary taxonomists have to talk about, that is, with animal affinities." (Simpson 1945:13.)

One can see from this overview that within the order Carnivora there has been specialization for a wide variety of different feeding niches. Whereas, for example, polar bears, wolves, lions, weasels, and otters are truly carnivorous, many of the other members of the order show adaptations for a broader diet. Given the variety of adaptations displayed by the extant Carnivora, what then characterizes these animals? They have all descended from miacids that were probably nocturnal, semi-arboreal small predators. The extant Carnivora all possess enlarged canines, but the shearing adaptation of the molars is not present in those forms that have adapted to a more omnivorous or herbivorous diet. Our early carnivores may have well been plantigrade, but many modern forms such as dogs and cheetahs that are specialized for cursorial pursuit of prey have become digitigrade.

The present-day small carnivores include some forms that are rather conservative in their morphology. The true civets of the Old World tropics probably occupy an ecological niche similar to that occupied by the miacids in the Miocene. These are nocturnal forms, many highly arboreal, that hunt small vertebrate prey and also feed on fruit. Civets rely on the tactile senses, vision, olfaction, and audition in locating prey and in orientation. Civets tend to be solitary except for mating or when rearing young.

The importance of the olfactory system as the main channel of gathering sensory information is retained by many of the extant carnivores but has declined in importance among the felines. Many modern carnivores are diurnal, and vision is extremely important among such diurnal hunters as the cheetah. Olfaction remains preeminent in pursuit hunters, such as many of the family Canidae. As might be expected, the aquatically adapted otters have a highly refined tactile system involving the vibrissae and the forepaws to assist in locating food beneath the surface of the water. Olfaction is much less important to aquatic forms.

The larger canids such as wolves are specialized for the pursuit of large prey and frequently hunt in packs. The modern cats are specialized for concealment and dispatch of prey after a short rush; in short, they are ambush hunters. The modern cats are among the most carnivorous of extant carnivores. Many members of the weasel family are also specialized predators, but in this case the specialization is for smaller prey, and the weasel's body form and size reflect specialization on tunneling rodents as prey.

In modern carnivores the relative brain size varies greatly. Rather large brains are characteristic of the canids, ursids, and procyonids. The exact adaptive significance of this interesting difference is poorly understood. The relatively large brains of the ursids are in conformity with their life history strategy in that they tend to be very long-lived and highly iteroparous (Gittleman 1986).

To be a hunter requires a considerable capacity to learn and a corresponding versatility in behavior. One can think of predator and prey coevolving through time as the prey becomes craftier at avoiding the predator and the predator still improves its techniques at ambush, capture, and dispatch of prey. Although we think of carnivores as solitary hunters, as previously noted, some species of carnivores hunt or forage in groups and have a rich social life. Group life demands an ability to recognize individuals and exhibit considerable behavioral plasticity (see Gittleman, this volume).

Great variety is seen in the reproductive adaptations of the extant Carnivora. Of particular interest is the phenomenon of delayed implantation found among the Mustelidae and Ursidae in the temperate-zone latitudes (Mead, this volume). Here, the timing of mating appears to be geared to the time when males are in optimal physical condition. Females may then exercise some choice and may possibly have promoted the timing of mating to favor optimum vitality in males. The significance of delayed implantation is discussed further by Mead. Whatever selective forces have been in operation to set the time of mating, the female does not implant the blastocist until a considerable time has elapsed. This permits an optimal season of birth, usually a time when the demands of lactation can be met and the young at dispersal will be confronted with an adequate food supply.

The evolution of social behavior by the Carnivora exhibits many interesting contrasts. Whereas most small, nocturnal carnivores are solitary except for the female-young unit and at the time of pairing, many carnivores have evolved behavioral mechanisms promoting sociality. We find an enduring trend among canids for a monogamous pair bond during the rearing of the young. Often the male assists in provisioning the female, and in some canid species such as the golden jackal (*Canis aureus*) and the black-backed jackal (*C. mesomelas*), the young of the previous year may remain with their parents and assist in various aspects of raising the next litter (Moehlman 1983). Lions are the only truly social felid. Here, the social system is based mainly on a group of females, probably related by descent, that cooperatively rear their young. Males protect their group of females from other males, thereby ensuring some exclusivity in mating. Cooperation among male lions in defense of prides of females has been an active area of research in recent years (Schaller 1972; Bertram 1975).

Among the Procyonidae the coatis exemplify diurnal, social carnivores. The females form bands during the later phases of rearing their young. The advantages of group foraging and the possible antipredator effects of group living have recently been the object of investigation (Kaufmann 1962; Smythe 1970;

Russell 1983). The Herpestidae show several species with advanced sociality, including the banded mongoose (*Mungos mungo*), the meerkat (*Suricata suricatta*), and the dwarf mongoose (*Helogale parvula*). Group living serves an important anti-predator function, and predator selection may well be the most potent selective force in enhancing group life among the diurnal mongooses (Rood 1983).

Archeologists concede that the domestication of the wolf was an early domestication event resulting in the modern breeds of the dog. Thus, a medium-sized social carnivore was the first species to be brought into intense contact with humans (Zeuner 1963). The larger carnivores have held the fascination of people for a long time. When humans were primarily hunters they were in direct competition with large carnivores, and in some cases might have found themselves a prey item. This has led to a duality in our attitudes toward large carnivores. On the one hand, they are admired for their beauty, strength, and efficiency in dispatching prey and, on the other hand, feared.

During the Neolithic when humans became domesticators of animal and plant life, large predators became an even greater menace to the livestock now under their control. This ultimately led to a constant war of atttrition between the farmers and the predators. It is no wonder that with the advent of modern fire arms large carnivores rapidly became exterminated over large parts of their range. In fact, in the history of the settlement of the North American continent by Europeans, the persecution of predators in favor of pastoral interests makes up a large part of North American folklore. Yet, at the same time, respect and intense curiosity concerning the lives of large carnivores remained high.

Small carnivores less often came into direct conflict with humans, although the weasel family has been persecuted for centuries because of its proclivity for raiding hen houses. On the other hand, many of the temperate-zone small carnivores, and especially otters, have been harvested for their pelts. Ultimately, overharvesting led to attempts to domesticate certain forms of small carnivores for a sustained yield of pelts. The mink (*Mustela vison*) is one classic example of this process.

Now we are at a crossroads with respect to the future of the earth's biomes. There has been much discussion of attempts to preserve ecosystems intact. If this course is to be followed, full recognition must be made that top carnivores play an important role in structuring communities. The removal of a top carnivore from an ecosystem can have an impact on the relative abundance of herbivore species within a guild. In the absence of predation, usually one or two species come to dominate the community. The consequence of this is often a direct alteration of the herbaceous vegetation fed on by the herbivore guild or assemblage. Top carnivores have an important role to play in the structuring of communities and, ultimately, of ecosystems. Thus, the preservation of carnivores becomes an important consideration in the discipline of conservation biology.

Because so much of the earth's surface is being vastly modified by humans, it

8   *John F. Eisenberg*

is only appropriate that recent field studies have focused on a wide variety of carnivores on different continents. Much of this literature is scattered and has not heretofore been brought together in a single volume. R. F. Ewer in 1973 published her classic, *The Carnivores,* in which she reviewed anatomy, behavior, reproduction, and aspects of natural history. The present work does not attempt to duplicate her standard reference but, rather, pulls together the threads on the behavior, ecology, and evolution of carnivores into one volume that may serve as a guide and reference to the conduct of future studies of carnivore biology. The time is indeed short, and we all hope that this volume provides a useful synthesis.

References

Bertram, B. C. R. 1975. Social factors influencing reproduction in wild lions. *J. Zool.* 177:463–482.
Dawson, M. R., and Krishtalka, L. 1984. Fossil history of the families of recent mammals. In: S. Anderson & J. Knox Jones, Jr., eds. *Orders and Families of Recent Mammals of the World,* pp. 11–58. New York: John Wiley & Sons.
Eisenberg, J. F. 1981. *The Mammalian Radiations.* Chicago: Univ. Chicago Press.
Eisenberg, J. F. 1986. Life history strategies of the Felidae: Variations on a common theme. In: S. Douglas Miller & D. D. Everett, eds. *Cats of the World: Biology, Conservation, and Management,* pp. 293–303. Washington, D.C.: National Wildlife Federation.
Ewer, R. F. 1973. *The Carnivores.* Ithaca, N.Y.: Cornell Univ. Press.
Gentry, R. L., and Kooyman, G. L., eds. 1986. *Fur Seals: Maternal Strategies on Land and at Sea.* Princeton: Princeton Univ. Press.
Gittleman, J. L. 1986. Carnivore brain size, behavioral ecology and phylogeny. *J. Mamm.* 67:23–36.
Kaufmann, J. 1962. Ecology and social behavior of the coati, *Nasua narica,* on Barro Colorado Island, Panama. *Univ. California Publ. Zool.* 60:95–222.
King, J. E. 1964. *Seals of the World.* London: British Museum.
Kruuk, H. 1972. *The Spotted Hyena.* Chicago: Univ. Chicago Press.
Kleiman, D. G., and Eisenberg, J. F. 1973. Comparisons of canid and felid social systems from an evolutionary perspective. *Anim. Behav.* 21:637–659.
Moehlman, P. 1983. Socioecology of silverbacked and golden jackals. In: J. F. Eisenberg & D. G. Kleiman, eds. *Advances in the Study of Mammalian Behavior,* pp. 423–453. Special Publication no. 7. Lawrence, Kans.: American Society of Mammalogists.
O'Brien, S. J., Nash, W. G., Wildt, D. E., Bush, M. E., and Benveniste, R. E. 1984. A molecular solution to the riddle of the giant pandas phylogeny. *Nature* 317:140–144.
Rood, J. 1983. The social system of the dwarf mongoose. In: J. F. Eisenberg & D. G. Kleiman, eds. *Advances in the Study of Mammalian Behavior,* pp. 454–488. Lawrence, Kan.: American Society of Mammalogists.
Russell, J. K. 1983. Altruism in coati bands: Nepotism or reciprocity: In: S. K. Wasser, ed. *Social Behavior in Female Vertebrates,* pp. 263–290. New York: Academic Press.
Sarich, V. M. 1969. Pinniped phylogeny. *Syst. Zool.* 18:286–295.
Schaller, G. 1972. *The Serengeti Lion.* Chicago: Univ. Chicago Press.

Simpson, G. G. 1945. The principles of classification and a classification of mammals. *Bull. Amer. Mus. Nat. Hist.* 85.

Smythe, N. 1970. The adaptive value of the social organization of the coati (*Nasua narica*). *J. Mamm.* 51:818–820.

Stains, H. J. 1984. Carnivores. In: S. Anderson & J. Knox Jones, Jr., eds. *Orders and Families of Recent Mammals of the World,* pp. 491–522. John Wiley & Sons.

Van Valen, L. 1966. Deltatheridia, a new order of mammals. *Bull. Amer. Mus. Nat. Hist.* 132:1–126 and 8 plates.

Zeuner, F. E. 1963. *A History of Domestic Animals.* London: Hutchinson.

PART I

BEHAVIOR

INTRODUCTION

Behavioral studies of carnivores form an integral part in the history and development of ethology, behavioral ecology, and other behaviorally oriented disciplines. For example, mechanistic approaches in motivation theory were guided by Leyhausen's (1973, 1979) classic work on the ontogeny and loco-motion of prey killing by felids and viverrids, as well as Eibl-Eibesfeldt's (1950, 1956) observations of the play behavior and aggression of European badgers and polecats. Further, Wilson's (1975) groundbreaking volume on sociobiology, which spurred the subdiscipline of behavioral ecology, used examples from gray wolves (*Canis lupus*), African hunting dogs (*Lycaon pictus*), and African lions (*Panthera leo*) to illustrate theoretical ideas of kin selection and reciprocal altruism. Future research of carnivore behavior should continue to provide important case studies for testing theory and revealing a more accurate understanding of behavioral variation and mating systems.

On the surface the chapters in this section are varied and eclectic. Indeed, with studies of carnivore behavior developing so rapidly, it is difficult to select behavioral problems and individual taxa that represent the full diversity and variation represented by carnivores. Nevertheless, these chapters reflect subject areas in behavior that have received such considerable attention that critical reviews are now appropriate to assess what is known and what directions should be taken in the future.

The first two chapters describe mechanisms of communication that allow carnivores to establish mating systems, modes of parental care, foraging patterns, and other behavioral features. Utilizing much of the obscure German literature, Peters and Wozencraft discuss the physical, physiological, developmental, and evolutionary aspects of vocalizations. As most studies of carnivore vocalizations are from captive animals, this chapter should provoke further analyses of wild populations. Gorman and Trowbridge consider the importance of olfaction, both in terms of anatomical properties and functional effects on reproduction and territory utilization. With many carnivore species using olfactory means to communicate reproductive status, it will be interesting to learn whether olfactory "fingerprints" designate individual fitness.

Moving from communication mechanisms to ontogeny, Bekoff critically reviews methodological problems and general trends in carnivore behavioral development. Most studies of carnivore development are on captive individuals, which unfortunately often produce spurious results. Bekoff describes how future studies may analyze carnivore development and life histories more rigorously and why solid developmental data will provide the answers to questions of individual dispersal patterns and subsequent mating strategies. At present, intense discussion in evolution theory surrounds the interrelationships of ontogeny, neoteny, and phylogeny of mammalian lineages. Data on carnivore behavioral development and life histories should have direct bearing on some of these theoretical issues.

The remaining chapters in this part all concern various mating systems and their ecology and evolution. Mills reviews recent field work comparing the diet, reproduction, social organization, communication and denning of brown and spotted hyenas (*Hyaena brunnea* and *Crocuta crocuta,* respectively); in these species, which live in facultative groups, comparisons of specific differences in diet lead to the evolution of dramatic differences in social behavior and reproduction. Clearly, the most well-studied family of carnivores is the Canidae. Moehlman discusses the degree and kind of intraspecific variation observed across nine species of canids and stresses the need to understand the ecological factors selecting for intraspecific variation in order to uncover the evolutionary forces influencing many interspecific trends. As multiple studies are now available on some species in the Canidae, such as wolves, coyotes (*Canis latrans*), silverbacked jackals (*C. mesomelas*), red foxes (*Vulpes vulpes*), and gray foxes (*Urocyon cinereoargenteus*), finer resolution of analyses may provide new insights into parallel or divergent trends of intra- and interspecific variation.

Finally, the chapters by Sandell and Gittleman consider comparative evidence for the evolution of solitary and group living, respectively. Although most carnivores are solitary, little attention has been given to different forms of solitary living and the ecological factors selecting for these forms. To date, discussion has centered around the influence of diet, specifically, the distribution and abundance of foods, on individual home range movements and day range patterns. Sandell reconsiders the evolution of solitary living in terms of mating strategy, particularly the spatial patterning of males and females during the breeding season. This perspective, which is more aligned with contemporary issues in evolution theory and behavioral ecology, should force us to reevaluate the supposed simplicity of solitary living. Gittleman analyzes the comparative evidence for what ecological factors are associated with group living. Previous discussion has emphasized that large carnivores (e.g., wolves, African lions, spotted hyenas) live in groups to aid in bringing down large prey whereas smaller species (e.g. dwarf mongoose, *Helogale parvula,* and banded mongoose, *Mungos mungo*) reside in groups to ward off potential predators. The comparative data do not support this size-dependent hypothesis; rather, anti-predatory factors *and* dietary characteristics are probably operating simultaneously to select for group living. As for primates (see Smuts et al. 1987), when specific data are available on the composition, sex ratio, and relatedness of individuals in carnivore groups, more detailed and accurate models of the evolution of carnivore sociality will be forthcoming.

John L. Gittleman

References

Eibl-Eibesfeldt, I. 1950. Über die Jugendentwicklung des Verhaltens eines männlichen Dachses (*Meles meles* L.) unter besonderer Berücksichtigung des Spieles. *Z. Tierpsychol.* 7:327–355.

Eibl-Eibesfeldt, I. 1956. Zur Biologie des Iltis (*Putorius putorius* L.). *Zool. Anz. Suppl.* 19:304 314.

Leyhausen, P. 1973. On the function of the relative hierarchy of moods (as exemplified by the phylogenetic and ontogenetic development of prey-catching in carnivores). In:K. Lorenz & P. Leyhausen, *Motivation of Animal and Human Behavior*, pp. 144–247. New York: Van Nostrand Reinhold.

Leyhausen, P. 1979. *Cat Behavior*. New York: Garland.

Smuts, B. B., Cheney, D. L., Seyfarth, R. M., Wrangham, R. W., and Struhsaker, T. T. eds. 1987. *Primate Societies*. Chicago: Univ. Chicago Press.

Wilson, E. O. 1975. *Sociobiology: The New Synthesis*. Cambridge: Harvard Univ. Press.

CHAPTER 1

Acoustic Communication
by Fissiped Carnivores

GUSTAV PETERS AND
W. CHRIS WOZENCRAFT

The domestic dog (*Canis lupus* f. *familiaris*) and cat (*Felis silvestris* f. *catus*), which are quite vocal by mammalian standards, are not good representatives of the acoustic activities of fissiped carnivores. Fissipeds are generally thought of as mammals that communicate with smell rather than with vocalizations (Gorman and Trowbridge, this volume). Nevertheless, several carnivore acoustic signals like the howling of gray wolves (*Canis lupus*), the whooping of spotted hyenas (*Crocuta crocuta*), and the roaring of African lions (*Panthera leo*) capture the human imagination as few other animal sounds do. It is probably no accident that wolf howling—unlike other acoustic signals of carnivores—is one of the best-studied mammalian vocalizations (Theberge and Falls 1967; Cohen and Fox 1976; Tembrock 1976a, 1976b; Fox and Cohen 1977; Shalter et al. 1977; Field 1978, 1979; Fox 1978; Harrington and Mech 1978a, 1978b, 1979, 1982, 1983; Schassburger 1978; Klinghammer and Laidlaw 1979; Filibeck et al. 1982; Harrington 1986, 1987; Nikolskii and Frommolt 1986).

Slater (1983) presented various concepts of the ways communication in animals can be defined. Studies of mammalian communication rarely deal with more than one of their four signaling modes (acoustic, olfactory, tactile, visual); only the first two are considered in this book (see Gorman and Trowbridge, this volume, for olfactory). Signals of different modes often occur together, especially in close-range communication, and are qualitatively and quantitatively interdependent. This has not been well studied nor will it be considered here, but ought to be kept in mind when statements on the functions of acoustic signals are evaluated.

Fissiped sound communication is covered in reviews by Tembrock (1963a, 1968, 1970), Fox and Cohen (1977), Pruitt and Burghardt (1977), and Wem-

This chapter is dedicated to Paul Leyhausen and Günter Tembrock, who contributed so substantially to the study of carnivore behavior and acoustic communication by mammals.

mer and Scow (1977). Ewer (1973) reviewed the subject on three pages, with only an occasional mention of vocalization in other contexts, and did not present structural data. This review summarizes acoustic communication by fissiped carnivores and puts it into a wider mammalian perspective, following Gould (1983). We have organized this topic around five basic ethological concepts: (1) *structure* of acoustic signals, (2) *motivation* of acoustic signaling behavior, (3) *functions* of acoustic signals, (4) *ontogeny* of vocalization, and (5) *evolution* of acoustic communication.

The terms "acoustic signal," "acoustic communication," and "vocalization" are used for any sound produced by animals, irrespective of structures and mechanisms generating the sound. Signals produced by oscillations of the vocal folds are designated as "vocal"; those generated in any other way, as "nonvocal" signals. "Tonal" signals show a distinct frequency band or a harmonic structure, whereas "atonal" or "noisy" have a broad frequency range without such bands.

The taxonomy applied in this review follows Honacki et al. (1982), as listed in the appendix of this volume. References on acoustic communication data for fissipeds, grouped taxonomically, are listed in Table 1.3. Because this review covers only the terrestrial fissiped carnivore families, the terms "carnivore" and "fissiped" will be used interchangeably throughout the text without any phylogenetic implication. The Phocidae and Otariidae are not included in this review.

Physical and Physiological Aspects of Vocalization by Carnivores

Sound Production

Carnivores, like most mammals, generate sound by oscillations of the vocal cords in the larynx. However, they can also produce signals in various other ways (Gould 1983; Müller-Preuss and Ploog 1983). The process of sound generation by the vocal folds or other structures in the upper respiratory tract and the modification of this sound in the oral and nasal cavity is poorly understood in nonhuman mammals. The only well-studied carnivore acoustic signal in this respect is felid purring (Denis 1969; Remmers and Gautier 1972).

A generally held assumption is that the tonal calls of mammals (the principal exception being some whistle-type sounds produced by cetaceans and rodents) are produced by oscillations of the vocal folds in the larynx. Atonal and partially atonal sounds may be generated by the same process or may involve contributions of other sound-producing sources to that of the vocal folds. Fully atonal sounds can also be produced by structures other than the vocal cords. It is not known whether the pure tonal, whistle-like calls of some carnivores, such as chirps in otters (Lutrinae) (Duplaix 1982) or "whistles" in jaguarundi (*Felis yagouaroundi*) (Hulley 1976), can be produced in a way other than by oscillations of the vocal folds.

Comparative anatomical studies of the mammalian larynx have included relatively few carnivores (Negus 1949; Kelemen 1963; Schneider 1964). Davis (1964) compared the gross laryngeal anatomy of the coyote (*Canis latrans*), black bear (*Ursus americanus*), Asiatic black bear (*U. thibetanus*), sloth bear (*Melursus ursinus*), giant panda (*Ailuropoda melanoleuca*), raccoon (*Procyon lotor*), coatimundi (*Nasua nasua*), and red panda (*Ailurus fulgens*). The ursid larynx probably represents a more primitive type among caniforms, whereas the canid and procyonid larynx are more derived, each with peculiar anatomical characteristics. The larynx of the giant panda is most similar to that of ursids (Davis 1964) in spite of differences in the acoustic repertoires and pitch of some vocalizations between bears and giant pandas (G. Peters 1982, 1985; Schaller et al. 1985). Structurally, the canid larynx is more complex than that of felids (Kelemen 1963); however, major qualitative differences between these families are lacking in the normal range of their acoustic signals. There does not appear to be a direct correlation between gross anatomy of a larynx and the range and quality of sounds it can produce (Kelemen 1963). The length, mass, and other physical dimensions and properties of sound-producing and sound-modifying structures may be crucial. A correlation proposed by Pocock (1916, 1917) between the degree of ossification of the hyoid apparatus and the presence of roaring or purring in a felid's acoustic repertoire was not verified by G. Peters (1981a). Morphologically, the vocal folds of *Panthera* species, with the exception of the snow leopard (*P. uncia*), differ from those of the other felids studied; the structure of the larynx in species of *Panthera* enables them to roar (Hast 1986).

Nonvocal sounds can be classified into three broad categories according to the mechanism(s) and structures involved in sound production (for a more detailed classification, see Tembrock 1977): (1) respiratory, (2) nonrespiratory, and (3) instrumental sounds.

Respiratory sounds used in communication are produced by stressed and stereotyped exhalation and/or inhalation through the mouth and/or nose. They differ from normal and increased breathing sounds by temporal patterning, duration, amplitude, and sound quality. Respiratory sounds may also be components of complex vocalizations involving sounds generated by other sources as well. Nasal exhalatory sounds are known in the giant panda (Kleiman 1983; G. Peters 1985; Schaller et al. 1985), raccoon (Sieber 1984), kinkajou (*Potos flavus*) (Poglayen-Neuwall 1962, 1976b), red panda (Roberts and Gittleman 1984), and the binturong (*Arctictis binturong*) (Wemmer and Murtaugh 1981). Nasal exhalatory sound as a component of a complex with different sources occurs in the Viverridae (Wemmer 1977) and Felidae (G. Peters 1978b, 1984a, 1984b). Oral exhalatory sounds may also occur together with noisy inhalation and nasal respiratory sounds. Examples are the chuffing of ursids (Wemmer et al. 1976; G. Peters 1978a, 1984b) or the huffing in the giant panda (Kleiman 1983; G. Peters 1985; Schaller et al. 1985). It has not been established whether the widespread carnivore hissing sound involves the vocal folds or whether it is an exhalatory sound. Reschke (1960) differentiates between "guttural" (laryngeal sound generation?) and "palatal" (without it?) hissing in felids, whereas Eisenberg (1981) considered hissing an unvoiced breathing sound.

Nonrespiratory sounds produced in the upper respiratory tract (e.g., by lips, tongue, teeth, and/or cheeks) often involve several structures and may be components of complex signals involving other sound-generating sources. There is an overlap with respiratory sounds in cases where a forceful exhalatory jet of air is accompanied by a sound generated by the lips releasing the air. Ursid chuffing, produced primarily by the lips (Meyer-Holzapfel 1957; Jordan 1976, 1979; Wemmer et al. 1976), has two such structural components (G. Peters 1978a, 1984b). The exact mechanism of sound production of jaw-clapping in the red panda has not been described in detail (Roberts and Gittleman 1984). "Chomping" in the giant panda probably involves both tooth clicking and lip smacking (Kleiman 1983; G. Peters 1985; Schaller et al. 1985).

Instrumental sounds, which animals generate by interaction of parts of their body with each other or their environment, are made by bears: during threat behavior they slap their front paws against the ground or other objects (Jordan 1976, 1979).

A major conceptual problem exists in comparative mammal vocalization studies because of the lack of data on the mechanisms of sound production (Eisenberg 1974; Eisenberg et al. 1975). The way an acoustic signal is produced is essential for any classification system. An example of a mammalian sound with structural uniformity that can be produced in different ways is the click. Despite structural similarity and possible functional equivalence, clicks produced by different structures must be classified as nonhomologous sound types.

The physical size of a species influences the range of sounds it can produce. As a general rule, large species can produce sounds of lower pitch than can smaller ones (Hutterer and Vogel 1977); the same should hold true for individuals of different size within the same species (Balph and Balph 1966; August and Anderson 1987). Although this phenomenon was substantiated in an interspecific comparison of many mammalian taxa (August and Anderson 1987), it is not true that signals of species within the same genus or family generally follow this rule. "Whistles" of the puma (*Felis concolor*) and jaguarundi are much higher in pitch than homologous calls of other similar or smaller-sized felids (G. Peters 1978b, pers. obs.). The giant panda's "chirp" is higher in pitch than any other sound known in a similar-sized carnivore (G. Peters 1982, 1985; Schaller et al. 1985). Frequency parameters in "whistles" of juvenile raccoons of the same age do not show a significant correlation with body weight (Sieber 1986).

Central nervous control of sound production by carnivores has been studied in the domestic dog (Skultety 1962) and domestic cat (Kanai and Wang 1962; de Molina and Hunsperger 1962; Skultety 1965). Lesions in the central grey and parabrachial area of the brainstem in both species destroy the ability to vocalize. In the domestic cat electrical stimulation of structures in the hypothalamus and the rostral portion of the midbrain evoke agonistic vocalizations such as screaming, growling, and spitting. Mews can be elicited by stimulation of the caudal portions of the midbrain and at different levels of the pons and the medulla (Kanai and Wang 1962).

Properties of Sounds and Sound Transmission

Carnivores can produce a wide range of tonal, atonal, and mixed vocal and nonvocal signals, with wide variation in all three structural dimensions: frequency, amplitude, and time. The lowest frequencies recorded in carnivore sounds are in the roar of lions at 50 Hz (Jarofke 1982), and the highest are up to 107 kHz in the distress calls of juvenile ferrets (*Mustela putorius* f. *furo*) (Solmsen and Apfelbach 1979). Most adult fissiped acoustic signals fall in the frequency range below 10 kHz, even in the smallest species, the least weasel (*Mustela nivalis*) (Huff and Price 1968; Gossow 1970; Heidt and Huff 1970). Romand and Ehret (1984) reported occasional frequency components up to 60 kHz in distress calls of domestic cat kittens during their first months of life. The only known pure ultrasonic signals in carnivores have been recorded in domestic cats around the time when the kittens start to leave the nest at an age of about one month. At that time the mother cat produces pure ultrasounds around 50 kHz, the kittens around 80 kHz (Härtel 1972). In a careful experimental study Romand and Ehret (1984) did not detect pure ultrasonic calls in kittens. Lehner (1978b) suggested that adult coyotes may be able to produce ultrasonic sounds, whereas Huff and Price (1968) expressly stated that a least weasel female and her four young at 3 weeks of age did not produce any calls with frequency components in the range from 20 to 180 kHz.

There are very few direct measurements of call amplitude in carnivore acoustic signals. Average sound pressure level (SPL) of isolation calls of domestic cat kittens stays rather constant, around 70 to 75 dB, between day 1 and 105 (Romand and Ehret 1984). Indirect measurements of amplitude in distress calls of juvenile ferrets reveal components with up to 90 dB (Solmsen and Apfelbach 1979). The highest amplitude recorded from lion roaring was 114 dB SPL for a male and 110 dB in a female (Jarofke 1982). Amplitude measurements of faint carnivore vocalizations like feline purring are not available.

Duration of signals may vary from between 10 and 20 ms, such as the felid spitting sound (G. Peters 1980) or tooth clicking in the giant panda (G. Peters 1985), to continuous loud calls of large felids of several seconds' length (G. Peters 1978b; Rieger and Peters 1981), to continuous sound production for minutes on end in felid or viverrid purring (Denis 1969; Wemmer 1977; G. Peters 1981a) or the ursid nursing sound (Schneider 1933; Meyer-Holzapfel 1957). Purring in the domestic cat may go on continuously for up to 2 h (Kiley-Worthington 1984). A solo wolf howl lasts for up to about 11 s (Theberge and Falls 1967; Harrington and Mech 1978b; Schassburger 1978). Composite vocal signals like the structured call sequence of a roaring lion may last for up to 40 s (Reschke 1960, 1966; Schaller 1972; G. Peters 1978b). Chorus roaring in lions or chorus howling in wolves can have a duration of more than 1 min. Call repetition rate is also an important temporal parameter in acoustic communication (Schleidt 1973), but it is relatively little studied in carnivores.

Although there are experimental data for anurans and birds on sound transmission in different habitats (Morton 1975; Wiley and Richards 1982; Gerhardt 1983), this aspect of carnivore vocalization has not been studied. Therefore, hypotheses about the adaptive significance of carnivore signals can only be inferred from evidence in other vertebrates. Fissipeds are found in nearly all types of terrestrial habitat, from arctic tundra to tropical rainforest. Most species are terrestrial, some are semi-aquatic, semi-arboreal, or arboreal. Carnivore families with species largely adapted to an arboreal way of life are the Procyonidae and the Viverridae. As a structural adaptation for optimum sound transmission, the long-distance calls of ground-living carnivores have their highest amplitude below 1 kHz, which are the frequencies that are transmitted best (Wiley and Richards 1982). Roaring by lions (Schaller 1972; G. Peters 1978b; Jarofke 1982) and wolf howling (Theberge and Falls 1967; Harrington and Mech 1978b; Schassburger 1978) both have their maximum intensity below 0.5 kHz, the frequency range least affected by absorption in open grassland (Morton 1975; Wallschläger 1981). Two arboreal viverrids, the masked palm civet (*Paguma larvata*) and the African palm civet (*Nandinia binotata*), and perhaps the common palm civet (*Paradoxurus hermaphroditus*), have relatively high-pitched, repetitive long-distance calls, a possible adaptation for optimum sound propagation in higher forest strata (Wemmer 1977). The same may hold for calls of the arboreal olingo (*Bassaricyon* sp.), with their highest amplitude near 4 kHz (Poglayen-Neuwall 1976a). Another adaptation for optimum transmission in long-range calls involves the daily temporal distribution of vocalizing activity. Lion roaring (Schaller 1972), wolf howling (Harrington and Mech 1978a; Schassburger 1978; Klinghammer and Laidlaw 1979), roar barking by maned wolves (*Chrysocyon brachyurus*) (Brady 1981), and coyote vocalization (Laundré 1981) are most frequent around dawn and dusk (or during the night), when sound propagation in open habitat is optimal (Wallschläger 1981). Seasonal variation of overall vocalization activity or in frequency of occurrence of specific signals, as in the cases of wolves (Harrington and Mech 1978a; Klinghammer and Laidlaw 1979; Nikolskii and Frommolt 1986; Nikolskii et al. 1987) or coyotes (Laundré 1981), has not been shown to be an adaptation to optimum sound propagation but is probably influenced by circannual physiological rhythms related to reproductive state or factors such as long-range mobility or developmental state of the young.

Sound Perception

Many studies of mammalian auditory physiology have used the domestic cat and were not done with animal sounds as auditory stimuli that are biologically significant to a cat, but with artificial sounds or with natural sounds whose meaning to the domestic cat has not been established. Thus, relatively little is known as to the specific perception and processing of the cat's own vocal signals or that of prey species.

Behavioral audiograms are published for domestic cat (Neff and Hind 1955; Heffner and Heffner 1985b), domestic dog (Heffner 1983), raccoon (Wollack 1965), and the least weasel (Heffner and Heffner 1985a). Data on the auditory response in a variety of carnivores published by Peterson et al. (1969) were based on the measurement of the cochlear microphonic potential, recorded at the round window membrane, and therefore cannot be equated with the behavioral audiograms.

The domestic cat and dog, the raccoon, and the least weasel all have a broad range of best sensitivity of hearing, from 1 to 16 kHz, with no prominent optimum (Heffner and Heffner 1985a). The cat's hearing limits at 60 dB SPL are 55 Hz and 78 kHz; the other three differ relatively little from this (Heffner and Heffner 1985a, 1985b). Special aspects of sound perception such as frequency discrimination (Elliott et al. 1960; Ehret 1977), temporal resolution (Gerken and Sandlin 1977), and masking effects (Watson 1963) have been studied in domestic cats.

Red foxes (*Vulpes vulpes*) were tested for localization performance with pure tones between 0.3 and 34 kHz broadcasted from speakers 35.5° apart. The animals showed the best performance for frequencies between 0.9 and 14 kHz, with an optimum at 3.5 kHz and a slight decrease at 8.5 kHz (Isley and Gysel 1975). According to Heffner (pers. comm., in Gourevitch 1980:363), domestic dogs can discriminate click sources about 4° apart. Minimum audible angle function in the domestic cat was tested with pure tones between 0.25 and 8 kHz. Best localization performance with angles smaller than 10° is for frequencies between 0.5 and 2 kHz; angular thresholds for wide band noise signals are about 5° (Casseday and Neff 1973). Locatability of carnivore signals has not been studied experimentally; however, structural characteristics of vocalizations have been discussed as adaptations to locatability (G. Peters 1984b; Sieber 1984), based on the mechanisms of sound source localization in other mammals (Gourevitch 1980).

Most mammals are able to vocalize within minutes after birth and, given an appropriate stimulus like cold, hunger, or pain, will do so frequently and at high intensity. Hearing, however, develops gradually in carnivores during the first weeks of life (Ehret 1983). Carnivore species studied in this respect are the domestic cat, domestic dog, and mink (*Mustela vison*) (Foss and Flottorp 1974; Olmstead and Villablanca 1980; Ehret and Romand 1981). Because of differences in experimental procedure, conclusions about the onset of hearing from the different studies are not directly comparable (Ehret 1983). In general, the auditory perception of juveniles begins to develop in the lower and middle hearing range of adults and then extends to even lower and then higher frequencies (Ehret 1983). At thresholds above 100 dB SPL, hearing in domestic cat kittens starts on the first or second day of life. It goes below this threshold after the sixth day. Hearing first develops in the range between 0.5–2 kHz, extending to 0.2–6 kHz until the sixth day, and full adult hearing range is established by about one month of age (Ehret and Romand 1981). Olmstead and Villablanca (1980) found first specific and differential directional auditory responses to kitten and mother cat calls (orientation toward stimulus) and cat growls (orientation away from it) in kittens at about 25 days. Kittens can be senders of diverse vocal signals from their first day of life but do not attain full receiver status until later, when auditory sensitivity and resolution have fully developed (Brown et al. 1978; Ehret 1983).

The adaptive significance of auditory perceptual performance in carnivores has been discussed only in a general context (Heffner and Heffner 1985a, 1985b). The complete vocal repertoire is well documented for all species in which a behavioral audiogram is established (domestic cat, domestic dog, raccoon, least weasel). The frequency range of acoustic signals in adults hardly goes beyond 10 kHz (with the exception of the possible occurrence of ultrasounds in some species). Only some calls of juveniles have frequency components in the ultrasonic range; however, in these calls the main energy is also below 10 kHz. Therefore, the hearing range of these four species by far exceeds the frequency range of the species' acoustic signals, which are well within the range of their best auditory sensitivity (1–16 kHz). Auditory perception by juveniles also starts to develop in this range, especially in its lower portion, where maternal vocalizations have their main energy. It has been argued that hearing in the high-frequency range in carnivores is an adaptation to the detection of small mammals, especially rodents, which have calls in the ultrasonic range. Juvenile and adult rodent ultrasonic vocalizations are in the range from 17 to 148 kHz, mainly below 80 kHz (Sales and Pye 1974), so most are within the hearing range of cats and weasels that specialize on this type of prey. At present, there is no experimental proof that hearing of these calls actually plays an important role in detection and capture of prey. High-frequency hearing sensitivity seems to be important in the perception of high components of neonate and juvenile vocalizations (Solmsen and Apfelbach 1979).

The Motivational Basis of Vocalization by Carnivores

The concept of motivation is still rather vague in ethology (Halliday 1983a), and therefore studies discussing the motivational basis of vocalization differ in

theoretical approaches. General motivational concepts were proposed by Andrew (1963), Tembrock (1971, 1975, 1977), Kiley (1972), Cohen and Fox
(1976), and Morton (1977, 1982); and Scherer (1985) integrated mammalian
models in a review of vocal affect signaling in humans. August and Anderson
(1987) tested the motivation-structural rules postulated by Morton (1977,
1982) in a large sample of mammalian acoustic signals.

Research on the motivational basis of vocalization can be grouped into five
basic questions: (1) Why does an animal vocalize at all in a given behavioral
context? (2) How context-specific are vocalizations? (3) Is there a correlation
between a vocalizing animal's motivation and basic structural characteristics
of its acoustic signals? (4) In what ways are changes in motivation reflected by
structural changes in the acoustic signals used? (5) Do vocalizations depend
only on the motivational state of the sender, or can they also refer to external
stimuli?

Some of these questions have been discussed in the literature, but there is no
holistic concept of the motivational basis of vocal behavior in mammals incorporating all of these areas. Questions (1) and (2) and to a lesser extent (5) were
discussed by Andrew (1964) for birds. His concept was adopted and detailed
for mammals by Kiley (1972). They argued that acoustic signals generally do
not convey information on the specific motivational state of the sender.
Rather, they convey the degree of interest attached by the sender to a stimulus
when there is a discrepancy between an observed and an expected pattern of
stimuli while the animal is prevented from obtaining its goal. This concept was
biased by its original formulation in a domesticated species (G. Peters 1981b).
Kiley-Worthington (1984) modified her concept of stimulus contrast (Kiley
1972) in relation to canid and felid vocalizations, which are slightly more
specific.

Questions (3) and (4) were discussed by Tembrock (1971, 1975, 1977) and
Cohen and Fox (1976). They suggested that a basic motivational and structural dichotomy exists between acoustic signals that promote approach and
those that cause increase of distance between sender and receiver. Within this
general framework Tembrock (1977) postulated basic structural parameters
for vocalizations in the behavioral contexts of friendly close-range interaction,
defensive and offensive threat at close distance, submission, and calls that
promote approach between sender and receiver over long distances.

Morton's (1977, 1982) motivation-structural rules also concern questions
(3) and (4); this is currently the most widely accepted model and refers to a
classification proposed by Collias (1960). Morton (1982:188) noted that "the
sounds used by aggressive birds and mammals are low in frequency, whereas
fearful or appeasing individuals use high-frequency sounds."

None of the motivational concepts published are sufficient to explain all
relevant phenomena described for carnivores (G. Peters 1978b, 1984b; Brady
1981; Sieber 1984). Acoustic appeasement signals of felids and ursids do not
fit into the model proposed by Tembrock (1977) because they are atonal and

have an abrupt onset (G. Peters 1984a, 1984b). Morton's (1977, 1982) motivation-structural hypothesis lumps fearful and appeasing states in the sender of a signal and therefore mixes different motivations. This is well substantiated by friendly close-range and appeasing vocalizations by fissipeds (G. Peters 1984a, 1984b; Sieber 1984) and by other mammals (August and Anderson 1987) that do not fit the structural scheme of this model.

Motivation-structural rules must incorporate all three basic dimensions (frequency, amplitude, time) in relation to the individual's motivation. Morton's (1977, 1982) model predicts that the structure of relevant vocal signals follows a code with two physical dimensions: sound quality (noisy/harsh versus tonal) and frequency range. This is a variation of two parameters within the frequency dimension. According to Morton's model, increasing aggressiveness of the sender would be encoded in increasing harshness and lower pitch of the signal. However, this also may be encoded in other ways, such as increasingly higher intensity, longer duration, or higher repetition rate, with or without change in sound quality or frequency.

Aggressive sounds described for various fissiped species (e.g., neotropical canids, Brady 1981; raccoon, Sieber 1984) fit into Morton's model. In wolves, close-approach aggressive howling is significantly deeper in pitch than howling of the same individual from a greater distance (Harrington 1987).

There is a growing literature on signals used to manipulate receivers, ways in which receivers can exploit signals a sender is emitting, and the evolution of such behavior (Wiley 1983). In a manipulating sender, this would mean that motivation is not encoded in the vocal signal and/or an exploitation of decoding mechanisms in the receivers. Selfishness of receivers can have two forms, making use of signals addressed to another receiver ("eavesdropping") or making use of any imperfection of deceit detected in the sender's signals (Wiley 1983). The encoding of aggressiveness in increasing harshness and lower pitch of the signals used (Morton 1982) is based on a deceit of the receiver by exploiting its decoding mechanisms that would correlate low pitch of a sound with the sender's size, size being an important factor in the outcome of aggressive interactions (cf. Harrington 1987).

Functional Aspects of Vocalization by Carnivores

Functional concepts of vocalization refer to the interpretation of the sender's and receiver's behavior associated with an acoustic signal and thought by the observer to be influenced by this signal in a specific way. The functions of an acoustic signal pertain to proximal causal relationships without regard to their ultimate adaptive significance in the species' evolutionary history. The interpretation of motivation in the vocalizing individual and the functional context of an acoustic signal are likely to be influenced by captive conditions, where most studies have been done.

Encoding and Decoding

In studying functional aspects of communication, one must consider the motivation of the sender and the effect of its signal on the addressee(s) and possible other receivers. W. J. Smith (1977) introduced the concepts of the message and the meaning of a signal to differentiate between these two sides from which one can view a communicatory act. Encoding of the message and the decoding of the meaning of a signal are central to an understanding of animal communication processes and the phylogeny of communication signals (see Green and Marler 1979; Slater 1983; Wiley 1983); however, encoding and decoding have not been well studied in carnivores. "Encoding" in this review is defined as the mapping of the sender's message(s) onto its vocal signal, and "decoding" as the translation process by the addressee(s) and any other receiver(s) of the signal on which the signal's meaning to them is based. The latter process incorporates more information about the sender and the behavioral context than that encoded in the respective acoustic signal.

Lactating ferrets react to playbacks of juvenile distress calls by approaching the loudspeaker, irrespective of whether these are the calls of their own young or not. Near the age of 21 days the juvenile calls have a frequency range from 0.1 to 55 kHz, with their main energy below 5 kHz. Females show the same reaction to a modified call missing all frequency components below 16 kHz, but not to calls missing all above 16 kHz (Solmsen and Apfelbach 1979). Therefore, the frequency range with the highest energy in these calls does not appear essential in eliciting the appropriate response in the addressee. This finding is a caveat that the most prominent structural characteristics of an acoustic signal should not be assumed to be those that encode its main message.

Indirect evidence indicates that various structural parameters may contribute to the encoding of messages in carnivore acoustic signals. For example, in coyotes an increase in frequency and amplitude modulation plus the addition of "yipping" is characteristic of group "yip-howling" as compared with group howling (Lehner 1978b). Experimental playbacks indicate that the latter call type primarily serves in localization of the sender and in group reunion, the former primarily in territory advertisement (Lehner 1982). These structural differences between the two call types contribute to the encoding of their different messages. Call repetition rate in ermine (*Mustela erminea*) "trilling," a friendly close-range call, is stereotyped and thus may also encode species identity (Gossow 1970; G. Peters 1984b). Individual identity of the sender of dwarf mongoose (*Helogale parvula*) "contact calls" is probably encoded in the call's pitch (Marquardt 1976; Rasa 1986). Differentiations of the same basic type are the "play" and the "moving out" calls, their messages probably being partially encoded in different repetition rates (Marquardt 1976; Maier et al. 1983). In meerkat (*Suricata suricatta*) alarm calls, the type of predator is partly encoded in call intensity (Moran 1984), whereas in dwarf mongoose alarm calls it is mainly in call duration, frequency modulation, and noisiness (Maier et al. 1983; Rasa 1983). Individual identity in mother raccoons' "chitter1" calls may be encoded in frequency characteristics and pulse repetition rate (Sieber 1986), the latter structural parameter probably also contributing to the encoding of the same message type in gurgles of serval (*Felis serval*) and the caracal (*Lynx caracal*) (G. Peters 1983). In "whistle" calls of juvenile raccoons, individual identity of the sender seems to be encoded in frequency parameters and call duration (Sieber 1986); in wolf howling the energy distribution within the calls' frequency range probably encodes the same message type (Theberge and Falls 1967; Harrington and Mech 1978b; Schassburger 1978; Filibeck et al. 1982).

The widespread occurrence of rapid rhythmical sound patterns in friendly close-contact situations in mammals (Eisenberg et al. 1975; G. Peters 1984b) indicates that this structural characteristic is important in encoding the message of these signals. Defensive threat vocalizations, such as spitting by felids or herpestids (Mulligan and Nellis 1975), often start abruptly at their full intensity, this parameter being important for the encoding of the message. The difference in the abruptness of onset may be important in the differentiation of two call types of the red fox (Tembrock 1976a). A strong correlation was found between the structural characteristics (frequency modulation, temporal parameters) of the whistles of human shepherds addressed to their herding dogs and messages of whistles that intended to stimulate or inhibit the dog's activity toward stock (McConnell and Baylis 1985). Close-approach howling by aggressive wolves is significantly deeper in pitch than howling by wolves that do not approach (Harrington 1987). There are currently no experimental data for carnivores on the decoding of the meaning(s) of conspecific acoustic signals, but structural characteristics that contribute to the encoding of the message must also play a role in the decoding process.

Message Systems and Message Types

R. Peters (1980) defined four message systems in mammal communication (including neonatal messages) and a total of 30 message types (Table 1.1). W. J. Smith (1977) differentiated between behavioral and nonbehavioral messages in animal communication, of which only the latter will be dealt with here (Table 1.2); another classification of message types was presented by Halliday (1983b).

A distinction can be made as to whether a message type is represented by a particular acoustic signal or several signals (Tables 1.1 and 1.2, column 1) or whether the message type is included additionally in one or several acoustic signals that primarily encode another message (Tables 1.1 and 1.2, column 2). An example of this distinction involves the individual identity of the sender.

This message type is not known to be represented by a particular vocalization in the fissipeds, but structural characteristics typical of the sender are present in various types of acoustic signals that represent other message types. The sex of the sender, on the other hand, may be encoded in a call peculiar to this sex (Table 1.1, column 1). Sex-specific structural parameters may also be present in various calls encoding other messages (Table 1.1, column 2).

One must realize that in most behavioral contexts in which animals vocalize, signals are used in close temporal association with other signaling modes (e.g., visual or olfactory) that also contribute to their meaning for the receiver. This classification of message types will not consider such interdependence here.

Synopsis of Acoustic Message Types

This section deals with message types as outlined by R. Peters (1980) and discusses their presence in acoustic communication of carnivores. The following survey cannot present a complete catalogue of each fissiped vocalization known to represent the respective message type but is restricted to one or a few examples in each relevant type. Neonatal message types are dealt with only as far as they require additional comments compared with the equivalent adult messages. Table 1.1 summarizes the presence of these message types in carnivore vocalization under the two different categories.

Integrative Message System

Play. Examples of species with calls to play are the ermine (Müller 1970) and the dwarf mongoose (Marquardt 1976; Maier et al. 1983; Rasa 1984). Play can also be encoded in other signals by variation in syntax, rate of emission, or other structural characteristics. A decrease in the call repetition rate by the dwarf mongoose indicates a decrease in motivation to play (Rasa 1984).

Contact. Contact calls are made by some social herpestids such as the dwarf mongoose (Marquardt 1976; Maier et al. 1983; Rasa 1986), the meerkat (Ewer 1963), and the banded mongoose (*Mungos mungo*) (Messeri et al. 1987). In many carnivores, contact calls of the mother and juveniles are especially frequent when the latter start to make their first excursions from the nest and try to follow the mother (e.g., Härtel 1975; Roeder 1984b; Sieber 1984, 1986). It is likely that contact can also be encoded as an additional message in other call types through increased or temporally stereotyped emission rate.

Affiliation. Not documented in carnivores as a specific acoustic signal, affiliation calls help to establish and maintain the affiliative bond between the individuals in a group (R. Peters 1980). They probably can be represented by signals that are used only toward certain group members or by the frequent use of a type of vocalization in this context that otherwise is rarely used, e.g., tonic communication (Schleidt 1973). In both types of acoustic affiliation messages, it may be difficult to discriminate them from acoustic signals encoding a contact message.

Assembly. Wolves aggregate in response to a "woof" or a bark when close to their den with cubs (Schassburger 1978). A message to assemble is probably encoded in communal howling of coyotes (Lehner 1978b) or wolves (Harrington and Mech 1978a, 1978b, 1979; Schassburger 1978) or the chorus roaring of lions (Schaller 1972) as an additional message to which stray members respond by joining the group.

Identity. Any signal peculiar to one species encodes species identity; furthermore, a species' identity call can be any vocalization that can be decoded only by conspecifics of the sender. The structural parameters encoding species identity are influenced by the habitat and the signals of sympatric species. Irrespective of other messages, many signals have structural characteristics that encode species identity as an additional message. However, experimental proof of which structural parameters of a fissiped acoustic signal encode species identity is not available.

Sexual identity can be encoded by signals restricted to either sex, like the "chitter2" and purring vocalizations in adult female raccoons (Sieber 1984), or additionally by differences in energy distribution within the frequency range of a call type (Jarofke 1982) or call sequence duration in lions (G. Peters 1978b). Age-specific messages are known in carnivores. Wolves can discriminate between pup and adult howls (Harrington 1986). A rank-specific identity message has not been documented for fissipeds; however, rank may be demonstrated in the role an individual takes in group vocalizations like howling by coyotes or roaring by lions. Among coyotes, the dominant individual often initiates group "yip-howls" (Lehner 1978b), and among lions the dominant male of the pride tends to start and terminate the chorus roaring with his calls (G. Peters 1978b). High-ranking wolves are joined in chorus howling more often than low-ranking individuals (Klinghammer and Laidlaw 1979). Aggressive howling is performed only by the alpha male (Harrington 1987).

Familiarization. Familiarization cannot be physically represented by an acoustic signal because it involves a signal for recognition in later encounters with the sender (R. Peters 1980).

Solicitation. There is some evidence that carnivores solicit a specific response from conspecifics by uttering a special type of vocalization. This may be what is represented by the dwarf mongoose's "moving out" call, in which the alpha female calls other individuals to start on the daily foraging tour (Maier et al. 1983; Rasa 1983, 1985). Distress calls of juveniles generally try to solicit caregiving behavior by the mother. A call of the small Indian mongoose (*Herpestes auropunctatus*) with which conspecifics are attracted to food may encode solicitation (Mulligan and Nellis 1975). The Malagasy ring-tailed mongoose (*Galidia elegans*) utters specific calls when small prey are found, which may solicit other individuals to approach (Albignac 1973). Solicitation may also be encoded as an additional message in parameters such as emission rate or intensity. This may be the case in the "chitter1" call uttered by young raccoons trying to suckle (Sieber 1984).

Alarm. Klump and Shalter's (1984) detailed classification of alarm calls should be consulted for further differentiation. Some species have generalized predator alarm calls, for example, the meerkat (Ewer 1963) or small Indian mongoose (Mulligan and Nellis 1975). However, Moran (1980, 1984) reported that the meerkat has different alarm calls for different predator types, as reported in the dwarf mongoose (Maier et al. 1983; Rasa 1983, 1985). Two types of alarm calls by the giant river otter (*Pteronura brasiliensis*) have been described, and it appears that other otters have a similar repertoire of alarm calls (Duplaix 1980, 1982). The olingo has one type of alarm call (Poglayen-Neuwall 1976a), whereas coyotes are known to have at least three different calls (Lehner 1978b). Different intensity levels of the same type of alarm call may represent different alarm messages, as in two forms of barking of the red fox (Tembrock 1976b). Lions do not appear to have specific alarm vocalizations (Schaller 1972).

Distress. Distress calls are made by all neonate and juvenile carnivores suffering from pain, hunger, cold, or isolation from mother or siblings (Ehret 1980). Adults in pain also utter such distress calls. Ewer (1963, 1973) stated that the meerkat does not have a specific vocalization when suffering from pain. Further differentiation of distress calls is necessary because they may encode different messages and accordingly the reaction of receivers will vary (Tembrock, in litt.).

Satisfaction. A satisfaction message is represented by the continuous, pulsed, low-intensity sounds like purring in felids (G. Peters 1981a), viverrids (Wemmer 1977), and procyonids (Sieber 1984) or the ursid nursing sound (Schneider 1933). Neonate mustelines (Channing and Rowe-Rowe 1977) and canids (Schassburger 1978) also have vocalizations that encode this message. The functional significance of purring by adult individuals is not yet fully clear (Leyhausen 1979; G. Peters 1981a).

Agonistic Message System

Territorial advertisement. Specific long-distance calls encoding territorial advertisement have been documented in several solitary carnivore species such as the leopard (*Panthera pardus*) (Eisenberg and Lockhart 1972) and tiger (*P. tigris*) (Schaller 1967), and in social species like the wolf (Field 1978; Har-

Table 1.1. Message systems and types in mammalian communication according to R. Peters (1980) as applied to vocalization in fissiped carnivores

System	Type	Specific acoustic signal	Included in other acoustic signal
Integrative	Play	+	+
	Contact	+	?
	Affiliation	?	?
	Assembly	+	+
	Identity, species	+	+
	Identity, sex	+	+
	Identity, age	+	+
	Identity, rank	?	+
	Familiarization	−	−
	Solicitation	+	?
	Alarm	+	?
	Distress	+	+
	Satisfaction	+	?
Agonistic	Territory advertisement	+	?
	Submission	+	?
	Defensive threat	+	?
	Offensive threat	+	?
	Dominance	?	+
	Fighting	?	?
Sexual	Male advertisement	+	+
	Female advertisement	+	+
	Courtship	?	?
	Synchronization	U	U
	Suppression	U	U
	Copulatory signal	+	?
Neonatal	Infant distress	+	+
	Infant identity, species	+	+
	Infant identity, sex	U	+
	Infant identity, age	+	+
	Infant identity, rank	?	+
	Infant affiliation	?	?
	Infant satisfaction	+	?
	Neonatal contact	?	?
	Maternal assembly	+	?
	Maternal identity	U	+
	Maternal alarm	+	?

Note. + = present, − = not present, ? = may be present, U = unlikely to be present.

rington and Mech 1978a, 1978b, 1979, 1983; Schassburger 1978) and lion (Schaller 1972).

Submission. Canid whines (Cohen and Fox 1976; Schassburger 1978) are one example of a submission vocalization.

Defensive threat. Examples of defensive threat calls are spitting of felids (Reschke 1960; Wemmer and Scow 1977) and herpestids, e.g., small Indian mongoose (Mulligan and Nellis 1975).

Offensive threat. Examples of offensive threat calls are widespread growl-like vocalizations made by a variety of carnivores. Hissing of felids encodes a mild offensive threat message (G. Peters 1983).

Dominance. Messages that encode dominance are related to the rank of an individual.

Fighting. Fighting messages have not been found to be encoded in specific carnivore acoustic signals, but they most certainly occur. Increased intensity and/or emission rate of acoustic signals may encode as an additional message that the sender is going to fight.

Sexual Message System

Male advertisement. A vocalization representing male advertisement seems to be present in a specific "rut call" in the red fox (Tembrock, in litt.). In most carnivores males that call to find a potential mate do so with signals encoding other messages; the specific message of advertisement may be encoded in increased calling rate, call intensity, or duration, and perhaps other structural parameters. There is a marked increase of vocal activity during the mating season by giant panda males (Kleiman et al. 1979; G. Peters 1982, 1985; Kleiman 1983, 1985; Schaller et al. 1985) and snow leopard males (G. Peters 1980; Rieger and Peters 1981).

Female advertisement. Some females produce specific vocal types; examples are pumas (G. Peters 1978a) and ringtails (*Bassariscus astutus*) (Willey and Richards 1981). In other taxa, female acoustic advertisement during estrus is encoded in the same signal form as in males. Females also increase vocalization rate during estrus (G. Peters 1978b; Kleiman et al. 1979; Kleiman 1983).

Courtship. Courtship messages probably exist, although they have not been demonstrated. Structural characteristics such as emission rate, regular temporal patterning, or pitch of calls may encode this type as an additional message.

Synchronization and suppression. Calls encoding synchronization or suppression are unlikely to be represented in acoustic communication of fissipeds.

Copulatory signal. In viverrids (Wemmer and Murtaugh 1981; Baumgarten 1985), herpestids (Albignac 1973), mustelines (Channing and Rowe-Rowe 1977), and procyonids (Poglayen-Neuwall 1976a, 1976b; Sieber 1984), females utter specific calls during copulation. Felid males and females also have specific vocalizations (G. Peters 1978b; Rieger and Peters 1981). It is likely that a special rhythmical emission of a signal encoding other messages includes the copulation message.

Neonatal Message System

Infant distress. Infant distress calls are made by all mammals. Slight distress felt by neonates and juveniles can be encoded in increased call intensity or rate of calls encoding other messages (Härtel 1975; Haskins 1977, 1979; Romand and Ehret 1984).

Infant identity. Juvenile females and males have not been found to have different call types during ontogeny. Their calls may, however, differ in various structural aspects, and these may change during ontogeny (Leschke 1969; Tembrock 1976b). Although dominance relationships may develop quite early in some species such as juvenile wolves (Mech 1970), there is no evidence what role acoustic signals play in this process.

Infant affiliation. See the affiliation discussion, above.

Infant satisfaction. Juvenile raccoon "churring", which is similar in structure to felid or viverrid purring, represents infant satisfaction. The continuous vibration of the body during vocalization may also be an important tactile signal, as the animals usually are in close body contact when this sound is emitted (Sieber 1984).

Neonatal contact. R. Peters (1980) does not present an explicit definition of neonatal contact calls. It can be assumed that a message of this type indicates that the young of the litter have body contact. This message may be represented by purring of felids and viverrids and equivalent sounds of other carnivores that also encode infant satisfaction. Blind young of zorillas (*Ictonyx striatus*) and white-naped weasels (*Poecilogale albinucha*) utter contact calls when the mother enters the nest (Channing and Rowe-Rowe 1977). This behavioral context probably represents a message type not specified by R. Peters (1980).

(The following maternal message types adopted from R. Peters (1980) break from the preceding types because the sender of these messages is the mother, and her young are the addressees. All other message types in the neonatal system are defined with the juveniles as senders.)

Maternal assembly. In response to the female raccoon's "chitter1" call, young aggregate and stay close to the mother (Sieber 1984). Felid mews and gurgles probably can encode this as an additional message (Härtel 1975; G. Peters 1983).

Maternal identity. An acoustic signal specific to the mother is not known in fissipeds. Her identity is encoded in various calls in individual-specific structural parameters. Domestic cat kittens at an age of 21 days respond significantly more often to calls of their mother than to calls of another cat (Härtel 1975).

Maternal alarm. The grunts of mother raccoons are an example of maternal alarm calls; the young react according to their developmental stage and their physical location at the moment the mother calls (Sieber 1984). As an additional message, maternal alarm may be encoded in increased calling rate.

Nearly all message types defined by R. Peters (1980) are represented in carnivore vocalization (Table 1.1). Moreover, there are vocalizations in these animals that seem to represent message types not specified by this author. R. Peters (1980) listed 122 signal forms distributed among 25 message types in the wolf and 43 forms and 22 types in the domestic cat. Table 1.1 shows the possible presence of 27 of the 30 defined message types in the acoustic signal repertoire of fissiped carnivores (according to R. Peters's procedure, the "identity" message type is counted as one type each in the integrative and neonatal systems).

SYNOPSIS OF ACOUSTIC NONBEHAVIORAL MESSAGE TYPES

Nonbehavioral message types found in fissiped acoustic signals are classified according to the system outlined by W. J. Smith (1977). Table 1.2 summarizes the presence or absence of Smith's message types and is analogous to Table 1.1. Some of the message types used by R. Peters (1980) were inclusive of the nonbehavioral messages originally defined by Smith, and these will not be discussed.

Population classes. Poglayen-Neuwall (1976a) believed that the subspecies message may be present in the olingo, although it is unlikely that the vocal repertoires of different subspecies would differ in the presence of certain call types that represent subspecies identity. Dialects have not been described in carnivores. Structural differences in pitch, call duration, and frequency modulation in the rutting calls of red deer subspecies (*Cervus elaphus*) (Tembrock 1965) suggest that similar differences may exist in different subspecies of geographically widespread carnivores such as the wolf, leopard, or brown bear (*Ursus arctos*). Tembrock (1965) found that in the barking sequences of the arctic and red fox, individual identity appears to be encoded in the number of calls per sequence, the duration of the barks, and their pitch. The call sequences of lions, leopards, and jaguars (*Panthera onca*) individually differ in several structural characteristics (G. Peters 1978b). Wolf pups can discriminate the howls of different adults (Shalter et al. 1977), illustrating that individually specific structural differences are registered by conspecific receivers of this signal. Individual identity is very unlikely to be encoded in a specific acoustic signal in fissipeds.

Physiological classes. W. J. Smith's (1977) message type "maturity" should be considered in R. Peters's (1980) age type, being represented by the vocalizations of mature individuals. The "breeding state" message is equivalent to R. Peters's (1980) female and male advertisement types.

Table 1.2. Message types in animal communication according to
W. J. Smith (1977) as applied to vocalization in fissiped carnivores

Type	Specific acoustic signal	Included in other acoustic signal
Identifying messages		
Population classes		
Species	+	+
Subspecies	U	?
Population	U	?
Individual	U	+
Physiological classes		
Maturity	+	+
Breeding state	+	+
Sex	+	+
Bonding classes		
Pair	+	?
Family	+	?
Troop	+	?
Location messages	U	+

Note. + = present, − = not present, ? = may be present, U = unlikely to be present.

Bonding classes. Bonding messages are represented by the individuals' joining together in the performance of communal vocalizations. This is seen with members of a pair or family, for example, golden jackals (*Canis aureus*) (Nikolskii and Poyarkov 1979), or with members of a group, like lions (Schaller 1972) or wolves (Harrington and Mech 1978a, 1978b, 1983; Schassburger 1978). It is not clear whether the specific bonding class is encoded as an additional message. This may happen in the coordination of the vocal utterances of the individuals as they join the group. The howling rate of wolves may be positively correlated with the number of adults in a pack (Harrington and Mech 1978a).

Location. Although not demonstrated in carnivores as a specific signal, the location message is present as part of other messages. Structural characteristics in vocalizations that make the sound source easy to localize define this message.

There are specific acoustic or additionally encoded messages not classifiable within the systems presented above. In the case of the dwarf mongoose, an individual on guard utters the contact call with increased intensity, thus informing the rest of the group (1) of its individual identity, (2) that it is on guard, and (3) of its location (Rasa 1986). In the system of behavioral messages defined by W. J. Smith (1977) this signal may be grouped (with one encoded message) as the "attention behavior" message because the vocalizing individual is monitoring the environment for predators. However, this call also encodes the individual identity of the guard and, for the time the animal vocalizes, it encodes that it is performing a specific role in the family group; thus, this complex message may represent a type not classified by W. J. Smith (1977) or R. Peters (1980). This may also hold for vocal signals made by the small Indian mongoose (Mulligan and Nellis 1975) and the ring-tailed mongoose (Albignac 1973) when they find small prey and for the "water call" of the banded mongoose (Messeri et al. 1987), all of which probably also include a location message. The individual-specific pitch of the adult dwarf mongooses' contact call is fixed by learning. When they become adults, individuals call at a pitch not yet occupied by others in the family group (Rasa 1985). The message encoded in the contact call's pitch in fitting into the family's pitch pattern may be classified into Smith's (1977) bonding classes (as an additional message) but could also represent a kinship message. It is likely that acoustic signals of carnivores encode more message types than have been listed thus far.

All nonbehavioral message types listed by Smith (1977) as occurring in animal communication may be represented by acoustic signals in fissiped carnivores (Table 1.2). This information, together with that in the preceding section, demonstrates that for carnivores vocalization is a highly versatile communication mode. Indeed, nearly any type of message defined in animal communication is represented by an acoustic signal of carnivores.

Ontogeny of Vocalization by Carnivores

Ehret's (1980) detailed and comprehensive review of the ontogeny of sound communication by mammals included data for domestic cat, some other felids (mainly *Panthera* species), and the domestic dog. General aspects of ontogeny of vocal communication by mammals were discussed in canid vocalization studies by Tembrock (1976a, 1976b) and Schassburger (1978).

Comprehensive statements on ontogeny are reasonable only if based on a large enough sample of individuals studied throughout their development from birth until adulthood. Neonatal and juvenile acoustic types must be defined within the same system as the acoustic signals of adults if hypotheses on the ontogenetic precursors of the adult sound forms and the schedule of the species' complete vocal repertoire are to be proposed. To a limited extent this situation is available only in a few species: wolf, domestic dog, red fox, raccoon, zorilla, ermine, least weasel, European polecat (*Mustela putorius*), white-naped weasel, puma, domestic cat, and *Panthera* species.

During the ontogenetic unfolding of a species' vocal repertoire the following basic patterns of occurrence of an acoustic signal type may be represented, irrespective of structural and/or functional changes during development: (1) those that persist throughout life; (2) those that occur later during ontogeny and then persist throughout life; (3) those that are restricted to certain juvenile developmental periods; (4) those that are present only in adults.

In their first days of life carnivores have a relatively small acoustic repertoire; in some *Mustela* species and in the domestic cat there is only a general distress call with various modifications (Gossow 1970; Härtel 1975). Neonatal zorillas, raccoons, and white-naped weasels have three basic message types represented by vocal signals: distress, contact, and satisfaction (Channing and Rowe-Rowe 1977; Sieber 1984). In some domestic dog breeds, the neonatal vocal repertoire consists of two differentiations of a call encoding the distress message and one vocal satisfaction message (Bleicher 1963). Chihuahua puppies have four different vocal signals at birth, all encoding distress (Cohen and Fox 1976). During days 1–4 German shepherd dog puppies have four basic vocalization types, one of them with two structural modifications, another with three. With the exception of the development of barking at about 14 days the repertoire does not change until about 52 days of age (Leschke 1969). The neonatal vocal repertoire of wolves comprises three distress messages and one satisfaction message (Schassburger 1978); according to Frommolt et al. (1988) there are four types of distress calls and one sound type with an unknown message. Neonatal distress calls in zorillas and white-naped weasels are replaced by a juvenile distress call at about 3 weeks of age, the latter type also not being present after about the third month of life (Channing and Rowe-Rowe 1977). The neonatal/juvenile distress call of raccoons persists until the cubs are about two and a half months old (Sieber 1984); in wolves the cry modification of the neonatal distress call system is not heard after about the first month (Schassburger 1978). The neonatal distress calls of pumas and lions are absent after about the end of the first month of life and then are replaced by other distress call types; in other *Panthera* species the neonatal/juvenile distress call type can probably persist until about six months of age (G. Peters 1978b).

Examples of vocalizations that persist from birth are whines in canids (Bleicher 1963; Cohen and Fox 1976; Schassburger 1978), whistles in raccoons (Sieber 1984), or purring in felids (G. Peters 1981a). Most adult vocal types, or their precursors, occur quite early in juveniles and undergo structural changes of various degree. The basic adult repertoire of canids is complete at about 4 weeks (Tembrock 1958, 1959b; Cohen and Fox 1976). In raccoons (Sieber 1984) and some *Mustela* species (Gossow 1970) most adult types are present by about the third month of life. The main vocal forms of adult wolves, including superimpositions and sequential combinations, develop between the third and ninth week (Schassburger 1978). In *Panthera* species major adult call types do not develop until after the first six months (G. Peters 1978b).

The neonatal and juvenile distress calls are restricted to a certain developmental period during ontogeny. Juveniles of several *Mustela* species produce different vocal types only for certain periods of their development (Gossow 1970; Solmsen and Apfelbach 1979). The purr and grunts of juvenile raccoons are heard only in cubs one and a half to three months old (and in adults only nursing females purr and grunt) (Sieber 1984). The scream of wolf pups is only found at an age of about one to two months (Schassburger 1978).

Examples of vocal types that do not develop before adulthood are the structured call sequences of the lion, the jaguar, and the leopard (G. Peters 1978b). The call types composing these sequences also develop late, some not before the second year of life. Specific vocal forms occurring during copulation in felids (G. Peters 1978b) and in female raccoons (Sieber 1984) are found only in adults. In some cases these probably have precursors in the vocal repertoire of juveniles. An exact temporal sequence of the ontogenetic unfolding of a carnivore's complete repertoire and the established derivation of adult types from their juvenile precursors has not been fully documented.

As a general rule, the ontogenetic development of the vocal repertoire is closely related to the physical and behavioral development of juveniles. In raccoons the major developmental steps in the unfolding of the vocal repertoire occur around the time when the cubs start to leave the nest and around the time of weaning (Sieber 1984). The main vocal developmental push in wolf pups is during the phase of socialization, between 3 and 12 weeks of age when pups are gradually weaned and social relationships are established between littermates (Schassburger 1978; Frommolt et al. 1988). During ontogeny the relative importance of the different communicatory channels changes, as well as rates of signaling and frequencies of the occurrence of specific types (Roeder 1984b).

Romand and Ehret (1984) correlate structural changes in kitten vocalizations with the growth of the sound-producing and -modifying apparatus, but this interdependence has not yet been quantified. As a general rule, fissipeds do not attain full adult hearing range until about the end of the first month of life (Ehret 1983). However, hearing starts earlier in the frequency range of signals between the mother and littermates. Specific responses of domestic cat kittens to mother and littermate calls begin at about 3 weeks (Härtel 1975; Olmstead and Villablanca 1980).

The domestic cat mother is especially responsive to juvenile distress calls for about 30 days after the birth of the kittens (Haskins 1977). Experimental data for ferrets (Solmsen and Apfelbach 1979) and raccoons (Sieber 1986) indicate that nursing females generally react to neonatal distress calls by approaching the sound source. The other main message type of neonates, satisfaction, is also addressed to the mother, informing her that the young ones are well and no immediate care-giving behavior is necessary.

Learning has not been shown to play a major role in the unfolding of the vocal repertoire of any fissiped. Numerous involuntary experiments of isolated hand-rearing by humans or nonconspecific nurse rearing of wild and domestic carnivores did not result in atypical acoustic signals. The only controlled experiment on the role of learning was carried out by Romand and Ehret (1984) on the domestic cat. They studied early ontogeny of certain call types made by kittens growing up under normal conditions and compared these with deafened kittens and kittens reared in isolation. The calls of kittens of the latter two groups differed in various quantitative parameters from those of normal kittens. The structural changes in deafened kittens document that feedback through the auditory system is necessary for full normal call development

(Romand and Ehret 1984). Indirect evidence for the genetic basis of call structure in certain vocalizations is available in *Panthera* hybrids (Tembrock 1977; G. Peters 1978b). The structured call sequences of these hybrids are intermediate in some parameters between the two parental species. However, certain aspects show similarity to one or the other species (G. Peters 1978b).

General structural changes of acoustic signals that persist during most of juvenile development (and throughout life) differ quantitatively in vocal (tonal, atonal) and nonvocal signals (with the exception of possible tonal, whistle-like sounds). Only tonal forms show considerable structural changes during ontogeny. These are much less pronounced in atonal vocal signals like purring of felids (G. Peters 1981a) and nonvocal ones like the snort of raccoons (Sieber 1984). Tonal signals, and the tonal component of mixed signals, with increasing age undergo a decrease in fundamental frequency, upper frequency limit, frequency range, and harmonic with highest intensity (Harrington and Mech 1978b; G. Peters 1978b; Solmsen and Apfelbach 1979; Romand and Ehret 1984; Sieber 1984). The frequencies with highest amplitude increase for some time after birth and only then decrease during further development (G. Peters 1978b; Romand and Ehret 1984); this is known in other mammals, for example, *Peromyscus maniculatus* (Hart and King 1966). In human baby cries the fundamental frequency shows a similar ontogenetic change (Kent 1976). During the ontogeny of kittens, further call parameters like frequency and amplitude modulation, duration, amplitude, and general structural variability also change (Romand and Ehret 1984). In acoustic signals that are performed in regular temporal sequences, the temporal emission pattern develops later than the call types themselves, as in the Mustelinae (Gossow 1970), Canidae (Tembrock 1976a, 1976b), or Felidae (G. Peters 1978b).

Vocalizations that persist over a considerable period may change the message encoded during ontogeny or in adults as compared with the message they encode in neonates or juveniles. The juvenile contact calls of the zorilla and the white-naped weasel in their adult derivation encode submission and, in another modified version, greeting (Channing and Rowe-Rowe 1977). R. Peters (1980) did not list a "greeting" message type, but it may be included in types such as affiliation or solicitation. The juvenile distress cry in wolves probably is the progenitor of various different signals in adults, including the howl, which can encode several messages (Schassburger 1978). In general, the vocal repertoire of adult carnivores is richer in acoustic signal types than that of juveniles and so is able to encode more message types. However, the precursors of many adult signals already occur early in juvenile development.

Phylogenetic Aspects of Vocalization by Carnivores

To a limited extent, a comprehensive data base is now available that can be used for intra- and interfamilial comparisons for all fissiped families except the Ursidae. However, because of different classification systems adopted for the acoustic types, repertoire size sometimes is difficult to compare. Within the Canidae it appears quite uniform, with about 10 to 12 basic types (Cohen and Fox 1976; Tembrock 1976a, 1976b; Schassburger 1978). Thirteen types are described in the raccoon (Sieber 1984), 15 in the domestic cat (McKinley 1981), and 12 in the Lutrinae (Duplaix 1982).

Within families there can be clear differences as to the presence of certain types of acoustic signals; usually, though, different types of vocalizations that are functionally equivalent replace each other in the repertoires of the different taxa. In the Canidae there is a difference in vocalization between the vulpine group and the canine group (Tembrock 1976a, 1976b; Schassburger 1978), and among felids, some *Panthera* species clearly differ in certain vocal forms from the other species (G. Peters 1978b, 1983). The giant panda has several vocal types not present in other ursids (G. Peters 1982, 1985; Schaller et al. 1985). Close contact calls of *Lutra* species differ from those of the other Lutrinae (Duplaix 1980). There are no signal types equivalent to spotted hyenas' "lowing" and "whooping" in the brown hyena (*Hyaena brunnea*) (Mills 1981; Henschel 1986).

Signals that carnivores use in interspecific communication are common to the repertoires of sender and receiver and encode an equivalent message in both. This is true of the widespread agonistic sounds like growling, hissing, and spitting, which have a similar basic structure in most carnivores. Another form of interspecific acoustic communication involving dwarf mongooses and hornbills (*Tockus* spp.) was described by Rasa (1983). The mongooses react to hornbill predator alarm calls that are given in response to predators that prey on the mongooses but not on them.

Various authors (e.g., Cohen and Fox 1976; Schassburger 1978; Kiley-Worthington 1984) have proposed hypotheses as to the influence of the following criteria on size and structure of vocal repertoires in carnivores (and mammals in general): (1) social structure, (2) habitat, (3) activity pattern (diurnal versus nocturnal), and (4) duration of dependence of young on parental care in a species. These hypotheses are that (generally compared between closely related species): (1) social species have a richer and more complex vocal repertoire than solitary species; (2) species living in forest habitats have a more diverse vocal repertoire than species of open habitats; (3) nocturnal species have a more complex vocal repertoire than diurnal species; and (4) species with a long dependence of the juveniles on parental care have a more complex vocal communication between young and parents (mother). On the basis of the carnivore vocalization data available, only preliminary statements as to the soundness of these hypotheses are possible.

Repertoire size is quite uniform in all carnivores irrespective of the species' social system. There is one established exception: the social spotted hyena's acoustic repertoire is larger than that of the solitary brown hyena (Mills 1981; Henschel 1986). A basic conceptual aspect in the study of animal acoustic communication is the graded versus discrete structure of the repertoire, in which structural and functional aspects must be differentiated from each other (Klingholz and Meynhardt 1979). Carnivore species differ quantitatively in this respect, but in general their acoustical system comprises discrete signals and graded portions. A classification system based on sonographic analysis is not automatically equivalent to the communication "potential" available in it to sender and receiver. There is some evidence that the vocalization of arboreal viverrid species differs from that of terrestrial species (Wemmer 1977). Wemmer argues that loud long-distance calls have likely evolved in arboreal forms because predation risk is less in trees than on the ground. Valid evidence for structural differences in vocal forms between forest and savannah species is lacking, with the exception of that for the high-frequency long-distance calls in some arboreal viverrids (Wemmer 1977). Differences in acoustic communication between closely related species with a diurnal versus a nocturnal mode of life have not been documented. Nearly all carnivores have altricial young, but it is not known in which way the duration of the juvenile dependence on parental care influences the complexity of acoustic communication during this period.

Summary

Fissipeds can produce a wide range of vocal and nonvocal acoustic signals. The process of sound generation and modification in the various vocalizations is not well understood. Irrespective of the species' size, most signals in adults are restricted to the frequency range below 8 kHz. The existence of pure ultrasonic sounds made by adult fissipeds and of ultrasonic components in audible signals is not well established. The hearing range of the carnivore species studied exceeds the main frequency range of conspecific vocalizations by at least three octaves. Current models of the motivational basis of acoustic signaling behavior and motivation-structural correlations in their vocalizations are not sufficient to explain all relevant aspects of fissiped behavior. In many species a fundamental structural dichotomy exists between vocalizations during friendly approach or close contact between sender and addressee and those that result in withdrawal. Nearly all message types defined in animal communication behavior are represented in fissiped vocalization, documenting the functional diversity of this signaling mode. Still lacking is an understanding of the way acoustic signals function with the other communication modes with which they are associated in given behavioral contexts.

The ontogeny of vocalization by carnivores proceeds in accordance with the general mammalian pattern. Neonatal and juvenile acoustic signals occupy a higher (and wider) frequency range than those of adult conspecifics. The acoustic repertoire of neonates and young juveniles is more restricted in number of signal types and messages encoded than is that of adults. During further juvenile development the repertoire unfolds through structural and functional

Table 1.3. References on vocalizations by fissiped carnivores

Taxon	Comprehensive[a]	Limited[a]	Nontechnical
Herpestidae[b]			Dücker 1965
Galidiinae			
Galidia elegans		Albignac 1973 G. Peters 1984b	
Mungotictis de-cemlineata		Albignac 1973	
Herpestinae			
Herpestes auro-punctatus	Mulligan and Nellis 1975		
H. ichneumon			Dücker 1960
H. sanguineus		Baker 1982	Jacobsen 1982
Mungotinae			
Cynicitis pen-icillata			Earlé 1981
Helogale par-vula	Marquardt 1976[c] Maier et al. 1983	Rasa 1984[c], 1986	Rasa 1983, 1985[c]
Mungos mungo	Messeri et al. 1987[c]		
Suricata suricat-ta	Moran 1980	Moran 1984	Dücker 1962[c] Ewer 1963
Viverridae	Wemmer 1977[c]		Dücker 1965
Cryptoproctinae			
Cryptoprocta ferox		Albignac 1973 G. Peters 1984b	Vosseler 1929
Fossa fossa		Albignac 1973 Wemmer 1977[c]	
Paradoxurinae			
Arctictis bin-turong		Wemmer and Murtaugh 1981	Huf 1965
Paguma larvata		Wemmer 1977	
Paradoxurus her-maphroditus		Baumgarten 1985	Wemmer 1977
Viverrinae			
Civettictis civet-ta		Wemmer 1977	Ewer and Wemmer 1974
Genetta genetta	Roeder 1984[b]		Gangloff and Ropartz 1972; Roeder 1984b
G. tigrina	Wemmer 1977[c]		
Nandinia bi-notata		Wemmer 1977	Dücker 1971
Viverra zibetha		G. Peters 1984b	
Viverricula indi-ca		Wemmer 1977[c]	
Felidae	Reschke 1960	Reschke 1966; G. Peters 1978a[c], 1981a[c], 1984a, 1984b	Tembrock 1962, 1970; Wemmer and Scow 1977
Felinae			
Felis aurata		G. Peters 1984b	
F. concolor		G. Peters 1978b[c], 1981a; Movchan and Opahova 1981	

Table 1.3. (*Continued*)

Taxon	Comprehensive[a]	Limited[a]	Nontechnical
F. margarita		G. Peters 1983[c]	
F. serval		G. Peters 1984b	
F. silvestris f. catus	McKinley 1981; Kiley-Worthington 1984	Denis 1969[c]; Härtel 1972[c], 1975[c]; Haskins 1977[c], 1979[c]; Brown et al. 1978[c]; G. Peters 1981a[c], 1983[c]; Romand and Ehret 1984[c]	Moelk 1944[c], 1979[c]
F. wiedii			Petersen 1979
F. yagouaroundi		G. Peters 1984b	Hulley 1976[c]
Lynx caracal	G. Peters 1983[c]		
L. lynx	G. Peters 1987[c]		
L. rufus		G. Peters 1981a[c], 1987[c]	
Pantherinae			
Neofelis nebulosa		G. Peters 1978b[c], 1984a[c], 1984b[c]	
Panthera leo		Schaller 1972[c]; G. Peters 1978b[c]; Movchan and Opahova 1981; Jarofke 1982	
P. onca		G. Peters 1978b[c], 1984a, 1984b; Movchan and Opahova 1981	
P. pardus		G. Peters 1978b[c]; Movchan and Opahova 1981	Schaller 1972
P. tigris		G. Peters 1978b[c], 1984a, 1984b; Movchan and Opahova 1981	Schaller 1967, 1972
P. uncia	G. Peters 1980	G. Peters 1978b[c], 1984a; Rieger and Peters 1981	
Incertae sedis			
Acinonyx jubatus		Schaller 1972; Movchan and Opahova 1981; G. Peters 1984b	Eaton 1974; Frame and Frame 1981
Hyaenidae			
Crocuta crocuta	Henschel 1986	Schaller 1972	Kruuk 1972
Hyaena brunnea		Mills 1981	Owens and Owens 1978; Mills 1982
H. hyaena		G. Peters 1984b	Rieger 1981
Proteles cristatus			Kingdon 1977
Ursidae		G. Peters 1978a, 1984b	Meyer-Holzapfel 1957[c]; Pruitt and Burghardt 1977

(*continued*)

Table 1.3. (*Continued*)

Taxon	Comprehensive[a]	Limited[a]	Nontechnical
Ailuropoda melanoleuca	G. Peters 1985; Schaller et al. 1985	G. Peters 1982, 1984b	Kleiman 1983
Melursus ursinus		Tembrock 1975	Laurie and Seidensticker 1977
Tremarctos ornatus		G. Peters 1978a, 1984b	Eck 1969
Ursus americanus	Jordan 1979[c]		Jordan 1976
U. arctos ssp.		G. Peters 1984b	Couturier 1954
U. maritimus		Wemmer et al. 1976; G. Peters 1978a, 1984b	Schneider 1933[c]
U. thibetanus		G. Peters 1978a	
Incertae sedis			
Ailurus fulgens	Roberts 1981[c]	G. Peters 1984b	Roberts 1975[c]; Roberts and Gittleman 1984
Canidae	Cohen and Fox 1976[c]; Tembrock 1976a[c], 1976b[c]; Fox and Cohen 1977[c]; Brady 1981[c]	Tembrock 1960a, 1961; Lehner 1978b	Tembrock 1970
Alopex lagopus		Cohen and Fox 1976[c]; Tembrock 1960a, 1976a, 1976b	
Canis aureus		Cohen and Fox 1976[c]; Tembrock 1976a, 1976b; Nikolskii and Poyarkov 1979[c]	Seitz 1959[c]
C. latrans	Lehner 1978a, 1978b	McCarley 1975; Cohen and Fox 1976[c]; McCarley and Carley 1976; Tembrock 1976a, 1976b; Bekoff 1978[c]; Laundré 1981; Lehner 1982	
C. lupus	Field 1978[c]; Harrington and Mech 1978b[c]; Schassburger 1978[c]	Theberge and Falls 1967; Cohen and Fox 1976[c]; Tembrock 1976a, 1976b; Fox and Cohen 1977; Shalter et al. 1977[c]; Field 1979; Harrington and Mech 1979, 1982; Filibeck et al. 1982;	Klinghammer and Laidlaw 1979; Nikolskii and Frommolt 1986

Table 1.3. (*Continued*)

Taxon	Comprehensive[a]	Limited[a]	Nontechnical
		Harrington 1986, 1987; Nikolskii et al. 1987; Frommolt et al. 1988	
C. lupus f. familiaris	Bleicher 1963[c]; Leschke 1969[c]	Cohen and Fox 1976[c]; Tembrock 1976a[c], 1976b[c]; Fox and Cohen 1977[c]	
C. rufus		McCarley and Carley 1976; McCarley 1978	
Chrysocyon brachyurus	Brady 1981[c]	Cohen and Fox 1976; Tembrock 1976a, 1976b	
Cuon alpinus		Tembrock 1976a, 1976b	Johnsingh 1982
Dusicyon culpaeus		Cohen and Fox 1976; Tembrock 1976a, 1976b	
D. thous	Brady 1981[c]	Tembrock 1976a, 1976b	
Lycaon pictus		Schaller 1972; Cohen and Fox 1976; Tembrock 1976a[c], 1976b[c]	Frame and Frame 1981
Nyctereutes procyonoides		Tembrock 1976b	Seitz 1955
Otocyon megalotis		Tembrock 1976a, 1976b; Lamprecht 1979[c]	Nel and Bester 1983[c]
Speothos venaticus	Brady 1981[c]	Cohen and Fox 1976; Drüwa 1976; Tembrock 1976b	
Urocyon cinereoargenteus		Cohen and Fox 1976[c]; Tembrock 1976b	
Vulpes chama		Tembrock 1960a, 1976a, 1976b	
V. corsac		Tembrock 1976a, 1976b	
V. macrotis			Egoscue 1962
V. vulpes		Tembrock 1958[c], 1959a[c], 1959b, 1960a, 1960b, 1961, 1963b, 1965, 1976a[c], 1976b[c]; Cohen and Fox 1976[c]	
V. zerda			Gauthier-Pilters 1962; Koenig 1970[c]

continued)

Table 1.3. (*Continued*)

Taxon	Comprehensive[a]	Limited[a]	Nontechnical
Procyonidae			
Bassaricyon sp.		Poglayen-Neuwall 1976a[c]	Poglayen-Neuwall and Poglayen-Neuwall 1965
Bassariscus astutus	Willey and Richards 1981[c]	Bailey 1974	Toweill and Toweill 1978[c]; Poglayen-Neuwall and Poglayen-Neuwall 1980
B. sumichrasti			Poglayen-Neuwall 1973
Nasua nasua			Kaufmann 1962; H. J. Smith 1980
Potos flavus		Poglayen-Neuwall 1976b[c]	Poglayen-Neuwall 1962
Procyon cancrivorus			Löhmer 1976[c]
P. lotor	Sieber 1984[c]	G. Peters 1984b; Sieber 1986[c]	
Mustelidae			
Lutrinae	Duplaix 1982[c]		
Aonyx cinerea		Duplaix 1982	
Enhydra lutris		Sandegren et al. 1973[c]; Konstantinov et al. 1980	
Lutra longicaudis		Duplaix 1982	
L. lutra	Rogoschik 1987[c]	Duplaix 1982; Scheffler 1985[c]	Goethe 1964
L. perspicillata		Duplaix 1982	
L. sumatrana		Duplaix 1982	
Pteronura brasiliensis	Duplaix 1980[c], 1982[c]	G. Peters 1984b	
Mephitinae			
Mephitis mephitis			Verts 1967
Melinae			
Meles meles			Goethe 1964; Neal 1977
Mustelinae	Gossow 1970[c]; Farley et al. 1987		Goethe 1964
Eira barbara		Poglayen-Neuwall and Poglayen-Neuwall 1976[c]	Poglayen-Neuwall 1975, 1978[c]
Ictonyx striatus	Channing and Rowe-Rowe 1977[c]		
Martes americana		Belan et al. 1978	
M. foina		Gossow 1970	
Mustela erminea	Gossow 1970[c]		Müller 1970[c]
M. eversmanni	Farley et al. 1987		

Table 1.3. (*Continued*)

Taxon	Comprehensive[a]	Limited[a]	Nontechnical
M. frenata		Svendsen 1976	
M. nivalis	Gossow 1970[c]	Huff and Price 1968[c]; Heidt and Huff 1970[c]	Heidt et al. 1968[c]
M. putorius	Gossow 1970[c]		
M. putorius f. furo		Solmsen 1978[c]; Solmsen and Apfelbach 1979[c]	
M. vison		Gilbert 1969	
Poelicogale albinucha	Channing and Rowe-Rowe 1977[c]		

[a]Comprehensive = nearly complete vocal repertoire; Limited = limited to a few vocalizations or limited in data presented.

[b]Order of taxa follows Wozencraft (this volume, appendix). Papers listed under a family or subfamily deal with three or more species of the respective taxon.

[c]Data on ontogeny of vocalizations included.

changes and the fission of existing forms and the appearance of new forms to encode additional more specific and new messages. Some acoustic signals are restricted to certain developmental periods. The important steps in the unfolding of a species' sound repertoire occur at the same time as decisive phases, such as when the juveniles start to leave the nest or at around the time of weaning. Hearing by juveniles, during the first week in the few species studied, is restricted to the lower frequency range, which in the mother's vocalizations has the highest amplitude. Learning is not known to play an important role in the unfolding of the acoustic repertoire of fissipeds or in the appropriate contextual use of specific signals.

Phylogenetic aspects of vocalization are not well known for fissipeds. Neither the adaptive significance of structural characteristics of certain acoustic signals, nor the composition and structure of a species' whole repertoire has been established. Acoustic signal repertoire size seems to be quite uniform for most fissipeds studied, but this finding is not equivalent to the communicatory potential available in it to sender and receivers.

Acknowledgments

We are grateful to L. R. Heaney, R. S. Hoffmann, and E. S. Morton for their careful reading of the manuscript and helpful comments. Peters gratefully acknowledges grants from Deutsche Forschungsgemeinschaft (Le 37/21, Le 37/23) and the Max-Planck-Gesellschaft, by which his early bioacoustic research in fissipeds was supported. For intermittent support during later parts of this research, he would like to thank the Smithsonian Institution and Friends of the National Zoo for grants kindly provided by D. G. Kleiman, and the New

York Zoological Society for a grant by G. B. Schaller. He is deeply indebted to his institute, Zoologisches Forschungsinstitut und Museum Alexander Koenig, for providing the setting for this research. Support for Wozencraft was provided by the National Museum of Natural History, Smithsonian Institution. Both authors gratefully acknowledge the kind permission given by various colleagues to cite unpublished data.

References

Albignac, R. 1973. *Mammifères carnivores*. Faune de Madagascar, vol. 36. Paris: O.R.S.T.O.M. and C.N.R.S.

Andrew, R. J. 1963. The origin and evolution of the calls and facial expressions of the primates. *Behaviour* 20:1–109.

Andrew, R. J. 1964. Vocalizations in chicks and the concept of "stimulus contrast." *Anim. Behav.* 12:64–76.

August, P. V., and Anderson, J. G. T. 1987. Mammal sounds and motivation-structural rules: A test of the hypothesis. *J. Mamm.* 68:1–9.

Bailey, E. P. 1974. Notes on the development, mating behavior, and vocalization of captive ringtails. *Southwestern Nat.* 19:117–119.

Baker, C. M. 1982. Methods of communication exhibited by captive slender mongooses *Herpestes sanguineus*. *South African J. Zool.* 17:143–146.

Balph, D. M., and Balph, D. F. 1966. Sound communication of Uinta ground squirrels. *J. Mamm.* 47:440–450.

Baumgarten, L. 1985. Beobachtungen zum Fortpflanzungsverhalten des Palmenrollers, *Paradoxurus hermaphroditus* (Pallas, 1777), im Zoo Halle. *Zool. Garten (N.F.)* 55:29–38.

Bekoff, M. 1978. Behavioral development in coyotes and Eastern coyotes. In: M. Bekoff, ed. *Coyotes: Biology, Behavior, and Management*, pp. 97–126. New York: Academic Press.

Belan, I., Lehner, P. N., and Clark, T. 1978. Vocalizations of the American pine marten, *Martes americana*. *J. Mamm.* 59:871–874.

Bleicher, N. 1963. Physical and behavioral analysis of dog vocalizations. *Amer. J. Vet. Res.* 24:415–427.

Brady, C. A. 1981. The vocal repertoires of the bush dog (*Speothos venaticus*), crab-eating fox (*Cerdocyon thous*), and maned wolf (*Chrysocyon brachyurus*). *Anim. Behav.* 29:649–669.

Brown, K. A., Buchwald, J. S., Johnson, J. R., and Mikolich, D. J. 1978. Vocalization in the cat and kitten. *Devel. Psychobiol.* 11:559–570.

Casseday, J. H., and Neff, W. D. 1973. Localization of pure tones. *J. Acoust. Soc. Amer.* 54:365–372.

Channing, A., and Rowe-Rowe, D. T. 1977. Vocalizations of South African mustelines. *Z. Tierpsychol.* 44:283–293.

Cohen, J. A., and Fox, M. W. 1976. Vocalizations of wild canids and possible effects of domestication. *Behav. Processes* 1:77–92.

Collias, N. E. 1960. An ecological and functional classification of animal sounds. In: W. E. Lanyon & W. N. Tavolga, eds. *Animal Sounds and Communication*, pp. 368–391. Washington, D.C.: American Institute of Biological Sciences, Publication no. 7.

Couturier, M. A. J. 1954. *L'ours brun* (Ursus arctos L.). Grenoble: published by the author.

Davis, D. D. 1964. The giant panda—A morphological study of evolutionary mechanisms. *Fieldiana Zool. Mem.* 3:1–339.

Denis, B. 1969. *Contribution à l'étude du ronronnement chez le chat domestique* (Felis catus L.) *et chez le chat sauvage* (Felis silvestris S.). *Aspects morpho-fonctionnels, acoustiques et éthologiques.* Alfort, France: published by the author.

Drüwa, P. 1976. Beobachtungen zum Verhalten des Waldhundes (*Speothos venaticus,* Lund 1842) in der Gefangenschaft. Ph.D. dissert., Univ. Bonn.

Dücker, G. 1960. Beobachtungen über das Paarungsverhalten des Ichneumons (*Herpestes ichneumon* L.). *Z. Säugetierk.* 25:47–51.

Dücker, G. 1962. Brutpflegeverhalten und Ontogenese des Verhaltens bei Surikaten (*Suricata suricatta* Schreb., Viverridae). *Behaviour* 19:305–340.

Dücker, G. 1965. Das Verhalten der Viverriden. *Handb. Zool.* 8 (10), 20a:1–48.

Dücker, G. 1971. Gefangenschaftsbeobachtungen an Pardelrollern *Nandinia binotata* (Reinwardt). *Z. Tierpsychol.* 28:77–89.

Duplaix, N. 1980. Observations on the ecology and behavior of the giant otter *Pteronura brasiliensis* in Suriname. *Rev. Ecol. (Terre Vie)* 34:495–620.

Duplaix, N. 1982. Contribution à l'écologie et à l'éthologie de *Pteronura brasiliensis* Gmelin 1788 (Carnivora, Lutrinae): implications évolutives. Ph.D. dissert., Univ. Paris.-Sud.

Earlé, R. A. 1981. Aspects of social and feeding behaviour of the yellow mongoose, *Cynictis penicillata* (G. Cuvier). *Mammalia* 45:143–152.

Eaton, R. L. 1974. *The Cheetah: The Biology, Ecology, and Behavior of an Endangered Species.* New York: Van Nostrand Reinhold.

Eck, S. 1969. Über das Verhalten eines im Dresdner Zoologischen Garten aufgezogenen Brillenbären (*Tremarctos ornatus* (Cuv.)). *Zool. Garten (N.F.)* 37:81–92.

Egoscue, H. J. 1962. Ecology and life history of the kit fox in Tooele County, Utah. *Ecology* 43:481–497.

Ehret, G. 1977. Comparative psychoacoustics: Perspectives of peripheral sound analysis in mammals. *Naturwissenschaften* 64:461–470.

Ehret, G. 1980. Development of sound communication in mammals. In: J. S. Rosenblatt, R. A. Hinde, C. Beer & M.-C. Busnel, eds. *Advances in the Study of Behavior,* 11:179–225. New York: Academic Press.

Ehret, G. 1983. Development of hearing and response behavior to sound stimuli: Behavioral studies. In: R. Romand, ed. *Development of the Auditory and Vestibular Systems,* pp. 211–237. New York: Academic Press.

Ehret, G., and Romand, R. 1981. Postnatal development of absolute auditory thresholds in kittens. *J. Comp. Physiol. Psychol.* 95:304–311.

Eisenberg, J. F. 1974. The function and motivational basis of hystricomorph vocalizations. *Symp. Zool. Soc. London* 34:211–247.

Eisenberg, J. F. 1981. *The Mammalian Radiations: An Analysis of Trends in Evolution, Adaptation, and Behavior.* Chicago: Univ. Chicago Press.

Eisenberg, J. F., Collins, L. R., and Wemmer, C. 1975. Communication in the Tasmanian devil (*Sarcophilus harrisii*) and a survey of auditory communication in the Marsupialia. *Z. Tierpsychol.* 37:379–399.

Eisenberg, J. F., and Lockhart, M. 1972. An ecological reconnaissance of Wilpattu National Park, Ceylon. *Smithsonian Contrib. Zool.* 101. 118 pp.

Elliott, D. N., Stein, L., and Harrison, M. J. 1960. Determination of absolute-intensity thresholds and frequency-difference thresholds in cats. *J. Acoust. Soc. Amer.* 32:380–384.

Ewer, R. F. 1963. The behaviour of the meerkat, *Suricata suricatta* (Schreber). *Z. Tierpsychol.* 20:570–607.

Ewer, R. F. 1973. *The Carnivores.* London: Weidenfeld and Nicolson.

Ewer, R. F., and Wemmer, C. 1974. The behaviour in captivity of the African civet (*Civettictis civetta*). *Z. Tierpsychol.* 34:359–394.

Farley, S. D., Lehner, P. N., Clark, T., and Trost, C. 1987. Vocalizations of the Siberian ferret (*Mustela eversmanni*) and comparisons with other mustelids. *J. Mamm.* 68:413–416.

Field, R. 1978. Vocal behavior of wolves (*Canis lupus*): Variability in structure, context, annual/diurnal patterns and ontogeny. Ph.D. dissert., Johns Hopkins Univ., Baltimore.

Field, R. 1979. A perspective on syntactics of wolf vocalizations. In: E. Klinghammer, ed. *The Behavior and Ecology of Wolves*, pp. 182–205. New York: Garland STPM Press.

Filibeck, U., Nicoli, M., Rossi, P., and Boscagli, G. 1982. Detection by frequency analyzer of individual wolves howling in a chorus: A preliminary report. *Boll. Zool.* 49:151–154.

Foss, I., and Flottorp, G. 1974. A comparative study of the development of hearing and vision in various species commonly used in experiments. *Acta Oto-lar.* 77:202–214.

Fox, M. W. 1978. *The Dog: Its Domestication and Behavior*. New York: Garland STPM Press.

Fox, M. W., and Cohen, J. A. 1977. Canid communication. In: T. A. Sebeok, ed. *How Animals Communicate*, pp. 728–748. Bloomington: Indiana Univ. Press.

Frame, G., and Frame, L. 1981. *Swift and Enduring: Cheetahs and Wild Dogs of the Serengeti*. New York: E. P. Dutton.

Frommolt, K.-H., Kaal, M. I., Paschina, N. M., and Nikolskii, A. A. 1988. Die Entwicklung der Lautgebung beim Wolf (*Canis lupus* L., Canidae L.) während der postnatalen Ontogenese. *Zool. Jahrb. Physiol.* 92:105–115.

Gangloff, B., and Ropartz, P. 1972. Le répertoire comportemental de la Genette *Genetta genetta* (Linné). *Terre Vie* 26:489–560.

Gauthier-Pilters, H. 1962. Beobachtungen an Feneks (*Fennecus zerda* Zimm.). *Z. Tierpsychol.* 19:440–464.

Gerhardt, H. C. 1983. Communication and the environment. In: T. R. Halliday & P. J. B. Slater, eds. *Animal Behaviour*, vol. 2: *Communication*, pp. 82–113. Oxford: Blackwell.

Gerken, G. M., and Sandlin, D. 1977. Auditory reaction time and absolute threshold in cat. *J. Acoust. Soc. Amer.* 61:602–607.

Gilbert, F. F. 1969. Analysis of basic vocalizations of the ranch mink. *J. Mamm.* 50:625–627.

Goethe, F. 1964. Das Verhalten der Musteliden. *Handb. Zool.* 8 (10), 19:1–80.

Gossow, H. 1970. Vergleichende Verhaltensstudien an Marderartigen. I. Über Lautäußerungen und zum Beuteverhalten. *Z. Tierpsychol.* 27:405–480.

Gould, E. 1983. Mechanisms of mammalian auditory communication. In: J. F. Eisenberg & D. G. Kleiman, eds. *Advances in the Study of Mammalian Behavior*, pp. 265–342. Special Publication no. 7. Lawrence, Kans.: American Society of Mammalogists.

Gourevitch, G. 1980. Directional hearing in terrestrial mammals. In: A. N. Popper & R. R. Fay, eds. *Comparative Studies of Hearing in Vertebrates*, pp. 357–373. New York: Springer Verlag.

Green, S., and Marler, P. 1979. The analysis of animal communication. In: P. Marier & J. G. Vandenbergh, eds. *Handbook of Behavioral Neurobiology.* 3:73–158. New York: Plenum Press.

Halliday, T. R. 1983a. Motivation. In: T. R. Halliday & P. J. B. Slater, eds. *Animal Behaviour*, vol. 1: *Causes and Effects*, pp. 100–133. Oxford: Blackwell.

Halliday, T. R. 1983b. Information and communication. In: T. R. Halliday & P. J. B. Slater, eds. *Animal Behaviour*, vol. 2: *Communication*, pp. 43–81. Oxford: Blackwell.

Harrington, F. H. 1986. Timber wolf howling playback studies: Discrimination of pup from adult howls. *Anim. Behav.* 34:1575–1577.

Harrington, F. H. 1987. Aggressive howling in wolves. *Anim. Behav.* 35:7–12.
Harrington, F. H., and Mech, L. D. 1978a. Howling at two Minnesota wolf pack summer homesites. *Canadian J. Zool.* 56:2024–2028.
Harrington, F. H. and Mech, L. D. 1978b. Wolf vocalizations. In: R. L. Hall & H. S. Sharp, eds. *Wolf and Man: Evolution in Parallel,* pp. 109–132. New York: Academic Press.
Harrington, F. H., and Mech, L. D. 1979. Wolf howling and its role in territory maintenance. *Behaviour* 68:207–249.
Harrington, F. H., and Mech, L. D. 1982. An analysis of howling response parameters useful for wolf pack censusing. *J. Wildl. Mgmt.* 46:686–693.
Harrington, F. H., and Mech, L. D. 1983. Wolf spacing: Howling as a territory-independent spacing mechanism in a territorial population. *Behav. Ecol. Sociobiol.* 12:161–168.
Hart, F. M., and King, J. A. 1966. Distress vocalizations of young in two subspecies of *Peromyscus maniculatus. J. Mamm.* 47:287–293.
Härtel, R. 1972. Frequenzspektrum und akustische Kommunikation der Hauskatze. *Wissenschaftliche Zeitschrift, Humboldt-Universität, Berlin, Mathematisch-Naturwissenschaftliche Reihe* 21:371–374.
Härtel, R. 1975. Zur Struktur und Funktion akustischer Signale im Pflegesystem der Hauskatze (*Felis catus* L.). *Biol. Zentralblatt* 94:187–204.
Haskins, R. 1977. Effect of kitten vocalizations on maternal behavior. *J. Comp. Physiol. Psychol.* 91:830–838.
Haskins, R. 1979. A causal analysis of kitten vocalizations: An observational and experimental study. *Anim. Behav.* 27:726–736.
Hast, M. H. 1986. The larynx of roaring and non-roaring cats. (Abstract.) *J. Anat.* 149:221–222.
Heffner, H. E. 1983. Hearing in large and small dogs: Absolute thresholds and size of tympanic membrane. *Behav. Neurosci.* 97:310–318.
Heffner, R. S., and Heffner, H. E. 1985a. Hearing in mammals: The least weasel. *J. Mamm.* 66:745–755.
Heffner, R. S., and Heffner, H. E. 1985b. Hearing range of the domestic cat. *Hearing Res.* 19:85–88.
Heidt, G. A., and Huff, J. N. 1970. Ontogeny of vocalization in the least weasel. *J. Mamm.* 51:385–386.
Heidt, G. A., Petersen, M. K., and Kirkland, G. L., Jr. 1968. Mating behavior and development of least weasels (*Mustela nivalis*) in captivity. *J. Mamm.* 49:413–419.
Henschel, J. R. 1986. The socio-ecology of a spotted hyaena (*Crocuta crocuta*) clan in the Kruger National Park. D.Sc. dissert., Univ. Pretoria.
Honacki, J. H., Kinman, K. E., and Koeppl, J. W., eds. 1982. *Mammal Species of the World: A Taxonomic and Geographic Reference.* Lawrence, Kan.: Allen Press and Association of Systematic Collections.
Huf, K. 1965. Über das Verhalten des Binturong (*Arctictis binturong* Raffl.). *Portugaliae Acta Biologica* 9:249–304.
Huff, J. N., and Price, E. O. 1968. Vocalizations of the least weasel, *Mustela nivalis. J. Mamm.* 49:548–550.
Hulley, J. T. 1976. Maintenance and breeding of captive jaguarundis, *Felis yagouaroundi,* at Chester Zoo and Toronto. *Internat. Zoo Yearb.* 16:120–122.
Hutterer, R., and Vogel, P. 1977. Abwehrlaute afrikanischer Spitzmäuse der Gattung *Crocidura* Wagler, 1832 und ihre systematische Bedeutung. *Bonner zoologische Beiträge* 28:218–227.
Isley, T. E., and Gysel, L. W. 1975. Sound localization in the red fox. *J. Mamm.* 56:397–404.

Jacobsen, N. H. G. 1982. Observations on the behaviour of slender mongooses *Herpestes sanguineus* in captivity. *Säugetierk. Mitt.* 30:168–183.

Jarofke, D. 1982. Messungen der Lautstärke des Chorgebrülls der Löwen (*Panthera leo*) im Zoo Berlin. *Bongo* 6:73–78.

Johnsingh, A. J. T. 1982. Reproductive and social behaviour of the dhole, *Cuon alpinus* (Canidae). *J. Zool. (Lond.)* 198:443–463.

Jordan, R. H. 1976. Threat behavior of the black bear (*Ursus americanus*). In: M. R. Pelton, J. W. Lentfer & G. E. Folk, eds. *Bears: Their Biology and Management*, pp. 57–63. Morges: IUCN Publications, new series no. 40.

Jordan, R. H. 1979. An observational study of the American black bear *(Ursus americanus)*. Ph.D. dissert., Univ. Tennessee, Knoxville.

Kanai, T., and Wang, S. C. 1962. Localization of the central vocalization mechanism in the brainstem of the cat. *Exp. Neurol.* 6:426–434.

Kaufmann, J. H. 1962. Ecology and social behavior of the coati, *Nasua narica*, on Barro Colorado Island, Panama. *Univ. California Publ. Zool.* 60:95–222.

Kelemen, G. 1963. Comparative anatomy and performance of the vocal organ in vertebrates. In: R.-G. Busnel, ed. *Acoustic Behaviour of Animals*, pp. 489–521. Amsterdam: Elsevier.

Kent, R. D. 1976. Anatomical and neuromuscular maturation of the speech mechanism: Evidence from acoustic studies. *J. Speech Hear. Res.* 19:421–447.

Kiley, M. 1972. The vocalizations of ungulates, their causation and function. *Z. Tierpsychol.* 31:171–222.

Kiley-Worthington, M. 1984. Animal language? Vocal communication of some ungulates, canids and felids. *Acta Zool. Fennica* 171:83–88.

Kingdon, J. 1977. *East African Mammals: An Atlas of Evolution in Africa*, vol. 3A. New York: Academic Press.

Kleiman, D. G. 1983. Ethology and reproduction of captive giant pandas (*Ailuropoda melanoleuca*). *Z. Tierpsychol.* 62:1–46.

Kleiman, D. G. 1985. Social and reproductive behavior of the giant panda *(Ailuropoda melanoleuca)*. *Bongo* 10:45–58.

Kleiman, D. G., Karesh, W. B., and Chu, P. R. 1979. Behavioural changes associated with oestrus in giant panda (*Ailuropoda melanoleuca*) with comments on female proceptive behaviour. *Internat. Zoo Yearb.* 19:217–223.

Klinghammer, E., and Laidlaw, L. 1979. Analysis of 23 months of daily howl records in a captive grey wolf pack (*Canis lupus*). In: E. Klinghammer, ed. *The Behavior and Ecology of Wolves*, pp. 153–181. New York: Garland STPM Press.

Klingholz, F., and Meynhardt, H. 1979. Lautinventare der Säugetiere—diskret oder kontinuierlich? *Z. Tierpsychol.* 50:250–264.

Klump, G. M., and Shalter, M. D. 1984. Acoustic behaviour of birds and mammals in the predator context. I. Factors affecting the structure of alarm signals. II. The functional significance and evolution of alarm signals. *Z. Tierpsychol.* 66:189–226.

Koenig, L. 1970. Zur Fortpflanzung und Jugendentwicklung des Wüstenfuchses (*Fennecus zerda* Zimm. 1780). *Z. Tierpsychol.* 27:205–246.

Konstantinov, A. I., Makarov, A. K., and Sokolov, B. V. 1980. [Acoustic signalisation of some species of pinnipeds (Otariidae, Phocidae) and the sea otter (*Enhydra lutris*) in open air] (English summary). *Zool. Zh.* 59:1397–1408.

Kruuk, H. 1972. *The Spotted Hyena: A Study of Predation and Social Behavior*. Chicago: Univ. Chicago Press.

Lamprecht, J. 1979. Field observations on the behaviour and social system of the bat-eared fox, *Otocyon megalotis* Desmarest. *Z. Tierpsychol.* 49:260–284.

Laundré, J. W. 1981. Temporal variation in coyote vocalization rates. *J. Wildl. Mgmt.* 45:767–769.

Laurie, A., and Seidensticker, J. 1977. Behavioural ecology of the sloth bear (*Melursus ursinus*). *J. Zool. (Lond.)* 182:187–204.

Lehner, P. N. 1978a. Coyote communication. In: M. Bekoff, ed. *Coyotes: Biology, Behavior, and Management,* pp. 127–162. New York: Academic Press.

Lehner, P. N. 1978b. Coyote vocalizations: A lexicon and comparisons with other canids. *Anim. Behav.* 26:712–722.

Lehner, P. N. 1982. Differential vocal response of coyotes to "group howl" and "group yip-howl" playbacks. *J. Mamm.* 63:675–679.

Leschke, M. 1969. Die Lautentwicklung bei Schäferhund-Welpen. Ph.D. dissert., Humboldt-Univ., Berlin.

Leyhausen, P. 1979. *Katzen, eine Verhaltenskunde.* Berlin: Paul Parey.

Löhmer, R. 1976. Zur Verhaltensontogenese bei *Procyon cancrivorus cancrivorus* (Procyonidae). *Z. Säugetierk.* 41:42–58.

McCarley, H. 1975. Long-distance vocalizations of coyotes (*Canis latrans*). *J. Mamm.* 56:847–856.

McCarley, H. 1978. Vocalizations of red wolves (*Canis rufus*). *J. Mamm.* 59:27–35.

McCarley, H., and Carley, C. J. 1976. *Canis latrans* and *Canis rufus* vocalization: A continuum. *Southwestern Nat.* 21:399–400.

McConnell, P. B., and Baylis, J. R. 1985. Interspecific communication in cooperative herding: Acoustic and visual signals from human shepherds and herding dogs. *Z. Tierpsychol.* 67:302–328.

McKinley, P. E. 1981. Cluster analysis of the domestic cat vocal repertoire. Ph.D. dissert., Univ. Maryland, College Park.

Maier, V., Rasa, O. A. E., and Scheich, H. 1983. Call-system similarity in a ground-living social bird and a mammal in the bush habitat. *Behav. Ecol. Sociobiol.* 12:5–9.

Marquardt, F.-R. 1976. Zur akustischen Kommunikation beim Zwergmungo (*Helogale undulata rufula*). Diplomarbeit, Univ. Marburg.

Mech, L. D. 1970. *The Wolf: The Ecology and Behavior of an Endangered Species.* New York: Natural History Press.

Messeri, P., Masi, E., Piazza, R., and Dessi'-Fulgheri, F. 1987. A study of the vocal repertoire of the banded mongoose (*Mungos mungo*). *Monitore Zoologico Italiano (N.S.) Suppl.* 22:341–373.

Meyer-Holzapfel, M. 1957. Das Verhalten der Bären (Ursiden). *Handb. Zool.* 8 (10), 17:1–28.

Mills, M. G. L. 1981. The socio-ecology and social behaviour of the brown hyaena *Hyaena brunnea* Thunberg, 1820 in the southern Kalahari. D.Sc. thesis, Univ. Pretoria.

Mills, M. G. L. 1982. *Hyaena brunnea. Mammalian Species* no. 194:1–5. Lawrence, Kans.: American Society of Mammalogists.

Moelk, M. 1944. Vocalizing in the house-cat: A phonetic and functional study. *Amer. J. Psychol.* 57:184–205.

Moelk, M. 1979. The development of friendly approach behavior in the cat: A study of kitten-mother relations and the cognitive development of the kitten from birth to eight weeks. In: J. S. Rosenblatt, R. A. Hinde, C. Beer & M.-C. Busnel, eds. *Advances in the Study of Behavior,* 10:163–224. New York: Academic Press.

Molina, A. F. de, and Hunsperger, R. W. 1962. Organization of the subcortical system concerning defence and flight reactions in the cat. *J. Physiol.* 160:200–213.

Moran, G. 1980. Behavioural studies of the meerkat (*Suricata suricatta*): An illustration of an alternative to "on-site" research in zoos. *Annual Proc. Amer. Assoc. Zoo Vet.* 119–128.

Moran, G. 1984. Vigilance behaviour and alarm calls in a captive group of meerkats, *Suricata suricatta. Z. Tierpsychol.* 65:228–240.

Morton, E. S. 1975. Ecological sources of selection on avian sounds. *Amer. Nat.* 109:17–34.

Morton, E. S. 1977. On the occurrence and significance of motivation-structural rules in some bird and mammal sounds. *Amer. Nat.* 111:855–869.

Morton, E. S. 1982. Grading, discreteness, redundancy, and motivation-structural rules. In: D. E. Kroodsma & E. H. Miller, eds. *Acoustic Communication in Birds,* vol. 1: *Production, Perception, and Design Features of Sounds,* pp. 183–212. New York: Academic Press.

Movchan, V. N., and Opahova, V. R. 1981. [Acoustic signals of cats (Felidae) living in the zoo] (English summary). *Zool. Zh.* 60:601–608.

Müller, H. 1970. Beiträge zur Biologie des Hermelins (*Mustela erminea* Linné, 1758). *Säugetierk. Mitt.* 18:293–380.

Müller-Preuss, P., and Ploog, D. 1983. Central control of sound production in mammals. In: B. Lewis, ed. *Bioacoustics: A Comparative Approach,* pp. 125–146. New York: Academic Press.

Mulligan, B. E., and Nellis, D. W. 1975. Vocal repertoire of the mongoose *Herpestes auropunctatus. Behaviour* 55:237–267.

Neal, E. G. 1977. *Badgers.* Poole, Dorset: Blanford Press.

Neff, W. D., and Hind, J. E. 1955. Auditory thresholds of the cat. *J. Acoust. Soc. Amer.* 27:480–483.

Negus, V. E. 1949. *The Comparative Anatomy and Physiology of the Larynx.* New York: Grune & Stratton.

Nel, J. A. J., and Bester, M. H. 1983. Communication in the southern bat-eared fox *Otocyon m. megalotis* (Desmarest, 1822). *Z. Säugetierk.* 48:277–290.

Nikolskii, A. A., and Frommolt, K.-H. 1986. [Sound activity of wolves during their breeding period] (English summary). *Zool. Zh.* 65:1589–1591.

Nikolskii, A. A., Frommolt, K.-H., Schniebs, K., and Wuntke, B. 1987. Rhythmen der Lautaktivität des Wolfes. In: J. Schuh, R. Gattermann & J. A. Romanov, eds. *Chronobiologie-Chronomedizin: Vorträge d. III. DDR-UdSSR-Symposiums Chronobiologie-Chronomedizin* (Halle/Saale, 1986), pp. 315–319. Halle: Kongreß- und Tagungsberichte der Martin-Luther-Universität Halle-Wittenberg.

Nikolskii, A. A., and Poyarkov, A. D. 1979. [Merging of individual characters in the group howling of jackals] (English summary). *Zh. Obshch. Biol.* 40:785–788.

Olmstead, C. E., and Villablanca, J. R. 1980. Development of behavioral audition in the kitten. *Physiol. Behav.* 24:705–712.

Owens, D. D., and Owens, M. J. 1978. Feeding ecology and its influence on social organization in brown hyaenas (*Hyaena brunnea*) in the central Kalahari desert. *East African Wildl. J.* 16:113–136.

Peters, G. 1978a. Einige Beobachtungen zur Lautgebung der Bären—Bioakustische Untersuchungen im zoologischen Garten. *Z. Kölner Zoo* 21:45–51.

Peters, G. 1978b. Vergleichende Untersuchung zur Lautgebung einiger Feliden (Mammalia, Felidae). *Spixiana, Suppl.* 1. 206 pp.

Peters, G. 1980. The vocal repertoire of the snow leopard (*Uncia uncia,* Schreber 1775). *International Pedigree Book of Snow Leopards,* 2:137–158. Helsinki: Helsinki Zoo.

Peters, G. 1981a. Das Schnurren der Katzen (Felidae). *Säugetierk. Mitt.* 29:30–37.

Peters, G. 1981b. Einige Anmerkungen zu domestikationsbedingten Veränderungen im Lautgebungsverhalten von Säugetieren. *Bonner zoologische Beiträge* 32:91–101.

Peters, G. 1982. A note on the vocal behaviour of the giant panda, *Ailuropoda melanoleuca* (David, 1869). *Z. Säugetierk.* 47:236–246.

Peters, G. 1983. Beobachtungen zum Lautgebungsverhalten des Karakal, *Caracal caracal* (Schreber, 1776) (Mammalia, Carnivora, Felidae). *Bonner zoologische Beiträge* 34:107–127.

Peters, G. 1984a. A special type of vocalization in the Felidae. *Acta Zool. Fennica* 171:89–92.

Peters, G. 1984b. On the structure of friendly close range vocalizations in terrestrial carnivores (Mammalia: Carnivora: Fissipedia). *Z. Säugetierk.* 49:157–182.

Peters, G. 1985. A comparative survey of vocalization in the giant panda (*Ailuropoda melanoleuca*, David, 1869). *Bongo* 10:197–208.

Peters, G. 1987. Acoustic communication in the genus *Lynx* (Mammalia: Felidae)—comparative survey and phylogenetic interpretation. *Bonner zoologische Beiträge* 38:315–330.

Peters, R. 1980. *Mammalian Communication: A Behavioral Analysis of Meaning.* Monterey, Cal.: Brooks/Cole.

Petersen, M. K. 1979. Behavior of the margay. *Carnivore* 2(2):69–76.

Peterson, F. A., Heaton, W. C., and Wruble, S. D. 1969. Levels of auditory response in fissiped carnivores. *J. Mamm.* 50:566–578.

Pocock, R. I. 1916. On the hyoidean apparatus of the lion (*F. leo*) and related species of the Felidae. *Ann. Mag. Nat. Hist.*, 8th ser., 18:222–229.

Pocock, R. I. 1917. The classification of existing Felidae. *Ann. Mag. Nat. Hist.*, 8th ser., 20:329–350.

Poglayen-Neuwall, I. 1962. Beiträge zu einem Ethogramm des Wickelbären (*Potos flavus* Schreber). *Z. Säugetierk.* 27:1–44.

Poglayen-Neuwall, I. 1973. Preliminary notes on maintenance and behaviour of the Central American cacomistle (*Bassariscus sumichrasti*) at Louisville Zoo. *Internat. Zoo Yearb.* 13: 207–211.

Poglayen-Neuwall, I. 1975. Copulatory behaviour, gestation and parturition of the tayra (*Eira barbara* L. 1758). *Z. Säugetierk.* 40:176–189.

Poglayen-Neuwall, I. 1976a. Fortpflanzung, Geburt und Aufzucht, nebst anderen Beobachtungen von Makibären (*Bassaricyon* Allen, 1876). *Zool. Beitr. (N.F.)* 22:179–233.

Poglayen-Neuwall, I. 1976b. Zur Fortpflanzungsbiologie und Jugendentwicklung von *Potos flavus* (Schreber, 1774). *Zool. Garten (N.F.)* 46:237–285.

Poglayen-Neuwall, I. 1978. Breeding, rearing and notes on the behaviour of tayras, *Eira barbara. Internat. Zoo Yearb.* 18:134–140.

Poglayen-Neuwall, I., and Poglayen-Neuwall, I. 1965. Gefangenschaftsbeobachtungen an Makibären (*Bassaricyon* Allen, 1876). *Z. Säugetierk.* 30:321–366.

Poglayen-Neuwall, I., and Poglayen-Neuwall, I. 1976. Postnatal development of tayras (Carnivora: *Eira barbara* L., 1758). *Zool. Beitr. (N.F.)* 22:345–405.

Poglayen-Neuwall, I., and Poglayen-Neuwall, I. 1980. Gestation period and parturition of the ringtail *Bassariscus astutus* (Liechtenstein, 1830). *Z. Säugetierk.* 45:73–81.

Pruitt, C. H., and Burghardt, G. M. 1977. Communication in terrestrial carnivores: Mustelidae, Procyonidae, and Ursidae. In: T. A. Sebeok, ed. *How Animals Communicate*, pp. 767–793. Bloomington: Indiana Univ. Press.

Rasa, O. A. E. 1983. Dwarf mongoose and hornbill mutualism in the Taru Desert, Kenya. *Behav. Ecol. Sociobiol.* 12:181–190.

Rasa, O. A. E. 1984. A motivational analysis of object play in juvenile dwarf mongooses (*Helogale undulata rufula*). *Anim. Behav.* 32:579–589.

Rasa, O. A. E. 1985. *Mongoose Watch: A Family Observed.* London: John Murray.

Rasa, O. A. E. 1986. Coordinated vigilance in dwarf mongoose family groups: The "watchman's song" hypothesis and the costs of guarding. *Ethology (Z. Tierpsychol.)* 71:340–344.

Remmers, J. E., and Gautier, H. 1972. Neural and mechanical mechanisms of feline purring. *Respiration Physiol.* 16:351–361.

Reschke, B. 1960. Untersuchungen zur Lautgebung der Feliden. Diplomarbeit, Humboldt-Univ., Berlin.

54   *Gustav Peters and W. Chris Wozencraft*

Reschke, B. 1966. Vergleichende Untersuchungen an Lautfolgen der Pantherinen. *Wiss. Z. Karl-Marx-Univ, Leipzig* 15:499–505.

Rieger, I. 1981. *Hyaena hyaena. Mammalian Species* no. 150:1–5. Lawrence, Kans.: American Society of Mammalogists.

Rieger, I., and Peters, G. 1981. Einige Beobachtungen zum Paarungs- und Lautgebungsverhalten von Irbissen (*Uncia uncia*) im Zoologischen Garten. *Z. Säugetierk.* 46:35–48.

Roberts, M. S. 1975. Growth and development of mother-reared red pandas, *Ailurus fulgens. Internat. Zoo Yearb.* 15:57–63.

Roberts, M. S. 1981. The reproductive biology of the red panda, *Ailurus fulgens,* in captivity. M.Sc. thesis, Univ. Maryland, College Park.

Roberts, M. S., and Gittleman, J. L. 1984. *Ailurus fulgens. Mammalian Species* no. 222:1–8. Lawrence, Kans.: American Society of Mammalogists.

Roeder, J.-J. 1984a. Contribution à l'étude de modalités de communication chez la genette: *Genetta genetta* L. Evolution ontogénétique et phylogénétique des systèmes de communication. Ph.D. dissert., Univ. Louis Pasteur, Strasbourg.

Roeder, J.-J. 1984b. Ontogenèse des systèmes de communication chez la genette (*Genetta genetta* L.). *Behaviour* 90:259–301.

Rogoschik, B. 1987. Vokalisation des Europäischen Fischotters (*Lutra lutra* Linné 1758) und Untersuchungen zur Lautentwicklung. Diplomarbeit, Univ. Göttingen.

Romand, R., and Ehret, G. 1984. Development of sound production in normal, isolated, and deafened kittens during the first postnatal months. *Devel. Psychobiol.* 17:629–649.

Sales, G., and Pye, D. 1974. *Ultrasonic Communication by Animals.* London: Chapman and Hall.

Sandegren, F. E., Chu, E. W., and Vandevere, J. E. 1973. Maternal behavior in the California sea otter. *J. Mamm.* 54:668–679.

Schaller, G. B. 1967. *The Deer and the Tiger: A Study of Wildlife in India.* Chicago: Univ. Chicago Press.

Schaller, G. B. 1972. *The Serengeti Lion: A Study of Predator-Prey Relations.* Chicago: Univ. Chicago Press.

Schaller, G. B., Hu Jinchu, Pan Wenshi, and Zhu Jing. 1985. *The Giant Pandas of Wolong.* Chicago: Univ. Chicago Press.

Schassburger, R. M. 1978. The vocal repertoire of the wolf: Structure, function, and ontogeny. Ph.D. dissert., Cornell Univ., Ithaca, N.Y.

Scheffler, E. 1985. Beobachtungen zum Verhalten des Europäischen Fischotters (*Lutra lutra* L. 1758) unter besonderer Berücksichtigung seiner Postembryonalentwicklung—Beobachtungen aus dem Alpenzoo Innsbruck. Diplomarbeit, Univ. Innsbruck.

Scherer, K. R. 1985. Vocal affect signalling: A comparative approach. In: J. S. Rosenblatt, C. Beer, M.-C. Busnel & P. J. B. Slater, eds. *Advances in the Study of Behavior,* 15:189–244. New York: Academic Press.

Schleidt, W. M. 1973. Tonic communication: Continual effects of discrete signs in animal communication systems. *J. Theor. Biol.* 42:359–386.

Schneider, K. M. 1933. Zur Jugendentwicklung eines Eisbären. II. Aus dem Verhalten: Lage, Bewegung, Saugen, stimmliche Äußerung. *Zool. Garten (N.F.)* 6:224–237.

Schneider, R. 1964. Der Larynx der Säugetiere. *Handb. Zool.* 5(7), 35:1–128.

Seitz, A. 1955. Untersuchungen über angeborene Verhaltensweisen bei Caniden. III. Beobachtungen an Marderhunden (*Nyctereutes procyonoides* Gray). *Z. Tierpsychol.* 12:463–489.

Seitz, A. 1959. Beobachtungen an handaufgezogenen Goldschakalen (*Canis aureus algirensis* Wagner 1843). *Z. Tierpsychol.* 16:747–771.

Shalter, M. D., Fentress, J. C., and Young, G. W. 1977. Determinants of response of wolf pups to auditory signals. *Behaviour* 60:98–114.

Sieber, O. J. 1984. Vocal communication in raccoons (*Procyon lotor*). *Behaviour* 90:80-113.

Sieber, O. J. 1986. Acoustic recognition between mother and cubs in raccoons (*Procyon lotor*). *Behaviour* 96:130–163.

Skultety, F. M. 1962. Experimental mutism in dog. *Arch. Neurol.* 6:235–241.

Skultety, F. M. 1965. Mutism in cats with rostral midbrain lesions. *Arch. Neurol.* 12:211–225.

Slater, P. J. B. 1983. The study of communication. In: T. R. Halliday & P. J. B. Slater, eds. *Animal Behaviour*, vol. 2: *Communication*. pp. 9–42. Oxford: Blackwell.

Smith, H. J. 1980. Behavior of the coati (*Nasua narica*) in captivity. *Carnivore* 3:88–136.

Smith, W. J. 1977. *The Behavior of Communicating: An Ethological Approach*. Cambridge: Harvard Univ. Press.

Solmsen, E. 1978. Untersuchungen zur Entwicklung und Bedeutung der Laute nestjunger Frettchen (*Mustela putorius* f. *furo*, Carnivora). Diplomarbeit, Univ. Tübingen.

Solmsen, E., and Apfelbach, R. 1979. Brutpflegewirksame Komponenten im Weinen neonater Frettchen (*Mustela putorius* f. *furo* L.). *Z. Tierpsychol.* 50:337–347.

Svendsen, G. E. 1976. Vocalizations of the long-tailed weasel (*Mustela frenata*) *J. Mamm.* 57:398–399.

Tembrock, G. 1958. Lautentwicklung beim Fuchs—sichtbar gemacht. *Umschau* 58:566–568.

Tembrock, G. 1959a. Beobachtungen zur Fuchsranz unter besonderer Berücksichtigung der Lautgebung. *Z. Tierpsychol.* 16:351–368.

Tembrock, G. 1959b. Zur Entwicklung rhythmischer Lautformen bei *Vulpes*. *Zool. Anzeiger. Suppl.* 22:194–202.

Tembrock, G. 1960a. Homologie-Forschung an Caniden-Lauten. *Zool. Anzeiger Suppl.* 23:320–326.

Tembrock, G. 1960b. Spezifische Lautformen beim Rotfuchs (*Vulpes vulpes*) und ihre Beziehungen zum Verhalten. *Säugetierk. Mitt.* 8:150–154.

Tembrock, G. 1961. Lautforschung an *Vulpes* und anderen Caniden. *Zool. Anzeiger Suppl.* 24:482–487.

Tembrock, G. 1962. Methoden der vergleichenden Lautforschung. In: J. Kratochvil & J. Pelikan, eds. *Symposium Theriologicum*. Proc. Internat. Symp. on Methods of Mammalogical Investigation, Brno 1960, pp. 329–338. Praha: Czechoslovak Acad. of Sciences.

Tembrock, G. 1963a. Acoustic behaviour of mammals. In: R.-G. Busnel, ed. *Acoustic Behaviour of Animals*, pp. 751–788. Amsterdam: Elsevier.

Tembrock, G. 1963b. Mischlaute beim Rotfuchs (*Vulpes vulpes* L.). *Z. Tierpsychol.* 20:616–623.

Tembrock, G. 1965. Untersuchungen zur intraspezifischen Variabilität von Lautäußerungen bei Säugetieren. *Z. Säugetierk.* 30:257–273.

Tembrock, G. 1968. Communication in land mammals. In: T. A. Sebeok, ed. *Animal Communication*, pp. 338–404. Bloomington: Indiana Univ. Press.

Tembrock, G. 1970. Bioakustische Untersuchungen an Säugetieren des Berliner Tierparkes. *Milu* 3:78–96.

Tembrock, G. 1971. *Biokommunikation: Informationsübertragung im biologischen Bereich*. Berlin (DDR): Akademie-Verlag.

Tembrock, G. 1975. Die Erforschung des tierlichen Stimmausdrucks (Bioakustik). In: F. Trojan, ed. *Biophonetik*, pp. 51–68. Mannheim: Bibliographisches Institut.

Tembrock, G. 1976a. Canid vocalizations. *Behav. Processes* 1:57–75.

Tembrock, G. 1976b. Die Lautgebung der Caniden—eine vergleichende Untersuchung. *Milu* 4:1–44.

Tembrock, G. 1977. *Tierstimmenforschung: Eine Einführung in die Bioakustik*, 2nd ed. Neue Brehm-Bücherei Bd. 250. Wittenberg Lutherstadt: A. Ziemsen Verlag.

Theberge, J. B., and Falls, J. B. 1967. Howling as a means of communication in wolves. *Amer. Zool.* 7:331–338.

Toweill, D. E., and Toweill, D. B. 1978. Growth and development of captive ringtails (*Bassariscus astutus flavus*). *Carnivore* 1(3):46–53.

Verts, B. J. 1967. *The Biology of the Striped Skunk*. Urbana: Univ. Illinois Press.

Vosseler, J. 1929. Beitrag zur Kenntnis der Fossa (*Cryptoprocta ferox* Benn.) und ihrer Fortpflanzung. *Zool. Garten (N.F.)* 2:1–9.

Wallschläger, D. 1981. Der Einfluß von Kanalparametern auf phonetische und syntaktische Eigenschaften akustischer Signale. *Nova Acta Leopoldina N.F.* 54(245):231–238.

Watson, C. S. 1963. Masking of tones by noise for the cat. *J. Acoust. Soc. Amer.* 35:167–172.

Wemmer, C. 1977. Comparative ethology of the large-spotted genet (*Genetta tigrina*) and some related viverrids. *Smithsonian Contrib. Zool.* 239. 93 pp.

Wemmer, C., Ebers, M. von, and Scow, K. 1976. An analysis of the chuffing vocalization in the polar bear (*Ursus maritimus*). *J. Zool. (Lond.)* 180:425–439.

Wemmer, C., and Murtaugh, J. 1981. Copulatory behavior and reproduction in the binturong, *Arctictis binturong. J. Mamm.* 62:342–352.

Wemmer, C., and Scow, K. 1977. Communication in the Felidae with emphasis on scent marking and contact patterns. In: T. A. Sebeok, ed. *How Animals Communicate*, pp. 749–766. Bloomington: Indiana Univ. Press.

Wiley, R. H. 1983. The evolution of communication: information and manipulation. In: T. R. Halliday & P. J. B. Slater, eds. *Animal Behaviour*, vol. 2: *Communication*, pp. 156–189. Oxford: Blackwell.

Wiley, R. H., and Richards, D. G. 1982. Adaptations for acoustic communication in birds: Sound transmission and signal detection. In: D. E. Kroodsma & E. H. Miller, eds. *Acoustic Communication in Birds*, vol. 1: *Production, Perception, and Design Features of Sounds*, pp. 131–181. New York: Academic Press.

Willey, R. B., and Richards, R. E. 1981. Vocalizations in the ringtail (*Bassariscus astutus*). *Southwestern Nat.* 26:23–30.

Wollack, C. H. 1965. Auditory thresholds in the raccoon (*Procyon lotor*). *J. Aud. Res.* 5:139–144.

CHAPTER 2

The Role of Odor in the Social Lives of Carnivores

MARTYN L. GORMAN AND
BEVERLEY J. TROWBRIDGE

Carnivores are complex creatures living complex social lives in which order is maintained by the transmission of information between individuals. Sometimes the signals are passed visually, sometimes by sound, and very often by odor.

Olfactory communication has a number of advantages over other forms of signaling. It can be used when visual or auditory signals are difficult to detect, for example at night, under the ground, or in dense vegetation. Odors can be deposited in the environment as scent marks and thus provide a spatial and historical record of an individual's movement and behavior. As signals, scent marks have the important property of remaining active for long periods, even in the absence of their producer.

The odors used as signals by mammals are not equivalent to the pheromones of lower animals. Mammalian odors are usually complex mixtures, not simple chemicals, and responses to them are not stereotyped but depend upon context, prior experience, and developmental status (Beauchamp et al. 1976). Brown (1979) suggested, therefore, that the term "social odor" would be more appropriate for mammalian chemical signals.

Carnivores are profligate in their use of social odors; they are equipped with a dazzling variety of odoriferous organs, and they make full use of the olfactory opportunities presented by their urine and feces (Gorman 1980; Macdonald 1980).

Sources of Social Odors

Odorous chemicals may be compounds derived from the diet, molecules synthesized by the animal itself, or the products of bacterial metabolism; many specialized scent organs are warm, moist, and anaerobic and provide ideal conditions for the proliferation of bacteria.

57

Figure 2.1. A brown hyena pasting (scent marking) a stem of grass with its anal pouch. (From Mills et al. 1980, courtesy of M. G. L. Mills.)

Urine and Feces

When urine and feces are used as scent marks, one is faced with the difficult problem of distinguishing between excretion and communication. One distinction is that signaling with urine and feces usually, but by no means always, involves small, token volumes placed at specific, and often prominent, objects that are reanointed frequently (Kleiman 1966; Macdonald 1985). Such token marking is common in all the carnivore families, with the exception, perhaps, of the Hyaenidae. In many species, including the hyenas, large quantities of feces can accumulate at discrete sites, known as latrines, over long periods.

Skin Glands

Many carnivores have evolved elaborate organs whose function is the production, storage, modification, and dissemination of odorous chemicals (Schaffer 1940; Quay 1977). Although diverse in structure, they are all derived from skin glands, whose primary roles are the maintenance of the pelage and thermoregulation. Basically, there are two types of glands involved, flask-shaped sebaceous glands and tubular sudoriferous glands. Scent organs con-

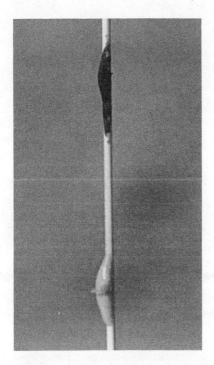

Figure 2.2. The scent mark left by a brown hyena consists of two distinct components. One is white, lipid-rich, and of sebaceous origin; the other is black, watery, and derived from apocrine glands.

taining sebaceous glands produce oily secretions that are long lasting, releasing their volatiles only slowly, and that are used to mark objects and conspecifics. Sudoriferous glands, in contrast, produce watery volatile secretions. Scents derived from them are generally involved in short-term signaling and may be applied to objects or released directly into the air. Many specialized scent-producing organs contain both types of glands, indicating complex functions.

A most striking example is the anal pouch of the brown hyena (*Hyaena brunnea*) (Mills et al. 1980). Brown hyenas normally live in small groups, the members of which share and defend a large territory, although they forage alone (see also Mills, this volume). As they move through their territory they pause two or three times in every kilometer and scent mark grass stems with their pouch (Figure 2.1). The paste they leave on the grass consists of two distinct components, a white fatty deposit produced by sebaceous glands and above it a smear of black watery, apocrine secretion (Figure 2.2). The white paste remains detectable for several weeks, even to the human nose, and Mills et al. argue that it functions as a signal to potential intruders that the area is already occupied. In contrast, the black apocrine paste is thought to convey information within the social group. Brown hyenas usually feed on small items that are only slowly replaced (Mills 1978). It is important, therefore, that each knows where other hyenas have foraged in the recent past and so avoids unproductive areas. The black paste loses its odor within a few hours and may indicate how long it has been since a hyena passed that way.

Scent-producing Organs

Carnivores display an array of scent-producing organs that range in complexity from simple, local increases in the density and size of skin glands to anatomically complex structures. In general, scent glands are associated with the face, the tail, the perineum, and in particular the anal region. The structures involved are briefly described here; for a fuller account the reader is referred to Schaffer (1940) and Macdonald (1985).

THE ANAL REGION

In most species the skin immediately surrounding the anus is richly endowed with a scattering of sebaceous and apocrine *anal glands*. There is no unequivocal evidence that these function in olfactory communication, but the attraction they engender in conspecifics suggests that they do.

Anal sacs are paired reservoirs lying lateral to the rectum and emptying by ducts within the anus, onto the anal skin, or into an anal pouch. There is much variation in the details of structure, but basically an anal sac is an invagination of the skin into which apocrine or sebaceous glands, or both, secrete. These secretions accumulate in the sac, together with sloughed epidermal cells, and support rich populations of anaerobic bacteria that may modify the original components (Gorman et al. 1974; Albone et al. 1978). The secretion is voluntarily expelled by contraction of the layers of muscle that surround the sac and may be deposited at defecation onto objects or onto conspecifics. In some species the secretion may be so violently ejected as to form a jet of material; the most notorious example is the striped skunk (*Mephitis mephitis*) (Blackman 1911). Anal sacs are commonly found among carnivores, being absent only in the Hyaenidae, although they are much reduced in most ursids.

Anal pouches are depressions of hairless skin, richly endowed with skin glands, into which opens the anus. They occur primarily in the hyenas, where the anus emerges toward the lower margin of the pouch, and in the viverrids, where it opens more centrally. Usually the pouch is retracted and all that can be seen is an oval or transverse slit. When it is to be used for scent marking, the pouch is everted by muscular action that exposes its secretion-covered surface and the openings of its skin glands. In the case of the brown hyena these glands open over the whole surface of the pouch, although the apocrine and sebaceous elements have different distributions (Figure 2.3). In the other species of hyenas, particularly the spotted hyena (*Crocuta crocuta*), the major secretory elements are a pair of grossly enlarged and lobulated sebaceous glands that superficially resemble anal sacs and each of which opens into the pouch by a duct (Matthews 1939). Hyenas use their anal pouches to mark, or paste, objects in their territories, particularly stems of grass.

In viverrids, sebaceous glands are generally dispersed over the whole of the pouch, although they may be particularly large and concentrated in certain

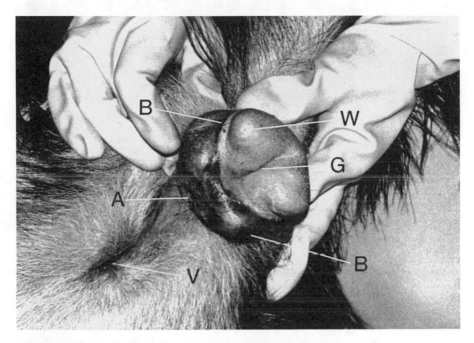

Figure 2.3. The everted anal pouch of a female brown hyena. The anus (A) opens on the ventral lip of the pouch, which is bisected by a groove (G) Myriads of sebaceous glands secrete a white secretion and open onto the central parts of the pouch (W) Lateral to the sebaceous areas, and separated from them by nonsecretory tissue, are two concentrations of apocrine tissue (B), which produce a black secretion. The white secretion had been removed prior to the taking of the photograph. V = vulva. (From Mills et al. 1980, courtesy of M. G. L. Mills.)

arcas, for example, in the swollen crescent of tissue lying above the anus of the small Indian mongoose (*Herpestes auropunctatus*) (Gorman et al. 1974). Viverrids use their anal pouches to mark objects in their range and, in the case of social species, their fellow group members.

THE TAIL

Supracaudal glands are elliptical masses of sebaceous and apocrine glands situated on the dorsal surface of the tail, toward its root. They are restricted to the Canidae but are widely distributed within the family, being absent only in the African hunting dog (*Lycaon pictus*). Their function, and mode of use, is unknown.

Subcaudal glands range from simple concentrations of skin glands on the ventral surface of the root of the tail in some felids and possibly canids to the complex subcaudal pocket of the European badger (*Meles meles*). In this badger the skin between the anus and tail is deeply invaginated to form a deep, relatively hairless pocket (Figure 2.4). The whole of the pocket is punctuated

A

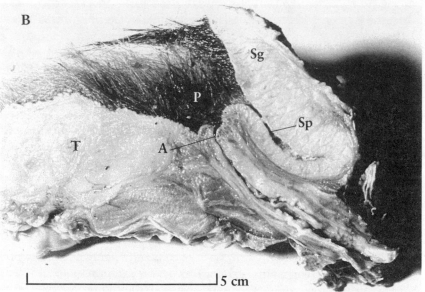

B

Sg

P

Sp

T

A

|⊢————————————⊣ 5 cm

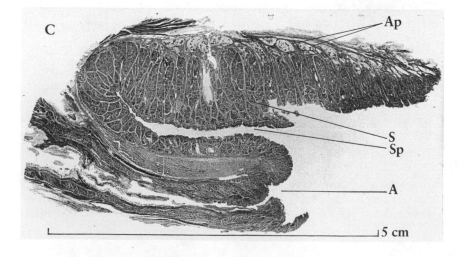

C

Ap

S

Sp

A

|⊢————————————⊣ 5 cm

with the openings of enlarged sebaceous and apocrine glands whose creamy white secretion accumulates within and which is used to mark the ground, bedding, and conspecifics. The anal sacs do not open into the subcaudal pocket, but inside the anus, as in all mustelids.

Perineal glands are found only in the Viverridae but are absent from the true mongooses, the Herpestinae. These organs vary enormously in structure from species to species but essentially consist of glandular tissue opening into a simple depression in the skin, or into a distinct storage chamber, situated between the anus and genital opening and sometimes extending more anteriorly. The glands appear to be used to mark objects in the environment. Many mustelid species have a diffuse concentration of enlarged skin glands in their ventral fur (*ventral glands*), forward of the genital opening. Also, in many species in the Canidae, Viverridae, and Felidae diffuse glandular tissue opens on to the chin, lips, and cheeks. The precise function of such *facial glands* is unknown.

The Chemistry of Carnivore Social Odors

Relatively little is known of the chemistry of social odors in carnivores (Albone 1984). A number of compounds have been isolated and identified (Table 2.1), but in no single case has a specific chemical been unequivocally associated with a particular behavioral function. Many proposed functions of social odors, for example, individual or group recognition, are unlikely to be achieved through the possession of single, unique chemical compounds, but rather by differences in the relative concentrations of the constituents of a complex chemical mixture. Such differences have been reported in a number of species including the stoat (*Mustela erminea*) (Brinck et al. 1983), brown hyena (Mills et al. 1980), red fox (*Vulpes vulpes*) (Albone and Perry 1976), small Indian mongoose and Egyptian mongoose (*Herpestes ichneumon*) (Gorman et al. 1974; Hefetz et al. 1984), European badger (Gorman et al. 1984), and European otter (*Lutra lutra*) (Trowbridge 1983).

Marking Behavior

Carnivores usually deploy social odors by one of three different methods: (1) Odors may be placed on the substrate or onto objects in the environ-

---

Figure 2.4. (*A*). A European badger squat marking the ground with its subcaudal pocket. (*B*). Vertical section through the subcaudal pocket of a male European badger. T = testis; A = anus; P = perianal surface; Sg = glandular tissue; Sp = subcaudal pocket. (*C*). Longitudinal section through the subcaudal pocket of a European badger. The 15-μm section was stained with Mallory's triple stain. A = anus; Sp = subcaudal pocket; S = sebaceous glands; Ap = apocrine glands. (All from Gorman et al. 1984, courtesy of the Zoological Society of London.)

Table 2.1. Chemicals identified in carnivore social odors

Anal sacs and other anal structures
  Volatile carboxylic acids
    acetic
    propionic
    butyric
    isobutyric
    valeric
    isovaleric
    isocaproic
    2-methylbutyric acids (*Vulpes vulpes*)[a,b]
    A similar range is found in *Canis familiaris* and *C. latrans*[y], *Chrysocyon brachyurus* and
      Speothos venaticus[d], *Panthera leo*[e], *P. tigris*[d], *Felis catus*[x], *Mustela vison*[aa], *Herpestes
      auropunctatus*[r].
    Longer chain fatty acids, with 10–22 carbon atoms are present in *Herpestes ichneumon*, with
      2,4,6,10-tetramethylundecanoic acid specific to males[t].
  Amino acids and amines
    5-aminovaleric acid (*Vulpes vulpes*)[f]
    ammonia (*V. vulpes*)[c]
    putrescine 1,4-diaminobutane (*V. vulpes, Panthera leo*)[c], (*Mustela vison*)[aa]
    cadaverine 1,5-diaminopentane (*V. vulpes, P. leo*)[c]
    2-phenylethylamine (*P. tigris*, possibly urinary in origin)[j]
  Organosulphur compounds
    5-thiomethylpentane-2,3-dione (*Hyaena hyaena*)[ff]
    3-methylbutane-1-thiol, trans-2-butene-1-thiol, trans-2-butenyl methyl disulphide, butane-1-
      thiol (*Mephitis mephitis*)[g,h]
    2,2-dimethylthietane (*Mustela vison*)[k,l]
    2-propylthietane, 2-ethylthietane, 3-ethyl-1,2-dithiolane (*Mustela erminea*)[m,n,p]
    2-propylthietane, 2-pentylthietane, trans- and cis-2,3-dimethylthietane, trans- and cis-3,4-
      dimethyl-1,2-dithiolane (*Mustela putorius*)[n,o,aa]
Perineal glands
  civetone, a macrocyclic ketone with 17 carbon atoms (*Viverra civetta*)[w,z]
  long-chain fatty acid esters of macrocyclic alcohols (*V. civetta*)[dd]
Vaginal secretions
  methyl-*p*-hydroxybenzoate (*Canis familiaris*)[q,v]
  C2-C9 and isoC4-isoC8 fatty acids, 2,2-dimethylbutanoic and 2,2-dimethylhexanoic acids,
    volatile amines (*Mustela vison*)[bb]
Urine
  isopent-3-enyl methyl sulphide
  2-phenylethyl methyl sulphide
  4-heptanone
  6-methylhept-5-en-2-one
  benzaldehyde
  acetophenone
  2-methylquinoline
  trans-geranylacetone (*Vulpes vulpes*)[i,u]
  felinine, cysteine-S-isopentanol (*Felis catus*, probably from kidney)[s,cc,ee]
  Urine contains a wide variety of potentially important odorants, including steroids, steroid
    metabolites and conjugated steroids

[a]Albone 1984.
[b]Albone and Fox 1971.
[c]Albone and Gronneberg 1977.
[d]Albone and Perry 1976.
[e]Albone et al. 1974.
[f]Albone et al. 1976.
[g]Anderson and Bernstein 1975.
[h]Anderson et al. 1982.
[i]Bailey et al. 1980.
[j]Brahmachary and Dutta 1979.
[k]Brinck et al. 1978.
[l]Brinck et al. 1983.
[m]Crump 1980b.
[n]Crump 1980a.
[o]Crump and Moors 1985.
[p]Erlinge et al. 1982.

[q]Goodwin et al. 1979.
[r]Gorman et al. 1974.
[s]Greaves and Scott 1960.
[t]Hefetz et al. 1984.
[u]Jorgenson et al. 1978.
[v]Kruse and Howard 1983.
[w]Lederer 1950.
[x]Michael et al. 1972.
[y]Preti et al. 1976.
[z]Ruzicka 1926.
[aa]Sokolov et al. 1980.
[bb]Sokolov and Khorlina 1976.
[cc]Tallan et al. 1954.
[dd]Van Dorp et al. 1973.
[ee]Westall 1953.
[ff]Wheeler et al. 1975.

ment, to be investigated immediately or at some future time. Feces, urine, and durable sebaceous secretions are all regularly used this way.

(2) Odors may be applied to the animal's own body or, more usually, to the bodies of other members of the social group. Skin gland secretions are usually involved in such behavior, and, again, the information remains available for future investigation.

(3) More rarely, odors may be released into the air to be detected by conspecifics, or by members of another species, at some distance. Volatile apocrine secretions are particularly important in this context.

Object Marking

Social odors are a limited resource, whether they be feces, urine, or glandular secretions. Scent marking may also involve a significant investment in terms of time and energy; for example, the white component of the secretion of the brown hyena is 97% lipid, and over the course of a year each individual deposits some 29,000 scent marks (Mills et al. 1980). One would predict, therefore, that scent marks should be distributed in a way that maximizes their chance of being discovered by the individuals for whom they are intended. This indeed seems to be the case; a recurring feature of object marking is that scent marks are placed not at random within the environment, but instead at visually conspicuous, often elevated, and traditionally used landmarks. Placing marks at such sites reduces the number of potential places to be searched, puts them at nose level, and helps disperse their odor. In the absence of suitably elevated sites, some species manufacture them. For example, along rivers in Greece, European otters scrape up mounds of sand on which to mark (Mason and Macdonald 1986). In many species, for example, gray wolf (*Canis lupus*) (Peters and Mech 1975), marks are frequently placed along well-used tracks, particularly at junctions, with the result that they are available for detection by animals arriving from several directions.

Feces, urine, and glandular secretions are widely used in object marking by all the carnivore families. The use of urine and feces is often, to use Macdonald's (1985) terminology, token, and involves small volumes placed at prominent and frequently revisited sites. In some species, for example, the red panda (*Ailurus fulgens*), these sites receive such intense attention that large accumulations of feces gather in a small area, resulting in the formation of latrines, or middens.

Canids scatter feces throughout their ranges, either singly or at latrines. Red foxes, for example, leave single tokens on top of conspicuous features such as grass tussocks, and mole hills and may even defecate from a handstand position in order to get them high (Murie 1936; Macdonald 1979). In the featureless tundra of the far North, arctic foxes (*Alopex lagopus*) defecate on rocks, eskers, and on cast caribou antlers (Muuller-Schwarze 1983). In the case of the domestic dog (*Canis familiaris*), and probably in other species, the feces receive a smear of anal sac secretion as they are passed (Ashdown 1968). Canids also token urinate throughout their ranges, and almost all species elevate the marks by cocking their leg to do so (raised leg urination, RLU). RLUs are, again, left at prominent sites, usually after much sniffing. In their classic study of the gray wolf, Peters and Mech (1975) tracked 13 packs as they moved around 240 km of trails. In doing so they came upon an RLU every 450 m on average, 40% of them at the junctions of the trails used by the wolves. The frequency with which the wolves stopped to urinate was very much higher on the borders of the territories than it was in the interiors. However, it seems that not all wolves enjoy the right to indulge in RLU since the rate of marking does not increase with pack size, indicating that RLU is the prerogative of the alpha pair. This is also the case in the African hunting dog; the alpha male routinely places his own RLU on top of that of the alpha female (van Lawick 1974; Frame and Frame 1976; Frame and Frame 1981).

Spotted hyenas living in large and stable social groups in the Ngorongoro defecate at latrines on the borders of their group territories (Kruuk 1972). They visit them en masse, and after sniffing and scratching, they add to the piles of visually striking white ordure. However, in areas with shifting populations such as the Serengeti, the feces are placed along tracks and at other landmarks scattered through the range. Brown hyenas living in the vastness of the Kalahari concentrate their latrines near the center of the territory, where food is at its most abundant, and along any border that is shared with neighbors (Mills et al. 1980). These hyenas go to extreme lengths to place their marks at predictable sites; not only are 75% of latrines placed at conspicuous Shepherd's trees (*Boscia albitrunca*), almost all of them are to be found on their south-facing sides!

All the extant species of hyena scent mark with their anal pouches, using them to paste stems of grass, but the pasting behavior of the brown hyena is particularly striking (Mills et al. 1980). During pasting a hyena bends a grass stalk forward by walking over it until the root of the grass comes to lie between its hind legs with the stalk running forward under its belly (Figure 2.1). Then the hyena, with its tail curved up over its back and with its back legs slightly bent, extrudes its anal pouch, which consists of two distinct regions (Figure 2.3). The large central area, which is normally covered in an accumulation of white sebaceous secretion, has a distinct, deep groove running vertically from top to bottom. Lying one to each side of the central area and separated from it by nonsecretory epithelium, are two circular areas onto which open apocrine glands producing a black secretion. Having extended its pouch, the hyena now feels for the grass stalk, sometimes for several seconds, and eventually succeeds in locating it in the central groove. The hyena then moves forward, pulling the anal pouch along the grass stalk and at the same time retracting the pouch. The effect is to smear a blob of white paste onto the grass. Then, as the pouch continues to retract, the nonsecretory parts of the pouch and the apocrine areas collapse in turn onto the stem. In this way a thin smear of apocrine secretion is deposited 1–2 cm above the white paste (Figure 2.2). The pasting behavior of other hyena species is essentially similar except that only a white sebaceous component is produced (Kruuk 1972, 1976; Kruuk and Sands 1972; Rieger 1977; Nel and Bothma 1983).

In the Ngorongoro spotted hyenas concentrate their pasting efforts along the borders and at kills (Kruuk 1972), but in the Kalahari they paste throughout their territory, as do brown hyenas (Gorman and Mills 1984; Mills and Gorman 1987). Pasting often appears to be stimulated by the presence of existing pastings, particularly those made by strange hyenas (Kruuk 1972; Mills et al. 1980). The significance of differences in the pattern of scent deployment is discussed in the section on the functions of scent marking.

African lions (*Panthera leo*) appear to defecate at random (Schaller 1972), but most other felids, for example, the bobcat (*Lynx rufus*) (Bailey 1974), defecate along trails and on top of elevated objects. Only within the core areas of their ranges do domestic cats (*Felis catus*) and Scottish wildcats (*F. silvestris*) bury their feces; elsewhere they are left prominently displayed (Corbett 1979; Panaman 1981; Macdonald 1985). The males of most felids token urinate, spraying backward between their legs onto rocks, termite mounds, trees, and other foci of attention (e.g., African lion: Schaller 1972; Cheetah, *Acinonyx jubatus:* Eaton 1973; puma, *Felis concolor:* Hornocker 1979; bobcat: Bailey 1974; tiger, *Panthera tigris:* Schaller 1967). In the case of both the lion and the tiger, anal sac secretions may be incorporated into the urine spray (Schaller 1967; Brahmachary 1979).

Relatively little is known about the marking behavior of wild viverrids. African civet cats (*Viverra* [*Civettictis*] *civetta*) use their perineal glands to mark at civetries, which are accumulations of compacted feces in depressions in the ground, and along the trails that run through their territories (Bearder and Randall 1978). The frugivorous African palm civet (*Nandinia binotata*) uses its gland to mark at territorial borders and around fruiting trees (Charles-Dominique 1978). The social dwarf mongoose (*Helogale undulata*) marks objects with a mixture of secretions from its cheek glands, anal sacs, and pouch, and with feces and urine (Rasa 1973). In captivity, they sniff the object to be marked for 3–4 s, grasp it in their forepaws, and stroke it a couple of times with each cheek; then they evert their anal pouch and drag it across the object, often from a handstand position, and finally sniff the object again. The cycle is then repeated up to 20 or 30 times, the whole marking episode lasting for 3–4 min.

Among the Mustelidae much of the available information on patterns of scent marking in the wild comes from studies made in the winter when the ground is covered in snow and it is relatively easy to follow the movements of animals and to find their scent-marking sites. For example, Pulliainen (1982) was able to show that in Finnish Lapland pine martens (*Martes martes*) leave anal sac marks throughout their ranges, males pausing to do so seven to eight times per kilometer and females three to four times.

The European badger places blobs of anal sac secretion onto feces at latrine pits (Kruuk 1978). In addition, it sports a subcaudal pocket whose secretion comes to cover the skin and hair of the perianal region, mixed together with anal sac secretions and often with fecal matter (Kruuk et al. 1984). This mixture is used in "squat-marking," a behavior in which the badger briefly presses its nether region onto the substrate, leaving a scent mark (Figure 2.4). Badgers frequently pause to squat-mark as they pass along the many paths that traverse their group territories, the same spots being repeatedly re-marked (Östborn 1976). Conspicuous objects such as hummocks and tussocks are also repeatedly marked by all passing badgers, sometimes from a handstand position. In the wild, much of this squat-marking takes place near the sett, or along the paths and at the latrines associated with the border of the territory (Kruuk 1978; Kruuk et al. 1984). Both sexes squat-mark, but males do so most frequently and dominant females more frequently than subdominant ones (Kruuk et al. 1984). Badgers also squat-mark on vegetation before taking it down into their sett to be used as bedding, rejecting any vegetation that has been marked by strange badgers (Kruuk et al. 1984).

European otters deposit feces (spraints) at nose height on top of prominent objects, such as large rocks and tussocks of grass, throughout their home ranges. Repeated sprainting and urination at these spraint piles by successive generations of otters can lead to the formation of distinct mounds and to the lush growth of nitrophilous grasses and algae, all of which makes them visually conspicuous (Figure 2.5). Around the Rhue peninsula in western Scotland, where otters forage exclusively in the sea, spraint piles are dispersed along the coast in an organized manner, with most being clumped together at distinct spraint stations (Trowbridge 1983). The organization can be clearly seen by comparing the frequency distribution of distances between spraint piles with the distribution that would result were the same number to be dispersed randomly around the coast (Figure 2.6). Typically, a spraint station consists of a number of spraint piles connected one to the other, and to the sea, by distinct trails through the vegetation (Figure 2.7). The great majority of stations also contain a relatively large pool of fresh water. Within stations, nearly 50% of spraint piles occur right on the edges of the freshwater pools, with the rest dispersed along the trails, many at junctions, thus ensuring their encounter whatever the direction of approach by an otter. The stations are distributed along the coast in a regular fashion, with a modal interstation distance of 50 m, and with very few stations closer together than 35 m or farther apart than 165 m (Figure 2.8).

With spraint stations spaced out in this way, at regular and frequent intervals, any otter landing from the sea will never be more than a short distance

Figure 2.5. (*A*). A European otter scent marking with feces (sprainting) (*arrow*). (*B*). Sprainting (defecating) pile made by European otters on the west coast of Scotland. The mound, 42 cm high and grass covered, formed as a result of otters' sprainting at the same site for several generations. (Photo *B* from Trowbridge 1983.)

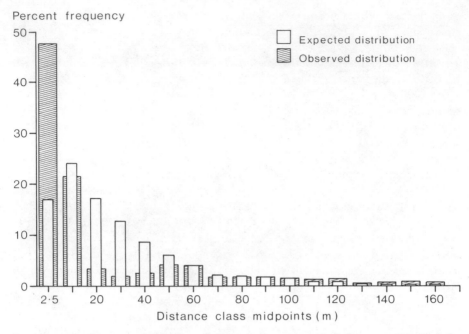

Figure 2.6. The frequency distribution of distances between spraint piles (n = 575) made by European otters on the coastal plateau around the Rhue peninsula in western Scotland. The expected distribution is that which would result were the same number of piles distributed at random around the 15.9 km of coastline and is the mean of ten computer simulations. The two distributions are significantly different at $P < 0.001$ (K-S test, D = 0.3051). (After Trowbridge 1983.)

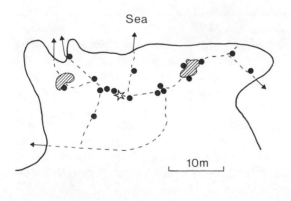

Spraint pile	•
Fresh water pool	🌀
Sleeping site	☆
Otter trail	---
Exit to the sea	⟶

Figure 2.7. A scale drawing of a spraint station showing the major environmental features with which they are associated on the Rhue peninsula. (After Trowbridge 1983.)

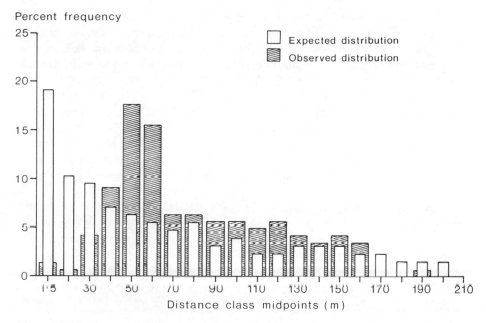

Figure 2.8. The frequency distribution of distances between spraint stations ($n = 143$) on the Rhue peninsula. The expected random distribution is the mean of ten computer simulations and is significantly different from the observed distribution (K-S test, $D = 0.2804$; $P < 0.001$). (After Trowbridge 1983.)

from the nearest scent marks. Coastal otters require fresh water for drinking and for washing salt from their pelage. By placing their spraints on prominent objects, around and on the trails leading to freshwater pools, otters increase yet further the likelihood of their being detected by other individuals. In other coastal populations spraints accumulate at spaced intervals in the same way, but the focus of attention may be different. Thus, Kruuk and Hewson (1978), working just a few miles away on the Applecross peninsula, found that over 50% of spraints were concentrated at dens (holts), which themselves were uniformly spaced approximately 1.1 km apart.

Social Marking

Sometimes, the object to be marked is another individual of the same species. Such allomarking is particularly common in those canids, felids, viverrids, and mustelids that live in organized social groups.

The behavior of the European badger is in many ways typical (Kruuk et al. 1984). When two badgers meet in the dead of night and away from their sett, they often sniff each others' flanks and rumps. These are just the regions that receive attention during social squat-marking, whereby one badger wipes its anal region on the flank of another badger. Most of our detailed information on social squat-marking comes from a study of captive animals living in a large outdoor enclosure, but more casual observations of wild animals indicate that they behave in a essentially similar manner (Gorman et al. 1984; Kruuk et al. 1984).

All the badgers in a clan squat-mark on each other, but not equally. In the captive group of six adults, the dominant male made 66% of the social marks, two other males 19%, and three females the remaining 15%. Of the marks made by the dominant male, 78% were directed toward the sows. In this way the flanks and rumps of each badger came to bear a mixture of the secretions of all the clan members, but a mixture dominated by the odor of the top-ranking boar.

During courtship the dominant boar remains close to the sow, continually sniffing her and attempting to mount. Most of these advances are met with a growl and snap of the teeth. Typically, the male responds by squat-marking the female.

There are close behavioral parallels between social marking in the badger and in other species, including the dwarf mongoose, which uses its anal sacs and pouch for allomarking (Rasa 1973). Here, too, the alpha male is responsible for most (65%) of the social marks, and these are applied to the alpha female. The female is marked on a regular basis but particularly so when she shows signs of being sexually receptive, with the frequency of marking reaching a peak on the first day of her estrus (Figure 2.9).

Sometimes European badgers indulge in mutual squat-marking whereby two individuals back into each other, with their tails raised and their subcaudal pockets open, and press their anal regions together. This behavior is quite rare but seems to take place when clan members have been separated for some days. It is always accompanied by intense sniffing of the flanks. The return to the group of temporarily absent members is a signal for intense allomarking in a number of other species also. For example, Kingdon (1977, pers. comm. in Macdonald 1985) describes how banded mongooses (*Mungos mungo*) indulge in a veritable orgy of allomarking on such occasions, piling into a ball of bodies and rubbing their anal pouches over each other. Rasa (pers. comm. in Macdonald 1985) reports that dwarf mongooses behave in a similar manner prior to combat with a pack of neighbors.

It is difficult to avoid the conclusion that allomarking, in all its different facets, leads to a similarity of odor among the members of a social group, allowing them to be recognized by their colleagues, even in the dead of night or in the heat of battle.

Releasing Odors into the Air

All scent organs release odorous molecules into the air, and these may be sampled by conspecifics during close encounters. Most carnivores, on meeting,

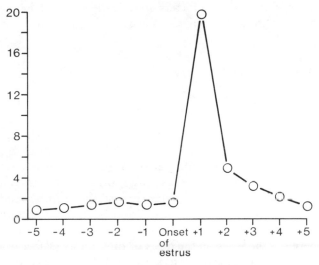

Figure 2.9. The frequency with which dwarf mongooses allomark before, during, and after estrus in the dominant female. The peak around estrus is due to the dominant male's marking the dominant female. (After Rasa 1973, courtesy of *Zeitschrift für Tierpsychologie.*)

sniff each other in the region of their scent organs or at places where secretions have been applied to the body. Badgers sniff each other's flanks, domestic dogs the anal region, and so on.

Some scent organs appear, however, to be specialized for the release of relatively large quantities of odor into the air, to be detected at greater distances by conspecifics, or by members of other species. Such organs usually have important apocrine elements and among carnivores are most often used when an individual is excited, aggressive, or perhaps frightened. Thus, the air around fighting mink (*Mustela vison*) is usually redolent with a sickly sweet odor released from the anal sacs. In this species massive blocks of apocrine tissue open into the base of the anal sac, near its duct (Sokolov et al. 1980).

In some species the anal sacs appear to function primarily in defense against predators. Most carnivores violently void the contents of their anal sacs when in great distress. This tendency is most pronounced among the mustelids, particularly in the genera *Mellivora, Galictis, Ictonyx, Mephitis,* and *Conepatus,* whose grossly enlarged anal sacs produce particularly disgusting secretions (Aldrich 1896; Pocock 1921; Anderson and Bernstein 1975). The anal sacs of the striped skunk are surrounded by powerful muscles contiguous with those of the tail (Blackman 1911). When threatened, the skunk stamps its feet in warning, arches its back, and throws up its hindquarters and tail toward the enemy. As a result, the everted openings of the anal sacs are exposed and the sacs themselves are placed under great pressure. Should the adversary persist, the skunk simply relaxes the sphincters around the ducts of the sacs, releasing jets of the most noxious of chemicals.

Functions of Social Odors

Identity

Many species appear to be able to determine an individual's gender using olfactory cues. In some cases this judgment may result because scent marks are deployed in different ways, or in different quantities, by the two sexes. Female bushdogs (*Speothos venaticus*), for example, urinate from a handstand and thus contrive to place their marks some 15 cm above those made by males (Kleiman 1972). In other cases the distinction is based on differences in the chemical composition of the odor. Both beagle dogs and domestic cats can distinguish between the sexes on the basis of their odors (Veberne and de Boer 1976; Dunbar 1977). The Egyptian mongoose is one of the very few examples for which an apparently sex-specific chemical has been identified; 2,4,6,10-tetramethylundecanoic acid is present in the anal sac secretion of males but is absent from females (Hefetz et al. 1984). Unfortunately, there is no evidence to show that the substance is used to determine the sex of the marker.

An ability to distinguish between individuals on the basis of odor (but not necessarily to recognize individuals) has been clearly demonstrated in the case of the dwarf mongoose (Rasa 1973), the small Indian mongoose (Gorman 1976), the brown hyena (Mills et al. 1980), and the European badger (Öst-born, 1976; Kruuk et al. 1984). In each of these species the relevent information is encoded in glandular secretions, from the anal sacs and anal pouch in the two mongooses and the hyena and from the subcaudal pocket in the badger.

The European otter can discriminate between the feces (spraints) of different individuals. Trowbridge (1983) has shown this by training an otter to discriminate between pairs of spraints, each member of any pair having been produced by a different individual. The otter was trained to associate one spraint from each pair with the reward of a cube of eel. The correct response after sniffing a rewarded spraint was to move to a feeding bowl, 1.5 m away, and to wait for the reward to be delivered. Following the presentation of an unrewarded spraint the correct response was to remain stationary at the feet of the trainer. In order to avoid any possibility of trainer bias, a number of spraints, the identities of which were unknown to the trainer, were introduced at random points in each trial. The results presented in Table 2.2 show that the otter could distinguish between his own spraints and those of other individuals, and between those of two other individuals, regardless of their sex. A comparison of the "blind" presentations of spraints with those made when the trainer was

Table 2.2. Discrimination by a captive otter between spraints from different individuals

Otters providing the spraints	Response to presentation			Yates χ^2
	Correct	Incorrect	% Correct	
Self vs. female A	81	2	97.5	71.6[a]
Self vs. male A	106	20	84.1	62.7[a]
Self vs. female B	68	12	85.0	38.8[a]
Female B vs. male A	79	10	88.8	53.3[a]
Female B vs. Female A	109	13	89.3	70.1[a]
Female A vs. male A	92	46	66.7	25.3[a]

Source. After Trowbridge 1983.
[a]$P < 0.001$.

aware of their identity showed that the otter required no help from the trainer (Table 2.3).

The existence of a group odor produced by all the members of a social group, and allowing an individual's group membership to be recognized, is a beguiling idea, but one that has yet to be conclusively demonstrated. What at first sight might appear to be supporting evidence can usually be explained in terms of the mixing of individual secretions during allomarking or as a result of repeated marking of the same sites. Such is the case for the subcaudal secretion of the European badger, a complex mixture of compounds that can be resolved by gas chromatography (Gorman et al. 1984). A statistical analysis of chromatograms of the secretions of 39 individuals from nine different clans in Gloucestershire, England, demonstrated significantly greater variation *between clans* than within clans in the relative proportions of 14 out of 20 components. However, this does not necessary imply that the members of a clan *produce* a similar odor; badgers rub their subcaudal pockets together during mutual squat-marking and all members of a group squat-mark the same objects. Both behaviors result in a physical mixing of the secretions of the different badgers, leading to a similarity of odor within the clan. More than just secretion may be passed during such encounters. It is now well established that many scent organs support rich populations of bacteria that may, by their metabolism, modify the animal's own secretions (Albone et al. 1974, 1978).

Table 2.3. The effect of the trainer's awareness of spraint identity on a captive otter's ability to discriminate between spraints

	Responses to presentation		
	Correct	Incorrect	% Correct
Identity of spraint known to trainer	535	103	84.3
Identity of spraint not known to trainer	107	14	88.4

Source. After Trowbridge 1983.
Note. $\chi^2 = 1.63$, d.f. $= 1$, $P < 0.3$. See Table 2.2 for data from trials.

Indeed, in the small Indian mongoose individual differences in odor are due, at least in part, to short-chain fatty acids produced by anaerobic bacteria living in the anal sacs (Gorman et al. 1974). Cross infection between individuals during allomarking, or while marking communal scent posts, may thus further increase any similarity of odor profiles among the members of a social group.

Reproduction and the Detection of Estrus

There are clear indications that social odors are involved during sexual interactions, but little evidence as to precisely what is being signaled, or by which particular odor.

In many species the frequency of scent marking increases markedly in the breeding season, and particularly during courtship as the female approaches estrus. Such increases involve marking with scent organs, as described above for the badger and dwarf mongoose, but in particular token marking with urine. Macdonald (1979, 1985) showed that the rate of token urination by a tame red fox vixen increased as the winter progressed, reaching a peak just before mating in late January. The pungency of the urine of both vixens and dog foxes increases during December to February, the breeding season, with the result that scent marks can be detected from much greater distances, even by humans (Jorgenson et al. 1978; Henry 1980). This coincides closely, in dog foxes, with an increase in the concentration of two urinary volatiles, 4-heptanone and 3-methylbutyl methyl sulphide, both of which peak in February (Figure 2.10) (Bailey et al. 1980). In coyotes *(Canis latrans)* the rate of RLU varies seasonally, with high rates during the breeding season, November to February, followed by a decline through April, when the cubs are born, to a minimum in May (Wells and Bekoff 1981).

As some owners of dogs suspect, male dogs can smell when bitches are estrus. In formal demonstrations of this ability, sexually experienced male beagles investigated urine and vaginal secretions collected during estrus for longer than those collected during diestrus (Beach and Gilmore 1949; Doty

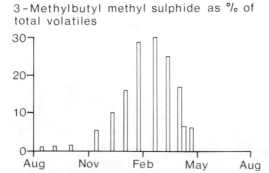

3-Methylbutyl methyl sulphide as % of total volatiles

Figure 2.10. Changes in the concentration (as a percentage of total volatiles) of 3-methylbutyl methyl sulphide in the urine of red foxes. The peak in January–February coincides with the mating period. (After Bailey et al. 1979, courtesy of Plenum Press.)

and Dunbar 1974; Dunbar 1977, 1978). There is some evidence that vaginal secretions may be involved in the recognition of the estrous state; methyl-*p*-hydroxybenzoate appears in the vagina during estrus and, if applied to the vulvae of diestrous bitches, causes males to attempt to mount, despite the adverse reactions of the females (Goodwin et al. 1979).

Urine is a potentially rich source of information concerning reproductive state. In female mammals blood titers of estradiol increase during the follicular phase of the estrous cycle and drop abruptly at ovulation. Estrogen levels are therefore an accurate indication of changes in female receptivity. Clearly, males cannot follow changes in blood levels, but they can monitor odorous free steroids in voided urine. The renal handling of steroids is very efficient, and urine levels reflect accurately the production by the ovaries (Baird 1976). The European otter, which is continually polyestrous if unmated (Gorman et al. 1978; Wayre 1979), is one of the few species of carnivores for which we have data on seasonal changes in levels of urinary estrogens. Trowbridge (1983) collected 24-hour urine samples from an unmated female on a daily basis over a period of two years. During that time estradiol levels peaked on 16 occasions, with a mean periodicity of 36 days (range 17–51). A sample of the data, collected in the summer of 1979, is shown in Figure 2.11.

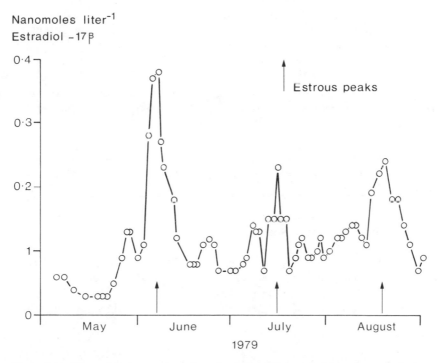

Figure 2.11. Changes in the concentration of estradiol-17β in daily 24-hour samples in the urine of a captive, unmated female European otter. (After Trowbridge 1983.)

Temporal Information

Most social odors consist of complex mixtures of compounds of widely differing volatility. Since these will evaporate at different rates, it is not inconceivable that scent marks may contain temporal information. In fact, at least one species, the dwarf mongoose, can discriminate between scent secretions of different ages (Rasa 1973).

Leyhausen (1965) has argued that in populations of domestic cats several individuals may make use of the same area, but with each doing so at different times. He envisaged such a time-sharing scheme being based on the ability of cats to tell the age of urine spray-marks, a fresh mark indicating that the area was currently in use, an older mark that it was free for the next cat to take over, adding its own mark as it did so (Leyhausen and Wolff 1959; Leyhausen 1971, 1979). In much the same way, cheetahs avoid moving along trails that have recently been urine marked (Eaton 1970).

Temporal information may also be important during foraging, allowing individuals to avoid areas in which other individuals have recently fed and which will, therefore, be unproductive. We described earlier how foraging brown hyenas repeatedly pause to paste and how one of the components of the mark rapidly wanes, possibly signaling the passage of time. European badgers squat-mark onto the grass as they forage for earthworms, and Neal (1977) has suggested that this informs the badger where it has been in the recent past.

Red foxes, coyotes, and wolves all seem to use urine in "bookkeeping," to help them keep a tab on where they have foraged or hidden food (Henry 1977; Herrington 1981, 1982). Henry found that red foxes placed 88% of their urine marks at sites where they had scavenged and unearthed cached food. There is strong evidence that such behavior informs the fox, when it next returns to the area, that although the smell of food may linger, there is so little present that it would do better to forage elsewhere. The evidence comes from a series of experiments in which Henry presented, to wild foxes, the smell of food either alone or in combination with urine; the foxes spent significantly less time investigating the sites when the urine was present.

Territoriality and Social Status

Individual animals may gain an advantage over others by denying them access to resources such as food and mates. They do so either by being territorial or by gaining high status within a social group. Scent marking is centrally involved in the advertisement of both land tenure and social dominance.

TERRITORIALITY

Fights over the possession of a territory are rare because individuals are generally reluctant to enter occupied areas. When intruder and resident do meet, the result is usually withdrawal by the intruder, without escalation to fighting. Almost all carnivore territories are scent marked, and the earliest explanation of such behavior was that the scent marks acted as an impenetrable, olfactory barrier, deterring the entry of intruders (Hediger 1949). In fact, animals do on occasion leave their own territories and trespass on those of others. During such excursions carnivores often cease to scent mark, indicating that they are well aware that they are outside their own property. Such a change in behavior has been observed in the cases of brown hyenas (Mills et al. 1980), coyotes (Bowen and McTaggart Cowan 1980), and red foxes (Macdonald 1979).

Gosling (1982), in reviewing a variety of published hypotheses relating to the function of territorial scent marking, found them all wanting in one way or another. He argues convincingly that all the available evidence supports a single hypothesis concerned with the assessment of the quality of potential competitors. His argument goes as follows:

The individuals resident in a territory have more to gain from retaining the territory than do intruders from taking it over. This is because residents will have invested a great deal of energy and time into getting to know their areas and resources, and may well have dependent young. Since a resident has more to lose, it will be more likely to escalate any encounter than will an intruder. In addition, a resident, by virtue of having gained and held a territory, is likely to be an animal of high quality and fighting ability. Since escalation brings the risk of injury and death to both animals, it is in the resident's interest to allow itself to be recognized as such, in a completely unambiguous manner and in a way that precludes any possibility of bluff. Territorial scent marking may be one way to do so; only a long-term resident can have had the opportunity to pepper an area with scent marks. So, if an intruder should meet an individual whose odor matches that of the majority of the scent marks in the area, then it can be reasonably sure that it has met the resident. Having identified the resident, by definition a quality individual quite likely to rapidly escalate conflict to horrible heights, the intruder would do well to withdraw as rapidly as possible. In essence, the scent marks in a territory act as a cue to potential fighting ability and willingness to fight, in an asymmetric contest between resident and intruder (Maynard Smith and Parker 1976).

A number of testable predictions follow naturally from such an interpretation of territorial scent marking (Gosling 1982, 1985). In particular, territorial owners should: (1) replace any scent marks that do not match their own odor; (2) ensure that they smell strongly of their own odor; (3) make themselves available for investigation; (4) deploy their scent marks in a manner that maximizes the chance of their being encountered by intruders. There is evidence in support of each prediction.

(1) A number of species scent mark at increased rates when they are near the borders of a territory. Barrette and Messier (1980), for example, discovered that coyotes marked at the highest rates at places where intrusion by neighbors was most common, and Mills et al. (1980) showed that although brown

Table 2.4. The reaction of brown hyenas to pastings made by members of their own social group and by members of alien groups.

Response	Pastings from own group	Pastings from alien group
Approached to within 1 m but did not sniff or paste	6	0
Approached to within 1 m and then sniffed and/or pasted	6	11

Source. After Mills et al. 1980.
Note. Fisher's exact probability = 0.0092.

hyenas place most marks in the interior of their territory, the rate of marking per kilometer traveled increases at the border. One possible interpretation of such behavior is that the animals are encountering, and attempting to obliterate, alien marks. In a series of fortuitous observations, and experimental transplantations of pasted grasses, Mills et al. were able to confirm that brown hyenas almost invariably overmark any foreign pastes that they encounter (Table 2.4).

(2) It is particularly important that the animals occupying a group territory should mark the same sites, and each other, so that the group and its environment comes to achieve some uniformity of odor. We described in an earlier section how it is often the dominant individuals who are mainly responsible for scent marking the territory and the group members. For example, the members of a badger group come to share a common olfactory identity dominated by the odor of high-status males (Gorman et al. 1984; Kruuk et al. 1984). These are just the animals most likely to get involved in possibly escalating conflicts over territory.

(3) When territorial carnivores meet an intruder, they may allow themselves to be sniffed, at least momentarily, and the focus of attention is usually a scent organ or a place that receives attention during allomarking. For example, spotted hyenas excitedly evert their anal pouches and also indulge in pasting during border conflicts (Kruuk 1972). By doing so, they provide the best possible opportunity for an immediate comparison of their odor with that of the scent marks in the surrounding area.

(4) A recurring phenomenon is the way in which carnivores place their marks at visually conspicuous sites, where they are most likely to be found. However, the best way to disperse a limited supply of marks within the territory depends upon the particular environmental circumstances in which an individual or group finds itself (Gosling 1981). The same species may opt for radically different solutions in different circumstances, as exemplified by hyenas (Gorman and Mills 1984).

Hyenas adopt one of two strategies for dispersing scent marks within their territories. In the Ngorongoro spotted hyenas form large (30–80) clans and live in small (30 km²) clan territories. In these territories the latrines and

Pastings per square km

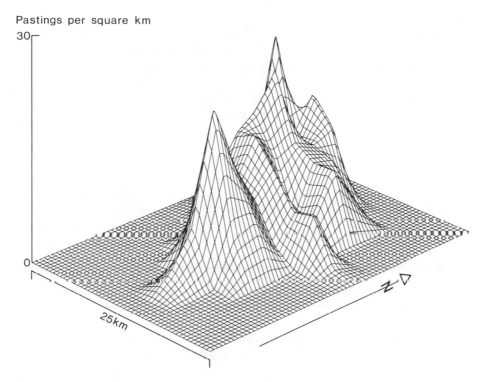

Figure 2.12. A three-dimensional representation of differences in the density of pastings (scent marks) throughout the territory of a group of brown hyenas in the southern Kalahari. The plot was produced by the SURFACE II graphics system (Kansas Geological Survey) from the numbers of pastings present in the elements of a matrix of 2.5-by-2.5-km squares. The resultant map is displayed as if seen from the southeast at an altitude of 35° above the horizontal. (Based on data in Gorman and Mills 1984.)

pasting sites are located along the border and are replenished during regular border patrols (Kruuk 1972). The second strategy is exemplified by brown hyenas living in the much less productive southern Kalahari, where small groups of one to nine individuals share large (300 km²) territories. These animals deposit pastings and latrines throughout their territories, but mainly in the interior, where they spend most of their time (Figure 2.12) (Mills et al. 1980; Mills 1982).

These two strategies are not genetically determined species differences, but adaptive responses to local conditions. Hinterland marking seems to be a solution to the problem of marking a very large territory, within a limited time budget, and with a finite supply of scent. Thus, we find that in the Kalahari spotted hyenas live in small (3–13) clans, occupy large (1100 km²) territories, and they too adopt the hinterland solution (Gorman and Mills 1984; Mills and Gorman 1987). Border marking may give the earliest possible warning of trespass, but it involves a single line of defense, which must be maintained intact if overt conflicts are to be avoided. As territory size increases, it becomes progressively more difficult to do this, and hinterland marking becomes a safer strategy. Intruders may penetrate some distance into the territory, but sooner or later they will encounter scent marks. In fact, Gorman and Miller (1984) have shown by computer simulations that despite the large territory of a brown hyena, intruders penetrate only 540 m on average before passing within 50 m of scent mark, not a great distance for a species that can detect carrion at a distance of 2 km (Mills 1978).

SOCIAL STATUS

Over a wide range of animals there is a correlation between the rates at which individuals scent mark and their dominance ranking (Ralls 1971). Indeed, in some of the more social carnivores, including the wolf and African wild dog, marking may be the prerogative of only high-ranking individuals (Peters and Mech 1975; Frame and Frame 1976). Thus, dominant individuals by their intense marking continually advertise their status to subordinates. However, there is no evidence to suggest that the marks of high-ranking individuals contain chemicals signifying their status.

Gosling (1985) argues that the concept of scent matching as a means of assessing the quality of competitors can be applied equally to dominance hierarchies. By marking at high rates, dominant individuals provide a mechanism by which subordinate animals can recognize them in encounters, by matching their odor with that of the majority of the scent marks in the vicinity. In hierarchial systems competition is at its most intense where access to mates is concerned. It is at this time that confrontation with subordinates is most likely to occur and correct assessment of status most important. It is not surprising, therefore, to find that males frequently mark during courtship. In some species, including the badger and dwarf mongoose (Figure 2.9), the female herself becomes the prime target for marking (Rasa 1973; Kruuk et al. 1984). By presenting his odor to the female in this way, the male may be reinforcing in the clearest possible manner that he is a dominant animal of proven competitive ability. In an earlier section we described how during the courtship of European badgers the early approaches of the males are met with hostility and that he responds by squat marking the female. The result of this is that the sow usually ceases her aggression (Table 2.5). This is consistent with her having received verification of his high social status and suitability as a mate.

What is unclear in all this is why subordinate animals mark at such low rates, or not at all. It could be that they are capable of doing so but are prevented by the more dominant individuals. There are snippets of information in support of such an idea; for example, Frame and Frame (1981) describe how a female African hunting dog began to token urinate for the very first time in her life shortly after leaving her natal group and the influence of her social superiors.

Table 2.5. The response of a female badger, during courtship, to being squat-marked by a male

Behavior of the female after being marked by the male	Observations
More aggressive	1
Less aggressive	36
More avoidance of the male	4
Less avoidance of the male	7

Source. After Kruuk et al. 1984, courtesy of *Animal Behaviour.*

Note. Binomial test for less aggressive or more avoidance, versus more aggressive or less avoidance: $z = -4.8$, $P < 0.001$

References

Albone, E. S. 1984. *Mammalian Semiochemistry: The Investigation of Chemical Signals between Mammals.* Chichester: Wiley & Sons.

Albone, E. S., Eglinton, G., Walker, J. M., and Ware, G. C. 1974. The anal gland secretion of the red fox, *Vulpes vulpes,* its chemistry and microbiology: A comparison with the anal sac secretion of the lion, *Panthera leo. Life Sci.* 14:387–400.

Albone, E. S., and Fox, M. W. 1971. Anal gland secretion of the red fox. *Nature* 233:569–570.

Albone, E. S., Gosden, P. E., Ware, G. C., Macdonald, D. W., and Hough, N. G. 1978. Bacterial action and chemical signalling in the red fox, *Vulpes vulpes,* and other mammals. In: R. W. Bullard, ed. *Flavor Chemistry of Animal Foods, Proc. Amer. Chem. Soc.* 67:78–91.

Albone, E. S., and Gronneberg, T. O. 1977. Lipids of the anal sac secretions of the red fox, *Vulpes vulpes,* and of the lion, *Panthera leo. J. Lipid Res.* 18:474–479.

Albone, E. S., and Perry, G. C. 1976. Anal sac secretion of the red fox, *Vulpes vulpes.* Volatile fatty acids and diamines. Implications for a fermentative hypothesis of chemical recognition. *J. Chem. Ecol.* 2:101–111.

Albone, E. S., Robins, S. P., and Patel, D. 1976. 5-Aminovaleric acid, a major free amino acid component of the anal sac secretion of the red fox, *Vulpes vulpes. Comp. Biochem. Physiol.* 55B:483–486.

Aldrich, T. B. 1896. A chemical study of the secretions of the anal glands of *Mephitis mephitis* (common skunk) with remarks on the physiological properties of this secretion. *J. Exp. Med.* 1:323.

Andersen, K. K., and Bernstein, D. T. 1975. Some chemical constituents of the scent of the striped skunk, *Mephitis mephitis. J. Chem. Ecol.* 1:493–499.

Andersen, K. K., Bernstein, D. T., Caret, R. L., and Romanczyk, L. J. 1982. Chemical constituents of the defensive secretion of the striped skunk, *Mephitis mephitis. Tetrahedron* 38:1965–1970.

Ashdown, R. R. 1968. Symposium on canine recto-anal disorders. I. Clinical anatomy. *J. Small Anim. Pract.* 9:315–322.

Bailey, S., Bunyan, P. J., and Page, J. M. J. 1979. Variation in the levels of some components of the volatile fraction of urine from captive red fox (*Vulpes vulpes*) and its relationships to the state of the animal. In: D. Müller-Schwarze & R. M. Silverstein, eds. *Chemical Signals: Vertebrates and Aquatic Invertebrates,* pp. 391–403. New York: Plenum Press.

Bailey, T. N. 1974. Social organization in a bobcat population. *J. Wildl. Mgmt.* 38:435–446.

Baird, D. T. 1976. Oestrogens in clinical practice. In: J. A. Loraine & E. T. Bell, eds. *Hormone Assays and Their Clinical Application.* New York: Churchill Livingstone.

Barrette, C., and Messier, F. 1980. Scent marking in free-ranging coyotes, *Canis latrans. Anim. Behav.* 28:814–819.

Beach, F. A., and Gilmore, R. N. 1949. Response of male dogs to urine from females in heat. *J. Mamm.* 30:391–392.

Bearder, S. K., and Randall, R. M. 1978. The use of faecal marking sites by spotted hyenas and civets. *Carnivore* 1:32–48.

Beauchamp, G. K., Doty, R. L., Moulton, D. G., and Mugford, R. A. 1976. The pheromone concept in mammalian chemical communication: A critique. In: Doty, R. L., ed. *Mammalian Olfaction, Reproductive Processes, and Behavior*, pp. 144–160. New York: Academic Press.

Blackman, M. W. 1911. The anal glands of *Mephitis mephitis. Anat. Rec.* 5:491–515.

Bowen, W. D., and McTaggart Cowan, I. 1980. Scent marking in coyotes. *Canadian J. Zool.* 58:473–480.

Brahmachary, R. L. 1979. The scent marking of tigers. *Tiger Paper* 6:19–20. FAO Regional Office for Asia and the Far East.

Brahmachary, R. L., and Dutta, J. 1979. Phenylethylamine as a biochemical in the tiger. *Z. Naturforsch.* 34C:632–633.

Brinck, C., Erlinge, S., and Sandell, M. 1983. Anal sac secretion in mustelids. A comparison. *J. Chem. Ecol.* 9:727–745.

Brinck, C., Gerell, R., and Odham, G. 1978. Anal pouch secretion in mink, *Mustela vison. Oikos* 30:68–75.

Brown, R. E. 1979. Mammalian social odours: A critical review. In: J. S. Rosenblatt, R. A. Hinde, C. Beer & M.-C. Busnel, eds. *Advances in the Study of Behaviour* 10:103–162. New York: Academic Press.

Charles-Dominique, P. 1978. Ecology and social behaviour of the African palm civet, *Nandinia binota,* in Gabon: With a comparison of sympatric prosimians. *Terre Vie* 32:477–528.

Corbett, L. C. 1979. Feeding ecology and social organisation of wildcats (*Felis sylvestris*) and domestic cats (*F. catus*) in Scotland. Ph.D. thesis, Univ. Aberdeen, Scotland. 296 pp.

Crump, D. R. 1980a. Anal gland secretion of the ferret (*Mustela putorius forma furo*). *J. Chem. Ecol.* 6:837–844.

Crump, D. R. 1980b. Thietanes and dithiolanes from the anal gland of the stoat, *Mustela erminea. J. Chem. Ecol.* 6:341–347.

Crump, D. R., and Moors, P. J. 1985. Anal gland secretions of the stoat *Mustela erminea* and the ferret *Mustela putorius* forma *furo.* Some additional thietane compounds. *J. Chem. Ecol.*11:1037–1043.

Doty, R. L., and Dunbar, I. F. 1974. Attraction of beagles to conspecific urine, vaginal and anal sac secretion odours. *Physiol. Behav.* 12:825–833.

Dunbar, I. F. 1977. Olfactory preferences in dogs: The response of male and female beagles to conspecific odours. *Behav. Biol.*20:471–481.

Dunbar, I. F. 1978. Olfactory preferences in dogs: The response of male and female beagles to conspecific urine. *Biol. Behav.* 3:273–286.

Eaton, R. L. 1970. Group interactions, spacing and territoriality in cheetahs. *Z. Tierpsychol.* 27:481–491.

Eaton, R. L. 1973. *The World's Cats*, vol. 1. Winston, Ore.: World Wildlife Safari.

Erlinge, S., Sandell, M., and Brinck, C. 1982. Scent marking and its territorial significance in stoats, *Mustela erminea. Anim. Behav.* 30:811–812.

Frame, G. W., and Frame, L. H. 1981. *Swift and Enduring: Cheetahs and Wild Dogs of the Serengeti*. New York: Dutton.

Frame, L. H., and Frame, G. W. 1976. Female African wild dogs emigrate. *Nature (Lond.)* 256:227–229.

Goodwin, M., Gooding, K. M., and Regnier, F. 1979. Sex pheromone in the dog. *Science* 203:559–561.

Gorman, M. L. 1976. A mechanism for individual recognition by odour in *Herpestes auropunctatus* (Carnivora: Viverridae). *Anim. Behav.* 24:141–145.

Gorman, M. L. 1980. Sweaty mongooses and other smelly carnivores. *Symp. Zool. Soc. London* 45:87–105.

Gorman, M. L., Jenkins, D., and Harper, R. J. 1978. The anal sacs of the otter *Lutra lutra*. *J. Zool. (Lond.)* 186:463–474.

Gorman, M. L., Kruuk, H., and Leitch, A. 1984. Social functions of the sub-caudal scent gland secretion of the European badger *Meles meles* (Carnivora: Mustelidae). *J. Zool. (Lond.)* 204:549–559.

Gorman, M. L., and Mills, M. G. L. 1984. Scent marking strategies in hyaenas (Mammalia). *J. Zool. (Lond.)* 202:535–547.

Gorman, M. L., Nedwell, D. B., and Smith, R. M. 1974. An analysis of the contents of the anal scent pockets of *Herpestes auropunctatus*. *J. Zool. (Lond.)* 172:388–389.

Gosling, L. M. 1981. Demarcation in a gerenuk territory: An economic approach. *Z. Tierpsychol.* 56:305–322.

Gosling, L. M. 1982. A reassessment of the function of scent marking in territories. *Z. Tierpsychol.* 60:89–118.

Gosling, L. M. 1985. The even-toed ungulates. In: R. E. Brown & D. W. Macdonald, eds. *Social Odours in Mammals*, pp. 550–618. Oxford: Clarendon Press.

Greaves, J. P., and Scott, P. P. 1960. Urinary amino acid pattern of cats on diets of varying protein content. *Nature* 187:242.

Hamilton, P. H. 1976. The movements of leopards in Tsavo National park, Kenya, as determined by radio-tracking. M.Sc. thesis, Univ. Nairobi, Kenya.

Harrington, F. H. 1981. Urine marking and caching behaviour in the wolf. *Behaviour* 76:280–288.

Harrington, F. H. 1982. Urine marking at food and caches in captive coyotes. *Canadian J. Zool.* 60:776–782.

Hediger, H. 1949. Säugetier-Territorien und ihre Markierung. *Bijdr. Dierkd.* 28:172–184.

Hefetz, A., Ben Yaacov, R., and Yom Tov, Y. 1984. Sex specificity in the anal gland secretion of the Egyptian mongoose *Herpestes ichneumon*. *J. Zool. (Lond.)* 203:205–209.

Henry, J. D. 1977. The use of urine marking in the scavenging behaviour of the red fox (*Vulpes vulpes*). *Behaviour* 61:82–105.

Henry, J. D. 1980. The urine marking behaviour and movement patterns of red foxes (*Vulpes vulpes*) during a breeding and post-breeding period. In: D. Müller-Schwarze & R. M. Silverstein, eds. *Chemical Signals: Vertebrates and Aquatic Invertebrates*, pp. 11–27. New York: Plenum.

Hornocker, M. G. 1969. Winter territoriality in mountain lions. *J. Wildl. Mgmt.* 33:457–464.

Jorgenson, J. W., Novotny, M., Carmack, M., Copland, G. B., Wilson, S. R., Katona, S., and Whitten, W. K. 1978. Chemical scent constituents in the urine of the red fox (*Vulpes vulpes*) during the winter season. *Science* 199:796–798.

Kingdon, J. 1977. *East African Mammals*, vol. 3. London: Academic Press.

Kleiman, D. G. 1966. Scent marking in the Canidae. *Symp. Zool. Soc. London* 18:167–177.

Kleiman, D. G. 1972. Social behavior of the maned wolf, *Chrysocyon brachyurus* and the bush dog *Speothos venaticus:* A study in contrasts. *J. Mamm.* 53:791–806.

Kruse, S. M., and Howard, W. E. 1983. Canid sex attractant studies. *J. Chem. Ecol.* 9:1503–1510.

Kruuk, H. 1972. *The Spotted Hyena, A Study of Predation and Social Behavior.* Chicago: Univ. Chicago Press.

Kruuk, H. 1976. Feeding and social behaviour of the striped hyaena (*Hyaena vulgaris* Desmarest). *East African Wildl. J.* 14:91–111.

Kruuk, H. 1978. Spatial organisation and territorial behaviour of the European badger *Meles meles. J. Zool. (Lond.)* 184:1–19.

Kruuk, H., Gorman, M. L., and Leitch, A. 1984. Scent marking with the subcaudal gland by the European badger, *Meles meles* L. *Anim.Behav.* 32:899–907.

Kruuk, H., and Hewson, R. 1978. Spacing and foraging of otters (*Lutra lutra*) in a marine habitat. *J. Zool. (Lond.)* 185:205–212.

Kruuk, H., and Sands, W. A. 1972. The aardwolf (*Proteles cristatus,* Sparrman 1783) as a predator of termites. *East African Wildl. J.* 10:211–227.

Lederer, E. 1950. Odeurs et parfums des animaux. *Fortschr. Chem. Organ. Naturst.* 6:87–153.

Leyhausen, P. 1965. The communal organisation of solitary mammals. *Symp. Zool. Soc. London* 14:249–263.

Leyhausen, P. 1971. Dominance and territoriality as complements in mammalian social structure. In: H. Esser, ed. *Behavior and Environment,* pp. 22–33. New York: Plenum Press.

Leyhausen, P. 1979. *Cat behavior: The Predatory and Social Behavior of Domestic and Wild Cats.* London: Garland Press.

Leyhausen, P., and Wolff, R. 1959. Das Revier einer Hauskatz. *Z. Tierpsychol.* 16:666–670.

Macdonald, D. W. 1979. Some observations and field experiments on the urine marking behaviour of the red fox, *Vulpes vulpes. Z. Tierpsychol.* 51:1–22.

Macdonald, D. W. 1980. Patterns of scent marking with urine and faeces among carnivore communities. *Symp. Zool. Soc. London* 45:107–139.

Macdonald, D. W. 1985. The carnivores: Order Carnivora. In: R. E. Brown & D. W. Macdonald, eds. *Social Odours in Mammals,* pp. 619–722. Oxford: Clarendon Press.

Mason, C. E., and Macdonald, S. M. 1986. *Otters: Ecology and Conservation.* Cambridge: Cambridge Univ. Press.

Matthews, L. H. 1939. Reproduction in the spotted hyaena, *Crocuta crocuta* (Erxleben). *Philos. Trans. Ser. B.* 230:1–78.

Maynard Smith, J., and Parker, G. A. 1976. The logic of asymmetric contests. *Anim. Behav.* 24:159–175.

Michael, R. P., Zumpe, D., Keverne, E. B., and Bonsall, R. W. 1972. Neuro-endrocine factors in the control of primate behaviour. *Rec. Progr. Horm. Res.* 28:665–704.

Mills, M. G. L. 1978. The comparative sociology of the Hyaenidae. *Carnivore* 1:1–7.

Mills, M. G. L. 1982. Factors affecting group size and territory size of the brown hyaena, *Hyaena brunnea,* in the Southern Kalahari. *Z. Tierpsychol.* 48:113–141.

Mills, M. G. L., and Gorman, M. L. 1987. The scent marking behaviour of the spotted hyaena, *Crocuta crocuta,* in the southern Kalahari. *J. Zool. (Lond.)* 212:483–497.

Mills, M. G. L., Gorman, M. L., and Mills, M. E. J. 1980. The scent marking behaviour of the brown hyaena, *Hyaena brunnea,* in the Southern Kalahari. *South African J. Zool.* 15:240–248.

Müller-Schwarze, D. 1983. Scent glands in mammals and their functions. In: J. F.

Eisenberg & D. G. Kleiman, eds. *Advances in the Study of Mammalian Behavior*, pp. 150–197. Special Publication no. 7. Lawrence, Kans.: American Society of Mammalogists.

Murie, A. 1936. Following fox trails. *Univ. Mich. Misc. Publ.* 32:7–45.

Neal, E. 1977. *Badgers*. Poole: Blandford Press.

Nel, J. A. J., and Bothma, J. du P. 1983. Scent marking and midden use by aardwolves *(Proteles cristatus)* in the Namib desert. *African J. Ecol.* 21:25–39.

Östborn, H. 1976. Doftmarkering hos graveling. *Zool. Revy.* 38:103–112.

Panaman, R. 1981. Behaviour and ecology of free-ranging female farm cats (*Felis catus* L.). *Z. Tierpsychol.* 56:59–73.

Peters, R. P., and Mech, L. D. 1975. Scent-marking in wolves. *Amer. Scient.* 63:628–637.

Pocock, R. I. 1921. On the external characters and classification of the Mustelidae. *Proc. Zool. Soc. London* 1921:803–807.

Preti, G., Muetterties, E. L., Furman, J. M., Kennelly, J. J., and Johns, B. E. 1976. Volatile constituents of dog (*Canis familiaris*) and coyote (*Canis latrans*) anal sacs. *J. Chem. Ecol.* 2:177–186.

Pulliainen, E. 1982. Scent marking in the pine marten *(Martes martes)* in Finnish Forest Lapland in winter. *Z. Säugetierk.* 47:91–99.

Quay, W. B. 1977. Structure and function of skin glands. In D. Müller-Schwarze & M. M. Mozell, eds. *Chemical Signals in Vertebrates*, pp. 1–17. New York: Plenum.

Ralls, K. 1971. Mammalian scent marking. *Science* 171:443–449.

Rasa, O. A. E. 1973. Marking behaviour and its significance in the African dwarf mongoose, *Helogale undulata rufula*. *Z. Tierpsychol.* 32:449–488.

Rieger, I. von 1977. Markierungsverhalten von Streifen-hyänen, *Hyaena hyaena*, im Zoologischen Garten Zürich. *Z. Säugetierk.* 42:307–317.

Ruzicka, L. 1926. Zur Kenntnis des Kohlenstoffringes. I. Über die Konstitution des Zibetons. *Helvetica Chimica Acta* 9:230–248.

Schaffer, J. 1940. *Die Hautdrusenorgane der Säugetiere*. Berlin: Urban & Schwarzenberg.

Schaller, G. B. 1967. *The Deer and the Tiger*. Chicago: Univ. Chicago Press.

Schaller, G. B. 1972. *The Serengeti Lion: A Study of Predator-Prey Relations*. Chicago: Univ. Chicago Press.

Sokolov, V. E., Albone, E. S., Flood, P. F., Heap, P. F., Kagan, M. Z., Vasilieva, V. S., Roznov, V. V., and Zinkevich, E. P. 1980. Secretion and secretory tissues of the anal sac of the mink, *Mustela vision:* Chemical and histological studies. *J. Chem. Ecol.* 6:805–825.

Sokolov, V. E., and Khorlina, I. M. 1976. Pheromones of mammals: study of the composition of volatile acids in vaginal discharges of minks (*Mustela vison*). *Doklady Akademii Nauk SSSR (Biochem.)* 228:225–227.

Tallan, H. H., Moore, S., and Stein, W. H. 1954. Studies on the free amino acids and related compounds in the tissues of the cat. *J. Biol. Chem.* 211:927–939.

Trowbridge, B. J. 1983. Olfactory communication in the European otter *Lutra l. lutra*. Ph.D. dissert., Univ. Aberdeen, Scotland. 201 pp.

Van Dorp, D. A., Klok, R., and Nugteren, D. H. 1973. New macrocyclic compounds from the secretion of the civet cat and the musk rat. *Recueil des Travaux Chimiques du Pays-Bas et de la Belgique* 92:915–924.

van Lawick, H. 1974. *Solo: The Story of an African Wild Dog*. Boston: Houghton Mifflin.

Veberne, G., and de Boer, J. N. 1976. Chemocommunication among domestic cats. *Z. Tierpsychol.* 42:86–109.

Wayre, P. 1979. *The Private Life of the Otter*. London: Batsford.

Wells, M. C., and Bekoff, M. 1981. An observational study of scent marking in coyotes, *Canis latrans. Anim. Behav.* 29:332–350.

Westall, R. G. 1953. The amino acids and other ampholytes of urine. 2. The isolation of a new sulphur-containing amino acid from cat urine. *Biochem. J.* 55:244–247.

Wheeler, J. W., Von Endt, D. W., and Wemmer, C. 1975. 5-Thiomethylpentane-2,3-dione. A unique natural product from the striped hyaena. *J. Amer. Chem. Soc.* 97:441.

CHAPTER 3

Behavioral Development
of Terrestrial Carnivores

MARC BEKOFF

Why Development?

The importance of fully understanding behavioral development cannot be emphasized too strongly. Without detailed knowledge of how the behavior of individuals unfolds throughout life, and not only during infancy, we can only guess at the supposed adaptive significance of various ontogenetic patterns and how they may be related to (1) the immediate situation in which a young animal finds itself and (2) its later reproductive activities and fitness (Tinbergen 1951, 1963; Bekoff 1977d, 1981a, 1981b; Gould 1977; Galef 1981; Wiley 1981; Mayr 1983; Calow 1984; Lee 1984; Bekoff and Byers 1985; Gray 1985a, 1985b; Maynard Smith et al. 1985; Brooks and Wiley 1986; Jamieson 1986; Buss 1987; Clark and Ehlinger 1987; Lomnicki 1988). Therefore, studies of adult behavior conducted in the absence of developmental data may make unwarranted assumptions.

Unfortunately, there are still only scanty data on the development of behavior for most members of the order Carnivora. Thus, even careful generalizing about (1) how proximate factors might influence development and (2) the comparative evolution of life-long ontogenetic trajectories (Wiley 1981) in this diverse group must be considered tentative at best. The large amount of behavioral (and morphological) variability among even the few extant carnivores that have been studied (Scott and Fuller 1965; Ewer 1973; Fox 1975; Bekoff et al. 1984; Gittleman 1986a, 1986b; Rabinowitz and Nottingham 1986) suggests that it would be premature to conclude either that unstudied species will conform to what is already known about close relatives or even conspecifics living in the same geographical area (Bekoff and Wells 1982, 1986; Bekoff et al. 1984; Lee 1984).

The purpose of this review is threefold. First, I briefly discuss some aspects of methods and sampling. Next, data are selectively reviewed for carnivores for which there is quantitative information stemming either from field studies or from systematic research on captive animals (Tables 3.1 and 3.2). Less

89

Table 3.1. Representative studies of carnivores that contain developmental data of various degrees of sophistication

Canidae

Afik and Alkon 1983; Allison 1971; Ashmead et al. 1986; Bekoff 1972a, 1972b, 1974, 1975b, 1975c, 1977a, 1977b, 1977c, 1978a, 1981a, 1981b, 1987; Bekoff and Wells 1982, 1986; Bekoff et al. 1981a, 1981b; Biben 1982a, 1982b, 1983; Brady 1979, 1981; Burrows 1968; Caley 1972; Camenzind 1978; Coppinger et al. 1987; Daniels 1987; Dietz 1984, 1987; Egoscue 1979; Feddersen-Petersen 1986a, 1986b; Fentress and Ryon 1982; Fentress et al. 1987; Fox 1969a, 1969b, 1970, 1971a, 1971b; Fox and Clark 1971; Frame et al. 1979; Garrott et al. 1984; Harrington 1986, 1987; Harrison and Gilbert 1985; Harrison and Harrison 1984; Havkin and Fentress 1985; Henry 1986; Hill and Bekoff 1977; Jean et al. 1986; Johnsingh 1982; Keith 1983; Knight 1978; Lamprecht 1979; Lindström 1983, 1986; Lloyd 1980; Lockwood 1976; Macdonald 1980, 1987; Macdonald and Moehlman 1982; McGrew 1979; Malcolm 1986; Malcolm and Marten 1982; Mech 1970, 1974, 1975, 1988; Mech and Seal 1987; Medjo and Mech 1976; Moehlman 1983, 1986, 1987; Oftedal 1984; Ortega 1988; Packard et al. 1985; Paradiso and Nowak 1972; Pyrah 1984; Reich 1978, 1981; Rowe-Rowe 1978, 1982, 1984; Ryon 1986; Scott and Fuller 1965; Scott and Marston 1950; Silver and Silver 1969; Snow 1967; Storm et al. 1976; Tullar and Berchielli 1980; van der Merwe 1953a and 1953b; van Lawick and van Lawick 1970; Vincent and Bekoff 1978; Wandrey 1975; Wayne 1986; Yamamoto 1984, 1987; Zimen 1976, 1981

Felidae

Baerends–van Roon and Baerends 1979; Bertram 1978; Caro 1979, 1981, 1987; Caro and Collins 1987; Cooper 1942; Eloff 1973; Fitzgerald and Karl 1986; Hanby and Bygott 1987; Hemmer 1972; Kuo 1931; C. Lawrence 1980; Leyhausen 1979; McVittie 1978; Martin 1984a, 1984b; Martin and Bateson 1985a, 1985b; Mendl 1988; Moelk 1979; Pusey 1987; Pusey and Packer 1987; Rudnai 1973; Schaller 1967, 1972; Schneirla et al. 1963; Seidensticker et al. 1973; Sunquist 1981; Tan and Counsilman 1985; West 1974

Mustelidae

Apfelbach 1986; Apfelbach and Weiler 1985a, 1985b; Biben 1982c; Diener 1984, 1985; Erlinge 1979; Estes 1980; Garshelis and Garshelis 1985; Garshelis et al. 1984; Neal 1986; Poole 1966; Powell 1982; Rosatte and Gunson 1984; Stockman et al. 1986

Ursidae

Alt and Beecham 1984; Alt and Gruttadauria 1984; Craighead and Craighead 1972; Dean et al. 1986; Glenn et al. 1976 (cited in Lunn 1986); Latour 1981a, 1981b; Lunn 1986; Pruitt 1974; Ramsey and Dunbrack 1986; Ramsey and Stirling 1986a, 1986b; Rogers 1987; Stirling and Latour 1978

Viverridae

Crawford et al. 1983; Ewer and Wemmer 1974; Hinton and Dunn 1967; Rasa 1973, 1977, 1984; Rood 1978, 1983, 1986; Vilijoen 1980; Wemmer 1977; Wemmer and Fleming 1974

Hyaenidae

Frank 1986a, 1986b; Glickman et al. 1987; Golding 1969; Henschel and Skinner 1987; Kruuk 1972; Kruuk and Parish 1987; Mills 1983, 1984, 1985; Owens and Owens 1978, 1979, 1984

Procyonidae

Fiero and Verts 1985; Fritzell 1977, 1978; Fritzell et al. 1985; Lotze and Anderson 1979; Schneider et al. 1971; Sieber 1986

Note. For general reviews and extensive bibliographies, see Fox 1971a, 1971b; Ewer 1968, 1973; Bekoff 1977a, 1978a; Bekoff and Byers 1981, 1985; Bekoff et al. 1981b, 1984; Fagen 1981; Chapman and Feldhamer 1982; Riedman 1982; Peters 1984; Martin and Caro 1985; Gittleman 1986a, 1986b; Lillegraven et al. 1987; and Burghardt 1988.

Table 3.2. Guide by subject to representative studies of carnivore development

Ethograms
Baerends–van Roon and Baerends 1979; Bekoff 1972a, 1979; Bertram 1978; Daniels 1987; Fox 1969a, 1969b, 1970, 1971a, 1971b; Hill and Bekoff 1977; Kruuk 1972; C. Lawrence 1980; Leyhausen 1979; Mech 1970; Poole 1966; Rasa 1977; Rudnai 1973; Schaller 1967, 1972; Scott and Fuller 1965; Sunquist 1981; Wandrey 1975; Zimen 1981, 1982

Play
Baerends–van Roon and Baerends 1979; Bekoff 1972a, 1972b, 1974, 1975a, 1975b, 1976, 1978c, 1984; Bekoff and Byers 1981, 1985; Biben 1982a, 1982b, 1982c, 1983; Caro 1979, 1981, 1987; Diener 1984; Feddersen-Petersen 1986b; Henry 1986; Hill and Bekoff 1977; Kruuk 1972; Lamprecht 1979; Leyhausen 1979; Martin 1984a, 1984b; Martin and Bateson 1985a, 1985b; Mech 1970; Mendl 1988; Ortega 1988; Poole 1966; Pruitt 1974; Rasa 1973, 1977, 1984; Rudnai 1973; Schaller 1967, 1972; Stockman et al. 1986; Sunquist 1981; Vincent and Bekoff 1978; Wemmer and Fleming 1974; West 1974; Zimen 1981, 1982

Aggression/dominance
Bekoff 1972a, 1977c, 1978a, 1981a; Bekoff et al. 1975, 1981a, 1984; Dean et al. 1986; Fox 1969a, 1971a; Fox and Clark 1971; Frank 1986b; Henry 1986; Knight 1978; Kruuk 1972; Leyhausen 1979; Moehlman 1983; Poole 1966; Rasa 1977; Schaller 1967, 1972; Silver and Silver 1969; van Lawick and van Lawick 1970; Wandrey 1975; Zimen 1976, 1981, 1982

Predation
Baerends–van Roon and Baerends 1979; Bekoff 1978a; Biben 1982a; Caley 1972; Caro 1979, 1981; Coppinger et al. 1987; Harrison and Harrison 1984; Henry 1986; Kruuk 1972; Kuo 1931; Lamprecht 1979; Leyhausen 1979; Macdonald 1980, 1987; Rasa 1973, 1977; Schaller 1967, 1972; Stirling and Latour 1978; Tan and Counsilman 1985; van Lawick and van Lawick 1970; Vincent and Bekoff 1978

Denning
Bekoff and Wells 1982, 1986; Camenzind 1978; Craighead and Craighead 1972; Fentress et al. 1987; Frame et al. 1979; Garrott et al. 1984; Harrison and Gilbert 1985; Henry 1986; Kruuk 1972; Lloyd 1980; Malcolm and Marten 1982; Mech 1970; Mills 1983; Ortega 1988; Owens and Owens 1979; Rogers 1987; Ryon 1986; Schaller 1967, 1972; Schneider et al. 1971; Seiden-sticker et al. 1973; Sunquist 1981; Tullar and Berchielli 1980

Dispersal, philopatry, and movement patterns
Allison 1971; Bekoff 1977b, 1978a; Bekoff and Wells 1982, 1986; Bertram 1978; Burrows 1968; Camenzind 1978; Caro and Collins 1987; Daniels 1987; Frame et al. 1979; Frank 1986a; Fritts and Mech 1981; Fritzell 1977, 1978; Garshelis and Garshelis 1984; Hanby and Bygott 1987; Harrington 1987; Henschel and Skinner 1987; Johnsingh 1982; Kruuk 1972; Kruuk and Parish 1987; Liberg and von Schantz 1985; Lindström 1986; Lloyd 1980; Macdonald 1980; Marks and Redmond 1987; Mech 1970; Moore and Ali 1984; Pusey 1987; Pusey and Packer 1987; Reich 1978, 1981; Rogers 1987; Schaller 1967, 1972; Schneider et al. 1971; Seidensticker et al. 1973; Shields 1982, 1983; Stirling and Latour 1978; Storm et al. 1976; Sunquist 1981; van Lawick and van Lawick 1970; Waser and Jones 1983; Zimen 1976, 1981, 1982

Life history analyses
Bekoff et al. 1981b; Bekoff and Conner 1987; Craighead and Mitchell 1982; Eloff 1973; Fiero and Verts 1985; Frame et al. 1979; Fritzell et al. 1985; Gittleman 1986a, 1986b; Gittleman and Harvey 1987; Lloyd 1980; Mech 1975; Moehlman 1986 (see text for discussion); Ramsey and Dunbrack 1986

Helping
Bekoff and Wells 1982, 1986; Camenzind 1978; Fentress and Ryon 1982; Frame et al. 1979;

(continued)

Bunge 1979; Thompson 1981; Eaton and DiDomenico 1985) relationships between ontogeny and individual reproductive fitness or success (Lee 1984; Bekoff and Byers 1985).

Many carnivores are also appealing for developmental studies because of their social organization. In a number of species "typical" or modal (Eisenberg 1981) social groups (packs, prides) are composed of related individuals or of individuals of varying degrees of relatedness (Bekoff et al. 1984; Bennett 1987). Therefore, the likelihood that various types of social behavior evolved via kin selection and/or reciprocity can be assessed (Reich 1978; Rood 1978, 1983, 1986; Bekoff and Wells 1982, 1986; Bekoff et al. 1984; Mills 1985). Also, the way in which young are integrated into their natal group can be studied, as can the question of why offspring retention (natal philopatry: Waser and Jones 1983) has evolved (Lindström 1986; Bekoff 1987). The addition of infants into a group and the emigration of juveniles (and adults) from their natal group may greatly influence the social dynamics among all individuals. For example, the importance of young animals as "social glue" has been stressed by Rasa (1977). She suggests that in the case of the dwarf mongoose (*Helogale undulata refula*), amicable and appeasing behavior of young animals is probably the most important factor in maintaining group stability.

A word of caution is necessary. Despite the appeal of such studies, the obvious difficulty of directly observing the development of *identified* young carnivores in or around dens and following them during forays and dispersal from the natal area must be accepted. Furthermore, assessing kin relationships in many carnivores may be impossible under field conditions because of the pooling of litters in communal dens (Owens and Owens 1979; Lloyd 1980; Rood 1980; Tullar and Berchielli 1980).

Methodological Issues

"In studies of . . . development it is clear that 'what you measure is what you get'" (Fentress 1985:8). The methods used to study any problem directly determine the results. Consequently, they have a profound influence on how data are interpreted and thus on the generation of subsequent hypotheses and future research programs (Magee 1973; Feyerabend 1975; Himsworth 1986; Murray 1986). To present useful data on development, we must be able to study identified individuals of known litter size, age, sex, and genetic relatedness beginning as early in life as possible and for as long a period of time as conditions permit. There simply is no alternative if we want to learn about individual patterns of development and make reasonable causal inferences about development up to some specific age and how events at that time were influenced by past experience. Of course, collecting these data on most carnivores under field conditions is extremely difficult, and even for studies of animals in captivity this would be a tall order.

The Study Site

Assuming we have accepted the known or identified limitations of our methods, how do we go about tackling a developmental problem? After it has been decided what questions are to be asked and which species is to be studied, the study site at which the required information can be best collected must be chosen with care. Detailed questions about mechanisms of development or about the ontogeny of individual motor patterns or early social interactions are usually best done using captive animals, for the loss of even a small amount of information can render a study meaningless.

Altthough detail is almost always sacrificed in the field, at some point such research is necessary. It is unlikely that general trends uncovered in the study of captive animals will be inapplicable to natural populations of the same species, although the rates at which various actions are performed may differ greatly due to differences in group structure or spatial limitations (Ryon 1986). For example, the manner in which captive canids communicate, fight, mate, play, or kill prey is very similar to the way in which their wild conspecifics do so; the form of the actions used and the sequences of motor patterns are almost identical (Fox 1969a, 1969b; 1971a; Mech 1970; Bekoff 1972a, 1978a; Zimen 1981, 1982; Bekoff and Wells 1982, 1986). But in many instances it would be impossible to collect such detailed information under field conditions. Furthermore, data from captive animals can, and should be, used to guide future field studies. For example, the relationship between early patterns of behavioral development and a species' typical pattern of social interaction remains unclear, but there have been some efforts to clarify the situation in different species (see below)

Other Difficulties: Observing and Estimating
Life History Variables

Developmental studies are also faced with the difficulty of (1) observing known individuals before they emerge from their den(s); (2) following youngsters that have emerged from the den but are difficult to see because they are small and cryptically colored; (3) marking young animals so that the tags or collars are not lost because of increases in body size or vigorous playful or other type of interactions; and (4) making sure that the devices used to mark the animals do not injure them (Sallaberry 1985). Expandable radio collars have proven to be useful in various studies (Bekoff and Wells 1980, 1982, 1986; J. Laundré, pers. comm.). Following young animals is usually not as difficult as tracking older individuals because youngsters' movements are typically not as wide-ranging until dispersal (Schaller 1972; Bertram 1978; Bekoff and Wells 1986). Thus, fewer locations are needed to determine reliably their home ranges and general patterns of movement (Bekoff and Mech 1984).

Reliable estimation of litter size (Eloff 1973; Bertram 1978; Lloyd 1980;

Bekoff and Wells 1986; Frank 1986b), sex ratio, and age of young is also difficult. For example, Bertram (1978) rarely knew the litter size of the African lions (*Panthera leo*) he studied until they were about six weeks of age. Lloyd (1980), studying red foxes (*Vulpes vulpes*), listed three reasons for the difficulty of estimating litter size under field conditions: (1) all animals may not be found in a den, (2) there may be pooling of litters, and (3) there may already have been mortality when individuals are first counted in a den or when they emerge. Eloff (1973) stressed that reporting an average litter size on the basis of individuals of different ages is not valid. The difficulties associated with determining differential mortality by sex and as a function of age (Ralls et al. 1980) and estimating how sex ratios may be influenced by population density and degree of exploitation (Mech 1975) also emphasize the importance of reliably estimating these parameters as early in life as possible. The fact that accurate estimation may be difficult does not mean that it is not worth doing these types of studies; we simply need to be aware of the possible limitations of these data.

Despite inherent difficulties, coming to terms with these types of data along with interspecific and intraspecific variation is essential for gaining a grasp on developmentally oriented life history analyses of carnivores (Bekoff et al. 1981b, 1984; Gittleman 1986a, 1986b). For example, Moehlman (1986:69, fig. 4.2) reported a significantly positive ($r = +0.85$, $P < 0.01$) relationship between mean litter size (MLS) and mean female body weight (MFBW) for 15 canids. (The results actually are based on *ln*-transformed data and not *log*-transformed data as the figures are labeled. The regression line drawn on figure 4.2 is incorrect: the y-intercept should be -1.24. This difference is important because an incorrect line-of-best-fit is used to analyze the relative position of species, including "outliers," on these axes.) Moehlman mentions that this is the first time that such a relationship has been recorded for any mammalian family; thus, select canids appear to be unique. A recalculation of information presented in Bekoff et al. (1981b) using ln-transformed data produces a positive but nonsignificant association ($r = +0.39$, $P > 0.05$) between MLS and *minimum* female weight (MINFW) for 13 canids. In her analysis Moehlman (1986) omitted three "outlying" species (arctic fox, *Alopex lagopus;* crab-eating fox, *Cerdocyon thous;* maned wolf, *Chrysocyon brachyurus*). When the arctic fox is left out of Bekoff et al.'s data set, the relationship between MLS and MINFW is stronger ($r = +0.52$) but still nonsignificant. When all 18 species for which Moehlman amassed data are considered, the association between MLS and MFBW is nonsignificant and strikingly similar ($r = +0.33$) to that found for the relationship between MLS and MINFW ($r = 0.39$). This finding is important because further reference to this relationship (Moehlman 1987:370) can be misleading. The use of slightly different indices of female weight does not seem to be a major issue.

Ewer's (1968:ix) admonition that there is no such thing as a typical mammal stresses that we must be aware that general statements about development may actually be limited in scope. For example, Gittleman and Harvey (1987) reported a negative correlation ($r = -0.71$, $P < 0.001$) between age at eye opening (EO) and age at independence (AI) for 35 diverse carnivores after removing the influence of body weight. They suggest that the association between EO and AI may be influenced by the amount of energy that is allocated to growth; species that show a delay in EO may remain relatively inactive inside dens and thus allocate more intensive energy toward early postnatal growth and reach independence earlier. However, for 11 species in the family Canidae there is a positive but nonsignificant relationship between EO and AI ($+0.46$) after the effects of body weight are removed (Bekoff and Conner 1987). The demonstration of different associations between these two variables does not mean that one analysis is right and the other wrong. Rather, one simply needs to remain aware of the amount of variation that may exist at different taxonomic levels (J. Gittleman, pers. comm.).

I have gone into some detail here not only because there is a lot of interest in developmental life history analyses but also because it is imperative that researchers (1) decide on when it is appropriate to discount interspecific variation (remove "outliers"), (2) determine how such a procedure *influences the formulation of reliable assessments of how different life history variables are related within different taxonomic levels,* and (3) come to terms with possible restrictions of broad generalizations among different levels of classification. It is also essential that objective guidelines be established for the selective exclusion of species when different variables are being considered; an "outlier" in one analysis may not be an "outlier" in another (Moehlman 1986). Solid developmental (and other) studies of life history strategies rely not only on the accurate collection of data but also on the correct analyses and interpretation of this information. One only needs to keep in mind that Cohen (1963) demonstrated that Alexander the Great did not exist and that he had an infinite number of limbs.

Finally, it should be noted that some multivariate techniques (e.g., discriminant function and principal components analysis) and other statistical methods may be useful for analyzing developmental data (Bekoff et al. 1975; Bekoff 1977c, 1978b; Rushen 1982; Stanislaw and Brain 1983; Martin and Bateson 1986). Many ontogenetic studies need to consider numerous variables simultaneously (age, sex, body size, litter size, litter sex ratio, mortality, the nature of food resources, presence or absence of helpers, experience of the mother), for which multivariate techniques may be particularly well suited.

Developmental Trends

In keeping with Ewer's (1968, 1973) broad view of behavioral biology, I will consider different "levels" of analysis and address some "big questions," recognizing fully that my choice is subjective and certainly debatable. However, space limitations require that I be selective and brief. Thus, the neural bases of motor behavior and sensory perception are not covered (see Fox 1971b; Panksepp et al. 1984; Sheppey and Bernard 1984; Apfelbach and Weiler 1985a, 1985b; Fentress 1985, 1986; Apfelbach 1986; Ashmead et al. 1986), nor are possible relationships between brain size, development, and metabolism (Bennett and Harvey 1985) considered. Studies at this level are very helpful in answering questions concerning (1) possible relationships among the development of various senses, the endocrine system, and ontogenetic changes in behavior and (2) differences between species varying in the degree of development of the central and peripheral nervous systems at, and shortly after, birth.

Motor Patterns and Sequences of Behavior

The first step in any behavioral study is the construction of an ethogram (behavioral repertoire) of motor patterns that are used in different contexts (see Hinde 1970; Bekoff 1979; Schleidt et al. 1984 for discussions of the description and classification of motor patterns). An ethogram is typically purely descriptive; causation, context, function, and motivation are not included in this inventory (Schleidt and Crawley 1980), in which pattern recognition is based on the most stereotyped features of a given movement (Schleidt 1982). The level of description varies according to the researcher's needs; it can be an extremely detailed analysis describing motor patterns in terms of movements on a series of spherical coordinates (Golani 1976; Havkin and Fentress 1985) or a more crude listing in which motor patterns are described simply in terms of form. Even a relatively gross level of description is useful for differentiating among closely related young canids (Bekoff et al. 1975, 1984; Bekoff 1978b). Many of the references listed in Tables 3.1 and 3.2 contain ethograms for different species.

From a developmental perspective, the precise description and classification of motor patterns is essential for a number of reasons. First, actions may change form as animals mature and grow. Also, gross and diffuse movements of the body may become individuated (Coghill 1929) into discrete, localized, and subtle movements of the limbs, trunk, head, and tail. Likewise, independent movements of different body parts may become integrated (Windle 1940; for review see A. Bekoff 1988) into coordinated motor patterns seen in activities such as predatory behavior (Fox 1969b; Mech 1970; Kruuk 1972; Schaller, 1972; Rasa 1973; Bekoff 1978a; Baerends–van Roon and Baerends 1979; Leyhausen 1979; Powell 1982; Bekoff and Wells 1986), play (Bekoff 1972a, 1972b, 1974, 1975b), agonistic behavior (Fox 1969a; Fox and Clark 1971; Bekoff 1978a; Knight 1978), or social communication (Fox 1970; Chevalier-Skolnikoff 1974; Lehner 1978). Generally, as an individual gains increasing experience in a wide variety of social and other types of encounters, its motor patterns become more refined, expanded, and perfected (Meier

1981) as different actions combine synergistically (Corning 1983; see Altmann 1986 and Levins and Lewontin 1985:267ff. for a discussion of relationships between wholes and parts that is relevant to the study of development). Finally, by studying discrete motor patterns, one can answer questions about the "innateness" of different actions (e.g., Rasa's 1973 and Leyhausen's 1979 studies of the development of predatory behavior).

Sequences of behavior are composed of individual acts, and the analysis of behavior sequences is one level above the study of single motor patterns (Brown and Colgan 1985; Bakeman and Gottman 1986). Typically, relationships between pairs of acts are studied and conditional probabilities for two-act transitions are calculated. For example, based on analyses of two-act transitions for predatory encounters by young coyotes (*Canis latrans*) under field conditions, Bekoff and Wells (1986) concluded that by the time coyotes are five to nine months old, predatory sequences resemble those of adults in structure and length.

There actually has been very little detailed analysis of the way sequences are constructed for different behavior patterns in carnivores. Data, mostly qualitative, are available for diverse topics, including mother-infant interactions (C. Lawrence 1980 and references therein), predation (Ewer 1968, 1973; Fox 1969b; Mech 1970; Caley 1972; Kruuk 1972; Schaller 1972; Rasa 1973; Bekoff 1978a; Leyhausen 1979), agonistic behavior (Poole 1966; Ewer 1968, 1973; Fox 1979a), and social play (Bekoff 1972a, 1972b, 1974, 1975a, 1976).

Agonistic Behavior and Dominance Relationships

Many comprehensive studies of carnivores include at least some information on the development of agonistic behavior and dominance relationships (Bekoff 1981a; Bernstein and Williams 1983). In coyotes (Fox 1969a, 1971a; Silver and Silver 1969; Bekoff 1974, 1978a; Knight 1978), golden jackals (*C. aureus*) (Wandrey 1975), at least some populations of red foxes (Burrows 1968; Henry 1986) and arctic foxes (MacPherson 1969), and occasionally wolves (*C. lupus*) (Mech 1970) and dholes (*Cuon alpinus*) (Johnsingh 1982), agonistic interactions may result in the formation of dominance hierarchies very early in life. L. Frank (pers. comm.) reports that early and severe aggression between neonatal spotted hyenas (*Crocuta crocuta*) of the same sex often results in the death of one individual (siblicide). Biben (1983) suggests that the observations of early intensive agonistic encounters in coyotes that result in the formation of dominance hierarchies may be an artifact of observation conditions. However, this is unlikely because animals reared in a wide variety of situations (and observed by different investigators) behave similarly early in life (Silver and Silver 1969; Fox and Clark 1971; Bekoff 1978a; Knight 1978; Bekoff et al. 1981a; Feddersen-Petersen 1986b). Also, social interaction patterns observed among captive young animals for whom dominance relationships were known and clear-cut (Bekoff 1978a; Knight 1978; Bekoff et al. 1981a) have been observed among youngsters in the field (pers. observ.).

The significance of the formation of early dominance relationships remains unclear for either the youngsters at the time they have formed their social relationships or with respect to long-term effects of differential social rank. Frank (1986b) reported that the maternal rank of spotted hyenas is inherited and cubs of higher ranking females are more able than other females' cubs to feed successfully at kills in competition with adults. Furthermore, he postulates that the highly aggressive sons of alpha females would be very successful competitors in the context of the spotted hyena's polygynous mating system.

Individuals of different ranks also may show varying predispositions to disperse from their natal groups for behavioral (Burrows 1968; Bekoff 1977c; Gaines and McClenaghan 1980) or energetic (Golightly 1981) reasons, but more data are needed to obtain a clear picture of developmental influences on dispersal. If young dispersing carnivores suffer higher mortality than do more sedentary littermates (Bekoff et al. 1984; Bekoff and Wells 1986), advantages (protection, access to a dependable food supply) of juvenile philopatry are obvious. A territory also may be inherited and used over a number of genera-tions by related individuals (Bekoff and Wells 1986; Lindström 1986).

Quantitative analyses of the proximate mechanisms of hierarchy formation are also essential (Lockwood 1976; Bekoff 1977c, 1978a; Knight 1978; Bekoff et al. 1981a; Chase 1982a, 1982b; Nelissen 1986). The applicability of Chase's (1982a, 1982b, 1986; Slater 1986) "jigsaw model" of hierarchy for-mation to the development of dominance relationships among young individu-als needs to be assessed. In this model, hierarchy formation basically is seen "as a dynamic process in which sequences of dominance relationship formation is [*sic*] smaller groups concatenate to form hierarchies in larger groups. It argues that the structural form of a hierarchy can be explained by regularities or 'building blocks' of interaction involving two individuals and a bystander" (Chase 1982b:230) involved in a triadic relationship.

Other questions that need to be addressed deal with the development of sex differences in dominance relationships, the stability of relationships formed early in life (Knight 1978; Bernstein and Williams 1983), and the influence of age, size, sex, general health, and prior win/loss record on future encounters. Patterns of initiation, escalation, and termination during agonistic interactions also should be studied (Bekoff et al. 1981a) to gain more information about the development of dominance hierarchies and how relative social rank influ-ences subsequent relationships among young individuals.

Social Play Behavior

Comparative reviews of social play behavior in carnivores and other mam-mals can be found in Bekoff (1978c, 1984), Bekoff and Byers (1981, 1985),

Fagen (1981), Martin and Caro (1985), Meaney et al. (1985), and Meaney (1988). Because the young of most species typically do (and should: Fagen 1977) play more than do adults, many discussions of play take on a strongly developmental perspective, although quantitative data are rare. Furthermore, detailed information on the taxonomic distribution of different types of play among carnivores is scanty. How play is distributed relative to different life history strategies (Bekoff and Byers 1981; Fagen 1981; Ortega and Bekoff 1987) also needs to be addressed in greater detail.

Functional analyses of social play, in which the questions being asked stress the evolution (Hinde 1975; Symons 1979; van Dongen and van den Bercken 1981; Jamieson 1986) of this behavioral phenotype, suggest that play in many species is important for at least physical training (see below), socialization, and/or sensorimotor/cognitive training (Brownlee 1954; Bekoff and Byers 1981, 1985; see also Fagen 1981 and Martin and Caro 1985). Baerends–van Roon and Baerends (1979) suggested that play in kittens is important for teaching them how to deal with the opposed tendencies of attacking and fleeing, an important prerequisite for socialization. Two kittens that were deprived of the opportunity for social play after 7 weeks of age showed "later signs of insufficient harmonious control of their attack and escape tendencies in agonistic, sexual, and parental encounters" (Baerends–van Roon and Baerends 1979:103).

However, possible relationships between early play experience and later behavior are not necessarily clear-cut (Bekoff and Byers 1981, 1985; Fagen 1981; Martin and Caro 1985). For example, Schaller (1972) observed in African lions that the most common act used in play, wrestling, was rarely used by adults, whereas stalking, an important component of hunting, was infrequently observed in play by cubs. Martin and Caro (1985) noted that data relevant to practice theories of play actually suggest that play improves later skills rather than being absolutely necessary for their development. Play also may enhance behavioral flexibility as a result of learning that takes place during these social encounters.

Despite a lot of interest in the phenomenon of social and other forms of play, it must be stressed that just about all suggestions about the possible functions of play are merely suggestions. It is currently impossible to make any hard-and-fast statements about "the" function of play because of (1) the lack of quantitative comparative data, (2) species differences in the predominant types of play in which individuals engage, and (3) the possibility that functions of play may change during ontogeny (Bekoff and Byers 1985; Martin and Caro 1985).

However, there are some data for a few species suggesting that variations in early social play experience may affect the strength and maintenance of social bonds formed within (and possibly between: Ortega 1988) litters and among young animals and adults. As a consequence, an individual's tendency to leave its natal group (and the group's social organization) may be affected (Bekoff

1977b, 1984; Englund 1980; Fagen 1981; Zimen 1981, 1982). However, detailed longitudinal field observations of identified individuals are lacking.

Whether or not variations in play experience are associated with different habitats and resources (Berger 1979; Fagen 1981; Lee 1984) is also unknown. Lee (1984) observed that the play of vervet monkeys (*Cercopithecus aethiops*) was influenced by time budgets, the energy available from the diet, and the overall abundance and distribution of food. Bekoff (1978c) suggested that in some cases mothers (or the primary care-giver) may restrict play if it placed undue energetic demands on them to provide food to active youngsters when food was scarce. This type of parent-offspring conflict has been observed in domestic cats (*Felis catus*) (C. Lawrence 1980) and other mammals (Lee 1984). Right whale (*Eubalaena australis*) mothers may quiet boisterous play to conserve energy when bringing a calf to weaning (Thomas and Taber 1984).

Quantitative cost analyses of play have also been attempted. Based on a study of play in captive domestic cats, Martin (1984a) concluded that the time and energy costs of this activity may be nominal and that play may not be costly, at least in this species under the conditions in which it was studied. There may be indirect costs to youngsters and hidden costs to parents, however, if play is associated with hunting failures (Caro 1987). The survivorship cost of play has not yet been determined in any species (Martin and Caro 1985).

In any study of relative costs and benefits, regardless of whether time and energy or individual survivorship and fitness are being considered, one must account for the specific situation in which the animals are being observed (Lee 1984; Bekoff and Byers 1985) and the type of play that is under scrutiny (Bekoff and Wells 1986). High-energy locomotor social play may require a lot of energy in a short period of time. In some situations even the minimal amount of time and energy spent in play may be costly and place unnecessarily high energetic stress on youngsters or care-givers (see Caro 1987). Also, this type of activity can result in tissue, tendon, muscle, bone, and psychological damage, the repair of which also requires energy that might otherwise be allocated to growth and maintenance (Calow 1984; for a discussion of possible genetic mechanisms concerning individual responses to stressful environmental variations, see Mitton and Koehn 1985). Furthermore, there might be costs associated with convalescing; a young individual might be unable to play and thus not gain any of the potential benefits associated with the activity. It might also be unable to feed or defend itself or to travel with its group while recuperating. Convalescence requiring inactivity also may induce "detraining" (see next section), resulting in changes in muscles' ability to do work and the loss of aerobic and anaerobic capacities. Thus, the costs associated with play or any other behavior are not limited to the performance of the activity itself.

More detailed comparative, developmental, ecological, and physiological information about play in carnivores is needed. Structural analyses of the forms of motor patterns used in carnivore play (Henry and Herrero 1974; Hill and Bekoff 1977) will undoubtedly shed some light on function (Bekoff and Byers 1981; Fagen 1981; Bekoff 1982). Experimental studies (Martin and Bateson 1985a) should be done whenever possible. Although it seems likely that variations in play experience would have some influence on later reproductive activities, there are no data that address this question.

Motor Training and Physical Fitness

For active and cursorial animals such as carnivores, being physically fit is probably essential for survival. However, very little attention has been devoted to the development of physical (aerobic and anaerobic) fitness (Bekoff 1988). As mentioned above, discussions of play often deal with possible physical training effects that may result from engaging in this activity (Fagen 1976, 1981; Bekoff and Byers 1981, 1985; Martin and Caro 1985). Of course, just about any type of vigorous and varied motor activity will result in some form of training.

There are four general physiological, anatomical, and behavioral effects of motor training: (1) bones thicken and become remodeled in response to the specific stresses associated with the activity; (2) muscles used in the activity, if stressed specifically and sufficiently, hypertrophy and show biochemical and cellular changes associated with an increased ability to do work; (3) cardiopulmonary capacity and efficiency increases; there are changes in maximum and minimum heart rates, cardiac output, stroke volume, and oxygen extraction from the blood (Blomqvist and Saltin 1983; Garland 1985) and there is an increase in endurance; and (4) the smoothness and economy of repeated movements increase. The first three effects of physical exercise are usually referred to as the "training response" by physiologists.

It is intriguing to think of how differential motor training during development may be linked to individual differences in physical fitness that may influence later behavior (Bekoff 1988). In addition to the possible effects that differential individual social experience may have on social behavior, there is no reason to think that early variations in motor activity and physical fitness do not influence the performance of many behaviors (predation, territory or den defense, courtship, play, aggression) that are associated with later reproductive activities (see Watt et al. 1986).

It also is known for humans (Malina and Bouchard 1986) and other animals (Ryan 1975; Powers et al. 1983; Garton et al. 1985; Danzmann et al. 1987, 1988; see Bekoff 1988 for additional references) that there are strong genetic influences on individual physical fitness, including glycogen metabolism (Brown 1977) and heart and breathing frequencies (Arieli et al. 1986). In many studies protein heterozygosity (Mitton and Grant 1984) appears to be strongly and positively associated with variations in oxygen consumption, metabolic efficiency, endurance, superior viability, greater fecundity, and growth

(Frelinger 1972; Serradilla and Ayala 1983; Garton 1984; Garton et al. 1985; Mitton and Koehn 1985). Enzyme heterozygosity may also be related to developmental stability (Leary et al. 1984; Mitton and Grant 1984; Mitton and Koehn 1985).

With respect to development, some individuals may be able to train more than others because they are (perhaps genetically) better able to withstand the stresses associated with growth, maintenance, vigorous motor activity early in life, *and* to recover faster from stressful situations. Mitton and Grant (1984) suggest a genetic mechanism by which individual differences in responses to stress may rise. They postulate that enzyme heterozygosity may enhance efficiency by decreasing energetic costs of standard metabolism, which leads to developmental stabilization.

Along these lines, Martin and Bateson (1985b) found consistent differences between litters of domestic cats with respect to locomotor activity during play. Individual differences also were apparent. Social rank, body size, sex, litter size, group composition, and nutritional state can all influence social interaction patterns that may have some effect on the development of physical fitness. Genetic differences also may be important to consider. Martin and Bateson (1985b:509) concluded that "the ontogenetic origins of behavioural differences such as these between individuals or litters, and their *functional significance* [my emphasis] (if any), are as yet largely unknown and present an important challenge to those studying behavioural development." Clearly, the future is wide open with respect to research opportunities concerning basic questions dealing with possible relationships among physical fitness, genetics, developmental stability, and individual variations in behavior (Bekoff 1988).

Dispersal

Despite the fact that it is often implied that we know a lot about dispersal by young (or adult) carnivores, there are few solid data concerning the diverse but interrelated factors that may influence the likelihood of an individual's moving away from, or remaining at, its birthplace past the age of independence. Thus, generalizations about sex differences in dispersal or conclusive explanations about proximate or ultimate influences (inbreeding avoidance, competition for mates or other resources; Greenwood 1980; Dobson 1982; Moore and Ali 1984; Dobson and Jones 1985; Liberg and von Schantz 1985; Waser 1985; Waser et al. 1986; Bekoff 1987; Marks and Redmond 1987; but see Pusey 1987; Pusey and Packer 1987) on movement patterns are currently speculative, especially for carnivores (Storm et al. 1976; Bekoff et al. 1984; see also Shields 1982, 1983; Zimen 1982; Theberge 1983; Chesser and Ryman 1986; Waser et al. 1986; Pusey 1987 for further discussions of inbreeding).

Likewise, ontogenetic influences on individual behavioral phenotypes and dispersal (Burrows 1968; Bekoff 1977b; Harcourt 1978; Harcourt and Stewart 1981; Lott 1984) remain unclear, although there is some evidence that dispersal in rodents may be a consequence of the development of individuality (Armitage 1986). Young animals may be less successful than older group members in competing for various resources; thus, they may be more disposed than older individuals to leave their natal group (Bertram 1973; Elliott and McT. Cowan 1978; Macdonald 1980; Fritts and Mech 1981).

It is a common observation that not all individuals in a litter are equally likely to disperse (Bekoff 1977b; Lloyd 1980; Downhower and Armitage 1981; Marks and Redmond 1987), independent of sex. Within litters, differential dispersal may be related to relative dominance rank (Bekoff 1977b, 1981b; Zimen 1976). Because of the risks associated with dispersal (Macdonald 1980; Bekoff and Wells 1982, 1986), dominant individuals may be more likely than subordinate animals to exercise their prerogative and drive out lower ranking littermates. (In some species, dominant individuals appear to have more freedom of movement [Scott and Fuller 1965; Knight 1978], may have priority of access to food [Frank 1986b], and are less vigilant than subordinate animals [Knight 1978]). However, there are few clear indications from field work that this logical explanation is generally valid.

It is also possible that both dominant and subordinate individuals may be predisposed to disperse from their natal site because of their inability to develop strong social bonds with siblings and other group members (Burrows 1968; Bekoff 1977b) or because of the stresses associated with being a high- or low-ranking animal (Golightly 1981). Dominant individuals may be avoided by other group members and subjected to regular challenges, whereas subordinate animals may actively avoid social encounters and be subjected to continued harasssment (Bekoff 1977b; Feddersen-Petersen 1986b).

Long-term studies of identified individuals are essential if we are to further our understanding of dispersal patterns in carnivores. We must pinpoint what conditions favor the retention of some individuals rather than others in their natal group and account for intraspecific variation in movement patterns.

Social Development, Sex Differences, and Social Organization

Possible relationships between the development of social behavior within a species and species-typical patterns of social organization are difficult to tease apart for a number of reasons. First, detailed developmental data for identified individuals are difficult to gather under most field conditions. Also, in some cases it is difficult to characterize a species as typically being solitary or social because of pronounced intraspecific variability in social organization among diverse carnivores (Bekoff et al. 1984; Bekoff and Wells 1986). And, it is not known whether these intraspecific differences are reflections of local variations in early social development, as they appear to be in bighorn sheep (*Ovis canadensis*) (Berger 1979) and lemurs (*Lemur* spp.) (Sussman 1977). However, in some species, at least sex differences in development seem to be

consistent and related to species-typical social organization (e.g., female philopatry and male dispersal in the case of lions; Schaller 1972; Bertram
1978; Pusey and Packer 1987), whereas in others, sex differences in behavior
seem to be habitat-specific. For example, movement patterns by male and
female African hunting dogs (*Lycaon pictus*) living in different habitats vary
(see Frame et al. 1979; Reich 1981). In their extensive review of sex differences
in social play, Meaney et al. (1985; see also Meaney 1988) concluded that for
species in which there are large sex differences in adult social roles, one would
expect to see sex differences in developmental strategies. Thus, "sex differences
in social play contribute to the socialization process by enhancing the appropriate forms of social interactions for each sex" (p. 46). For behaviorally
(species in which males and females serve as helpers, for example) and physically monomorphic carnivores, one would predict minimal sex differences in
early development (see Biben 1982c; Meaney et al. 1985). Indeed, among
mammals, paternal care (and monogamy) are usually associated with a reduction in sexual dimorphism (Kleiman 1977; Ralls 1977).

Despite the difficulties of studying the relationship between early development and social organization, there are some data from work on captive
animals that suggest that interspecific differences in social organization may be
associated with species differences in social development. For example, the
ontogeny of behavior, especially agonistic encounters and dominance relationships, in what are typically social species is different from patterns seen in
typically less social species (Bekoff 1974, 1977b; Henry 1986); rank-related
agonistic encounters often precede the emergence of social play. In the case of
solitary red pandas (*Ailurus fulgens*), most early play activities are directed
toward bamboo stalks that are batted, manipulated, and bitten rather than
toward littermates (J. Gittleman, pers. comm.). Furthermore, variations in the
way food is handled by young individuals may be associated with species
differences in food-related behavior (Biben 1982a, 1982b; for a discussion of
Biben's 1982a data as they relate to the practice theory of play, see Caro and
Alawi 1985; Martin and Caro 1985). Along these lines, Rasa (1973) suggested
that the slow development of predatory behavior may favor sociality. Although this does not seem to be the case for some canids (Fox 1969b), there are
too few data to make any generalizations about this interesting possibility.

Biben (1983) suggested that differences between highly social bush dogs
(*Speothos venaticus*) and less social crab-eating foxes and maned wolves do
not follow conventional ideas about the way sociality is likely to develop. She
postulated that the young of more social species should (1) show more complex behavior, (2) be less aggressive, and (3) have a well-defined dominance
hierarchy. Reliable assessments of "levels of complexity" are difficult to establish, especially because of the lack of comparative field data, the subjective
nature of this type of analysis, and the absence of complete data sets even
within species. It is not at all clear that more social species do possess more
complex behavioral repertoires or that their sequences of behavior are more

elaborate or complex than those performed by less social species. Agonistic displays (Baker 1980) and facial expressions (Fox 1970) of some solitary carnivores may be less well developed than those of more social species, but this does not mean that as a whole the species' behavioral repertoire is less complex.

Likewise, questions dealing with rates of aggressiveness and the rigidity (and stability) of dominance hierarchies have not been studied in any detail in the field. For example, depending on the situation, highly social wolves can be very aggressive early in life, but stable dominance relationships may not form until a later age (Mech 1970; Zimen 1981, 1982). Furthermore, there may be seasonal patterns of alternating stability and instability in dominance and other types of social relationships that are influenced by food supply and group composition.

Because development is a dynamic process and what happens at one age may not be a good indicator of what happens at a later age (Bateson 1976; Knight 1978), it would be wrong to assume that there will be absolute rules by which sociality will develop, even within a species. Byers's (1983) study of the social development of collared peccaries *(Tayassu tajacu)* and Lee's (1984) data on the development of vervet monkeys clearly illustrate that there may be alternative developmental pathways that result in sociality, especially under varying field conditions. Martin and Caro's (1985) concept of "equifinality," which states that in an open system the same steady state in development can be reached from different initial conditions and in different ways, also is relevant here. In summary, we simply need more data in this area to see if, indeed, any strong comparative statements can be made relating "typical" patterns of development with "typical" patterns of social organization (see Lindstrom 1986).

Other Areas of Interest

Other relevant areas are predatory behavior, ontogenetic perspectives on helping, learning, recognition, relationships between evolution and development, and the importance of rare events during development (Weatherhead 1986a, 1986b). Additionally, questions about behavioral (Bookstaber and Langsam 1985) and neural (Dumont and Robertson 1986) optimality are important to any study of development. Space limitations preclude even a brief consideration of these areas, but many of the papers listed in Tables 3.1 and 3.2 contain information on these topics. General surveys and theoretical treatments that are useful for carnivore research can be found in Gould (1977), Eisenberg (1981), Fink (1982), Johnston (1982a, 1982b), McFarland (1982), Riedman (1982), Colgan (1983), Corning (1983), Byers and Bekoff (1986), Emlen et al. (1986), Jamieson (1986), Kortmulder (1986), Peck and Feldman (1986), Toulouse et al. (1986), and Blaustein et al. (1987).

Conclusions

More data are needed in almost all areas of behavioral development. Necessary and sufficient conditions need to be specified for the evolution and organization of various developmental pathways for specific behavior patterns and across species. We need to consider whether or not big developmental questions are simply too large to answer (Slobodkin 1986) and also whether or not we are cursed by our mental ability to create explanations for just about anything we care to (Rowell 1979). At this stage we need simple models (Boyd and Richerson 1985) and should avoid unnecessary complexity (Williams 1985; Watson 1986). As Sternberg (1985:1117) warned, "Models should be our servants rather than our masters." They should be formulated with the intention of being rigorously tested. Jamieson's (1986) cautious approach to stretching selectionist theory, especially in developmental studies, needs to be taken to heart, especially when it is so very difficult to demonstrate causal relationships between ontogenetic events and later behavior.

The absence of developmental data basically means that any study is incomplete (Bekoff and Byers 1985). The importance of studying animals in as close to natural conditions as possible and attempting to fit in patterns of development with a species' natural history (R. D. Lawrence 1980; Bartholomew 1986) cannot be emphasized too strongly. The usefulness of comparative ontogenetic data and the excitement of observing behavior unfold throughout life makes these types of endeavors worthwhile and enjoyable. We should take advantage of the supportive atmosphere within which most of us work to tackle a wide variety of questions dealing with development.

Acknowledgments

I thank the John Simon Guggenheim Memorial Foundation for providing a fellowship and the University of Colorado for permitting a sabbatical leave, during which time some of the ideas contained herein were developed. In particular, this time allowed me to read papers and books that otherwise would have been shelved for eternity. John Gittleman, Beth Bennett, Douglas Conner, Thomas Daniels, David Manry, Allen Moore, Joseph Ortega, and an anonymous reviewer offered useful and insightful comments on an earlier draft of this chapter. Jeffry Mitton helped with the analyses of life history data, and Muriel Sharp and Bay Roberts helped in procuring and organizing many of the papers that are cited herein. J. Mitton, Ted Garland, and Todd Gleeson provided information on the genetics of aerobic performance, and Andrew Pruitt discussed many of these issues with me. Of course, I remain liable for all shortcomings, omitted references, and flights of fancy. This chapter was completed in February 1987 and updated in March 1988.

References

Afik, D., and Alkon, P. U. 1983. Movements of a radio-collared wolf (*Canis lupus pallipes*) in the Negev Highlands, Israel. *Israel J. Zool.* 32:138–146.

Allison, L. 1971. Activity and behaviour of red foxes in central Alaska. M.Sc. thesis, Univ. Toronto. 92 pp.

Alt, G. L., and Beecham, J. J. 1984. Reintroduction of orphaned black bear cubs into the wild. *Wildl. Soc. Bull.* 12:169–174.

Alt, G. L., and Gruttadauria, J. M. 1984. Reuse of black bear dens in northeastern Pennsylvania. *J. Wildl. Mgmt.* 48:236–239.

Altmann, S. A. 1986. [Book review, Levins and Lewontin 1985.] *Ethology* 71:85–87.

Apfelbach, R. 1986. Imprinting on prey odours in ferrets *(Mustela putorius f. furo)* and its neural correlates. *Behav. Proc.* 12:363–381.

Apfelbach, R., and Weiler, E. 1985a. Is there a neural basis for olfactory imprinting in ferrets? *Naturwissenschaften* 72:106–107.

Apfelbach, R,. and Weiler, E. 1985b. Olfactory deprivation enhances normal spine loss in the olfactory bulbs of developing ferrets. *Neurosci. Letters* 62:169–173.

Arieli, R., Heth, G., Nevo, E., Zamir, Y., and Neutra, O. 1986. Adaptive heart and breathing frequencies in 4 ecologically differentiating chromosomal species of mole rats in Israel. *Experientia* 42:131–133.

Armitage, K. B. 1986. Individual differences in the behavior of juvenile yellow-bellied marmots. *Behav. Ecol. Sociobiol.* 18:419–424.

Ashmead, D. H., Clifton, R. K., and Reese, E. P. 1986. Development of auditory localization on dogs: Single source and precedence effect sounds. *Devel. Psychobiol.* 19:91–103.

Baerends–van Roon, J. M., and Baerends, G. P. 1979. *The Morphogenesis of the Behaviour of the Domestic Cat.* Amsterdam: North-Holland.

Bakeman, R., and Gottman, J. M. 1986. *Observing Interaction: An Introduction to Sequential Analysis.* New York: Cambridge Univ. Press.

Baker, C. M. 1980. Agonistic behaviour patterns of the slender mongoose, *Herpestes sanguineus. South African J. Zool.* 16:262–265.

Barlow, G. B. 1977. Modal action patterns. In: T. A. Sebeok, ed. *How Animals Communicate*, pp. 98–134. Bloomington: Indiana Univ. Press.

Bartholomew, G. A. 1986. The role of natural history in contemporary biology. *Bio-Science* 36:324–329.

Bateson, P. P. G. 1976. Specificity and the origins of behavior. *Adv. Study Behav.* 6:1–20.

Bekoff, A. 1988. The neural basis for the ontogeny of behavior in vertebrates. In: W. M. Cowan, ed. *Handbook of Physiology: Volume on Developmental Neurobiology.* Bethesda, Md.: American Physiological Society. In press.

Bekoff, M. 1972a. An ethological study of the development of social interaction in the genus *Canis:* A dyadic analysis. Ph.D. dissert., Washington Univ., St. Louis. 164 pp.

Bekoff, M. 1972b. The development of social interaction, play, and metacommunication in mammals: An ethological perspective. *Quart. Rev. Biol.* 47:412–434.

Bekoff, M. 1974. Social play and play-soliciting by infant canids. *Amer. Zool.* 14:323–340.

Bekoff, M. 1975a. Animal play and behavioral diversity. *Amer. Nat.* 109:601–603.

Bekoff, M. 1975b. The communication of play intention: Are play signals functional? *Semiotica* 15:231–239.

Bekoff, M. 1975c. Social behavior and ecology of the African Canidae: A review. In: M. W. Fox, ed. *The Wild Canids: Their Systematics, Behavioral Ecology and Evolution*, pp. 120–142. New York: Van Nostrand Reinhold.

Bekoff, M. 1976. Animal play: Problems and perspectives. *Persp. Ethol.* 2:165–188.
Bekoff, M. 1977a. The coyote, *Canis latrans* Say. *Mamm. Species* 79:1–9.
Bekoff, M. 1977b. Mammalian dispersal and the ontogeny of individual behavioral phenotypes. *Amer. Nat.* 111:715–732.
Bekoff, M. 1977c. Quantitative studies of three areas of classical ethology: Social dominance, behavioral taxonomy, and behavioral variability. In: B. A. Hazlett, ed. *Quantitative Methods in the Study of Animal Behavior,* pp. 1–46. New York: Academic Press.
Bekoff, M. 1977d. Socialization in mammals with an emphasis on nonprimates. In: S. Chevalier-Skolnikoff & F. E. Poirier, eds. *Primate Bio-Social Development: Biological, Social, and Ecological Determinants,* pp. 603–636. New York: Garland.
Bekoff, M. 1978a. Behavioral development in coyotes and eastern coyotes. In: M. Bekoff, ed. *Coyotes: Biology, Behavior, and Management,* pp. 97–126. New York: Academic Press.
Bekoff, M. 1978b. A field study of the development of behavior in Adelie penguins: Univariate and numerical taxonomic approaches. In: G. Burghardt & M. Bekoff, eds. *The Development of Behavior: Comparative and Evolutionary Aspects,* pp. 177–202. New York: Garland.
Bekoff, M. 1978c. Social play: Structure, function, and the evolution of a cooperative social behavior. In: G. Burghardt & M. Bekoff, eds. *The Development of Behavior: Comparative and Evolutionary Aspects,* pp. 367–383. New York: Garland.
Bekoff, M. 1979. Behavioral acts: Description, classification, ethogram analysis, and measurement. In: R. B. Cairns, ed. *The Analysis of Social Interactions: Methods, Issues, and Illustrations,* pp. 67–80. Hillsdale, N.J.: Lawrence Erlbaum.
Bekoff, M. 1981a. Development of agonistic behavior: Ethological and ecological perspectives. In: P. F. Brain & D. Benton, eds. *Multidisciplinary Approaches to Aggression Research,* pp. 161–178. New York: Elsevier/North Holland.
Bekoff, M. 1981b. Mammalian sibling interactions: Genes, facilitative environments, and the coefficient of familiarity. In: D. J. Gubernick & P. H. Klopfer, eds. *Parental Care in Mammals,* pp. 307–346. New York: Plenum.
Bekoff, M. 1982. Functional aspects of play as revealed by structural components and social interaction patterns. *Behav. Brain Sci.* 5:156–157.
Bekoff, M. 1984. Social play behavior. *BioScience* 34:228–234.
Bekoff, M. 1987. Group living, natal philopatry, and Lindström's lottery: It's all in the family. *Trends Ecol. Evol.* 2:115–116.
Bekoff, M. 1988. Motor training and physical fitness: Possible Short- and long-term influences on the development of individual differences in behavior. *Dev. Psychobiol.* 21:601–612.
Bekoff, M., and Byers, J. A. 1981. A critical reanalysis of the ontogeny and phylogeny of mammalian social play and locomotor play: An ethological hornet's nest. In: K. Immelmann, G. W. Barlow, L. Petrinovich & M. Main, eds. *Behavioral Development: The Bielefeld Interdisciplinary Conference,* pp. 296–337. New York: Cambridge Univ. Press.
Bekoff, M., and Byers, J. A. 1985. The development of behavior from evolutionary and ecological perspectives in mammals and birds. *Evol. Biol.* 19:215–286.
Bekoff, M. and Conner, D. 1987. Relationships between age at eye-opening and age at independence in canids. Unpublished data.
Bekoff, M., Daniels, T. J., and Gittleman, J. L. 1984. Life history patterns and the comparative social ecology of carnivores. *Ann. Rev. Ecol. Syst.* 15:191–232.
Bekoff, M., Diamond, J., and Mitton, J. B. 1981b. Life history patterns and sociality in canids: Body size, reproduction, and behavior. *Oecologia* 50:386–390.
Bekoff, M., Hill, H. L., and Mitton, J. B. 1975. Behavioral taxonomy in canids in discriminant function analysis. *Science* 190:1223–1225.

Bekoff, M., and Mech, L. D. 1984. Simulation analyses of space use: Home range estimates, variability, and sample size. *Behav. Res. Methods Instruments Computers* 16:32–37.

Bekoff, M., Tyrrell, M., Lipetz, V. E., and Jamieson, R. 1981a. Fighting patterns in young coyotes: Initiation, escalation, and assessment. *Aggressive Behav.* 7:225–244.

Bekoff, M., and Wells, M. C. 1980. The social ecology of coyotes. *Sci. Amer.* 242:130–148.

Bekoff, M., and Wells, M. C. 1982. Behavioral ecology of coyotes: Social organization, rearing patterns, space use, and resource defense. *Z. Tierpsychol.* 60:281–305.

Bekoff, M., and Wells, M. C. 1986. Social ecology and behavior of coyotes. *Adv. Study Behav.* 16:251–338.

Bennett, B. 1987. Measures of relatedness. *Ethology* 74:219–236.

Bennett, P. M., and Harvey, P. H. 1985. Brain size, development and metabolism in birds and mammals. *J. Zool.* 207:491–509.

Berger, J. 1979. Social ontogeny and behavioural diversity: Consequences for bighorn sheep, *Ovis canadensis*, inhabiting desert and mountain environments. *J. Zool.* 118:252–266.

Bernstein, I. S., and Williams, L. E. 1983. Ontogenetic changes and the stability of rhesus monkey dominance relationships. *Behav. Processes* 8:379–392.

Bertram, B. C. R. 1973. Lion population regulation. *East African Wildl. J.* 11:215–225.

Bertram, B. C. R. 1978. *Pride of Lions*. New York: Scribners.

Biben, M. 1982a. Object play and social treatment of prey in bush dogs and crab eating foxes. *Behaviour* 79:210–211.

Biben, M. 1982b. Ontogeny of social behaviour related to feeding in the crab-eating fox (*Cerdocyon thous*) and the bush dog (*Speothos venaticus*). *J. Zool.* 196:207–216.

Biben, M. 1982c. Sex differences in the play of young ferrets. *Biol. Behav.* 7:303–308.

Biben, M. 1983. Comparative ontogeny of social behaviour in three South American canids, the maned wolf, crab eating fox and bush dog: Implications for sociality. *Anim. Behav.* 31:814–826.

Blaustein, A. R., Bekoff, M., and Daniels, T. J. 1987. Kin recognition in vertebrates (excluding primates): Empirical evidence; Mechanisms, functions and future research. In: D. J. C. Fletcher & C. D. Michener, eds. *Kin Recognition in Animals*, pp. 287–331; 333–357. New York: John Wiley & Sons.

Blomqvist, C. G., and Saltin, B. 1983. Cardiovascular adaptations to physical training. *Ann. Rev. Physiol.* 45:169–189.

Bookstaber, R., and Langsam, J. 1985. On the optimality of coarse behavior rules. *J. Theor. Biol.* 116:161–193.

Boyd, R., and Richerson, P. J. 1985. *Culture and the Evolutionary Process*. Chicago: Univ. Chicago Press.

Brady, C. A. 1979. Observations on the behavior and ecology of the crab-eating fox (*Cerdocyon thous*). In: J. F. Eisenberg, ed. *Vertebrate Ecology in the Northern Neotropics*, pp. 161–171. Washington, D.C.: Smithsonian Institution Press.

Brady, C. A. 1981. The vocal repertoires of the bush dog (*Speothos venaticus*), crab-eating fox (*Cerdocyon thous*), and maned wolf (*Chrysocyon brachyurus*). *Anim. Behav.* 29:649–669.

Brooks, D. R., and Wiley, E. O. 1986. *Evolution as Entropy: Toward a Unified Theory of Biology*. Chicago: Univ. Chicago Press.

Brown, A. J. L. 1977. Physiological correlates of enzyme polymorphism. *Nature* 269:803–804.

Brown, J. A., and Colgan, P. W. 1985. The ontogeny of social behaviour in four species of centrarchid fish. *Behaviour* 92:254–276.

Brownlee, A. 1954. Play in domestic cattle in Britain: An analysis of its nature. *British Vet. J.* 110:46–68.

Bunge, M. 1979. *Causality and Modern Science.* New York: Dover.

Burghardt, G. M. 1988. Precocity, play, and the ecotherm-endotherm transition: Profound reorganization or superficial adaptation? In: E. M. Blass, ed. *Handbook of Behavioral Neurobiology,* 9:107–148. New York: Plenum.

Burrows, R. 1968. *Wild Fox.* New York: Taplinger.

Buss, L. W. 1987. *The Evolution of Individuality.* Princeton, N.J.: Princeton Univ. Press.

Byers, J. A. 1983. Social interactions of juvenile collared peccaries, *Tayassu tajacu* (Mammalia: Artidactyla). *J. Zool.* 201:83–96.

Byers, J. A., and Bekoff, M. 1986. What does "kin recognition" mean? *Ethology* 72:342–345.

Caley, M. T. 1972. The ontogeny of predatory behavior in captive arctic foxes. M.Sc. thesis, Univ. Alaska, Fairbanks.

Calow, P. 1984. Economics of ontogeny—adaptational aspects. In: B. Shorrocks, ed. *Evolutionary Ecology,* pp. 81–104. London: Blackwell Scientific Publications.

Camenzind, F. J. 1978. Behavioral ecology of coyotes on the National Elk Rufuge, Jackson, Wyoming. In: M. Bekoff, ed. *Coyotes: Biology, Behavior, and Management,* pp. 267–294. New York: Academic Press.

Caro, T. M. 1979. Relations between kitten behaviour and adult predation. *Z. Tierpsychol.* 51:158–168.

Caro, T. M. 1981. Predatory behaviour and social play in kittens. *Behaviour* 76:1–24.

Caro, T. M. 1987. Indirect costs of play: Cheetah cubs reduce maternal hunting success. *Anim. Behav.* 35:295–297.

Caro, T. M., and Alawi, R. M. 1985. Comparative aspects of behavioural development in two species of free-living hyrax. *Behaviour* 95:87–109.

Caro, T. M., and Collins, D. A. 1987. Male cheetah social organization and territoriality. *Ethology* 74:52–64.

Chapman, J., and Feldhamer, G., eds. 1982. *Wild Mammals of North America: Biology, Economics, and Management.* Baltimore: Johns Hopkins Univ. Press.

Chase, I. D. 1982a. Behavioral sequences during dominance hierarchy formation in chickens. *Science* 216:439–440.

Chase, I. D. 1982b. Dynamics of hierarchy formation: The sequential development of dominance relationships. *Behaviour* 80:218–240.

Chase, I. D. 1986. Explanations of hierarchy structure. *Anim. Behav.* 34:1265–1267.

Chesser, R. K., and Ryman, N. 1986. Inbreeding as a strategy in subdivided populations. *Evolution* 40:616–624.

Chevalier-Skolnikoff, S. 1974. The ontogeny of communication in the stumptail macaque (*Macaca arctoides*). *Contrib. Primatology* 2:1–174.

Clark, A. B., and Ehlinger, T. J. 1987. Pattern and adaptation in individual behavioral differences. *Perspect. Ethol.* 7:1–47.

Coghill, G. F. 1929. *Anatomy and the Problem of Behavior.* New York: Hafner.

Cohen, J. 1963. On the nature of mathematical proofs. In: R. A. Baker, ed. *Stress Analysis of a Strapless Evening Gown,* pp. 84–90. Englewood Cliffs, N.J.: Prentice-Hall.

Colgan, P. 1983. *Comparative Social Recognition.* New York: John Wiley & Sons.

Cooper, J. 1942. An exploratory study on African lions. *Comp. Psychol. Monogr.* 17:1–48.

Coppinger, R., Glendinning, J., Torop, E., Matthay, C., Sutherland, M., and Smith, C. 1987. Degree of behavioral neoteny differentiates canid polymorphs. *Ethology* 75:89–108.

Corning, P. A. 1983. *The Synergism Hypothesis: A Theory of Progressive Evolution.* New York: McGraw-Hill.

Craighead, F. C., and Craighead, J. J. 1972. Grizzly bear prehibernation and denning activities as determined by radiotracking. *Wildl. Monogr.* 32:1–35.

Craighead, J. J., and Mitchell, J. A. 1982. Grizzly bear. In: J. W. Chapman & G. A. Feldhamer, eds. *Wild Mammals of North America: Biology, Management, and Economics,* pp. 515–556. Baltimore, Md.: Johns Hopkins Univ. Press.

Crawford, P. B., Crawford, S. A. H., and Crawford, R. J. M. 1983. Some observations on cape gray mongooses, *Herpestes pulverulentus,* in the Tsitsikamma National Parks. *South African J. Wildl. res.* 13:35–40.

Daniels, T. J. 1987. The social ecology and behavior of free-ranging dogs. Ph.D. dissert., Univ. Colorado, Boulder. 303 pp.

Danzmann, R. G., Ferguson, M. M., and Allendorf, F. W. 1987. Heterozygosity and oxygen-consumption rate as predictors of growth and developmental rate in rainbow trout. *Physiol. Zool.* 60:211–220.

Danzmann, R. G., Ferguson, M. M., and Allendorf, F. W. 1988. Heterozygosity and components of fitness in a strain of rainbow trout. *Biol. J. Linn. Soc.* 33:285–304.

Dean, F. C., Darling, L. M., and Lierhaus, A. G. 1986. Observations of intraspecific killing by brown bears, *Ursus arctos. Canadian Field-Nat.* 100:208–211.

Diener, A. 1984. Hormonelle Einfluesse auf das geschlechtsspezifische Sozialspiel bei Iltisfrettchen (*Mustela putorius* f. *furro*). *Z. Säugetierk.* 49:242 246.

Diener, A. 1985. Verhaltsanalysen zum Sozialspiel von Iltisfrettchen (*Muestela putorious* f. *furo*). *Z. Tierpsychol.* 67:179 197.

Dietz, J. M. 1984. Ecology and social organization of the maned wolf (*Chrysocyon brachyurus*). *Smithsonian Contrib. Zool.* no. 392:1–51.

Dietz, J. M. 1987. Grass roots of the maned wolf. *Nat. Hist.* 96:52–59.

Dobson, F. S. 1982. Competition for mates and predominant juvenile male dispersal in mammals. *Anim. Behav.* 30:1183–1192.

Dobson, F. S., and Jones, W. T. 1985. Multiple causes of dispersal. *Amer. Nat.* 126:855–858.

Downhower, J. F., and Armitage, K. B. 1981. Dispersal of yearling yellow-bellied marmots (*Marmota flaviventris*). *Anim. Behav.* 29:1064–1069.

Dumont, J. P. C., and Robertson, R. M. 1986. Neuronal circuits: An evolutionary perspective. *Science* 233:849–853.

Eaton, R. C., and DiDomenico, R. 1985. Command and the neural causation of behavior: A theoretical analysis of the necessity and sufficiency paradigm. *Brain Behav. Evol.* 27:132–164.

Eco, U. 1986. *Art and Beauty in the Middle Ages.* New Haven, Conn.: Yale Univ. Press.

Egoscue, H. J. 1979. The swift fox, *Vulpes velox* Say. *Mamm. Species* 122.1–5.

Eisenberg, J. F. 1981. *The Mammalian Radiations: An Analysis of Trends in Evolution, Adaptation, and Behavior.* Chicago: Univ. Chicago Press.

Elliott, J. P., and McT. Cowan, I. 1978. Territoriality, density, and prey of the lion in Ngorongoro Crater, Tanzania. *Canadian J. Zool.* 56:1726–1734.

Eloff, F. C. 1973. Ecology and behavior of the Kalahari lion. In: R. L. Eaton, ed. *The World's Cats,* 1:90–126. Winston, Ore.: World Wildlife Safari.

Emlen, S. T., Emlen, J. M., and Levin, S. A. 1986. Sex-ratio selection in species with helpers-at-the-nest. *Amer. Nat.* 127:1–8.

Englund, J. 1980. Population dynamics of the red fox (*Vulpes vulpes* L., 1758). In: E. Zimen, ed. *The Red Fox: Symposium on Behaviour and Ecology,* pp. 107–121. The Hague: Junk.

Erlinge, S. 1979. Adaptive significance of sexual dimorphism in weasels. *Oikos* 33:233–245.

Estes, J. A. 1980. Sea otter, *Enhydra lutris. Mamm. Species* 133:1–8.

Ewer, R. F. 1968. *Ethology of Mammals.* New York: Plenum.

Ewer, R. F. 1973. *The Carnivores.* Ithaca, N.Y.: Cornell Univ. Press.

Ewer, R. F., and C. Wemmer, 1974. The behaviour in captivity of the African civet, *Civettictis civetta* (Schreber). *Z. Tierpsychol.* 34:359–394.

Fagen, R. 1976. Exercise, play, and physical training in animals. *Persp. Ethol.* 2:189–219.

Fagen, R. 1977. Selection for optimal age-dependent schedules of play behavior. *Amer. Nat.* 111:395–414.

Fagen, R. 1981. *Animal Play Behavior.* New York: Oxford Univ. Press.

Feddersen-Petersen, D. 1986a. *Hundepsychologie: Wesen und Sozialverhalten.* Stuttgart: Franckh'sche Verlagshandlung, W. Keller and Co.

Feddersen-Petersen, D. 1986b. Observations on social play in some species of Canidae. *Zoologischer Anzeiger* 217:130–144.

Fentress, J. C. 1985. Development of coordinated movement: Dynamic, relational and multileveled perspectives. In: M. G. Wade & H. T. A. Whiting, eds. *Motor Skill Acquisition in Children: Aspects of Coordination and Control,* pp. 1–26. Maastricht, Netherlands: NATO Advanced Study Institute.

Fentress, J. C. 1986. Ethology and the neural sciences. In: R. Campan & R. Zavan, eds. *Relevance of Models and Theory in Ethology,* pp. 77–107. Toulouse, France: Privat IEC.

Fentress, J. C., and Ryon, J. 1982. A long-term study of distributed pup feeding in captive wolves. In: F. H. Harrington & P. C. Paquet, eds. *Wolves of the World: Perspectives of Behavior, Ecology, and Conservation,* pp. 238–261. Park Ridge, N.J.: Noyes.

Fentress, J. C., Ryon, J., and McLeod, P. J. 1987. Coyote adult-pup interactions in the first three months. *Canadian J. Zool.* 65:760–763.

Feyerabend, P. 1975. *Against Method.* New York: Schocken Books.

Fiero, B. C., and Verts, B. J. 1985. Age-specific reproduction in raccoons in northwestern Oregon. *J. Mamm.* 67:169–172.

Fink, W. L. 1982. The conceptual relationship between ontogeny and phylogeny. *Paleobiology* 8:254–264.

Fitzgerald, B. M., and Karl, B. J. 1986. Home range of feral house cats (*Felis catus* L.) in forest of the Orongorongo Valley, Wallington, New Zealand. *New Zealand J. Ecol.* 9:71–81.

Fox, M.W. 1969a. The anatomy of aggression and its ritualization in canidae: A developmental and comparative study. *Behaviour* 35:242–258.

Fox, M. W. 1969b. Ontogeny of prey killing behaviour in canidae. *Behaviour* 35:259–271.

Fox, M. W. 1970. A comparative study of the development of facial expressions in canids, wolf, coyote, and foxes. *Behaviour* 36:49–73.

Fox, M. W. 1971a. *Behaviour of Wolves, Dogs and Related Canids.* New York: Harper & Row.

Fox, M. W. 1971b. *Integrative Development of the Brain and Behavior of the Dog.* Chicago: Univ. Chicago Press.

Fox, M. W., ed. 1975. *The Wild Canids: Their Systematics, Behavioral Ecology and Evolution.* New York: Van Nostrand Reinhold.

Fox, M. W., and Clark, A. L. 1971. The development and temporal sequencing of agonistic behavior in the coyotes (*Canis latrans*). *Z. Tierpsychol.* 28:262–278.

Frame, L. H., Malcolm, J. R., Frame, G., and van Lawick, H. 1979. Social organization of African wild dogs (*Lycaon pictus*) on the Serengeti Plains, Tanzania, 1967–1978. *Z. Tierpsychol.* 50:225–249.

Frank, L. G. 1986a. Social organization of the spotted hyena (*Crocuta crocuta*). I. Demography. *Anim. Behav.* 34:1500–1509.

Frank, L. G. 1986b. Social organization of the spotted hyena (*Crocuta crocuta*). II. Dominance and reproduction. *Anim. Behav.* 34:1510–1527.

Frank, L. G., Davidson, J. M., and Smith, E. R. 1985. Androgen levels in the spotted hyaena *Crocuta crocuta:* The influence of social factors. *J. Zool.* 206:525–531.

Frelinger, J. A. 1972. The maintenance of transferring polymorphism in pigeons. *Proc. Nat. Acad. Sci.* 69:326–329.

Fritts, S. H., and Mech, L. D. 1981. Dynamics, movements, and feeding ecology of a newly protected wolf population in northwestern Minnesota. *Wildl. Monogr.* 80:1–79.

Fritzell, E. K. 1977. Dissolution of raccoon sibling bonds. *J. Mamm.* 58:427–428.

Fritzell, E. K. 1978. Aspects of raccoon (*Procyon lotor*) social organization. *Canadian J. Zool.* 56:260–271.

Fritzell, E. K., Hubert, G. F., Jr., Meyen, B. E., and Sanderson, G. C. 1985. Age-specific reproduction in Illinois and Missouri raccoons. *J. Wildl. Mgmt.* 49:901–905.

Gaines, M. S., and McClenaghan, L. R. 1980. Dispersal in small mammals. *Ann. Rev. Ecol. Syst.* 11:163–196.

Galef, B. G. 1981. The ecology of weaning: Parasitism and the achievement of independence by altricial mammals. In: D. J. Gubernick & P. H. Klopfer, eds. *Parental Care in Mammals*, pp. 211–241. New York: Plenum.

Garland, T. 1985. Ontogenetic and individual variation in size, shape and speed in the Australian agamid lizard *Amphiborlurus nuchalis. J. Zool.* 207:425–439.

Garrott, R. A., Eberhardt, L. E., and Hanson, W. C. 1984. Arctic fox denning behavior in Alaska. *Canadian J. Zool.* 62:1636–1640.

Garshelis, D. L, and Garshelis, J. A. 1984. Movements and management of sea otters in Alaska. *J. Wildl. Mgmt.* 48:665–678.

Garshelis, D. L., Johnson, A. M., and Garshelis, J. A. 1984. Social organization of sea otters in Prince William Sound, Alaska. *Canadian J. Zool.* 62:2648–2658.

Garton, D. W. 1984. Relationship between multiple locus heterozygosity and physiological correlates of growth in the estuarine gastropod *Thais haemastoma. Physiol. Zool.* 57:530–543.

Garton, D. W., Koehn, R. K., and Scott, T. M. 1985. The physiological energetics of growth in the clam, *Mulinia lateralis:* An explanation for the relationship between growth rate and individual heterozygosity. In: P. E. Gibbs, ed. *Proceedings XIX European Marine Biology Symposium*, pp. 455–464. New York: Cambridge Univ. Press.

Gittleman, J. L. 1986a. Carnivore brain size, behavioral ecology, and phylogeny. *J. Mamm.* 67:23–36.

Gittleman, J. L. 1986b. Carnivore life history patterns: Allometric, phylogenetic, and ecological associations. *Amer. Nat.* 127:744–771.

Gittleman, J. L., and Harvey, P. H. 1987. Compensatory life histories and natal denning patterns in carnivores. Unpublished manuscript.

Glickman, S. E., Frank, L. G., Davidson, J. M., Smith, E. R., and Siiteri, P. K. 1987. Androstenedione may organize or activate sex-reversed traits in female spotted hyenas. *Proc. Natl. Acad. Sci.* 84:3444–3447.

Golani, I. 1976. Homeostatic motor processes in mammalian interactions: A choreography of display. *Persp. Ethol.* 2:69–134.

Golding, R. R. 1969. Birth and the development of spotted hyenas, *Crocuta crocuta*, at the University of Ibadan Zoo, Nigeria. *Internat. Zoo Yearb.* 9:93–95.

Golightly, R. 1981. The comparative energetics of two desert canids: The coyote (*Canis latrans*) and the kit fox (*Vulpes macrotis*). Ph.D. dissert., Arizona State Univ., Tempe.

Gould, S. J. 1977. *Ontogeny and Phylogeny*. Cambridge: Harvard Univ. Press.

Gray, J. P., ed. 1985a. *A Guide to Primate Sociobiological Theory and Research*. New Haven, Conn.: Human Relations Area Files Press.

Gray, J. P. 1985b. *Primate Sociobiology*. New Haven, Conn.: Human Relations Area Files Press.

Greenwood, P. J. 1980. Mating systems, philopatry and dispersal in birds and mammals. *Anim. Behav*. 28:1140–1162.

Hanby, J. P., and Bygott, J. D. 1987. Emigration of subadult lions. *Anim. Behav*. 35:161–169.

Harcourt, A. H. 1978. Strategies of emigration and transfer by primates, with particular reference to gorillas. *Z. Tierpsychol*. 48:401–420.

Harcourt, A. H., and Stewart, K. J. 1981. Gorilla male relationships: Can differences during immaturity lead to contrasting reproductive tactics in adulthood? *Anim. Behav*. 29:206–210.

Harrington, F. H. 1986. Timber wolf howling playback studies: Discrimination of pup from adult howls. *Anim. Behav*. 34:1575–1577.

Harrington, F. H. 1987. The man who cries wolf. *Nat. Hist*. 96:22, 24–26.

Harrington, F. H., Mech, L. D., and Fritts, S. H. 1983. Pack size and wolf pup survival: Their relationship under varying ecological conditions. *Behav. Ecol. Sociobiol*. 13:19–26.

Harrison, D. J., and Gilbert, J. R. 1985. Denning ecology and movements of coyotes in Maine during pup rearing. *J. Mamm*. 66:712–719.

Harrison, D. J., and Harrison, J. A. 1984. Foods of adult Maine coyotes and their known-aged pups. *J. Wildl. Mgmt*. 48:922–926.

Havkin, Z., and Fentress, J. C. 1985. The form of combative strategy in interactions among wolf pups (*Canis lupus*). *Z. Tierpsychol*. 68:177–200.

Hemmer, H. 1972. Snow leopard, Ounce *Uncia uncia* (Schreber, 1775). *Mamm. Species* 20:1–5.

Henry, J. D. 1986. *Red Fox: The Catlike Canine*. Washington, D.C.: Smithsonian Institution Pres.

Henry, J. D., and Herrero, S. M. 1974. Social play in the American black bear: Its similarity to canid social play and examination of its identifying characteristics. *Amer. Zool*. 14:371–389.

Henschel, J. R., and Skinner, J. D. 1987. Social relationships and dispersal patterns in a clan of spotted hyaenas, *Crocuta crocuta*, in the Kruger National Park. *South African J. Zool*. 22:18–24.

Hill, H. L., and Bekoff, M. 1977. The variability of some motor components of social play and agonistic behaviour in infant Eastern coyotes, *Canis latrans* var. *Anim. Behav*. 25:907–909.

Himsworth, H. 1986. *Scientific Knowledge and Philosophic Thought*. Baltimore, Md.: Johns Hopkins Univ. Press.

Hinde, R. A. 1970. *Animal Behaviour*. New York: McGraw-Hill.

Hinde, R. A. 1975. The concept of function. In: G. Baerends, C. G. Beer & A. Manning, eds. *Function and Evolution in Behaviour*, pp. 3–15. New York: Oxford Univ. Press.

Hinton, H., and Dunn, A. 1967. *Mongooses: Their Natural History and Behaviour*. London: Oliver & Boyd.

Jamieson, I. G. 1986. The functional approach to behavior: Is it useful? *Amer. Nat*. 127:195–208.

Jean, Y., Bergeron, J.-M., Bisson, S., and Larocque, B. 1986. Relative age determination of coyotes, *Canis latrans*, from southern Quebec. *Canadian Field-Nat*. 100:483–487.

Johnsingh, A. J. T. 1982. Reproductive and social behaviour of the dhole, *Cuon alpinus* (Canidae). *J. Zool*. 198:443–463.

Johnston, T. D. 1982a. Learning and the evolution of developmental systems. In: H. C. Plotkin, ed. *Learning, Development, and Culture*, pp. 411–442. New York: John Wiley & Sons.

Johnston, T. D. 1982b. Selective costs and benefits in the evolution of learning. *Adv. Study Behav.* 12:65–106.

Keith, L. B. 1983. Population dynamics of wolves: In: L. N. Carbyn, ed. *Wolves in Canada and Alaska*, pp. 66–77. Canadian Wildlife Service Reports Series no 45. Edmonton.

Kleiman, D. G. 1977. Monogamy in mammals. *Q. Rev. Biol.* 52:39–69.

Knight, S. W. 1978. Dominance hierarchies in captive coyote litters. M.Sc. thesis, Utah State Univ., Logan. 142 pp.

Kortmulder, K. 1986. The congener: A neglected area in the study of behaviour. *Acta Biotheoretica* 35:39–67.

Kovacs, K. M., and Lavigne, D. M. 1986. Maternal investment and neonatal growth in phocid seals. *J. Anim. Ecol.* 55:1035–1051.

Kruuk, H. 1972. *The Spotted Hyena*. Chicago: Univ. Chicago Press.

Kruuk, H. H., and Parish, T. 1987. Changes in the size of groups and ranges of the European badger (*Meles meles* L.) in an area in Scotland. *J. Anim. Ecol.* 56:351–364.

Kuo, Z. Y. 1931. The genesis of the cat's response to the rat. *J. Comp. Psychol.* 11:1–35.

Lamprecht, J. 1979. Field observations on the behaviour and social system of the bat-eared fox, *Otocyon megalotis* Desmarest. *Z. Tierpsychol.* 49:260–284.

Latour, P. B. 1981a. Interactions between free-ranging, adult male polar bears (*Ursus maritimus* Phipps): A case of adult social play. *Canadian J. Zool.* 59:1775–1783.

Latour, P. B. 1981b. Spatial relationships and behaviour of polar bears (*Ursus maritimus* Phipps) concentrated on land during the ice-free season of Hudson Bay. *Canadian J. Zool.* 59:1763–1774.

Lawrence, C. 1980. Individual differences in the mother-kitten relationship in the domestic cat (*Felis catus*). Ph.D. dissert., Univ. Edinburgh, Scotland. 248 pp.

Lawrence, R. D. 1980. The study of life: A naturalist's view. *Proc. The Myrin Institute* 35:1–43.

Leary, R. F., Allendorf, F. W., and Knudsen, K. L. 1984. Superior developmental stability of heterozygotes at enzyme loci in salmonid fishes. *Amer. Nat.* 124:540–551.

Lee, P. C. 1984. Ecological constraints on the social development of vervet monkeys. *Behaviour* 91:245–262.

Lehner, P. N. 1978. Coyote communication. In: M. Bekoff, ed. *Coyotes: Biology, Behavior, and Management*, pp. 127–162. New York: Academic Press.

Levins, R., and Lewontin, R. 1985. *The Dialectical Biologist*. Cambridge: Harvard Univ. Press.

Leyhausen, P. 1979. *Cat Behavior*. New York: Garland STPM Press.

Liberg, O., and von Schantz, T. 1985. Sex-biased philopatry and dispersal in birds and mammals: The Oedipus hypothesis. *Amer. Nat.* 126:129–135.

Lillegraven, J. A., Thompson, S. D., McNab, B. K., and Patton, J. L. 1987. The origin of eutherian mammals. *Biol. J. Linnean Soc.* 32:281–336.

Lindström, E. 1983. Condition and growth of red foxes (*Vulpes vulpes*) in relation to food supply. *J. Zool.* 199:117–122.

Lindström, E. 1986. Territory inheritance and the evolution of group living in carnivores. *Anim. Behav.* 34:1825–1835.

Lloyd, H. G. 1980. *The Red Fox*. London: Batsford.

Lockwood, R. 1976. An ethological analysis of social structure and affiliation in captive wolves. Ph.D. dissert., Washington Univ., St. Louis. 362 pp.

Łomnicki, A. 1988. *Population Ecology of Individuals.* Princeton, N.J.: Princeton Univ. Press.

Lott, D. F. 1984. Intraspecific variation in social systems of wild vertebrates. *Behaviour* 88:266–325.

Lotze, J.-H., and Anderson, S. 1979. Raccoon, *Procyon lotor. Mamm. Species* 119:1–8.

Lunn, N. J. 1986. Observations of nonaggressive behavior between polar bear family groups. *Canadian J. Zool.* 64:2035–2037.

Macdonald, D. W. 1980. Social factors affecting reproduction amongst red foxes. In: E. Zimen, ed. *The Red Fox: Symposium on Behaviour and Ecology,* pp. 123–175. The Hague: Junk.

Macdonald, D. 1987. *Running with the Fox.* New York: Facts on File.

Macdonald, D. W., and Moehlman, P. D. 1982. Cooperation, altruism, and restraint in the reproduction of carnivores. *Persp. Ethol.* 5:433–467.

McFarland, D., ed. 1982. *Functional Ontogeny.* Boston: Pitman Advanced Publishing Program.

McGrew, J. C. 1979. Kit fox, *Vulpes macrotis* Merriam. *Mamm. Species* 123:1–6.

Macpherson, A. H. 1969. *The Dynamics of Canadian Arctic Fox Populations.* Canadian Wildlife Service Reports Series, no. 8. Edmonton.

McVittie, R. 1978. Nursing behavior of snow leopard cubs. *Appl. Anim. Ethol.* 4:159–168.

Magee, B. 1973. *Karl Popper.* New York: Viking Press.

Malcolm, J. 1986. Socio-ecology of bat-eared foxes (*Otocyon megalotis*). *J. Zool.* 208:457–467.

Malcolm, J., and Marten, K. 1982. Natural selection and the communal rearing of pups in African wild dogs (*Lycaon pictus*). *Behav. Eco. Sociobiol.* 10:1–13.

Malina, R. M., and Bouchard, C. eds. 1986. *Sport and Human Genetics.* Champaign, Ill.: Human Kinetics.

Marks, J. S., and Redmond, R. L. 1987. Parental-offspring conflict and natal dispersal in birds and mammals: Comments on the Oedipus hypothesis. *Amer. Nat.* 129:158–164.

Martin, P. 1984a. The (four) whys and wherefores of play in cats: A review of functional, evolutionary, developmental and causal issues. In: P. K.Smith, ed. *Play in Animals and Humans,* pp. 71–94. New York: Basil Blackwell.

Martin, P. 1984b. The time and energy costs of play behaviour in the cat. *Z. Tierpsychol.* 64:298–312.

Martin, P., and Bateson, P. 1985a. The influence of experimentally manipulating a component of weaning on the development of play in domestic cats. *Anim. Behav.* 33:511–518.

Martin, P., and Bateson, P. 1985b. The ontogeny of locomotor play in the domestic cat. *Anim. Behav.* 33:502–510.

Martin, P., and Bateson, P. 1986. *Measuring Behaviour: An Introductory Guide.* New York: Cambridge Univ. Press.

Martin, P., and Caro, T. M. 1985. On the functions of play and its role in behavioral development. *Adv. Study Behav.* 15:59–103.

Martin, R. D., and MacLarnon, A. M. 1985. Gestation period, neonatal size and maternal investment in placental mammals. *Nature* 313:220–223.

Mason, W. 1979. Ontogeny of social behavior. In: P. Marler & J. G. Vandenbergh, eds. *Handbook of Behavioral Neurobiology,* vol. 3: *Social Behavior and Communication,* pp. 1–28. New York: Plenum.

Maynard Smith, J., Burian, R., Kauffman, S., Alberch, P., Campbell, J., Goodwin, B., Lande, R., Raup, D., and Wolpert, L. 1985. Developmental constraints and evolution. *Q. Rev. Biol.* 60:265–287.

Mayr, E. 1961. Cause and effect in biology. *Science* 134:1501–1506.

Mayr, E. 1974. Behavior programs and evolutionary strategies. *Amer. Sci.* 62:650–659.

Mayr, E. 1983. How to carry our the adaptationist program? *Amer. Nat.* 121:324–334.

Meaney, M. J. 1988. The sexual differentiation of play. *Trends Neurosci.* 11:54–58.

Meaney, M. J., Stewart, J., and Beatty, W. W. 1985. Sex differences in social play: The socialization of sex roles. *Adv. Study Behav.* 15:1–58.

Mech, L. D. 1970. *The Wolf.* New York: Natural History Press, Doubleday.

Mech, L. D. 1974. Gray wolf, *Canis lupus* Linnaeus. *Mamm. Species* 37:1–6.

Mech, L. D. 1975. Disproportionate sex ratios of wolf pups. *J. Wildl. Mgmt.* 39:737–740.

Mech, L. D. 1988. Longevity in wild wolves. *J. Mamm.* 69:197–198.

Mech, L. D., and Seal, U. S. 1987. Premature reproductive activity in wild wolves. *J. Mamm.* 68:871–873.

Medjo, D. C., and Mech, L. D. 1976. Reproductive activity in nine- and ten-month old wolves. *J. Mamm.* 57:406–408.

Meier, G. W. 1981. The bounty of behavior. *Devel. Psychobiol.* 14:173–175.

Mendl, M. 1988. The effects of litter-size variation on the development of play behaviour in the domestic cat: Litters of one and two. *Anim. Behav.* 36:20–34.

Millar, J. S., Burkholder, D. A. L., and Lang, T. L. 1986. Estimating age at independence in small mammals. *Canadian J. Zool.* 64:910–913.

Mills, M. G. L. 1983. Mating and denning behaviour of the brown hyaena *Hyaena brunnea* and comparisons with other Hyaenidae. *Z. Tierpsychol.* 63:331–342.

Mills, M. G. L. 1984. The comparative behavioural ecology of the brown hyaena *Hyaena brunnea* and the spotted hyaena *Crocuta crocuta* in the southern Kalahari. *Koedoe Suppl.* 1984:237–247.

Mills, M. G. L. 1985. Related spotted hyaenas forage together but do not cooperate in rearing young. *Nature* 316:61–62.

Mitton, J. B., and Grant, M. C. 1984. Associations among protein heterozygosity, growth rate, and developmental homeostasis. *Ann. Rev. Ecol. Syst.* 15:479–499.

Mitton, J. B., and Koehn, R. K. 1985. Shell shape variation in the blue mussel, *Mytilus edulis* L., and its association with enzyme heterozygosity. *J. Exp. Marine Biol. Ecol.* 90:73–80.

Moehlman, P. D. 1983. Socioecology of silverbacked and golden jackals (*Canis mesomelas* and *Canis aureus*). In: J. F. Eisenberg & D. G. Kleiman, eds. *Advances in the Study of Mammalian Behavior,* pp. 423–453. American Society of Mammalogists Special Publication no. 7. Lawrence, Kan.: American Society of Mammalogists.

Moehlman, P. D. 1986. Ecology of cooperation in canids. In: D. I. Rubenstein & R. W. Wrangham, eds. *Ecological Aspects of Social Evolution: Birds and Mammals,* pp. 64–86. Princeton, N.J.: Princeton Univ. Press.

Moehlman, P. D. 1987. Social organization in jackals. *Amer. Sci.* 75:366–375.

Moelk, M. 1979. The development of friendly approach behavior in the cat: A study of kitten-mother relations and the cognitive development of the kitten from birth to eight weeks. *Adv. Study Behav.* 10:163–224.

Moore, J., and Ali, R. 1984. Are dispersal and inbreeding avoidance related? *Anim. Behav.* 32:94–112.

Murray, B. G. 1986. The influence of philosophy on the interpretation of interspecific aggression. *Condor* 88:543.

Neal, E. 1986. *The Natural History of Badgers.* London: Croom Helm.

Nelissen, M. H. J. 1986. The effect of tied rank numbers on the linearity of dominance hierarchies. *Behav. Processes* 12:159–168.

Oftedal, O. T. 1984. Lactation in the dog: Milk composition and intake by puppies. *J. Nutr.* 114:803–812.

Ortega, J. C. 1988. Activity patterns of different-aged coyote (*Canis latrans*) pups in southeastern Arizona. *J. Mamm.* In press.

Ortega, J. C., and Bekoff, M. 1987. Avian play: Comparative evolutionary and developmental trends. *Auk* 104:338–341.

Owens, D. D., and Owens, M. J. 1979. Communal denning and clan associations in brown hyenas (*Hyaena brunnea* Thunberg) of the central Kalahari Desert. *African J. Ecol.* 17:35–44.

Owens, D. D., and Owens, M. J. 1984. Helping behavior in brown hyenas. *Nature* 308:843–845.

Owens, M. J., and Owens, D. D. 1978. Feeding ecology and its influence on social organization of brown hyenas (*Hyaena brunnea*) of the central Kalahari. *East African Wildl. J.* 16:113–135.

Packard, J. M,. Seal, U. S., Mech, L. D., and Plotka, E. D. 1985. Causes of reproductive failure in two family groups of wolves (*Canis lupus*). *Z. Tierpsychol.* 68:24–40.

Panksepp, J., Siviy, S., and Normansell, L. 1984. The psychobiology of play: Theoretical and methodological perspectives. *Neurosci. & Biobehav. Rev.* 8:465–492.

Paradiso, J. L., and Nowak, R. M. 1972. Red wolf *Canis rufus* Audubon and Bachman. *Mamm. Species* 22:1–4.

Peck, J. R., and Feldman, M. W. 1986. The evolution of helping behavior in large, randomly mixed populations. *Amer. Nat.* 127:209–221.

Peters, G. 1984. On the structure of friendly close range vocalizations in terrestrial carnivores (Mammalia: Carnivora: Fissipedia). *Z. Säugetierk.* 49:157–182.

Poole, T. B. 1966. Aggressive play in polecats. *Symp. Zool. Soc. London* 18:23–44.

Powell, R. A. 1982. *The Fisher: Life History, Ecology, and Behavior.* Minneapolis: Univ. Minnesota Press.

Powers, D. A., DiMichele, L., and Place, A. R. 1983. The use of enzyme kinetics to predict cellular metabolism, developmental rate, and swimming peformance between LDH-B genotypes of the fish, *Fundulus heteroclitus. Isoenzymes: Current Topics Biol. Med. Res.* 10:147–170.

Pruitt, C. H. 1974. Social behavior of young captive black bears. Ph.D. dissert., Univ. Tennessee, Knoxville.

Pusey, A. E. 1987. Sex-biased dispersal and inbreeding avoidance in birds and mammals. *Trends Ecol. Evol.* 10:295–299.

Pusey, A. E., and Packer, C. 1987. The evolution of sex-biased dispersal in lions. *Behaviour* 101:275–310.

Pyrah, D. 1984. Social distribution and population estimates of coyotes in north-central Montana. *J. Wildl. Mgmt.* 48:679–690.

Rabinowitz, A. R., and Nottingham, B. G. 1986. Ecology and behaviour of the jaguar (*Panthera onca*) in Belize, Central America. *J. Zool.* 210:149–159.

Ralls, K. 1977. Sexual dimorphism in mammals: Avian models and unanswered questions. *Amer. Nat.* 111:917–938.

Ralls, K., Brownell, R. L., Jr., and Ballou, J. 1980. Differential mortality by sex and age in mammals, with specific reference to the sperm whale. *Rep. Internat. Whale Comm.* 2:233–243.

Ramsey, M. A., and Dunbrack, R. L. 1986. Physiological constraints on life history phenomena: The example of small bear cubs at birth. *Amer. Nat.* 127:735–743.

Ramsey, M. A., and Stirling, I. 1986a. Long-term effects of drugging and handling free-ranging polar bears. *J. Wildl. Mgmt.* 50:619–626.

Ramsey, M. A., and Stirling, I. 1986b. On the mating system of polar bears. *Canadian J. Zool.* 64:2142–2151.

Rasa, O. A. E. 1973. Prey capture, feeding techniques, and their ontogeny in the African dwarf mongoose, *Helogale undulata refula*. *Z. Tierpsychol.* 32:449–488.

Rasa, O. A. E. 1977. The ethology and sociology of the dwarf mongoose (*Helogale undulata rufula*). *Anim. Behav.* 32:579–589.

Rasa, O. A. E. 1984. A motivational analysis of object play in juvenile dwarf mongooses (Helogale undulata rufula). *Anim. Behav.* 32:579–589.

Reich, A. 1978. A case of inbreeding in the African wild dog, *Lycaon pictus,* in the Kruger National Park. *Koedoe* 21:119–123.

Reich, A. 1981. The behavior and ecology of the African wild dog (*Lycaon pictus*) in the Kruger National Park. Ph.D. dissert., Yale Univ., New Haven, Conn. 425 pp.

Riedman, M. L. 1982. The evolution of alloparental care and adoption in mammals and birds. *Quart Rev. Biol.* 57:405–435.

Rogers, L. L. 1987. Effects of food supply and kinship on social behavior, movements, and population growth of black bears in northeastern Minnesota. *Wildl. Monogr.* 97:1–72.

Rood, J. P. 1978. Dwarf mongoose helpers at the den. *Z. Tierpsychol.* 48:277–287.

Rood, J. P. 1980. Mating relationships and breeding suppression in the dwarf mongoose. *Anim. Behav.* 28:143–150.

Rood, J. P. 1983. The social system of the dwarf mongoose. In: J. F. Eisenberg & D. G. Kleiman, eds. *Advances in the Study of Mammalian Behavior,* pp. 454–488. American society of Mammalogists Special Publication no. 7. Lawrence, Kan.: American Society of Mammalogists.

Rood, J. P. 1986. Ecology and social evolution in the mongooses. In: D. I. Rubenstein & R. W. Wrangham, eds. *Ecological Aspects of Social Evolution: Birds and Mammals,* pp. 131–152. Princeton, N.J.: Princeton Univ. Press.

Rosatte, R. C., and Gunson, J. R. 1984. Dispersal and home range of striped skunks, *Mephitis mephitis,* in an area of population reduction in southern Alberta. *Canadian Field-Nat.* 98:315–319.

Rowell, T. 1979. How would we know if social organization were *not* adaptive? In: I. S. Bernstein & E. O. Smith, eds. *Primate Ecology and Human Origins: Ecological Influences on Social Organization,* pp. 1–22. New York: Garland.

Rowe-Rowe, D. T. 1978. The small carnivores of Natal. *Lammergeyer* 25:1–48.

Rowe-Rowe, D. T. 1982. Home range movements of black-backed jackals in an African montane region. *South African J. Wildl. Res.* 12:79–84.

Rowe-Rowe, D. T. 1984. Black-backed jackal population structure in the Natal Drakensberg. *Lammergeyer* 32:1–7.

Rudnai, J. A. 1973. *The Social Life of the Lion.* Wallingford, Pa.: Washington Square.

Rushen, J. 1982. Development of social behaviour in chickens: A factor analysis. *Behav. Processes* 7:319–333.

Ryan, J. E. 1975. The inheritance of track performance in greyhounds. M.Sc. thesis, Trinity College, Dublin, Ireland.

Ryon, J. 1986. Den digging and pup care in captive coyotes (*Canis latrans*). *Canadian J. Zool.* 64:1582–1585.

Sallaberry, M. 1985. Wounds due to flipper bands on penguins. *J. Field Ornithol.* 56:275–277.

Schaller, G. B. 1967. *The Deer and the Tiger.* Chicago: Univ. Chicago Press.

Schaller, G. B. 1972. *The Serengeti Lion.* Chicago: Univ. Chicago Press.

Schleidt, W. 1982. Stereotyped feature variables are essential constituents of behaviour patterns. *Behaviour* 79:230–238.

Schleidt, W., and Crawley, J. N. 1980. Patterns in the behaviour of organisms. *J. Social Biol. Struct.* 3:1–15.

Schleidt, W., Yakalis, G., Donnelly, M., and McGarry, J. 1984. A proposal for a standard ethogram, exemplified by an ethogram of the bluebreasted quail *(Coturnix chinensis)*. *Z. Tierpsychol.* 64:193–220.

Schneider, D. G., Mech, L. D., and Tester, J. R. 1971. Movements of female raccoons and their young as determined by radio-tracking. *Anim. Behav. Monogr.* 4:1–43.

Schneirla, T. C., Rosenblatt, J. S., and Tobach, E. 1963. Maternal behavior in the cat. In: H. Rheingold, ed. *Maternal Behavior in Mammals,* pp. 122–168. New York: J. Wiley & Sons.

Scott, J. P., and Fuller, J. L. 1965. *Genetics and the Social Behavior of the Dog.* Chicago: Univ. Chicago Press.

Scott, J. P., and Marston, M. V. 1950. Critical periods affecting normal and maladjustive social behavior in puppies. *J. Genet. Psychol.* 77:25–60.

Seidensticker, J. C., Hornocker, M. G., Miles, W. V., and Messick, J. P. 1973. Mountain lion social organization in the Idaho Primitive Area. *Wildl. Monogr.* 35:1–60.

Seradilla, J. M., and Ayala, F. J. 1983. Alloprocoptic selection: A mode of natural selection promoting polymorphism. *Proc. Natl. Acad. Sci.* 80:2022–2025.

Sheppey, K., and Bernard, R. T. F. 1984. Relative brain size in the mammalian carnivores of the Cape Province of South Africa. *South African J. Zool.* 19:305–308.

Shields, W. M. 1982. *Philopatry, Inbreeding, and the Evolution of Sex.* Albany: State Univ. New York Press.

Shields, W. M. 1983. Genetic considerations in the management of the wolf and other large vertebrates: An alternative view. In: L. N. Carbyn, ed. *Wolves in Canada and Alaska,* pp. 90–92. Canadian Wildlife Service Reports Series no. 45. Edmonton.

Sieber, O. J. 1986. Acoustic recognition between mother and cubs in raccoons *(Procyon lotor). Behaviour* 96:130–163.

Silver, H., and Silver, W. T. 1969. Growth and behavior of the coyote-like canid of northern New England with observations on canid hybrids. *Wildl. Monogr.* 17:1–41.

Slater, P. J. B. 1986. Individual differences and dominance hierarchies. *Anim. Behav.* 34:1264–1265.

Slobodkin, L. B. 1986. The role of minimalism in art and science. *Amer. Nat.* 127:257–265.

Snow, C. J. 1967. Some observations on the behavioral and morphological development of coyote pups. *Amer. Zool.* 7:353–355.

Stanislaw, H., and Brain, P. F. 1983. The systematic response of male mice to differential housing: A path-analytical approach. *Behav. Processes* 8:165–175.

Sternberg, R. J. 1985. Human intelligence: The model is the message. *Science* 230:1111–1118.

Stirling, I., and Latour, P. B. 1978. Comparative hunting abilities of polar bear cubs of different ages. *Canadian J. Zool.* 56:1768–1772.

Stockman, E. R., Callagnan, R. S., Gallagher, C. A., and Baum, M. J. 1986. Sexual differentiation of play behavior in the ferret. *Behav. Neurosci.* 100:563–568.

Storm, G. L., Andrews, R. D., Phillips, R. L., Bishop, R. A., Siniff, D. B., and Tester, J. R. 1976. Morphology, reproduction, dispersal, and mortality of midwestern red fox populations. *Wildl. Monogr.* 49:1–82.

Sunquist, M. E. 1981. The social organization of tigers *(Panthera tigris)* in Royal Chitawan National Park, Nepal. *Smithsonian Contrib. Zool.* 336:1–98.

Sussman, R. W. 1977. Socialization, social structure, and ecology of two sympatric species of *Lemur.* In: S. Chevalier-Skolnikoff & F. E. Poirier, eds. *Primate Bio-Social Development,* pp. 515–528. New York: Garland.

Symons, D. 1979. *The Evolution of Human Sexuality.* New York: Oxford Univ. Press.

Tan, P. L., and Counsilman, J. J. 1985. The influence of weaning on prey-catching behaviour in kittens. *Z. Tierpsychol.* 70:148–164.

Theberge, J B. 1983. Considerations in wolf management related to genetic variability and adaptive change. In: L. N. Carbyn, ed. *Wolves in Canada and Alaska*, pp. 86–89. Canadian Wildlife Service Reports Series no. 45. Edmonton.

Thomas, P. O., and Taber, S. M. 1984. Mother-infant interaction and behavioral development in southern right whales, *Eubalaena australis*. *Behaviour* 88:42–60.

Thompson, N. S. 1981. Toward a falsifiable theory of evolution. *Persp. Ethol.* 4:51–73.

Tinbergen, N. 1951. *The Study of Instinct.* New York: Oxford Univ. Press.

Tinbergen, N. 1963. On aims and methods of ethology. *Z. Tierpsychol.* 20:410–433.

Toulouse, G., Dehaene, S., and Changeux, J.-P. 1986. Spin glass model of learning by selection. *Proc. Nat. Acad. Sci. USA* 83:1695–1698.

Tullar, B. F., Jr., and Berchielli, L. T., Jr. 1980. Movement of the red fox in central New York. *N.Y. Fish and Game J.* 27:179–204.

van der Merwe, N. J. 1953a. The coyote and the black-backed jackal. *Flora and Fauna* 3:45–51.

van der Merwe, N. J. 1953b. The jackal. *Flora and Fauna* 4:4–80.

van der Molen, P. P. 1984. Bi-stability of emotions and motivations: An evolutionary consequence of the open-ended capacity for learning. *Acta Biotheor.* 33:227–251.

van Dongen, P. A. M. and van den Bercken, J. H. L. 1981. Structure and function in neurobiology: A conceptual framework and the localization of functions. *Internat. J. Neurosci.* 15:49–68.

van Lawick, H., and van Lawick, J. 1970. *Innocent Killers.* Boston: Houghton Mifflin.

Vilijoen, S. 1980. Early postnatal development, parental care and interaction in the banded mongoose *Mungos mungo*. *South African J. Zool.* 15:119–120.

Vincent, L. E., and Bekoff, M. 1978. Quantitative analyses of the ontogeny of predatory behaviour in coyotes, *Canis latrans*. *Anim. Behav.* 26:225–231.

Wandrey, R. 1975. Contribution to the study of social behaviour of golden jackals (*Canis aureus* L.). *Z. Tierpsychol.* 39:365–402.

Waser, P. M. 1985. Does competition drive dispersal? *Ecology* 66:1170–1175.

Waser, P. M., and Jones, W. T. 1983. Natal philopatry among solitary mammals. *Quart. Rev. Biol.* 58:355–390.

Watson, J. D. 1986. Foreward. In: H. Himsworth, *Scientific Knowledge and Philosophic Thought*. Baltimore: Johns Hopkins Univ. Press.

Watt, W. B., Carter, P. A., and Donohue, K. 1986. Females' choice of "good genotypes" as mates is promoted by an insect mating system. *Science* 233:1187–1190.

Wayne, R. K. 1986. Cranial morphology of domestic and wild canids: The influence of development on morphological change. *Evolution* 40:243–261.

Weatherhead, P. J. 1986a. How unusual are unusual events? *Amer. Nat.* 128:150–154.

Weatherhead, P. J. 1986b. Erratum. How unusual are unusual events? *Amer. Nat.* 128:942.

Wemmer, C. M. 1977. Comparative ethology of the large-spotted genet (*Genetta tigrina*) and some related viverrids. *Smithson. Contrib. Zool.* 239:1–93.

Wemmer, C. M., and Fleming, M. J. 1974. Ontogeny of playful contact in a social mongoose, the meerkat, *Suricata suricatta*. *Amer. Zool.* 14:415–426.

West, M. 1974. Social play in the domestic cat. *Amer. Zool.* 14:427–436.

Wiley, R. H. 1981. Social structure and individual ontogenies: Problems of description, mechanism, and evolution. *Persp. Ethol.* 4:105–133.

Williams, G. C. 1985. A defense of reductionism in evolutionary biology. In: R. Dawkins & M. Ridley, eds. *Oxford Surveys in Evolutionary Biology*, pp. 1–27. New York: Oxford Univ. Press.

Wilson, E. O. 1975. *Sociobiology: The New Synthesis.* Cambridge: Harvard Univ. Press.

Windle, W. F. 1940. *Physiology of the Fetus: Origin and Extent of Function in Prenatal Life.* Philadelphia: W. B. Saunders.

Yamamoto, I. 1984. Latrine utilization and feces recognition in the raccoon dog. *J. Ethol.* 2:47–54.

Yamamoto, I. 1987. Male parental care in the raccoon dog, *Nyctereutes procyonoides,* during the early rearing period. In J. L. Brown & K. Kikkawa, eds. *Animal Societies: Theories and Facts,* pp. 189–195. Tokyo: Japan Scientific Societies Press.

Zimen, E. 1976. On the regulation of pack size in wolves. *Z. Tierpsychol.* 40:300–341.

Zimen, E. 1981. *The Wolf: A Species in Danger.* New York: Delacorte.

Zimen, E. 1982. A wolf pack sociogram. In: F. H. Harrington & P. C. Paquet, eds. *Wolves of the World: Perspectives of Behavior, Ecology, and Conservation,* pp. 282–322. Park Ridge, N.J.: Noyes.

CHAPTER 4

The Comparative Behavioral Ecology of Hyenas:
The Importance of Diet and Food Dispersion

M. G. L. MILLS

A close relationship between diet and food-dispersion patterns, on the one hand, and behavior and social organization, on the other, was first recorded in birds by Crook (1965). Subsequently, this relationship has been studied in a range of mammals: in bats (Bradbury and Vehrencamp 1976), in antelope (Jarman 1974), in primates as reviewed by Clutton-Brock and Harvey (1977), and in carnivores as reviewed by Macdonald (1983) and Bekoff et al. (1984). The hyaenids are highly suited for studies of this nature; they show a wide range of ecological and behavioral adaptations and social organizations (Kruuk 1975; Mills 1978a, 1984) and constitute only four extant species of three genera.

Two hyaenids, the brown hyena (*Hyaena brunnea*) and the spotted hyena (*Crocuta crocuta*), have been extensively studied in the southern Kalahari. In this chapter I review and discuss their diet, foraging behavior, social organization, and social behavior to show how the two species are able to inhabit the same region by exploiting rather different food sources and how these differences in diet have lead to the evolution of different foraging and social behavior patterns, as well as differences in social organization. I also compare my findings with those of other workers who have studied the same species in other locations, discuss hyena social ecology in comparison with other carnivores, and suggest areas for further research.

The area referred to as the southern Kalahari comprises the adjacent Kalahari Gemsbok (South Africa) and Gemsbok (Botswana) national parks, which together cover an area of about 36,000 km². It is an arid region with an irregular rainfall ($\overline{X}$ 220 mm annually) and experiences large temperature fluctuations both daily and seasonally. The area is covered with sand dunes broken by pans and two fossil riverbeds, the dunes and the riverbeds providing two distinct habitats. The vegetation is an extremely open shrub or tree savanna (Leistner 1967). The larger ungulates are mainly nomadic, concentrating along the riverbeds during the rains and dispersing into the dunes during the dry times (Mills and Retief 1984). Gemsbok (*Oryx gazella*), blue wildebeest

125

(*Connochaetes taurinus*), red hartebeest (*Alcelaphus buselaphus*), eland (*Taurotragus oryx*), springbok (*Antidorcas marsupialis*), and steenbok (*Raphicerus campestris*) are the most common antelope species. In addition to the hyenas, African lion (*Panthera leo*), leopard (*P. pardus*), cheetah (*Acinonyx jubatus*), black-backed jackal (*Canis mesomelas*), and caracal (*Felis caracal*), as well as eight smaller carnivores, are resident.

The brown hyena is the most common large carnivore in the area, occurring at a density of 1.8 per 100 km², whereas the spotted hyena occurs at a lower density of 0.9 per 100 km² (Mills 1989).

Feeding Habits

Figure 4.1 shows the diets of the two species as determined from direct observations made when following individuals by vehicle at night, revealing large differences in their diets. Brown hyenas feed on a wide variety of mainly small food items such as small mammals, bones, wild fruits, and insects. Spotted hyenas have a far more specialized diet, consuming mainly large and medium-sized mammals; 64.3% of spotted hyenas observed feeding on food items that could be identified as to the species were feeding on either gemsbok or

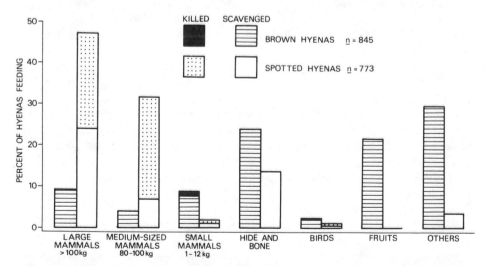

Figure 4.1. The diets of the brown hyena and the spotted hyena in the southern Kalahari, as determined from direct observations. The number of hyenas observed feeding on each food item is expressed as the percentage of the total number of hyenas of each species observed feeding. The data for mammals have been analyzed on a mass rather than a species basis, for example, gemsbok and wildebeest calves less than a year old have been recorded as medium-sized mammals, whereas those older than one year are recorded as large mammals. The proportions killed:scavenged are the percentages of the number of each species observed feeding on each food item. The heading "Others" mainly includes insects, but also reptiles, birds' eggs, and small unidentifiable pieces of food. (From Mills 1984, used by permission of National Parks Board, Republic of South Africa.)

wildebeest. Thus, whereas much of the brown hyena's food consists of small food items that provide a meal for only one hyena, most of the spotted hyena's food consists of large food items that simultaneously provide a meal for several hyenas.

The differences in diet are further accentuated by differences in the manner in which the two species procure their food. Brown hyenas are predominantly scavengers. They killed only 5.8% of the biomass of food they were observed to eat, and their kills comprised only small animals (Mills 1978b). Spotted hyenas kill much of their food themselves; 72.6% of the biomass of food they were observed to eat came from their kills, comprising 49.2% of the large mammals they consumed and 77.5% of the medium-sized mammals (Mills 1989). Food competition between the two species, therefore, is limited; the most important competition occurs when both scavenge from large ungulate carcasses.

These differences in feeding habits have led to the evolution of many differences in the behavior and social systems of the two species.

Foraging Behavior

The brown hyena, the spotted hyena, and the striped hyena (*Hyaena hyaena*) are predominantly nocturnal throughout their range (Kruuk 1972, 1976; Bearder 1977; Mills 1978b, 1989; Owens and Owens 1978; Tilson and Hamilton 1984; Goss 1986). Apart from being an important water-conservation strategy, nocturnal behavior may have evolved in the Hyaenidae as a means of reducing competition with the other dominant scavengers in African ecosystems, the vultures, which are exclusively diurnal (Houston 1979).

In the southern Kalahari brown hyenas were active for an average of 80.2% of the hours of darkness (1800h–0600h) (Mills 1978b), whereas spotted hyenas were active only for an average of 55.3% of this time (Mills 1989), a significant difference (Mann-Whitney test, U = 15.5, P < 0.0001, one-tailed). However, the distances moved per night by the two species were similar; brown hyenas moved an average of 31.1 km ± SE 2.1 per night and spotted hyenas moved an average of 27.1 km ± SE 1.4 per night (t = 1.437, d.f. = 180, P > 0.05) (Mills 1978b, 1989). Because spotted hyenas frequently feed on large food items that satiate them, once they have found their food they can afford to rest for some time. Brown hyenas, on the other hand, typically move from one small food item to the next and therefore spend more time foraging than spotted hyenas do. However, the actual distances covered during foraging by the two species are similiar. Spotted hyenas move far more quickly, approximately half of the time at a lope of 10 km per hour, looking for large and medium-sized ungulates (Mills 1989), whereas brown hyenas walk at a speed of about 4 km per hour, spending time investigating a wider range of smaller potential feeding opportunities (Mills 1978b).

Brown hyenas are exclusively solitary foragers (Mills 1978b). As most of their food is carrion, the olfactory sense is most important for locating food, and they repeatedly make upwind sniffs, locating much of their food from downwind. Hunting behavior is unspecialized and opportunistic and comprises short chases of 10–1100 m after small animals such as springbok lambs, springhares (*Pedetes capensis*), and bat-eared foxes (*Otocyon megalotis*), as well as ground-nesting birds such as korhaans (*Eupodotis* sp.). Prey are encountered at close quarters mostly by chance. Of 104 hunting attempts observed, only ten (9.6%) were successful (Mills 1978b). Whenever a large amount of food is found by a brown hyena and providing no other brown hyenas also find the food, the excess is usually stored close by in a clump of tall grass, under a bush, or rarely down a hole. Some of the food may also be carried back to the den for the cubs (Mills 1978b, 1982b).

Brown hyenas would not seem to gain by foraging in a group, as this would not enhance their efficiency at finding food. One hyena can probably detect the smell of carrion as well as several can. Furthermore, as most food items provide a meal for only one hyena, group foraging could lead to unnecessary aggression between individuals.

Spotted hyenas may forage solitarily or more often in a group. The mean foraging group size was $3.0 \pm$ SE 0.1 ($n = 566$), with 29.9% comprising one animal, 19.3% two, 18.7% three, 13.3% four, and 19.2% five or more animals (Mills 1985). In addition to sight, olfactory and auditory cues are used for locating prey. On 24 occasions the prey were smelled from a mean distance of 1.1 km $\pm$ SE 0.1, and on 11 occasions scent trails of prey were followed for a mean of 0.9 km $\pm$ SE 0.2. On another nine occasions prey were heard—that is, horns clashing or young bleating—from a mean distance of 2.4 km $\pm$ SE 0.7 (Mills 1989). Once the prey have been encountered, sight is important in prey selection. Carrion is located through the olfactory sense, and auditory cues often lead other spotted hyenas to a carcass on which conspecifics are feeding. Hunting behavior is directed mainly at large and medium-sized ungulates, which are run down at speeds of up to 50 km per hour, usually over distances of 0.5–2.5 km. Gemsbok less than a year old made up 43% of their kills, followed by wildebeest of all ages (15%), and gemsbok subadults and adults (10%) (Mills 1985, 1988). Food storing is rare because most carcasses are consumed rapidly by several hyenas feeding simultaneously.

Several selective pressures may have caused group foraging in carnivores. It has been suggested that group foraging allows animals to overcome larger prey and increases hunting success (Kruuk 1972; Ewer 1973). Subadult and adult gemsbok were observed to be killed only by two and by four spotted hyenas, respectively, lending weight to this hypothesis. However, hunting success on gemsbok calves and wildebeest of all ages was not correlated with hunting group size. Single spotted hyenas were particularly adept at catching gemsbok calves, 65% ($n = 17$) of hunts by solitary spotted hyenas on gemsbok calves being successful, compared with an overall hunting success rate on gemsbok calves of 64% (Mills 1985).

Group foraging may also lead to a more successful defense of the food from competitors (Caraco and Wolf 1975; Lamprecht 1978). This does not seem to be important in the southern Kalahari, as the density of large carnivores is so low that at only 2% of the carcasses on which spotted hyenas fed were they seriously challenged by other carnivores (Mills 1985, 1989). Group foraging, by Kalahari spotted hyenas at any rate, may also be related to kin selection. Foraging and feeding groups comprised more closely related individuals than did random groups from the same clans (Mills 1985). The large size of much of the food ensures that several individuals may feed on a prey item even if they do not all cooperate in catching the prey. This may be of particular benefit to younger animals that have not yet acquired the skills to hunt successfully themselves.

Social Organization

Although solitary foragers, most brown hyenas live in small social groups of varying sizes called clans. The members of a clan share and defend a common territory, feed together on large food items, and carry food back to the den for the cubs. The number of brown hyenas inhabiting a territory depends on the quality of food in the territory (Mills 1982a). In a territory where wildebeest were dying because of drought, there were nine adult and subadult brown hyenas, whereas in another where most of the food consisted of small scattered pieces of bone and some wild fruits, only an adult female and her litter of three cubs lived. The mean number of adults and subadults in six brown hyena clans was 3.7 ± SE 0.8 (range 1–9) (Mills 1982a). No dominance hierarchy was apparent in these social groups (Mills 1983a), probably because of their mainly solitary feeding.

Some male and female brown hyenas leave their natal clans at subadulthood, but others stay longer, at least some females doing so for life. The majority of the members of a clan are related, although two males were known to join groups that they were not born into (Mills 1982b, 1989). The average degree of relatedness (*r*) between the members of an intensively studied brown hyena clan was 0.26 (Mills 1989).

Approximately 8% of the brown hyena population in the southern Kalahari (33% of the adult male segment of the population) were found to be nomadic males. These animals apparently range widely and do not belong to any clan. They do, however, perform an important function, as they were the males that were observed mating with the group-living females. Group-living male brown hyenas were never observed to mate with their own or any other females (Mills 1982b, 1983b), although recent evidence suggests that immigrant males may do so (Mills 1989; see also Owens and Owens 1984).

The mean number of adults and subadults in six spotted hyena clans was 8.0 ± SE 1.6 (range 3–12) (Mills 1989). Although the quality of the food resources in the southern Kalahari is responsible for the small clan sizes, in comparison with, for example, the Ngorongoro Crater (Kruuk 1972), the differences within southern Kalahari clan sizes may not have been entirely due to differences in food quality; rabies may also have been implicated (Mills 1989).

Spotted hyena males leave their natal clans at the attainment of adulthood and become nomadic. Some may eventually join another clan, but before doing so have to undergo a prolonged period of assimilation, during which they are repeatedly chased away by established clan members. Southern Kalahari clans usually only have one immigrant male, and the indications are that these are the males that do the mating (Mills 1985, 1989; Frank 1986b). In contrast, female spotted hyenas show more fidelity to their natal clans than do males, although this is not rigid, and cases of females' transferring clans have been recorded (Mills 1989). A dominance hierarchy among the adult females exists, and they and their offspring are dominant to the generally smaller immigrant adult males. Dominance and reproductive success of females have been found to be related (Mills 1989). As with brown hyenas, the average degree of relatedness between clan members is high, being 0.29 and 0.33 for two intensively studied clans (Mills 1985, 1989).

The two species have similar basic patterns of social organization, matrilineal breeding groups. Despite this, fluctuations in group size within a group appear to be considerably greater for brown hyenas than for spotted hyenas. One clan from each species was intensively studied for a six-year period. Taking the mean group size for each year for each clan, I determined that the coefficient of variation for the brown hyena clan was 42%, whereas for the spotted hyena clan it was 13% (Mills 1989). Brown hyenas with their more catholic feeding habits may be more sensitive to changes in food availability in the southern Kalahari than are spotted hyenas, particularly in their ability to capitalize on favorable conditions.

The mean size of six brown hyena group territories was 308 km² ± SE 39, and the mean of six spotted hyena territories was 1095 km² ± SE 177 (Mills 1989). In neither species were group size and territory size correlated, but there was a significant correlation between territory size and the average distance moved between food items. These correlations suggest that territory size is influenced by the way food is distributed in the territory (Mills 1982a, 1989). Furthermore, the average distance moved between locations of meal's (one meal is defined here as a large vertebrate food item or ten wild fruits for brown hyenas and a kill or meaty carcass for spotted hyenas) was 32.7 km for spotted hyenas and only 9.2 km for brown hyenas. Spotted hyenas need to travel longer distances between food items than brown hyenas do because of their more specialized feeding habits and because they normally feed off larger food items. Consequently, they forage over a much larger territory (see Gittleman and Harvey 1982).

Denning Behavior

Both species keep their cubs in holes in the ground, and although the entrances to these dens may be large, the dens quickly narrow down into tunnels 30–50 cm high and 50–60 cm wide (Mills 1989), which are only large enough for the cubs to enter. The dens, therefore, provide ideal refuges for the cubs during the long periods that adults are absent. In both species there is an extended period of 15 months that cubs are attached to the den, and cubs are weaned as late as 12–15 months of age (Mills 1983b).

Other denning behaviors of the two species are different. Only one litter of cubs was found at most southern Kalahari brown hyena dens, although occasionally (three out of 12 dens observed) two females raised their cubs simultaneously at the same den (Mills 1983b). Spotted hyena dens, on the other hand, were usually communal, being used simultaneously by several females with cubs of varying ages. The modal size of 15 brown hyena litters was 3 (range 1–4), whereas no spotted hyena female was ever observed to have more than two cubs (Mills 1989). Brown hyena females occasionally suckled each other's cubs, but spotted hyena females were never observed to do so (Kruuk 1972; Mills 1983b, 1989).

The most marked difference in the denning behavior of the two species is that brown hyenas regularly carry carcasses with meat back to the den for the cubs to eat (Mills 1982b), whereas spotted hyenas do not do so. The milk diet of brown hyena cubs, therefore, is substituted from about 12 weeks of age with meat, whereas spotted hyena cubs obtain a substantial amount of meat only when they are nine to 12 months old and able to accompany foraging adults (Mills 1989).

Again, differences in feeding habits seem to be the main selective pressures for these differences in denning behavior. Because brown hyenas so often feed alone, an individual that finds a suitable food item for cubs can usually eat some of it and then carry the rest back to the den, as there are unlikely to be any other hyenas competing with it for the food. At the den there are normally few cubs and competition for the food is likely to be spread among cubs of equal age.

In contrast, the competitive feeding behavior of spotted hyenas makes it important for each individual to eat as much as it can as quickly as possible. There is, therefore, seldom any meat left over to take back to the cubs. The large size and dominance of the females gives them priority at carcasses, so that they can meet the increased demands of lactation and quickly satiate themselves before returning to suckle their cubs. It is possible that the comparatively small litter size of spotted hyenas has evolved as a result of the heavy dependence of spotted hyena cubs on milk. By providing additional nourishment for their cubs, brown hyenas can raise larger litters. Additionally, the total dependence of the spotted hyena cubs on their mother's milk may mean that females can afford to suckle only their own cubs, thus explaining the lack of communal suckling among closely related females, as is found among brown hyenas (Mills 1983b, 1985, 1989).

Communication

Although vocal systems of hyenas are difficult to classify because they are graded and linked by intermediates, the spotted hyena has an obviously larger vocal repertoire than the brown hyena. Eight vocalizations have been identified for the brown hyena: a yell, a grunt-laugh, two whines, and four growls (Mills 1989). All are short-distance vocalizations aimed at conspecifics or competitors in sight of the vocalizing animal and, except for the yell, can be heard only over a few meters.

Twelve vocalizations—two whoops, two groans, a giggle, a yell, four growls, a grunt-laugh, and two whines—have been identified for the spotted hyena (Kruuk 1972; Mills 1989). Some of the vocalizations, particularly the whoop, often function as long-distance calls, which may be directed at conspecifics not in sight of the vocalizing animal. In addition to their larger vocal repertoire, spotted hyenas vocalize far more frequently and, even where equivalent calls occur, louder than brown hyenas.

It is often important for spotted hyenas to know where the other members of their clan are so that they can come together to form a hunting group, or to interact with major competitors such as lions and spotted hyenas from other clans. They have, therefore, evolved a long-range call (the whoop), one of the functions of which is to help accomplish this (Mills 1989). They have also evolved a number of other vocalizations such as the low groan and the giggle, which are important in communal antagonistic situations at food, in territorial defense, and against lions. Brown hyenas do not need to quickly join up with fellow group members, nor do they need a complex repertoire of group-orientated vocalizations, as most of their feeding and foraging is solitary. The long-distance and communal vocalizations of the spotted hyena are mainly responsible for the differences in its vocal repertoire from that of the brown hyena. The other vocalizations are similiar, further testimony to the close phylogenetic relationship between the two species.

Chemical communication is found in both species (Mills et al. 1980; Gorman and Mills 1984; Mills and Gorman 1987). This occurs chiefly by means of their pasting anal gland secretions onto grass stalks, but also by their defecating at latrines, both of which are often accompanied by scratching the ground with the forefeet. Pasting is unique to the Hyaenidae. Pasting is performed by brown hyenas at a far higher frequency ($\overline{X}$ = 2.64 pastings per km) than by spotted hyaenas ($\overline{X}$ = 0.13 pastings per km). Moreover, the paste of a brown hyena consists of two distinct components; a long-lived, lipid-rich white secretion and a short-lived, watery black one, whereas the paste of a spotted hyena consists of a long-lived component only. Brown hyenas secrete individually unique pastes and are able to recognize each other's pastes (Mills et al. 1980), and the same may be true for spotted hyenas (M. Gorman, pers. comm.).

The long-acting component of pastes and defecating at latrines are seen to have an intergroup function in both species, and are the means by which individuals communicate an unambiguous cue to territorial ownership, thus allowing a conventional settlement of conflicts (Gorman and Mills 1984). In the southern Kalahari both species employ a hinterland method of marking their territories rather than border marking, probably because the length of border to be marked is so large that border marking is impractical.

The long-acting paste may also have an intragroup function. It may help to reinforce an individual's presence in the group. This may be particularly important for brown hyenas, as they do not meet up with their fellow group members as often as spotted hyenas do.

The short-acting paste of the brown hyena is also believed to have an intragroup function (Mills et al. 1980); it may communicate to other members of the group where a brown hyena has recently been foraging. This message could be important in preventing individuals from wasting time and energy foraging in areas that are likely to be unproductive or where competition may be increased, as so much of the brown hyena's food consists of small items with a slow renewal rate. Pasting, therefore, appears to be of more importance to brown hyenas than to spotted hyenas. This is reflected in the higher frequency at which they paste, the larger size of the anal gland, and in the more complex secretions they deposit. Again, the selective pressures accounting for these differences seem to be chiefly related to feeding habits.

Visual communication patterns are better developed in the spotted hyena than in the brown hyena (Kruuk 1972; Mills 1989). For example, the dark tail of the spotted hyena contrasts markedly with its light-colored body, thus enhancing the signaling function of the tail, which is raised or curled over the back whenever the hyenas are excited. The brown hyena's dark tail does not contrast with the rest of its body color, and so the signaling function of the tail is less striking. Furthermore, the spotted hyena uses a greater variety of head movements in communication than the brown hyena does. Spotted hyenas also indulge in communal social activities such as social sniffing, female baiting (Kruuk 1972), and communal scent marking, which are not found in the brown hyena (Mills 1989).

The spotted hyena has also evolved an elaborate meeting ceremony (Kruuk 1972). Two animals greeting stand head to tail, lift the leg nearest to the other, and mutually sniff and lick at each other's erected sexual organs, which in the case of the female have become virilized. This ritual occurs irrespective of the sexes or ages of the animals concerned, and there is no sexual connotation to this behavior. It is seen as a way in which the social bonds in animals that are at times solitary and at others very social are reestablished (Kruuk 1972). Brown hyenas greet far less elaborately. One animal presents its anal region to the other, protruding its anal pouch, which the other then sniffs at. Then the two animals often switch roles. There is no modification of the sexual organs in the brown hyena (Mills 1983a).

These generally more complex visual communication patterns of the spotted hyena, particularly the involved meeting ceremony and greater emphasis on head and tail movements, are further manifestations of the higher degree of sociality of this species.

The most striking visual display of the brown hyena is pilo-erection of the long hair over the back and neck, which occurs in any situation where there is a tendency to either attack or flee. This display is also prominent in the striped hyena (Rieger 1978) and the fourth member of the Hyaenidae, the aardwolf (*Proteles cristatus*) (Kingdon 1977). Spotted hyenas have much shorter hair, and although they erect the hairs on their necks during aggression, it is not nearly so striking a display as in the other hyaenids. The fact that pilo-erection is so well developed in the brown hyena may seem inconsistent with the preceding argument. It is perhaps significant that this posture, which at least under some conditions makes the animal performing it appear to be larger, has evolved in the smaller and less aggressive members of the family. Among other things, it is used in defense against larger and more aggressive competitors. Alternatively, spotted hyenas frequently get covered in blood, particularly around their heads, necks, and chests, when killing and feeding. As long hair would be difficult to keep clean, short hair in this species might also be an adaption to feeding (Mills 1989).

Relationships between Brown Hyenas and Spotted Hyenas

Brown hyenas and spotted hyenas compete for food mainly when they are scavenging from large and medium-sized mammal carcasses (Figure 4.1). This competition may be particularly important for brown hyenas, as they stand to lose a significant amount of food should they lose such a carcass, either directly or indirectly, to spotted hyenas. This, in fact, happened in a certain part of the study area well frequented by spotted hyenas (Mills and Mills 1982; Mills 1989).

Whenever the two species meet, even if no food is present, spotted hyenas are clearly dominant. Such an encounter usually results in an unpleasant experience for the brown hyena. These interactions may escalate to physical combat between the two species, with a spotted hyena grabbing a brown hyena by the side of the neck and vigorously shaking it, even on occasion killing it (Mills and Mills 1982, Mills 1988). Yet there is often a measure of attraction shown by the one species toward the other; they often approach closer when they sense each other and sometimes use the same latrines and paste on the same grass stalks (Mills and Mills 1982).

Brown hyenas tend to avoid areas well frequented by spotted hyenas in the southern Kalahari (Mills and Mills 1982). Although the influence of spotted hyenas on brown hyenas in the southern Kalahari is small, a major influence on the distribution and abundance of brown hyenas throughout their range may be the regional density of spotted hyenas (Mills and Mills 1982, Mills 1989). Several other studies on closely related sympatric carnivores have also suggested that the smaller species tend to be absent from areas well frequented by the larger ones (Schaller 1967; Kruuk 1976; Seidensticker 1976; Berg and Chessness 1978; Fuller and Keith 1981; Skinner and Van Aarde 1981).

Comparisons with Other Studies of Hyenas

Both brown and spotted hyenas, particularly the latter, live in a variety of habitats in Africa, ranging from areas with very high prey densities such as the Ngorongoro Crater to areas of extremely low densities such as the interior of the Namib Desert. Studies of these two species in different habitats have shown how differences in the food supply can also affect the diet and foraging behavior of each species, as well as aspects of their social organization and behavior. Some examples are discussed below.

Kruuk's (1972) study of spotted hyenas in east Africa was the first to show that they are efficient hunters of large and medium-sized antelope as well as efficient scavengers. In the Ngorongoro Crater 90.6% of feeding hyenas observed were feeding from their own kills, compared with 54.8% in the Serengeti and 56.0% in the southern Kalahari. Because of the far higher density of ungulates in the Ngorongoro Crater, clan and territory sizes there are very different from those of the southern Kalahari. Ngorongoro clans contained 30–80 individuals living in territories of 10–40 km². Because the small territories were easy to demarcate, scent marking was confined to territorial boundaries.

In the Serengeti the ungulates migrate over large distances, leaving areas without any prey for long periods of the year. In contrast to spotted hyenas in both the southern Kalahari and Ngorongoro, Serengeti spotted hyenas do not usually form clans of fixed membership defending a particular area. Rather, individuals originating from different areas come together to form a temporary clan in an area where there happens to be a concentration of ungulates. Serengeti, like Kalahari spotted hyenas, make far more use of carrion and smaller prey such as Thompson's gazelle (*Gazella thomsonii*) and springbok than Ngorongoro hyenas do. Consequently, foraging group sizes of spotted hyenas are usually smaller in the Serengeti than they are in Ngorongoro, more like those in the southern Kalahari. Similarly, scent marking in Serengeti spotted hyenas is carried out along the main hyena pathways and not around territory boundaries, more in the Kalahari manner.

In the neighboring Masai-Mara National Reserve the social organization of a clan of 60–80 spotted hyenas has been intensively studied by Frank (1986a, 1986b). This revealed a similar system of social dynamics to that of southern

Kalahari spotted hyenas, albeit with far more animals; that is, male emigration from the natal clan, immigration and nomadism, and female philopatry to the natal territory and the formation of dominant matrilines were the central features. Territories in this area are not contiguous, there being unproductive areas between the prey-rich shortgrass areas. Consequently, contact between clans was rare, and boundary-maintenance behaviors such as scent marking were uncommon. Only two latrines were known of on the clan's boundaries.

Henschel (1986) studied a clan of ten spotted hyaenas in the Kruger National-al Park in an open woodland with a high biomass and diversity of ungulates and other mammals, but without brown hyenas. The spotted hyenas scavenged most of their food from drought victims and lion kills and had a less specialized diet than in other regions, inclining to that of the brown hyena in the Kalahari. They mainly foraged in small groups ($\overline{X} = 1.8$). Territory size was intermediate (130 km²), but the social dynamics of the clan was similar to that of southern Kalahari spotted hyenas. Latrines were mainly situated on territory boundaries, and when on boundary patrols, the hyenas tended to be in larger groups ($\overline{X} = 2.6$) than when foraging.

In the Namib Desert spotted hyenas studied by Tilson et al. (1980), Tilson and Hamilton (1984), and Tilson and Henschel (1986) live at even lower densities than in the southern Kalahari. They feed mainly off large and medium-sized ungulates but appear to kill a larger proportion of adult gemsbok than southern Kalahari spotted hyenas do, perhaps because in the Namib Desert there is no competition with lions for this food source (Mills 1989). Clan size was 3–8, and ranges were not contiguous, measured as being between 383 and 816 km² (Tilson and Henschel 1986). Namib spotted hyenas are reported to feed at a far more leisurely pace than their counterparts in east Africa (Tilson and Hamilton 1984) and in the southern Kalahari. This is partly because of the relatively fewer hyenas feeding at the carcass per kilogram of meat available. However, Tilson and Hamilton (1984) never observed a kill being made when the chances that a scramble competition might occur were highest. Thus they may have missed observing some form of scramble competition.

The brown hyena has been studied in the central Kalahari by Owens and Owens (1978), and Owens and Owens (1979a, 1979b, 1984). This area is similar to the southern Kalahari except that with about twice the annual rainfall it is more productive, and large ungulate species absent from or only rarely found in the southern Kalahari, such as giraffe (*Giraffa camelopardalis*) and kudu (*Tragelaphus strepsiceros*), occur there. The feeding and foraging habits of the brown hyena are similar in the two areas, as is much of their social behavior. Central Kalahari brown hyenas are, however, apparently more sociable, with a higher frequency of social contacts than southern Kalahari ones. This is possibly because there are more opportunities for them to feed together from large carcasses.

In certain areas along the Namib Desert coast the brown hyena has become a feeding specialist on Cape fur seal (*Arctocephalus pusillus*) pups (Goss 1986). These appear to provide an abundant and highly concentrated food source. Contrary to expectations, the brown hyenas maintained an excessively large territory, spending time moving through areas that were unlikely to be productive, visiting some mining towns that had been abandoned sometime during the last fifty years. Goss hypothesized that the towns were once food sources that the hyenas' ancestors used and that the relatively recent disappearance of this food supply has not yet changed the hyenas' lifestyle. If this is so, this would appear to be uncharacteristic of the high degree of flexibility hyenas are able to exhibit to changes in food dispersion patterns.

In spite of the flexibility in certain aspects of hyena socioecology, there are certain aspects that seem to be less labile and that may be limited by phylogenetic constraints. The basic structure of brown and spotted hyena social systems is similar: female bonded groups with various types of multi-male associations. Although it is not well studied, the striped hyena, it appears, has a similar social system (Kruuk 1976; Bouskila 1984). The fourth member of the Hyaenidae, the termite-eating aardwolf, however, is socially monogamous, with an adult pair inhabiting a territory with only their most recent offspring (Kruuk and Sands 1972; Richardson 1985). Aardwolves do not feed on large food items, so rarely utilize a food source that could feed several individuals simultaneously. This may have prevented the evolution of larger groups in this species.

The mating systems of the brown hyena and spotted hyena are broadly similar. All females in a group breed; they are polygynous; and the mating males originate from outside the group, although with brown hyenas it is usually nomadic males that mate and with spotted hyenas it is immigrant males. Even the socially monogamous aardwolf tends toward a polygynous mating system, with less aggressive males being cuckolded by neighbors (Richardson 1985). Unusual for polygamous animals is the lack of striking sexual dimorphism and secondary sexual characteristics in the Hyaenid males.

Conclusions

I have shown how two closely related carnivores are able to survive in an area by exploiting different ecological niches and have discussed their adaptations to accomplish this. These findings are summarized in Table 4.1. Of the two species, the brown hyena is the more successful one in the arid southern Kalahari, where the larger ungulates are erratically distributed. The brown hyena is the most common and widespread, and seemingly the best adapted, of the larger carnivores in the area. Its ability to survive on small, sparsely distributed food items of many kinds and to take advantage of changes in food availability, together with its suitably flexible social system, make it a success-

Table 4.1. The behavioral ecology of hyenas in the southern Kalahari

	Brown hyena	Spotted hyena
Diet		
Main food	All kinds of vertebrate remains, wild fruits, insects, birds' eggs	Large and medium-sized ungulates, mainly gemsbok and wildebeest
Manner in which food acquired	Scavenged	Killed and scavenged
Foraging		
Mean and range of foraging group size	1	3(1–15)
Percentage of hours of darkness active	55.3%	80.2%
Mean distance moved per night	31.1 km	27.1 km
Social organization		
Basis of social group	Matrilineal	Matrilineal
Mean and range of social group size	3.7(1–9)	8(3–12)
Sexual dimorphism	None	Females larger than males, mimic males' reproductive organs
Dominance hierarchy	Not present	Females dominant
Average degree of relatedness (r) of group	0.26	0.31
Males responsible for mating	Nomadic	Immigrants
Coefficient of variation in group size	42%	13%
Mean and range of territory size	330 km² (235–481 km²)	1095 km² (553–1776 km²)
Mean distance moved per meal	9.2 km	32.7 km
Denning behavior		
Suckling period	12 months	12 months
Denning period	15 months	15 months
Solitary/communal	Usually solitary, occasionally communal	Communal
Litter size	1–4	1–2
Suckle each other's cubs	Occasionally	Not observed
Method of feeding cubs	Suckling and carrying food to den	Suckling only
Communication		
Vocal	8 vocalizations	12 vocalizations
	Short distance only	Short and long distance and group oriented
Chemical	Pasting and latrines	Pasting and latrines
	2.64 pastings/km	0.13 pastings/km
	Short- and long-acting paste	Long-acting paste only
	Hinterland marking	Hinterland marking
Visual	Tail movements not accentuated	Striking tail movements
	Simple meeting ceremony	Involved meeting ceremony
	No communal social activities	Communal social activities present
	Striking pilo-erection	Pilo-erection less obvious

ful inhabitant of the southern Kalahari ecosystem. On the other hand, the spotted hyena with its greater dependence on large and medium-sized prey animals, is better adapted to the more productive areas of Africa.

Although these two species have been extensively studied in several habitats, there are still gaps in our understanding of many aspects of their socioecology and even more so in the family Hyaenidae as a whole. The mating systems in particular are imperfectly understood; in spite of the many hours members of both species have been observed, few matings by known individuals have been documented. Furthermore, there appear to be several options individuals can choose, for example, males can be nomadic or belong to a group, and it is not known under which conditions these options are chosen. For this, long-term studies on known individuals are essential.

Studies in more habitats would be valuable for learning the extent of behavioral flexibility of these species, although there are severe observational problems to be overcome in thick bush areas. Of the four members of the Hyaenidae the striped hyena is the least well known even though it has the widest distribution (Rieger 1979). Studies on this species are urgently needed, both to help in its conservation and to further investigate the effects of diet and food dispersion on behavior and social organization.

The influence of resource dispersion on the behavior and social organization of other carnivores is as strong as it is on the Hyaenidae (Bekoff et al. 1984; Kruuk and Macdonald 1985). Nowhere is the flexibility in social organization in the Carnivora more conspicuous than in fluctuations in group and territory sizes; gray wolves (*Canis lupus*), red foxes (*Vulpes vulpes*), raccoons (*Procyon lotor*), brown bears (*Ursus arctos*), and European badgers (*Meles meles*), to mention a few examples, all have fluctuations in group and/or territory size comparable in magnitude to those of spotted hyenas (Macdonald 1983). As in the case of the hyaenids, the benefits accruing to individuals living in groups both between and within species vary. Lions, for example, may cooperate to overcome large prey (Schaller 1972), and females in groups reduce the frequency of infanticide (Packer 1986). Dwarf mongooses (*Helogale parvula*) collectively ward off predators (Rood 1983), and African hunting dogs (*Lycaon pictus*) cooperate to feed their young (Malcolm and Marten 1982). These differences are largely determined by different feeding and other ecological pressures. Overriding these ecological pressures are certain phylogenetic constraints that seem to limit the range of flexibility in social system and behavior.

Acknowledgments

I thank the National Parks Board, South Africa, and the Department of Wildlife and National Parks, Botswana, for allowing the study to be conducted in their parks, the National Parks Board, South Africa, for facilities; Hans Kruuk for inspiration and ideas; Margie Mills for help in the field and in data analysis; and Philip Richardson for comments on the manuscript.

References

Bearder, S. K. 1977. Feeding habits of spotted hyaenas in a woodland habitat. *East African Wildl. J.* 15:263–280.

Bekoff, M., Daniels, T. J., and Gittleman, J. L. 1984. Life history patterns and the comparative social ecology of carnivores. *Ann. Rev. Ecol.Syst.* 15:191–232.

Berg, W. E., and Chessness, R. A. 1978. Ecology of coyotes in northern Minnesota. In: M. Bekoff, ed. *Coyotes: Biology, Behaviour and Management,* pp. 229–247. New York: Academic Press.

Bouskila, Y. 1984. The foraging groups of the striped hyaena, *Hyaena hyaena syriaca. Carnivore* 7:2–12.

Bradbury, J. W., and Vehrencamp, S. L. 1976. Social organization and foraging in emballonurid bats. II. A model for the determination of group size. *Behav. Ecol. Sociobiol.* 1:384–404.

Caraco, T., and Wolf, L. L. 1975. Ecological determinants of group size of foraging lions. *Amer. Nat.* 109:343–352.

Clutton-Brock, T. H., and Harvey, P. H. 1977. Primate ecology and social organization. *J. Zool. (Lond.)* 183:1–39.

Crook, J. H. 1965. The adaptive significance of avian social organisations. *Symp. Zool. Soc., London* 14:181–218.

Ewer, R. F. 1973. *The Carnivores.* London: Weidenfield and Nicholson.

Frank, L. G. 1986a. Social organization of the spotted hyaena, *Crocuta crocuta.* I. Demography. *Anim. Behav.* 35:1500–1509.

Frank, L. G. 1986b. Social organization of the spotted hyaena, *Crocuta crocuta.* II. Dominance and reproduction. *Anim. Behav.* 35:1510–1527.

Fuller, T. K., and Keith, L. B. 1981. Non-overlapping ranges of coyotes and wolves in northeastern Alberta. *J. Mamm.* 62:403–405.

Gittleman, J. L., and Harvey, P. H. 1982. Carnivore home-range size, metabolic needs and ecology. *Behav. Ecol. Sociobiol.* 10:57–63.

Gorman, M. L., and Mills, M. G. L. 1984. Scent marking strategies in hyaenas (Mammalia). *J. Zool.* 202:535–547.

Goss, R. 1986. The influence of food source on the behavioural ecology of brown hyaenas, *Hyaena brunnea,* in the Namib Desert. M.Sc. thesis, Univ. Pretoria. 166 pp.

Henschel, J. R. 1986. The socio-ecology of a spotted hyaena (*Crocuta crocuta*) clan in the Kruger National Park. D.Sc. thesis, Univ. Pretoria. 215 pp.

Houston, D. C. 1979. The adaptations of scavengers. In: A. R. E. Sinclair & M. Norton-Griffiths, eds. *Serengeti: Dynamics of an Ecosystem,* pp. 263–286. Chicago: Univ. Chicago Press.

Jarman, P. J. 1974. The social organization of antelope in relation to their ecology. *Behaviour* 48:215–267.

Kingdon, J. 1977. *East African Mammals,* vol. 3. London: Academic Press.

Kruuk, H. 1972. *The Spotted Hyena.* Chicago: Univ. Chicago Press.

Kruuk, H. 1975. Functional aspects of social hunting by carnivores. In: G. Baerends, C. Beer & A. Manning, eds. *Function and Evolution in Behavior,* pp. 119–141. Oxford: Clarendon Press.

Kruuk, H. 1976. Feeding and social behavior of the striped hyaena (*Hyaena vulgaris* Desmarest). *East African Wild J.* 14:91–111.

Kruuk, H., and Macdonald, D. W. 1985. Group territories of carnivores: Empires and enclaves. In: R. M. Sibly & R. H. Smith, eds. *Behavioral Ecology: Ecological Consequences of Adaptive Behavior,* pp. 521–536. 25th Symp. British Ecol. Soc., Oxford: Blackwell.

Kruuk, H., and Sands, W. A. 1972. The aardwolf (*Proteles cristatus* Sparrman) 1783 as predator on termites. *East African Wildl. J.* 10:211–227.

Lamprecht, J. 1978. The relationship between food competition and foraging group size in some larger carnivores: A hypothesis. *Z. Tierpsychol.* 46:337–343.

Leistner, O. A. 1967. The plant ecology of the southern Kalahari. *Mem. Bot. Survey South Africa* 38:1–172.

Macdonald, D. W. 1983. The ecology of carnivore social behavior. *Nature* 301:379–384.

Malcolm, J. R., and Marten, K. 1982. Natural selection and the communal rearing of pups in African wild dogs. *Behav. Ecol. Sociobiol.* 10:1–13.

Mills, M. G. L. 1978a. The comparative socio-ecology of the Hyaenidae. *Carnivore* 1:1–7.

Mills, M. G. L. 1978b. Foraging behavior of the brown hyaena (*Hyaena brunnea* Thunberg, 1820) in the southern Kalahari. *Z. Tierpsychol.* 48: 113–141.

Mills, M. G. L. 1982a. Factors affecting group size and territory size of the brown hyaena, *Hyaena brunnea*, in the southern Kalahari. *J. Zool.(Lond.)* 198:39–51.

Mills, M. G. L. 1982b. The mating system of the brown hyaena, *Hyaena brunnea*, in the southern Kalahari. *Behav. Ecol. Sociobiol.* 10:131–136.

Mills, M. G. L. 1983a. Behavioural mechanisms in territory and group maintenance of the brown hyaena, *Hyaena brunnea*, in the southern Kalahari. *Anim. Behav.* 31:503–510.

Mills, M. G. L. 1983b. Mating and denning behavior of the brown hyaena, *Hyaena brunnea*, and comparisons with other Hyaenidae. *Z. Tierpsychol.* 63:331–342.

Mills, M. G. L. 1984. The comparative behavioural ecology of the brown hyaena, *Hyaena brunnea*, and the spotted hyaena, *Crocuta crocuta*, in the southern Kalahari. *Koedoe* 27 (supplement): 237–247.

Mills, M. G. L. 1985. Related spotted hyaenas forage together but do not cooperate in rearing young. *Nature* 316:61–62.

Mills, M. G. L. 1989. *Kalahari Hyaenas: The Behavioural Ecology of Two Species.* London: Unwin Hyman. In press.

Mills, M. G. L., and Gorman, M. L. 1987. The scent marking behavior of the spotted hyaena in the southern Kalahari. *J. Zool. (Lond.)* 212:483–497.

Mills, M. G. L., Gorman, M. L., and Mills, M. E. J. 1980. The scent marking behavior of the brown hyaena, *Hyaena brunnea*. *South African J. Zool.* 15:240–248.

Mills, M. G. L., and Mills, M. E. J. 1978. The diet of the brown hyaena, *Hyaena brunnea*, in the southern Kalahari. *Koedoe* 21:125–149.

Mills, M. G. L., and Mills, M. E. J. 1982. Factors affecting the movement patterns of brown hyaenas, *Hyaena brunnea*, in the southern Kalahari. *South African J. Wildl. Res.* 12:111–117.

Mills, M. G. L., and Retief, P. F. 1984. The response of ungulates to rainfall along the riverbeds of the southern Kalahari, 1972–1982. *Koedoe* 27 (supplement):129–142.

Owens, D. D., and Owens, M. J. 1979a. Communal denning and clan associations in brown hyaenas (*Hyaena brunnea*, Thunberg) of the central Kalahari Desert. *African J. Ecol.* 17:35–44.

Owens, D. D., and Owens, M. J. 1979b. Notes on social organization and behavior in brown hyaenas (*Hyaena brunnea*) *J. Mamm.* 60:405–408.

Owens, D. D., and Owens, M. J. 1984. Helping behavior in brown hyaenas. *Nature* 308:843–845.

Owens, M. J., and Owens, D. D. 1978. Feeding ecology and its influence on social organization in brown hyaenas (*Hyaena brunnea*, Thunberg) of the central Kalahari Desert. *East African Wildl. J.* 16:113–135.

Packer, C. 1986. The ecology of sociality in felids. In: D. I. Rubenstein & R. W. Wrangham, eds. *Ecological Aspects of Social Evolution: Birds and Mammals*, pp. 429–451. Princeton, N.J.: Princeton Univ. Press.

Richardson, P. R. K. 1985. The social behaviour and ecology of the aardwolf, *Proteles cristatus* (Sparrman, 1783) in relation to its food resources. D. Phil. thesis, Univ. Oxford. 288 pp.

Rieger, I. 1978. Social behavior of striped hyaenas at the Zurich Zoo. *Carnivore* 1:49–60.

Rieger, I. 1979. A review of the biology of striped hyaenas, *Hyaena hyaena* (Linné, 1758). *Säugetierk. Mitt.* 27:81–95.

Rood, J. P. 1983. The social system of the dwarf mongoose. In: J. F. Eisenberg & D. G. Kleiman, eds. *Advances in the Study of Mammalian Behavior*, pp. 454–488. Special Publication no. 7. Lawrence, Kans.: American Society of Mammalogists.

Schaller, G. B. 1967. *The Deer and the Tiger*. Chicago: Univ. Chicago Press.

Schaller, G. B. 1972. *The Serengeti Lion*. Chicago: Univ. Chicago Press.

Seidensticker, J. C. 1976. On the ecological separation between tigers and leopards. *Biotropica* 8:225–234.

Skinner, J. D., and van Aarde, R. J. 1981. The distribution and ecology of the brown hyaena, *Hyaena brunnea,* and spotted hyaena, *Crocuta crocuta,* in the central Namib Desert. *Madoqua* 12:231–239.

Tilson, R. L., and Hamilton, W. J. 1984. Social dominance and feeding patterns of spotted hyaenas. *Anim. Behav.* 32:715–724.

Tilson, R. L., and Henschel, J. R. 1986. Spatial arrangement of spotted hyaena groups in a desert environment, Namibia. *African J. Ecol.* 24:173–180.

Tilson, R. L., von Blottnitz, F., and Henschel, J. 1980. Prey selection by spotted hyaenas, *Crocuta crocuta,* in the Namib Desert. *Madoqua* 12:16–30.

CHAPTER 5

Intraspecific Variation in
Canid Social Systems

PATRICIA D. MOEHLMAN

The family Canidae is composed of approximately 37 species that are categorized into 10–13 genera (Clutton-Brock et al. 1976; Macdonald 1984). Canids typically are lithe, muscular runners possessing the ability to travel at speeds of up to 30 km/h for extended periods. They are diverse in body weight (1.5–31.1 kg), diet, and habitat (Gittleman 1984; Macdonald 1984). They usually breed once a year and initially raise their litters in ground dens. Compared with most mammals, they have a large litter size and a long period of infant dependency (Kleiman and Eisenberg 1973). The pervasive mating system among canids is obligatory monogamy, a trait that is rare in mammals (Kleiman 1977). Canids are also unusual in that family members share food and provide care for sick adults and dependent young. The larger canid species regurgitate food to family members, which allows greater efficiency in and opportunity for sharing food.

Interspecific Variation among Canidae

Among the canids there are general behavioral trends that correlate with body size (Macdonald and Moehlman 1983; Moehlman 1986). The smaller canids (<6.0 kg) like red foxes (*Vulpes vulpes*) and bat-eared foxes (*Otocyon megalotis*) are usually monogamous but are on occasion polygynous, and they tend to have a sex ratio biased toward females, female helpers, and male dispersal. Medium-sized canids (6.0–13.0 kg) like jackals (silverbacked jackal, *Canis mesomelas*, and golden jackal, *C. aureus*) and coyotes (*C. latrans*) appear to be strictly monogamous; their adult sex ratios are equal, and their male and female helping behavior and dispersal are equivalent. The largest canids (>13.0 kg) like the African hunting dog (*Lycaon pictus*) have a monogamous mating system with a tendency toward polyandry and an adult sex ratio skewed toward males, male helpers, and female emigration. Feeding ecology also shows a body size trend: smaller canids tend to be solitary hunters, and medium-sized canids sometimes hunt cooperatively. Among most large canids cooperative hunting is an important if not critical method of obtaining food.

Allometric analyses indicate that there are strong correlations between mean female body weight and a number of important life history traits (Bekoff et al. 1981; Gittleman 1984, 1985; Moehlman 1986). In particular there are strong correlations when (1) natural log median birth weight is regressed against natural log mean female body weight ($r^2 = 0.97$, slope $= 0.76 \pm 0.08$ at $t_{.05}$), (2) natural log mean litter size is regressed against natural log mean female body weight ($r^2 = 0.72$, slope $= 0.33 \pm 0.12$), and (3) natural log litter weight is regressed against natural log mean female body weight ($r^2 = 0.89$, slope $= 1.14 \pm 0.24$) (Moehlman 1986). The correlations indicate that as canid females increase in body weight, they tend to have relatively smaller and potentially more altricial young. Unlike most mammals (Eisenberg 1981) and carnivores (Gittleman 1984), canids have a positive correlation between litter size and female body weight. Thus, larger females will not only be producing increasingly altricial young, but more of them. Concurrently, their prepartum investment will remain high and may even increase with the larger females. As maternal weight increases, the trend is toward more prepartum investment in gestation of larger litters composed of proportionally smaller neonates. Correspondingly, more postpartum investment may be needed to rear these larger litters to the age of independence.

This allometric and essentially physiological scenario is consistent with the general interspecific behavioral pattern observed in Canidae. Smaller females will produce fewer, more developed neonates that will potentially require less postpartum investment. Parental investment and sexual selection theory predicts that as males contribute less, there will be reduced competition by females for males, there will be a tendency toward polygyny, the adult sex ratio will skew toward females, and males will disperse (Trivers 1972). This suite of behaviors has been observed in small canids like kit foxes (*Vulpes macrotis*), arctic foxes (*Alopex lagopus*), bat-eared foxes, and red foxes (Storm and Ables 1966; Ables 1975; Storm et al. 1976; Brady 1978, 1979; Nel 1978; Egoscue 1979; Lamprecht 1979; Macdonald 1979b, 1980, 1981; Hersteinsson 1984; Nel et al. 1984; Moehlman 1986).

By contrast, large canid females produce larger litters of relatively less developed neonates. They appear to be making a larger prepartum investment that will potentially require substantial male investment in the rearing of these offspring. Females cannot afford to share this investment with other females (e.g., polygyny is unlikely), and competition for males could be intense. In the cases of both African hunting dogs and gray wolves (*Canis lupus*), males tend to provide more food than do females to pups (Malcolm 1980; Fentress and Ryon 1982), there is fierce competition between females for males, and there are limited observations of polyandrous matings (van Lawick 1973; Davidar 1975; Reich 1981; Harrington and Mech 1982; Harrington et al. 1982, 1983; M. Rabb, pers. comm.). African hunting dogs do exhibit a significant pup and adult sex ratio bias toward males, female emigration, and male helpers (Dekker 1968; Frame and Frame 1976; Frame et al. 1979; Malcolm 1980; Malcolm and Marten 1982; Heerden and Kuhn 1985). The dhole (*Cuon alpinus*) displays similar life history allometry, mating and rearing strategies, and feeding ecology (Johnsingh 1982). These large canids tend to be more dependent on cooperative hunting, and improved hunting success and defense of prey by larger groups can interact positively with the investment needs for cooperatively rearing large litters (Ewer 1973; Kleiman and Eisenberg 1973; Kruuk 1975; Lamprecht 1981). When obligatory cooperative hunting is not linked to pup rearing, litters tend to be smaller. Gray wolves are cooperative hunters in the winter, but in the spring and summer their feeding ecology may shift to solitary hunting of small prey and groups may disperse (Mech 1970; Peterson et al. 1984). The major exception among large canids is the maned wolf (*Chrysocyon brachyurus*), a solitary forager that feeds on rodents and fruit, has a monogamous mating system, and produces only two pups on average.

Allometric analyses are useful for delineating life history strategies that are related to size, but they are just one step in understanding the variation in canid social systems. Ecological constraints and their effect on feeding, spacing, and reproductive strategies must be understood if one is to elucidate both inter- and intraspecific variation among Canidae. Research on the correlation of food habits and basal rate in eutherian mammals (McNab 1986, this volume) offers important insights into the potential effects of diet on ecology and behavior. Although scaling of basal rate generally follows Kleiber's (1961) curve, different diets—for example, frugivory versus carnivory—are correlated with lower- or higher-than-expected basal rates. Climate is also a contributing factor to basal rate, but recent research indicates that diet may dominate climate with regard to influencing basal rate (Hennemann et al. 1983). The strong correlation of diet with basal rate could be causative since food type attributable to (1) digestibility, (2) toxicity, (3) availability, and (4) energetic cost of acquisition may limit the rate at which a mammal can acquire energy and hence expend energy (McNab 1986). Since strong correlations do exist between basal rate of metabolism and eutherian mammal reproductive traits such as gestation period, postnatal growth constant, and fecundity, the combined effects of body size, food habits, and climate can have a major impact on maximum intrinsic population growth rate (Hennemann et al. 1983; McNab 1986) and social behavior (Gittleman 1985; Moehlman 1986).

The potential effect of such ecological contraints can be seen in the maned wolf (22.7 kg), the only large canid that has a mixed diet and feeds exclusively on small food items (rodents and fruit). Canids with mixed diets generally have basal rates intermediate to fruit-eating specialists (low) and vertebrate-eating specialists (high). The costs and rewards of foraging for many small food packets can impose energetic constraints on the ability of adults to invest in reproduction, and maned wolves have the lowest mean litter size recorded for canids, $\bar{X} = 2.0$ (Acosta 1972; Brady and Dinen 1979; Dietz 1984). Another example is arctic fox populations in the Northwest Territory of Canada which periodically have access to very abundant populations of lemmings (*Dicrostonyx torquatus* and *Lemmus sibiricus*) during the whelping season. These vertebrate-eating specialists have very large litter sizes ($\bar{X}$ = 10.1) compared both with other populations of arctic foxes and with canids of similar weight (Macpherson 1969; Hersteinsson 1984). By exploiting periodically abundant vertebrate prey, the arctic fox may dramatically increase its reproductive rate.

Single Species Studies

Interspecific analyses of reproductive traits and maternal body weight indicate that "first-order" or "physiologically based" strategies are strongly correlated with size (Western 1979). Such analyses are useful in extracting what components of a species' life history strategy may depend on size and what components reflect environmental and ecological selection ("second-order strategies"). Single species studies under different or varying ecological circumstances offer the best opportunity for understanding linkages between ecology and behavior. The majority of such canid studies are descriptive and examine the significance of correlations rather than determine causality.

Field studies indicate that canids can exhibit an impressive degree of intraspecific variation both between populations and within the same population seasonally and from year to year. A plethora of data exist on variability in diet composition, home range size, group size and sex composition, and litter size; but few studies provide concurrent detailed information on (1) food availability and energetics of acquisition, (2) predation/parasite/disease pressure (3) territory/home range availability and utilization, and link these factors to (1) group size, social organization, and mating behavior (see Gittleman, this volume); (2) care of young, pup development, and survival; and (3) spacing behavior and dispersal (see Bekoff, this volume). In species where there are adequate data, I will examine the potential effects of diet and feeding ecology, habitat and climate, and predation and disease on spacing systems, social groups, mating systems, cooperative rearing of young, and dispersal.

Bat-eared Fox

Bat-eared foxes are insectivores; harvester termites (*Hodotermes*) and dung beetles (Scarabaeidae) make up most of their diet (Nel 1978; Lamprecht 1979; Malcolm 1985). In all locales where they have been studied, bat-eared foxes are nonterritorial and social groups forage together. Given the ephemeral nature of their food resources, individuals would have little negative effect on another's foraging success (Lamprecht 1979; Waser 1980), and in fact bat-eared foxes call each other to food resources, thereby enhancing group members' access to food (Nel et al. 1984). In the Kalahari Desert bat-eared fox group size fluctuates seasonally and is strongly correlated with rainfall, which in turn is a major correlate of invertebrate abundance (Nel et al. 1984).

Bat-eared foxes are monogamous, though polygynous groups in which both females nurse all the pups have been observed (Nel et al. 1984). Given the nature of this fox's feeding ecology, the addition of another breeding female to the group might not detract from available food resources and would potentially provide additional food and guarding for the pups as well as a substitute source of milk if one of the females should die. Bat-eared foxes bear small litters of pups that are more developed than other canid young, and females potentially can sustain a majority of the needed parental investment. On the Serengeti, when the male of a pair died, the female succeeded in gestating and rearing five pups (B. Mass, pers. comm.). Given a female's potential to provide most of her pups' nutritional requirements, a male could be polygynous and attempt to improve his reproductive success. The division of his paternal care might have little effect on pup survival. Because of the relative ease of dispersal by nonterritorial species, the major selective force for larger social groups of bat-eared foxes would be anti-predator vigilance and defense and shortage of available dens.

Crab-eating Fox

Crab-eating foxes (*Cerdocyon thous*) are monogamous. Brady's (1979) study in Venezuela determined that their diet and spacing system had a marked seasonal change. In the wet season insects and fruit were their predominant food (72%), and there was home range overlap and tolerance of adjacent pairs (e.g., they were nonterritorial). Apparently there was relatively little problem with resource depletion. By contrast, during the dry season vertebrates and crabs composed 79% of their diet, and the foxes were territorial and intolerant of nonfamily conspecifics. The type of food resources (insects versus vertebrates) is a possible explanation for the change in spacing systems.

Arctic Fox

Detailed field studies have been done on arctic foxes in northwest Canada (Macpherson 1969) and coastal and inland habitats of Iceland (Hersteinsson 1984). There were important differences in food type and availability between the inland tundra of northwest Canada and the habitats in Iceland. On the coast of Iceland food resources are composed of seal carcasses, seabirds, fish, invertebrates, and berries; their spatiotemporal availability was variable and patchy. By contrast, food availability in the inland Iceland habitat is seasonally more steady and reliable, and the resources are composed of ptarmigan and sheep carcasses in the winter and migrant birds in the summer. In a comparison of these habitats (where fox hunting was not a major cause of mortality), litter sizes were significantly larger in the coastal area (coastal = 4.53 ± 1.47, n = 57; inland = 4.00 ± 1.46, n = 129; $p \leqq 0.02$) (Hersteinsson 1984).

There also was a significant difference in group size in the two locales. On the coast groups were composed of the breeding male and female plus a nonbreeding yearling female. In some families the nonreproductive females were helpers who brought food to the pups and guarded them. Coastal habitats also had smaller territories and a higher density of breeding dens. There were observations of groups in which there was one male and two lactating females, but no detailed behavior data were taken. By contrast, the inland area groups rarely had adults in addition to the breeding pair. The Iceland populations exhibited group sizes consistent with the predictions of Macdonald's (1983) resource dispersion hypothesis and Lindström's (1986) territory inheritance hypothesis.

In the northwestern Canadian tundra lemmings composed 50–90% of arctic fox diets. During a five-year study of 203 dens, data were collected on lemming abundance, the composition of scats, and arctic fox reproduction. The adult sex ratio was 1:1; $\frac{1}{3}$ of the one- and two-year-old females bred, and $\frac{5}{6}$ of the 3+-year-old females bred. It was determined, based on placental scars, that mean litter size at birth did not vary year to year, but litter size was comparatively large ($\bar{X}$ = 10.6 ± 0.28, n = 118). Mean size of weaned litters did vary yearly and correlated positively with lemming abundance. Macpherson (1969) attributed the large litter sizes to the heightened seasonal contrast in food resources in this northern latitude and the large relative food surplus during the breeding season. Braestrup (1941, in Macpherson 1969) had similar results in Greenland, observing that arctic foxes in the interior that were dependent on lemmings raised twice as many pups per litter as foxes living on the coast that preyed mainly on marine animals and birds. Arctic foxes are opportunistic in utilizing seasonally abundant food resources and can dramatically increase their reproductive rate. They are morphologically equipped to have relatively large litters in that they have twice as many teats as other canids of similar body size (Ewer 1973).

Red Fox

Red foxes typically have a monogamous mating system but are also found in single male groups with two to five females (Macdonald 1979b, 1980, 1981). The female component of the group can range from (1) one reproductive female, (2) one reproductive female plus nonbreeding female helper(s), (3) one reproductive female plus additional breeding female(s) that lose their pups due to reabsorption, abortion, or negligence, and (4) two reproductive females that den and nurse communally.

As long as the subordinate females do not breed, the alpha female derives benefits of (1) additional food to her pups, (2) increased anti-predator surveillance and defense, (3) increased time spent foraging while the helper(s) do guard duty, (4) an additional female that can "substitute" if the mother dies (Macdonald 1983), and (5) assurance that the additional female (kin) might inherit her territory (Lindstrom 1986). For the subordinate female(s) the benefits of (1) staying on a known territory, (2) potentially inheriting the territory and reproductive status, and (3) investing in close kin and deriving inclusive fitness benefits must be weighed against the costs of (1) dispersing, (2) acquiring a territory and mate, and (3) delaying reproduction (Emlen 1982a).

But why don't additional males stay? In all fox populations studied there is a sex bias toward males' emigrating. Small canid females have fewer and heavier newborns, which potentially require less paternal investment postpartum, and if a female can provide most of the offsprings' needs, then the male can invest in more than one female, and polygyny is possible. Females invest more than males and are the limiting sex. Sexual selection theory would then predict a polygynous mating system, with males more likely to disperse (Trivers 1972).

Food availability and dispersion can affect the spacing of a social group. In Macdonald's (1980, 1981) main study area territories were small (0.19–0.72 km², $n = 7$), individuals within the group were frequently in contact, and suitable denning sites may have been limited. Subordinate females did not exhibit reproductive behavior and may have suffered endocrine suppression because of the alpha female's behavior.

Von Schantz's (1981, 1984) study area also had groups composed of one male and several females. But in this locale territories were larger, food was more dispersed, and subordinate females occupied smaller (2.7 km²) and suboptimal areas within the territory (5.3 km²). They presumably were able to avoid the dominant female and had a reduced frequency of stressful encounters. They did breed, but they did not successfully raise litters. When the alpha female died, a subordinate female provisioned her litter and took over her portion of the territory. Although subordinate females did not breed successfully, they were positioned to inherit a better part of the territory. One male could control an area in which several females bred but additional pups did not survive, presumably because of inadequate resources and provisioning.

Communal denning is rarely observed, and there are no direct data on the factors that might select for it. Macdonald (1980) has postulated that communal denning would be mutualistic and allow females to share maternal duties; for example, if one female died, the other could raise both litters. In addition, the male would be more likely to provision both litters if they were in the same den.

Silverbacked Jackal

Silverbacked jackals are medium-sized canids that are monogamous, territorial, and have equal sex ratios in social group composition, helpers, and dispersal (Moehlman 1983, 1986). These canids are both solitary foragers and facultative cooperative hunters. When hunting cooperatively, they have a higher success rate in killing Thomson's gazelle fawns (*Gazella thomsoni*) (Wyman 1967; Lamprecht 1978). Cooperative groups of jackals are also more successful in defending and feeding on carcasses (Lamprecht 1978; Moehlman 1983).

Long-term pair bonding (six to eight years) in silverbacked jackals reflects both physiological and ecological constraints. The female bears large litters ($\bar{X}$ = 5.7, range = 1–9) that have a long period of dependency (4+ months). In the Serengeti study area (Moehlman 1983, 1986) they forage on rodents (~60 gm) and fruit (5 gm) that are abundant but energetically costly since they involve foraging trips of 6–8 km. Paternal investment is critical to pup survival, and pairs on average raise only 1.3 pups (n = 6, range = 0–2). When male parents have died, whole litters have been lost and the females have disappeared (n = 2).

Of known surviving pups, 24% stayed and helped (n = 20, male:female = 1:1) by feeding the pups and lactating female, guarding the pups, and socializing with the pups. With the addition of helpers there was a positive and significant correlation with rates of regurgitation (n = 7 litters, 472 h observation, r_s = 0.90, $P \leq 0.01$), and pups were seldom left unguarded. Pup survival at 14 weeks of age had a significant correlation with the number of adults in the family (r_s = 0.89, $P \leq 0.01$) (Moehlman 1986).

Parents could improve their reproductive success by allowing offspring to remain on the natal territory and invest in the new litter of pups. Pup survival was increased, the female's future reproductive success potentially was improved by the provisioning that she received, and offspring were in place to inherit the territory if the parents died. The benefits to parents of retaining helpers were limited by the available resources on the territory and the energetics of provisioning the pups.

A year-old jackal had the option of staying and helping or of dispersing, attempting to acquire mate and territory and trying to raise a litter of its own (Moehlman 1979; Emlen 1982a, 1982b). Ecological and demographic constraints would determine the costs and benefits of this choice. In this study area jackals had an abundant food supply throughout the year (3000–13,000 rats/km^2) (Senzota 1978), and conditions were favorable for the retention of offspring in the natal territory (Macdonald 1983). However, the food resource situation that made it possible for young adults to stay also made it easier for them to disperse and reproduce. Because they live in a brush woodland habitat with poor visibility, individuals could establish residence at the edges of existing territories.

Silverbacked helpers derived significant inclusive fitness benefits since on average a helper contributed to the survival of 1.74 pups and their average relatedness was r = ½ (Moehlman 1981, 1983). In addition they gained extended experience in familiar terrain that might increase their survivorship and the quality of their future parental care (see Gittleman 1985). Potential dispersers would have constraints on rearing their first litter since pairs on average can raise only one pup and even experienced parents lose whole litters. Even with this pattern of inclusive fitness benefits and ecological constraints on rearing offspring, most one-year-old silverbacked jackals opted for dispersing and attempting to reproduce at an earlier age. The real costs and benefits of this choice cannot be evaluated until there are data on one-year-olds' versus two-year-olds' success at emigrating, acquiring a territory and mate, and successfully rearing pups.

Present data indicate that silverbacked jackals and other medium-sized canids are exclusively monogamous. It seems that the paternal investment of the silverbacked jackal is critical to pup survival and hence to both male and female reproductive success. The female cannot raise a litter on her own and requires substantial male investment. Their relative investment in the pups is comparable, and hence intrasexual competition for access to the opposite sex is equal and selects for a monogamous mating system (Trivers 1972). Territorial defense is almost exclusively between the resident and intruders of the same sex (Moehlman 1979, 1983). A monogamous mating system in turn would select for equal sexual selection for helpers and hence equal sexual dispersal (Trivers 1972; Emlen et al. 1986). However, if food availability changed such that it was energetically possible for the female to invest much more than the male, hence energetically feasible for the male to invest in several litters, then the balance could tip toward polygyny. On the other hand, if the female's ability to invest became more limited and more male investment was required, then males would become the limiting resource and the tendency would be toward polyandry. Anecdotal data indicate that silverbacked jackals sometimes have multiple litters (Ferguson et al. 1983). Thus, in some locales silverbacked jackals might either have multiple monogamous pairs or a polygynous group on a territory.

Golden Jackal

Golden jackals are also monogamous, territorial, and have families in which some pups stay and help raise the next year's litter. The male provisions his mate during her pregnancy, and the male and the helpers feed the lactating female and the pups. During whelping season in the Serengeti golden jackals feed on larger prey and carcasses. Cooperative hunting and defense of carcasses are a more important component of their feeding ecology, and they are more carnivorous than the neighboring silverbacked jackals. Golden jackal pairs feed their pups at more than twice the rate that silverbacked jackals do, and pup survival does not appear to be limited by food provisioning. Rainfall and flooded dens, density-independent factors, are a leading cause of pup mortality. Pairs on average raise 1.8 pups ($n = 12$ litters, range $= 0-4$), and although there is a significant correlation between number of adults in the family and pup survival, it is weaker ($r_s = 0.36$, $P \leqq 0.05$) and more variable than the correlation for silverbacked jackals. In this population the nature of the food resources and their availability make it possible for the female to provision food at a higher rate, and male investment could be less critical. In particular, golden females nurse their pups at three to five times the rate of silverbacked jackals (Moehlman 1986). Correspondingly, the golden jackal pair bond is not as strong, and mate changes do occur. Under these circumstances it might be possible for golden jackal males to successfully provision more than one litter, allowing the possibility of polygyny.

Young golden jackals leave their natal territory during the dry season when a food bottleneck occurs, and they are transient in the woodlands. But during the wet season 70% return and help to raise the next year's litter. The higher proportion of young golden jackals' returning and helping versus dispersing may reflect a high cost in obtaining a territory. Golden jackal territories are small (<1 km²) and form a tight mosaic on the open shortgrass plains.

Macdonald's study (1979a) in Israel illustrates the behavioral variability possible in golden jackals. The jackals in this study population obtained 92% of their food from a large provisioning site and a garbage dump. These sites constituted a highly clumped and defendable food resource that was available throughout the year. Social group size was large, with two groups of ten and 20 individuals each. Territory size was quite small (0.1 km²), and territory boundaries were marked with fecal piles (middens). Macdonald suggested that the large groups were possible because the nature of the food resources made it economically defendable (Bradbury and Vehrencamp 1976; Emlen and Oring 1977).

Coyote

Coyotes are medium-sized canids that typically have stable pair bonds (three to four years), territories, and some offspring of both sexes that stay and help (Knowlton 1972; Gize 1975; Camenzind 1978; Bekoff and Wells 1980). Relative prey size may be an important determinant of coyote group size. In habitats where mule deer (*Odocoileus hemionus*) and elk (*Cervus elaphus*) were important food items, there was a correlation with delayed pup dispersal and larger group sizes (Bowen 1978, 1981; Bekoff and Wells 1980, 1982). Bowen found a strong correlation between pack size, territory size, and the percentage of mule deer in the winter diet. He attributed this relationship to an increase in individual fitness with an increase in food acquisition efficiency that resulted from a combination of searching, capture, and defense of food. Bekoff and Wells (1986) attributed the larger size of one of their two study groups to the presence of elk carcasses, a large, abundant, defendable food resource. This larger group had delayed dispersal of offspring, and some individuals stayed and helped. There was a positive correlation with the presence of helpers and pup survival, but it was not significant. Yearlings that dispersed did suffer a higher mortality than their age peers that stayed at home. Data on reproductive rates in yearling coyotes (an indication of early and successful dispersion rate) varies from southern to northern latitudes. Data from the more northern latitudes shows a lower rate of pregnancy in yearling females. This may reflect delayed maturation and dispersal and/or higher costs of dispersal.

In locales where coyote diets are mainly composed of small prey items, e.g., rodents, group size tends to be smaller and dispersal is earlier (Bekoff and Wells 1980). However, a scenario in which prey size determines group size is confounded by studies finding that when coyotes feed on small prey but the coyote population density is high, dispersal presumably is difficult and groups are relatively large (Andelt 1982). Once again the costs and benefits of staying versus leaving must be assessed if one is to determine the selective factors for determining group size. Messier and Barrette (1982) concluded from their study at a northern latitude (46°N) that large prey facilitated group living but that the major selective force for larger social groups was delayed dispersal of juveniles due to (1) saturation of available territories and (2) later age of maturity. Thus in their study, population density and nutrition were critical to dispersal time and rate, and group size.

Coyote field studies indicate that sex ratios are typically equal for group composition and emigration. There are data from radio-tracking studies in which sex ratios are skewed toward males or females and studies in which male territories contain several female territories (Nellis and Keith 1976; Berg and Chesness 1978). These data are intriguing but lack resolution as to the age and reproductive status of individuals, and hence the mating system.

Camenzind (1978) studied groups of coyotes ($n = 3$) in Wyoming that ranged in size from four to seven adults. In several cases there was circumstantial evidence that groups contained multiple breeding pairs. In these groups several females suckled all the pups, and the total number of pups exceeded average litter size. In one group three females appeared to have borne a total of 16 pups that were kept in two separate dens. At 14 weeks all pups were merged into one den and communally reared by what appeared to be three breeding pairs. In this study population a large group size may have been selected for by improved foraging efficiency (better defense of elk carcasses) and improved defense of offspring. Camenzind had circumstantial evidence of infanticide by trespassing conspecifics.

Gray Wolf

The basic component of wolf social organization is the breeding pair. The typical pack composition observed in the wild (Rausch 1967; Mech 1970) is that of a mated pair and its offspring. The pair bond may persist for several years, but direct long-term observations are limited. Wild packs occasionally have two pregnant females (Rausch 1967, $n = 3$), and there is one report of two females' raising their young in a communal den (Mech 1970). The literature generally refers to the pregnant females as being members of mated pairs. The work of Jordan et al. (1967) on the Isle Royale wolf population reported a main pack of 11–22 wolves (1961–66) with three breeding pairs. These data are interesting as they resemble the multiple mated pairs recorded in a pack of coyotes (Camenzind 1978). The pervasive mating system in the genus *Canis* is one of long-term bonded pairs, and when multiple pregnant females occur in a group, it appears to be the result of multiple monogamy rather than polygyny.

The norm for most wolf packs (wild and captive) is one breeding pair, which by agonistic behavior prevents subordinates from mating (Packard et al. 1983). As Packard et al. (1985) have stressed, it is important to distinguish between reproductive failure due to (1) suppression of endocrine cycles and (2) suppression of reproductive behavior. In female wolves age at first ovulation ranges from ten to 22 months, and both social and environmental factors may delay the age of puberty. However, once a female has cycled, anestrus is rarely observed and most reproductive failure in adult females is attributed to lack of copulation. Failure to copulate correlates positively with (1) high rate of aggression received, (2) low rate of sexual behavior received, (3) levels of preovulatory progesterone and cortisol, and (4) low discharge duration (Packard et al. 1985). Subordinate females in a pack exhibit normal estrous cycles and ovulation; and a study of captive packs found that four nonpregnant females had serum hormone concentrations through the luteal stage similar to those in females that produced litters (Seal et al. 1987). Not only do subordinate females not experience endocrine suppression, but they are clearly "primed" to produce and provide milk for another female's offspring. If a dominant reproductive female produces pups and then dies, a subordinate female can provide milk. This "help" raises the issue of whether all cases of observed communal nursing involve strictly birth mothers. Although some subordinate females do breed and produce litters, this normally occurs when there is no parental pair in the pack or when a member of the parental pair is very old or dies (Packard et al. 1983). When subordinate females breed, they tend to be less successful in raising pups (Zimen 1976). Dominance hierarchy is well established in wolf social groups (Mech 1970; Zimen 1976), and the pack is territorial, highly integrated, and aggressive toward nonmembers of the pack. Immigration appears to be almost nonexistent, but, once again, the field data that would clarify this situation are not available.

The review by Harrington et al. of mating systems in wolves (1982) presents evidence for flexibility based on a male or female's ability to control reproductive activity in the group. Subordinate animals can be prevented from breeding through direct threats and aggression. In groups where cooperative hunting and defense of prey is common and group members are always associated, the potential for subjugation and control of low-ranking members increases. It is important to note that although during the winter wolves are primarily cooperative hunters, their foraging pattern may shift to solitary foraging and small prey in the spring and summer, when groups tend to disperse (Mech 1970; Peterson et al. 1984). Since pups are born and reared in spring and summer, wolf social groups may be less cohesive during the reproductive season and subordinate members of the pack could temporarily split away and raise a litter of pups. Food type and availability, population density, and individual pack histories all can affect pack size and cohesiveness, the mating system, and the cooperative rearing of pups.

African Hunting Dog

African hunting dogs are large canids that cooperatively hunt prey as heavy as zebra (*Equus burchelli*, 200 kg). Among the canids, they are the most obligatory cooperative hunters and achieve a high degree of hunting success, 50–70% (Frame et al. 1979). A long-term study found that average pack size in the Serengeti was 9.8, with a range of 1–26 ($n = 12$) (Frame et al. 1979). A pack unit is typically composed of one adult female, one to ten adult males, the yearlings, and the pups. Some packs contain a subordinate female, but she rarely reproduces successfully. In 26 observed natal dens, the dominant female was mother of 20 of the litters and a subordinate female whelped six litters. Of the latter, only one litter survived. Dominant females will prevent the feeding and care of a subordinate female's litter, and dominant females with litters have been observed killing a subordinate female's litter. African hunting dogs have relatively large litters with a mean of 10.1 and a range of 1–16. In allometric analyses, hunting dogs have relatively small (altricial) neonates and very large litters. Thus the female is incurring large prepartum investment costs and large postpartum investment needs (Moehlman 1986). Two pairs that tried to breed without helpers had no pup survival. Four pairs that had yearlings but no adult helpers also had no surviving pups. Adult hunting dogs allow yearlings to feed first at a kill. Thus, if food is abundant, there will be enough for the yearlings and the pups, but if it is scarce, priority goes to the yearlings and the adults will have little to regurgitate to the pups. In the Serengeti if any pups survived, there was a positive correlation between number of helpers and pup survival at one year of age ($r_s = 0.85$, $p < 0.05$) (Malcolm and Marten 1982). Parental investment and sexual selection theory would predict intense competition by females for males and a tendency toward polyandry. This is consistent with observations in the Serengeti population (Frame et al. 1979; Frame 1986). In addition, sex ratio at birth is skewed toward males (Heerden and Kuhn 1985). Malcolm (1980) proposed that this skewed sex ratio was an evolved response to a social organization in which sons were more likely than daughters to contribute to the raising of subsequent litters (Trivers and Hare 1976; Emlen et al. 1986).

Hunting dogs in the Serengeti have very large overlapping home ranges (1500–2000 km²). In this locale males are recruited into the pack and females emigrate. No female remained in her natal pack during ten years of observation (Frame and Frame 1976).

Hunting dogs were also studied in Kruger National Park (Reich 1981). Average pack size was 11 ($n = 27$), and both pup and adult sex ratios were skewed toward males. Polyandrous matings did occur, although monogamy was the norm. However, pack range size was much smaller (500 km²) and density was higher. Female emigration was much less frequent than in the Serengeti, and Reich attributed this difference to (1) the high density of packs, (2) the fact that packs with breeding vacancies did not exist, and (3) the fact that subordinates instead of emigrating were remaining within their natal packs and passively awaiting the death of a dominant or actively fighting for dominant breeding status. In one pack a young female supplanted her mother as dominant, and she bred with her father. When packs were large and the potential for achieving dominant status was low, pack fission instead of emigration occurred.

Conclusions

Among canids the pervasive theme of obligatory monogamy appears to be closely linked to a critical need for male investment in the rearing and survival of offspring. Allometric analyses indicate that a positive and significant relationship exists between neonate weight, litter size, litter weight, and maternal body weight (Moehlman 1986). This trend of larger litters of less developed pups with increasing maternal weight is unique among mammals and may affect the relative investment needed from individuals other than the mother for the successful rearing of a litter. Species in the order Carnivora that communally raise young tend to have relatively heavier litter weights than those that raise young without "helpers", and canids as a family have the heaviest relative litter weights (Gittleman 1984, 1985).

A general relationship between body weight and behavioral trends occurs among canids. Smaller female canids tend to have fewer and heavier pups, require less paternal investment, and are the limiting sex. There is a concurrent tendency toward polygyny and/or female helpers, with males dispersing. The availability of food and the energetics of nutritive input to the mother and pups can alter group size, litter size, and mating system (monogamy ↔ polygyny). Species at the heavy end of the scale have larger litters of relatively more altricial pups and require substantial postpartum investment in pups. Males are the limiting sex, and there is a tendency toward polyandry. Within this trend there are anomalous species (e.g., arctic fox, maned wolf) that emphasize the important role of ecological factors. In particular, the availability and energetics of food acquisition and utilization can affect the ability of individuals to control resources, their access to the opposite sex, and nutritional input to pups (Emlen and Oring 1977; Davies and Lundberg 1984). These scenarios are complicated further by population demography.

The type and size of food resource tends to correlate with feeding and spacing systems. Hence, small canids that feed primarily on invertebrates and fruit (crab-eating foxes and bat-eared foxes) have a spacing system of overlapping home ranges and may have little impact on one another's foraging success. In the case of crab-eating foxes, when the diet changed from exclusively invertebrates to crabs and vertebrates, there was a concurrent change in the spacing system and pairs were territorial. Presumably this change was related to defense of a food resource and to optimization of feeding efficiency, but the energetics of food acquisition was not examined.

Among territorial solitary foragers the distribution (patchiness) of food resources may determine the size of the territory, and the richness of those patches might allow additional group members (Kruuk 1978; Macdonald 1983; Kruuk and Macdonald 1985). This appears to be an important factor for group size in some populations of red foxes (Macdonald 1981) and arctic foxes (Hersteinsson and Macdonald 1982; Hersteinsson 1984). In populations that experience strong seasonal and/or yearly fluctuations, territory size might be determined by the minimal food resource conditions and additional individuals could remain during "good" times. Such resource fluctuations play a role in group size in some populations of red foxes (von Schantz 1981, 1984) and golden jackals (Moehlman 1983).

As body size increases in canids, there is a concurrent tendency toward cooperative hunting and defense of prey. It has long been postulated that social groups in carnivores have evolved in response to increased hunting success that resulted from cooperative foraging and defense of prey (Kleiman and Eisenberg 1973; Kruuk 1975; Lamprecht 1978). But Messier and Barrette (1982) made the important point that it is necessary to demonstrate a per capita increase in food intake with increased group size for cooperative foraging to be a strong selective force. Among coyotes there is intraspecific variation in prey size and group size. However, group size across all populations studied does not correspond strictly to prey size. Although several populations that have larger prey (e.g., mule deer, elk) also have larger groups (Bekoff and Wells 1980, 1982; Bowen 1981), coyotes that prey primarily on rodents may have large groups when coyote population density is high (Andelt 1982). Wild dogs appear to be the only canids that specialize in large prey throughout the year and the only canids for which cooperative foraging may have been the predominant force for the evolution of social groups.

Brown (1982) has made the salient point that all group territories documented involve breeding groups. These breeding groups may be composed of two or more reproductive individuals and/or nonreproductive helpers. Brown then incorporates the contributions that helpers can make to the original territory holder's reproductive fitness through care of the young and anti-predator behavior. These models focus attention on the dynamics of costs incurred via resource depletion versus the benefits gained by having more group members to share the burden of territorial defense and care of the young.

Both the spatial and temporal patterning of key resources and population demography can affect the distribution of potential mates and the costs and benefits of one sex's monopolizing reproductive access to the opposite sex. The mating system will be the outcome of a conflict of interest between males and females concerning their individual reproductive success and will involve complicated individual behavior that is conditioned by operational sex ratio, population density, and food resources (Davies and Lundberg 1984).

In some studies of red foxes the spatial availability of food correlated with the spacing of group members and affected their reproductive status and strategies. In smaller territories with clumped food resources, subordinate females did not reproduce, possibly because of stress induced by frequent encounters with the dominant female. In larger territories with more dispersed resources subordinate females utilized separate but suboptimal territories and presumably avoided contact with the dominant female; and there was a polygynous mating system. Polygynous mating systems with communal denning have also been observed, but the role of food resources and population demography has not been examined.

Variance in group size also occurs in medium-sized canids, but in all populations studied sexual roles have been symmetrical. The relationship between body weight and the associated weight and number of neonates may be at a fulcrum point in the balance between female investment at birth and the relative investment needed from the male to ensure survival of their offspring. Thus, when multiple litters have been observed, they appear to be the result of multiple monogamous pairings. This sexual balance could tip toward polyandry or polygyny if the availability of food were significantly different and changed the parental investment needed from the male. There are suggestions in the literature that this might occur, but the documentation is inadequate. Field studies need to determine genealogies and reproductive status and to quantify individual time and energy budgets.

Among the larger canids mating systems vary intraspecifically from monogamy to polyandry. In gray wolves polyandrous matings have been observed only in captive situations, and observations on free-ranging wolves are very limited. In allometric analyses wolf litter size and weight lie below the line of regression (Moehlman 1986). Thus, physiological constraints and postpartum investment needs (e.g., fewer pups) may not select as strongly for increased male investment. In addition, wolves are flexible in their feeding ecology and typically do not hunt cooperatively during the reproductive season, thus allowing the opportunity for pairs to disperse and successfully raise pups.

African wild dogs are obligatory cooperative hunters, and monogamy with a low frequency of poyandry is well documented for this species in the wild. Females produce very large litters of altricial pups, and there are no observations of a pair of wild dogs' successfully raising a litter of pups on its own. Additional adults (not yearlings) are critical to pup survival. Females compete for male investment and dominant females can prevent subordinates from breeding or kill subordinate females' pups. Subordinate females tend to emigrate in low-density populations (Frame and Frame 1976), but in higher density populations (×3, Reich 1981) subordinate females remain in their natal packs and attempt to become dominant, or pack fission occurs. Physiological constraints in the case of both of these large canids would presumably preclude the viability of polygynous mating system for provisioning and successfully rearing pups.

Among territorial canids, as these models suggest, ecological constraints determine whether it is possible for additional individuals to remain with the parental pair. But it is also necessary to examine when it is worthwhile for a subordinate individual to stay (Emlen 1982a, 1982b). Only when environmental constraints are severe and it is difficult for an individual to breed independently (because of lack of available territories or high cost of successfully rearing young), will it remain on the natal territory as a nonbreeder. Individuals that stay potentially will accrue such benefits as (1) avoiding the high risks of dispersal, (2) acquiring experience in the care of young, (3) increasing inclusive fitness, (4) inheriting a portion of the natal territory, and (5) eventually achieving a reproductive status. It is among the large canids that cooperative hunting and pup rearing are often necessary for survival and reproduction. Some adults within a group may remain nonreproductive for years, and if so, the development of behavioral conflict is often alleviated by shared paternity, communal maternity, and reciprocity (Emlen 1982b).

Variation in food size and temporal and spatial availability of food can greatly affect canid spacing and mating systems. Analyses of resource availability and depletion in terms of individual energy budgets and fitness are critical to understanding optimum group size and social systems in canids. Single species studies do illustrate the ecological variability that populations can contend with and their concurrent behavioral flexibility. Better quantitative data on genealogy, individual time and energy budgets, reproductive success, and survivorship are needed to determine how ecological factors affect intraspecific variation in social systems.

Acknowledgments

I am grateful to the Tanzanian National Scientific Research Council for permission to conduct research and for the support of Karim Hirji (Coordinator, SWRI), Juma Kayera (Conservator, NCAA), and David Babu (Director, TANAPA). This work was financially supported by Wildlife Conservation International, the Harry Frank Guggenheim Foundation, the Muskiwinni Foundation, and the National Geographic Society. I am also grateful to friends and colleagues for their assistance and helpful critiques of my work—in particular, John Gittleman, Bjorn Figenschou, Dan Rubenstein, Devra Kleiman, Jane Packer, and David Macdonald.

References

Ables, E. D. 1975. Ecology of the red fox in America. In: M. W. Fox, ed. *The Wild Canids*, pp. 216–236. New York: Van Nostrand Reinhold.

Acosta, A. L. 1972. Hand-rearing a litter of maned wolves, *Chrysocyon brachyurus*, at Los Angeles Zoo. *Internat. Zoo Yearb.* 12:170–174.

Andelt, W. F. 1982. Behavioral ecology of coyotes on Welder Wildlife Refuge, south Texas. Ph.D. dissert., Colorado State Univ., Fort Collins. 169 pp.

Bekoff, M., Diamond, J., and Mitton, J. B. 1981. Life-history patterns and sociality in canids: Body size, reproduction, and behavior. *Oecologia (Berl.)* 50:386–390.

Bekoff, M., and Wells, M. 1980. The social ecology of coyotes. *Sci. Amer.* 242:130–151.

Bekoff, M., and Wells, M. 1982. Behavioral ecology of coyotes: Social organization, rearing patterns, space use, and resource defense. *Z. Tierpsychol.* 60:281–305.

Bekoff, M., and Wells, M. 1986. Social ecology and behavior of coyotes. In: J. S. Rosenblatt, C. Beer, M.-C. Busnel & P. Slater, eds. *Advances in the Study of Behavior*, 16:251–338. New York: Academic Press.

Berg, W. E., and Chesness, R. A. 1978. Ecology of coyotes in northern Minnesota. In M. Bekoff, ed. *Coyotes: Biology, Behavior, and Management*, pp. 229–247. New York: Academic Press.

Bowen, W. D. 1978. Social organization of the coyote in relation to prey size. Ph.D. dissert., Univ. British Columbia, Vancouver. 230 pp.

Bowen, W. D. 1981. Coyote social organization and prey size. *Canadian J. Zool.* 59:639–652.

Bradbury, J. W., and S. L. Vehrencamp. 1976. Social organization and foraging in emballonurid bats. II. A model for the determination of group size. *Behav. Ecol. Sociobiol.* 1:383–404.

Brady, C. A. 1978. Reproduction, growth and parental care in crab-eating foxes (*Cerdocyon thous*) at the National Zoological Park, Washington. *Internat. Zoo Yearb.* 18:130–134.

Brady, C. A. 1979. Observations on the behavior and ecology of the crab-eating fox, *Cerdocyon thous*. In: J. F. Eisenberg, ed. *Studies of Vertebrate Ecology in the Northern Neotropics*, pp. 161–172. Washington, D.C.: Smithsonian Institution Press.

Brady, C., and Ditton, M. K. 1979. Management and breeding of maned wolves (*Chrysocyon brachyurus*) at the National Zoological Park, Washington, D.C. *Internat. Zoo Yearb.* 19:171–176.

Brown, J. L. 1982. Optimal group size in territorial animals. *J. Theor. Biol.* 95:793–810.

Camenzind, F. J. 1978. Behavioral ecology of coyotes (*Canis latrans*) on the National Elk Refuge, Jackson, Wyoming. Ph.D. dissert., Univ. Wyoming, Laramie. 97 pp.

Clutton-Brock, J., Corbett, G. B., and Hills, M. 1976. A review of the family Canidae, with a classification by numerical methods. *Bull. Amer. Mus. Nat. Hist.* 29:117–199.

Davidar, E. R. E. 1975. Ecology and behavior of the dhole or Indian wild dog (*Cuon alpinus* Pallas). In: M. W. Fox, ed. *The Wild Canids*, pp. 109–119. New York: Van Nostrand Reinhold.

Davies, N. B., and Lundberg, A. 1984. Food distribution and the variable mating system in the dunnock (*Prunella modularis*). *J. Anim. Ecol.* 53:895–912.

Dekker, D. 1968. Breeding the Cape hunting dog at Amsterdam Zoo. *Internat. Zoo Yearb.* 8:27–30.

Dietz, J. M. 1984. Ecology and social organization of the maned wolf (*Chrysocyon brachyurus*). *Smithsonian Contrib. Zool.* 392:1–51.

Egoscue, H. J. 1962. Ecology and life history of the kit fox in Toole County, Utah. *Ecology* 43:481–497.

Egoscue, H. J. 1979. *Vulpes velox. Mamm. Species* 122:1–5.

Eisenberg, J. F. 1981. *The Mammalian Radiations.* Chicago: Univ. of Chicago Press.

Emlen, S. T. 1982a. The evolution of helping. I. An ecological restraints model. *Amer. Nat.* 119:29–39.

Emlen, S. T. 1982b. The evolution of helping. II. The role of behavioral conflict. *Amer. Nat.* 119:40–53.

Emlen, S. T., Emlen, J. M., and Levin, S. A. 1986. Sex-ratio selection in species with helpers-at-the-nest. *Amer. Nat.* 127:1–8.

Emlen, S. T., and Oring, L. W. 1977. Ecology, sexual selection, and evolution of mating systems. *Science* 197:215–223.

Ewer, R. F. 1973. *The Carnivores.* Ithaca, N.Y.: Cornell University Press.

Fentress, J. C., and Ryon, J. 1982. A long-term study of distributed pup feeding in captive wolves. In: F. H. Harrington & P. C. Paquet, eds. *Wolves of the World,* pp. 238–261. Park Ridge, N.J.: Noyes Publications.

Ferguson, J. W. H., Nel, J. A. J., and DeWet, M. J. 1983. Social organization and movement patterns of black-backed jackals, *Canis mesomelas,* in South Africa. *J. Zool. (Lond.)* 199:487–502.

Frame, L. H. 1986. Social dynamics and female dispersion in African wild dogs. M.S. thesis. Utah State Univ., Logan. 169 pp.

Frame, L. H., and Frame, G. W. 1976. Female African wild dogs emigrate. *Nature* 263:227–229.

Frame, L. H., Malcolm, J. R., Frame, G. W., and Van Lawick, H. 1979. Social organization of African wild dogs (*Lycaon pictus*) on the Serengeti Plains, Tanzania, 1967–1978. *Z. Tierpsychol.* 50:225–249.

Gier, H. T. 1975. Ecology and social behavior of the coyote. In: M. W. Fox, ed. *The Wild Canids,* pp. 247–262. New York: Van Nostrand Reinhold.

Gittleman, J. L. 1984. The behavioral ecology of carnivores. Ph.D. dissert. Univ. Sussex, Brighton, England.

Gittleman, J. L. 1985. Functions of communal care in mammals. In: P. J. Greenwood, P. H. Harvey & M. Slatkin, eds. *Evolution: Essays in Honor of John Maynard Smith,* pp. 187–205. Cambridge: Cambridge Univ. Press.

Harrington, F. H., and Mech, L. D. 1982. Patterns of homesite attendance in two Minnesota wolf packs. In: F. H. Harrington & P. C. Paquet, eds. *Wolves of the World,* pp. 81–105. Park Ridge, N.J.: Noyes Publications.

Harrington, F. H., Mech, L. D., and Fritts, S. H. 1983. Pack size and wolf pup survival: Their relationship under varying ecological conditions. *Behav. Ecol. Sociobiol.* 13:19–26.

Harrington, F. H., Paquet, P. C., Ryon, J., and Fentress, J. C. 1982. Monogamy in wolves: A review of the evidence. In: F. H. Harrington & P. C. Paquet, eds. *Wolves of the World,* pp. 209–222. Park Ridge, N.J.: Noyes Publications.

Heerden J., and Kuhn, F. 1985. Reproduction in captive hunting dogs, *Lycaon pictus. South African J. Wildl. Res.* 15:80–84.

Hennemann, W. W. III, Thompson, S. D., Konecny, M. J. 1983. Metabolism of crab-eating foxes, *Cerdocyon thous*: Ecological influences on the energetics of canids. *Physiol. Zool.* 56:319–324.

Hersteinsson, P. 1984. The behavioral ecology of the artic fox (*Alopex lagopus*) in Iceland. Ph.D. dissert., Oxford Univ., Oxford. 305 pp.

Hersteinsson, P., and Macdonald, D. W. 1982. Some comparisons between red and artic foxes, *Vulpes vulpes* and *Alopex lagopus,* as revealed by radio tracking. *Symp. Zool. Soc. London* 49:259–289.

Johnsingh, A. J. T. 1982. Reproductive and social behavior of the dhole, *Cuon alpinus* (Canidae). *J. Zool. (Lond.)* 198:443–463.

Jordan, P. A., Shelton, P. C., and Allen, D. L. 1967. Numbers, turnover, and social structure of the Isle Royale wolf population. *Amer. Zool.* 7:233–252.

Kleiber, M. 1961. *The Fire of Life.* New York: Wiley.

Kleiman, D. G. 1977. Monogamy in mammals. *Quart. Rev. Biol.* 52:39–69.

Kleiman, D. G., and Eisenberg, J. F. 1973. Comparisons of canid and felid social systems from an evolutionary perspective. *Anim. Behav.* 21:637–659.

Knowlton, F. F. 1972. Preliminary interpretations of coyote population mechanics with some management implications. *J. Wildl. Mgmt.* 36(2):269–382.

Kruuk, H. 1975. Functional aspects of social hunting in carnivores. In: G. Baerends, C. Beer & A. Manning, eds. *Function and Evolution in Behavior: Essays in Honor of Professor Niko Tinbergen,* pp. 119–141. New York: Oxford Univ. Press.

Kruuk, H. 1978. Foraging and spatial organization of the European badger, *Meles meles* L. *Behav. Ecol. Sociobiol.* 4:75–89.

Kruuk, H., and MacDonald, D. W. 1985. Group territories of carnivores: Empires and enclaves. In: R. M. S. Sibly & R. H. Smith, eds. *Behavioural Ecology: Ecological Consequences of Adaptive Behaviour,* pp. 521–536. 25th Symp. Brit. Ecol. Soc. Oxford: Blackwell.

Lamprecht, J. 1978. On diet, foraging behavior and interspecific food competition of jackals in the Serengeti National Park, East Africa. *Z. Säugetierk.* 43:210–223.

Lamprecht, J. 1979. Field observations on the behavior and social system of the bat-eared fox (*Otocyon megalotis* Demarest). *Z. Tierpsychol.* 49:260–284.

Lamprecht, J. 1981. The function of social hunting in larger terrestrial carnivores. *Mamm. Rev.* 11:169–179.

Lindström, E. 1986. Territory inheritance and the evolution of group living in carnivores. *Anim. Behav.* 34:1825–1835.

Macdonald, D. W. 1979a. Flexibility of the social organization of the golden jackal, *Canis aureus. Behav. Ecol. Sociobiol.* 5:17–38.

Macdonald, D. W. 1979b. Helpers in fox society. *Nature* 282:69–71.

Macdonald, D. W. 1980. Social factors affecting reproduction amongst red foxes. In: E. Zimen, ed. *Biogeographica,* vol. 18: *The Red Fox,* pp. 123–175. The Hague: Dr. W. Junk Publishers.

Macdonald, D. W. 1981. Resource dispersion and the social organization of the red fox, *Vulpes vulpes.* In: J. A. Chapman & D. Ursley, eds. *Proceedings of the Worldwide Furbearer Conference,* pp. 918–949. Frostburg, Md.

Macdonald, D. W. 1983. The ecology of carnivore social behavior. *Nature* 301:379–384.

Macdonald, D. W. 1984. *The Encyclopedia of Mammals.* New York: Facts on File.

Macdonald, D. W., and Moehlman, P. D. 1983. Cooperation, altruism, and restraint in the reproduction of carnivores. In: P. Bateson & P. Klopfer, eds. *Perspectives in Ethology,* 5:433–467. New York: Plenum Press.

McNab, B. 1986. The influence of food habits on the energetics of eutherian mammals. *Ecol. Monogr.* 56:1–19.

Macpherson, A. H. 1969. The dynamics of Canadian arctic fox populations. *Canadian Wildl. Serv. Rept. Ser.* 8:1–49.

Malcolm, J. R. 1980. Social organization and communal rearing of pups in African wild dogs (*Lycaon pictus*). Ph.D. dissert., Harvard Univ., Cambridge.

Malcolm, J. R. 1985. Socioecology of bat-eared foxes (*Otocyon megalotis*). *J. Zool. (Lond.)* 208:457–467.

Malcolm, J. R., and Marten, K. 1982. Natural selection and the communal rearing of pups in African wild dogs (*Lycaon pictus*). *Behav. Ecol. Sociobiol.* 10:1–13.

Mech, L. D. 1970. *The Wolf: Ecology and Social Behavior of an Endangered Species.* New York: Natural History Press.

Messier, F., and C. Barrette. 1982. The social system of the coyote (*Canis latrans*) in a forested habitat. *Canadian J. Zool.* 60:1743–1753.

Moehlman, P. D. 1979. Jackal helpers and pup survival. *Nature* 277:382–383.

Moehlman, P. D. 1981. Why do jackals help their parents? Reply to Montgomerie. *Nature* 289:824–825.

Moehlman, P. D. 1983. Socioecology of silverbacked and golden jackals (*Canis mesomelas, C. aureus*). In: J. F. Eisenberg & D. G. Kleiman, eds. *Recent Advances in the Study of Mammalian Behavior*, pp. 423–453. Special Publication no. 7. Lawrence, Kans.: American Society of Mammalogists.

Moehlman, P. D. 1986. Ecology of cooperation in canids. In: D. I. Rubenstein & R. W. Wrangham, eds. *Ecological Aspects of Social Evolution*, pp. 64–86. Princeton, N.J.: Princeton Univ. Press.

Nel, J. A. J. 1978. Notes on the food and foraging behavior of the bat-eared fox (*Otocyon megalotis*). *Bull. Carnegie Mus. Nat. Hist.* 6:132–137.

Nel, J. A. J., Mills, M. G. L., and van Aarde, R. J. 1984. Fluctuating group size in bat-eared foxes (*Otocyon megalotis*), in the south-western Kalahari. *J. Zool. (Lond.)* 203:294–298.

Nellis, C. H., and L. B. Keith. 1976. Population dynamics of coyotes in Central Alberta. *J. Wildl. Mgmt.* 40:389–399.

Packard, J. M., Mech, L. D., Seal, U. S. 1983. Social influences on reproduction in wolves. In: L. Garbyn, ed. *Wolves in Canada*, pp. 78–85. Canadian Wildl. Serv. Rep. no. 88.

Packard, J. M., Seal, U. S., Mech, L. D., Plotka, E. D. 1985. Causes of reproductive failure in two families of wolves (*Canis lupus*). *Z. Tierpsychol.* 68:24–40.

Peterson, R. D., Woolington, J. D., and Bailey, T. N. 1984. Wolves of the Kenai Peninsula, Alaska. *Wildl. Monogr.* 88:1–52.

Rausch, R. A. 1967. Some aspects of the population ecology of wolves, Alaska. *Amer. Zool.* 7:253–265.

Reich, A. 1981. The behavior and ecology of the African wild dog in Kruger National Park. Ph.D. dissert. Yale Univ., New Haven, Conn. 425 pp.

Seal, U. S., Plotka, E. D., Mech, D., Packard, J. M. 1987. Seasonal metabolic and reproductive cycles in wolves. In: H. Frank, ed. *Man and Wolf*, pp. 109–125. Dordrecht, The Netherlands: Dr. W. Junk Publishers.

Senzota, R. B. M. 1978. Some aspects of the ecology of two dominant rodents in the Serengeti ecosystem. M.S. thesis, Univ. Dar-es-Salaam, Tanzania.

Storm, G. L., and E. D. Ables. 1966. Notes on newborn and full-term wild red foxes. *J. Mamm.* 47:116–118.

Storm, G. L., R. D. Andrews, R. L. Phillips, R. A. Bishop, D. B. Siniff, and J. R. Tester. 1976. Morphology, reproduction, dispersal, and mortality, of midwestern red fox populations. *Wildl. Monogr.* 49:1–82.

Trivers, R. L. 1972. Parental investment and sexual selection. In: B. Campbell, ed., *Sexual Selection and the Descent of Man*, pp. 136–179. Chicago: Aldine Press.

Trivers, R. L., and H. Hare. 1976. Haplodiploidy and the evolution of the social insects. *Science* 191:249–263.

van Lawick, H. 1973. *Solo*. London: Collins Publications.

von Schantz, T. 1981. Female cooperation, male competition, and dispersal in red fox, *Vulpes vulpes*. *Oikos* 37:63–68.

von Schantz, T. 1984. Carnivore social behavior—Does it need patches? *Nature* 307:388–390.

Waser, P. M. 1980. Small nocturnal carnivores: Ecological studies in the Serengeti. *African J. Ecol.* 18:167–185.

Western, D. 1979. Size, life history, and ecology in mammals. *African J. Ecol.* 17:185–204.

Wyman, J. 1967. The jackals of the Serengeti. *Animals* 10:79–83.

Zimen, E. 1976. On the regulation of pack size in wolves. *Z. Tierpsychol.* 40:300–341.

CHAPTER 6

The Mating Tactics and Spacing Patterns of Solitary Carnivores

MIKAEL SANDELL

A majority of the carnivore species are primarily solitary, having very little contact with conspecifics (Gittleman, this volume). These solitary species have received less attention than the group-living species, which have attracted much interest (see reviews in Macdonald and Moehlman 1982; Macdonald 1983; Bekoff et al. 1984).

This chapter focuses on the spacing patterns and mating systems of solitary carnivores. Because these two characteristics are closely interrelated (cf. Clutton-Brock and Harvey 1978), an analysis of one must also include the other. I assume that food determines the distribution of females, whereas spacing in males, at least during the mating season, is determined by the distribution of females (Erlinge and Sandell 1986). From this I make a number of predictions about the spacing patterns in solitary carnivores and test them with available data.

Most analyses of mating systems have been classifications (e.g., Eisenberg 1966, 1981; Emlen and Oring 1977; Wittenberger 1979, 1981), which are not easy to use for making testable predictions. In this chapter another approach is taken; it centers on the individual male and the tactics used to maximize reproductive success. This approach provides a number of testable predictions, some which are tested with data from the literature.

Solitary Life—What Is It and Who Lives It?

All mammalian species are more or less social and regularly interact with conspecifics, so "solitary" is not contrary to "social" (see Leyhausen 1965). Instead, solitary behavior is contrasted with cooperative behavior. A carnivore is solitary if it never, except when mating, cooperates with conspecifics; that is, if two or more animals of any given species cooperate to rear young, forage, achieve matings, or defend against predators, the species is classified as cooperative (which, so defined, resembles group living, as in Gittleman, this volume).

164

The defense of common area has not been included as a criterion of coopera-tion. For animals moving solitarily, it is very difficult to separate group defense of an area from individual defense of overlapping ranges where the residents tolerate each other. The European badger (*Meles meles*) is described as living in groups that defend territories (Kruuk 1978a, 1978b). The ranges of group members are, however, different (Kruuk 1978b; Harris 1982), and it is diffi-cult to evaluate whether it is a group territory or simply individual ranges that are defended. Since group members have not been shown to cooperate in any other way, this species is defined as noncooperative. The sea otter (*Enhydra lutris*) is highly gregarious and spends a large proportion of its time together with conspecifics (e.g., Loughlin 1980; Garshelis et al. 1984; Estes, this vol-ume), but since no cooperative activities have been reported, it is classified as noncooperative.

Many species show a large variation in social structure between populations and may cross the demarcation line between solitary and cooperative. The red fox (*Vulpes vulpes*) has been reported as highly cooperative, with nonbreeding helpers in some places (Macdonald 1979), as monogamous in other areas (Sargeant 1972), and as solitary in still other places (Ables 1969; von Schantz 1981). A species shown to exhibit cooperative behavior in one or more popula-tions will be classified as cooperative, although it may be solitary over large parts of its distribution. The analyses and discussions can probably also be applied to cooperative species in populations where environmental conditions give rise to solitary living, but they are not included in the analyses.

In some species the sexes behave differently—for example, cheetah (*Acinonyx jubatus*) males form coalitions whereas females are solitary (Pettifer 1981; Caro and Collins 1987), and coati (*Nasua nasua*) males are solitary but coati females form tight social groups (Russell 1981, 1983). These species are classified as cooperative.

A solitary lifestyle is widespread among the carnivores, occurring in five of the seven families.

Canidae

All species that have been studied reasonably well in the wild have shown cooperative behavior (Moehlman 1986, this volume). The males of most spe-cies join in the rearing of the young, for example, the maned wolf (*Chrysocyon brachyurus*) (Dietz 1984), the coyote (*Canis latrans*) (e.g., Bowen 1982; Mes-sier and Barrette 1982), the gray wolf (*Canis lupus*) (e.g., Harrington et al. 1983), the raccoon dog (*Nyctereutes procyonides*) (Ikeda 1986), and the bat-eared fox (*Otocyon megalotis*) (Lamprecht 1979; Malcolm 1986). Several species also include nonbreeding helpers, for example, the arctic fox (*Alopex lagopus*) (Hersteinsson and Macdonald 1982), the coyote (Bekoff and Wells 1982), the gray wolf (e.g., Harrington and Mech 1982), and the black-backed jackal (*Canis mesomelas*) (Moehlman 1979; Ferguson et al. 1983). And a few species form packs, for example, the gray wolf, (Harrington et al. 1982), the Indian dhole (*Cuon alpinus*) (Johnsingh 1982), and the African hunting dog (*Lycaon pictus*) (Frame et al. 1979; Malcolm and Marten 1982). There are, however, species that live solitarily in some parts of their distribution (e.g., red fox, refs. above).

Ursidae

No form of cooperative behavior has been reported for bears, but good data are available for only three species: the giant panda (*Ailuropoda melanoleuca*) (Schaller et al. 1985), the brown (grizzly) bear (*Ursus arctos*) (e.g., Ballard et al. 1982; Servheen 1983), and the American black bear (*Ursus americanus*) (e.g., Amstrup and Beecham 1976; Lindzey and Meslow 1977; Garshelis and Pelton 1981; Young and Ruff 1982).

Procyonidae

Data are available from very few species of procyonids. In the case of one species, the coati, the females are cooperative (see above), but the other species studied are solitary, namely, the raccoon (*Procyon lotor*) (Fritzell 1978a, 1978b), and the ringtail (*Bassariscus astutus*) (Trapp 1978; Toweill and Teer 1981).

Mustelidae

One mustelid species, the giant otter (*Pteronura brasiliensis*), has been reported as cooperative (Duplaix 1980); all other species for which data are available are solitary, for example, the American marten (*Martes americana*) (Steventon and Major 1982; Wynne and Sherburne 1984), the beech marten (*Martes foina*), (Skirnisson 1986), the stoat (*Mustela erminea*) (Erlinge 1977; Erlinge and Sandell 1986), the American mink (*Mustela vison*) (Gerell 1970; Linn and Birks 1981; Dunstone and Birks 1985), the Euroasian otter (*Lutra lutra*) (Green et al. 1984), the sea otter (Loughlin 1980; Ribic 1982), the wolverine (*Gulo gulo*) (Hornocker and Hash 1981; Whitman et al. 1986), the American badger (*Taxidea taxus*) (Lindzey 1978; Messick and Hornocker 1981), and the striped skunk (*Mephitis mephitis*) (Storm 1972).

Viverridae

In spite of the large number of viverrid species, very few have been studied. Several of them are cooperative (Rood 1986; Gittleman, this volume), and good data are only available for two solitary species: the white-tailed mongoose (*Ichneumia albicauda*) (Waser and Waser 1985) and the African palm civet (*Nandinia binotata*) (Charles-Dominique 1978).

Hyaenidae

All hyaenid species studied show cooperative behavior, for example, the spotted hyena (*Crocuta crocuta*) (Kruuk 1972; Mills 1985) and the brown hyena (*Hyaena brunnea*), (Owens and Owens 1979, 1984; Mills 1982). Also, the aardwolf (*Proteles cristatus*) male joins in the rearing of the young (P. Richardson, pers. comm.).

Felidae

With the exception of the African lion (*Panthera leo*), (Schaller 1972) and the cheetah (see above), all wild felids for which data are available live a solitary life, for example, the Canadian lynx (*Lynx canadensis*) (Mech 1980; Bailey et al. 1986), the European lynx (*Lynx lynx*) (Haller and Breitenmoser 1986), the bobcat (*Lynx rufus*) (review in McCord and Cardoza 1982), the mountain lion (*Felis concolor*) (e.g., Seidensticker et al. 1973; Hemker et al. 1984), and the tiger (*Panthera tigris*) (Sunquist 1981).

Why Live a Solitary Life?

There is no simple answer to this question. Many factors are involved in shaping the social structure of a population. Individuals of all species are simultaneously exposed to several counteracting selection pressures, namely, those favoring cooperative living and those favoring solitary life. The behavior observed is the realized compromise between these selection pressures. Thus, solitary living by an animal indicates both the absence of strong selection pressures for cooperation and the presence of factors promoting solitariness.

Many factors have independently generated cooperative behaviors in carnivores: increased foraging efficiency, improved young production, more successful predator defense, higher mating success (Macdonald 1983; Gittleman 1984; Gittleman, this volume; Mills, this volume; Moehlman, this volume).

The main factors promoting solitary living probably are prey characteristics and hunting mode. Predators that generally take prey much smaller than themselves can almost always subdue the prey alone and consume the whole prey rather quickly. In this situation, which applies to most carnivores, the presence of conspecifics in the immediate surroundings almost always has a negative effect on foraging efficiency, either through disturbance of prey or through depletion of local food sources. A second cause for noncooperative living is the absence of male parental investment. The factors that determine whether or not a male assists in the rearing of the young are to a large extent unclear (Kleiman and Malcolm 1981), and it is not possible to make any predictions.

My conclusion is that solitary living is mainly, though not only, the result of an absence of selection pressures for cooperation, and at present it seems that the social structure of almost every population requires a unique explanation.

Data on Solitary Carnivores

I have been very critical of the data on solitary carnivores used in the following analyses. The requirements for inclusion in the comparative data table (Table 6.1) are that (1) the data come from free-living animals; (2) the study covers most of the year, including information collected both during and outside the mating season; (3) data concerning both sexes are available, and at least one sex is represented by more than one individual; and (4) accounts on the movement and behavior of individually marked animals is reasonably detailed. For almost all carnivore species this type of information can be acquired only by radiotelemetry. My literature search has not been exhaustive, and I am aware of studies reported in publications not accessible to me; however, Table 6.1 gives a representative picture of the data available on spatial organization in solitary carnivores. The use of this critical approach has led to the discovery that these data are needed also from many of the most common and widespread species.

The main problem encountered in comparisons of data from different studies is the large diversity in methods, sampling design, and data analysis. Even when only radio-tracking studies are considered, calculations of home range size are based on data of widely variable quality: continuous tracking over long periods versus sporadic tracking with less than one position per week; positions of active animals recorded around the clock versus one point for the daytime retreat place; tracking times per animal of some weeks versus several years. When range size has been calculated from these data, several methods have been used. Many of these methods begin with the subjective exclusion of "nontypical" or "excursion" positions. In the present analysis the value arrived at by the "convex polygon" (or "minimum area") method (Mohr and Stumpf 1966) has been used whenever possible, mainly because most studies present only this value. Where separate ranges for the mating and nonmating seasons were presented, these have been used, in all other cases annual ranges are employed.

Whereas overlapping ranges are easy to detect, it is more difficult to prove that ranges are exclusive. Either there must be a high level of confidence that all animals within an area are radio collared, or data must be acquired on several animals with adjacent ranges. The latter criterion has been used here, and three to four animals with adjacent ranges and a mean overlap of less than

Table 6.1. The spacing of solitary carnivores

Species	Body weight (kg)		Density (N/km²)	Range size (km²), females				Range size (km²), males				References
	Male	Female		X	Range	N	Overlap	X	Range	N	Overlap	
Giant panda (Ailuropoda melanoleuca)	102	89	0.35	4.1	3.9–4.3	2	O	6.2	—	1	—	Schaller et al. 1985
Brown (grizzly) bear (Ursus arctos)	200	177	0.02	285	112–458	2	O	1403	293–3029	3	O	Servheen 1983
	—	—	—	408	194–734	12	—	769	313–1382	10	—	Ballard et al. 1982
American black bear (Ursus americanus)	—	—	0.37	19.6	—	15	(E)	119	—	25	O	Young and Ruff 1982
	—	—	—	49	17–130	7	O	112	109–115	2	O	Amstrup and Beecham 1976
	—	—	—	8	2–23	14	O	21	13–28	10	O	Garshelis and Pelton 1981
	—	—	1.1–1.5	2.4	1.4–3.8	6	O	5.0	1.6–12.5	5	O	Lindzey and Meslow 1977
Raccoon (Procyon lotor)	—	—	0.5–1.0	8.1	2.3–16.3	7	O	25.6	6.7–49.5	9	(E)	Fritzell 1978a, 1978b
Ringtail (Bassariscus astutus)	—	—	2.2	0.20	0.16–0.28	3	—	0.43	0.35–0.52	2	O	Toweill and Teer 1980
	—	—	1.5	1.29	0.52–2.03	4	—	1.39	0.49–2.35	9	—	Trapp 1978
American marten (Martes americana)	—	—	—	2.9	—	3	—	5.6	—	2	—	Wynne and Sherburne 1984
	—	—	—	2.5	—	1	—	8.1	5.0–10.0	3	—	Steventon and Major 1982
Beech marten (Martes foina)	1.83	1.45	—	2.0	1.2–2.8	2	—	3.6	3.2–4.0	2	—	Skirnisson 1986
Stoat (Mustela erminea)	0.23	0.13	0.25–1.0	0.04	0.02–0.07	5	E	0.16[b]	0.11–0.26	11	E	Erlinge 1977, 1983; Erlinge and Sandell 1986
	—	—	—					7.3[c]	1.05–26.4	8	O	
American mink (Mustela vison)	—	—	—	2.8[d]	—	1	—	4.8[d]	2.8–5.9	4	(E)	Linn and Birks 1981
	—	—	—	1.8[d]	1.0–2.8	2	(E)	2.6[d]	1.8–5.0	4	E	Gerell 1970
	—	—	—	1.1[d]	—	4	E	1.5[d]	—	4	E	Dunstone and Birks 1985
Euroasian otter (Lutra lutra)	9.0	5.2	—	19.2[d]	16.0–22.4	2	O	39.1[d]	—	1	—	Green et al. 1984
Sea otter (Enhydra lutris)	27.6	19.9	—	0.8	0.3–2.0	8	O	0.4	0.2–1.4	11	E,O	Loughlin 1980
	—	—	—	5.4	—	12	O	3.2	—	8	O	Ribic 1982

(continued)

Table 6.1. (Continued)

Species	Body weight (kg)		Density (N/km²)	Range size (km²), females				Range size (km²), males				References
	Male	Female		X	Range	N	Overlap[a]	X	Range	N	Overlap[a]	
Wolverine (Gulo gulo)	—	—	0.02	388	—	11	O	422	—	9	O	Hornocker and Hash 1981
	—	—	—	105	—	3	—	535	—	4	—	Whitman et al. 1986
American badger (Taxidea taxus)	7.6	6.3	0.4	2.4	1.4–3.0	3	O	5.8	5.4–6.3	2	—	Lindzey 1978
			3–5	1.2	0.4–3.8	7	—	1.4	0.8–3.4	3	—	Messick and Hornocker 1981
Striped skunk (Mephitis mephitis)	—	—	—	3.8	2.5–6.7	5	—	5.1	4.0–6.2	2	—	Storm 1972
White-tailed mongoose (Ichneumia albicauda)	3.58	3.44	4.3	0.64	0.39–1.18	9	O	0.97	0.80–1.23	4	E	Waser 1980; Waser and Waser 1985
African palm civet (Nandinia binotata)	2.97	2.32	5	0.45	0.29–0.50	7	E	0.85	0.34–1.53	11	E	Charles-Dominique 1978
Canadian lynx (Lynx canadensis)	10.6	9.1	—	—	51–122	—	O	783	145–243	—	(E)	Mech 1980
	—	—	0.01	70	51–89	2	—	—	—	1	—	Bailey et al. 1986
European lynx (Lynx lynx)	24.0	17.2	0.01	116	96–135	2	—	362	275–450	2	—	Haller and Breitenmoser 1986
Bobcat (Lynx rufus)	—	—	0.05	19.3	9.1–45.3	8	E	42.1	6.5–107.9	4	E	Bailey 1974
	13.0	9.2	0.05	38	15–92	6	E	62	13–201	16[b]	O	Berg 1981
	7.7	5.9	0.77–1.16	1.1	—	6	E	2.6	—	6	E	Miller and Speak 1981
	12.4	8.1	0.05	43	26–59	4	O	73	39–95	3	O	Zezulak and Schwab 1981; McCord and Cardoza 1982
	10.2	7.0	—	26	—	3	O	77	—	2	O	Kitchings and Story 1984
Mountain lion (Felis concolor)	12.6	9.0	—	33	18–41	8	—	96	21–200	10	—	Litvaitis et al. 1986
	—	—	0.01	268	173–373	4	O	453	—	1	O	Seidensticker et al. 1973
	—	—	0.04	94	54–119	3	O	178	78–277	5	O	Sitton and Wallen 1976
	—	—	0.03	66	57–74	2	—	152	109–238	4	O	Kutilek et al. 1980, cited in Hemker et al. 1984
Tiger (Panthera tigris)	—	—	0.004	685	396–1454	4	O	826	—	1	—	Hemker et al. 1984
	214	138	0.03	16.9	16.4–17.7	3	E	66	60–72	2	(E)	Sunquist 1981

aE = exclusive ranges (defined in text); O = overlapping ranges.
bNonmating season.
cMating season.

10% (measured on "convex polygons") is considered to be a strong indication of exclusivity.

The sea otter has not been included in the analyses since this aquatic species shows a pattern totally different from those of the terrestrial and semi-aquatic species (see below). Also, the giant panda has been excluded; as a strict herbivore it deviates from the general carnivore pattern concerning range size (cf. Gittleman and Harvey 1982).

Spatial Organization

The spacing pattern in a population is the result of the tactics chosen by the individual animals in their attempts to survive and maximize reproductive success. It is assumed here that female spacing patterns are determined by the abundance and dispersion of food, whereas male spatial organization, at least during the mating season, is determined by the distribution of females (see Erlinge and Sandell 1986).

Females

Because females in noncooperative species must rear young by themselves, their reproductive success is closely correlated with the amount of energy they can allocate to reproduction. In turn, this amount mainly depends on the food resources available during the rearing period. Thus, for solitary females food is the most important resource, and females should follow a behavioral tactic that maximizes their chances of securing food resources for reproduction and survival.

Range size is expected to be adjusted so that a female retains enough resources also when resources are low. Thus, range size should be determined by food availability during the most critical period, though food dispersion also may have some influence. When two or more different food sources are used during the year, the dispersion of these resources in relation to each other may influence range size. Black bears in the Great Smoky Mountains National Park feed on berries and fruits during summer, whereas their staple food during autumn is the acorn. Many of the radio-tracked bears showed fall ranges completely disjunct from their summer ranges (Garshelis and Pelton 1981), and thus their annual ranges are determined more by the distribution of the two resources than by their abundance. Generally, when one food source is utilized throughout the year, its dispersion probably has relatively little influence on range size. A patchily distributed resource also usually has a lower total abundance when calculated over the whole area. One can therefore predict that female range size is correlated with food abundance, and especially with food biomass during the most critical period of the year.

A correlation between prey density and range size has been found in the cases of the bobcat (Litvaitis et al. 1986) and the Canadian lynx (Ward and Krebs 1985). Van Orsdol et al. (1985) found that for the African lion the range size was correlated with lean-season prey biomass, but not with good-season prey biomass. Since most studies do not present data on food abundance, the prediction cannot be tested directly. There is, however, a way to test it indirectly. Density is expected to be directly correlated with food abundance. For the lion three measurements of density (overall mean, adult females, and pride members) were correlated with lean-season prey biomass (Van Orsdol et al. 1985). A correlation between density and home range size would indicate the dependence of both variables on food abundance, although causality is not clear. A strong correlation between density and female range size is apparent in the data on solitary carnivores in Table 6.1 ($r_{22} = -0.9380$, $P < 0.001$). Thus, range size in females is mainly determined by food abundance.

One of the main characteristics of the spacing pattern in a population is the extent of range overlap between individuals. For ranges to be exclusive, the food resource must be so evenly distributed and stable that an area just large enough to support the animal during the most critical period contains food enough throughout the year. If the food resource varies in space and time, the range must be larger to provide for the animal at all times. This larger area may contain a surplus of food for most of the year; thus, several animals can utilize the same area, and a system of overlapping ranges develops. Essentially the same explanation has been proposed by Macdonald (1983) for the evolution of group living in some carnivores. The same scenario may, however, lead to a system of solitary animals with overlapping ranges, indicating that this hypothesis is not sufficient to explain the evolution of group living in these species.

From the discussion above it follows that exclusive ranges are expected when food resources are stable and evenly distributed, whereas a system of overlapping ranges is likely when the timing and spacing of available food varies.

So long as there are no methods to measure resource distribution, this prediction cannot be satisfactorily tested. Exclusive ranges are, however, not common among female solitary carnivores; they are found in only seven out of 24 studies ($P = 0.032$, binomial test), and mainly when ranges are small (range size for exclusive and overlapping ranges; $U = 17$, $n_1 = 6$, $n_2 = 14$, $P < 0.05$, Mann-Whitney U-test).

If exclusive ranges include only food enough for one animal, whereas overlapping ranges contain enough food for several individuals, the latter should accordingly be relatively larger than the former. Hence the prediction is that exclusive ranges are relatively smaller than overlapping ranges.

A way to measure relative range size is to examine the deviations from the regression line for female range size on density. For the studies where both range overlap and density are available ($n = 17$, Table 6.1) all ten points above

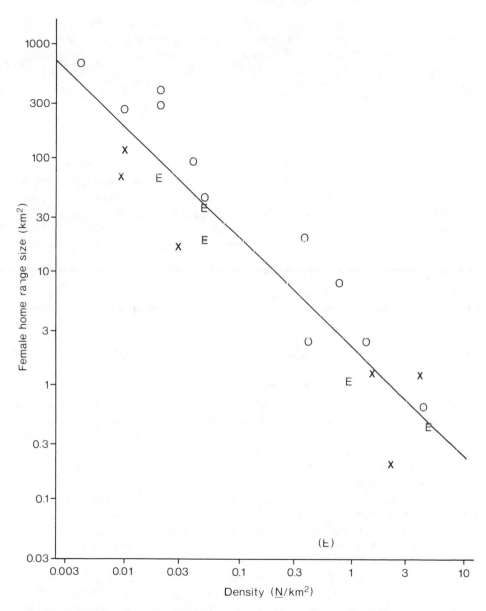

Figure 6.1. The relation between density and female home range size in solitary carnivores. The line is the regression line for all points except (E). O = studies where females have overlapping ranges; E = studies where females have exclusive ranges; and X = studies without data on overlap. Exclusive ranges are relatively small, that is, below the regression line, whereas overlapping ranges are relatively large (see text).

the line, namely, relatively large ranges, are from studies with overlapping ranges, whereas only one out of seven of the points below the line are from a study with overlapping ranges (Figure 6.1). This difference is statistically significant ($P = 0.0006$, Fisher exact test). Thus, overlapping ranges are relatively larger than exclusive ones.

Males

According to the basic hypothesis male spatial organization is influenced by two resources: food outside the mating season and receptive females during the mating period. It follows that during a substantial part of the year male and female spacing patterns are determined by different factors, and range size in males should be a function not only of food requirements but also of female distribution. Since food ranges are minimized whereas mating ranges are expected to be maximized, it follows that male ranges should be larger than predicted by energy requirements.

If it is assumed that female range sizes are determined by food abundance, the body weight of the two sexes can be used to predict the male range size required from an energy standpoint:

$$\text{male range size} = \frac{\text{female range size} \times (\text{male weight})^{0.75}}{(\text{female weight})^{0.75}}$$

(data in Table 6.1). Predicted ranges are 1.22 ± 0.10 ($\bar{X} \pm SD$) times the female range size, whereas the observed ranges are 2.47 ± 1.06 times larger than corresponding female ranges, that is, observed male ranges are significantly larger than expected on the basis of energy requirements ($T = 0$, $N = 14$, $P < 0.01$, Wilcoxon matched-pairs test). The stoat was not included in the means because of the extreme value; male mating ranges are on average 180 times larger than female ranges (Table 6.1). Thus, male ranges are determined by factors other than food requirements.

Males may adopt one of two alternative ways to achieve matings; either they stay and try to monopolize a number of females, or they roam and compete over access to each single female that comes into heat.

In general, it is assumed that the resident males achieve almost all matings in carnivore populations. Evidence has, however, started to accumulate that many carnivore males range widely during the mating season, and that these wandering males are not only young transients. In the cases of the brown hyena and the stoat there are indications that roaming males sequester the majority of matings (Mills 1982; Sandell 1986). Wide-ranging adult males have been reported from many carnivore species, for example, an eight-year-old brown bear moved over more than 3000 km² during one year (Servheen 1983), and an adult male Canadian lynx roamed over 783 km² during one year and visited two widely separated females during the breeding season (Bailey et al. 1986). A study of martens in Ontario showed a large turnover of males during the mating season; individuals stayed only a short time at each place and then moved on (Taylor and Abrey 1982). Thus, during the mating season males of many carnivore species roam over large areas. Generally, the roaming patterns probably are less extensive than these examples, but the patterns shows a continuum, and I define a male as roaming when his mating range overlaps with other males' ranges (range overlap defined as above). Thus, in a population of staying males ranges are exclusive.

When the goal is to maximize the number of matings, the defendability of a resource should not be the main factor influencing the system. Even if a dominant male may be able to monopolize a number of females, a roaming tactic, though much more costly, may lead to higher reproductive success and is probably the tactic followed. The spatial pattern of a population should be determined mainly by the tactic chosen by the dominant fraction of the population. A roaming tactic is more profitable the more dominant a male is, since the probability to take over other females will be related to the male's social status. If the most rewarding tactic for the dominant males is to roam, it is impossible for other males to have exclusive ranges. The dominant males will move in and take over the females. (Because the most dominant male present will take over the female, the result of male competition will coincide with female choice.) A system with exclusive ranges is possible only when it is the best tactic for all dominant males. In that situation there is a mutual interest in exclusivity among all contestants. Thus, exclusive ranges are maintained through the mutual interest of all participants in the system, and when a roaming tactic becomes the best option for the dominant fraction of the population, the system with exclusive ranges should break down. Subordinate males are more or less harassed by dominants in both exclusive and overlapping systems, and in most cases their best mating tactic probably is different from the one employed by the dominant males (Sandell 1986; Liberg and Sandell, 1988).

The maintenance of exclusive ranges should be the best tactic when females are dense and evenly distributed. In this situation the male can control a number of females and secure matings with them. At lower densities there is a decrease in the number of females that a male can control, and at some threshold density it pays more to move around in search of receptive females over a larger area than to stay and secure matings from a few females. Thus, when females are evenly distributed a threshold density should exist above which the system shifts from roaming males to exclusive ranges. This threshold of course differs among species, but I predict that for each population the shift should occur within a rather narrow density interval. Unfortunately, there are no data available to test this prediction.

When females are concentrated in patches, a dominant male can double the number of potential matings by including a second patch within his range, probably losing only a small number of matings in the first group. A dominant male should continue to incorporate patches until he reaches a level where the addition of another patch results in a loss in matings in the other patches within his range equal to the number he gains in the new patch. Other males are able to explore the same patches, and a system of overlapping ranges should develop. It can be predicted that male ranges will be exclusive when females are dense and evenly distributed, whereas overlapping ranges are expected in all other situations.

There are no data on female density and distribution available to test this prediction, but some indications can be derived from Table 6.1. Relative density for a population can be determined by the position in relation to the regression line for density on female metabolic body weight ($W^{0.75}$). Points above the line indicate a relatively high density and points below, a relatively low density. Of the studies with data on body weight, density, and overlap ($n = 8$), all three studies with values above the line, that is, with relatively high densities, are studies where males have exclusive ranges, whereas four out of five studies with values below the line, that is, with relatively low densities, are studies where males have overlapping ranges ($P = 0.071$, Fisher exact test). Thus, although not significant, this small sample indicates a relation between relative density and male mating tactic, with roaming males at relatively low densities and staying males at relatively high densities. Further data are needed to test the prediction.

When the mating season is restricted to one part of the year the decisive resource for males is different during the mating and nonmating seasons. As the two resources (receptive females and available food) in almost all cases have different characteristics, a change in tactics used to exploit the different resources is expected. The behavior shown and the area needed to secure necessary food is different from the tactic employed and the range covered when the goal is to maximize the number of matings. Thus, the spatial organization of the male population will differ between the mating and the nonmating season (Erlinge and Sandell 1986). When mating ranges are exclusive, competition over access to areas of high female density probably go on for most of the year, since it is easier to maintain an exclusive area than to establish one. A male that has acquired a range is expected to show his presence in that range throughout the year. Thus, for species with a restricted mating season it can be predicted that if mating ranges overlap, there is a change in range size between the mating and the nonmating season, with mating ranges being the largest. In contrast, exclusive mating ranges should show little variation in size during the year.

Since very few references provide separate data for the mating and the nonmating seasons, the prediction cannot be properly tested. In the case of male stoats there is a drastic change in range size between the two periods, with overlapping mating ranges that are on average 50 times larger then the exclusive nonmating ranges (Table 6.1, Erlinge and Sandell 1986). The same pattern of increased range size during the mating season has been reported for black bear (Lindzey and Meslow 1977) and domestic cat (*Felis catus*) (Liberg and Sandell 1988), both of which have overlapping mating ranges.

The sea otter deviates from the general pattern. The male mating ranges in this aquatic species are exclusive and smaller than female ranges (Loughlin 1980; Ribic 1982; Garshelis et al. 1984). The cheetah shows the same pattern: the females wander over large areas, whereas certain males (sometimes in coalition) defend small, exclusive areas (Caro and Collins 1987). With this staying tactic females are not included within the range, but instead the male range includes areas that females will pass through or visit for other reasons. An analysis according to the "hotspot hypothesis" (cf. Bradbury et al. 1986) might give insight into these systems.

There is an almost total lack of data and analytical work on most aspects of spatial organization in solitary carnivores. The predictions in this paper present one approach that promises to increase our understanding in this area, and they also indicate the types of data and analyses that are needed.

Summary and Conclusions

Solitary living (here defined as noncooperative living) is mainly an effect of the absence of selection pressures for cooperation.

From the basic assumption that spatial organization in female solitary carnivores is determined by the characteristics of the food resource, it can be predicted that: (1) female range size should be correlated with food abundance, and especially with food biomass during the most critical period of the year; (2) exclusive ranges should be expected when food resources are stable and evenly distributed, whereas temporal and spatial variation in food availability should lead to a system of overlapping ranges; and (3) exclusive ranges should be relatively smaller than overlapping ranges. Available data, although scarce, support the basic assumption.

Spatial organization in male solitary carnivores is influenced by two resources: food outside the mating season and receptive females during the mating period. From this it can be predicted that: (1) male ranges should be larger than predicted on the basis of energy requirements, which is supported by data; (2) when females are evenly distributed, a threshold density should exist above which the system shifts from roaming males to exclusive ranges; (3) male ranges are exclusive when females are dense and evenly distributed, whereas overlapping ranges are to be expected in all other situations; (4) if mating ranges overlap there is a change in range size between the mating and the nonmating season, with mating ranges being the largest. In contrast, exclusive mating ranges should show little variation in size during the year. The latter three predictions cannot be properly tested, since field data are not available.

To explain the patterns found in solitary carnivore populations we need detailed information on individuals with known social status that are followed during both the mating and the nonmating season. Such data are still lacking for almost all solitary species.

Acknowledgments

I thank S. Erlinge, O. Liberg, J.-Å. Nilsson, and H. Smith for discussions and comments during the development of this paper; J. L. Gittleman for his patience and for constructive comments; D. Tilles for correcting the English; and E. Karlsson for drawing the figure.

References

Ables, E. D. 1969. Home-range studies of red foxes (*Vulpes vulpes*). *J. Mamm.* 50:108–120.

Amstrup, S. C., and Beecham, J. 1976. Activity patterns of radio-collared black bears in Idaho. *J. Wildl. Mgmt.* 40:340–348.

Bailey, T. N. 1974. Social organization in a bobcat population. *J. Wildl. Mgmt.* 38:435–446.

Bailey, T. N., Bangs, E. E., Portner, M. F., Malloy, J. C., and McAvinchey, R. J. 1986. An apparent overexploited lynx population on the Kenai peninsula, Alaska. *J. Wildl. Mgmt.* 50:279–290.

Ballard, W. B., Miller, S. D., and Spraker, T. H. 1982. Home range, daily movements, and reproductive biology of brown bear in southcentral Alaksa. *Canadian Field-Nat.* 96:1–5.

Bekoff, M., Daniels, T. J., and Gittleman, J. L. 19845. Life history patterns and the comparative social ecology of carnivores. *Ann. Rev. Ecol. Syst.* 15:191–232.

Bekoff, M., and Wells, M. C. 1982. Behavioral ecology of coyotes: Social organization, rearing patterns, space use, and resource defence. *Z. Tierpsychol.* 60:281–305.

Berg, W. E. 1981. Ecology of bobcats in northern Minnesota. *Natl. Wildl. Fedn. Sci. Tech. Ser.* 6:55–61.

Bowen, D. W. 1982. Home range and spatial organization of coyotes in Jasper National Park, Alberta. *J. Wildl. Mgmt.* 46:201–216.

Bradbury, J., Gibson, R., and Tsai, I. M. 1986. Hotspots and the dispersion of leks. *Anim. Behav.* 34:1694–1709.

Caro, T. M., and Collins, D. A. 1987. Male cheetah social organization and territoriality. *Ethology* 74:52–64.

Charles-Dominique, P. 1978. Ecologie et vie sociale de *Nandinia binotata* (Carnivores, Viverrides): Comparaison avec les prosimiens sympatriques du Garbon. *La Terre et la Vie* 32:477–528.

Clutton-Brock, T. H., and Harvey, P. H. 1978. Mammals, resources and reproductive strategies. *Nature* 273:191–195.

Dietz, J. M. 1984. Ecology and social organization of the maned wolf (*Chrysocyon brachyurus*). *Smithsonian Contrib. Zool.* 392:1–51.

Dunstone, N., and Birks, J. D. S. 1985. The comparative ecology of coastal, riverine and lacustrine mink *Mustela vison* in Britain. *Z. Angewandte Zool.* 72:60–70.

Duplaix, N. 1980. Observations on the ecology and behavior of the giant river otter, *Pteronura brasiliensis,* in Suriname. *Rev. Ecol.* 34:495–620.

Eisenberg, J. F. 1966. The social organization of mammals. *Handb. Zool.,* Band 8, Lieferung 39; 10(7):1–92.

Eisenberg, J. F. 1981. *The Mammalian Radiations.* London: Athlone Press.

Emlen, S. T., and Oring, L. W. 1977. Ecology, sexual selection, and the evolution of mating systems. *Science* 197:215–223.

Erlinge, S. 1977. Spacing strategy in stoat, *Mustela erminea. Oikos* 28:32–42.

Erlinge, S. 1983. Demography and dynamics of a stoat (*Mustela erminea*) population in a diverse community of vertebrates. *J. Anim. Ecol.* 52:705–726.

Erlinge, S., and Sandell, M. 1986. Seasonal changes in the social organization of male stoats, *Mustela erminea:* An effect of shifts between two decisive resources. *Oikos* 47:57–62.

Ferguson, J. W. H., Nel, J. A. J., and de Wet, M. J. 1983. Social organization and movement patterns of black-backed jackals, *Canis mesomelas,* in South Africa. *J. Zool.* 199:487–502.

Frame, L. H., Malcolm, J. R., Frame, G. W., and van Lawick, H. 1979. Social organization of African wild dogs (*Lycaon pictus*) on the Serengeti Plains, Tanzania, 1967–1978. *Z. Tierpsychol.* 50:225–249.

Fritzell, E. K. 1978a. Aspects of raccoon (*Procyon lotor*) social organization. *Canadian J. Zool.* 56:260–271.

Fritzell, E. K. 1978b. Habitat use by prairie raccoons during the waterfowl breeding season. *J. Wildl. Mgmt.* 42:118–127.

Garshelis, D. L., Johnson, A. M., and Garshelis, J. A. 1984. Social organization of sea otters in Prince William sound, Alaska. *Canadian J. Zool.* 62:2648–2658.

Garshelis, D. L., and Pelton, M. R. 1981. Movements of black bears in the Great Smoky Mountains National Park. *J. Wildl. Mgmt.* 45:912–925.

Gerell, R. 1970. Home ranges and movements of the mink *Mustela vision* Schreber in southern Sweden. *Oikos* 21:160–173.

Gittleman, J. L. 1984. The behavioural ecology of carnivores. Ph.D. dissert., Univ. Sussex, U.K.

Gittleman, J. L., and Harvey, P. H. 1982. Carnivore home-range size, metabolic needs and ecology. *Behav. Ecol. Sociobiol.* 10:57–63.

Green, J., Green, R., and Jefferies, D. J. 1984. A radio-tracking survey of otters *Lutra lutra* on a Perthshire river system. *Lutra* 27:85–145.

Haller, H., and Breitenmoser, U. 1986. Zur Raumorganisation der in den Schweizer Alpen wiederangesiedelten Population des Luchses (*Lynx lynx*). *Z. Saugetierk.* 51:289–311.

Harrington, F. H., and Mech, L. D. 1982. Patterns of homesite attendance in two Minnesota wolf packs. In: F. H. Harrington & P. C. Paquet, eds. *Wolves of the World: Perspectives of Behavior, Ecology, and Conservation,* pp. 81–109. Park Ridge, N.J.: Noyes.

Harrington, F. H., Mech, L. D., and Fritts, S. H. 1983. Pack size and wolf pup survival: Their relationship under varying ecological conditions. *Behav. Ecol. Sociobiol.* 13:19–26.

Harrington, F. H., Paquet, P. C., Ryon, J., and Fentress, J. C. 1982. Monogamy in wolves: A review of the evidence. In: F. H. Harrington & P. C. Paquet, eds. *Wolves of the World: Perspectives of Behavior, Ecology, and Conservation,* pp. 209–222. Park Ridge, N.J.: Noyes.

Harris, S. 1982. Activity patterns and habitat utilization of badgers (*Meles meles*) in suburban Bristol: A radio tracking study. *Symp. Zool. Soc. London* 49:301–323.

Hemker, T. P., Lindzey, F. G., and Ackerman, B. B. 1984. Population characteristics and movement patterns of cougars in southern Utah. *J. Wildl. Mgmt.* 48:1275–1284.

Hersteinsson, P., and Macdonald, D.W. 1982. Some comparisons between red and arctic foxes, *Vulpes vulpes* and *Alopex lagopus,* as revealed by radiotracking. *Symp. Zool. Soc. London* 49:259–289.

Hornocker, M. G., and Hash, H. S. 1981. Ecology of the wolverine in northwestern Montana. *Canadian J. Zool.* 59:1286–1301.

Ikeda, H. 1986. Old dogs, new treks. *Nat. Hist.* 1986(8):38–44.

Johnsingh, A. J. T. 1982. Reproductive and social behaviour of the dhole, *Cuon alpinus* (Canidae). *J. Zool.* 198:443–463.

Kitchings, J. T., and Story, J. D. 1984. Movements and dispersal of bobcats in east Tennessee. *J. Wildl. Mgmt.* 48:957–961.

Kleiman, D. G., and Malcolm, J. R. 1981. The evolution of male parental investment in mammals. In: D. J. Gubernick & P. H. Klopfer, eds. *Parental Care in Mammals,* pp. 347–387. New York: Plenum Press.

Kruuk, H. 1972. *The Spotted Hyena: A Study of Predation and Social Behaviour.* Chicago: Univ. Chicago Press.

Kruuk, H. 1978a. Foraging and spatial organization of the European badger, *Meles meles* L. *Behav. Ecol. Sociobiol.* 4:75–89.

Kruuk, H. 1978b. Spatial organization and territorial behaviour of the European badger, *Meles meles. J. Zool.* 184:1–19.

Lamprecht, J. 1979. Field observations on the behaviour and social system of the bat-eared fox (*Otocyon megalotis* Desmarest). *Z. Tierpsychol.* 49:260–284.

Leyhausen, P. 1965. The communal organization of solitary mammals. *Symp. Zool. Soc. London* 14:249–263.

Liberg, O., and Sandell, M. 1988. Spatial organization and reproductive tactics in the domestic cat and other felids. In: P. P. G. Bateson & D. C. Turner, eds. *The Domestic Cat: The Biology of Its Behaviour.* Cambridge: Cambridge Univ. Press.

Lindzey, F. G. 1978. Movement patterns of badgers in northwestern Utah. *J. Wildl. Mgmt.* 42:418–422.

Lindzey, F. G., and Meslow, E. C. 1977. Home range and habitat use by black bears in southwestern Washington. *J. Wildl. Mgmt.* 41:413–425.

Linn, I. J., and Birks, J. D. S. 1981. Observations on the home ranges of feral American mink (*Mustela vison*) in Devon, England, as revealed by radio-tracking. In: J. A. Chapman & D. Pursley, eds. *Proceedings of the Worldwide Furbearer Conference,* pp. 1088–1102. Frostburg: Univ. Maryland Press.

Litvaitis, J. A., Sherburne, J. A., and Bissonette, J. A. 1986. Bobcat habitat use and home range size in relation to prey density. *J. Wildl. Mgmt.* 50:110–117.

Loughlin, T. R. 1980. Home range and territoriality of sea otters near Monterey, California. *J. Wildl. Mgmt.* 44:576–582.

McCord, C. M. and Cardoza, J. E. 1982. Bobcat and lynx. In: J. A. Chapman & G. A. Feldhamer, eds. *Wild Mammals of North America: Biology, Management, and Economics,* pp. 728–766. Baltimore: Johns Hopkins Univ. Press.

Macdonald, D. W. 1979. Helpers in fox society. *Nature* 282:69–71.

Macdonald, D. W. 1983. The ecology of carnivore social behaviour. *Nature* 301:379–384.

Macdonald, D. W., and Moehlman, P. D. 1982. Cooperation, altruism, and restraint in the reproduction of carnivores. In: P. P. G. Bateson & P. H. Klopfer, eds. *Perspectives in Ethology,* 5:433–467. New York: Plenum Press.

Malcolm, J. R. 1986. Socio-ecology of bat-eared foxes (*Otocyon megalotis*). *J. Zool.* 208:457–467.

Malcolm, J. R., and Marten, K. 1982. Natural selection and the communal rearing of pups in African wild dogs (*Lycaon pictus*). *Behav. Ecol. Sociobiol.* 10:1–13.

Mech, L. D. 1980. Age, sex, reproduction, and spatial organization of lynxes colonizing northeastern Minnesota. *J. Mamm.* 61:261–267.

Messick, J. P., and Hornocker, M. G. 1981. Ecology of the badger in southwestern Idaho. *Wildl. Monogr.* 76:1–51.

Messier, F., and Barrette, C. 1982. The social system of the coyote (*Canis latrans*) in a forested habitat. *Canadian J. Zool.* 60:1743–1753.

Miller, S. D., and Speak, D. W. 1981. Progress report: Demography and home range of the bobcat in south Alabama. *Natl. Wildl. Fedn. Sci. Tech. Ser.* 6:123–124.

Mills, M. G. L. 1982. The mating system of the brown hyaena, *Hyaena brunnea,* in the southern Kalahari. *Behav. Ecol. Sociobiol.* 10:131–136.

Mills, M. G. L. 1985. Related spotted hyaenas forage together but do not cooperate in rearing young. *Nature* 316:61–62.

Moehlman, P. D. 1979. Jackal helpers and pup survival. *Nature* 277:382–383.

Mochlman, P. D. 1986. Ecology of cooperation in canids. In: D. I. Rubenstein & R. W. Wrangham, eds. *Ecological Aspects of Social Evolution,* pp. 64–86. Princeton, N.J.: Princeton Univ. Press.

Mohr, C. O., and Stumpf, W. A. 1966. Comparison of methods for calculating areas on animal activity. *J. Wildl. Mgmt.* 30:293–304.

Owens, D. D., and Owens, M. J. 1979. Communal denning and clan associations in brown hyenas of the central Kalahari desert. *African J. Ecol.* 17:35–44.

Owens, D. D., and Owens, M. J. 1984. Helping behaviour in brown hyenas. *Nature* 308:843–845.

Pettifer, H. L. 1981. Aspects on the ecology of cheetahs (*Acinonyx jubatus*) on the Suikerbosrand Nature Reserve. In: J. A. Chapman & D. Pursley, eds. *Proceedings of the Worldwide Furbearer Conf.,* pp. 1121–1142. Frostburg: Univ. Maryland Press.

Ribic, C. A. 1982. Autumn movement and home range of sea otters in California. *J. Wildl. Mgmt.* 46:795–801.

Rood, J. P. 1986. Ecology and social evolution in the mongooses. In: D. I. Rubenstein & R. W. Wrangham, eds. *Ecological Aspects of Social Evolution,* pp. 131–152. Princeton, N.J.: Princeton Univ. Press.

Russell, J. K. 1981. Exclusion of adult male coatis from social groups: Protection from predation. *J. Mamm.* 62:206–208.

Russell, J. K. 1983. Altruism in coati bands: Nepotism or reciprocity? In: S. K. Wasser, ed. *Social Behavior of Female Vertebrates,* pp. 263–290. New York: Academic Press.

Sandell M. 1986. Movement patterns of male stoats, *Mustela erminea,* during the mating season: Differences in relation to social status. *Oikos* 47:63–70.

Sargeant, A. B. 1972. Red fox spatial characteristics in relation to waterfowl predation. *J. Wildl. Mgmt.* 36:225–236.

Schaller, G. B. 1972. *The Serengeti Lion: A Study of Predator-Prey Relations.* Chicago: Univ. Chicago Press.

Schaller, G. B., Jinchu, H., Wenshi, P., and Jing, Z. 1985. *The Giant Pandas of Wolong.* Chicago: Univ. Chicago Press.

Seidensticker, J. C., Hornocker, M. G., Wiles, W. V., and Messick, J. P. 1973. Mountain lion social organization in the Idaho Primitive Area. *Wildl. Monogr.* 35:1–60.

Servheen, C. 1983. Grizzly bear food habits, movements, and habitat selection in the Mission Mountains, Montana. *J. Wildl. Mgmt.* 47:1026–1035.

Sitton, L. W., and Wallen, S. 1976. *California Mountain Lion Study.* Sacramento: Dept. of Fish and Game, California.

Skirnisson, K. 1986. Untersuchungen zum Raum-Zeit-System freilebender Steinmarder (*Martes foina* Erxleben, 1777). *Beiträge zur Wildbiologie* 6:1–200.

Steventon, J. D., and Major, J. T. 1982. Marten use of habitat in a commercially clear-cut forest. *J. Wildl. Mgmt.* 46:175–182.

Storm, G. L. 1972. Daytime retreats and movements of skunks on farmlands in Illinois. *J. Wildl. Mgmt.* 36:31–45.

Sunquist, M. E. 1981. The social organization of tigers (*Panthera tigris*) in Royal Chitawan National Park, Nepal. *Smithsonian Contrib. Zool.* 336:1–98.

Taylor, M. E., and Abrey, N. 1982. Marten, *Martes americana,* movements and habitat use in Algonquin Provincial Park, Ontario. *Canadian Field-Nat.* 96:439–447.

Toweill, D. E., and Teer, J. G. 1981. Home range and den habits of Texas ringtails (*Bassariscus astutus flavus*). In: J. A. Chapman & D. Pursley, eds. *Proceedings of the Worldwide Furbearer Conference,* pp. 1103–1120. Frostburg: Univ. Maryland Press.

Trapp, G. R. 1978. Comparative behavioral ecology of the ringtail and gray fox in southwestern Utah. *Carnivore* 1(2):3–32.

Van Orsdol, K. G., Hanby, J. P., and Bygott, J. D. 1985. Ecological correlates of lion social organization (*Panthera leo*). *J. Zool. (Lond.)* 206:97–112.

von Schantz, T. 1981. Evolution of group living, and the importance of food and social organization in population regulation: A study on the red fox (*Vulpes vulpes*). Ph.D. dissert., Univ. Lund, Sweden.

Ward, R. M. P., and Krebs, C. J. 1985. Behavioural responses of lynx to declining snowshoe hare abundance. *Canadian J. Zool.* 63:2817–2824.

Waser, P. M. 1980. Small nocturnal carnivores: Ecological studies in the Serengeti. *African J. Ecol.* 18:167–185.

Waser, P. M., and Waser, M. S. 1985. *Ichneumia albicauda* and the evolution of viverrid gregariousness. *Z. Tierpsychol.* 68:137–151.

Whitman, J. S., Ballard, W. B., and Gardner, C. L. 1986. Home range and habitat use by wolverines in southcentral Alaska. *J. Wildl. Mgmt.* 50:460–463.

Wittenberger, J. F. 1979. The evolution of mating systems in birds and mammals. In: P. Marler & J. G. Vandenbergh, eds. *Handbook of Behavioral Neurobiology,* 3:271–349. New York: Plenum Press.

Wittenberger, J. F. 1981. *Animal Social Behavior.* Boston: Duxbury Press.

Wynne, K. M., and Sherburne, J. A. 1984. Summer home range use by adult marten in northwestern Maine. *Canadian J. Zool.* 62:941–943.

Young, B. F., and Ruff, R. L. 1982. Population dynamics and movements of black bears in east central Alberta. *J. Wildl. Mgmt.* 46:845–860.

Zezulak, D. S., and Schwab, R. G. 1981. A comparison of density, home range and habitat utilization of bobcat populations at Lava Beds and Joshua Tree National Monuments, California. *Natl. Wildl. Fedn. Sci. Tech. Ser.* 6:74–79.

CHAPTER 7

Carnivore Group Living:
Comparative Trends

JOHN L. GITTLEMAN

In contrast to some other mammalian orders, members of the Carnivora do not commonly live in groups: only about 10–15% of all species aggregate at some period outside of the breeding season (Bekoff et al. 1984; Gittleman 1984). Because most carnivores reside in dense habitats and are solitary, dangerous, and nocturnal, little information existed on their social behavior until recently. Now, more comprehensive and comparative data are available to examine functional explanations of interspecific variation in grouping patterns across carnivores (for previous qualitative comparisons, see Ewer 1973; Kleiman and Eisenberg 1973; Kruuk 1975; Bertram 1979; Macdonald 1983). In this chapter I briefly review selected hypotheses for the evolution and maintenance of grouping in carnivores, focusing on those that are broadly applicable across the order and are testable from the available comparative data. I then analyze quantitative measures of interspecific variation in social behavior with respect to differences in morphology, physiology, and ecology. The analysis differs from previous cross-species comparisons of carnivore social ecology (Ewer 1973; Kleiman and Eisenberg 1973; Kruuk 1975; Bertram 1979; Macdonald and Moehlman 1982; Macdonald 1983; Bekoff et al. 1984; Kruuk and Macdonald 1985) by being more quantitative, by accounting for morphological and metabolic constraints, and by deriving general trends across the order as a whole rather than in particular taxonomic families.

To analyze carnivore social behavior, one must first classify functional aspects of grouping in terms of what general behaviors are being performed. At least four types of grouping may be distinguished: population groups—individuals sharing a common home range area; feeding groups—individuals utilizing the same food resource at a given time; foraging groups—individuals banding together while searching for food or hunting; and breeding groups—individuals forming a reproductive unit. Population, feeding, and foraging group sizes are similar, at least with regard to the number of adult individuals in the group, for most carnivores. However, for species that remain in groups outside of the breeding season, different individuals are included in various

183

activities. For example, banded mongooses (*Mungos mungo*) live in packs of approximately 16 individuals that forage on invertebrates independently of other pack members (Rood 1975, 1986), whereas dwarf mongooses (*Helogale parvula*) live in multi-male packs of up to 24 individuals, in which about four adults are reproductively active (Rood 1978, 1980, 1983). In the case of the African lion (*Panthera leo*), approximately eight individuals live in a pride encompassing the same home range area, but only three members of the pride hunt for the entire group (Schaller 1972; Bertram 1979). Perhaps the most extreme example of a division of labor within a group is provided by the spotted hyena (*Crocuta crocuta*); in the Ngrorongoro Crater as many as 55 individuals make up a clan that divides into hunting groups of about seven adults and feeding groups of roughly 19 individuals (Kruuk 1972; see also Mills, this volume). Thus, from these examples, it is clear that functional explanations of group living must distinguish each form of grouping (see also Kleiman and Brady 1978; Van Orsdol et al. 1985).

Many functional explanations have been suggested for grouping in eutherian mammals (for reviews, see Alexander 1974; Wilson 1975; Bertram 1978; Eisenberg 1981; Harvey and Greene 1981; Pulliam and Caraco 1984; Clark and Mangel 1986). Those pertaining to carnivores fall in two categories: anti-predator defense and exploitation of food.

Hypotheses for the Evolution of Group Living

Anti-Predator Defense

Carnivores that compete with other species for food or sometimes serve as food may benefit from group vigilance, whereby encroachers are detected more effectively. Among dwarf mongooses subordinate males are found on the periphery of the pack, where they keep lookout for threatening predators (Rasa 1977, 1986); further, groups with few vigilant guards are preyed upon more frequently (Rasa 1986). Grouping may also help minimize a predator's effect on the group: if by clustering together the members of a group cause a predator to catch only one individual while the rest are able to escape, then gregariousness may evolve (Hamilton 1971). Banded mongoose packs respond to raptors or terrestrial predators by immediately aggregating into a tight bunch, approaching the predator collectively (Rood 1975), and "with mouths pointed in all directions, giving the appearance of one large organism defending itself " (Kruuk 1975).

Group defense is expected to be more common in smaller species that are not able individually to ward off larger species (Ewer 1973; Kruuk 1975; Rood 1986). Also, species living in open habitats (e.g., grassland plains) are more vulnerable to predators (or competitors) and therefore more likely to form groups (Lamprecht 1981; Rood 1986). Many of these general associations of size and ecology with group living are also found in primates (Crook 1970; Clutton-Brock and Harvey 1977), ungulates (Jarman 1974; Jarman and Jarman 1979), sciurids (Hoogland 1981), and marsupials (Kaufman 1974; Lee and Cockburn 1985).

Exploitation of Food

In general, group living may be advantageous for locating food resources (Ward and Zahavi 1973), improving chances of finding and catching prey (Schaller 1972; Kruuk 1975), increasing the diversity and size of prey (Kruuk 1972, 1975; Schaller 1972; Caraco and Wolf 1975), and competing successfully for food (Lamprecht 1978, 1981). Not all of these factors have been considered for carnivores, mainly because of the methodological difficulties in carrying out detailed field experiments necessary for teasing apart hypotheses (see Bekoff et al. 1984).

In the search for food it is obvious that many pairs of eyes (or ears) are better than one. Yet, it is difficult actually to test whether, once food is located, information is being passed on among members of a group. Only a few experimental studies (e.g., Menzel 1971; Krebs et al. 1972) have shown that individuals forage more successfully by learning from one another; nevertheless, descriptive studies show that contact calls by smaller carnivores (e.g., dwarf mongoose; slender mongoose, *Herpestes sanguineus*; white-tailed mongoose, *Ichneumia albicauda*) foraging in groups for invertebrates may communicate the location of new food resources (Ewer 1973; Kingdon 1977).

Whether they find new food resources, predators hunting in groups may be more successful at taking down prey. For example, Schaller (1972) found that African lions had a higher success rate in capturing Thomson's gazelle (*Gazella thomsoni*), zebra (*Equus burchelli*), and wildebeest (*Connochaetes taurinus*) when two or more lionesses hunted together (see Caraco and Wolf 1975; Van Orsdol 1984). More extensive data, since collected by Bertram (1975, 1976, 1979) and Packer and Pusey (1982, 1983a, 1983b), and recently analyzed by Packer (1986), indicate that group hunting lions may not increase hunting success; data are not conclusive on either the average biomass of kills made by groups of different sizes or the hunting rates of different sized groups. The African lion story is a classic case of the paradox wherein the more information we have, the less we seem to know. Nevertheless, other carnivores do tend to support the association of grouping with hunting success: Wyman (1967) observed that golden jackals (*Canis aureus*) and black-backed jackals (*C. mesomelas*) were successful at catching Thomson's gazelle fawns only when hunting in pairs. And for spotted hyenas hunting wildebeest, 15% of 74 attempts were successful when a single hyena pursued a calf in contrast to 74% of 34 attempts when two or more hyenas attacked (Kruuk 1972). Even though within species variation of hunting methods seems to indicate benefits from grouping, it is difficult to compare hunting success rates across species because: (1) definitions of a hunting attempt vary among observers (Schaller 1972; Bertram 1979), (2) hunting success may depend on hunger level or hunting technique (e.g., ambush versus cursorial hunting: Van Orsdol 1984; Van Valkenburgh 1985; Taylor, this volume) and not grouping, and (3) various ecological constraints such as vegetation, habitat density, or time of day are confounding factors (Bertram 1979; Van Orsdol 1984).

The most common explanation for grouping in larger predators is that concerted effort permits a wider selection of prey in terms of amount, diversity, and size. Schaller (1972) and Bertram (1979) observed that African lions living in groups frequently hunted adult buffalos (*Syncerus caffer*) whereas single lions rarely even attempted an attack at buffalo. However, as Packer (1986) cautions, even though grouping lions take down and prefer larger prey than solitaries, this does not prove that lion sociality evolved as a consequence of the advantages of cooperative hunting: cooperative hunting may only be an adaptation to group living, rather than the evolutionary force resulting in group living (see Alexander 1974). Nevertheless, cooperative hunting is certainly an important benefit of grouping, and similar accounts of coordinated hunting have been reported for the African hunting dog (*Lycaon pictus*) (Estes and Goddard 1967; Malcolm and van Lawick 1975; Frame et al. 1979; Malcolm 1979), golden jackal (Lamprecht 1978), gray wolf (*Canis lupus*) (Mech 1966, 1970), coyote (*C. latrans*) (Bekoff 1978; Bekoff and Wells 1978; Bowen 1981; Wells and Bekoff 1982), spotted hyena (Kruuk 1972, 1975; Mills, this volume), Indian dhole (*Cuon alpinus*) (Davidar 1975; Johnsingh 1982), and cheetah (*Acinonyx jubatus*) (Caro and Collins 1986; Ashwood and Gittleman, 1989).

There are a number of carnivores that do not fit these generalizations (see also Packer 1986). Among larger species the mountain lion (*Puma concolor*), leopard (*Panthera pardus*), jaguar (*P. onca*), and tiger (*P. tigris*) exploit larger prey than themselves while hunting solitarily. In the case of some of the smaller Mustelidae (especially species of *Mustela*) individuals regularly kill prey of larger size than themselves (King, this volume). Even so, these examples do not deny the fact that group living is an important benefit, either direct or indirect, for catching large prey.

Finally, group living may carry advantages in defending kills or other food resources from neighboring predators (or other groups). For many carnivores, particularly medium-sized species, protecting kills is difficult: black-backed jackals in the Serengeti lose up to 30% of their Thomson gazelle and hare kills to spotted hyenas (Lamprecht 1978, 1981); spotted hyenas and African lions frequently scavenge from each other (Kruuk 1972); and, both hyenas and lions steal kills from cheetahs, leopards, and African hunting dogs (Estes and Goddard 1967; Kruuk 1972; Schaller 1972; Bertram 1979; Frame et al. 1979; Packer 1986). In each case, species feeding in groups will usually stand their ground against a scavenger or competitor and will retreat only when feeding individually or in pairs. The exception is that smaller species such as the African hunting dog may give way to a considerably larger species (such as the African lion or the spotted hyena) even when feeding in a group (see Frame et al. 1979).

An inherent difficulty in assessing the importance of competition or defense of kills for the evolution of grouping in carnivores is that there are many other behaviors that reduce losses in competitive situations. Carnivores will (1) make kills inaccessible to competitors (e.g., leopards hide carcasses in trees), (2) reduce exploitation time by fast feeding or group feeding, or (3) cache food (Macdonald 1976). Furthermore, the advantages of group living mentioned above (increased hunting success, prey size, and prey diversity) potentially are associated with a confounding variable, body size. Both population group size and prey size frequently increase with body mass (Clutton-Brock and Harvey 1977, 1983; Gittleman 1985a). All of these variables may be closely linked because they are influenced by similar energetic constraints (see McNab 1980, this volume). Therefore, size-related effects must be considered in searching for comparative trends in the functions of carnivore grouping.

Related Hypotheses

Other advantages have also been shown to result from grouping. These include reproductive access to members of the other sex (Wrangham 1975; Bygott et al. 1979; Packer and Pusey 1982), facilitation of learning (e.g., teaching young to hunt: Kleiman and Eisenberg 1973), and collective resistance against harsh environments (Eisenberg 1981; Gittleman 1985b). These additional factors are less well documented than those mentioned above, and are generally considered to be secondarily important, at least for carnivores.

Even though some authors (e.g., Hoogland 1979; Harvey and Greene 1981) suggest that variation in group living may best be explained by the disadvantages of grouping, these have not received as much attention as the beneficial factors. Undoubtedly this is because many disadvantages are more subtle, are difficult to observe, and depend on mechanisms within the group. Four general disadvantages are likely: group living increases the chances of being detected by potential predators (Jarman 1974; Clutton-Brock and Harvey 1977; Underwood 1982; Rasa 1986), decreases the amount of food intake to individuals (Jarman 1974; Wrangham 1977), increases transmission of disease or parasites (Hoogland 1979; Gittleman 1985b), and increases the possibility of aggression or injury. Because few studies have assessed these factors in carnivores (but see Rood 1983; Packer 1986), they will not be considered in the comparative analyses presented here.

Methods

All of the data, except for some body weight values (Gittleman 1985a) and life history information (Gittleman 1986b), were taken from studies of natural populations. Because species within a genus often share similar ecological and behavioral characteristics, thus biasing analyses by not representing independent sample points (see Clutton-Brock and Harvey 1984), statistical tests were performed on congeneric data. These data were calculated from mean values for species within a genus which share the same ecological type and social system (see definitions below; for further discussion of data calculations, see Gittleman 1985a, 1986a, 1986b).

Average figures were calculated, or descriptive categories were assigned, for each of the following variables (see Table 7.1):

1. Body weight: average weight (kg) of adult male and female.

2. Feeding group size: the number of individuals usually found feeding together at a kill or at a primary food source.

3. Foraging group size: the number of individuals hunting or foraging for the most common prey (see "Diet," below) in the diet.

4. Population group size: the number of individuals that regularly associate together and share a common home range.

5. Group metabolic rate: many behavioral and ecological factors are related to metabolic rate (McNab 1980, this volume; Eisenberg 1981; Martin 1981; Gittleman and Harvey 1982; Mace et al. 1983). In the present analysis metabolic requirements of group sizes were approximated by: body weight$^{0.75}$ (Kleiber's Value) multiplied by each group size variable (population, feeding, or foraging group size, respectively).

6. Litter size: average number of offspring at birth (for more complete definitions and data sources of life history traits, see Gittleman 1986a).

7. Age of independence: age when juvenile disperses from natal territory or is independent of parental care (days).

8. Prey size: size of most common prey in the diet. Categories are: very small (<1 kg); small (1–10 kg); medium (10–100 kg); large (100–400 kg).

9. Vegetation: forest, woodland, dense brush or scrub, open grassland, aquatic. Occasionally species could not be accurately described by one category and types were combined (e.g., American black bear (*Ursus americanus*): open grassland and woodland; small Indian civet (*Viverricula indica*): open grassland and forest).

10. Activity pattern: nocturnal, diurnal, crepuscular, arhythmic, nocturnal, and crepuscular.

11. Diet: type of food constituting at least 60% of the diet. Those species that do not feed on any single type making up 60% of the diet were classified as omnivores. Also, species that are primarily scavengers (e.g., wolverine, *Gulo gulo*) or frugivores/invertebrate feeders (e.g., coati) were not included in the dietetic analyses. Categories are: carnivores (flesh eaters), insectivores (this

Table 7.1. Species and data used for analysis

Species	Activity pattern[a]	Zonation[b]	Diet[c]	Vegetation[d]	Prey size[e]	Body weight (kg)	Litter size	Age of independence (days)	Group size		
									Population	Feeding	Foraging
Canidae											
Canis lupus	A	T	M	N	M	33.12	6	—	7.0	7.0	7.0
C. latrans	C	T	M	N	S	10.59	6	—	2.0	1.5	1.5
C. aureus	C	T	O	O	V	8.76	3	—	3.0	—	1.8
C. adustus	N	T	O	S	S	11.25	4	—	2.0	1.5	1.5
C. mesomelas	C	T	O	W	M	7.69	4	270	2.0	—	1.8
Lycaon pictus	D	T	M	O	M	21.98	9	392	8.0	8.0	8.0
Cuon alpinus	D	T	M	R	M	15.80	4	—	6.0	—	—
Alopex lagopus	A	T	M	O	—	5.19	7	166	—	—	1.0
Vulpes vulpes	N	T	M	R	V	6.14	5	226	3.0	1.0	1.0
V. bengalensis	O	T	O	S	—	2.39	4	—	2.0	—	—
V. ruppelli	O	T	M	T	—	3.60	2	—	4.0	—	2.0
V. chama	N	T	O	O	V	3.10	4	—	2.0	—	—
V. velox	N	T	O	O	V	2.20	5	—	1.0	—	1.0
Otocyon megalotis	O	T	I	S	V	3.94	4	—	2.0	1.0	1.0
Urocyon cinereoargenteus	O	T	O	D	V	3.63	4	—	1.0	—	—
Dusicyon gymnocercus	N	T	M	O	—	4.44	4	—	—	—	1.0
Ursidae											
Ursus arctos	A	T	V	N	—	292.87	2	645	1.0	1.0	1.0
U. americanus	D	S	V	S	—	111.05	3	483	1.0	1.0	1.0
Thalarctos maritimus	D	T	—	—	—	365.04	2	821	1.0	—	—
Selenarctos thibetanus	A	S	V	R	—	103.54	2	337	1.0	—	—
Melursus ursinus	A	B	V	N	—	101.49	2	916	1.0	—	1.0
Procyonidae											
Bassariscus astutus	N	B	O	W	V	0.95	3	—	1.0	—	—
Nasua narica	D	S	O	R	—	5.00	4	—	8.0	—	—
Procyon lotor	O	T	O	S	—	5.35	4	—	1.0	—	—
Ailuridae											
Ailuropoda melanoleuca	C	S	V	R	—	135.64	2	—	1.0	—	—
Mustelidae											
Mustela erminea	A	T	M	R	M	0.95	5	—	1.0	—	—
M. sibirica	O	S	M	F	V	0.57	5	—	1.0	—	—
M. vison	N	—	M	F	V	1.09	5	—	1.0	—	—
Martes martes	N	B	M	R	V	1.20	3	—	1.0	—	—

(continued)

Table 7.1. (Continued)

Species	Activity pattern[a]	Zonation[b]	Diet[c]	Vegetation[d]	Prey size[e]	Body weight (kg)	Litter size	Age of independence (days)	Group size		
									Population	Feeding	Foraging
M. americana	N	B	M	F	V	0.87	3	—	1.0	—	—
M. pennanti	A	—	—	F	V	3.74	3	—	1.0	—	—
Gulo gulo	A	T	S	N	—	11.59	3	—	1.0	—	—
Ictonyx striatus	N	T	O	S	V	0.77	2	—	1.0	—	—
Poecilogale albinucha	O	T	M	D	—	0.30	3	—	1.0	—	—
Mellivora capensis	O	T	I	N	V	8.08	2	—	1.0	—	—
Meles meles	A	T	I	S	V	11.59	3	211	7.0	1.0	1.0
Taxidea taxus	A	T	M	O	S	4.07	4	—	1.0	—	—
Mephitis mephitis	A	T	M	O	S	2.41	6	84	1.0	—	—
Lutra maculicollis	A	Q	F	A	—	4.07	3	—	2.0	—	—
Lutrogale perspicillata	—	Q	—	Q	—	8.77	—	—	2.0	—	—
Aonyx capensis	O	Q	O	A	—	18.92	3	—	2.0	—	—
Viverridae											
Civettictus civetta	N	T	O	D	V	12.06	3	—	1.0	—	—
Viverricula indica	N	B	O	N	—	2.66	4	—	2.0	—	—
Genetta tigrina	N	B	O	R	V	2.05	4	—	1.0	—	—
Nandinia binotata	N	B	V	R	—	1.16	2	—	1.0	—	—
Paguma larvata	N	A	—	R	—	4.71	3	—	1.0	—	—
Fossa fossa	N	S	O	R	—	1.80	1	365	2.0	—	—
Herpestes sanguineus	D	T	O	N	V	0.49	3	—	2.0	—	1.5
H. auropunctatus	D	—	I	N	V	0.78	3	—	1.0	1.0	1.0
H. smithi	N	T	O	R	—	1.70	3	—	2.0	—	—
H. fuscus	A	—	—	—	—	1.19	4	—	3.0	—	—
Mungos mungo	D	T	I	S	V	1.26	4	—	17.0	1.0	1.0
Helogale parvula	D	T	I	W	V	0.27	4	—	10.0	1.0	1.0
Ichneumia albicauda	N	T	O	S	V	3.90	3	—	1.0	—	—

Atilax paludinosus	N	T	O	R	V	3.71	3	—	1.0	—	—
Cynictus penicillata	D	T	I	O	V	0.60	3	—	5.0	—	—
Paracynictus selousi	N	T	I	S	V	1.69	2	—	2.0	—	—
Hyaenidae											
Hyaena brunnea	N	T	O	S	—	43.38	2	898	6.0	3.0	1.0
Crocuta crocuta	N	T	M	O	L	51.94	2	916	55.0	18.5	6.4
Proteles cristatus	N	T	I	O	V	8.33	3	—	1.0	—	—
Felidae											
Felis silvestris	A	T	M	F	V	4.66	3	140	1.0	1.0	1.0
F. libyca	N	T	O	S	V	4.31	3	140	1.0	—	1.5
Leptailurus serval	A	T	M	N	V	11.70	2	—	1.0	—	1.5
Prionailurus bengalensis	A	B	M	R	—	6.47	3	—	2.0	—	—
Profelis aurata	O	B	M	R	—	7.61	—	—	1.0	—	—
Caracal caracal	N	S	O	S	V	11.59	3	365	1.0	1.0	1.5
Puma concolor	N	T	M	W	L	51.94	2	420	1.0	1.0	1.0
Leopardus pardalis	—	B	—	R	—	11.82	3	—	1.0	—	—
Lynx lynx	A	T	M	R	S	15.30	2	240	1.0	1.0	1.0
L. rufus	A	T	M	R	S	6.17	3	365	1.0	1.0	—
Panthera leo	N	T	M	S	L	166.02	3	1075	9.0	6.5	2.5
P. tigris	N	T	M	F	M	160.77	3	572	1.0	1.0	1.0
P. pardus	O	S	M	W	M	52.46	3	602	1.0	1.0	1.0
P. uncia	A	T	—	—	—	32.46	3	—	1.0	—	—
Acinonyx jubatus	D	T	M	S	M	58.56	4	464	1.0	1.0	1.0

Note. Taxonomy follows Ewer 1973 except for Ailuropoda, which was placed in a separate family because of conflicting evidence; see Corbet and Hill 1980, Eisenberg 1981, and Schaller et al. 1985.

[a]D = diurnal, N = nocturnal, A = arhythmic, C = crepuscular, O = nocturnal and crepuscular.

[b]T = terrestrial, S = terrestrial and occasionally arboreal, B = arboreal and terrestrial, A = arboreal, Q = aquatic.

[c]M = carnivorous (flesh eater), O = omnivorous, I = insectivorous, F = piscivorous, F = frugivorous and folivorous.

[d]N = open grassland and forest, O = open grassland, R = forest, S = open grassland and woodland, D = dense brush and scrub, T = desert, W = woodland, Q = aquatic.

[e]V = very small, S = small, M = medium, L = large (see "Methods" for measurement values).

includes other invertebrate prey such as earthworms because of similar availability and distribution), folivores/frugivores, piscivores, omnivores.

12. Zonation: terrestrial, terrestrial and occasionally arboreal (primarily ground living but also adept at tree climbing), arboreal and terrestrial (both ground and tree living), aquatic.

Results

Across the order there is no relationship between body weight and population group size ($r_{72} = 0.02$), feeding group size ($r_{21} = 0.06$), or foraging group size ($r_{36} = 0.02$). At the family level, there only is a correlation between population group size and body weight in the Canidae ($r_{14} = 0.56, P < 0.05$). Because of these results, body weight was not incorporated in further analyses on group sizes and ecology. Furthermore, there were no consistent differences in group sizes among taxonomic families, and therefore phylogenetic effects were unlikely (see Harvey and Mace 1982).

Population group size is correlated with feeding group size ($r_{21} = 0.79, P < 0.01$) but not with foraging group size; and, foraging group size and feeding group size are significantly correlated ($r_{21} = 0.65, P < 0.01$).

For the discrete ecological categories examined (activity pattern, zonation, vegetation, diet), population group size varies only with vegetation ($F_{6,47} = 2.41, P < 0.05$; Figure 7.1): pair-wise comparisons reveal that population group size is smaller in forest-living species than open grassland species ($t_6 = 2.95, P < 0.05$) and open grassland and woodland species ($t_{12} = 2.18, P < 0.05$). Feeding group size and foraging group size are not significantly different between species with different ecologies; this is perhaps due to smaller sample sizes than with population group size. However, heterogeneity among vegetational types was in the same direction as that found with population group size.

Among predatory carnivores (those species that include some meat in the diet), population group size differs in relation to prey size ($F_{3,34} = 5.86, P < 0.005$; see Figure 7.2); and, at the 10% level of significance, foraging group size ($F_{3,15} = 2.65$) and foraging group size ($F_{3,12} = 2.90$) vary with prey size. Population group sizes of species feeding on very small and small prey, respectively, are smaller than those eating medium ($t_4 = 2.89, P < 0.05; t_4 = 2.63, P < 0.05$) and large prey ($t_4 = 2.79, P < 0.05; t_2 = 4.83, P < 0.05$).

Across the order and at the family level none of the group size variables are correlated with age of independence or litter size, either with or without accounting for maternal body size.

Differences in group metabolic needs (see Methods section) were examined only in relation to dietetic types because of the close relationship between metabolic rate and diet (McNab 1980, this volume). Heterogeneity of foraging

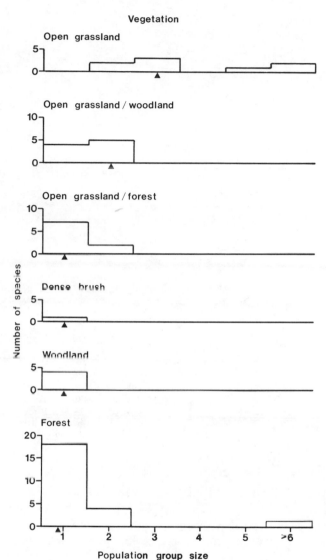

Figure 7.1. Distribution of population group sizes across vegetational types. Arrowhead indicates median value for population group size within each vegetational type.

group metabolic needs is related to diet ($F_{3,26}$ = 3.87, P < 0.025): strict carnivores have higher foraging group metabolic needs than omnivores (t_{12} = 2.68, P < 0.02) and insectivores (t_{18} = 3.00, P < 0.01); herbivores/frugivores have higher foraging group metabolic needs than omnivores (t_9 = 5.70, P < 0.001) and insectivores (t_5 = 2.80, P < 0.05).

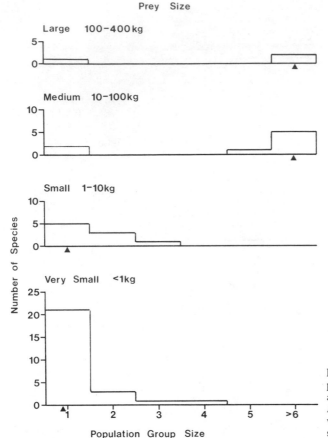

Figure 7.2. Distribution of population group sizes across prey size categories. Arrowhead indicates median value for population group size within each prey type.

Discussion

The quantitative analyses of ecological associations with carnivore grouping patterns reveal that two factors may be influential: exploitation of food resources and habitat. Both of these factors, as well as many others that could not be examined because of inadequate data across the order, are undoubtedly interrelated. A multivariate analysis would be necessary to partition relative effects to each factor; however, this was not possible because of small sample sizes and numerous empty cells in across-variable comparisons; see Clutton-Brock and Harvey, (1984) Harvey and Clutton-Brock (1985), and Gittleman (1988) for discussion of problems in using multivariate techniques for comparative studies. The following discussion, therefore, centers on grouping trends for each of the salient ecological factors. First it is necessary, though, to mention briefly the relationship between body size and grouping.

In primates (Clutton-Brock and Harvey 1977), ungulates (Estes 1974; Jarman 1974; Jarman and Jarman 1979), and some marsupials (Kaufmann 1974; Lee and Cockburn 1985), population group size increases with body size (weight). The function of this association may lie with similar energetic constraints, mediated through energy expenditure and food availability, on both variables (Clutton-Brock and Harvey 1977, 1983; McNab 1980). Body size and grouping were not found to be related in carnivores, although in Canidae, the family having the highest frequency of grouping, there appears to be an increase in grouping with body size (see also Bekoff et al. 1981). The general trend across the order is exemplified by the large ursids being primarily solitary and some of the small viverrids (e.g., yellow mongoose, *Cynictis penicillata*; dwarf mongoose; meerkat, *Suricata suricatta*) living in large packs. The failure to observe a relationship between group size and body weight across carnivores suggests that additional factors, other than energetic constraints, influence grouping patterns. Even so, this should not be construed as a statement that energetics is unimportant in the evolution or maintenance of grouping; rather, energetics, combined with an unusual diet, may simply operate in another fashion for carnivores: after all, the largest carnivores, the Ursidae, are among the most solitary and herbivorous/folivorous species in the order (see Herrero 1978).

Dietetic Correlates

Previous studies have shown that intraspecific variation in carnivore grouping patterns is related to food availability and distribution (Kruuk 1975; Bertram 1979; Macdonald 1983; Bekoff et al. 1984; Kruuk and Macdonald 1985; Lindström 1986). The results of this study are the first quantitative demonstration of this trend interspecifically. To assess the underlying functional reasons for this relationship, it is necessary to distinguish between the various grouping types and their likely associated causes (see also Mills 1978; von Schantz 1984).

Foraging group size is linked directly to food acquisition in terms of location, pursuit, and kill, whereas population group size is more a function of internal mechanisms of the group (i.e., kin-related effects) as well as ecological factors. Last, feeding group size is somewhat related to both population group size and foraging group size in the sense that, after foraging, a group of individuals is likely to remain together while feeding, and population groups may arise from these feeding congregations. The comparative data analyzed here, coupled with single species studies, tend to bear out these distinctions. Population group size is related to feeding group size but not to foraging group size: population groups frequently break-up into smaller foraging groups (e.g., gray wolf, spotted hyena, African lion) and then regroup while feeding. Thus, individual members of a group which are likely to share a kill or a clump of insects (as a feeding group) tend to remain together. In Israel, where striped hyenas (*Hyaena hyaena*) and golden jackals are artificially fed large quantities of food, individuals congregate while eating and remain together afterward even though they inhabit extensive areas. In east Africa, where these species forage independently of humans, both are primarily solitary (or occasionally seen in pairs; see Kingdon 1977; Moehlman 1983) whether considered in feeding or population groups (Macdonald 1978, 1979). Similar patterns of intraspecific differences in grouping have been observed in coyotes (Bekoff and Wells 1980, 1986; Bowen 1981), gray wolves (Mech 1970; Messier 1985a, 1985b), red foxes (*Vulpes vulpes*) (Zimen 1980), and African lions (Schaller 1972; Eloff 1973; Bertram 1978; Packer 1986). One interesting exception is the spotted hyena, which feeds in a smaller group than a clan (population group). Originally, it was reported that relatedness within spotted hyena clans is unusually low for a social carnivore (Bertram 1979). However, recent evidence indicates that there is a high degree of relatedness (Mills 1985); therefore, kin-related effects cannot explain the observed loose population structure. Mills (1985, this volume) suggests that food carrying, an important function of many carnivore groups, is difficult for spotted hyenas to accomplish because adults travel long distances for prey and the social nature of their feeding would provide little prey left over to take back to a den site; furthermore, Frank (1986a, 1986b) shows that individuals of high rank within the clan maintain close association and support each other effectively in competition for food. Thus, the evolution of clans among spotted hyenas may be derived from ecological factors different than those operating in many other social carnivores.

Among predatory species, population group size increases with prey size (Figure 7.2), and there is also a trend (at the 10% level of significance) for feeding group size and foraging group size to increase with prey size—lack of significance may be due to small samples sizes (see Results section). This association has been shown in a number of single species studies (e.g., coyote: Bekoff and Wells 1980; Bowen 1981; black-backed jackal: Wyman 1967; African lion: Schaller 1972) and in descriptive comparisons across particular taxonomic families (Canidae: Ewer 1973; Kleiman and Eisenberg 1973; Macdonald 1983. Hyaenidae: Kruuk 1975. Felidae: Kleiman and Eisenberg 1973; Bertram 1979; Packer 1986). This study quantitatively demonstrates the relationship between prey size and group size across the order.

As mentioned previously, functional explanations for each group size variable may be different even though various grouping patterns are interrelated. Therefore, caution must be taken when ascribing the same functional causes to different grouping characteristics on the basis of a common trend with one particular variable. The relationship between population group size and prey size is probably due to larger (either in size or distribution) food resources supporting a greater number of individuals and *perhaps* maximizing energetic returns and foraging efficiency (Caraco and Wolf 1975; Nudds 1978; also see

Packer 1986 for a discussion of the difficulties of testing these hypotheses). Species feeding on very small prey such as *Peromyscus* or *Apodemus* species are all solitary, with the exceptions of the red fox, Cape clawless otter (*Aonyx capensis*), slender mongoose, marsh mongoose (*Atilax paludinosus*), and yellow mongoose, which are occasionally observed in pairs or small family groups (see Kingdon 1977; Gorman 1979; Lynch 1980; Macdonald 1983; Arden-Clarke 1986). Species feeding on small prey are of size classes similar to those feeding on very small prey; again, there are some exceptions such as the black-backed jackal, side-striped jackal (*Canis adustus*), Malagasay civet (*Fossa fossa*), and perhaps Bengal cat (*Felis [Prionailurus] bengalensis*) (see Albignac 1972, 1973; Guggisburg 1975; Moehlman 1983). These exceptions are those species that tend to have more omnivorous feeding habits by including vegetation, fruit, and insects in their diet along with meat. Thus, increased food availability and more evenly distributed foods may select for occasional small groups among species exploiting very small or small prey (i.e., prey less than 10 kg). It is important to recognize that selection for grouping operates by the distribution and quality of prey in a given foraging area (Bradbury and Vehrencamp 1976; Kruuk 1978; Kruuk and Parish 1982; Macdonald 1983; Kruuk and Macdonald 1985) and not by benefits accrued in grouping when acquiring or hunting for food, as is the case with larger carnivores.

Predatory species living in larger groups, consisting of four or more adults, prey on animals of medium and large size categories. Undoubtedly, this association is partly related to the relative abundance and distribution of prey. For example, the gray wolf, coyote, African lion, spotted hyena, and perhaps cheetah vary their group sizes in response to local prey fluctuations (Kruuk 1972, 1975; Schaller 1972; Bertram 1979; Bowen 1981; Caro and Collins 1986, 1987; Ashwood and Gittleman, 1989). More importantly, though, with species living in larger groups there is the additional factor that cooperative hunting is necessary to bring down larger prey. Accounts of cooperative hunting among individuals may be found for wolves (Mech 1966, 1970), spotted hyenas (Kruuk 1972), and African lions (Schaller 1972; Bertram 1978, 1979).

The relationship between group size and prey size is confounded by another variable: habitat. All of the species living in larger groups are found in open vegetation (see Table 7.1; Figure 7.1). To hunt cooperatively the animals must fan out, maintain contact with other individuals, and adjust positions during pursuits. Such behaviors could hardly occur in a dense habitat. Previous discussions have ignored this point in the context of prey characteristics and group size in carnivores (e.g. Kruuk 1975; Macdonald 1978, 1983; Bertram 1979; Lindström 1986; but see Sunquist 1981) despite its importance in other mammalian groups (ungulates: Jarman 1974; Jarman and Jarman 1979; primates: Clutton-Brock and Harvey 1977). Interestingly, the only solitary species preying on larger animals than themselves (the leopard, the jaguar, the tiger, and the mountain lion) live in dense vegetation (see also Gittleman

1984). The only known exceptions where vegetation does not seem to be associated with grouping are the cheetah and the coati. Cheetahs are occasionally observed in small groups of related males (Bertram 1979; Caro and Collins 1986, 1987). Selection against more permanent, increased grouping may be due to their hunting technique of an inconspicuous approach, quick rush, straight long distance pursuit, and lack of maneuverability as well as their phylogenetic heritage (see Ashwood and Gittleman 1989). In the case of the coati, an unusual invertebrate eater/frugivore of the family Procyonidae, females form bands consisting of between four and 20 individuals (Kaufmann 1962; Russell 1983). These bands reside in relatively dense woodland/forest regions, thereby contradicting the general trend for grouping and open habitats. Although specifics of the social system of this species have yet to be studied, it is presumed that interactions among females in a band, particularly during the intensive rearing and lactation period, do not establish tightly knit groups. Thus, even though grouping is observed, the amount of direct or continuous contact among individuals requires further study in order to test the influence of habitat on grouping in this species.

Finally, the capacity for grouping that is closely tied to the exploitation of prey may also be associated with various morphological characters (see Taylor, this volume). Canids and spotted hyenas are limited in lateral movement (because of their restrictive ankle joint) and have heavily built skeletons with nonretractile claws. Such characteristics are well suited for long-range tracking of prey (Ewer 1973), which eventually tires a potential prey victim. In this mode of hunting, cooperation allows for longer and faster pursuit, with more effective closing in on prey. By contrast, felids rely more on an elaborate sequence of stalking, use of retractile claws for pulling down quarry, and truncated jaws used for a precisely oriented killing bite (Van Valkenburgh, this volume). All of these characteristics add up to a solitary hunt "with the predator in full control of the situation" (Ewer 1973:226). Thus, in addition to various ecological factors, morphological constraints may at least maintain, if not contribute to the origin of, group living in some carnivores and solitariness in others (see also Eisenberg and Leyhausen 1972).

Anti-predator/Competitor Correlates

Even though carnivores are usually considered to be threatening to noncarnivorous animals, they are also harmful to each other. Dietetic analyses of most medium- and large-sized species indicate that they will eat an infant or juvenile of another carnivore species (see Kruuk 1972; Schaller 1972). Clearly, the risk of predation would be greater for smaller species, but also species living in more open habitats would be more vulnerable, irrespective of size. Gorman (1979) and Rood (1986) qualitatively compared discrete categories of social versus solitary viverrids and found that social species generally live in more open habitats. The results of comparing group sizes among vegetational types reveal that this pattern holds across the order: open grassland and open grassland/woodland species have larger population group sizes than do forest dwellers (see Figure 7.1); foraging and feeding group sizes show the same trend at the 10% level of significance.

In smaller species (e.g., dwarf mongoose, banded mongoose, African hunting dog), specific group defense mechanisms serve as direct protection of young and other members of the group. For example, Rood (1975) observes in banded mongoose that "bunching was . . . used in offensive contexts to drive off other species, some of them potential predators and competitors. A pack sighting a raptor such as fish eagle on the ground invariably responded by aggregating and approaching en masse. Animals toward the front of the group would frequently stand up giving the appearance of a single large animal in continuous motion. This spectacle always caused the raptor to fly off" (p.108). Other small carnivores display similar forms of anti-predator behavioral defense (e.g., meerkat; common kusimanse, *Crossarchus obscurus*) or warning vocalizations (dwarf mongoose. Rasa 1986.)

Just as it is necessary to consider habitat effects in the context of the exploitation of food with sociality (as discussed previously), it is important to include the effects of food availability in those species for which anti-predatory defense is probably the principle force behind sociality. The relevant question is whether those species displaying group vigilance feed on readily available and/or evenly distributed foods that allow for sociality. For smaller species the answer to this question is probably yes. Most of the smaller carnivores displaying group defense are insectivores or omnivores, for whom food is relatively abundant. Waser (1981) produced a quantitative model that suggests that insects commonly found in the diet of carnivores are distributed in highly renewable patches and that "the cost of social tolerance is very small; a mongoose excluding a single competitor from its foraging range will gain only a 1% increase in prey density" (p.234). Similar conclusions are reached after evaluating the feeding ecology of omnivorous canids and the earthworm feeder, the Eurasian badger (*Meles meles*) (Macdonald 1983; Kruuk and Macdonald 1985). Thus, distribution and abundance of food may be a precondition for sociality in the context of group vigilance (Gittleman 1984; Waser and Waser 1985).

For larger species group defense probably serves two functions: protection of young (Bekoff and Wells 1982; Macdonald and Moehlman 1982; Moehlman 1983, 1986) and defense of valuable food resources (Lamprecht 1978, 1981; Bowen 1981). Schaller (1972) reported that African lion cubs were attacked and sometimes killed by leopards, spotted hyenas, African hunting dogs, elephants, buffalos and other African lions. Numerous studies have shown the extent of parental care and guarding of young by the gray wolf (Mech 1970), coyote (Bekoff and Wells 1982, 1986), brown hyena (*Hyaena brunnea*) (Owens and Owens 1984; Mills, this volume), spotted hyena (Kruuk 1972; Mills, this volume), and African lion (Schaller 1972; Packer 1986). However, it is difficult to tease apart the relative influence of group protection from resource defense because these are also the species that commonly feed on larger, scarcer prey animals, the type of food that would likely select for grouping to defend carcasses. Individuals grouping around a kill are often more effective at warding off scavengers (Lamprecht 1978, 1981).

Finally, it should be mentioned that some studies (e.g., gray wolf: Messier 1985a, 1985b; bat-eared fox, *Otocyon megalotis*: Nel et al. 1984) are finding that the effects of a food resource on grouping are mediated via other factors such as climate, territory size and availability, age distribution, and sex ratio of group members. Further studies should incorporate these factors after controlling for variation in food resources.

Phylogeny

As with most other mammalian traits (see Eisenberg 1981; Gittleman 1988), phylogeny is an influential but yet vexing factor in evolutionary explanations. In the case of carnivores, for example, after removing allometric effects, one sees that life history traits are significantly correlated with some phylogenetic component at the family level (Gittleman 1986a). Similarly, as one considers the evolution of group living, it appears that certain taxonomic groups (e.g., canids; herpestids) have a greater tendency for grouping than others. Nevertheless, as shown in the general phylogenetic tree of Figure 7.3, some form of social behavior (either in groups or in pair formation) has evolved in each major taxonomic family across the order. It appears that carnivores' grouping behaviors evolved independently many times. Although a specific phylogenetic methodology is currently not available to assess these patterns (Felsenstein 1985), it would be useful for future analysis to take into account phylogeny in the evolution of carnivore grouping and perhaps to use contemporary genetic and morphological studies (see Wayne et al., this volume) for testable hypotheses.

Acknowledgments

I thank John Eisenberg, Paul Harvey, Hans Kruuk, and Craig Packer for comments on previous versions of this chapter. During the preparation of this work I received support from the Department of Zoology and Graduate Programs in Ecology and Ethology, University of Tennessee; the Scholarly Studies Program, Smithsonian Institution; and an NICHD Training Grant (T32-HD-07303).

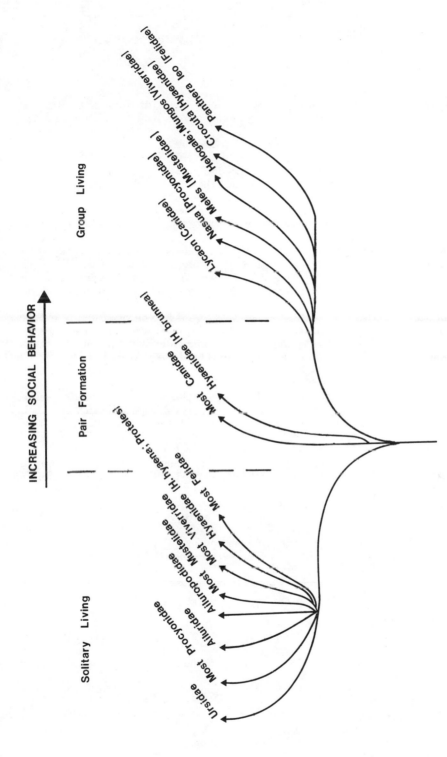

Figure 7.3 General phylogenetic tree of social behavior in Carnivora.

References

Albignac, R. 1972. The carnivores of Madagascar. In: R. Battistini & G. Richard-Vindard, eds. *Biogeography and Ecology of Madagascar*, pp. 21–35. The Hague: Dr. W. Junk.

Albignac, R. 1973. Mammifères carnivores. *La Terre et la Vie* 36:1–206.

Albignac, R. 1984. The carnivores. In: A. Jolly, P. Oberle & R. Albignac, eds. *Madagascar*, pp. 167–181. Oxford: Pergamon Press.

Alexander, R. D. 1974. The evolution of social behavior. *Ann. Rev. Ecol. Syst.* 5:325–383.

Arden-Clarke, C. H. G. 1986. Population density, home range size and spatial organization of the Cape clawless otter, *Aonyx capensis*, in a marine habitat. *J. Zool.* 209:201–211.

Ashwood, T. L., and Gittleman, J. L. 1989. Behavioral and ecological factors influencing population density in cheetah (*Acinonyx jubatus*): A critical review. In: H. G. Genoways, ed. *Current Mammalogy*, vol. 3. New York: Plenum. In press.

Bekoff, M., ed. 1978. *Coyotes: Their Biology, Behavior and Management* New York: Academic Press.

Bekoff, M., Daniels, T. J., and Gittleman, J. L. 1984. Life history patterns and the comparative social ecology of carnivores. *Ann. Rev. Ecol. Syst.* 15:191–232.

Bekoff, M., Diamond, J., and Mitton, J. B. 1981. Life history patterns and sociality in canids: Body size, reproduction and behavior. *Oecologia* 50:386–390.

Bekoff, M., and Wells, M. C. 1980. The social ecology of coyotes. *Sci. Amer.* 242:130–148.

Bekoff, M., and Wells, M. C. 1982. The behavioral ecology of coyotes: Social organization, rearing patterns, space use, and resource defense. *Z. Tierpsychol.* 60:281–305.

Bekoff, M., and Wells, M. C. 1986. Social ecology and behavior of coyotes. *Adv. Study Behav.* 16:251–338.

Bertram, B. C. R. 1975. The social systems of lions. *Sci. Amer.* 177:463–482.

Bertram, B. C. R. 1976. Kin selection in lions and in evolution. In: P. P. G. Bateson & R. A. Hinde, eds. *Growing Points in Ethology*, pp. 281–301. Cambridge: Cambridge Univ. Press.

Bertram, B. C. R. 1978. Living in groups: Predators and prey. In: J. R. Krebs & N. B. Davies, eds. *Behavioural Ecology: An Evolutionary Approach*, pp. 281–301. Sunderland, Mass.: Sinauer.

Bertram, B. C. R. 1979. Serengeti predators and their social systems. In: A. R. E. Sinclair & M. Norton-Griffiths, eds. *Serengeti: Dynamics of an Ecosystem*, pp. 159–179. Chicago: Univ. Chicago Press.

Bowen, W. D. 1981. Variation in coyote social organization: The influence of prey size. *Canadian J. Zool.* 59:539–559.

Bradbury, J. W., and Vehrencamp, S. L. 1976. Social organization and foraging in Emballonurid bats. I. field studies. *Behav. Ecol. Sociobiol.* 2:1–17.

Bygott, J. D., Bertram, B. C. R., and Hanby, J. P. 1979. Male lions in large coalitions gain reproductive advantages. *Nature* 282:838–840.

Caraco, T., and Wolf, L. L. 1975. Ecological determinants of group sizes of foraging lions. *Amer. Nat.* 109:343–352.

Caro, T. M., and Collins, D. A. 1986. Male cheetahs of the Serengeti. *Natl. Geogr. Rep.* 2:75–86.

Caro, T. M., and Collins, D. A. 1987. Ecological characteristics of territories of male cheetahs (*Acinonyx jubatus*). *J. Zool.* 211:89–105

Clark, C. W., and Mangel, M. 1986. The evolutionary advantages of group foraging. *Theor. Pop. Biology.* 30:45–75.

Clutton-Brock, T. H., and Harvey, P. H. 1977. Primate ecology and social organisation. *J. Zool.* 183:1–39.

Clutton-Brock, T. H., and Harvey, P. H. 1978. Mammals, resources and reproductive strategies. *Nature* 273:191–195.

Clutton-Brock, T. H., and Harvey, P. H. 1983. The functional significance of variation in body size among mammals. In: J. F. Eisenberg and D. G. Kleiman, eds. *Advances in the Study of Mammalian Behavior*, pp. 632–663. Special Publication no. 7. Lawrence, Kans.: American Society of Mammalogists.

Clutton-Brock, T. H., and Harvey, P. H. 1984. Comparative approaches to investigating adaptation. In: J. R. Krebs & N. B. Davies, eds. *Behavioral Ecology: An Evolutionary Approach* (2nd ed.), pp. 7–29. Sunderland, Mass.: Sinauer.

Corbet, G. B., and Hill, J. E. 1980. *A World List of Mammalian Species*. London: British Museum (Natural History).

Crook, J. H. 1970. The socio-ecology of primates. In: J. H. Crook, ed. *Social Behavior in Birds and Mammals*, pp. 103–159. London: Academic Press.

Davidar, E. R. C. 1975. Ecology and behavior of the dhole or Indian wild dog (*Cuon alpinus*). In: M. W. Fox, ed. *The Wild Canids*, pp. 109–119. New York: Van Nostrand Reinhold.

Eisenberg, J. F. 1981. *The Mammalian Radiations*. Chicago: Univ. Chicago Press.

Eisenberg, J. F., and Leyhausen. P. 1972. The phylogenesis of predatory behaviour. *Z. Tierpsychol.* 30:59–83.

Eloff, F. C. 1973. Ecology and behavior of the Kalahari lion. In: R. Eaton, ed. *The World's Cats* 1: pp. 90–126. Winston, Ore.: World Wildlife Safari.

Estes, R. D. 1974. Social organization of the African Bovidae. In: V. Geist & V. Walther, eds. *The Behavior of Ungulates and Its Relation to Management*, pp. 166–255. Morges, Switzerland: IUCN.

Estes, R.D., and J. Goddard. 1967. Prey selection and hunting behavior of the African wild dog. *J. Wild. Mgmt.* 31:52–70.

Ewer, R. F. 1973. *The Carnivores*. Ithaca, N.Y.: Cornell Univ. Press.

Felsenstein, J. 1985. Phylogenies and the comparative method. *Amer. Nat.* 125:1–15.

Frame, L. H., Malcolm, J. R., Frame, G. W., and van Lawick, H., 1979. Social organization of African wild dogs (*Lycaon pictus*) on the Serengeti Plains. *Z. Tierpsychol.* 50:225–249.

Frank, L. G. 1986a. Social organization of the spotted hyaena (*Crocuta crocuta*). I. demography. *Anim. Behav.* 34:1500–1509.

Frank, L. G. 1986b. Social organization of the spotted hyaena (*Crocuta crocuta*). II. dominance. *Anim. Behav.* 34:1510–1527.

Gittleman, J. L. 1984. *The Behavioural Ecology of Carnivores*. Ph.D. thesis, Univ. Sussex, England.

Gittleman, J. L. 1985a. Carnivore body size: Ecological and taxonomic correlates. *Oecologia* 67:540–554.

Gittleman, J. L. 1985b. Functions of communal care in mammals. In: P. J. Greenwood, P. H. Harvey & M. Slatkin, eds. *Evolution: Essays in Honour of John Maynard Smith*, pp. 187–205. Cambridge: Cambridge Univ. Press.

Gittleman, J. L. 1986a. Carnivore brain size, behavioral ecology, and phylogeny. *J. Mamm.* 67:23–36.

Gittleman, J. L. 1986b. Carnivore life history patterns: Allometric, phylogenetic, and ecological associations. *Amer. Nat.* 127:744–771.

Gittleman, J. L. 1988. The comparative approach in ethology: Aims and limitations. In: P. P. G. Bateson and P. H. Klopfer, eds. *Perspectives in Ethology*, 8:55–83. New York: Plenum.

Gittleman, J. L., and Harvey, P. H. 1982. Carnivore home-range size, metabolic needs, and ecology. *Behav. Ecol. Sociobiol.* 10:57–63.

204   *John L. Gittleman*

Gorman, M. L. 1979. Dispersion and foraging of the small Indian mongoose, *Herpestes auropunctatus* (Carnivora: Viverridae) relative to the evolution of social viverrids. *J. Zool.* 187:65–73.
Guggisburg, C. A. W. 1975. *Wild Cats of the World.* New York: Taplinger.
Hamilton, W. D. 1971. Geometry for the selfish herd. *J. Theor. Biol.* 31:295–311.
Harvey, P. H., and Clutton-Brock, T. H. 1985. Life history variation in primates. *Evolution* 39:559–581.
Harvey, P. H., and Greene, P. J. 1981. Group composition: An evolutionary perspective. In: H. Kellerman, ed. *Group Cohesion: Theoretical and Clinical Perspectives,* pp. 148–169. New York: Grune & Stratton.
Harvey, P. H., and Mace, G. M. 1982. Comparisons between taxa and adaptive trends: Problems of methodology. In: King's College Sociobiology Group, eds. *Current Problems in Sociobiology,* pp. 343–361. Cambridge: Cambridge Univ. Press.
Herrero, S. 1978. A comparison of some features of the evolution, ecology, and behavior of black and grizzly/brown bears. *Carnivore* 1:7–17.
Hoogland, J. L. 1979. Aggression, ectoparasitism, and other costs of prairie dog (Sciuridae: *Cenomys* spp.) coloniality. *Behaviour* 69:1–35.
Hoogland, J. L. 1981. The evolution of coloniality in white-tailed and black-tailed prairie dogs (Sciuridae: *Cenomys leucrus* and *C. ludocianus*). *Ecology* 62:252–272.
Jarman, P. J. 1974. The social organization of antelope in relation to their ecology. *Behaviour* 48:215–256.
Jarman, P. J., and Jarman, M. V. 1979. The dynamics of ungulate social organization. In: A. R. E. Sinclair and M. Norton-Griffiths, eds. *Serengeti: Dynamics of an Ecosystem,* pp. 185–220. Chicago: Univ. Chicago Press.
Johnsingh, A. J. T. 1982. Reproductive and social behaviour of the dhole, *Cuon alpinus* (Canidae). *J. Zool.* 198:443–464.
Kaufmann, J. H. 1962. Ecology and social behavior of the coati, *Nasua narica,* on Barro Colorado Island, Panama. *Univ. California Publications in Zoology* 60:95–222.
Kaufmann, J. H. 1974. The ecology and evolution of social organization in the kangaroo family (Macopodidae). *Am. Zool.* 14:51–62.
Kingdon, J. 1977. *East African Mammals.* 3A. *Carnivores.* New York: Academic Press.
Kleiman, D. G., and Brady, C. 1978. Coyote behavior in the context of recent canid research: Problems and perspectives. In: M. Bekoff, ed. *Coyotes,* pp. 163–186. New York: Academic Press.
Kleiman, D. G., and Eisenberg, J. F. 1973. Comparisons of canid and felid social systems from an evolutionary perspective. *Anim. Behav.* 21:637–659.
Krebs, J. R., MacRoberts, M. H., and Cullen, J. M. 1972. Flocking and feeding in the great tit *Parus major*—an experimental study. *Ibis* 114:507–530.
Kruuk, H. 1972. *The Spotted Hyena: A Study of Predation and Social Behavior.* Chicago: Univ. Chicago Press.
Kruuk, H. 1975. Functional aspects of social hunting in carnivores. In: G. Baerands, C. Beer, and A. Manning, eds. *Function and Evolution in Behaviour,* pp. 119–141. Oxford: Oxford Univ. Press.
Kruuk, H. 1978. Spatial organization and territorial behaviour of the European badger, *Meles meles. J. Zool.* 184:1–19.
Kruuk, H., and Macdonald, D. 1985. Group territories of carnivores: Empires and enclaves. In: R. M. Sibly and R. H. Smith, eds. *Behavioural Ecology,* pp. 521–526. Oxford: Blackwell Scientific Publications.
Kruuk, H., and Parish, T. 1982. Factors affecting population density, group size and territory size of the European badger. *J. Zool.* 196:31–39.
Lamprecht, J. 1978. On diet, foraging behaviour and interspecific food competition of jackals in the Serengeti National Park. *Z. Säugetierk.* 43:210–233.

Lamprecht, J. 1981. The function of social hunting in larger terrestrial carnivores. *Mamm. Rev.* 11:169–179.

Lee, A. T., and Cockburn, A. 1985. *Evolutionary Ecology of Marsupials.* Cambridge: Cambridge Univ. Press.

Lindström, E. 1986. Territory inheritance and the evolution of group-living in carnivores. *Anim. Behav.* 34:1825–1835.

Lynch, C. D. 1980. Ecology of the suricate, *Suricata suricatta* and yellow mongoose, *Cynictis penicillata* with special reference to reproduction. *Memoirs Van Die Nasionale Museum.* 14:1–145.

Macdonald, D. W. 1976. Food caching by red foxes and some other carnivores. *Z. Tierpsychol.* 42:170–185.

Macdonald, D. W. 1978. Observations on the ecology and behaviour of the striped hyaena, *Hyaena hyaena*, in Israel. *Israel J. Zool.* 27:189–198.

Macdonald, D. W. 1979. The flexible social system of the golden jackal, *Canis aureus. Behav. Ecol. Sociobiol.* 5:17–38.

Macdonald, D. W. 1983. The ecology of carnivore social behaviour. *Nature* 301:379–384.

Macdonald, D. W., and Moehlman, P. D. 1982. Cooperation, altruism, and restraint in the reproduction of carnivores. *Persp. Ethol.* 5:433–467.

Mace, G. M., Harvey, P. H., and Clutton-Brock, T. H. 1983. Vertebrate home range size and metabolic requirements. In. I. Swingland & P. J. Greenwood, eds. *The Ecology of Animal Movement*, pp. 32–53. Oxford: Clarendon Press.

McNab, B. K. 1980. Food habits, energetics, and the population biology of mammals. *Amer. Nat.* 116:106–124.

Malcolm, J. 1979. Social organization and the communal rearing of pups in African wild dogs (*Lycaon pictus*). Ph.D. dissert., Harvard Univ., Cambridge, Mass.

Malcolm, J., and van Lawick, H. 1975. Notes on wild dogs hunting zebras. *Mammalia* 39:231–240.

Martin, R. D. 1981. Field studies of primate behaviour. *Symp. Zool. Soc. London.* 46:287–331.

Mech, L. D. 1966. *The Wolves of Isle Royale.* Fauna Series 7. Washington, D. C.: Government Printing Office.

Mech, L. D. 1970. *The Wolf: the Ecology and Behavior of an Endangered Species.* New York: Natural History Press.

Menzel, E. W. 1971. Communication about the environment in a group of young chimpanzees. *Folia Primatol.* 15:220–232.

Messier, F. 1985a. Social organization, spatial distribution, and population density of wolves in relation to moose density. *Canadian J. Zool.* 63:1068–1077.

Messier, F. 1985b. Solitary living and extraterritorial movements of wolves in relation to social status and prey abundance. *Canadian J. Zool.* 63:239–245.

Mills, M. G. L. 1978. The comparative socio-ecology of the Hyaenidae. *Carnivore* 1:1–7.

Mills, M. G. L. 1985. Related spotted hyaenas forage together but do not cooperate in rearing young. *Nature* 316:61–62.

Moehlman, P. D. 1983. Socioecology of silverbacked and golden jackals (*Canis mesomelas* and *Canis aureus*). In: J. F. Eisenberg & D. G. Kleiman, eds. *Advances in the Study of Mammalian Behavior*, pp. 423–453. Special Publication no. 7. Lawrence, Kans., American Society of Mammalogists.

Moehlman, P. D. 1986. Ecology of cooperation in canids. In: D. I. Rubenstein & R. W. Wrangham, eds. *Ecological Aspects of Social Evolution*, pp. 64–86. Princeton, N.J.: Princeton Univ. Press.

Nel, J. A. J., Mills, M. G. L., Van Aarde, R. J. 1984. Fluctuating group size in bat-eared foxes (*Otocyon megalotis*) in the south-western Kalahari. *J. Zool.* 203:294–298.

Nudds, T. D. 1978. Convergence of group size strategies by mammalian social carnivores. *Amer. Nat.* 112:957–960.

Owens, D. D., and Owens, M. J. 1984. Helping behaviour in brown hyaenas. *Nature* 308:843–845.

Packer, C. 1986. The ecology of sociality in felids. In: D. I. Rubenstein & R. W. Wrangham, eds. *Ecological Aspects of Social Evolution*, pp. 429–451. Princeton, N.J.: Princeton Univ. Press.

Packer, C., and Pusey, A. E. 1982. Cooperation and competition within coalitions of male lions: Kin selection or game theory? *Nature* 296:740–742.

Packer, C., and Pusey, A. E. 1983a. Adaptations of female lions to infanticide by incoming males. *Amer. Nat.* 121:716–728.

Packer, C., and Pusey, A. E. 1983b. Male takeovers and female reproductive parameters: A simulation of oestrous synchrony in lions (*Panthera leo*). *Anim. Behav.* 31:334–340.

Pulliam, H. R., and Caraco, T. 1984. Living in groups: Is there an optimal group size? In: J. R. Krebs & N. B. Davies, eds. *Behavioural Ecology* (2nd ed.), pp. 122–147. Sunderland, Mass.: Sinauer.

Rasa, O. A. E. 1977. The ethology and sociology of the dwarf mongoose, *Helogale undulata rufula*. *Z. Tierpsychol.* 43:337–407.

Rasa, O. A. E. 1986. Coordinated vigilance in dwarf mongoose family groups: The "watchman's song" hypothesis and the costs of guarding. *Z. Tierpsychol.* 71:340–344.

Rood, J. P. 1974. Banded mongoose males guard young. *Nature* 248:176.

Rood, J. P. 1975. Population dynamics and food habits of the banded mongoose. *East African Wildl. J.* 13:89–111.

Rood, J. P. 1978. Dwarf mongoose helpers at the den. *Z. Tierpsychol.* 48:277–287.

Rood, J. P. 1980. Mating relationships and breeding suppression in the dwarf mongoose. *Anim. Behav.* 28:143–150.

Rood, J. P. 1983. The social system of the dwarf mongoose. In: J. F. Eisenberg & D. G. Kleiman, eds. *Advances in the Study of Mammalian Behavior*, pp. 454–488. Special Publication no. 7. Lawrence, Kans.: American Society of Mammalogists.

Rood, J. P. 1986. Ecology and social evolution in the mongooses. In: D. I. Rubenstein & R. W. Wrangham, eds. *Ecological Aspects of Social Evolution*, pp. 131–152. Princeton, N.J.: Princeton Univ. Press.

Russell, J. K. 1983. Altruism in coati bands: Nepotism or reciprocity? In: S. K. Wasser, ed. *Social Behavior of Female Vertebrates*, pp. 263–290. New York: Academic Press.

Schaller, G. B. 1972. *The Serengeti Lion: A Study of Predator-Prey Relations*. Chicago: Univ. Chicago Press.

Schaller, G. B., Jinchu, H., Wenshi, P., and Jing, Z. 1985. *The Giant Pandas of Wolong*. Chicago: Univ. Chicago Press.

Sunquist, M. E. 1981. The social organization of tigers (*Panthera tigris*) in Royal Chitwan National Park, Nepal. *Smithsonian Contrib. Zool.* 336.

Underwood, R. 1982. Vigilance behaviour in grazing African ungulates. *Behaviour* 79:81–107.

Van Orsdol, K. G. 1984. Foraging behaviour and hunting success of lions in Queen Elizabeth National Park, Uganda. *African J. Ecol.* 22:79–99.

Van Orsdol, K. G., Hanby, J. P. and Bygott, J. D. 1985. Ecological correlates of lion social organization (*Panthera leo*). *J. Zool.* 206:97–112.

Van Valkenburgh, B. 1985. Locomotor diversity within past and present guilds of large predatory mammals. *Paleobiology* 11:406–428.

von Schantz, T. 1984. Carnivore social behaviour—does it need patches? *Nature* 307:389–390.

Ward, P., and Zahavi, A. 1973. The importance of certain assemblages of birds as "information-centres" for food finding. *Ibis* 115:517–534.

Waser, P. M. 1981. Sociality or tarritorial defense? The influence of resource renewal. *Behav. Ecol. Sociobiol.* 8:231–237.

Waser, P. M., and Waser, M. S. 1985. *Ichneumia albicauda* and the evolution of viverrid gregariousness. *Z. Tierpsychol.* 68:137–151.

Wells, M. C., and Bekoff, M. 1982. Predation by wild coyotes: Behavioral and ecological analyses. *J. Mamm.* 63:118–127.

Wilson, E. O. 1975. *Sociobiology: The New Synthesis.* Cambridge: Harvard Univ. Press.

Wrangham, R. W. 1975. The behavioural ecology of chimpanzees in Gombe National Park. PhD. thesis, Cambridge Univ., Cambridge, England.

Wrangham, R. W. 1977. Feeding behaviour of chimpanzees in Gombe National Park, Tanzania. In: T. H. Clutton-Brock, ed. *Primate Ecology*, pp. 504–538. London: Academic Press.

Wyman, J. 1967. The jackals of the Serengeti. *Animals* 10:79–83.

Zimen, E. (ed.) 1980. *The Red Fox.* The Hague: Dr. W. Junk.

PART II

ECOLOGY

INTRODUCTION

Of the three parts in this volume, this, on ecology, is unique in that it not only conveys conceptual advances but also revolutionary techniques for the study of carnivores. Technological improvements in live trapping, radio tagging, aerial radiotelemetry, and spotting scopes have allowed more accurate investigations of home range movements, territoriality, denning habits, hunting behavior, social interactions, and a wide range of other features fundamental to carnivore ecology (see Mech 1974, 1983; Amlaner and Macdonald 1980). Further development in merging telemetric techniques with physiological methods will vastly increase our knowledge of the physiological capacity of carnivore species, the ability of carnivores to metabolically adjust to new habitat conditions and, most important, the effectiveness of conservation and management strategies.

Carnivores, as their name implies, are closely tied to dietary effects, and most of the chapters in this part examine the variety of such effects. Schaller, Qitao, Johnson, Xiaoming, Heming, and Jinchu, in the first comparative study of the giant panda (*Ailuropoda melanoleuca*) and Asiatic black bear (*Ursus thibetanus*) show the ecological influences of feeding on a nutritionally limited diet of bamboo, as in the case of the panda, versus feeding on a more diverse herbivorous diet, as in the case of the black bear. Although this chapter appears more specialized than others, it makes two important general points. First, it illustrates that to analyze carefully the feeding ecology, population dynamics, and home range movements of a carnivore species, researchers will find it instructive to use a comparative field approach by looking at sympatric species. Second, it shows that the comparative dietary efficiency of foraging on different foods may govern the activity cycle, movements, and reproduction of a carnivore.

Estes continues with a physiological theme, but rather than stressing dietary effects, he considers how a "terrestrial" carnivore deals with the aquatic en-

209

vironment. In his review Estes critically synthesizes information on the perception, physiology, locomotion, life histories, and behavior of the otters, drawing conclusions and raising questions that should encourage other workers to examine this ecologically rich yet relatively unstudied group of carnivores. These studies improve our understanding of why particular species successfully radiated into aquatic environments and may perhaps suggest new insights into the vexing comparative phylogeny of carnivores and pinnipeds.

Sunquist and Sunquist consider how prey characteristics, habitat, scavengers, and predatory behavior influence predation in large felids. This chapter also uses a comparative approach to predict interactive effects among feeding ecology, habitat utilization, and social behavior.

King evaluates the unusual niche filled by small carnivores, primarily species of *Mustela*. Because of their small size, weasels face different ecological constraints than do larger carnivores. The relatively high population density, wide geographical distribution, and comparative ease of doing research on captive animals make smaller mustelids an ideal group for future studies.

McNab pulls together the dietary effects discussed in previous chapters by showing that, on the basis of measurements of basal metabolic rate and body size, correlates of metabolism and diet may serve as powerful predictors of carnivore reproduction and population biology. Given these predictions, the final chapter by Oftedal and Gittleman includes a speculative discussion on the reproductive output of carnivores. Specifically, the authors present theoretical analysis and some empirical support for the idea that in carnivores various reproductive parameters such as growth rate, litter size, and birth weight are closely tied to milk quality, which in turn is influenced by dietary efficiency.

Although the contributions by McNab and by Oftedal and Gittleman establish the need to merge physiological and ecological perspectives, all of the data in this area are from captive animals. The next major advance in understanding the ecology of carnivores may lie with yet another methodological technique for pulling together physiological and dietary features of natural populations of carnivores. Perhaps application of the doubly labeled water technique, already being ardently used in studies of bat and rodent ecology (Kenagy 1987; Gittleman and Thompson 1988; Kunz and Nagy 1988), will prove useful for future studies.

JOHN L. GITTLEMAN

References

Amlaner, C. J. Jr., and Macdonald, D. W. eds. 1980. *A Handbook on Biotelemetry and Radio-Tracking*. Oxford: Pergamon Press.

Gittleman, J. L., and Thompson, S. D. 1988. Energy allocation in mammalian reproduction. *Amer. Zool.* In press.

Kenagy, G. J. 1987. Energy allocation for reproduction in the golden-mantled ground squirrel. In: A. Loudon & P. Racey, eds. *Reproductive Energetics in Mammals,* pp. 259–273. Oxford: Oxford Univ. Press.

Kunz, T. H., and Nagy, K. A. 1988. Methods of energy budget analysis. In: T. H. Kunz, ed. *Ecological and Behavioral Methods for the Study of Bats.* Washington, D.C.: Smithsonian Institution Press.

Mech, L. D. 1974. Current techniques in the study of elusive wilderness carnivores. *Proc. XIth Internat. Congress Game Biol.* 11:315–322.

Mech, L. D. 1983. *A Handbook of Animal Radio-Tracking.* Minneapolis: Univ. Minnesota Press.

CHAPTER 8

The Feeding Ecology of Giant Pandas and Asiatic Black Bears in the Tangjiahe Reserve, China

GEORGE B. SCHALLER, TENG QITAO, KENNETH G. JOHNSON, WANG XIAOMING, SHEN HEMING, AND HU JINCHU

The Asiatic black bear (*Ursus thibetanus*) has a wide though patchy distribution from Iran, Afghanistan, and Pakistan eastward along the Himalayas to Indochina and across China to northeastern Russia. By contrast, the giant panda (*Ailuropoda melanoleuca*) survives only along the mountainous eastern edge of the Tibetan plateau, confined to an area totaling about 29,500 km² mainly in China's Sichuan province, but also southern Gansu and Shaanxi provinces (Figure 8.1). In these mountain forests pandas and black bears of the subspecies *U.t. mupinensis* (Ma 1983) are sympatric.

The giant panda and the Asiatic black bear are much alike, being solitary carnivores of similar body build and size. Both are mainly herbivores (Bromlei 1973; Schaller et al. 1985), in spite of the fact that they have the short, relatively unspecialized digestive tract of carnivores. Lacking the microbial digestion in rumen or caecum typical of most herbivores, they are unable to break down cellulose and other structural carbohydrates composing the cell walls of plants. Most plants consist primarily of cell walls and water; therefore, the animals derive their nutrition principally from cell solubles (sugars, starches, lipids, protein). Because solubles represent only a small fraction of a plant, much bulk must be consumed to fulfill daily nutritional requirements. This chapter examines the strategies of panda and black bear for living not only as herbivores but also for doing so sympatrically.

In the Wolong Natural Reserve, where we first conducted panda studies, we noted that both bears and pandas seasonally forage on shoots of one bamboo (*Fargesia spathacea*). (Yi [1985] renamed this bamboo *Fargesia robusta,* but we retain the old name to avoid confusion with our previous publications.) But because bears were only sporadic visitors to our study area, we obtained no data on the amount of ecological overlap or possible competition between these two similar species. Such overlap might consist of spatial use of an area, of a similar daily activity schedule within that area, and most important, of the same food habits. The panda subsists primarily on bamboo and, in fact, has evolved two specializations for processing this plant efficiently: the forepaws

212

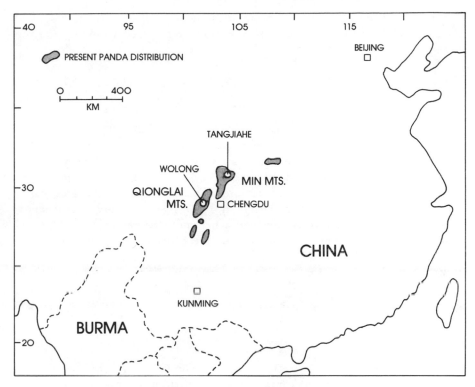

Figure 8.1. The present distribution of the giant panda. (Adapted from Chu and Long 1983.)

are adapted for grasping bamboo stems through the addition of a sixth digit or "thumb"—an enlarged wrist bone—and the posterior premolars and molars are broad and flat, modified for crushing bamboo (Davis 1964). The black bear lacks such morphological specializations. Its diet is more varied than the panda's, if data from Russia (Bromlei 1973), India (Schaller 1977), and China (Wu 1983) are indicative, consisting of forbs, fruits, and nuts. Feeding adaptations and strategies determine how an animal meets its nutritional requirements for maintenance, growth, and reproduction, and these, in turn, affect movements, activity cycles, and other aspects of existence. To what extent do the panda and the black bear overlap ecologically within their area of sympatry?

We studied both species in the Tangjiahe Natural Reserve of northern Sichuan (Figure 8.1). The research continued through 1987; this chapter, based on work conducted between March 1984 and March 1985, is limited to a preliminary discussion of the variety, abundance, dispersion, seasonality, and nutritional quality of panda and black bear foods, and to the effects of such variation in food supply on the behavior of these sympatric species. The Tangjiahe study is an extension of the cooperative China–World Wildlife Fund panda project, initiated in 1980 in the Wolong Natural Reserve, as described by Schaller et al. (1985). All data from Wolong in this report were taken from that publication.

The Study Area

The Tangjiahe Natural Reserve, established in 1978, lies in the Min Mountains of northern Sichuan bordering Gansu province. It extends over about 300 km² of rugged ridges and narrow valleys at elevations of from 1200 to 3800 m. Of the two main drainages in the reserve, we selected about 75 km² of the upper Beilu valley as our general study area. The eastern end of this area was at Maoxiangba, the reserve headquarters at 1420 m. From there a road winds westward for 14 km up the Beilu valley, where near the mouth of the Hongshi valley it becomes impassable to vehicles. We concentrated our activities in about 17.5 km² of the upper Beilu and lower Hongshi valleys between elevations of 1520 and 2300 m; our research base was located there at 1760 m (Figure 8.2).

Vegetation

The vegetation shows a vertical zonation similar to that of the Wolong Reserve, but being over 200 km farther north, each of the three zones extends about 300 m lower on the slopes. (1) An evergreen and deciduous broadleafed forest occurs below 1700 m. The evergreen trees *Lindera communis* and *Cyclobalanopsis oxyodon* are prominent, as are the deciduous beech (*Fargus longipetiolata*) and oak (*Quercus glandulifera*). (2) Between 1700 and 2100 m a mixed coniferous and deciduous broadleaved forest predominates, although various species from the previous zone persist, especially on south-facing slopes. Several species of maple (*Acer*), *Litsea, Hydrangea,* and *Viburnum* are common, as is birch (*Betula utilis, B. alba-sinensis*) and cherry (*Prunus sericea, P. brachypoda*); there are evergreen rhododendrons, ranging in size from low shrubby species to trees; and, among the conifers, pine (*Pinus armandii*) favors dry, southern exposures, and hemlock (*Tsuga chinensis*) and spruce (*Picea brachytyla*) moist, northern ones. In valley flats and on lower slopes, where humus is deep, lush forb meadows thrive, providing an important source of bear food. (3) A subalpine coniferous zone, with hemlock and spruce at lower elevations and fir (*Abies faxoniana*) higher up, begins at 2100–2300 m, depending on exposure, and extends to timberline; rhododendron and birch are the main broadleafed trees. On some slopes forest gives way to tussock grassland at only 2500 m, but generally the upper limit of tree growth is approximately 3200–3300 m and appears to be edaphically determined.

Bamboo is a critical resource for pandas. Since the taxonomy of bamboo in the Min Mountains remains unsettled, we follow the terminology of Yi (1985),

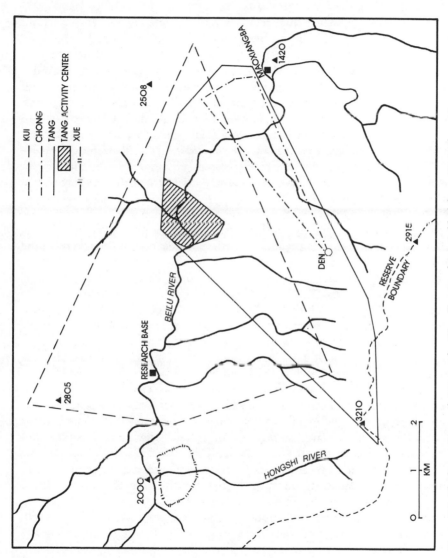

Figure 8.2. The home ranges of giant pandas (Xue, Tang) and Asiatic black bears (Kui, Chong) in the Tangjiahe Natural Reserve. The numbers refer to elevations in meters.

who examined material from Tangjiahe. There is little bamboo below 1600–1700 m in our study area, except for patches of *Bashania (Indocalamus) fargesii* (E. G. Camus) Keng f. et Yi around the mouth of the Shiqiao valley at 1580 m. The dominant bamboo from 1600–1700 m to 2200–2300 m is *Fargesia scabrida* Yi, which may cover whole hillsides in the Hongshi and some other valleys. This bamboo has flowered and died in patches between 1972 and the present, with a peak die-off in 1975. The die-off affected the bamboo primarily on the lower slopes of some valleys, leaving ridge tops and other valleys, such as the Hongshi, almost unaffected. This bamboo, therefore, exists as a mosaic, ranging in height from tiny seedlings to decade-old growth averaging less than 1 m tall to adult stands. From about 2100–2200 m and upward to timberline is a third bamboo, a small-leafed species closely resembling *F. nitida*, which has been named *F. denudata* Yi. It flowered so extensively in the mid-1970s that only some stands on ridges above 2600 m remained unaffected; pandas have been largely deprived of this resource since the die-off. At least two other species—*F. rufa* Yi and *Phyllostachys* species— occur locally in Tangjiahe, neither of importance to our study.

Human activity has greatly modified the vegetation in many areas below 2200 m, especially in the main valleys. The impact has been more severe on the relatively gentle southern and western exposures than on the more precipitous northern and eastern slopes. Cultivation was once extensive (and persisted downstream of Maoxiangba until 1986). Long-abandoned fields are now overgrown with trees (*Populus* spp.) and shrubs (species of *Salix, Spiraea, Deutzia, Rosa, Rubus*); recent fields are covered densely with forbs, conspicuously the tall *Artemisia subdigitata. Buddleia davidii*, willow, and other shrubs have replaced forest in the main valleys and some side ones. Roads were built into several valleys between 1965 and 1978 and timber extracted from all accessible parts, slopes being either clear-felled or selectively logged for conifers. Some slopes have been replanted with pines. This habitat destruction has affected both pandas and black bears. For pandas the impact has been entirely negative. Formerly cultivated slopes remain virtually devoid of bamboo, and clear-felling of timber has resulted in dense bamboo thickets without tree canopy which are little favored by pandas. Logging also removes conifers 1 m or more in diameter that, if hollow at the base, could serve as maternity dens; we saw no potential den trees in our study area. The impact on black bears has been mixed. Stands of oak, an important food source, have been decimated, often leaving only fringes of trees along ridge crests; however, the secondary growth on disturbed sites provides bears with *Rubus* species and other fruit.

Climate

Winter lasts from November to March, months during which temperatures dropped below freezing at our camp. The coldest month was January, with an average daily maximum of 1.8°C and minimum of −5.3°C; the absolute maximum was 8°C and minimum −11°C. The first flower of spring—a primula—was seen on March 18, and the last snow of winter fell on March 23. During April and early May forbs grew rapidly, trees leafed out, and rhododendrons blossomed. It rained on at least 15 days every month between May and October; 93.5% of the total 12-month precipitation of 1130 mm fell during this rainy season. June to August were the warmest months, with an average daily maximum of 22.6–24.7°C and minimum of 13.1–14.2°C; the absolute maximum was 30°C and minimum 10°C. The first yellow leaves were evident in mid-September, and a month later autumn coloring was at its peak. Most deciduous trees had shed their leaves by mid-November. December 14 brought the first heavy snowfall to the valley.

Methods

Since we observed black bears only eight times and pandas 26 times, most of our data are based on examinations of feeding sites, droppings, and other spoor. Monthly samples of panda droppings were analyzed to determine food selection. *Fargesia scabrida* bamboo samples were collected monthly at two sites for nutritional analysis. Both sites were at 2000 m, one of mature bamboo and the other of seedlings about nine years old. (Positive species identification of seedlings has not yet been made.) Collecting methods and analyses of feeds and droppings follow those described in Schaller et al. (1985). Bear droppings were given an ocular examination, and the percentage of each major component estimated. To determine bamboo shoot and stem mortality rates, 48 plots (2 m² and 4 m²), totaling 112 m², were established in unflowered *F. scabrida* in and near the Hongshi valley. Sites differed in degree of slope, altitude, exposure, and percentage of canopy. The plots were established in March–April 1984; at that time all new stems (shoots of 1983) were marked and all old stems (two years old and older) were counted. From July to September, during the shoot-growing season, plots were visited at least once every month and shoot mortality noted. Dead stems were tallied in October and again in March 1985, completing one annual cycle.

In 1984 we captured two pandas and two black bears in traps baited with goat meat, sedated the animals either with CI-744 or ketamine hydrochloride, radio-collared them, and subsequently monitored their movement and activity using equipment and techniques as reported in Schaller et al. (1985). One panda, Xue, a middle-aged female without infant weighing 67.3 kg, was caught on 14 December in one of the seven boxtraps of logs we had built in the Hongshi valley. The other panda, Tang, an adult male, was collared on 8 June, having been captured in a cave into which he retreated when we surrounded his bamboo thicket; on 1 June 1985, when we replaced his collar with a new one, he weighed 67.7 kg. Both pandas were considerably lighter than adults in Wolong, where two females weighed 86 and 89 kg and two males, 97 and 107 kg. Both bears were caught in Aldrich foot snares. Kui, an adult male of undetermined weight—Bromlei (1973) found that adults generally range from 100 to 150 kg—was captured on 23 July. Chong, a subadult male estimated to be almost three years old, was snared on 5 November; he weighed 70.5 kg. In contrast to the placid pandas, both bears were aggressive, lunging at us and roaring.

Radio-collared pandas were usually contacted daily, except that we seldom found Tang after he moved out of his usual haunts during the summer. Chong's signal was often received during the month between his capture and hibernation, whereas Kui's was sometimes lost for weeks when he traveled out of the reserve. We monitored activity on a 24-h basis every 15 min, 96 signal readings per day. Tang was monitored for 19 days, Xue for 9, Chong for 7, and Kui for 3.

Giant Panda

We estimated that about ten pandas frequented our 7 by 2.5 km main study area, some animals only part time. (Quoting other sources, Schaller et al. [1985] gave a population estimate of 100–140 pandas for Tangjiahe; in 1985 a census coordinated by Hu Jinchu revealed 50–60 pandas, a figure that agrees with our impressions formed in 1984.) Of these, Tang spent much of the year at 1500–1600 m in *Bashania fargesii* bamboo, whereas Xue and others were in *F. scabrida*, usually above 2000 m, and some seasonally high up in *F. denudata*. The three bamboo species differ in their annual cycle of shoot production; this in turn affects the movements, food habits, and nutrient intake of pandas.

B. fargesii is a large-leaved bamboo, 2–3 m tall, with stems up to 1–1.5 cm thick at the base. New shoots appear in mid-April, and these are almost fully grown by mid-June. *F. scabrida* averages about 1.8 m high, with some stems 3–4 m tall, although stands on dry, logged slopes may average only 1 m. Stems rarely reach a basal thickness of 1 cm, most being about half that. There was a mean of 27.5 stems/m^2 on our plots, and the mean above-ground biomass (fresh weight)—based on sampling seven plots of 1 m^2—was 1479 g/m^2. New shoots appear in mid-July and reach full height in late September. Pandas also eat *F. scabrida* seedlings after they reach a height of 40 cm or more. A 1 m^2 plot of seedlings, estimated to be nine years old, yielded 172 stems with a biomass of 609 g. Mean stem height was 58 cm, with the tallest stem 157 cm; mean stem diameter was 0.2 cm. *F. denudata* resembles *F. scabrida* in stem height and thickness. Although it grows at higher elevations than *F. scabrida*, its shoots appear from mid-June through July.

Food Habits

Pandas may consume plants other than bamboo, and they also eat meat when available (Hu 1981; Schaller et al. 1985). At Tangjiahe an animal once

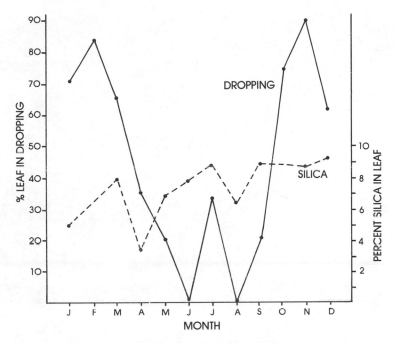

Figure 8.3. The percentage of bamboo (*Fargesia scabrida*) leaves in panda drop-pings (*n* = 333) each month compared with silica content of fresh leaves. We lack data for February silica. In August and September pandas ate mainly shoots.

ate the juicy stems of wild parsnip (*Heracleum moellendorffii*)—a favored bear food—and once chewed on old skin and leg bones of a tufted deer (*Elaphodus cephalophus*). *F. scabrida* bamboo was the principal food of most pandas. However, animals showed marked seasonal preferences for certain parts of the plant (Figure 8.3). They preferred leaves over stems from October to March in a ratio of 2:1. By mid-April animals had decreased their leaf consumption markedly, and this trend continued until June, when they avoided leaves. They again ate leaves during July. Several of our study animals also ascended into *F. denudata* that month. On 30 July 45 droppings of *F. denudata* at 2800 m were composed of 21% new shoot and 79% leaf. During August and September animals consumed primarily new *F. scabrida* shoots.

Tang, living mainly in *B. fargesii* bamboo, displayed a different pattern of dietary preference. From September to April he ate leaves almost exclusively. On 26 October we observed him for 1 h (Figure 8.4). He either bent stems with a forepaw toward his muzzle and ate the leaves off, or he detached the stem with a bite and, holding it upright with one forepaw, pushed the leafy branches into his mouth with the other. He did not select stems of particular age: after 2.5 days in the observed bamboo patch, he had eaten at least some leaves from 68% of the stems. During May and early June, Tang foraged on *B. fargesii* shoots. After shoots had grown tall and hard, he abandoned the valley for

Figure 8.4. The male panda Tang feeds in *Bashania fargesii* bamboo.

ridges, first to the east and then south. We found him on 4 July at 1990 m in *F. scabrida*. His droppings revealed a diet of 23% leaf and 77% stem. In early September he returned to his usual *B. fargesii* haunts. (D. Reid told us that Tang also spent June–September 1985 at high elevations.)

Wolong pandas usually peel the enveloping sheaths off *F. spathacea* shoots before eating them, perhaps because sheaths of this species are unusually hairy. By contrast, Tangjiahe animals often consume the shoots of all species without peeling them. In Wolong pandas also select thick shoots, ≥ 1 cm in diameter. Shoots are seldom that thick in Tangjiahe. Taking the *F. scabrida* shoots in our plots as a sample, we found that most shoots eaten were 0.7–0.8 cm thick, a size Wolong pandas seldom consumed. There was, however, selection against thinner shoots. As in Wolong, insect predators took a significantly greater proportion of thin shoots (diameter ≤ 0.7 cm) than did pandas (Figure 8.5). Rodents selected much like pandas. Tang displayed a similar preference for thick *B. fargesii* shoots, selecting for shoots ≥ 0.9 cm and against those ≤ 0.8 cm. It appears that pandas merely chose the thickest shoots in an area, with a lower limit of about 0.6 cm.

There was much shoot predation. One or more shoots were destroyed in 96% of our plots by insects, in 42% by pandas, and in 35% by rodents. Of the 996 shoots produced, 43.5% were destroyed between July and September, a mortality figure similar to that in Wolong (Table 8.1). Nevertheless, the num-

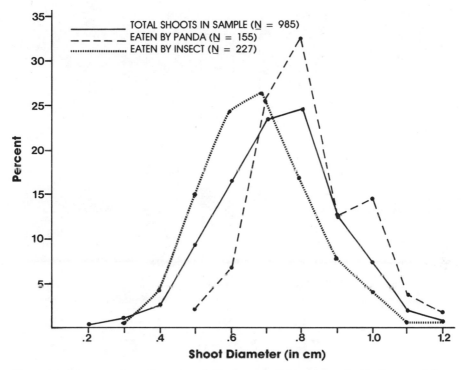

Figure 8.5. Diameter of new *Fargesia scabrida* shoots available compared with diameter of shoots eaten by pandas and by insects. The Vanderploeg and Scavia's selectivity coefficient and Chi-square analysis indicated that pandas prefer shoots from 0.61 to 1.20 cm in diameter and insects prefer shoots from 0.31 to 0.70 cm in diameter ($p \leq 0.001$). (See Schaller et al. 1985.)

ber of bamboo stems increased during the year. There were 3028 stems in the plots in April 1984, of which 256 (8.5%) died during the following 12 months. Of the shoots produced in 1984, 535 survived, an increment of 279 new stems (9.2%). This increase appeared due to a large number of 1984 shoots: only 3 shoots per m² of the 1983 crop survived to the age of one year, whereas 4.8 shoots per m² of the 1984 crop did so.

Table 8.1. Destruction of new bamboo shoots by predators in Wolong and Tangjiahe

	Wolong: *Fargesia spathacea*		Tangjiahe: *F. scabrida*
Predator	1982 n = 724 shoots	1983 n = 209 shoots	1984 n = 996 shoots
Insects	15.6%	12.0%	22.8%
Giant pandas	12.0	23.9	15.6
Rodents and *Ochotona* sp.	3.7	6.2	4.5
Others	4.0	0.5	0.6
Total	35.3%	42.6%	43.5%

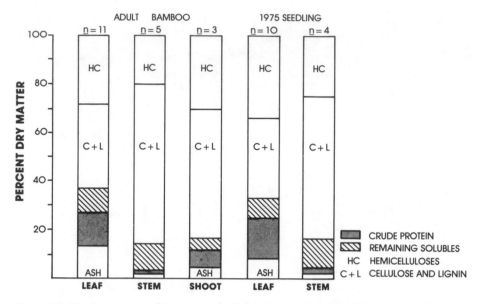

Figure 8.6. Nutrient content of *Fargesia scabrida* bamboo (adult and seedling), expressed as percentage of dry matter. Shoot data are based on shoots collected in July, August, and September at different growth stages.

Nutritional Content of Bamboo

Approximately 90% of protein, carbohydrates, and other cell solubles are nutritionally available to pandas (Dierenfeld et al. 1982). Of the cell wall components, cellulose and lignin are indigestible; however, pandas can break down a fraction of the hemicelluloses—digesting 18–26%, depending on season (Schaller et al. 1985).

F. scabrida leaves from both adult bamboo and seedlings have more protein, ash (minerals and salts), and hemicelluloses and less cellulose and lignin than do stems (Figure 8.6). *F. denudata* and *B. fargesii* leaves are chemically similar to *F. scabrida* (Table 8.2). New *F. scabrida* shoots are of lower average nutritional quality between July and September than are leaves; the percentage of total cell solubles in shoots is similar to that in stems (Figure 8.6). A tall *B. fargesii* shoot, collected on June 1, had 8.7% crude protein and 5.5% other cell content, similar to *F. scabrida* shoots.

All bamboo species in Tangjiahe retain green leaves throughout the year. Monthly samples show that the chemical composition of bamboo leaves and stems remains quite constant at all seasons, as illustrated for crude protein in *F. scabrida* leaves (Figure 8.7). The protein level in new *F. scabrida* shoots decreases, however, as shoots grow and harden until it is similar to that in stems; there is a concomitant increase in the percentage of cellulose.

Table 8.2. Chemical composition of bamboo leaves eaten by pandas and forbs eaten by bears

	Monthly samples (*n*)	Ash	Hemicelluloses	Cellulose and lignin	Crude protein	Remaining solubles
			(% of dry matter)			
Bamboo leaves						
Fargesia scabrida	11	13.4	29.2	34.4	12.6	10.5
F. denudata	1	10.8	35.1	27.3	15.8	11.0
Bashania fargesii	3	12.0	29.1	34.5	16.1	8.3
Forbs						
Mean of 12 species	—	16.3 ± 5.4	5.7 ± 3.0	28.9 ± 11.3	16.8 ± 8.7	32.3 ± 6.5

Wolong pandas seldom ate the leaves of *Sinarundinaria fangiana* bamboo between April and June. Silica (SiO_2) levels in leaves reached their highest levels (4–5%) during those months, then dropped to low levels from July to October, a period when pandas selected for leaves. In an attempt to explain this change in food selection, Schaller et al. (1985) suggested that silica—which can inhibit digestion (Van Soest 1982)—may be implicated. Our data from Tangjiahe do not support this idea. Although Tangjiahe pandas selected against *F. scabrida* leaves also from April to June, silica levels fluctuated little during the year and were always higher than at Wolong, even during the months when animals favored leaves (Figure 8.3). The average annual silica level was 7.4% in leaves and 0.1% in stems. The reason why pandas avoid leaves in spring remains unexplained.

The water content of *F. scabrida* leaves and stems is 40–60%, and in shoots

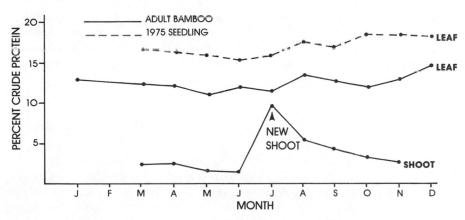

Figure 8.7. The percentage of crude protein in *Fargesia scabrida* bamboo leaves and shoots, by month.

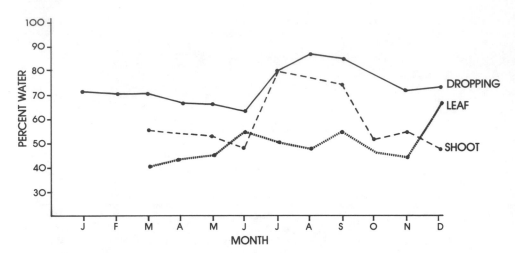

Figure 8.8. Monthly water content of panda droppings ($n = 315$) compared with water content of *Fargesia scabrida* leaves and shoots.

it is at least 75%; water content of droppings is about 70% when animals forage on leaves and stems, and 75–85% or more when they are on a shoot diet (Figure 8.8). Pandas thus eliminate more water in their droppings than they obtain from their food, except for a month or two each year when they eat shoots. Moist droppings may be essential for smooth and rapid passage of the coarse forage through the digestive tract. Oxidation of feeds forms metabolic water, but, in addition, pandas may need to drink at least once a day (Ruan and Yong 1983).

In sum, bamboo contains high levels of indigestible cellulose and lignin (35–65%) and partially digestible hemicelluloses (20–35%), and low levels of readily available nutrients as part of the cell contents (12–24%), making it a food of poor quality. But since nutritive content remains constant all year, bamboo represents a predictable food source. Analyses of two Wolong bamboo species gave results similar to those from Tangjiahe, as did analysis of two introduced species from Washington, D.C. (Dierenfeld 1981).

Activity

Most of a panda's day is devoted either to resting or to collecting, preparing, and eating bamboo; other activities, such as traveling and grooming, consume only about 4% of the day. In Wolong 300 days of 24-h activity monitoring of several individuals showed that animals may be active or inactive at any time of day or night. Pandas were, on the average, inactive for 9.8 h (41.6%) of the day. Part of this time was devoted to one or two long rest periods lasting 2–4 h or more. Pandas were active for 14.2 h (58.4%) of the day, a figure that remained relatively constant throughout the year. Daily activity reached its lowest level between 0800 and 0900h and after 1900h, and its highest level between 0400 and 0600h and between 1600 and 1900h. These activity peaks near dawn and at dusk were similar all year, regardless of amount of daylight. Although individuals showed no significant differences in activity levels, they often displayed idiosyncratic patterns seemingly unrelated to age, sex, or other obvious factors.

In Tangjiahe winter data for Xue showed that her average probability of activity in January was 0.53, in February 0.68 and in March 0.36 for a mean of 0.52, similar to Wolong animals (0.58). She had two activity peaks, one between 0200 and 0300h, and the other between 1900 and 2100h; her low activity between 0700 and 0900h was similar to that of Wolong animals (Figure 8.9).

Tang was less active than any panda we have monitored in Wolong and Tangjiahe: his average probability of activity was only 0.43. There was little variation between September and March (December excluded because of lack of data), average probability of activity ranging from 0.37 to 0.46; in June, the only other month during which we monitored Tang, the figure was an exceptionally active 0.79 from only one day's data. Tang's sedentary habits for most of the year and his preference for leaves—leaves are less time-consuming to eat than stems—probably account in part for his restful existence. His 24-h pattern was also unusual in that he tended to be inactive between 2100 and 0800h and showed only one prolonged daily peak between 1200 and 1800h (Figure 8.9). He sometimes rested for long periods. On 14 October, for instance, he began a rest at 1830h that lasted until 1215h on 15 October (nearly 18 h), so long that we became concerned about his health; after feeding, he also slept during the following night from 2100 to 0730h (10.5 h), yet appeared healthy.

Land Tenure

Home range sizes of six Wolong pandas varied from 3.9 to 6.4 km². Ranges of males were only as large as, or slightly larger than, those of females. Even though its range was small, an animal visited some parts only rarely; the amount of total range used each month seldom exceeded 25%. Ranges were stable and shared all or in part with other pandas. Land tenure appeared to be different in males and females: males occupied greatly overlapping ranges, whereas each female spent most of her time in a discrete core area of only 30–40 ha.

We lack detailed data on land tenure in Tangjiahe pandas. Xue confined her activities to 1.3 km² from mid-December to March (Figure 8.2), and within this small area she moved little (Table 8.3). However, that summer she shifted at least 3 km southeast to some high ridges (D. Reid, pers. comm.).

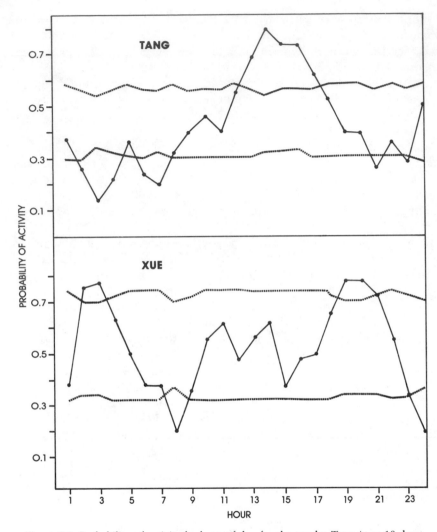

Figure 8.9. Probability of activity by hour of day for the pandas Tang (*n* = 18 days, September–March) and Xue (*n* = 9 days, January–March). The confidence intervals are indicated by broken lines. Values differing significantly (*P* ≤ 0.01) from the average probability of each animal lie outside these lines.

Tang had a range of at least 23.1 km². an unusually large area. However, within that range he had a center of activity, 1.1 km² large, in which he remained for nine months from September to June (Figure 8.2), using generally less than 1% of his total range each month (Table 8.3). And within his center of activity he was usually found in one of four bamboo patches whose total area comprised only about 5 ha; between September and March he was radio-located on 134 of 182 days (74%) in these patches. His sedentary habits

Table 8.3. Monthly variations in known home range used by two pandas

	Xue	Tang
September 1984	—	0.03%
October	—	0.07
November	—	0.03
December	0.20%	0.41
January 1985	0.30	0.32
February	0.40	0.63
March	1.20%	1.83%

Note. The total range for Xue, a female, was 1.3 km²; for Tang, a male, 23.1 km².

during the mating season from mid-March to early May surprised us, for no other pandas were in his center of activity, to our knowledge. Tang traveled widely during summer. In June, for example, he moved over 3 km down-valley and over a ridge to the east, where we lost contact with him; he had returned by July and remained south of his usual haunts, for a while outside of the reserve. Although some Wolong pandas shifted seasonally to lower elevations to feed on bamboo shoots, they remained within their small ranges, showing a pattern quite different from that of Tang and apparently also Xue. Tang's shift away from *B. fargesii* occurred at a time when that bamboo seemed to become less palatable. The unusual extent of Tang's movements cannot be explained solely on the basis of a food search, although much travel was necessary to reach the patchy bamboo remnants on the ridges.

Asiatic Black Bear

Bears were not abundant in Tangjiahe, though the many branches they broke while feeding in trees were conspicuous reminders of their presence. Only an estimated 10–12 bears frequented the slopes of the Beilu valley up-river from Maoxiangba, and none seemed to remain there permanently.

Food Habits

The feeding of Tangjiahe bears falls into three periods: from April to mid-July the bears eat mainly forbs and leaves from shrubs, from mid-July to mid-September they add fruits to their diet as soon as these ripen, and from mid-September to November they harvest primarily acorns and other nuts.

Spring provides bears with a variety of succulent forb species. Of 15 spring foods listed in Table 8.4, bears particularly favored the thick stalks of wild parsnip (*Anthriscus sylvestris, Heracleum* sp., *Angelica* sp.), *Petasites tricholobus* leaves, new *Hydrangea* growth, and *Rubus coreanus* shoots. Bears seldom lingered in a valley at this season, ignoring all but a few plant species in

228 *George B. Schaller et al.*

Table 8.4. Wild food plants eaten by black bears
in the Tangjiahe Reserve

Species	Part eaten
Acanthopanax henryi	New stem growth, leaf
Actinidia chinensis	Fruit
Angelica sp.	Succulent stalk
Anthriscus sylvestris	Succulent stalk
Arisaema lobatum	Succulent stalk
Aster ageratoides	Leaf
Cacalia tangutica	Leaf
Caraya sp.	Fruit
Celtis biondii	Fruit
Cnidium sp.	Succulent stalk
Cornus chinensis	Fruit
Corylus sp.	Nut
Cyclobalanopsis oxyodon	Acorn
Fargesia scabrida	Shoot
Heracleum moellendorffii	Succulent stalk
H. scabridum	Succulent stalk
Hydrangea sp.	New stem growth, leaf
Juglans cathayensis	Nut
Lunathyrium giraldii	Young frond
Petasites tricholobus	Leaf
Phlomis sp.	Leaf
Prunus brachypoda	Fruit
P. sericea	Fruit
Quercus aliena	Acorn
Q. glandulifera	Acorn
Q. spinosa	Acorn
Rubus coreanus	Fruit, new shoot
Salvia umbratica	Leaf

their travels. On 22 May one bear stopped to feed 18 times while moving 1 km along the base of a slope. It plucked some leaves and branch tops from one spiny *Acanthopanax henryi* and two *Hydrangea* shrubs; and it consumed jack-in-the-pulpit (*Arisaema lobatum*) at one site, *Heracleum* species at a second, and *Angelica* species at a third, each time removing the leafy tops before eating the stalks. However, this bear seemed to prefer *R. coreanus* shoots about 1 cm thick, which sprout much like bamboo shoots. The bear broke or bit off shoots, removed the leafy tassel at the top, and ate the juicy stalk, sometimes after peeling off the densely haired skin. Another bear angled down to the base of a long slope and foraged in a patch of *Heracleum* plants before continuing across the valley and up the other side, with only a brief halt to eat one *Arisaema lobatum* stalk, a total of two feeding stops in 1.5 km.

Forbs remained an important food all summer, sometimes augmented with bamboo (*F. scabrida*) shoots in August (D. Reid, pers. comm.). From mid-July to late August *R. coreanus* berries were frequently eaten (Table 8.5). The wild cherry (*Prunus* sp.) crop failed in 1984; judging by broken branches, bears had harvested fruits in previous years.

Table 8.5. Seasonal variation in the diet of black bears, as estimated from contents of droppings

Month	Droppings (n)	Type of food (%)			
		Leaves and stalks	*Rubus* berries	Other fruits	Acorns and other nuts
April to mid-July	9	100	—	—	—
Mid-July to mid-September	25	75	25	—	—
Mid-September to November	46	5	—	5	90

Bears had left the forb and berry patches of the upper valleys by mid-September and concentrated below 1800 m in oak stands on ridges and slopes. Acorns (*Quercus* sp.) supplemented by hazelnuts (*Corylus* sp.), butternuts (*Juglans cathayensis*), and fruits (*Celtis biondii, Actinidia chinensis*) composed the diet until hibernation. On 24 October, at 1615h, we watched a bear feeding in an oak. Squatting or standing in a fork, the bear pulled small branches toward itself with a forepaw, sometimes breaking one with a bite, and plucked the acorns directly with its mouth. Large branches required more effort. The bear pulled them with both forepaws, occasionally using mouth as well, until they broke or snapped off. The animal usually pushed discarded branches into the fork beneath its feet and stepped on them, creating a crude platform (Figure 8.10). Bears in India harvest the cherry-like *Celtis* fruits in a similar manner (Schaller 1969).

Almost every oak in Tangjiahe has broken branches, some so damaged that little beyond the tree trunk and branch stumps remain. Bears have seriously reduced the acorn supply of future years. By late October most acorns have fallen, and bears search for them beneath trees.

Bears no doubt ate more than the 28 species listed in Table 8.4. Grapes (*Vitis* sp.) and rosehips (*Rosa* sp.) are eaten in India (Schaller 1977), beech buds in Japan (Hazumi 1985), and *Pinus* seeds, *Ribes* berries, *Carex* and *Lilium* leaves and stalks in Russia (Bromlei 1973). These genera are also present in Tangjiahe. Wu (1983) listed nine genera of fruits and nuts as bear food in the Qinling Mountains of Shaanxi province, 300 km northeast of our study area. Among these were *Fragaria, Schisandra, Coriaria, Flaeagnus, Rhus* and *Castanea*, most, if not all, of which occur in Tangjiahe too.

Domesticated plants were also consumed in Tangjiahe. In late August and September bears took apples from orchards in the reserve and maize and walnuts from farms.

Meat, either killed or scavenged, forms a small percentage of a bear's total diet, and may include mammals, birds, fish, mollusks, and insects (Schaller 1969; Bromlei 1973; Wu 1983). One dropping in Tangjiahe contained takin (*Budorcas taxicolor*) hair, and one in nearby Wanglang Reserve an infant bamboo rat (*Rhizomys sinense*).

Bears in Tangjiahe and northeastern Russia have similar food habits and show similar seasonal changes in selecting forbs, fruits, and nuts. Bromlei

Figure 8.10. Wang Xiaoming sits on a crude platform of broken oak branches made by an Asiatic black bear while feeding on acorns. The panda Tang centered his activity in the valley below.

(1973) listed 27 genera of wild food plants for bears in Russia, and ten of these were also eaten in Tangjiahe.

Nutritional Content of Plants

The indigestible cellulose and lignin values were moderately high (mean 29%) in the various forbs, whereas the partially digestible hemicelluloses were low (6%). Ash content was high (16%). Protein varied from 2.6% in *Angelica* species to 35.2% in *Hydrangea* species, with an average of 17% for all forbs tested. The remaining cell contents (sugars, lipids) comprised a mean of 31% (Table 8.2). *Rubus* and *Actinidia* fruits were chemically similar to the other plants except that they were low in ash (Figure 8.11). Bears shell acorns with their mouths and eat only the kernels; our nutritional analyses were conducted on kernels only. The two acorn species differed in nutrient content from forbs and berries: they were much lower in cellulose and lignin (mean 6%) and

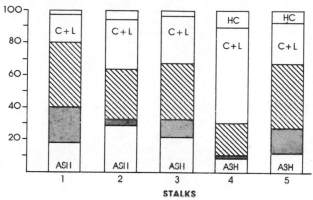

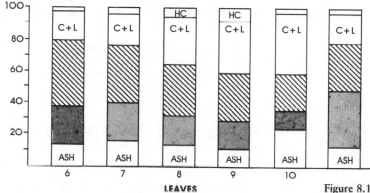

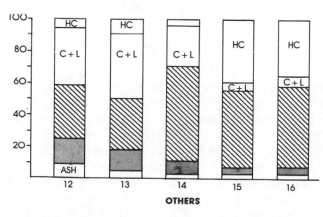

Figure 8.11. Nutrient content of bear food plants, expressed in percentage of dry matter. *Top row:* Stalks—(1) *Heracleum scabridum,* (2) *H. moellendorffii,* (3) *Anthriscus sylvestris,* (4) *Angelica* sp., (5) *Arisaema lobatum. Middle row:* Leaves—(6) *Petasites tricholobus,* (7) *Phlomis* sp., (8) *Cacalia tangutica,* (9) *Acanthopanax henryi,* (10) *Salvia umbratica,* (11) *Hydrangea* sp. *Bottom row:* Others— (12) *Rubus Loreanus* shoot, (13) *R. coreanus* berry, (14) *Actinidia chinensis* fruit, (15) Acorn (*Quercus aliena*), (16) Acorn (*Q. glandulifera*). Most forbs were collected in May, except *H. moellendorffii* on June 18, and *S. umbratica* on 14 July.

Table 8.6. Essential amino acid content
(in mg/100 g) of 11 spring plants
eaten by black bears

Amino acid	Content (*mean*)
Threonine	0.64 ± 0.41
Valine	0.71 ± 0.35
Methionine	0.15 ± 0.08
Isoleucine	0.60 ± 0.29
Leucine	1.06 ± 0.58
Phenylalanine	0.60 ± 0.35
Lysine	0.69 ± 0.46

Note. Tryptophan not tested. The plants tested were species of *Acanthopanax, Anthriscus, Arisaema, Aster, Cacalia, Cnidium, Hydrangea, Lunathyrium, Petasites, Phlomis,* and *Heracleum scabridum.*

moderately high in hemicelluloses (38%). Ash was low (3%). Protein was low as well (5%), whereas the remaining cell contents were higher (49%). Cell contents of acorns, as that of many seeds, contain much fat, 9% in *Quercus ilex* from northern Pakistan (Schaller 1977). Acorns thus represent a high-calorie, digestible plant food for bears.

Silica content of foods was low, averaging 0.2% for leaves and fruits and none in acorns.

The essential amino acid content of 11 spring food plants was analyzed. Only methionine showed a conspicuously low level (Table 8.6). This amino acid is usually deficient in plants, including bamboo (Schaller et al. 1985).

Forbs had a mean water content of 90% and *Rubus* berries of 86%; droppings from April to mid-September were soft, with a water content of 87%. During these months bears probably had no need to drink. The water balance changed in autumn. Acorns contained 52% water and droppings 73%—figures similar to those of pandas foraging on bamboo leaves and stems.

Activity

Kui's activity was monitored for 3 days in July, and Chong's for 6 days in November, the month prior to his hibernation. The two bears showed similar daily cycles (Figure 8.12): they became inactive in the evening, Chong at 1900h and Kui at 2100h, and spent much of the night at rest. Kui's probability of activity was similar to Chong's (0.48 vs. 0.46) even though their diet differed, the former foraging for forbs and fruit and the latter for acorns.

Bears usually rested on the ground, in the manner of pandas, but occasionally an animal built a bed. Three such beds, constructed of bamboo, were observed in Wolong. In Tangjiahe a bear climbed a spur in late afternoon to a

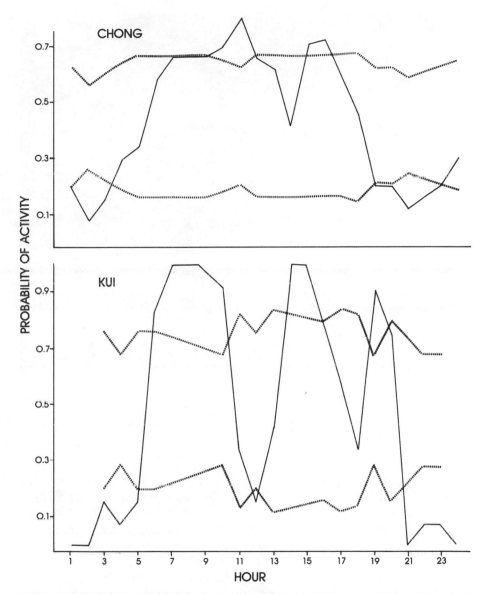

Figure 8.12. Probability of activity by hour of day for the bears Chong (n = 6 days, November) and Kui (n = 3 days, July). The confidence intervals are indicated by broken lines.

small, level spot. There it bent in a total of 51 bamboo stems and 13 beech saplings up to 3 cm thick and tucked them around and under its body; it either broke long stems to fit the rim, or, in the case of nine bamboos and 12 saplings, bit off the tops. The result of the effort was a springy, circular bed 130 cm in diameter (Figure 8.13). A total of 5.5 kg of droppings, consisting solely of forbs, were just outside the rim.

Figure 8.13. Teng Qitao examines a bed of bamboo and saplings constructed on the ground by an Asiatic black bear.

Bears in Tangjiahe, as in other temperate climates (Bromlei 1973; Maita 1985), hibernate during winter. Chong traveled last on 28 November, and he was active for part of the day, apparently in and around a den, from 29 November to 6 December. He became almost inactive on December 8; 24-h monitoring on that date revealed a probability of activity of 0.13. On 13 December we monitored his signal from 0915 to 1500h without recording any activity. On 26 March his signal indicated movement within the den—a rock cleft—and he emerged during the first week of April. In Wolong we found a hibernating yearling in the hollow base of a fir at 2600 m as late as 12 April. Late March to mid-April is the general emergence period in the region, according to local people, indicating a hibernation period of about four months.

Land Tenure

During the month before hibernation Chong meandered in search of acorns and then entered a den for four months, all within an area of 6 km². Kui roamed widely and often rapidly, covering at least 29 km² between July and September, excluding one or more trips outside the reserve to the south, where we did not track him (Figure 8.2); in October and November we could not find him. Our fragmentary data give no indication of the whole range of these males. In Japan the annual range of a subadult female was 6 km² and the autumn range of an adult male was 26 km² (Maita 1985).

Discussion

Giant pandas and Asiatic black bears showed moderate ecological overlap in activity schedules and use of space. Both were active in daytime; at night the bears tended to rest and pandas to feed intermittently. Ranges of the two species overlapped. However, animals were spatially separated in autumn, bears usually foraging below 1800 m and pandas above that elevation. And in winter the bears hibernated. There was almost no overlap in food habits. A panda occasionally sampled *Heracleum* species, a major bear food, and bears ate bamboo shoots. No competition existed for these resources, since pandas seldom ate forbs and bamboo shoots were seasonally abundant, providing ample forage for both species.

The panda has an extraordinarily limited diet, usually only one or two kinds of bamboo. But it has specialized on a plant resource that is available in unlimited amounts at all seasons. Furthermore, the nutritional quality of bamboo remains fairly constant throughout the year, an unusual situation for a plant growing in an environment with marked seasonal changes. Forage in temperate climates usually declines in nutritive value during winter, and in arid climates during the dry season. This lack of fluctuation in both amount and quality has enabled the panda to subsist on bamboo in spite of a low nutritive content.

Unable to digest cellulose and lignin in the cell wall, a panda must obtain nutrients from cell contents and from the fraction (av. 22%) of the hemi-celluloses it can break down. Total dry matter digestibility is only 17%, calculated from Wolong data, as compared with at least 60% from an ungulate eating green grass (Van Soest 1982). Such low digestibility implies that the panda must quickly move much bulk through its digestive tract to obtain needed nutrients. Indeed, passage time of bamboo is less than 14 h (Dierenfeld 1981) and as little as 5 h on a diet of bamboo shoots. The panda also needs to harvest bamboo selectively for the most nutritious parts. Stepwise discriminant analysis of bamboo parts eaten and not eaten in Wolong showed that cellulose was the best single variable for discriminating favored from other parts. Bamboo leaves are higher in protein, minerals, and hemicelluloses and lower in cellulose and lignin than are stems. They can also be collected and eaten quickly. One would, therefore, expect pandas to favor leaves. Animals feeding on *F. scabrida* selected leaves over stems (57% to 43%) on an annual basis.

Some stems were eaten at all seasons, possibly because the amount of soluble carbohydrates in the cell contents is similar in leaves and stems; for unknown reasons, leaves seemed to be avoided from April to June. Shoots were an almost exclusive feed in season, even though they rank nutritionally lower than leaves. Hemicellulose digestibility in young, growing tissue is higher than in mature tissue, and pandas perhaps select for the additional energy provided by these hemicelluloses.

Wolong pandas subsisting on *Sinarundinaria fangiana* bamboo obtained an estimated 4354 kcal/day of digestible energy in spring, 5488 kcal/day in summer and autumn, and 5542 kcal/day in winter. On one occasion in October, Tang foraged in a small *B. fargesii* patch for 2.5 days, permitting us to calculate his caloric intake. He defecated an average of 12.9 kg (wet weight) or 3.3 kg (dry weight) of droppings per day, the droppings composed wholly of leaves. His daily dry matter intake was, therefore, 3.3(100)/100 − 23.25 (where 23.25% is the autumn dry matter digestibility in Wolong), or 4.3 kg/day. Converted to fresh weight of leaves eaten, the figure is 4.3/100 − 45.5 (where 45.5% is water content of leaves), or 7.9 kg/day. Dierenfeld et al. (1982) found that gross energy of *Phyllostachys* species leaf was 4800 kcal/kg, and that of *B. fargesii* leaf is probably similar. On a dry matter intake of 4.3 kg, Tang ate 20,640 kcal, but with a 23.25% digestibility he obtained only 4799 kcal/day of digestible energy. This figure is somewhat lower than the autumn one calculated for Wolong. Tangjiahe adults, however, are a quarter to a third lighter than those in Wolong, if our small weight sample is characteristic. At 68 kg Tang has, for example, an energy expenditure for basal metabolism of 1658 kcal/day, as compared to 2214 kcal/day for a 100-kg Wolong male.

The total average energy expenditure of a 100-kg Wolong panda was calculated at approximately 3132 kcal/day, a figure that does not include energy invested in growth and reproduction. With an intake of only approximately 4300−5500 kcal/day, the nutritional margin of safety is fairly small. Although pandas have reduced energy expenditures to a minimum by resting much and, when active, concentrating on feeding, they are still constrained by low food quality and limited in food intake by the capacity of the digestive tract. Consequently, a panda can obtain only enough digestible energy to store a small amount of fat. But with a stable food source of constant quality and no hibernation period, a panda has no need for large fat deposits.

Conditions for the animals change, however, when at long intervals (40−80 or more years, depending on species) bamboo in an area flowers and dies synchronously. Pandas then are forced to forage in patchy remnants or switch to other bamboo species at different elevations for 5−10 years until new seedlings have grown tall enough to provide food. If alternative bamboo species are not available, the animals may starve. This happened in parts of the Min Mountains during the mid-1970s when at least three species flowered, and in certain areas of the Qionglai Mountains in 1983 when *S. fangiana* flowered.

Asiatic black bears differ markedly from pandas in their feeding strategy. Though selective in the plants they eat, they choose a moderately large variety—over 28 species in Tangjiahe. Emphasis on type of food changes with seasonal availability: forbs in spring, fruit in summer as soon as available, and acorns in autumn. The chemical compositions of these feeds are different from those of bamboo. Forbs average three times more cell contents than bamboo, and they are low in hemicelluloses (Table 8.2). *Rubus* berries are similar to forbs in nutritive levels (Figure 8.11) but presumably have higher concentrations of soluble carbohydrates, thus making it nutritionally beneficial for bears to pluck these small items. Bunnell and Hamilton (1985) found a 88.4% dry matter digestibility of blueberries in brown bear (*Ursus arctos*), and, judging by constituents (Figure 8.11), *Rubus* berries probably have a digestibility of 50–60%. Forbs and berries present bears with nutritional problems also faced by pandas when eating bamboo shoots. Succulent foods have a low retention time, and the animal must extract the maximum amount of energy within a few hours. Furthermore, it must eat much bulk to obtain few nutrients because of the high water content.

Acorns contain less cellulose and lignin than any other food analyzed; most of the kernel provides digestible energy (Figure 8.11). One method of measuring dry matter digestibility is by using an internal marker to compare concentrations of an indigestible constituent in both food and feces. Lignin is such a marker in mature plants. Lignin in two acorn samples averaged 2.55% and in three droppings 8.5%, indicating a digestibility of 70.2%. Acorns require time to collect. And about 30% of an acorn consists of shell, which bears discard, leaving kernels averaging 0.4 to 1.3 g each, depending on species. To eat 5 kg, a bear would require about 3850 large acorns or 12,500 small ones. During its 11-h daily active period in November, a bear would have to gather 5.8 to 18.9 acorns per minute to reach these numbers, a feasible effort. Time constraint is probably the reason why bears rarely eat the tiny *Cyclobalanopsis oxyodon* acorns.

Gross energy of bear foods is similar to that of bamboo (4400–4800 kcal/kg), with *Rubus coreanus* shoots 4200 kcal/kg, *R. coreanus* berries 5600, and acorns 4400. Unfortunately, we lack data on daily food intake of bears. Nelson (1980) calculated that the American black bear (*Ursus americanus*) may assimilate as many as 20,000 kcal in a day. Most aspects of bear digestion—passage rates, efficiency, and so forth—appear to be similar to the panda's. However, bear foods generally contain at least three times more digestible energy than does bamboo. Thus, although the gross energies of feeds are much alike, bears obtain at least three times more digestible energy per kilogram eaten than do pandas. A bear's resources are patchy and its food items small, requiring extra energy for searching and feeding, but the animals still obtain such a daily surplus of calories that they can store enough fat for hibernation.

With digestive capabilities of the two species so similar, the question still remains why bears do not eat bamboo leaves and especially why pandas only sample some nutritious bear foods on occasion. The theory of optimum foraging predicts that an animal should attempt to obtain the best balance and amount of nutrients for the least amount of time and effort (Pianka 1978). Yet we have noted pandas walking through stands of parsnip without halting in this favored bear food, only to forage on low-quality bamboo nearby. For unknown reasons, pandas do not always select forage of the highest available quality even though they seem to have retained the ability to assimilate it, as well as to store fat. Meat, with a gross energy digestibility of about 90% in bears (Bunnell and Hamilton 1985), is, however, eaten readily by both species. Both have conserved their digestive adaptations for carnivory in spite of some different morphological and possibly physiological adaptation for herbivory.

The panda's emphasis on bamboo, even when other forage is available, is surprising, for a low-quality diet has affected many facets of the animal's life. With a diet that provides little more than subsistence, a panda must keep its digestive tract filled by foraging at frequent intervals both day and night. By contrast, bears rest much of the night. In spite of abundant and concentrated resources, a panda's time budget resembles that of a nonruminant herbivore, with 50% or more of the day usually spent foraging. The bears Chong and Kui required only 46–48% of the day to fulfill energy needs, store fat, and then hibernate four months. Pandas can subsist for months within 1 km^2, whereas bears require large ranges and much travel—Kui used over 29 km^2—to forage on their seasonal, patchy, and small food resources. During November, for example, Tang remained in 1 ha while Chong roamed over 600 ha. Yet their total activity levels were alike.

A panda's need to conserve energy extends to reproduction. Females do not conceive until 5.5 or 6.5 years of age, implantation is delayed one and a half to four months, and, after a true gestation period of about one and a half to two months, one or two young are born in such altricial state that they are proportionately the smallest of eutherian mammals, about 1/900 the weight of the mother. If there are twins, one young soon dies, the female raising just one offspring to an early independence at one and a half years. The American black bear is the ecological counterpart of the Asiatic one, judging by similar body size, habitat, denning activity, range size, and other aspects of behavior (Jonkel and Cowan 1971; Garshelis and Pelton 1980, 1981; Johnson and Pelton 1980; Pelton 1982). For example, it consumes primarily grasses and forbs in spring, fruits in summer, and a mixture of fruits and acorns and other hard mast in autumn (Pelton 1982). With high-quality food available in summer and autumn, the mean monthly probability of activity of bears in the Great Smoky Mountains of Tennessee—a habitat similar to that of Tangjiahe—reaches a high level (0.5–0.6) only from June to October, and animals hibernate from late December to early April (Quigley 1982). The bears are primarily diurnal, with daily activity peaks around dawn and dusk (Garshelis and Pelton 1980; Pelton 1982). American black bears may reproduce as early as 2.5 or 3.5 years of age in areas with a good nutritional base (Alt et al. 1980), newborns are about 1/200–1/300 the weight of their mother, and litters average 1.7–2.9 young, depending on area, a female often raising all her offspring. A high-quality diet gives the American black bear, and no doubt also the Asiatic black bear, greater reproductive flexibility and potential than the panda can achieve, at less expense in energy.

Although a difference in feeding ecology influences various types of behavior in bear and panda, it apparently has not had an impact on the basic land tenure system, if data from studies of American black bears are valid for comparison. Adults of both species share ranges, overlap between females' ranges is less than between males', and a female has a more or less discrete core area in which other females are not tolerated (Rogers 1977). The ranges of male bears are generally two to three times larger than those of females. For example, in Tennessee males' ranges averaged 42 km² and females' 15 km² (Garshelis and Pelton 1981); in Pennsylvania males' ranges averaged 173 km² and females' 41 km² (Alt et al. 1980); and in Idaho males' ranges averaged 112 km² and females' 50 km² (Amstrup and Beecham 1976). Both sexes of pandas have ranges of approximately similar size, at least in Wolong.

In conclusion, differences in feeding strategy have had a marked impact on the lifestyles of the Asiatic black bear and the panda. The bear has opted for a nutritional boom-or-bust economy: it stuffs itself on a variety of high-energy seasonal foods, storing excess calories as fat, and then hibernates during times of scarcity. Opportunistic and adaptable, Asiatic black bears are widely distributed. "The panda, by contrast, has become a specialist; dependent on a low-quality but constant and abundant food source, it has chosen security over uncertainty. Its mode of life gives the impression of being a durable triumph of evolution. But by losing the sense of struggle, its curiosity, its need to explore and be observant and try something new, by tying itself to a fate without horizon, it has become defenseless, it has lost the adaptability that it now must have to survive" (Schaller et al. 1985:224).

Acknowledgments

This project was financed through an agreement between China's Ministry of Forestry and the World Wildlife Fund. Wildlife Conservation International, a division of the New York Zoological Society, also provided financial support. We received guidance and encouragement from many officials, including Wang Menghu (Ministry of Forestry, Beijing), Hu Tieqin, Fu Chengjun, Gung Tongyang, and Bi Fengzhou (Forest Bureau, Chengdu), and Yue Zhishun and Jiang Mingdao (Tangjiahe Natural Reserve). A. Taylor (University of Colorado) assisted with the bamboo work and provided us with data, Qin Zisheng (Nanchong Normal College) identified the plants, D. Reid (University of Calgary) supplied some recent information, and K. Schaller, Qiu Mingjiang, and

240   *George B. Schaller et al.*

Wang Fulin assisted with field work. Nutritional analyses were done by J. Robertson, Department of Animal Sciences, Cornell University, and the amino acid determinations by Pan Wenshi, Department of Zoology, Beijing University. The Department of Forestry, Wildlife and Fisheries of the University of Tennessee provided computer facilities through the courtesy of M. Pelton. A. Taylor and E. Dierenfeld read the manuscript critically. We are deeply grateful to these individuals and institutions for this assistance.

References

Alt, G., Matula, G., Alt, F., and Lindzey, J. 1980. Dynamics of home range and movements of adult black bears in northeastern Pennsylvania. In: C. Martinka & K. McArthur, eds. *Bears—Their Biology and Management,* pp. 131–136. Washington, D.C.: Government Printing Office.
Amstrup, S., and Beecham, J. 1976. Activity patterns of radio-collared black bears in Idaho. *J. Wildl. Mgmt.* 33:340–348.
Bromlei, G. 1973. *Bears of the South Far-eastern USSR.* New Delhi: Indian National Scientific Documentation Centre.
Bunnell, F., and Hamilton, T. 1985. Forage digestibility and fitness in grizzly bears. *Internat. Conf. Bear Res. and Mgmt.* 5:179–185.
Chu Ching and Long Zhi. 1983. [The vicissitudes of the giant panda.] *Acta Zool. Sinica.* 29:93–104. (In Chinese.)
Davis, D. 1964. The giant panda: A morphological study of evolutionary mechanisms. *Fieldiana. Zoology Mem.* 3:1–339.
Dierenfeld, E. 1981. The nutritional composition of bamboo and its utilization by the giant panda. M.Sc. thesis, Cornell Univ., Ithaca, N.Y.
Dierenfeld, E., Hintz, H., Robertson, J., Van Soest, P., and Oftedal, O. 1982. Utilization of bamboo by the giant panda. *J. Nutr.* 112:636–641.
Garshelis, D., and Pelton, M. 1980. Activity of black bears in the Great Smoky Mountains National Park. *J. Mamm.* 61:8–19.
Garshelis, D., and Pelton, M. 1981. Movements of black bears in the Great Smoky Mountains National Park. *J. Wildl. Mgmt.* 45:912–925.
Hazumi, Toshihiro. 1985. Habitat selection of Japanese black bears in Nikko. Abstracts distributed at First Asiatic Bear Conference, Utsunomiya and Nikko, Tochigi Prefecture, Japan.
Hu Jinchu. 1981. [Ecology and biology of the giant panda, golden monkey, and takin.] Chengdu: Sichuan People's Publishing House. (In Chinese.)
Johnson, K., and Pelton, M. 1980. Environmental relationships and the denning period of black bears in Tennessee. *J. Mamm.* 61:653–660.
Jonkel, C., and Cowan, I. 1971. The black bear in the spruce-fir forest. *Wildl. Monogr.* 27:1–57.
Ma Yiching. 1983. The status of bears in China. *Acta Zool. Fennica.* 174:165–166.
Maita, Kazuhiko. 1985. Movements of Japanese black bears (*Selenarctos thibetanus japonicus*) in Taiheizan Mountain region, Akita Prefecture. Abstracts distributed at First Asiatic Bear Conference, Usunomiya and Nikko, Tochigi Prefecture, Japan.
Nelson, R. 1980. Protein and fat metabolism in hibernating bears. *Amer. Physiol. Soc. Fed. Proc.* 39:2955–2958.
Pelton, M. 1982. Black bear, *Ursus americanus.* In: J. Chapman & G. Feldhamer, eds. *Wild Mammals of North America,* pp. 504–514. Baltimore: Johns Hopkins Univ. Press.

Pianka, E. 1978. *Evolutionary Ecology*. New York: Harper & Row.

Quigley, H. 1982. Activity patterns, movement ecology, and habitat utilization of black bears in the Great Smoky Mountains National Park, Tennessee. M.Sc. thesis, Univ. Tennessee, Knoxville. 140 pp.

Rogers, L. 1977. Social relationships, movements, and population dynamics of black bears in northeastern Minnesota. Ph.D. dissert., Univ. Minnesota, St. Paul. 194 pp.

Ruan Shiju and Yong Yange. 1983. [Observations on feeding and search for food of giant panda in the wild.] *Wildlife* 1:5–8. (In Chinese.)

Schaller, G. 1969. Food habits of the Himalayan black bear (*Selenarctos thibetanus*) in the Dachigam Sanctuary, Kashmir. *J. Bombay Nat. Hist. Soc.* 66:156–159.

Schaller, G. 1977. *Mountain Monarchs: Wild Sheep and Goats of the Himalaya*. Chicago: Univ. Chicago Press.

Schaller, G., Hu Jinchu, Pan Wenshi, and Zhu Jing. 1985. *The Giant Pandas of Wolong*. Chicago: Univ. Chicago Press.

Van Soest, P. 1982. *Nutritional Ecology of the Ruminant*. Corvallis, Ore.: O & B Books.

Wu Jiayian. 1983. [Preliminary observations on the food specialization of black bear from Qinling.] *Chinese J. Zool.* 4:47–51. (In Chinese.)

Yi Tongpei. 1985. Classification and distribution of the food bamboos of the giant panda. *J. Bamboo Res.* 4:11–27; (2):20–45. (In Chinese.)

CHAPTER 9

Adaptations for Aquatic
Living by Carnivores

James A. Estes

Before considering the carnivores' adaptations for aquatic living, one must define what is meant by an "adaptation" as well as identify those species that are aquatic. Neither task is simple.

Nowak and Paradiso (1983) recognized 238 extant species among seven families in the order Carnivora, yet there is no clear and unambiguous distinction between aquatic and nonaquatic forms. Nearly all carnivores are excellent swimmers, but no one would consider them all to be aquatic on that basis alone. The extent to which the carnivores live in aquatic habitats ranges from the strictly terrestrial life style of some species to the clearly aquatic habits of others. Most species fall into the former category, although many of these (such as bears; foxes; raccoons; jaguars, *Panthera onca*) may feed on aquatic prey in some circumstances. Others, such as mink (*Mustela vison*) and water mongooses (*Atilax paludinosus*), typically associate with aquatic habitats, although since these associations are not obligatory, few mammalogists would consider them to be strictly aquatic forms. A few other species, such as the otters, are strictly aquatic in the sense that they are inseparably tied to aquatic habits. However, in contrast with the cetaceans and sirenians, all of which live entirely in aquatic habitats, even the most "highly adapted" aquatic carnivores, including the pinnipeds (see Preface), retain some dependence on land (or a suitable substitute, such as floating ice) during at least part of their life cycles. Typically, dependence on solid substrata is for reproduction, since even the pinnipeds do not give birth in the water. Among the carnivores there is but one species, the sea otter (*Enhydra lutris*), that can conduct all life functions in the aquatic environment. Although sea otters haul out on land to rest, this does not appear to be an obligate function and in some areas animals may spend their entire lives at sea.

A precise definition of "adaptations" to aquatic living is even more troublesome because this term requires that the characteristic or function in question resulted primarily from selection imposed by life in the aquatic environment. But evolutionary cause-and-effect relationships are difficult to demonstrate (Gould and Lewontin 1979). To show causation, one must exclude the possibility that some presumed "adaptation" did not originate for other purposes, or perhaps for no purpose at all. Gould and Vrba (1982) urged a revision in terminology to reflect the distinction between two avenues by which some characteristic of a species or group of organisms has come to serve to its advantage. They defined all beneficial characteristics of organisms as "aptations": "adaptations" if selection for the same purpose resulted in their evolution, and "exaptations" if it did not. Some modifications that typify aquatic carnivores and serve clearly to their advantage in aquatic living must be exaptations.

My approach to the problem of distinguishing between adaptations and exaptations for aquatic living in carnivores will be comparative since recurrent patterns across taxa often can be used to infer cause-and-effect relationships on evolutionary time scales. Such comparisons can be made two ways. One is through study of convergence by distantly related forms in the same environment. Obligatory aquatic living in eutherian mammals has arisen independently in the insectivores, rodents, otters (the Lutrinae), pinnipeds (possibly between phocid and otarioid pinnipeds as well), cetaceans, and sirenians. Recurrent patterns across these taxa likely represent adaptations for aquatic living. This is because splitting events are comparatively old, making it unlikely that similarities across taxa had similar origins. Another approach is the study of divergence by closely related forms in different environments. In this case splitting times are comparatively recent, so that differences among species are more likely to be adaptations to different habitats than would be the case if the same comparison were made among more distantly related species. The Mustelidae family is suited for such treatment because it contains numerous fully terrestrial species, one that is semi-aquatic (the mink), and a number that are more fully aquatic (the Lutrinae).

I shall arbitrarily define aquatic carnivores as those species with an obligate link to aquatic habitats. By this criterion aquatic carnivores consist exclusively of the otters and the pinnipeds (seals, sea lions, and walruses). My focus will be largely on the otters, although I draw from other groups of aquatic mammals to illustrate recurrent themes related to life in aquatic habitats. My discussion generally follows Repenning's (1976a) organizational scheme for the pinnipeds, in which he recognized eight categories of adaptations for living in the aquatic environment: vision, hearing, olfaction, feeding, oxygen conservation, heat conservation, locomotion, and behavior. To these, I add water balance and life history. Since function and form in aquatic mammals (especially the pinnipeds and cetaceans) have been extensively discussed before (e.g., Harrison 1974; Ridgway and Harrison 1981), I will give more attention to life history and behavior.

Vision

Underwater vision presents three fundamental problems to the mammalian eye. One of these, the need for increased light-gathering capacity, is encountered by species that forage in murky water or the much-reduced light intensities encountered at depth. A second is the need to accommodate the spectral shift in light quality toward the blue-green wavelengths that occurs underwater, especially in scotopic (dim light) vision encountered at night or at depth. A third is a need to modify the eye's light-focusing capacity underwater because of refractive differences that occur at the water-corneal compared with the air-corneal interface.

The eye could accommodate structurally in three possible ways to focus light on the retina: (1) by increasing corneal convexity, (2) by increasing the focusing capacity of the lens, and (3) by increasing the length of the eye. Data on visual acuity in air and in water from a number of terrestrial, semi-aquatic, and aquatic mammals have been summarized by Sinclair et al. (1974) and Schusterman (1981). Piscivorous cetaceans and pinnipeds have high visual acuity in water (Table 9.1), which is not surprising considering the visual requirements for capturing fish with the mouth while swimming at a high rate of speed. Visual acuity in water is somewhat reduced in the Oriental small-clawed otter (*Aonyx cinerea*) (Schusterman and Barrett 1973). This species apparently uses tactile sensitivity in the fore limbs to locate and capture invertebrate prey; thus both the nature of its prey and the principal mode of prey capture likely require less visual acuity underwater than is needed by piscivorous lutrines, such as *Lutra* or *Pteronura* species. Similarly, the visual acuity of the sea otter might be expected to be comparable to that of the Oriental small-clawed otter because the former feeds principally on invertebrates and depends largely on fore limb sensitivity for detection and capture of prey. Gentry and Peterson (1967) re-

Table 9.1. Visual acuities of mammals in air and water

	Visual angle (min.)	
Species	Air	Water
Cetaceans and Pinnipeds		
California sea lion (*Zalophus californianus*)	5.0–7.4	4.8–6.4
Harbor seal (*Phoca vitulina*)		8.3
Killer whale (*Orcinus orca*)		5.5
Pacific white-sided dolphin (*Lagenorhyncus obliguidens*)		6.0
Steller sea lion (*Eumetopias jubata*)		7.1
Mustelids		
Asian clawless otter (*Aonyx cinerea*)	13.6–15.6	14.7–14.9
Ferret (*M. furo*)	16.2	
Mink (*Mustela vison*)	15.1	31.4
Other terrestrial mammals		
Cat (*Felis domesticus*)	5.5	
Rat (*Rattus norvegicus*)	26.0	

Sources. Modified from Sinclair et al. 1974; Schusterman 1981.

ported that the underwater vision of the sea otter was poorer than that of the California sea lion (*Zalophus californianus*) and the harbor seal (*Phoca vitulina*), although they presented no data on the minimum angle of visual acuity.

Judging from the data for the California sea lion and the Oriental small-clawed otter, in bright light both pinnipeds and lutrines have comparable visual acuities in air and water, thus indicating that structural modifications to the eye have occurred in these groups which allow a high degree of flexibility in their focusing ability. It is likely that selection has favored the retention of high visual acuity on land in the pinnipeds and lutrines because important activities requiring vision occur in both terrestrial and aquatic habitats. If the need for sharp visual acuity on land were not required, the eye might have more readily changed to accommodate to the aquatic environment without having to maintain a high degree of flexibility. The visual acuity of the semi-aquatic mink in air is comparable to that of the strictly terrestrial congeneric domestic ferret (*Mustela furo*) and the Oriental small-clawed otter. However, the mink's visual acuity is much poorer in water than on land (Table 9.1).

Collectively, these data indicate that adaptations in visual acuity to the aquatic environment have developed rapidly in mammals, and that at least in the case of the pinnipeds and lutrines a high degree of flexibility for comparable visual acuity in air and water also has developed. However, this ability apparently has been achieved in different ways. In bright light pinnipeds have comparable visual acuities in air and in water. As light intensity is progressively reduced, visual acuity in water remains nearly unchanged until background luminance becomes very low (Schusterman 1972). However, visual acuity in air declines more rapidly over the same range of background luminance, thus demonstrating that the pinniped eye is primarily adapted for vision in water. High visual acuity in air is achieved by virtue of the eye's extreme sensitivity to light (Lavigne et al. 1975). Thus, at high light intensities (i.e., normal photopic levels), the iris is closed to a small aperture, which allows the eye to function as a pin-hole lens (Walls 1963; Schusterman 1972). Different patterns occur in otters, judging from work on Oriental small-clawed otter by Schusterman and Barrett (1973). In this species constant acuity is not maintained over a range of declining background luminance—both decline in air and water. This is likely because otter eyes are less sensitive to light than are pinniped eyes. Furthermore, in contrast to the pinnipeds, the otter experiences a rate of decline with reduced luminance that is greater in water than in air, which indicates that otter eyes function better in air than in water. The underwater focusing mechanisms of *Aonyx* species are unknown; however, in *Lutra* species the lens is distorted by well-developed sphincter and ciliary muscles (Gentry and Peterson 1967).

Visual requirements under reduced light intensity, as would occur in deep or murky water, or at night, might be accommodated by increased orbital size or increased retinal sensitivity. Pinnipeds, which often feed at night and in deep water, have enlarged orbits (Repenning 1976a). Northern elephant seals (*Mirounga angustirostris*), which are known to dive to depths beyond 800 m (Le Bouef et al. 1986; Le Boeuf, et al. in press), have exceedingly large eyes. However, the orbits of lutrines are not enlarged. This may be because they are typically shallow divers and thus have little need for increased light gathering, or it may be because of trade-offs resulting from a limited anterior skull space, required by large olfactory fossae associated with the maintenance of an acute sense of smell. Retinal sensitivity may be increased by the presence of a reflective tapetum lucidum posterior to the retina (highly developed in cetaceans and pinnipeds [Jamieson and Fisher 1972]) and by increased rod density. A shift in scotopic sensitivity correlated with spectral distributions of radiant energy in the environment is well known in fish (Lythgoe and Dartnall 1970; McFarland 1971; Lythgoe 1972) and is thought to occur in pinnipeds (Schusterman 1981). Otters' responses to change in light quality are unstudied.

Vision is important in aquatic mammals because other sensory modalities used by terrestrial species (most notably, olfaction) do not function underwater. Three general problems would seem to be fruitful areas for further research. One is the variation in magnitude of visual acuity required by piscivorous as opposed to nonpiscivorous forms. Comparative studies among the otters would be especially interesting in this regard since there is a dichotomy between piscivorous and nonpiscivorous forms, and the nonpiscivorous forms have twice arisen and are represented by extant species (i.e., sea otter and *Aonyx* species) (Berta and Morgan, 1986). The mechanisms by which aquatic mammals maintain adequate visual acuity in both air and water is a second problem. The development of such flexibility would seem to be a difficult task for the vertebrate eye. Further comparative studies between otters and pinnipeds should prove interesting since this ability has arisen independently in the two groups. The rodents, which seem to have poor visual acuity, may also be important subjects for comparative study since some aquatic species (e.g., the cricetines *Icthyomys* species; aquatic rat, *Anotomys leander;* *Rheomys* species, fish eating rat, *Neusticomys monticolus;* and the hydromurine *Hydromys* species) have developed piscivorous habits (Eisenberg 1981). The spectral sensitivity of visual pigments in otters and other aquatic mammals is a third problem.

Hearing

Underwater hearing presents three difficulties to the mammalian ear, which has been designed to perceive variations in tone, intensity, and direction of airborne sound. First, waterborne sounds are attenuated and altered at the water-air interface since the ear canal is not flooded when aquatic mammals are submerged. Second, deep diving subjects the ear to substantial and rapid pressure changes. Third, because waterborne sound is virtually unattentuated by soft tissues and bone, the normal mechanism of directional sound perception in the aerial environment, through directional orientation of the ear pinnae and differential stimulation of the tympana on either side of the head, is precluded in water.

Modifications for underwater hearing are best known from studies of pinnipeds and cetaceans but are largely unknown for other groups of aquatic mammals. Pinnipeds have several evident modifications that protect hearing structures from pressure changes. Most species have small tympanic membrane–oval window ratios (Repenning 1972), which are thought to protect the cochlea from very high hydrosonic pressure. There is a recess in the petrosal bone—the round window fossa—which is thought to protect the round window membrane from damage by the cavernous tissue that lines the middle ear in pinnipeds and distends with blood in response to increasing hydrostatic pressure. These modifications occurred in fossil enaliarctids, the earliest known pinnipeds (Repenning 1976a).

Directional perception of waterborne sounds is achieved by specifically oriented bone surfaces of the skull, since sound transmission to the inner ear is via bone. In addition, the petrosal apex is enlarged for greater sensitivity to bone-conducted sound, and the incomplete fusion between the inner ear and the bones of the skull may act as a muting device for bone-conducted sound from other parts of the skull.

Hearing modifications for deep diving and direction perception underwater apparently have resulted in reduced hearing ability by pinnipeds in the aerial environment (however, see Schusterman 1981). One might thus expect that unless underwater sound perception is a substantial advantage to the less highly modified aquatic mammals, adaptations for hearing waterborne sounds in these species would be undeveloped because of their need to retain acute hearing in the aerial environment. Little is known of the importance, sensitivity, and mechanism of hearing by otters, in either the aquatic or aerial environment.

Olfaction

Olfaction is an important sense in mammals (Ewer 1973), serving many species as the primary means of detecting predators, locating prey, and communicating with conspecifics. However, since olfactory sensory tissues are located in the nasal fossae, mammals can "smell" only airborne chemicals. It would thus be expected that because fully aquatic mammals have little need for highly developed olfactory abilities, these abilities would be reduced or modified. Patterns in production and perception of scent across increasingly aquatic carnivores support this view.

The mustelids produce scents that are highly perceptible even to humans. These scents are produced by the anal and proctodeal glands (Gorman et al.

1978; see Gorman and Trowbridge, this volume), which are modified apocrine sebaceous glands. According to Stubbe (1970), scent glands are less specialized in the European otter (*Lutra lutra*) than they are in other mustelids, and Tarasoff et al. (1972) reported anal glands to be absent in the sea otter. However, Trowbridge (1983) identified nearly 100 compounds in the scent of the European otter consisting of proteins, mucopolysaccharides, and lipids (Gorman et al. 1978). Some compounds, which may function in species-level recognition, were common to all individuals. Others, which were variable among individuals, may function in individual-level recognition. Scent is deposited in the feces, or "spraints," as they are more delicately referred to.

Scent marking serves important social functions for the European otter. Spraints are deposited by males at territory boundaries and by females around den sites (Hillegaart et al. 1981; Green et al. 1984). Sprainting, although more poorly studied in other species, has been reported for the Cape clawless otter (*Aonyx capensis*) and the spotted-necked otter (*Lutra maculicollus*) (Mason and Macdonald 1986), the river otter (*L. canadensis*) (Melquist and Hornocker 1983), and the giant river otter (*Pteronura brasiliensis*) (Duplaix 1980). Gorman et al. (1978) reported that sprainting and scent production was highly concordant among captive male and female European otters and was closely correlated with the female estrous cycle.

The significance of sprainting and scent production by the sea otter is unclear. Sea otter spraints have a distinctly mustelid odor (Kenyon 1969; J. Estes, pers. observ.), although measurements of the concentration and composition of odoriferous material have not been made. In contrast with findings about the European otter and perhaps most other species of freshwater otters, there is no evidence that sea otters deposit spraints strategically; they seem simply to deposit spraints whenever the need to defecate arises, which usually occurs in the water, where the spraints either sink or rapidly degrade. Male sea otters often nose the anogenital region of females, presumably to determine reproductive condition. The sea otter's olfaction thus appears to function over short distances and brief times, whereas other lutrines use sprainting to extend olfactory functions both spatially and temporally.

Little is known of the olfactory perceptive abilities of aquatic mammals, largely because of the technical difficulties in making rigorous measurements. All otter species have large nasal fossae and well-developed turbinates, suggesting a keen sense of olfaction; however, the otters' olfactory lobes are small relative to other mustelids (Radinsky 1981a, 1981b; J. Gittleman, pers. comm.). Duplaix (1980), on the basis of field observations, concluded that the giant river otter has an excellent sense of smell. Sea otters appear to scent humans from long distances, and in areas where they are not accustomed to human odors, sea otters are difficult to approach except from down-wind (J. Estes, pers. observ.). They also seem to be able to detect males and estrous females by moving to a down-wind position of other otters (C. Deutsch, pers. comm.). Observations by M. Riedman and C. Deutsch (pers. comm.) suggest that male sea otters may locate estrous females by following waterborne scents across the ocean's surface. In addition, sea otters entering a group typically perform a ritualized greeting with other group members that appears to involve some form of scent recognition (M. Riedman and J. Estes, unpubl. ms.). Mason and Macdonald (1986) demonstrated the ability of the European otter to recognize individuals through olfaction by documenting behavioral changes of resident otters following the deposition of spraints from an unfamiliar otter at their sprainting sites.

To summarize: olfaction has been retained as an important sensory modality for aquatic mustelids, largely but not exclusively in support of their activities on land. There is evidence that otters have less complex scent-production capacities than do terrestrial mustelids, and that scent production by sea otters may be more poorly developed and less important than it is for other otter species. Pinnipeds, in contrast, have reduced nasal fossae due to their enlarged orbits, with a resultant loss in olfactory sensory ability (Repenning 1976a). These changes probably have resulted from the increased importance of vision and the reduced importance of olfaction in the aquatic environment. Thus, although there has been a modest reduction in the use of olfaction from terrestrial mustelids to freshwater otters to sea otter, the major break in loss of olfactory ability probably occurs between sea otters and pinnipeds. However, some pinniped species use olfaction to recognize their young (Schusterman 1981) and readily detect humans by scent (C. Heath, pers. comm.), thus suggesting their olfactory abilities may be better than has been thought.

Heat Conservation

The aquatic environment is indeed a cold place for homeotherms to live because the thermal conductivity of water is roughly an order of magnitude greater than that of air (Irving and Hart 1957). Even immersion in warm tropical water may extract more heat from an uninsulated mammal than can be compensated for by metabolic increases. Yet many aquatic species live in the cold waters of lakes and streams at high latitudes or high elevations, or in upwelled coastal marine habitats. Aquatic mammals may accommodate the potentially immense heat loss imposed by their cold environment in three ways: by increased insulation, increased metabolic heat production, and decreased surface-to-volume ratios (Scholander et al. 1950).

Insulation

Aquatic mammals are insulated with fur and/or blubber. Blubber is used exclusively by pinnipeds and cetaceans, and although it has a lower insulative value than fur (Costa and Kooyman 1982), it has several distinct advantages over fur. First, the air layer in fur compresses with depth; blubber does not. Thus, during deep dives furred animals are effectively uninsulated. Another advantage of blubber is that it permits a degree of insulative flexibility in aquatic mammals, by controlling blood flow to the body surface, that is not possible in furred species. Blubber also may serve as an energy reserve, permitting the animal to store energy in food-rich environments and thus providing opportunities for extensive behavioral and ecological flexibility. The disadvantage of blubber is that it is an encumbrance when the animal is out of the water. Furthermore, since skin must be maintained above freezing, a sharp thermal gradient, and thus substantial heat loss, cannot be avoided in a bare-skinned animal in extremely cold air. However, the fur of most pinnipeds has some insulative value in both air and water (Frisch et al. 1974).

Dense underfur may also insulate aquatic mammals by preventing water penetration to the skin. The insulative air layer surrounding the body of furred aquatic mammals secondarily provides flotation. The extent to which these functions are aquatic adaptations is arguable, since all mammals have fur and many terrestrial mammals, including nearly all terrestrial mustelids, have dense fur. However, the dense underfur of the primitive mustelids may have "pre-adapted" them to develop aquatic forms, in which there is a tendency toward increased fur density. Mink are the most aquatic of their extant congeners and they also have one of the highest underfur densities. Sea otters are the most aquatic lutrine and have the highest underfur density of any mustelid (Kenyon 1969). Aquatic rodents, such as beavers (*Castor canadensis*) and muskrats (*Ondatra zibethicus*), and duck-billed platypus (*Ornithorhynchus anatinus*) also have dense fur, indicating a functional convergence among these distantly related taxa.

The patterns of molt of aquatic and terrestrial mammals are distinctly different (Ling 1970). Terrestrial mammals, especially those living where there are extreme seasonal temperature fluctuations, tend to have two distinct molts, one during spring and one during autumn. Excluding pinnipeds, aquatic mammals typically molt more gradually and only once each year. Muskrats and sea otters, for example, molt gradually throughout the year; beavers have a prolonged summer molt; and river otters molt once during autumn (Ling 1970).

Furred aquatic mammals groom their fur to keep it clean and maintain its insulative air layer (Kenyon 1969; Tarasoff 1974). Sea otters may spend an average of 10% of their time grooming, which proceeds as a stereotyped five-step sequence (Loughlin 1977), and during which water is squeezed out of and air is blown into the pelage. Fur seals and other lutrines groom, but probably less extensively than do sea otters.

Fur seals and sea otters lack arrector pili muscles, which are present in all terrestrial mammals and serve to increase the insulative thickness of fur by holding the hair shaft vertical to the skin. Their absence in aquatic mammals allows the hair to lie flat against the body, thus presumably streamlining the body for more efficient aquatic locomotion (Ling 1970). Arrector pili are present in the muskrat; their presence is uncertain in other aquatic rodents and lutrines.

Because fur is an inflexible insulator, furred aquatic mammals require some means of controlling heat flux, especially during exercise. In the case of the sea otter most heat flux is conducted through the enlarged rear flippers (Iverson and Krog 1973; Morrison et al. 1974; Costa and Kooyman 1982), which are more sparsely furred than the rest of the body and extensively vascularized on the dorsal and plantar surfaces (Tarasoff 1972). The sea otter's flippers may also be used as solar panels to absorb heat (Tarasoff 1972), although the extent to which this serves a warming function is unknown. According to Chanin (1985), other otter species do not use the limbs to regulate heat flux. This probably is true of mink as well.

Metabolism

Increased basal metabolism, beyond that predicted from the Kleiber curve (i.e., the linear log-log relationship of basal metabolic rate vs. body mass— Kleiber 1975; see also Brody 1945), may be characteristic of aquatic carnivores (Morrison et al. 1974; McNab, this volume) and otariids (Iverson and Krog 1973). However, mustelids typically have basal metabolic rates about 20% above the standard curve (Iverson 1972), and basal rates in phocids are similar to predicted values (Lavigne et al. 1986), indicating that elevated metabolism is neither universally nor exclusively an aquatic function.

Surface-Volume Ratio

Aquatic mammals may reduce surface-volume ratios by modifying their shape toward the optimum spherical form or by becoming larger. Otters are somewhat less elongate than most other mustelids, and they tend to have smaller extremities. However, these may be modifications for increased streamlining as well as for heat conservation. One of the most evident characteristics of aquatic mammals is that they are large, especially when contrasted with terrestrial species from the same taxa. All of the extant pinnipeds and cetaceans are large, and there has been a trend toward increasing size in the pinnipeds over evolutionary time (Repenning 1976a). The largest rodents and shrews are aquatic (Eisenberg 1981). Among their congeners, mink are second in size only to the polecat (*Mustela putorius*). The extinct sea mink (*M. macrodon*) was apparently huge in comparison; its estimated head and body length of 914 mm (Nowak and Paradiso 1983) was twice that of the polecat. Otters are by far the largest mustelids, as a group ranging from about 4 to more than 45 kg (Gittleman 1985). The largest otter species, the giant river otter and the sea otter, are more than twice as large as the largest terrestrial mustelid species, the European badger (*Meles meles*). Terrestrial mustelids range in body weight from approximately 0.05 to 27 kg (Nowak and Paradiso 1983; Gittleman 1985). Sea otters are the largest and the most highly aquatic of lutrines, with males and females exceeding 45 and 30 kg, respectively (Kenyon 1969). However, there are exceptions that cast doubt on the invariability of the relationship of greater size and aquatic life style. For example, the chungungo, *Lutra felina*, which as the smallest of its congeners, inhabits the cold-temperate marine environment of Chile and Peru (Castilla and Bahamondes 1979; Sielfeld 1983).

Water Conservation

Vertebrates living in marine habitats must conserve water, especially if they feed on invertebrates that are isotonic to their sea water environment (Costa 1982). Conserving water is a lesser problem for piscivores because marine fish are hypotonic to sea water.

Sea otters are able to maintain water balance while feeding exclusively on marine invertebrate and without drinking fresh water (Costa 1982). They achieve this because their relatively large kidney (Barabash-Nikiforov 1947; Kenyon 1969) has an unusually high capacity for concentrating and processing large quantities of electrolytes compared with the kidneys of other mammals (Costa 1982). Sea otters (Costa 1982) and fasting otariids (Gentry 1981) have been observed to drink sea water, but apparently feeding pinnipeds (Pilson 1970; Depocas et al. 1971) and fasting phocids (Ortiz et al. 1978) do not need to do so. Nothing seems to be known of the water balance of other otter species that feed in marine habitats. This might prove especially interesting to study in the case of the Cape clawless otter, which feeds on marine invertebrates in the coastal habitat of South Africa (Van der Zee 1982; Arden-Clarke 1986).

Oxygen Conservation

Most aquatic mammals dive, usually to obtain food, although they may do so for other purposes as well. Diving poses two problems: (1) oxygen debts and associated chemical changes are incurred while the animal is holding its breath; and (2) rapid and extreme pressure increases are experienced en route to and from the surface. Aquatic mammals possess a broad range of diving abilities with which are associated structural, functional, and chemical modifications. Detailed discussions of these topics are given in Harrison and Tomlinson (1956, 1963), Harrison et al. (1968), and Kooyman (1973).

Among species of mustelids and pinnipeds that typically dive to feed, diving abilities (as measured by dive depth and duration) vary extensively (Table 9.2),

Table 9.2. Average and maximum dive times in selected mustelids and pinnipeds

Species	Dive times (s)		Depth (m)		Source
	$\bar{X}$	MAX	$\bar{X}$	MAX	
Mustelids					
European otter	13–16	49	1–2	15	Hewson (1973),
(*Lutra lutra*)					Kruuk and Hewson
					(1978), Watson
					(1978), Conroy and
					Jenkins (1986)
Giant river otter	25	72	—	—	Duplaix (1980)
(*Pteronura brasiliensis*)					
Mink	9.95	—	—	—	Dunstone and O'Con-
(*Mustela vison*)					nor (1979)
Sea otter	39–60	200	—	—	Estes et al. (1981)
(*Enhydra lutris*)	37–114	205	—	—	Garshelis (1983)
	—		—	100	Newby (1975)
Pinnipeds					
Antarctic fur seal	48–186	294	30	101	Kooyman et al.
(*Arctocephalus gazella*)					(1986)
Galapagos fur seal	—	462	26	112	Kooyman and Trill-
(*A. galapagoensis*)					mich (1986a)
Galapagos sea lion	—	360	37–38	186	Kooyman and Trill-
(*Zalophus californianus*)					mich (1986b)
Grey seal	—	1080	—	—	Harrison and Kooy-
(*Halichoerus grypus*)					man (1981)
Northern elephant seal	1260	2880	333	894	Le Boeuf et al.
(*Mirounga angustirostris*)					(1986), Le Boeuf et
					al. (1988)
Northern fur seal	130	460	68	207	Gentry et al. (1986)
(*Callhorinus ursinus*)					
Ringed seal	—	1020	—	—	Ferren and Elsner
(*Phoca hispida*)					(1979)
South African fur seal	102–150	450	41–49	204	Kooyman and Gentry
(*A. pusillus*)					(1986)
South American fur seal	138–198	426	27–63	170	Trillmich et al. (1986)
(*A. australis*)					
Walrus	—	600	—	—	Harrison and Kooy-
(*Odobenus rosmarus*)					man (1981)
Weddell seal	600	4320	—	600	Kooyman et al.
(*Leptonychotes weddelli*)					(1980)

from mink, which dive to only several meters' depth and typically remain underwater for less than 15 s, to large phocid seals, which dive to depths of 894 m (northern elephant seals) (Le Boeuf et al. 1988) and may remain submerged for nearly an hour (weddell seals [*Leptonychotes weddelli*]) (Kooyman 1981). Among the otters, which are better divers than mink, sea otters appear to be capable of the longest and deepest dives. Wright and Alton (1971) reported a forced breath-holding capacity in sea otters of about 5 min, and Newby (1975) established that they can dive to depths of at least 100 m.

Diving vertebrates may achieve extended periods of apnea by the so-called diving reflex, in which heart rate is reduced, oxygenated blood is diverted to the heart and brain, and energy for swimming and tissue maintenance is obtained from the anaerobic metabolism of glucose (Scholander 1940). However, such long dives probably are rare in the wild because lactate buildup, which occurs during anaerobic metabolism, requires a long recovery phase (Kooyman et al. 1980, 1981). This in turn constrains the proportion of an animal's time that can be spent diving and foraging. Thus, although sea otters are capable of remaining submerged for more than 200 s, their dives seldom last more than 100 s, which is probably near their aerobic dive limit (G. Kooyman, pers. comm.). In general, the average dive times of aquatic mammals are far below their physiological limits (Table 9.2).

Aquatic mammals possess structural modifications in their lungs and bronchial trees, which vary according to dive depth and duration. The tracheal length-width ratio decreases from river otters (*Lutra canadensis*) to sea otters to phocid seals (Tarasoff and Kooyman 1973a), presumably permitting more rapid and complete air exchange with the lungs before and after diving in the more highly adapted aquatic forms. The tracheal rings of river otters and sea otters are partially calcified, whereas those of phocid seals are entirely cartilaginous, thus permitting flexibility under the pressure of deep diving (Tarasoff and Kooyman 1973b). In the case of sea otters and pinnipeds, cartilaginous airways empty directly into the alveoli, thus ensuring patency until compression collapse during deep dives (Kooyman 1973). This modification is absent in river otters and in terrestrial mammals. More highly adapted aquatic mammals tend to have less lobulation of the lungs, which Tarasoff and Kooyman (1973a) speculated may facilitate gas exchange in a weightless environment where the normal support function of the lobular form of the lungs is unnecessary. Sea otters have remarkably large lungs, which are more than three times greater per unit of body weight than are those of either river otters or phocid seals, providing buoyancy (which is needed to support their young, food, and tools on the surface) and increased oxygen storage (Kooyman 1973).

Phocid seals have high hematocrits and blood hemoglobin levels, whereas sea otters and otariid seals have values that are similar to those of most terrestrial mammals (Lenfant et al. 1970). However, oxygen-hemoglobin affinities are higher in sea otters than they are in terrestrial mammals (Lenfant et al. 1970). These modifications increase blood-oxygen storage capacity. I know of no blood chemistry data for mink and river otters.

Locomotion

Fundamental locomotor changes required for swimming are among the most evident modifications in aquatic mammals. In addition, movement in an aquatic medium places a premium on the reduction of drag, which is achieved by modifications of the integument and general body form (see Taylor, this volume). Extensive limb modifications have occurred in many aquatic mammals. In general, there has been a tendency for limb shortening and an increase of surface area at the extremities. River otters have shortened limbs and webbed toes, but otherwise their limbs are largely unmodified from those of terrestrial mustelids (Ewer 1973), except that they have muscles in the feet that regulate interdigital web tension, which assists their propulsion through the water (Burton 1979). Sea otters are more highly modified for aquatic propulsion; they have greatly enlarged, flipper-like hind limbs and an extended fifth digit; except for its being even shorter than the river otter's, the sea otter's fore limb is unmodified. Fore limb modification in sea otter may have been prevented by the fact that a high degree of tactile sensitivity in this structure is required for the capture and consumption of food, for the sea otter, in contrast with piscivorous otter species and pinnipeds, does not capture prey with its mouth and thus does not require their increased sensitivity and motor function of the facial region (see Feeding). Piscivorous feeding may have freed the fore limbs of primitive phocids and otariids to become flipper-like, thus further assisting with aquatic locomotion.

Because of their reduced need for support on land, river otters, sea otters, and pinnipeds exhibit parallel modifications to the skeleton that permit the increased flexibility necessary for grooming and feeding in the aquatic environment. Greater flexibility has been achieved by numerous anatomical changes (see Taylor, this volume). Consequently, mobility on land is increasingly reduced in the more highly adapted aquatic forms. River otters, because of their shortened limbs and webbed feet, are generally less mobile on land than are most terrestrial mustelids (Tarasoff et al. 1972). Sea otters are even less mobile on land because of their shortened fore limbs and enlarged, flipper-like hind limbs. Tarasoff et al. (1972) suggested that the terrestrial mobility of sea otters and pinnipeds was further limited by heat-retention problems caused by their dense pelage or blubber.

River otters use both fore and hind limbs for swimming and must actively swim to maintain themselves on the surface. Sea otters and most pinnipeds can float passively. Sea otters do not use their fore limbs in aquatic locomotion, and there is a graded tendency among these forms for body area movement to function in aquatic locomotion (Howell 1930; Sokolov and Sokolov 1970; Tarasoff et al. 1972). There is a similar graded trend for increased foot surface area, and for the margin of the hind limb to form a perfect lunate border for maximum propulsive efficiency (Tarasoff et al. 1972), as there is a trend toward increased locomotor importance of the hind limb and decreased importance of the tail (Tarasoff et al. 1972).

Feeding

The otters have evolved two rather distinct foraging modes—piscivory and invertebrate feeding—although few species adhere strictly to either of these.

Piscivory is the primitive foraging mode in otters, apparently having twice given rise to invertebrate feeders (Berta and Morgan 1986). One lineage of invertebrate feeders led to sea otters and the other, to clawless and small-clawed otters (*Aonyx* spp.). The diets of mink and the extant otter species fit this pattern. Mink are catholic feeders, but wherever they have been studied, their diet contains aquatic organisms (Sealander 1943; Guilday 1949; Wilson 1954; Korschgen 1958; Hamilton 1959; Wise et al. 1981). In addition to various fish species, they eat mice, rats, squirrels, rabbits, and ralliform and anseriform birds. Piscivorous otters are represented by *Lutra* species and the giant river otter (including spotted-necked otters, *Hydrictis maculicollus,* and smooth-coated otters, *Lutrogale perspicillata,* by Davis's [1978] classification). Dietary studies of these species, summarized by Mason and Macdonald (1986), show that whereas many kinds of prey are consumed, fish are most commonly eaten and probably are most highly preferred. Diets of the invertebrate feeders are best known from studies of the sea otter (Estes et al. 1981) and the Cape clawless otter (Rowe-Rowe 1977b). Although fish are important prey under some circumstances (Van der Zee 1981; Estes et al. 1982), mollusks, crustaceans, and echinoderms probably are the preferred species.

Related to this dichotomy of foraging modes are differences in brain structure, dentition, search and handling strategies, and the incidence of tool use. The piscivorous otters capture prey with their mouths. Correspondingly, the medial cortical region of the posterior sigmoid gyrus of these species is expanded, a change that is associated with increased facial sensitivity (Radinsky 1968). Acute vision probably is an important accessory sensory modality. In contrast, the invertebrate feeding otters capture prey with their fore limbs. The sigmoid gyrus of these species is expanded laterally, a change associated with increased tactile sensity of the forepaws; acute vision probably is a less important accessory sensory modality than it is for the piscivores.

The number of incisors is reduced in sea otters (Ewer 1973) and pinnipeds (King 1983) from the primitive I3 condition of generalized carnivores, an alteration, Ewer (1973) suggests, related to the reduced importance of the incisors for tearing flesh from the prey. Furthermore, there is a distinct dichotomy in dental morphology between the piscivorous and invertebrate feeding otters. Typical of most terrestrial carnivores, the piscivores have carnassial molars and premolars for shearing the soft flesh of fish. The sea otter and *Aonyx* species, however, have broadened bunodont molars for crushing the exoskeletons of their invertebrate prey.

Tool use has developed in the case of the sea otter (Hall and Schaller 1964) and the Cape clawless otter (Donnelly and Crobler 1976) as a further aid to crushing invertebrate exoskeletons. These animals pound prey against a rock tool or other hard object. Tool use is unknown in other mammals, excluding primates (Alcock 1972). The origin of tool use by these otters was probably facilitated by their highly dexterous fore limbs. This behavior undoubtedly benefited the foraging economics of invertebrate feeding otters by making available to them a diverse array of prey species that would otherwise have retained a refuge behind their exoskeletons. Marine mollusks have evolved heavy or highly sculptured shells in tropical and subtropical environments, which Vermeij (1977) has shown to be closely associated in time with a Mesozoic radiation of crushing invertebrate and fish predators. This apparently resulted in the gradient of reduced shell thickness seen today from tropical to polar marine habitats because the crushing invertebrates and fishes have been absent from cold seas (Vermeij 1978; Palmer 1979). Early sea otters, having radiated into the marine environment late in the Cenozoic (Repenning 1976b; Berta and Morgan 1986), encountered an abundance of thick-shelled prey, particularly in warm-temperate seas, and thus to some extent the evolution of tool use in sea otters was likely a product of the "Mesozoic Marine Revolution" (Vermeij 1977). Correlated with the latitudinal gradient in molluscan shell architecture is the finding that tool use by sea otters is more common in California than it is in the Aleutian Islands (J. Estes, unpubl. data).

The searching patterns for prey of mink and otters vary, as do those of piscivorous and invertebrate feeding otters. Mink locate prey from the surface and then dive after them (Poole and Dunstone 1976; Dunstone 1979). Otters search for and locate prey after submergence (Erlinge 1968a). Piscivorous otters use vision in clear water, supplemented by mechanical wave perception by their mustacial vibrissae in murky water, to locate and pursue their prey. This was demonstrated in the case of the European otter by Green (1977), who found that when the vibrissae were intact, water clarity had no influence on capture time, whereas when the vibrissae were removed, it took the otter 20 times as long to capture fish in darkened than in clear water. The invertebrate feeding otters use vision and fore limb tactile sensitivity to locate their prey (Kenyon 1969; Rowe-Rowe 1977a, 1977c; Shimek 1977). Sea otters actively explore the sea floor with their forepaws when searching for food, sometimes seeming not to use vision at all (J. Estes, pers. observ.).

Studies of the free-ranging dive patterns of aquatic mammals show that metabolism during most dives is aerobic, which puts unusual constraints on foraging behavior. One obvious consequence of limited dive duration is a forced negative correlation between search and pursuit times. Although the effects of such constraints remain poorly known, Dunstone and O'Connor (1979) concluded (from studies of mink) that limited dive duration requires diving mammals to optimize search effort (which would include travel time to and from the surface for the deeper divers) while retaining a residual time for pursuit that should also be optimized. The result was a predicted maximum search time at intermediate encounter rates (prey abundances).

Activity budgets and patterns have been studied in several otter species, and in general up to half or more of their time is spent feeding (Duplaix 1980; Estes et al. 1982; Mason and Macdonald 1986). This finding supports the view that food is an important limiting resource to otters. The time sea otters spend foraging is correlated with population status and food availability (Estes et al.

1982; Garshelis et al. 1986; Estes et al. 1986). Often, their foraging activity increases near dawn and dusk, which may occur for several reasons. Their foraging efficiency on fish may be greater at these times. Mammalian piscivores, whose dilating pupils accommodate rapidly to light level changes, may fish near dawn and dusk when fish are visually impaired by the transformation between photopic and scotopic vision (Munz and McFarland 1973). Also, when they fish at night, pinnipeds can increase their visual sensitivity by adapting to the dark (Lavigne et al. 1975). Estes et al. (1982) reported that all observed fish captures during their study of sea otters at Amchitka Island occurred in the morning and evening hours. Similar studies of foraging success and efficiency in other piscivorous otters are needed.

Most otter species feed exclusively in freshwater habitats. One, the sea otter, is strictly marine (Estes 1980). Another, the chungungo, seems to be largely marine (Sielfeld 1983), whereas several others, the European otter, the river otter, the Cape clawless otter, and the southern river otter (*Lutra provocax*), feed on occasion or at some locations in the marine environment (Chanin 1985; Mason and Macdonald 1986). The distribution of otters in freshwater habitats ranges from tropical to subpolar regions (Mason and Macdonald 1986). However, only temperate and boreal marine regions are inhabited by otters. This latitudinal disparity in otter distribution between freshwater and marine environments is perplexing, but related patterns occur in other vertebrate taxa. Indeed, Gross et al. (1988) explained the evolution of diadromy in fishes by latitudinal variation in aquatic productivity between freshwater and marine habitats, anadramous species being most common at high latitudes where marine productivity exceeds that of freshwater habitats and catadramous species being most common at low latitudes where this pattern is reversed. Perhaps fresh water was the primitive environment of otters, and only because of the greatly increased food availability, related to high production of temperate and boreal coastal zones, have otters been drawn from freshwater to marine habitats. This idea will be discussed further in the following section.

Life History and Behavior

Life history and behavioral adaptations for aquatic living in mammals have been largely ignored. One important exception is the wealth of fine studies that followed Bartholomew's 1970 paper on the evolution of polygyny in pinnipeds. Yet, a key element in Bartholomew's model was life on land during the breeding season. The model does not work very well otherwise, and indeed we would know much less about pinniped mating systems if the animals did not come ashore to breed where they could be observed easily. In this final section I explore some of the modifications to life history and behavior that may have resulted from life in aquatic environments. Although these topics are separated for the sake of organization, they are so closely associated that often they cannot be discussed separately. To this point I have generally adhered to a simple accounting of information; however, now I will speculate more broadly, emphasizing (1) trends or patterns among terrestrial mustelids, lutrines, and pinnipeds, and (2) differences between freshwater and marine habitats.

Life History

Many factors may have promoted the evolution of large body size in aquatic mammals. Whatever the exact cause, large body size is related to several life history characters that in themselves appear to serve no particular advantage in aquatic habitats. One of these is longevity. Within particular taxa, large mammals generally live longer than small ones (Gould 1977; Eisenberg 1981; Gittleman 1986b). The river otter may live for as long as 15 years in the wild (Stephenson 1977; Tabor and Wight 1977), as compared with weasels, which live only several years. However, this generalization does not apply to all taxa. For example, the large-bodied elephant seals (the largest pinnipeds) are short-lived (Laws 1953; Ling and Bryden 1981) compared with such small-bodied species as the ringed seal (*Phoca hispida*) (McLaren 1958; Smith 1973). Other life history traits, such as age of first reproduction, also are likely correlated with body size and perhaps aquatic living (Gittleman 1986b).

More interesting, perhaps, is the contrast in life histories between freshwater and marine species. The only truly marine otter is the sea otter; all others that inhabit marine systems are expatriates from freshwater habitats.

The litter size characteristic of sea otters and other lutrines is remarkably different (Table 9.3). Sea otters bear single young (twins are rarely conceived, but they probably are never successfully weaned). This pattern is consistent among all other marine mammal species in the Pinnipedia, Cetacea, and Sirenia (Estes 1979). Freshwater otters, in contrast, all have multiple-young pregnancies, and in all species for which data are available, maximum litter size is 4–6 (Gittleman 1986b). Jameson and Bodkin (1986) provide an interesting account of events following a rare successful twin birth of sea otters in nature. Although the female attempted to care for both young, she was not able to do so, and about 24 h following the birth she abandoned one of them.

There are remarkable differences in pup development between freshwater and marine species (Table 9.4). Despite the fact that sea otters are the largest lutrine species, their young are more precocial than those of either *Lutra* species or the giant river otter. This condition is taken to an extreme in phocid seals, some species of which may wean and abandon their young less than 2 weeks following birth (Bonner 1984), and one species, the hooded seal (*Cystophora cristata*), lactates for only 3–5 days (Bowen et al. 1985)!

Other life history characteristics are less divergent between marine and terrestrial species. The age of first reproduction for the European otter and the

Table 9.3. Litter sizes of selected otter species

Species	Litter size		Source
	Range or mean	Maximum	
European otter (*Lutra lutra*)	1.1–2.4	4	Erlinge 1967a; Stubbe 1977; Jenkins 1980; Wayre 1981
River otter (*L. canadensis*)	2.0–3.4	4	Liers 1951; Hamilton and Eadie 1964; Tabor and Wight 1977; Mowbray et al. 1979
Chungungo (*L. felina*)	2.0	5	Sielfeld et al. 1977; Cabello 1983
Oriental small-clawed otter (*Aonyx cinerea*)	0.8–4.4	6	Wayre 1981
Cape clawless otter (*A. capensis*)	2–3	?	Rowe-Rowe 1978b
Giant river otter (*Peteronura brasiliensis*)	2.1	5	Duplaix 1980
Sea otter (*Enhydra lutris*)	1.0	1	Kenyon 1969; M. Riedman and J. Estes, unpubl. ms.
Pinnipeds (all species)	1.0	1	
Cetaceans (all species)	1.0	1	
Sirenians (all species)	1.0	1	

river otter is 2 years for both sexes (Hamilton and Eadie 1964; Tabor and Wight 1977; Wayre 1980). Although extensive data are still lacking for sea otters, females apparently first reproduce at 3–5 years, whereas males, which reach sexual maturity at 5–6 years, do not become active breeders until they are even older (R. Jameson and A. Johnson, unpubl. ms.; M. Riedman and J. Estes, unpubl. ms.). Sea otters are polygynous, and in polygynous species males typically undergo sexual maturity later than females (Jarman 1983). Polygyny appears to be more weakly expressed in freshwater otters (see below). Possible environmental reasons for this is discussed in the section on behavior.

Most otters are aseasonal breeders (Duplaix-Hall 1975), although this character is highly plastic within and among species. For example, the European

Table 9.4. Development of pups in three otter species (*time in weeks*)

	European otter (*Lutra lutra*)[a]	Giant river otter (*Pteronura brasiliensis*)[b]	Sea otter (*Enhydra lutris*)[c]
Eyes open	4–5	4	?—near birth
First solid food	7	?	<4
First swimming	12	10	3–5
Fully weaned	14	?	<8?
First dive	?	?	6
First prey capture	16	15	6
Length of ♀/young association	32–50	>50	26

[a]Wayre 1979b; Melquist and Hornocker 1983.
[b]Duplaix 1980; Autuori and Deutsch 1977.
[c]Payne and Jameson 1984.

otter breeds aseasonally in Britain (Harris 1968); aseasonally in Scotland, but with lower survival of young born during winter (Jenkins 1980); and mainly during spring in Sweden (Erlinge 1967b). The river otter, in contrast, is a highly seasonal breeder in North America, independent of latitude (Harris 1968; Tabor and Wight 1977; Melquist and Hornocker 1983; Chanin 1985). Reasons for this difference between species are unknown, although King (1984) suggested that similar differences between least weasels (*Mustela nivalis*) and ermine (*M. erminea*) were historical artifacts rather than divergent reproductive strategies with contemporary function, and van Zyll de Jong (1972) concluded, on the basis of this reproductive difference, that the European otter and the river otter were less closely related than had been previously supposed. The sea otter breeds and gives birth throughout the year (Kenyon 1969; Schneider 1972a, 1973a), and although there are seasonal peaks in both mating and pupping, timing of these peaks varies among different parts of the species' range. The timing of the peaks also can change over time at particular locations, as occurred in California following anomalous weather and oceanographic conditions caused by the El Niño Southern Oscillation (ENSO) event in 1982–83 (M. Riedman and J. Estes, unpubl. ms.; R. Jameson and J. Estes, unpubl. data).

Related to breeding seasonality in otters is the cyclicity of estrus and length of gestation. The aseasonal breeders tend to come into estrus every month or so (Leslie 1970), to have short gestation periods, and to forego delayed implantation. For example, the length of gestation in the European otter, the smooth-coated otter, the spotted-necked otter, the Oriental small-clawed otter, the Cape clawless otter, and the giant river otter all range from about 60–70 days (Desai 1974; Duplaix-Hall 1975; Wayre 1979b; Duplaix 1980; Davis 1981). The sea otter's gestation typically ranges from approximately 120 to 180 days (Wendell et al. 1984), whereas the river otter's ranges from approximately 280 to 370 days (Liers 1951), although the period of implanted pregnancy in this latter species is approximately 56 days (see Mead, this volume).

One interpretation of these patterns is that factors that select for strongly seasonal reproduction, so typical of many terrestrial mammals (but see Kleiman and Eisenberg 1973), have been relaxed in the otters. Terrestrial mustelids, in particular, tend to be seasonal breeders with long periods of unimplanted pregnancy and extended lengths of gestation, scaled appropriately for body size. In contrast with most terrestrial habitats, in which strong seasonal changes undoubtedly select for reproductive seasonality of species living within them, aquatic environments are buffered from extreme changes by the capacity of water to store heat. This, along with correlated differences such as prey abundance, may be one reason for the general lack of seasonality in otter reproduction. It should be noted, however, that sea otters are distinctly seasonal breeders (Kenyon 1969; Siniff and Ralls 1988), and most pinnipeds are highly synchronous seasonal breeders (Stirling 1983). Reasons for this difference in seasonality of reproduction between freshwater and marine species are likely varied. Seasonal events such as storms, food production, and the formation and breakup of sea ice undoubtedly influence reproductive success of mammals in the marine environment. Further, in highly polygynous species the "marginal male effect" (Bartholomew 1970) may have acted to aggregate breeding in time as well as space.

Whether copulations and births occur on land or in water, they vary in interesting ways among otter and pinniped species. Freshwater otters copulate both on land and in water. Most lutrine species apparently do both. Pinnipeds copulate on land and in water as well, but the tendency to do one or the other is more nearly invariant within species (Stirling 1983). Those pinnipeds that copulate on land tend to be highly polygynous and highly sexually dimorphic, whereas those that copulate in water tend to be less polygynous and more nearly monomorphic (Stirling 1983). The sea otter probably copulates exclusively in water (Kenyon 1969; Estes 1980; M. Riedman and J. Estes, unpubl. ms.). Birth locations show more inflexible relationships among species. All pinnipeds (Stirling 1983) and all otter species except the sea otter (Chanin 1985; Mason and Macdonald 1986) give birth exclusively on land or other solid substrates. The sea otter may give birth in water or on land (Barabash-Nikiforov 1947; Sandegren et al. 1973; Woodward 1981; Jameson 1983), although most births probably occur in water. These patterns may not be adaptive; however, they may have influenced the evolution of mating systems, as well as the opportunity for further radiations into aquatic environments.

Milk Composition

The composition of milk from various terrestrial and aquatic taxa of mammals indicates correlations associated with aquatic living (Table 9.5). The milks of terrestrial species generally have low milk-fat content; those of aquatic species, in contrast, have high levels of milk fat. The fat content of pinniped and cetacean milks was high in all species for which data were found. Within the mustelids, the milks of *Mustela* species, the North American badger (*Taxidea taxus*), and the striped skunk (*Mephitis mephitis*) are not notably different from that of other carnivores or terrestrial mammals in general. However, both river otter and the sea otter, the only two otter species for which data were found, have milks with high milk-fat contents (see Gittleman and Oftedal 1987). Within the Rodentia, eight terrestrial species showed milk-fat contents within the normal range of terrestrial mammals. Two aquatic species, the beaver (*Castor canadensis*) and the nutria (*Myocastor coypus*), had milks with high fat contents. Within the Insectivora, the water shrews (*Neomys* spp.) had higher than average milk-fat content, although the terrestrial white-toothed shrews (*Crocidura* spp.) was higher. Data on carbohydrate levels of milk are less complete, but generally show opposite trends to milk-fat levels across the same taxa. It is likely that milk composition has a number of ecological and

Table 9.5. Milk-fat and carbohydrate content of terrestrial and aquatic mammals

Group	n (species)	Fat $\bar{X}$	Fat $\tilde{r}$	Carbohydrate $\bar{X}$	Carbohydrate $\tilde{r}$
Artiodactyla	42	8.2	0.3–20.7	4.4	2.4–6.7
Carnivora	18	10.0	3.0–18.6	3.8	2.3–4.7
European otter (*Lutra lutra*)[a]		24.0	—	0.1	—
Mink (*Mustela vison*)[a]		3.5	—	9.3	—
North American badger (*Taxidea taxus*)[b]		9.5	—	—	—
Polecat (*Mustela putorius*)[c]		8.0	—	—	—
Sea otter (*Enhydra lutris*)[b]		23.0	—	—	—
Striped skunk (*Mephitis mephitis*)		13.8		3.0	—
Pinnipedia[a]	8	43.2	30.7–53.2	0.7	0.1–2.6
Cetacea[a]	9	29.1	16.7–45.8	1.8	0.6–5.6
Edentata		20.0	—	0.3	—
Insectivora	3	16.6	6.5–31.9	1.7	0–3.2
Neomys[a]		20.0	—	0.1	—
Lagomorpha	4	15.9	14.4–19.3	1.8	0.9–2.7
Perissodactyla	5	1.4	0.2–4.8	6.4	5.3–7.2
Primates	8	3.4	2.2–10.6	6.8	5.9–7.2
Proboscidea	2	6.2	5.0–7.3	5.3	5.2–5.3
Rodentia	8	10.2	4.9–13.2	3.3	1.7–4.9
Beaver (*Castor canadensis*)[a]		19.0	—	1.7	—
Nutria (*Myocaster coypus*)[a]		27.9	—	0.6	—

[a]Aquatic or semi-aquatic forms.
[b]Jenness et al. 1981.
[c]Jenness and Sloan 1970.
Sources. Data from Ben Shaul 1962 or Oftedal 1984, except where noted.

phylogenetic correlates. However, the recurrent pattern of high milk fat and low lactose across several aquatic taxa suggests that these changes are adaptations for aquatic living in mammals.

Relative Brain Weight

Brain weights of aquatic mammals may have been subjected to several conflicting selective influences. Compared with closely related terrestrial species, aquatic mammals might be expected to have large brains because of the information-processing needs imposed by their complex three-dimensional habitat (Eisenberg 1981). Robin (1973) and Hofman (1983) have argued conversely, that because of the high sustained energy consumption of neural tissue, large brains are a detriment to diving mammals. Worthy and Hickie (1986), however, demonstrated that marine mammals varied considerably around a linear function equating body mass and brain mass for mammals in general. The brain weights of some species, including most odontecete cetaceans, were larger than expected; others, including sirenians and mysticete cetaceans, were smaller than expected; and still others, including the pinnipeds, were about the same as expected. Worthy and Hickie (1986) rejected the Robin-Hofman hypothesis by pointing out that the variation in actual to predicted brain-mass ratios among these species is uncorrelated with dive duration.

Gittleman's (1986a) data and general approach were used to determine whether brain weights of lutrines are significantly different from those of other mustelids. Brain weight versus body weight was regressed for all mustelids as in brain wt. (g) = 2.4697 + 0.7038 ln body wt. (kg), providing a good linear fit ($r^2 = 0.93$). Of the 11 otter species for which Gittleman presented data, eight had observed brain weights greater than expected. However, on average, the otters had brain weights slightly less than predicted ($\bar{X} = -1.37$g). In view of the imprecision in weight measures (due to variation among individuals resulting from age, sex, and body condition differences), this analysis indicates that comparative brain sizes (Gittleman 1986a) of otters are not significantly different from those of mustelids in general. Gittleman (1986a) showed that relative brain size for mustelids is somewhat less than for the carnivores overall.

Sexual Dimorphism

Data on skull length of adult individuals, from Harris (1968), have been used in the analysis (Table 9.6). Since body weight is a power function of skull length, greater differences in weight dimorphism would be expected than are indicated from the skull length data in the table. More extensive analyses of size dimorphism could no doubt be done for most otter species using available museum collections. I regard this analysis only as an initial indication of the presence or absence of sexual dimorphism among otter species.

The European otter and the sea otter are the only species of otters that show highly significant sexual dimorphism. All other species, except perhaps *Aonyx* species, indicate weak and statistically insignificant sexual dimorphism. The European otter is an especially interesting species, since the analysis indicates that dimorphism is highly significant in specimens from northern and western Eurasia, whereas it is insignificant in specimens from southern and eastern Eurasia. More data on this possible difference would be interesting, especially if they could be combined with information on natural history and social behavior. I found no skull length data for the giant river otter, and although Duplaix (1980) contended that males are slightly larger than females in this species, that conclusion was not substantiated.

Table 9.6. Analyses of sexual dimorphism in skull length for selected otter species.[a]

Species	Mean skull length (mm)			d.f.	t[b]	P
	♀	♂	♂/♀			
European otter (*Lutra lutra*)						
All samples	106.1	112.9	1.06	52	3.582	<0.001
USSR, western Europe only	108.2	119.4	1.10	24	8.113	<0.001
India, China, S.E. Asia only	103.7	107.5	1.04	26	1.613	0.2 > P > 0.1
Europe, N. Africa (basal length)	102.3	112.8	1.10	13	6.155	<0.001
River otter	109.9	113.8	1.04	23	1.580	0.2 > P > 0.1
(*L. canadensis*)						
Smooth-coated otter	117.2	120.9	1.03	13	1.345	0.2 > P > 0.1
(*L. [Lutrogale] perspicillata*)						
Spotted-necked otter	93.6	100.8	1.08	9	1.960	0.1 > P > 0.05
(*L. [Hydrictis] maculicollis*)						
Oriental small-clawed otter	84.8	84.4	1.00	15	0.211	>0.5
(*Aonyx cinerea*)						
Cape clawless otter	119.8	122.5	1.02	17	0.815	0.5 > P > 0.4
(*A. capensis*)						
Sea otter	125.7	136.4	1.09	6	6.380	<0.001
(*Enhydra lutris*)						

[a]Basilocondylar lengths from Harris 1968.
[b]Student's *t*-test.

Social Behavior

This section examines the extent to which aquatic carnivores, particularly the lutrines, are polygynous, and why. Following a review of what is known about the social behavior of otters, the discussion centers on the way behavior may have been influenced by aquatic living, and in particular, how it differs between freshwater and marine systems.

Mating Systems

Although most species have not been well studied, and in no case are data available to rigorously characterize mating systems, among the otters mating systems appear to vary from polygynous to monogamous. The sea otter probably is the most strongly polygynous species, judging from its territorial system (Calkins and Lent 1975; Loughlin 1980; Garshelis et al. 1986) and female density (M. Riedman and J. Estes, unpubl. ms.; C. Deutsch, pers. comm.). The European otter and the river otter also may be polygynous (or promiscuous), as indicated by the lack of male association with females, except for mating. Most other species for which information is available tend more toward monogamy, as indicated by observations of frequent or extended male-female pairs. The Cape clawless otters live as male and female pairs plus young in inland habitats (Rowe-Rowe 1978a), but males usually are absent from family groups in coastal populations (Van der Zee 1982). Pair bonding has been reported for the giant river otter (Duplaix 1980) and the smooth-coated otter (Wayre 1974).

Male parental care is absent in the behavior of the sea otter, has not been reported for and is probably absent in the behavior of the European and river otters, and has been reported in the behavior of the giant river otter (Duplaix 1980), the smooth-coated otter (Desai 1974; Wayre 1974), the Cape clawless otter, and the spotted-necked otter (Mason and Macdonald 1986). Infanticide by adult male sea otters may occur in captivity (M. Riedman and J. Estes, unpubl. ms.).

Territoriality and Sexual Segregation

Territoriality is a somewhat obscure behavior because (1) various definitions have been used (Kaufmann 1983) and (2) for any particular definition, variation among species is graded such that extreme cases may be easily categorized whereas intermediate cases are less clear. Male sea otters apparently defend small, contiguous territories (Calkins and Lent 1975; Loughlin 1977; Garshelis 1983) to which they have a high degree of interannual fidelity (R. Jameson, unpubl. ms.). Exclusively male groups form in this species (Schneider 1972b; 1973b), presumably because juvenile or subordinate males are displaced by territorial males, although territorial males may also join these groups during the nonbreeding season in some areas (Loughlin 1980; R. Jameson, unpubl. ms.). Trowbridge (1983) suggested that most otter species are not strongly territorial because they range over areas that are too large to be economically defensible. Male European otters and river otters display mutual avoidance, but they do not appear to actively maintain territories by direct interactions or contests (Erlinge 1968b; Hornocker et al. 1983). Other wide-ranging mustelids such as wolverines (*Gulo gulo*) also do not defend territories, whereas the more narrowly ranging North American badger does (Hornocker et al. 1983), thus supporting Trowbridge's (1983) suggestion. In the case of other mustelids, higher population densities tend to be correlated with male territorial defense (e.g., European badger) (Kruuk 1978). Except for the sea otter, strongly developed sexual segregation is unknown in lutrines, although it is common behavior for many pinnipeds. Sexual segregation is a common behavior among many seasonally territorial species; for example, numerous ungulate species form so-called bachelor herds. Displacement by dominant males seems to be an important mechanism in some cases (Sinclair 1977).

Group Size

Otter species rarely form large groups, with the exception of the sea otter. Adult European otters and river otters are solitary except for females with their most recently born young (Erlinge 1967b; Melquist and Hornocker 1983; Chanin 1985; Mason and Macdonald 1986). Groups of giant river otters consist of an adult male and female plus their two most recent cohorts, typically containing three to eight individuals (Duplaix 1980), and occasionally in this species family groups may join (Laidler and Laidler 1983). The spotted-necked otter also occurs in family groups consisting of a male and female plus their most recently born young (Rowe-Rowe 1978a), which may join to form groups of up to about 20 individuals (Proctor 1963). Oriental small-clawed otters have been reported in groups of up to 15 individuals on the Malay Peninsula (Mason and Macdonald 1986) and form four to eight individuals in Sabah (Furuya 1976). Cape clawless otters and spotted-necked otters have been reported in groups containing up to five individuals (Arden-Clarke 1986; Mason and Macdonald 1986). Although sea otters forage alone (Estes and Jameson 1988), they commonly rest in groups that may contain hundreds and even more than a thousand individuals in some areas (K. Schneider, pers. comm.). The composition of these groups is poorly known, although often they appear to consist largely or exclusively of a single sex (except for dependent young). In sum, most otter species are solitary or occur in pairs, except for family groups with young. These occasionally coalesce into larger groups of apparently unrelated individuals. The sea otter is exceptional in that it typically rests in groups of unrelated individuals that are sometimes very large.

Population Density

Erlinge (1967b) reported European otter densities of 1.7–5.6 individuals/10 km in Sweden. Similar densities have been reported for this species on Scottish streams (0.75 breeding female/10 km) (Green et al. 1984) and for the river otter in Idaho (1.7–3.8 individuals/10 km) (Melquist and Hornocker 1983). Species that typically inhabit fresh water tend to live at increased densities in the marine environment. For example, Van der Zee (1982) and Arden-Clarke (1986) reported Cape clawless otter densities of 4–10 individuals/10 km along the coast of South Africa; Watson (1978) reported European otter densities of 4 females plus cubs/10 km (= 12–24 individuals/10 km) along the Shetland coast; and information from Kruuk and Hewson (1978), Jenkins and Burrows (1980), and Twelves (unpub., cited in Mason and Macdonald 1986) show that European otter holts are 1.3-1.9 times closer together in marine than in freshwater habitats. Home ranges of the Cape clawless otter and the European otter tend to be smaller in marine than in freshwater habitats (Watson 1978; Van der Zee 1982; Twelves, cited in Mason and Macdonald 1986). Sea otter densities have been reported to be 216–324/10 km at Amchitka Island (Estes 1977) and 50–60/10 km in central California (R. Jameson and J. Estes, unpubl. data). Furthermore, data suggest that otters forage more profitably in marine habitats. Conroy and Jenkins (1986) reported that the European otter in Scotland captured more prey per hunt, spent less time catching prey, and had higher proportions of successful hunts and dives in marine habitats than in freshwater lochs. Studies of the river otter in locations where the species feeds in closely associated freshwater and marine habitats would provide further tests of these hypotheses.

Evolution of Mating Systems

Aquatic living by carnivores probably arose first as the animals radiated from terrestrial to freshwater habitats, and from there to the sea (Stirling 1983). Comparisons among these habitats, and among the life histories and behaviors of species that occupy them, provide a broad view of mating systems from the perspective of selection and evolution (figure 9.1). If allowance is made for often poor life history and behavioral data and possible phylogenetic constraints, terrestrial mustelids and otters provide contrasts perhaps indicative of initial adaptations for aquatic living; differences among freshwater otters, marine otters, and pinnipeds suggest modifications associated with further radiation into the sea. Possible mechanisms become apparent when the patterns among taxa are viewed in the light of existing theory on the evolution of mating systems and sexual dimorphism (Table 9.7), although the imprecise nature of this theory adds further uncertainty to any evolutionary scenario.

It is generally true for mammals that males are larger than females (Ralls 1976) and that larger species are more sexually dimorphic than smaller ones (Ralls 1977). Three unrelated explanations have been offered for the reason that species are sexually dimorphic. One (the "resource partitioning model") is that it reduces intersexual food competition (Brown and Lasiewski 1972; Shubin and Shubin 1975; Powell 1979). Since the expected direction of dimorphism between sexes should be the same among dimorphic species, this model is an unlikely explanation for the principal cause of sexual dimorphism in mammals. A second explanation (the "small female model") holds that selection has been for small female size (Powell 1979; Moors 1980). This model predicts a negative correlation between species body size and the extent of sexual dimorphism. Ralls and Harvey (1985) criticized this model on several grounds. A third explanation (the "large male model") is that sexual selection acts to increase male size (Ralls 1977). The model holds that, among polygynous species without male parental care, large males should be selected for because big males are best able to compete for or attract females. The later model, which I follow here, is thought to be the most likely explanation for

Figure 9.1. Behavior and life history characteristics among terrestrial mustelids, freshwater otters, marine otters, and pinnipeds. Breaks in horizontal lines indicate points in the evolution of increasingly more "highly adapted" aquatic forms where major changes in function or form apparently have occurred.

Life History	Taxa/Habitat			
	Terrestrial Mustelids	Freshwater Otters	Sea Otters	Pinnipeds
Milk fat	Low		High	High
Pup development	Altricial		Precocial	Precocial to strongly precocial
Reproductive synchrony	typically seasonal	Plastic; often seasonal	Seasonal, but births throughout year; plastic	Seasonal; usually highly synchronous
Litter size	Multiple		Single	
Sexual dimorphism	Modest but typical	Modest when present; absent in some species	Modest	Extreme in those that copulate on land absent or reverse in water copulators
Bimaturation	?Absent	Absent in some species	Present	Present in dimorphic species
Behavior				
Copulation site	Land	Land and water	Water	Land or water
Birth site	Land	Land	Water	Land
Group size	Typically solitary or small groups of related individuals		Often form large groups of unrelated individuals	
Population density	Usually low		Usually high to very high	
Territoriality	Intrasexual; all ♀'s and some ♂'s			♂'s only
Sexual segregation	Weak or absent		Strong	Strong for many species
Mating system	Promiscuous/weakly polygynous?	Monogamous/promiscuous weakly polygynous	Polygynous	Serially monogamous; promiscuous; highly polygynous
Male parental care	Unknown; probably rare or absent	Common in some species	Absent; ♂ infanticide known	Absent; facultative ♂ infanticide known
Number of Dichotomies:		3–5	12	4

Table 9.7. Predictions from theory on the evolution of mating systems and sexual dimorphism

1. Intersexual variation in reproductive success should be greater in males than females in poly-gymous species and similar in sexes in monogamous ones.
2. Variation in reproductive success should be greater in males of polygynous species than among males of monogamous ones.
3. Competition for mates should be greater among males of polygynous species than among males of monogamous ones.
4. Sexual dimorphism should be most highly developed in strongly polygynous species.
5. Sexual dimorphism arises through competition by males for females (the greater investing sex).
6. Competition occurs if some males have the opportunity to monopolize matings.
7. Environmental factors, mainly dispersion for form of resources, will, by influencing distribution and reproductive timing of females, determine whether males can limit access to females.
8. Delayed maturation of males relative to females (bimaturism) is necessary for the evolution of polygyny and extreme sexual dimorphism.

Sources. From Clutton-Brock 1983; Jarman 1983.

sexual dimorphism in the mustelids (Ralls and Harvey 1985) and the pinnipeds (Stirling 1983).

Several influences of aquatic living on the social organization of carnivores are suggested by this theory and information. A dichotomy of conditions between freshwater and marine habitats may have influenced the extent to which polygynous mating systems, and selection by the "large male model" for sexual dimorphism, occur in aquatic carnivores (Table 9.8). The key variable in this dichotomy may be food availability, which I suggest is low in fresh water compared with coastal marine habitats. If food is an important limiting resource to aquatic carnivores, then population densities and foraging behavior should reflect the proposed disparity in food abundance between freshwater and marine habitats. This seems to occur. Low-density populations in freshwater habitats should favor monogamous mating systems (and therefore male parental care), because (1) male territories large enough to hold more than one or two females would not be economically defensible, and (2) if food were scarce, a male may increase the survival of his offspring by assisting the female, especially if she is the only female with whom he had sired a litter (Kleiman 1977; Wittenberger and Tilson 1980; but see Wickler and Seibt 1983). Although the mating and parental care systems of most otters have not been well studied, many of them appear to fit the predictions of this model, especially when viewed among the mustelids in general, which Moors (1980) contended all have polygynous mating systems. Similarly, there would be no

Table 9.8. Dichotomous predictions from two models of selection for sexual dimorphism

	Small ♀	Large ♂
Female density	Independent	Strongly dependent
Species size	Strong inverse correlation	Independent
Polygynous mating system	Correlated, but only because parental care is not provided by ♂	Strongly dependent

selection for sexual dimorphism by the "large male model" under these circumstances, a conclusion that appears to be supported by the data from most freshwater otters. In contrast, high-density populations in marine habitats should favor polygynous mating systems without male parental care, because (1) increased female densities permit males to defend smaller territories, within which they would expect to copulate with multiple females, and (2) if food is abundant, a female may not need the male's assistance to care for her young. This situation would be further reinforced if females aggregated, which would allow a dominant male to attain even more copulations.

The observed patterns seem to support these predictions, although more data are needed from all otter species, especially those that forage in both freshwater and marine habitats. The Cape clawless otter may provide male parental care in freshwater inland systems (Rowe-Rowe 1978a), whereas in coastal habitats they appear not to (Van der Zee 1982). The extent to which there is a corresponding shift from monogamous to polygynous mating systems in this species is unknown. Sea otter densities are far higher than those reported for other otter species, and as predicted, this species shows a stronger tendency toward polygyny, with an associated social structure that appears to be unique among the lutrines, being more typical of some ungulates that rely on relatively abundant food resources. However, selection for sexual dimorphism by the "large male model" is also predicted under these circumstances, and whereas sea otters are significantly sexually dimorphic in size, they are no more so than European otters in western Europe, and they fall far short of many of the pinnipeds. Reasons for these patterns remain unclear, although there are several possibilities. One is that although female sea otters are aggregated in space, the onset of estrus may be too asynchronous to create a strong environmental potential for polygyny. A second is that males of species that copulate in the water may benefit more from agility (and small size) than from large size. This argument was advanced by Stirling (1983) to explain the modest reverse sexual dimorphism in Weddell seals, which are polygynous but which maintain aquatic territories beneath the land-fast ice. Indeed, none of the aquatic-copulating pinniped species are strongly sexually dimorphic, whereas all species that copulate on land are. A third possibility is that females may not be sufficiently aggregated in space to permit selection for extreme sexual dimorphism. Even though female sea otters occur at higher densities and are more aggregated than other lutrine females, they fall far short of the colonial breeding pinnipeds in these characteristics, and accordingly the ratio of females to breeding males is probably modest in comparison.

In sum, available data on otters indicate trends in social behavior related to food availability, although these trends are not absolute, and more information is needed from all species to confirm relationships proposed herein. Freshwater species seem to have either monogamous or weakly polygynous mating systems, male parental care in several cases, and male territoriality that is weakly defined or absent. In contrast, marine species, or marine-living populations of typically freshwater species, tend to have more strongly polygynous mating systems, no male parental care, and more strongly defined male territoriality. I suggest that this apparent dichotomy is related to differences in food availability and to corresponding differences in adult distribution between marine and freshwater environments. Primary production probably is much higher in temperate coastal marine systems than it is in even the most productive lakes and streams. In marine systems otters appear to live at high densities and (in the case of the sea otter) to aggregate, whereas in freshwater systems otter densities are lower and distributions more uniform. Thus it would seem that coastal marine systems have a high environmental potential for polygyny (Emlen and Oring 1977). Marine otters, in this regard, may be "intermediate" between the monogamous or weakly polygynous freshwater otters and the strongly polygynous land-breeding pinnipeds.

The social systems of a wide range of vertebrates, including many carnivores, are known to be related to the distribution and abundance of food (see Carr and MacDonald 1986; Packer 1986; Moehlman 1986; Gittleman, this volume, and Moehlman, this volume, for discussions). In one remarkable example, Zabel (1986) demonstrated a shift from bigamy to monogamy in the red fox (*Vulpes vulpes*) on a Bering Sea island following a decline of their prey, shore-nesting sea birds, resulting from strong ENSO conditions in the early 1980s. The diversity and flexibility of vertebrate mating systems in relation to variation in food resources make patterns, such as those I suggest for otters and pinnipeds, not only reasonable but expected.

It is remarkable that sexual dimorphism is insignificant in about half the otter species (Table 9.6), although this may be due to small sample sizes. Insofar as patterns exist, sexual dimorphism in otters appears to be largely independent of species size and weakly related to polygynous mating systems, as predicted by the "large male model" (Table 9.8). However, extreme sexual dimorphism has not evolved in any otter species, nor does there appear to be a relationship between the extent of sexual dimorphism and the extent to which otter species are polygynous. Similar patterns occur among aquatic breeding pinniped species. These observations suggest that territorial defense and copulation in an aquatic medium may have constrained the evolution of sexual dimorphism in body size, even under conditions in which the environmental potential for polygyny was otherwise high.

Acknowledgments

I thank D. Costa, S. Feldcamp, J. Gittleman, M. Haley, C. Heath, D. Lavigne, B. Le Boeuf, K. Ralls, G. Rathbun, M. Riedman, R. Schusterman, J. Taggart, and C. Zabel for comments on earlier drafts of this paper. I am grateful to P. Himlan for typing the manuscript.

References

Alcock, J. 1972. The evolution of the use of tools by feeding animals. *Evolution* 26:464–473.

Arden-Clarke, C. H. G. 1986. Population density, home range size and spatial organization of the Cape clawless otter, *Aonyx capensis,* in a marine habitat. *J. Zool. (Lond.)* 209:201–211.

Autuori, M. P., and Deutsch, L. A. 1977. Contribution to the knowledge of the Brazilian otter, *Pteronura brasiliensis* (Gmelin, 1788), Carnivora, Mustelidae: Rearing in captivity. *Zool. Garten* 47:1–8.

Barabash-Nikiforov, I. I. 1947. *Kalan (The sea otter).* Soviet Ministry RSFSR. Trans. from Russian by Israel Program for Scientific Translation. Jerusalem, Israel, 1962.

Bartholomew, G. A. 1970. A model for the evolution of pinniped polygyny. *Evolution* 24:546–559.

Ben Shaul, D. M. 1962. The composition of milk of wild animals. *Internat. Zoo Yearb.* 4:333–342.

Berta, A., and Morgan, G. S. 1986. A new sea otter (Carnivora: Mustelidae) from the late Miocene and early Pliocene (Hemphilian) of North America. *J. Paleontol.* 59:809–819.

Bonner, W. N. 1984. Lactation strategies in pinnipeds: Problems for a marine mammalian group. In: M. Peaker, R. G. Vernon & C. H. Knight, eds. *Physiological Strategies in Lactation,* Symposia of the Zoological Society of London no. 51, pp. 253–270. London: Academic Press.

Bowen, W. D., Oftedal, O. T., and Boness, D. J. 1985. Birth to weaning in 4 days: Remarkable growth in the hooded seal. *Canadian J. Zool.* 63:2841–2846.

Brody, S. 1945. *Bioenergetics and Growth.* London: Hafner Press.

Brown, J. C., and Lasiewski, R. C. 1972. Metabolism of weasels: The cost of being long and thin. *Ecology* 53:939–943.

Burton, R. 1979. *Carnivores of Europe.* London: Batsford.

Cabello, C. C. 1983. La nutria de mar en la Isla de Chiloe. Corporación National Forestal. *Boletin Tecnico* 6:1–37.

Calkins, D. G., and Lent, P. C. 1975. Territoriality and mating behavior in Prince William Sound sea otters. *J. Mamm.* 56:528–529.

Carr, G. M., and MacDonald, D. W. 1986. The sociality of solitary foragers: A model based on resource dispersion. *Anim. Behav.* 34:1540–1549.

Castilla, J. C., and Bahamondes, I. 1979. Observaciones conductales y ecológicas sobre *Lutra felina* (Molina) 1782 (Carnivora: Mustelidae) en las zonas Central y Centro-Norte de Chile. *Archivos de Biologia y Medicina Experimentales* 12:119–132.

Chanin, P. 1985. *The Natural History of Otters.* New York: Facts on File.

Clutton-Brock, T. H. 1983. Selection in relation to sex. In: D. S. Bendall, ed. *Evolution from Molecules to Men,* pp. 457–481. Cambridge: Cambridge Univ. Press.

Conroy, J. W. H., and Jenkins, D. 1986. Ecology of otters in northern Scotland. VI. Diving times and hunting success of otters (*Lutra lutra*) at Dinnet Lochs, Aberdeenshire and in Yell Sound, Shetland. *J. Zool. (Lond.)* 209:341–346.

Costa, D. P. 1982. Energy, nitrogen and electrolyte flux and seawater drinking in the sea otter *Enhydra lutris. Physiol. Zool.* 55:35–44.

Costa, D. P., and Kooyman, G. L. 1982. Oxygen consumption, thermoregulation, and the effect of fur oiling and washing on the sea otter, *Enhydra lutris. Canadian J. Zool.* 60:2761–2767.

Davis, J. A. 1978. A classification of otters. In: N. Duplaix, ed. *Otters: Proceedings of the First Meeting of the Otter Specialist Group,* pp. 14–33. Morges, Switzerland: International Union for the Conservation of Nature.

Davis, J. A. 1981. Breeding the spot-necked otter (Abstract.) Distributed at Second International Otter Colloquium, Norwich, England, September 1981.

Depocas, F., Hart, J. S., and Fisher, H. D. 1971. Sea water drinking and water flux in starved and fed harbor seals, *Phoca vitulina. Canadian J. Physiol. Pharmacol.* 49:53–62.

Desai, J. H. 1974. Observations on the breeding habits of the Indian smooth otter *Lutrogale perspicillata* in captivity. *Internat. Zoo Yearb.* 14:123–124.

Donnelly, B. G., and Crobler, J. H. 1976. Notes on food and anvil using behaviour by the Cape clawless otter, *Aonyx capensis* in the Rhodes Matopos National Park, Rhodesia. *Arnoldia* 7:1–8.

Dunstone, N. 1979. The fishing strategy of the mink (*Mustela vison*): Time-budgeting of hunting effort? *Behaviour* 67:157–177.

Dunstone, N., and O'Connor, R. J. 1979. Optimal foraging in an amphibious mammal. I. The aqualung effect. *Anim. Behav.* 27:1182–1194.

Duplaix, N. 1980. Observations on the ecology and behavior of the giant river otter *Pteronura brasiliensis* in Surinam. *Revue d'écologie: La terre et la vie* 34:495–620.

Duplaix-Hall, N. 1975. River otters in captivity: A review. In: R. D. Martin, ed. *Breeding Endangered Species in Captivity*, pp. 315–327. London: Academic Press.

Eisenberg, J. F. 1981. *The Mammalian Radiations*. Chicago: Univ. Chicago Press.

Emlen, S. T., and Oring, L. W. 1977. Ecology, sexual selection and the evolution of mating systems. *Science* 197:215–223.

Erlinge, S. 1967a. Food habits of the fish-otter *Lutra lutra* L. in South Swedish habitats. *Viltrevy* 4:371–443.

Erlinge, S. 1967b. Home range of the otter *Lutra lutra* L. in southern Sweden. *Oikos* 18:186–209.

Erlinge, S. 1968a. Food studies on captive otters (*Lutra lutra* L.). *Oikos* 19:259–270.

Erlinge, S. 1968b. Territoriality of the otter *Lutra lutra* L. *Oikos* 19:81–98.

Estes, J. A. 1977. Population estimates and feeding behavior of sea otters. In: M. C. Merritt & R. G. Fuller, eds. *The Environment of Amchitka Island, Alaska*, pp. 511–526. Springfield, Va.: National Technical Information Service.

Estes, J. A. 1979. Exploitation of marine mammals: R-selection of k-strategists? *J. Fish. Res. Board Canada* 36:1009–1117.

Estes, J. A. 1980. *Enhydra lutris*. Mammalian Speices no. 133. Lawrence, Kan.: American Society of Mammalogists.

Estes, J. A. 1981. Notes on feeding behavior. Unpubl. data. Available from J. A. Estes, Institute of Marine Sciences, UCSC.

Estes, J. A., and Jameson, R. J. 1988. A double-survey estimate for sighting probability of sea otters in California. *J. Wildl. Mgmt.* 52:70–76.

Estes, J. A., Jameson, R. J., and Johnson, A. M. 1981. Food selection and some foraging tactics of sea otters. In: J. A. Chapman & D. Pursley, eds. *Proceedings of the Worldwide Furbearers Conference*, pp. 606–641. Frostburg, Md.: Worldwide Furbearers Conference, Inc.

Estes, J. A., Jameson, R. J., and Rhode, E. B. 1982. Activity and prey selection in the sea otter: Influence of population status on community structure. *Amer. Nat.* 120:242–258.

Estes, J. A., Underwood, K. E., and Karmann, M. J. 1986. Activity-time budgets of sea otters in California. *J. Wildl. Mgmt.* 50:626–637.

Ewer, R. F. 1973. *The Carnivores*. Ithaca, N.Y.: Cornell Univ. Press.

Ferren, H., and Elsner, R. 1979. Diving physiology of the ringed seal: Adaptations and implications. *Proc. Alaska Sci. Conf.* 29:379–387.

Frisch, J., Øritsland, N. A., and Krog, J. 1974. Insulation of furs in water. *Comp. Biochem. Physiol.* 47A:403–410.

Furuya, Y. 1976. Otters in Padas Bay, Sabah, Malaysia. *J. Mamm. Soc. Japan* 7:34–43.

Garshelis, D. L. 1983. Ecology of sea otters in Prince William Sound, Alaska. Ph.D. dissert., Univ. Minnesota, Minneapolis. 321 pp.

Garshelis, D. L., Garshelis, J. A., and Kimker, A. T. 1986. Sea otter time budgets and prey relationships in Alaska. *J. Wildl. Mgmt.*50:637–647.

Garshelis, D. L., Johnson, A. M., and Garshelis, J. A. 1984. Social organization of sea otters in Prince William Sound, Alaska. *Canadian J. Zool.* 62:2648–2658.

Gentry, R. 1981. Seawater drinking in eared seals. *Comp. Biochem. Physiol.* 68A:81–86.

Gentry, R. L., Kooyman, G. L., and Goebel, M. E. 1986. Feeding and diving behavior of northern fur seals. In: R. L. Gentry & G. L. Kooyman, eds. *Fur Seals: Maternal Strategies on Land and at Sea,* pp. 61–78. Princeton, N.J.: Princeton Univ. Press.

Gentry, R. L., and Peterson, R. S. 1967. Underwater vision of the sea otter. *Nature* 216:435–436.

Gittleman, J. L. 1985. Carnivore body size: Ecological and taxonomic correlates. *Oecologia* 67:540–554.

Gittleman, J. L. 1986a. Carnivore brain size, behavioral ecology, and phylogeny. *J. Mamm.* 67:23–36.

Gittleman, J. L. 1986b. Carnivore life history patterns: Allometric, ecological and phylogenetic associations. *Amer. Nat.* 127:744–771.

Gittleman, J. L., and Oftedal, O. T. 1987. Comparative growth and lactation energetics in carnivores. In: A. S. I. Loudon & P. Racey, eds. *Reproductive Energetics in Mammals,* pp. 41–77. Oxford: Oxford Univ. Press.

Gorman, M. L., Jenkins, D., and Harper, R. J. 1978. The anal scent sacs of the otter (*Lutra lutra*). *J. Zool. (Lond.)* 186:463–474.

Gould, S. J. 1977. *Ontogeny and Phylogeny.* Cambridge: Harvard Univ. Press.

Gould, S. J., and Lewontin, R. L. 1979. The spandrels of San Marcos and the Panglossian paradigm: A critique of the adaptationist programme. *Proc. Royal Soc. London* ser. B. 205:581–598.

Gould, S. J., and Vrba, E. 1982. Exaptation—a missing term in the science of form. *Paleobiology* 8:4–15.

Green, J. 1977. Sensory perception in hunting otters. *Lutra lutra* L. *J. Otter Trust* 1977:13–16.

Green, J., Green, R., and Jeffries, D. J. 1984. A radio-tracking survey of otters *Lutra lutra* on a Perthshire river system. *Lutra* 27:85–145.

Gross, M. E., Coleman, R. M., and McDowall, R. M. 1988. Aquatic productivity and the evolution of diadromous fish migration. *Science* 239:1291–1293.

Guilday, J. E. 1949. Winter foods of Pennsylvania mink. *Penn. Game News* 20:32.

Hall, K. R. L., and Schaller, G. B. 1964. Tool-using behavior of the California sea otter. *J. Mamm.* 45:287–298.

Hamilton, W. J. 1959. Foods of mink in New York. *New York Fish Game J.* 1959:77–85.

Hamilton, W. J., and Eadie, W. R. 1964. Reproduction in the otter. *Lutra canadensis. J. Mamm.* 45:242–252.

Harris, C. J. 1968. *Otters: A Study of the Recent Lutrinae.* London: Weidenfeld and Nicolson.

Harrison, R. J. 1974. *Functional Anatomy of Marine Mammals,* vol. 2. New York: Academic Press.

Harrison, R. J., Hubbard, R. C., Peterson, R. S., Rice, C. E., and Shusterman, R. J., eds. 1968. *The Behavior and Physiology of Pinnipeds.* New York: Appleton-Century-Crofts.

Harrison, R. J., and Kooyman, G. L. 1981. *Diving in Marine Mammals*. Carolina Biology Readers. Burlington, N.C.: Carolina Biology Supply.

Harrison, R. J., and Tomlinson, J. D. W. 1956. Observations on the venous system in certain pinnipedia and cetacea. *Proc. Zool. Soc. London* 126:205–233.

Harrison, R. J., and Tomlinson, J. D. W. 1963. Anatomical and physiological adaptations in diving mammals. In: J. D. Carthy & C. L. Duddington, eds. *Viewpoints in Biology*, pp. 115–162. London: Butterworths.

Hewson, R. 1973. Food and feeding habits of otters *Lutra lutra* at Loch Park, northeast Scotland. *J. Zool. (Lond.)* 170:143–162.

Hillegaart, V., Ostman, J., and Sandegren, F. 1981. Area utilization and marking behaviour among two captive otter (*Lutra lutra* L.) pairs. (Abstract.) Distributed at Second International Otter Colloquium, Norwich, England. September 1981.

Hofman, M. A. 1983. Energy metabolism, brain size and longevity in mammals. *Quart. Rev. Biol.* 58:496–512.

Hornocker, M. G., Messick, J. P., and Melquist, W. E. 1983. Spatial strategies in three species of Mustelidae. *Acta Zoologica Fennica* 174:185–188.

Howell, A. B. 1930. *Aquatic Mammals: Their Adaptations to Life in the Water*. Springfield, Ill.: Charles C Thomas.

Irving, L., and Hart, J. S. 1957. The metabolism and insulation of seals as bare-skinned mammals in cold water. *Canadian J. Zool.* 35:497–511.

Iversen, J. A. 1972. Basal energy metabolism of mustelids. *J. Comp. Physiol.* 81:341–344.

Iversen, J. A., and Krog, J. 1973. Heat production and body surface area in seals and sea otters. *Norwegian J. Zool.* 21:51–54.

Jameson, R. J. 1983. Evidence of birth of a sea otter on land in central California. *California Fish Game* 69:122–123.

Jameson, R. J. 1987. Movements, home range, and territories of male sea otters in central California. Unpubl. ms. Available from R. J. Jameson, U.S. Fish and Wildlife Service, San Simeon, Calif.

Jameson, R. J., and Bodkin, J. L. 1986. An incidence of twinning in the sea otter (*Enhydra lutris*). *Marine Mammal Science* 2:305–309.

Jameson, R. J., and Estes, J. A. 1988. Unpublished data available from R. J. Jameson, U.S. Fish and Wildlife Service, San Simeon, Calif.

Jameson, R. J., and Johnson, A. M. 1988. Sea otter reproduction. Unpublished ms. Available from R. J. Jameson, U.S. Fish and Wildlife Service, San Simeon, Calif.

Jamieson, G. S., and Fisher, H. D. 1972. The pinniped eye: A review. In R. J. Harrison, ed. *Functional Anatomy of Marine Mammals, pp. 245–261*. London: Academic Press.

Jarman, P. 1983. Mating system and sexual dimorphism in large, terrestrial, mammalian herbivores. *Biol. Rev.* 58:485–520.

Jenkins, D. 1980. Ecology of otters in northern Scotland. 1. Otter (*Lutra lutra*) breeding and dispersion in mid-Deeside, Aberdeenshire in 1974–79. *J. Anim. Ecol.* 49:713–735.

Jenkins, D., and Burrows, G. O. 1980. Ecology of otters in northern Scotland. III. The use of faeces as indicators of otter (*Lutra lutra*) density and distribution. *J. Anim. Ecol.* 49:755–774.

Jenness, R., and Sloan, R. E. 1970. The composition of milks of various species: A review. *Dairy Sci. Abstr.* 32:599–612.

Jenness, R., Williams, T. D., and Mullin, R. J. 1981. Composition of milk of the sea otter (*Enhydra lutris*). *Comp. Biochem. Physiol.* 70A:375–379.

Kaufman, J. H. 1983. On the definitions and functions of dominance and territoriality. *Biol. Rev.* 58:1–20.

Kenyon, K. W. 1969. *The Sea Otter in the Eastern Pacific Ocean.* U.S. Fish and Wildlife Service, North American Fauna, no. 69. Washington, D.C.: Government Printing Office.

King, C. M. 1984. The origin and adaptive advantages of delayed implantation in *Mustela erminea. Oikos* 42:126–128.

King, J. E. 1983. *Seals of the World* Ithaca, N.Y.: Comstock Publishing Associates.

Kleiber, M. 1975. *The Fire of Life: An Introduction to Animal Energetics.* Huntington, N.Y.: R. E. Kreiger.

Kleiman, D. G. 1977. Monogamy in mammals. *Quart. Rev. Biol.* 52:39–68.

Kleiman, D. G., and Eisenberg, J. F. 1973. Comparisons of canid and felid social systems from an evolutionary perspective. *Anim. Behav.* 21:637–659.

Kooyman, G. L. 1973. Respiratory adaptations in marine mammals. *Amer. Zool.* 13:457–468.

Kooyman, G. L. 1981. Weddell seal: Consummate diver. London: Cambridge Univ. Press.

Kooyman, G. L., Castellini, M. A., and Davis, R. W. 1981. Physiology of diving in marine mammals. *Ann. Rev. Physiol.* 43:343–356.

Kooyman, G. L., Davis, R. W., and Croxall, J. P. 1986. Diving behavior of Antarctic fur seals. In: R. L. Gentry & G. L. Kooyman, eds. *Fur Seals: Maternal Strategies on Land and at Sea,* pp. 115–125. Princeton, N.J.: Princeton Univ. Press.

Kooyman, G. L., and Gentry, R. L. 1986. Diving behavior of South African fur seals. In: R. L. Gentry & G. L. Kooyman, eds. *Fur Seals: Maternal Strategies on Land and at Sea,* pp. 142–152. Princeton, N.J.: Princeton Univ. Press.

Kooyman, G. L., and Trillmich, F. 1986a. Diving behavior of Galapagos fur seals. In: R. L. Gentry & G. L. Kooyman, eds. *Fur Seals: Maternal Strategies on Land and at Sea,* pp. 186–195. Princeton, N.J.: Princeton Univ. Press.

Kooyman, G. L., and Trillmich, F. 1986b. Diving behavior of Galapagos sea lions. In: R. L. Gentry & G. L. Kooyman, eds. *Fur Seals: Maternal Strategies on Land and at Sea,* pp. 209–219. Princeton, N.J.: Princeton Univ. Press.

Kooyman, G. L., Wahrenbrock, E. A., Castellini, M. A., Davis, R. W., and Sinnett, E. E. 1980. Aerobic and anaerobic metabolism during voluntary diving in Weddell seals: Evidence of preferred pathways from blood chemistry and behavior. *J. Comp. Physiol.* 138:325–346.

Korschgen, L. J. 1958. December food habits of mink in Missouri. *J. Mamm.* 39:521–527.

Kruuk, H. 1978. Spatial organization and territorial behaviour of the European badger *Meles meles. J. Zool. (Lond.)* 184:1–19.

Kruuk, H., and Hewson, R. 1978. Spacing and foraging of otters (*Lutra lutra*) in a marine habitat. *J. Zool. (Lond.)* 185:205–212.

Laidler, K., and Laidler, E. 1983. *The River Wolf.* London: George Allen and Unwin.

Lavigne, D. M., Bernholz, C. D., and Ronald, K. 1975. Functional aspects of the marine mammal retina. In: R. J. Harrison, ed. *Functional Anatomy of Marine Mammals,* vol. 3. London: Academic Press.

Lavigne, D. M., Innes, S., Worthy, G. A. J., Kovacs, K. M., Schmitz, O. J., and Hickie, J. P. 1986. Metabolic rates of seals and whales. *Canadian J. Zool.* 64:279–284.

Laws, R. M. 1953. The elephant seal (*Mirounga leonina* Linn.). I. Growth and age. Falkland Islands Dependencies Surveys, Scientific Reports no. 8. N.p., n.p.

Le Boeuf, B. J., Costa, D. P., Huntley, A. C., and Feldcamp, S. D. 1988. Continuous, deep diving in female northern elephant seals, *Mirounga angustirostris. Canadian J. Zool.* In press.

Le Boeuf, B. J., Costa, D. P., Huntley, A. C., Kooyman, G. L., and Davis, R. W. 1986.

Pattern and depth of dives in Northern elephant seals, *Mirounga angustirostris. J. Zool. (Lond.)* 208:1–7.

Lenfant, C., Johansen, K., and Torrence, J. D. 1970. Gas transport and oxygen storage capacity in some pinnipeds and the sea otter. *Respiration Physiol.* 9:277–286.

Leslie, G. 1970. Observations on the Oriental short-clawed otter, *Aonyx cinerea,* at Aberdeen Zoo. *Internat. Zoo Yearb.* 10:79–81.

Liers, E. E. 1951. Notes on the river otter (*Lutra canadensis*). *J. Mamm.* 32:1–9.

Ling, J. K. 1970. Pelage and molting in wild mammals with special reference to aquatic forms. *Quart. Rev. Biol.* 45:16–54.

Ling, J. K., and Bryden, M. M. 1981. Southern elephant seal. In: S. H. Ridgeway & R. J. Harrison, eds. *Handbook of Marine Mammals,* 2:297–327. London: Academic Press.

Loughlin, T. R. 1977. Activity patterns, habitat partitioning, and grooming behavior of the sea otter, *Enhydra lutris,* in California. Ph.D. dissert., Univ. California, Los Angeles. 110 pp.

Loughlin, T. R. 1980. Home range and territoriality of sea otters near Monterey, California. *J. Wildl. Mgmt.* 44:576–582.

Lythgoe, J. N. 1972. The adaptation of visual pigments to the photic environment. In: H. J. A. Dartnall, ed. *The Handbook of Sensory Physiology,* VII. I. *The Photochemistry of Vision.* Hamburg: Springer-Verlag.

Lythgoe, J. N., and Dartnall, H. J. A. 1970. A "deep sea rhodopsin" in a marine mammal. *Nature* 227:995–996.

McFarland, W. N. 1971. Cetacean visual pigments. *Vision Research* 11:1065–1076.

McLaren, I. A. 1958. *The Biology of the Ringed Seal,* Phoca hispida, *in the Eastern Canadian Arctic.* Fisheries Research Board of Canada Bulletin, no. 118. N.p., n.p.

Mason, C. F., and Macdonald, S. M. 1986. *Otters: Ecology and Conservation.* Cambridge: Cambridge Univ. Press.

Melquist, W. E., and Hornocker, M. G. 1983. Ecology of river otters in west central Idaho. *Wildl. Monogr.* 83:1–60.

Moehlman, P. D. 1986. Ecology of cooperation in canids. In: D. I. Rubenstein & R. W. Wrangham, eds. *Ecological Aspects of Social Evolution,* pp. 64–86. Princeton, N.J.: Princeton Univ. Press.

Moors, P. J. 1980. Sexual dimorphism in the body size of mustelids (Mammalia: Carnivora): The role of food habits and breeding systems. *Oikos* 34:147–158.

Morrison, P., Rosenmann, M., and Estes, J. A. 1974. Metabolism and thermoregulation in the sea otter. *Physiol. Zool.* 47:218–229.

Mowbray, E. E., Pursley, D., and Chapman, J.A. 1979. *The Status, Population Characteristics and Harvest of the River Otter in Maryland.* Publ. Wildl. Ecol. no. 2. [Annapolis:] Maryland Wildlife Administration.

Munz, F. W., and McFarland, W. N. 1973. The significance of spectral position in the rhodopsins of tropical marine fishes. *Vision Res.* 13:1829–1874.

Newby, T. C. 1975. A sea otter (*Enhydra lutris*) food dive record. *Murrelet* 56:7.

Nowak, R. M., and Paradiso, J. L. 1983. *Walker's mammals of the world* (4th ed.). Baltimore: Johns Hopkins Univ. Press.

Oftedal, O. T. 1984. Milk composition, milk yield and energy output at peak lactation: A comparative review. *Symp. Zool. Soc. London* 51:33–85.

Ortiz, C. L., Costa, D., and LeBoeuf, B. J. 1978. Water and energy flux in elephant seal pups fasting under natural conditions. *Physiol. Zool.* 51:166–178.

Packer, C. 1986. The ecology of sociality in felids. In: D. I. Rubenstein & R. W. Wrangham, eds. *Ecological Aspects of Social Evolution,* pp. 429–451. Princeton, N.J.: Princeton Univ. Press.

Palmer, A. R. 1979. Fish predation and the evolution of gastropod shell structure: Experimental and geographic evidence. *Evolution* 33:697–713.

Payne, S. F., and Jameson, R. J. 1984. Early behavioral development of the sea otter, *Enhydra lutris. J. Mamm.* 65:527–531.

Pilson, M. E. W. 1970. Water balance in California sea lions. *Physiol. Zool.* 43:257–269.

Poole, T. B., and Dunstone, N. 1976. Underwater predatory behavior of the American mink (*Mustela vison*). *J. Zool. (Lond.)* 178:395–412.

Powell, R. A. 1979. Mustelid spacing patterns: Variations on a theme by *Mustela. Z. Tierpsychol.* 50:153–165.

Proctor, J. 1963. A contribution to the natural history of the spotted-necked otter (*Lutra maculicollis* Lichtenstein) in Tanganyika. *East African Wildl. J.* 1:93–102.

Radinsky, L. B. 1968. Evolution of somatic sensory specialization in otter brains. *J. Comp. Neurol.* 134:495–506.

Radinsky, L. B. 1981a. Evolution of skull shape in carnivores. 1. Representative modern carnivores. *Biol. J. Linnean Soc.* 15:369–388.

Radinsky, L. B. 1981b. Evolution of skull shape in carnivores. 2. Additional modern carnivores. *Biol. J. Linnean Soc.* 16:337–355.

Ralls, K. 1976. Mammals in which females are larger than males. *Quart. Rev. Biol.* 51:245–276.

Ralls, K. 1977. Sexual dimorphism in mammals: Avian models and some unanswered questions. *Amer. Nat.* 111:917–938.

Ralls, K., and Harvey, P. H. 1985. Geographic variation in size and sexual dimorphism of North American weasels. *Biol. J. Linnean Soc.* 25:119–167.

Repenning, C. A. 1972. Underwater hearing in seals: Functional morphology. In: R. J. Harrison, ed. *Functional Anatomy of Marine Mammals*, 1:307–331. New York: Academic Press.

Repenning, C. A. 1976a. Adaptive evolution of the sea lions and walruses. *Syst. Zool.* 25:375–390.

Repenning, C. A. 1976b. *Enhydra* and *Enhydriodon* from the Pacific coast of North America. *J. Res. U.S. Geol. Survey* 4:305–315.

Ridgway, S. H., and Harrison, R. J. 1981. *Handbook of Marine Mammals*, vols. 1 and 2. New York: Academic Press.

Riedman, M., and Estes, J. A. 1988. Biology of the sea otter: A review. Unpublished ms. Available from J. A. Estes, Institute of Marine Sciences, UCSC.

Robin, E. D. 1973. The evolutionary advantages of being stupid. *Perspect. Biol. Med.* 16:369–380.

Rowe-Rowe, D. T. 1977a. Food ecology of otters in Natal, South Africa. *Oikos* 28:210–219.

Rowe-Rowe, D. T. 1977b. Prey capture and feeding behaviour of South African otters. *Lammergeyer* 23:13–21.

Rowe-Rowe, D. T. 1977c. Variations in the predatory behaviour of clawless otter. *Lammergeyer* 23:22–27.

Rowe-Rowe, D. T. 1978a. Biology of two otter species in South Africa. In: N. Duplaix, ed. *Otters: Proceedings of the First Meeting of the Otter Specialists Group*, pp. 130–139. Morges, Switzerland: International Union for the Conservation of Nature.

Rowe-Rowe, D. T. 1978b. The small carnivores of Natal. *Lammergeyer* 25:1–48.

Sandegren, F. E., Chu, E. W., and Vandevere, J. E. 1973. Maternal behavior in the California sea otter. *J. Mamm.* 54:668–679.

Schneider, K. B. 1972a. *Reproduction in the Female Sea Otter.* Federal Aid in Wildlife Restoration Project W-17-4, Project Progress Report. Anchorage: Alaska Department of Fish and Game.

Schneider, K. B. 1972b. *Sex and Age Segregation of Sea Otters.* Federal Aid in Wildlife Restoration Project W-17-4, Project Progress Report. Anchorage: Alaska Department of Fish and Game.

Schneider, K. B. 1973a. *Reproduction in the Female Sea Otter*. Federal Aid in Wildlife Restoration Project W-17-5, Project Progress Report. Anchorage: Alaska Department of Fish and Game.

Schneider, K. B. 1973b. *Sex and Age Segregation of Sea Otters*. Federal Aid in Wildlife Restoration Project W-17-5, Project Progress Report. Anchorage: Alaska Department of Fish and Game.

Scholander, P. F. 1940. Experimental investigations on the respiratory function in diving birds and mammals. *Hvalradets Skrifter* 22:1–131.

Scholander, P. F., Hock, R., Walters, V., and Irving, L. 1950. Adaptation to cold in arctic and tropical mammals and birds in relation to body temperature, insulation, and basal metabolic rate. *Biol. Bull.* 99:259–271.

Schusterman, R. J. 1972. Visual acuity in pinnipeds. In: H. E. Winn & B. L. Olla, eds. *Behavior of Marine Mammals*, pp. 469–492. New York: Plenum.

Schusterman, R. J. 1981. Behavioral capabilities of seals and sea lions: A review of their hearing, visual, learning and diving skills. *Psych. Rec.* 31:125–143.

Schusterman, R. J., and Barrett, B. 1973. Amphibious nature of visual acuity in the Asian "clawless" otter. *Nature* 244:518–519.

Sealander, J. A. 1943. Winter food habits of mink in Michigan. *J. Wildl. Mgmt.* 7:411–417.

Shimek, S. J. 1977. The underwater foraging habits of the sea otter, *Enhydra lutris*. *California Fish Game* 63:120–122.

Shubin, I. G., and Shubin, H. G. 1975. [Sexual dimorphism in mustelids (Mustelidae, Carnivora).] *Zhurnal Obschei biologii* 36:283–290. (Trans. from Russian in C. M. King, ed. *Biology of Mustelids: Some Soviet Research*. Wellington, N.Z.: Science Information Division, Department of Scientific and Industrial Research.)

Sielfeld, W. K. 1983. *Mamiferos marinos de Chile*. Santiago: Edificiones de la Universidad de Chile.

Sielfeld, W., Venegas, C., and Atalah, A. 1977. Consideraciones acerca del estado de los mamiferos marinos en Chile. *Anales. Instituto de la Patagonia* 8:297–315.

Sinclair, A. R. E. 1977. *The African Buffalo*. Chicago: Univ. Chicago Press.

Sinclair, W., Dunstone, N., and Poole, T. B. 1974. Aerial and underwater visual acuity in the mink *Mustela vison*. *Anim. Behav.* 22:965–974.

Siniff, D. B., and Ralls, K. 1988. Population status of California sea otters. Final Report to Pacific Outer Continental Shelf Region of Minerals Management Service, U.S. Department of the Interior, Los Angeles, California under Contract No. 14-12-001-3003.

Smith, T. G. 1973. *Population Dynamics of the Ringed Seal in the Canadian Eastern Arctic*. Fisheries Research Board of Canada Bulletin no. 181. N.p., n.p.

Sokolov, A. S., and Sokolov, I. I. 1970. Some special features of the locomotory organs of the river and sea otters associated with their mode of life. *Byulleten' Moskovskogo Obshchestua Ispytalelei Prirody. Otdel Biologicheskii* 75:5–17.

Stephenson, A. B. 1977. Age determination and morphological variation of Ontario otters. *Canadian J. Zool.* 55:1577–1583.

Stirling, I. 1983. The evolution of mating systems in pinnipeds. In: J. F. Eisenberg & D. G. Kleiman, eds. *Recent Advances in the Study of Mammalian Behavior*, pp. 489–527. Special Publication no. 7. Lawrence, Kans.: American Society of Mammalogists.

Stubbe, M. 1970. Zur Evolution der analen Markierungsorgane bei Musteliden. *Biologisches Zentralblatt* 89:213–223.

Stubbe, M. 1977. Der Fischotter *Lutra lutra* (L. 1758) in der DDR. *Zoologischer Anzeiger* 199:265–285.

Tabor, J. E., and Wight, H. M. 1977. Population status of river otter in western Oregon. *J. Wildl. Mgmt.* 41:692–699.

Tarasoff, F. J. 1972. Comparative aspects of the hindlimbs of the river otter, sea otter, and harp seal. In: R. J. Harrison, ed. *Functional Anatomy of Marine Mammals,* 1:333–359. New York: Academic Press.

Tarasoff, F. J. 1974. Anatomical adaptations in the river otter, sea otter, and harp seal. In: R. J. Harrison, ed. *Functional Anatomy of Marine Mammals,* 1:111–141. New York: Academic Press.

Tarasoff, F. J., Bisaillon, A., Pierard, J., and Whitt, A. P. 1972. Locomotory patterns and external morphology of the river otter, sea otter, and harp seal (Mammalia). *Canadian J. Zool.* 50:915–927.

Tarasoff, F. J., and G. L. Kooyman. 1973a. Observations on the anatomy of the respiratory systems of the river otter, sea otter, and harp seal. 1. The topography, weight, and measurements of the lungs. *Canadian J. Zool.* 51:163–170.

Tarasoff, F. J., and Kooyman, G. L. 1973b. Observations on the anatomy of the respiratory system of the river otter, sea otter, and harp seal. 2. The trachea and bronchial tree. *Canadian J. Zool.* 51:171–177.

Trillmich, F., Kooyman, G. L., Mailuf, P., and Sanchez-Grinan, M. 1986. Attendance and diving behavior of South American fur seals during El Nino in 1983. In: R. L. Gentry & G. L. Kooyman, eds. *Fur Seals: Maternal Strategies on Land and at Sea,* pp. 153–167. Princeton, N.J.: Princeton Univ. Press.

Trowbridge, B. J. 1983. Olfactory communication in the European otter (*Lutra lutra*). Ph.D. dissert., University of Aberdeen, Aberdeen, Scotland.

Van der Zee, D. 1981. Prey of the Cape clawless otter (*Aonyx capensis*) in the Tsitsikama Coastal National Park, South Africa. *J. Zool. (Lond.)* 194:467–483.

Van der Zee, D. 1982. Density of Cape clawless otters *Aonyx capensis* (Schinz, 1821) in the Tsitsikama Coastal National Park. *South African J. Wildl. Res.* 12:8–13.

van Zyll de Jong, C. G. 1972. A systematic review of the Nearctic and Neotropical river otters (Genus *Lutra,* Mustelidae, Carnivora). *Life Sciences Contributions, Royal Ontario Museum* 80:1–104.

Vermeij, G. J. 1977. The mesozoic marine revolution: Evidence from snails, predators and grazers. *Paleobiology* 3:245–258.

Vermeij, G. J. 1978. *Biogeography and Adaptation: Patterns of Marine Life.* Cambridge: Harvard Univ. Press.

Walls, G. L. 1963. *The Vertebrate Eye.* New York: Hafner.

Watson, H. 1978. *Coastal Otters (Lutra lutra L.) in Shetland.* London: Vincent Wildlife Trust.

Wayre, P. 1974. Otters in western Malaysia. *Otter Trust Annual Report* 1974:16–38.

Wayre, P. 1979a. Otter havens in Norfolk and Suffolk, England. *Biol. Conserv.* 16:73–81.

Wayre, P. 1979b. *The Private Life of the Otter.* London: Batsford.

Wayre, P. 1980. Report of Council 1980—Breeding. *Otters: J. Otter Trust* 1980:6–8.

Wayre, P. 1981. Report of Council 1981—Breeding. *Otters: J. Otter Trust* 1981:6.

Wendell, F. E., Ames, J. A., and Hardy, R. A. 1984. Pup dependency period and length of reproductive cycle: Estimates from observations of tagged sea otters, *Enhydra lutris,* in California. *California Fish Game* 70:89–100.

Wickler, W., and Seibt, L. 1983. Monogamy: An ambiguous concept. In: P. Bateson, ed. *Mate Choice,* pp. 33–50. Cambridge: Cambridge Univ. Press.

Wilson, K. 1954. The role of mink and otter as muskrat predators in northwest North Carolina. *J. Wildl. Mgmt.* 18:199–207.

Wise, M. H., Linn, I. J., and Kennedy, C. R. 1981. A comparison of the feeding biology of mink *Mustela vison* and otter *Lutra lutra. J. Zool. (Lond.)* 195:181–213.

Wittenberger, J., and Tilson, R. 1980. The evolution of monogamy: Hypotheses and evidence. *Ann. Rev. Ecol. Syst.* 11:197–232.

Woodward, R. 1981. *On-the-rocks Birth of a Sea Otter Pup*. The Otter Raft no. 25. Carmel, Calif.: Friends of the Sea Otter.

Worthy, G. A. J., and Hickie, J. P. 1986. Relative brain size in marine mammals. *Amer. Nat.* 128:445–459.

Wright, R. A., and Allton, W. H. 1971. Sea otter studies in the vicinity of Amchitka Island. *Bioscience* 21:673–677.

Zabel, C. J. 1986. Reproductive behavior of the red fox (*Vulpes vulpes*): A longitudinal study of an island population. Ph.D. dissert., Univ. California, Santa Cruz. 98 pp.

CHAPTER 10

Ecological Constraints on
Predation by Large Felids

MEL E. SUNQUIST AND
FIONA C. SUNQUIST

The evolutionary fitness of any predator, whether it is a spider catching insects or a lion hunting buffalo, depends largely on the quality and quantity of its diet. Predatory strategies are shaped and refined by natural selection to maximize nutrient intake within the bounds of a wide range of ecological constraints (e.g., prey density, habitat) that may differ dramatically for the same species at the extremes of its geographical distribution. The basic task of finding and gathering food under these constraints fundamentally affects a species' spacing patterns and the structure of its social systems.

Any general discussion of ecological constraints on predation is bound to be complicated by definitions. The constraints must be biologically relevant (Clutton-Brock and Harvey 1983), as it is important that the variables reflect the animal's experience of its environment rather than the observer's (Jarman 1982). The problem is exacerbated when dealing with secretive, nocturnal predators because there may be factors that we as humans can neither perceive nor quantify. Certainly, we are a long way from being able to build realistic optimal foraging models for large predators.

For the purposes of this review we have defined a "large felid" as any adult cat with a body weight normally exceeding 36 kg (Guggisberg 1975; Gittleman 1985). Accordingly, the large felids include five species of *Panthera* (African lion, *P. leo;* tiger, *P. tigris;* jaguar, *P. onca;* leopard, *P. pardus;* snow leopard, *P. uncia*), one species of *Felis* (=*Puma*) (mountain lion, *F. concolor*), and one species of *Acinonyx* (cheetah, *A. jubatus*). The nomenclature follows that presented in Ewer (1973). The smaller cats were excluded from this review because of the general paucity of information on many aspects of their ecology and behavior.

Before considering the ecological constraints on predation by large felids, one must keep in mind those features or characteristics that are common to these species by virtue of phylogenetic inheritance (Martin, this volume) expressed as morphological (Van Valkenburgh, this volume) and physiological specializations.

283

The basic "phenotype set" is large, long-lived, carnivorous, and typically captures terrestrial prey at least half its own body weight (Gittleman 1985; Packer 1986). It forages over a large area and often exists at low densities (Gittleman and Harvey 1982; Robinson and Redford 1986). Females reproduce every two to three years and young spend a year to 18 months dependant on their mother while perfecting their hunting skills (Gittleman 1986). Prey are typically captured from ambush and/or a stalk and short rush or chase, and are dispatched with a swift killing bite delivered by the large bladelike canine teeth (Ewer 1973; Leyhausen 1979).

Only the cheetah differs radically from this generalized body plan and technique of prey capture. Cheetahs are specialized for high-speed pursuit of smaller prey and consequently exhibit a variety of morphological specializations (Ewer 1973).

Behaviorally, a solitary existence is compatible with felid specializations, and, with the exception of the African lion, the large felids are solitary hunters and feeders (Kleiman and Eisenberg 1973).

All felids are highly specialized carnivores, and their feeding habits are influenced by a number of ecological constraints that are to some degree measurable. Characteristics of the prey species such as their abundance, temporal and spatial distribution, size, defenses, and anti-predator tactics may to varying degrees represent ecological constraints on predation. In addition, the distribution and abundance of hunting cover, climatic conditions, and the presence and abundance of congeners and other potential competitors can act as constraints. However, an ecological constraint for one species may not equally influence a different species. In this chapter we review the most important ecological factors that affect foraging by large felids and draw examples from a variety of field studies that illustrate the effects of these factors. We then look at the ways these ecological constraints may have operated to shape felid social systems.

Prey Characteristics

Prey Density

Within any given area there is a measurable amount of energy that can be considered as potential food for carnivores. The density of prey species can be estimated using a variety of methods (e.g., transects, sample area counts, quadrats), and density is commonly expressed as total biomass, which is calculated by multiplying average density by the average individual weight.

Prey density and biomass figures for an area are useful because high herbivore densities usually mean more food for predators and thus higher predator densities. However, prey densities do not necessarily represent the amount of food available to predators (Bertram 1973). To take an extreme example, herbivore biomass in an area may be high, but if it consists largely of elephants and rhinos, then the adult prey are inaccessible to large felids because the prey are too large to kill. Similarly, humid, tropical evergreen forest supports a high mammalian biomass, but a large percentage of the species are arboreal (Eisenberg et al. 1972; Eisenberg 1980) and generally unavailable to the large felids.

Prey Distribution in Space

The amount of energy available in an area is determined by prey density, but a critical component is how this energy is spatially distributed. Whether prey are distributed randomly, evenly, or clumped influences predators' search time and has important energetic, spatial, and social consequences for them (Davies and Houston 1984). Also, if prey are arboreal or fossorial, they are not likely to be vulnerable to predation by large felids. African lions may occasionally dig warthogs (*Phacochoerus aethiopicus*) out of burrows, but this method is probably not often effective (Van Orsdol 1984). Similarly, young and adult female leopards may be more successful than males at hunting in trees (Muckenhirn and Eisenberg 1973), but the frequency of capture is probably low.

Prey Distribution in Time

Whether prey are migratory, sedentary, or renewed at some interval or rate has important consequences for predators. If prey make large seasonal movements in response to environmental changes, such as is seen in the Serengeti (Maddock 1979), Idaho (Seidensticker et al. 1973), and the Kalahari (Owens and Owens 1984), then predators either have to abandon or expand their ranges and follow the migratory prey or utilize alternative prey species (Schaller 1972; Seidensticker et al. 1973; Matjushkin et al. 1977; Hanby and Bygott 1979; Owens and Owens 1984; Van Orsdol et al. 1985). For example, in Idaho mule deer (*Odocoileus hemionus*) and elk (*Cervus canadensis*) are concentrated during the winter but are widely dispersed in the summer. The ranges of several radio-tagged resident female mountain lions in the same area were almost twice as large in summer as in winter (Seidensticker et al. 1973). During a period of drought in the Kalahari Desert large prey species were widely dispersed and the range of a resident pride of African lions increased from 702 km^2 to more than 3900 km^2 (Owens and Owens 1984). In the Soviet Far East tigers prey mainly on red deer (*C. elaphus*), wild pig (*Sus scrofa*), and moose (*Alces alces*) (Matjushkin et al. 1977). Some of these prey make large seasonal movements and tiger ranges are large; adult female ranges are 200–400 km^2 and those of adult males are 800–1000 km^2 (Matjushkin et al. 1977; Bragin 1986). In contrast, sedentary, predictable prey resources are often associated with small, exclusive predator ranges or territories (Brown 1964; Gill and Wolf 1975, 1977; Davies and Houston 1984). In the Himalayas snow leopards are typically found above timberline in dry, rocky, alpine steppe (Jackson and Ahlborn 1984). Given the nature of the terrain, range sizes would be expected to be large. However, the only radio-tracking study of snow leopards to date found that the ranges of several adults were relatively small (ca. 30 km²) and overlapped extensively (Jackson and Hillard 1986). Clearly, Jackson's study site contained excellent snow leopard habitat. Leopards in Sri Lanka maintain small (8–10 km²), exclusive ranges that are related to an abundant but localized prey base (Eisenberg and Lockhart 1972; Muckenhirn and Eisenberg 1973). Similarly, an abundant and seasonally stable prey base (i.e., nonmigratory) on the floodplain at Chitwan National Park, Nepal, is related to small (10–51 km²), exclusive ranges of tigresses (Sunquist 1981; Smith et al. 1987).

Prey distribution in space and time is a critical factor for maternal females because they are initially confined to an area near the den and are thus restricted in their foraging radius. Field data on leopards, tigers, mountain lions, and ocelots (*Felis pardalis*) show a dramatic reduction in the maternal home range immediately following the birth of cubs. Under these conditions females must locate and kill prey, feed, and return to the den every 24–36 h (Seidensticker et al. 1973; Seidensticker 1977; M. Sunquist, pers. observ.). Thus, prey distribution and "catachability" (Bertram 1973) are vital to females in the first two months of maternal dependency when cubs are largely immobile. The young of large felids are nutritionally dependent on their mother for one to two years.

Temporal or spatial separation of predators may also be related to the rate or interval at which prey are renewed. Waser (1980) suggested that a high rate of renewal of insect prey was related to high population densities, complete overlap of ranges, and low levels of competition among small, nocturnal carnivores in the Serengeti. For predators feeding on large ungulates the renewal rate may be affected by other conspecifics and competitors foraging in the same area, as prey are likely to be less vulnerable in an area that has recently been hunted (Hornocker 1970; Charnov et al. 1976).

Prey Quality and/or Size

Optimal foraging theory predicts that predators ought to choose the most "profitable" prey (MacArthur and Pianka 1966; Schoener 1971; Pulliam 1974; Werner and Hall 1974; Charnov 1976). Elegant experiments on shore crabs (*Carcinus maenas*) (Elner and Hughes 1978), sticklebacks (*Spinachia spinachia*) (Kislalioglu and Gibson 1976), pied wagtails (*Motacilla alba*) (Davies 1977), bluegill sunfish (*Lepomis macrochirus*) (Werner and Hall 1974), great tits (*Parus major*) (Krebs et al. 1972), and redshanks (*Tringa totanus*) (Goss-Custard 1977) demonstrated that these predators did indeed select the most profitable prey. For large felids the most profitable prey type would seem to be the largest available prey that could safely be killed. However, the importance of search time, encounter rates, and the energetic costs of capture for various prey types also need to be taken into account. Furthermore, if large prey are uncommon, search time may be increased to a point where it becomes too energetically costly. Smaller but more abundant prey might then be more profitable. Although differences in prey digestibility and nutrient quality may be selected for by some predators, their importance to large felids is not likely to be significant because they possess constant and efficient digestive systems (Ewer 1973).

The large felids are known occasionally to take very large prey, but the modal prey size is usually less than their body weight (Packer 1986). However, modal prey size can vary considerably between different geographic areas for the same species. The two interrelated factors most strongly influencing modal prey size of large felids appear to be availability and vulnerability. Mountain lions in Idaho killed equal numbers of elk (175 kg) and mule deer (64 kg), though mule deer were more abundant. Elk were apparently more vulnerable in the winter when they were forced into terrain offering ideal hunting conditions for mountain lions (Hornocker 1970). Mountain lions also took small prey, in addition to elk and mule deer. One female with young subsisted for a summer feeding primarily on ground squirrels (*Spermophilus columbianus*) (1 kg) and an occasional elk or deer (Seidensticker et al. 1973).

In south Florida mountain lions prey mainly on feral hogs (25 kg), white-tailed deer (*Odocoileus virginianus*) (40 kg), and raccoons (*Procyon lotor*) (5 kg) (Belden and Maehr 1986). Small (1–10 kg) prey are also important in the diet of mountain lions in southeastern Peru (Emmons 1987) but, as Emmons indicated, the sample size is too small for confident inference.

Jaguars in southeastern Peru preyed mainly on larger prey, including deer, capybara (*Hydrochoerus hydrochaeris*), and peccary (*Tayassu tajacu*) (Emmons 1987). Although jaguar took agouti (*Dasyprocta variegata*), paca (*Agouti paca*), deer, and capybara in proportion to their estimated abundance, they killed peccary more often than was predicted on the basis of availability, probably because they were more vulnerable (Emmons 1987). Jaguars in Belize preyed mainly on the small (5–6 kg) but abundant and particularly vulnerable armadillo (*Dasypus novemcinctus*) (54% of scats contained armadillo); relative densities of larger prey such as the paca, collared anteater (*Tamandua mexicana*), brocket deer (*Mazama americana*), and peccary appeared to be much less than that of armadillo, and large prey species occurred much less frequently in scats (Rabinowitz and Nottingham 1986).

Tigers in Chitwan killed sambar (*Cervus unicolor*) more often than predicted on the basis of availability, suggesting that this large (150–250 kg) deer was selected for or that they were more vulnerable than the smaller but more abundant chital (*Axis axis*) and hog deer (*A. porcinus*) (Sunquist 1981). The gaur (*Bos gaurus*), an extremely large (450–900 kg) bovid, was also present in

Chitwan, but it was rare and mainly confined to the hills. There was no evidence from kills or feces that tigers killed gaur at Chitwan, and it was surmised that this species was invulnerable to predation largely by virtue of its size (Sunquist 1981). However, Schaller (1967) reported that tigers at Kanha National Park, India, occasionally killed adult male gaur. In Nagarahole National Park in south India gaur densities are high, and this species makes up a substantial portion of the tiger's diet (K. Ullas Karanth, pers. comm.). Thus, the low density of gaur was probably the reason tigers at Chitwan did not prey on this large bovid.

At Lake Manyara National Park, Tanzania, buffalos (*Syncerus caffer*) were the most abundant of the large mammals and constituted 62% of African lion kills (Makacha and Schaller 1969; Schaller 1972). Of all buffalo kills, 81% were adult males; males were often separated from the herds and thereby apparently were more vulnerable to predation. At Manyara the modal prey size of lions would be in excess of 400 kg.

In most parts of the Serengeti African lions preyed mainly on wildebeests (*Connochaetes taurinus*) and zebras (*Equus burchelli*) when they were abundant during the annual migration; at other times buffalos and topi (*Damaliscus korrigum*) were the lion's main prey (Schaller 1972). All of these species are large, the modal prey size being approximately 150 kg (Packer 1986). Lions in several other African national parks, including Kafue, Kruger, Nairobi, and Albert, also prey mainly on the more numerically abundant large prey (see Schaller 1972). In Rwenzori National Park, Uganda, the modal prey size of lions in one area was 40–50 kg, whereas in an adjoining area of the park where prey densities were five times higher, the modal prey size was 65–100 kg (Van Orsdol 1982). In both areas lions preyed mainly on the larger, numerically abundant species.

Eloff (1973) found that in areas where large prey are relatively scarce, as in the Kalahari Desert, small mammals and juveniles constituted more than 50% of the lion's diet, and porcupines (*Hystrix africaeaustralis*) alone accounted for almost 26% of the kills. Gemsbok (*Oryx gazella*), particularly calves, formed the major part of the lion's diet. Under these circumstances small mammals (<50 kg) seem to be the most profitable prey.

The Kalahari Desert leopard, being considerably smaller than the lion, seems to be able to coexist with the lion by taking more small prey (Bothma and LeRiche 1986). A small sample of kills by both male and female leopards in the Kalahari indicated that most of the prey killed weighed less than 30 kg, and many kills were of animals weighing not much above 5 kg (Bothma and LeRiche 1986).

Similar observations were reported by Hoppe (1984) for leopards in the Ivory Coast, where 39% of scats contained small prey (<5 kg), 58% contained medium-sized prey (5–45 kg), and only 3% contained large prey (>45 kg). Small bovids, principally duikers (*Cephalophus* spp.), were the most common prey. Interestingly, almost 40% of the prey were arboreal, including at least seven primate species. This is the only report that suggests that leopards are significant predators of aboreal primates (reviewed by Cheney and Wrangham 1986).

Food habit studies of leopards in Chitwan (Seidensticker 1976; Sunquist 1981), Rhodesia (Smith 1978), Kenya (Hamilton 1976), and the northern Serengeti (Bertram 1982) also indicate that small animals (2–40 kg) can be the principal prey of adult leopards. In all of these studies leopards preyed mainly on those prey species that were abundant. However, studies of leopards in a variety of other areas (e.g., Kruger, Serengeti, Wilpattu, Kafue parks) suggest that leopards regularly take somewhat larger prey (Mitchell et al. 1965; Kruuk and Turner 1967; Pienaar 1969; Schaller 1972; Muckenhirn and Eisenberg 1973). Leopards are, however, found in a broad range of geographical locations, and their diet appears to be more varied than that of any other large felid.

Though even the largest felids can subsist and rear young on small (5–10 kg), abundant prey (Seidensticker et al. 1973; Owens and Owens 1984; Rabinowitz and Nottingham 1986), they are morphologically specialized to kill prey as large or larger than themselves and readily do so when the opportunity arises. In this sense the large felids forage optimally, and differences in food habits within a species seems largely to reflect differences in availability and vulnerability to various prey species.

Prey Defenses

It is difficult to measure the extent to which prey characteristics influence predator behavior. In open habitats many ungulate species form large herds, a behavior that may make prey less vulnerable to predators. Taylor's (1976) model of predation predicts that prey clumping almost always benefits the prey and hinders the predator. However, larger group size does not necessarily mean more effective predator detection. Van Orsdol (1984) reported that Africa lions hunting at night were detected significantly more often by kob (*Adenota kob*) than by topi even though the mean group size of kob was half that of topi. However, Schaller (1972) states that topi are the most vigilant and least vulnerable prey species in the Serengeti. In the case of a small sample of tiger and leopard kills in Chitwan, Mishra (1982) observed that no radio-collared chital were killed when the chital congregated (mean group size = 16.7) on cut and burned grasslands; all kills of collared chital occurred between April and December, when the vegetation was highest and chital group size was smaller (mean group size = 6.6).

Prey species may also have other defense tactics. Primates may climb, warthogs and armadillos may take refuge in burrows, whereas others may defend themselves with horns, antlers, tusks, and spines. Serious injuries and fatalities associated with prey capture have been noted for lions, mountain lions, and tigers, (Corbett 1944; Hornocker 1970; Packer 1986), but no data are available to indicate the frequency of injury by prey type. However, any injury that incapacitates a solitary predator may have serious consequences, as it cannot rely on other conspecifics to provide food and thus could easily starve to death.

Prey may also alter their distribution in response to the presence of predators. Temporary refuges can provide protection for prey at certain times of the day, year, or season. For example, chital deer in many national parks and reserves in India exhibit a nightly yarding behavior (Johnsingh 1983; pers. observ.). At dusk hundreds of chital collect in the open grounds of the park headquarters near staff houses; they spend the night on lawns close to buildings and return to the forest at dawn. These nightly aggregations are obviously not for the purpose of grazing, as there is very little forage available close to the buildings. Tiger and leopard, two of the most important predators of chital, are primarily nocturnal, whereas dhole (*Cuon alpinus*) are predominantly diurnal, suggesting that these chital aggregations are primarily to avoid felid predation. However, Johnsingh (1983) found that 23% of dhole kills, which included a few chital, were made before sunrise or after sunset, so it would seem that the yarding behavior of chital would also offer some protection against dhole predation. By spending the night in a relatively open area close to human habitation, the chital reduce their chances of being captured by all three of their most common predators. The fact that this yarding behavior does not occur throughout the chital's distribution may reflect differences in predation pressure and/or a cultural transmission of the behavior itself.

Mech (1977) also hypothesized that deer yarding behavior provided significant anti-predator benefits. In a study conducted in northern Minnesota, Nelson and Mech (1981) found that deer increased their chances of winter survival by migrating to large yards, especially those near human habitation or in buffer zones between wolf (*Canis lupus*) pack ranges.

Hunting Cover

All felids rely extensively on physical features in their environment, using almost any type of cover to get as close as possible to prey before making the final attack. Despite the importance of this maneuver, few studies have quantitatively analyzed its effects on felids' hunting success. The studies that have addressed this variable indicate that the minimum distance covered in the final charge is strongly correlated with a successful kill. Using data from actual stalks and computer simulations, Elliott et al. (1977) showed that African lions in Ngorongoro Crater had a high probability (0.8) of catching Thomson's gazelle (*Gazella thomsoni*) when the attack was launched at distances of 7.6 m or less, but at 15.2 m the probability was zero. Elliott et al. calculated that lions hunting the larger wildebeestes and zebras had a 50% chance of success at distances of 15.2 m. Similarly, in Queen Elizabeth National Park, Uganda, lions that were able to approach within about 20 m of prey were much more likely to succeed in making a kill (Van Orsdol 1984). Matjushkin et al. (1977), interpreting tracks in snow, concluded that tigers in the Soviet Far East sometimes launched the attack from as far away as 30 m, but more commonly from 10–15 m. They estimated that tigers were successful only once every three to five attempts but gave no success rates for varying distances.

The presence of stalking cover directly influences the distance traversed in the final charge. Elliott et al. (1977) suggest that 0.4 m of grass cover is necessary for successful daytime lion hunts. Van Orsdol (1984) stated that lion hunting success increased with grass height up to 0.8 m, and Schaller (1972) found that lions hunting in grass 0.3–0.6 m high were twice as successful as those hunting in grass less than 0.3 m high.

In many cases it is difficult to classify habitat constraints categorically. For example, in Chitwan National Park, Nepal, tigers infrequently hunted in the tall grassland portions of their ranges after the area had been burned and cover was reduced to almost zero (Seidensticker 1976; Sunquist 1981). Did the tigers alter their hunting patterns in response to the lack of stalking cover or was it because the usually dispersed prey species formed large groups (Mishra 1982) on the burned-over grasslands and became less vulnerable to predation? The importance of stalking cover was implicated by the observation that leopards frequently hunted the burns (Seidensticker 1976), suggesting there was enough stalking cover for them, and at this time tigers killed more sambar, a forest dwelling deer, than at any other time of the year (Sunquist 1981). Sambar are, however, essentially solitary and thus may be more vulnerable than group-living deer, especially when stalking cover is reduced (Mishra 1982).

As most felids are nocturnal (Gittleman 1985) and do much of their hunting under the cover of darkness, it seems logical to assume that lunar phase might also affect hunting success. Logistic problems have hampered most field scientists from gathering such information, but L. Emmons et al. (unpubl. ms.) report that ocelots in southeastern Peru spent equal time foraging on moonlit and dark nights, but on moonlit nights they confined their hunting to dense cover. Van Orsdol (1984) also found that lunar phase influenced lions' hunting success; they were almost twice as successful catching prey during moonless hours than during moonlit ones; at one of Van Orsdol's study site (Ishasha), no successful hunts were recorded on moonlit nights compared with 30.4% success on moonless nights.

Climatic Conditions

Climatic factors may dictate the time available for foraging and even affect the susceptibility of prey, or the effectiveness of the predator. Most large predators encounter periods of intense heat, cold, flooding, and storms, and many field studies suggest that climatic conditions affect hunting strategies and sometimes hunting success. Hornocker (1970) found that mountain lions were able to kill more elk in the winter when snow depth forced the elk into steep terrain where they were apparently more vulnerable. In Nepal tigers hunted less during the daytime in the hot season, and during periods of flooding they shifted their activities to higher, drier ground (Sunquist 1981). Van Orsdol (1984) found that African lions tended to initiate hunts more often when storms were imminent, and he hypothesized that this was related to the prey's reduced ability to detect the predator. Matjushkin et al. (1977) reported that in the Soviet Far East tigers have difficulty traveling for a two- to four-month period in the winter when snow cover is soft and deep (35–50 cm).

Scavengers and Other Competitors

Whether scavengers represent a constraint on predation by large felids depends on the abundance of scavengers and the frequency and amount of losses. Furthermore, losses to scavengers may not be energetically significant if ample prey are available, and some losses may be offset by felids' appropriating kills from other predators. Not surprisingly, few studies have documented these complex interactions. In the Serengeti, Schaller (1972) reported that of 238 cheetah kills, 32 (12%) were lost to other predators, primarily lions, before the cheetah had finished eating. Lions can usually appropriate kills from other predators because they are larger and live in groups (Schaller 1972). Similarly, spotted hyena (*Crocuta crocuta*) groups may displace single lions from carcasses (Kruuk 1972). From a sample of 23 large kills (i.e., eland, *Taurotragus oryx*; wildebeest; zebra) made by lions in the Serengeti, 44% were appropriated by hyenas before the lions had finished the carcass (Schaller 1972). However, Packer (1986) concluded that the amount of meat lost to hyenas is negligible as lions usually surrender only the remnants. He suggests that loss of meat to conspecifics, both pridemates and others, is more important, especially with medium-sized (100–250 kg) and large (>300 kg) carcasses. Packer further suggests that the openness of the lion's habitat coupled with high lion density and large modal prey size contribute to greater losses of meat by lions than by other large felids.

Leopards and other large felids probably suffer less significant losses to nonfelid scavengers either because prey are small and consumed rapidly or there are few scavengers in the area capable of supplanting the cat at a kill (Bertram 1979; Houston 1979). Leopards occasionally lose kills to hyenas and lions (Kruuk 1972; Schaller 1972; Owens and Owens 1984), and the Indian wild dog (dhole) has been known to drive tigers off kills (Schaller 1967), but this probably occurs infrequently. In some areas leopards take their kills into trees, which effectively eliminates scavenging by lions, tigers, or hyenas (Houston 1979). Additionally, tigers, leopards, and jaguars typically drag their kills into dense cover before feeding (Schaller 1967; Eisenberg and Lockhart

1972; Hamilton 1976; Schaller and Vasconcelos 1978; Sunquist 1981), a behavior that reduces scavenging even by vultures.

Intraspecific competition for kills among solitary felids seems to be rare, and the possibility of such felids' meeting at a kill is also reduced because of their temporally and spatially dispersed social system. Like-sexed ranges tend not to overlap, although each male's range usually encompasses the ranges of several females (Hamilton 1976; Seidensticker et al. 1973; Sunquist 1981; Caro and Collins 1986; Rabinowitz and Nottingham 1986). Whether the larger males compete with females for prey is not known, but the male's presence probably reduces food losses to females by preventing other males from settling in an area. When associations among solitary felids do occur at kills, they usually consist of a female with her young, although Seidensticker et al. (1973) observed mountain lions in large, temporary associations at large kills. Similar behavior is seen in tigers at bait sites (Schaller 1967; McDougal 1977), but few large associations have been recorded at natural kills (Sunquist 1981).

The effect of human scavenging of kills made by large felids seems to be insignificant, as judged by the infrequent reference to its occurrence. In Nepal, Sunquist (1979) reported that a tigress lost ten kills (seven sambar, two wild pigs, one hog deer) to villagers in an eight-month period. In most cases the tigress had fed for one or two days before the kill was scavenged. Although these losses appear insignificant in the short term, repeated losses can have dire consequences. An example is seen in Joslin's (1973) study of the Asian lion in the Gir forest, India. At that time lions in the Gir were feeding mainly on domestic livestock, and local herdsmen and hide and meat collectors regularly appropriated lion kills. Lions had not even begun to feed on 22% of their kills before the carcass was removed by people, and lions were displaced from more than 50% of their kills. These losses coupled with losses to vultures forced the lions to kill more often to secure food, which only increased the conflicts with herdsmen. Joslin (1973) concluded that competition between lions and humans for kills was a major factor in the decline of lions in the Gir.

Social Dominance and Other Behavioral Attributes

Throughout their geographic range the number of large felid species in an area ranges from one to three, with two being common. Sympatric felid congeners usually differ in size by a factor of two to four (Stanley et al. 1983; Van Valkenberg 1985), but the extent of competition between species is largely unknown. Few studies have been carried out on two large cats residing in the same place at the same time. In Chitwan National Park, Nepal, Seidensticker (1976) reported that tigers and leopards differed in size of prey killed, habitats used, and activity times. Leopards in the same area infrequently traveled on roads, whereas tigers commonly did (Sunquist 1981). That tigers in Chitwan act as a constraint on leopards is also inferred from the greater dietary diversity of leopards in Sri Lanka, where tigers are absent, though the same prey species are available in both areas (Seidensticker 1976). Seidensticker (1976) suggests that coexistence by tigers and leopards in Chitwan is facilitated by an abundance of prey, much of which is small, and by dense vegetation, which restricts opportunities for interaction.

In the northern Serengeti, Bertram (1982) found that African lions and leopards used the same area, were active at the same times, but differed in that lions tended to kill larger prey than did leopards. The only large prey killed by leopards were young animals. Leopards killed a wider variety of prey than lions did, but there was little overlap between the cats in terms of the prey species killed. Bertram (1982) concluded that there was little ecological competition and that the presence of retreats allowed leopards to coexist with the larger, dominant lions.

Jaguars and mountain lions live sympatrically in portions of Central and South America, but little information is available to assess the extent of competition between these cats. Both mountain lions and jaguars are capable of killing large prey (Hornocker 1970; Schaller and Crawshaw 1980; Ackerman et al. 1986; Anderson 1983), although one might expect the larger jaguar to be socially dominant and to kill larger prey than does the mountain lion (see Gittleman 1985). In Brazil mountain lions have been observed to avoid the larger jaguar (Schaller and Crawshaw 1980), though little dietary separation was noted. In Peru, Emmons (1987) observed jaguars to use riparian areas more than did mountain lions, suggesting that the two species differed in habitat use and food habits. In Belize both species apparently prey extensively on small mammals (Rabinowitz and Nottingham 1986), suggesting that both cats are killing the most readily available prey.

Presumably subordinate conspecifics may also be constrained to hunt at different times of the day or in different habitats, although the limited available data on solitary felids suggest that subadults forage in a manner similar to adults (Schaller 1972; Seidensticker et al. 1973; Hamilton 1976; McDougal 1977; Sunquist 1981; Bertram 1982; Rabinowitz and Nottingham 1986).

Conclusions

Most authors agree that food dispersion is one of the major ecological factors that influence mammalian social organization (Alexander 1974; Jarman 1974; Bradbury and Vehrencamp 1976, 1977; Clutton-Brock and Harvey 1977; see also Gittleman, this volume). Similarly, the distribution and abundance of prey exerts an importance influence on carnivore spatial organization (Kruuk 1972; Kleiman and Eisenberg 1973; Macdonald 1983). Studies of European badgers (*Meles meles*) (Kruuk 1978), golden jackal (*Canis aureus*) (Macdonald 1979), and other canids (see Moehlman 1986) have convincingly tied social groupings to resource dispersion. However, it has proved more difficult to measure the influence of prey availability and distribution as conclusively for other carnivores.

The openness of the habitat is also strongly correlated with the existence and size of groups (Clutton-Brock and Harvey 1977; Gittleman, this volume). And, more recently, Packer (1986) has suggested that other constraints such as the level of intraspecific competition may also profoundly influence the formation of groups by African lions.

Obviously, these ecological pressures vary in their impact throughout a species' geographical range. Each may or may not be independent of the other, and in many cases they act differently on different sex and age classes. Unraveling the effects of this web of ecological constraints on predatory and social behavior becomes easier when an animal's behavior is compared at two different study sites, as Van Orsdol (1982, 1984) and Van Orsdol et al. (1985) showed in comparative studies of lions.

However, ecological constraints and phylogenetic history are only part of the biological complex within which the observed social system exists. Kinship is another of the important factors influencing social relationships, and with the notable exception of the African lion, the influence of kinship on felid social systems has not been investigated.

In solitary mammals, females tend to be philopatric and have tolerant relationships with female kin (Waser and Jones 1983). If cooperative traits are to develop, they should, as suggested by Armitage (1986), occur among members of the sedentary sex. Though data do not yet exist for most felid species, we expect that solitary felids will show similar female philopatry. Recent results from a long-term study of tigers in Chitwan show that daughters commonly establish ranges next to their mothers, creating pockets of related females. The coefficient of relatedness within the study area is presently about 0.35 (Smith et al. 1987), which is similar to that for females in a lion pride.

The basic pattern of felid social organization is one in which males occupy larger, exclusive ranges that encompass several female ranges. The pattern of female ranges is keyed to resource distribution. Several females may use a large area but have separate core areas, or they may use a common space at different times. Their ranges may overlap partially or not at all, but regardless of the spatial arrangement, females (excluding lionesses) do not hunt or rear young cooperatively.

However, Packer (1986) has shown that the advantages of cooperative hunting fail to explain sociality in the lion. Instead, he argues that the ecological constraints that favor felid sociality are large prey size, high lion density, and open habitat. Under these conditions a kill quickly attracts conspecific scavengers. Packer hypothesizes that females allow their daughters to share their ranges because the costs of sharing food with relatives is low, and at high population densities dispersing subadult females would have difficulty establishing ranges. Packer's data are extensive and convincing, but he has yet to demonstrate that the costs of dispersal are high or would be if lions were not social. Indeed, as predator density reflects resource distribution and abundance, the costs of dispersal for lions may be no higher than for any other felid.

Information on the other large felids is patchy at best and provides only glimpses of the total picture. However, a recent observation of a group of tigers in Ranthambhore National Park in India (Thapar 1986) lends support to the hypothesis that where the habitat is open and prey are large, group living might be favored by the high costs of solitary life rather than by the high costs of dispersal (Wrangham 1982; Wrangham and Rubenstein 1986).

Ranthambhore is the only area that we know of where tigers live in comparatively open habitat. The vegetation is dry and deciduous, and prey concentrate around waterholes during the dry season. It is also the only place we know of where one of the tiger's most important prey species, the large (>250 kg), normally solitary sambar deer, lives in large herds. Some tigers in Ranthambhore have been identified by their facial markings, and a few genealogies are known. One observation of an aggregation of tigers at a natural kill confirms the importance of Packer's criteria of open habitat and intraspecific scavenging in the evolution of groups, and illustrates that other felids occasionally encounter situations where there are high costs associated with remaining solitary.

A tigress with three large cubs killed a 250-kg nilgai bull (*Boselaphus tragocamelus*) and was joined at the carcass by several neighboring tigers. They included two of the tigress's adult daughters from two previous litters, an adult son, an unrelated female, and an unidentified animal. No fights were observed and the animals fed, one at a time, respecting the owner's priority to the kill. The carcass was consumed in 24 hours and the tigers went their separate ways. By contrast, in the closed habitat of Chitwan neighbors rarely discovered one another's kills, and a 250-kg prey would feed a tigress and cubs for three days (Sunquist 1981).

It is unlikely that the volume and quality of information on the other large felids will ever approach that available for the lion. But improved radiotelemetry techniques and long-term studies of the "solitary" felids may reveal a new level of complexity in their spatially dispersed social systems. It would, for example, be interesting to quantify the frequency of natural kills shared by neighboring tigresses as a function of prey size and degree of relatedness. Related females sharing adjoining ranges probably interact far more than previously suspected, and the potential for interaction by females whose ranges overlap would be even higher.

L. Emmons (pers. comm.), for example, followed radio-tagged ocelots continuously during nocturnal hunting periods. She found that though both males and females hunted alone, they frequently encountered each other during the night. Meetings were either a brief encounter, lasting only a few seconds, or longer, lasting for several hours. The ocelots occasionally stopped and rested a few meters apart, or one followed another for several hundred meters.

Although more detailed studied of the social organization of solitary felids will almost certainly reveal new levels of complexity, for most felids the costs of remaining solitary are not outweighed by the benefits of living in a group. The frequent anecdotal references to groups of five to 10 individuals among tigers (see Schaller 1967; Bragin 1986) and other solitary felids—mountain lions (Seidensticker et al. 1973), jaguarundis (*F. yagouaroundi*) (Guggisberg 1975)—most likely represent occasions when the costs of maintaining a solitary lifestyle was too high. It is interesting to note that some of these aggregations occurred in species that occupy open habitats. Though these gatherings illustrate the flexibility of felid social systems, the phylogenetic characteristics of the family and ecological constraints of the environment rarely combine in a situation where the costs of solitary life are high enough to promote the formation of permanent social groups.

Acknowledgments

We thank John Eisenberg and Tim Caro for their comments on an earlier draft of this paper. The patience of John Gittleman is also deeply appreciated.

References

Ackerman, B. B., Lindzey, F. G., and Hemker, T. P. 1986. Predictive energetics model for cougars. In: S. D. Miller & D. D. Everett, eds. *Cats of the World: Biology, Conservation and Management,* pp. 333–352. Washington, D.C., National Wildlife Federation.

Alexander, R. D. 1974. The evolution of social behavior. *Ann. Rev. Ecol. Syst.* 5:325–383.

Anderson, A. E. 1983. *A Critical Review of Literature on Puma* (Felis concolor). Special Report no. 54. Denver: Colorado Division of Wildlife.

Armitage, K. B. 1986. Marmot polygyny revisited: Determinants of male and female reproductive strategies. In: D. I. Rubenstein & R. W. Wrangham, eds. *Ecological Aspects of Social Evolution,* pp. 303–331. Princeton, N.J.: Princeton Univ. Press.

Belden, R. C., and Maehr, D. S. 1986. *Florida Panther Food Habits.* Annual Performance Report. Tallahassee: Florida Game and Fresh Water Fish Commission.

Bertram, B. C. R. 1973. Lion population regulation. *East African Wildl. J.* 11:215–225.

Bertram, B. C. R. 1979. Serengeti predators and their social systems. In: A. R. E. Sinclair & M. Norton-Griffiths, eds. *Serengeti: Dynamics of an Ecosystem,* pp. 221–248. Chicago: Univ. Chicago Press.

Bertram, B. C. R. 1982. Leopard ecology as studied by radio tracking. *Symp. Zool. Soc. London* 49:341–352.

Bothma, J. Du P., and LeRiche, E. A. N. 1986. Prey preference and hunting efficiency of the Kalahari Desert leopard. In: S. D. Miller & D. D. Everett, eds. *Cats of the World: Biology, Conservation and Management,* pp. 389–414. Washington, D.C., National Wildlife Federation.

Bradbury, J. W., and Vehrencamp, S. L. 1976. Social organization and foraging in emballonurid bats. II. A model for determination of group size. *Behav. Ecol. Sociobiol.* 1:383–404.

Bradbury, J. W., and Vehrencamp, S. L. 1977. Social organization and foraging in emballonurid bats. III. Mating systems. *Behav. Ecol. Sociobiol.* 2:1–17.
Bragin, A. P. 1986. Territorial behaviour and possible regulatory mechanisms of population density in the Amur tiger (*Panthera tigris altaica*). *Zoologicheskii Zhurhal* 65:272–282.
Brown, J. L. 1964. The evolution of diversity in avian territorial systems. *Wilson Bull.* 76:160–169.
Caro, T. M., and Collins, D. A. 1986. Male cheetahs of the Serengeti. *Natl. Geogr. Rep.* 2:75–86.
Charnov, E. L. 1976. Optimal foraging: The marginal value theorem. *Theoretical Population Biol.* 9:129–136.
Charnov, E. L., Orians, G. H., and Hyatt, K. 1976. Ecological implications of resource depression. *Amer. Nat.* 110:247–259.
Cheney, D., and Wrangham, R. 1986. Predation. In: B. B. Smuts, D. L. Cheney, R. M. Seyfarth, R. W. Wrangham, & T. T. Struhsaker, eds. *Primate Societies,* pp. 227–239. Chicago: Univ. Chicago Press.
Clutton-Brock, T. H., and Harvey, P. H. 1977. Primate ecology and social organization. *J. Zool. (Lond.)* 183:1–39.
Clutton-Brock, T. H., and Harvey, P. H. 1983. The functional significance of variation in body size among mammals. In: J. F. Eisenberg & D. G. Kleiman, eds. *Advances in the Study of Mammalian Behavior,* pp. 632–658. Special Publication no. 7. Lawrence, Kans.: American Society of Mammalogists.
Corbett, J. 1944. *The Man-Eaters of Kumaon.* London: Penguin Books.
Davies, N. B. 1977. Prey selection and social behavior in wagtails (*Aves: Motacillidae*). *J. Anim. Ecol.* 46:37–57.
Davies, N. B., and Houston, A. I. 1984. Territory economics. In: J. R. Krebs & N. B. Davies, eds. *Behavioural Ecology: An Evolutionary Approach* (2nd ed.), pp. 148–169. Oxford: Blackwell Scientific Publications.
Eisenberg, J. F. 1980. The density and biomass of tropical mammals. In: M. E. Soulé & B. A. Wilcox, eds. *Conservation Biology,* pp. 35–55. Sunderland, Mass.: Sinauer.
Eisenberg, J. F., and Lockhart, M. 1972. An ecological reconnaissance of Wilpattu National Park. *Smithsonian Contrib. Zool.* 101:1–118.
Eisenberg, J. F., Muckenhirn, N. A., and Rudran, R. A. 1972. The relation between ecology and social structure in primates. *Science* 176:863–874.
Elliott, J. P., Cowan, I. M., and Holling, C. S. 1977. Prey capture in the African lion. *Canadian J. Zool.* 55:1811–1828.
Elner, R. W., and Hughes, R. N. 1978. Energy maximisation in the diet of the shore crab, *Carcinus maenas* (L.). *J. Anim. Ecol.* 47:103–116.
Eloff, F. C. 1973. Lion predation in the Kalahari Gemsbok National Park. *J. South African Wildl. Mgt. Assoc.* 3:59–63.
Emmons, L. H. 1987. Comparative feeding ecology of felids in a neotropical rainforest. *Behav. Ecol. Sociobiol.* 20:271–283.
Emmons, L. H., Sherman, P., Bolster, D., Goldizen, A., and Terborg, J. Unpublished ms. Ocelot behavior in moonlight. Available from L. H. Emmons, Museum of Natural History, Smithsonian Institution, Washington, D.C.
Ewer, R. F. 1973. *The Carnivores.* Ithaca, N.Y.: Cornell Univ. Press.
Gill, F. B., and Wolf, L. L. 1975. Economics of feeding territoriality in the golden-winged sunbird. *Ecology* 56:333–345.
Gill, F. B., and Wolf, L. L. 1977. Non-random foraging by sunbirds in a patchy environment. *Ecology* 58:1284–1296.
Gittleman, J. L. 1985. Carnivore body size: Ecological and taxonomic correlates. *Oecologia* 67:540–554.

Gittleman, J. L. 1986. Carnivore life history patterns: Allometric, phylogenetic, and ecological associations. *Amer. Nat.* 127:744–771.
Gittleman, J. L., and Harvey, P. H. 1982. Carnivore home-range size, metabolic needs and ecology. *Behav. Ecol. Sociobiol.* 10:57–63.
Goss-Custard, J. D. 1977. Optimal foraging and the size selection of worms by redshank *Tringa totanus*. *Anim. Behav.* 25:10–29.
Guggisburg, C. A. W. 1975. *Wild Cats of the World*. New York: Taplinger.
Hamilton, P. H. 1976. The movements of leopards in Tsavo National Park, Kenya, as determined by radio-tracking. M. S. thesis, University of Nairobi, Nairobi, Kenya. 256 pp.
Hanby, J. P., Bygott, J. D. 1979. Population changes in lions and other predators. In: A. R. E. Sinclair & M. Norton-Griffiths, eds. *Serengeti: Dynamics of an Ecosystem*, pp. 249–262. Chicago: Univ. Chicago Press.
Hoppe, B. 1984. Étude du spectre des proies de la panthère, *Panthera pardus*, dans le Parc National de Tai en Côte d'Ivoire. *Mammalia* 48:477–487.
Hornocker, M. G. 1970. An analysis of mountain lion predation upon mule deer and elk in the Idaho Primitive Area. *Wildl. Monogr.* 21:1–39.
Houston, D. C. 1979. The adaptations of scavengers. In: A. R. E. Sinclair & M. Norton-Griffiths, eds. *Serengeti: Dynamics of an Ecosystem*, pp. 263–286. Chicago: Univ. Chicago Press.
Jackson, R., and Ahlborn, G. 1984. A preliminary habitat suitability model for the snow leopard, *Panthera uncia*, in west Nepal. *International Pedigree Book of Snow Leopards* 4:43–52.
Jackson, R., and Hillard, D. 1986. Tracking the elusive snow leopard. *Natl. Geogr.* 169:793–809.
Jarman, P. 1974. The social organization of antelope in relation to their ecology. *Behaviour* 48:215–267.
Jarman, P. 1982. Prospects for interspecific comparison in sociobiology. In: King's College Sociobiology Group, eds. *Current Problems in Sociobiology*, pp. 323–342. Cambridge: Cambridge Univ. Press.
Johnsingh, A. J. T. 1983. Large mammalian prey-predators in Bandipur. *J. Bombay Nat. Hist. Soc.* 80:1–57.
Joslin, P. 1973. Factors associated with decline of the Asiatic lion. In: R. L. Eaton, ed. *The World's Cats*, 1:127–141. Winston, Ore.: World Wildlife Safari.
Kislalioglu, M., and Gibson, R. N. 1976. Prey handling time and its importance in food selection by the 15-spined stickleback *Spinachia spinachia* (L.). *J. Exp. Mar. Biol. Ecol.* 25:151–158.
Kleiman, D. G., and J. F. Eisenberg. 1973. Comparisons of canid and felid social systems from an evolutionary perspective. *Anim. Behav.* 21:637–659.
Krebs, J. R., MacRoberts, M. H., and Cullen, J. M. 1972. Flocking and feeding in the Great Tit *Parus major*—An experimental study. *Ibis* 114:507–530.
Kruuk, H. 1972. *The Spotted Hyena*. Chicago: Univ. Chicago Press.
Kruuk, H. 1978. Foraging and spatial organization of the European Badger, *Meles meles* L. *Behav. Ecol. Sociobiol.* 4:75–89.
Kruuk, H., and Turner, M. 1967. Comparative notes on predation by lion, leopard, cheetah, and wild dog in the Serengeti Area, East Africa. *Mammalia* 31:1–27.
Leyhausen, P. 1979. *Cat Behavior: The Predatory and Social Behavior of Domestic and Wild Cats*. New York: Garland STPM Press.
MacArthur, R. H., and Pianka, E. R. 1966. On the optimal use of a patchy environment. *Amer. Nat.* 100:603–609.
Macdonald, D. W. 1979. Flexibility of the social organization of the golden jackal, *Canis aureus*. *Behav. Ecol. Sociobiol.* 5:17–38.

Macdonald, D. W. 1983. The ecology of carnivore social behavior. *Nature* 301:379–384.

McDougal, C. 1977. *The Face of the Tiger.* London: Rivington Books.

Maddock, L. 1979. The "Migration" and grazing succession. In: A. R. E. Sinclair & M. Norton-Griffiths, eds. *Serengeti: Dynamics of an Ecosystem,* pp. 104–129.Chicago: Univ. Chicago Press.

Makacha, S., and Schaller, G. B. 1969. Observations on lions in the Lake Manyara National Park, Tanzania. *East African Wildl. J.* 7:99–103.

Matjushkin, E. N., Zhivotchenko, V. I., and Smirnov, E. N. 1977. The Amur tiger in the USSR. Unpublished report of the International Union for Conservation of Nature, Morges, Switzerland.

Mech, L. D. 1977. Wolf-pack buffer zones as prey reservoirs. *Science* 198:320–321.

Mishra, H. R. 1982. The ecology and behaviour of chital (*Axis axis*) in royal Chitwan National Park, Nepal. Ph.D. dissert., Univ. Edinburgh. 239 pp.

Mitchell, B. L., Shenton, J. B., and Uys, J. C. M. 1965. Predation on large mammals in the Kafue National Park. *Zool. Africana* 1:297–318.

Moehlman, P. D. 1986. Ecology of cooperation in canids. In: D. I. Rubenstein & R. W. Wrangham, eds. *Ecological Aspects of Social Evolution,* pp. 64–86. Princeton, N.J.: Princeton Univ. Press.

Muckenhirn, N. A., and Eisenberg, J. F. 1973. Home ranges and predation of the Ceylon leopard. In: R. L. Eaton, ed. *The World's Cats* 1:142–175. Winston, Ore.: Winston Wildlife Safari.

Nelson, M. E., and Mech, L. D. 1981. Deer social organization and wolf predation in northeastern Minnesota. *Wildl. Monogr.* 77:1–53.

Owens, M., and Owens, D. 1984. *Cry of the Kalahari.* Boston: Houghton Mifflin.

Packer, C. 1986. The ecology of sociality in felids. In: D. I. Rubenstein & R. W. Wrangham, eds. *Ecological Aspects of Social Evolution,* pp. 429–451. Princeton: Princeton Univ. Press.

Pienaar, U. de V. 1969. Predator-prey relationships amongst the larger mammals of the Kruger National Park. *Koedoe* 12:108–176.

Pulliam, H. R. 1974. On the theory of optimal diets. *Amer. Nat.* 108:59–75.

Rabinowitz, A. R., and Nottingham, B. 1986. Ecology and behaviour of the Jaguar, *Panthera onca,* in Belize, Central America. *J. Zool.* 210:149–159.

Robinson, J. G., and Redford, K. H. 1986. Body size, diet and population density of Neotropical forest mammals. *Amer. Nat.* 128:665–680.

Schaller, G. B. 1967. *The Deer and the Tiger: A Study of Wildlife in India.* Chicago: Univ. Chicago Press.

Schaller, G. B. 1972. *The Serengeti Lion: A Study of Predator-Prey Relations.* Chicago: Univ. Chicago Press.

Schaller, G. B., and Crawshaw, P. G. 1980. Movement patterns of jaguar. *Biotropica* 12:161–168.

Schaller, G. B., and Vasconelos, J. M. C. 1978. Jaguar predation on capybara. *Z. Säugetierk.* 43:296–301.

Schoener, T. W. 1971. Theory of feeding strategies. *Ann. Rev. Ecol. Syst.* 2:369–404.

Seidensticker, J. 1976. On the ecological separation between tigers and leopards. *Biotropica* 8:225–234.

Seidensticker, J. 1977. Notes on early maternal behavior of the leopard. *Mammalia* 41:111–113.

Seidensticker, J., Hornocker, M. G., Wiles, W. V., and Messick, J. P. 1973. Mountain lion social organization in the Idaho Primitive Area. *Wildl. Monogr.* 35:1–60.

Smith, J. L. D., McDougal, C. and Sunquist, M. E. 1987. Female land tenure system in tigers. In: R. L. Tilson & U. S. Seal, eds. *Tigers of the World: The Biology, Manage-*

ment, and Biopolitics of Conservation of an Endangered Species, pp. 97–109. Park Ridge, N.J.: Noyes Publications.

Smith, R. M. 1978. Movement patterns and feeding behavior of the leopard in Rhodes Matopos National Park, Rhodesia. *Carnivore* 1:58–69.

Stanley, S. M., Van Valkenburg, B., and Steneck, R. S. 1983. Co-evolution and the fossil record. In: D. J. Futuyma & M. Slatkin, eds. *Co-Evolution,* pp. 328–349. Sunderland, Mass.: Sinauer.

Sunquist, M. E. 1979. The movements and activities of tigers (*Panthera tigris tigris*) in Royal Chitawan National Park, Nepal. Ph.D. dissert., Univ. Minnesota, St. Paul. 170 pp.

Sunquist, M. E. 1981. The social organization of tigers (*Panthera tigris*) in Royal Chitawan National Park, Nepal. *Smithsonian Contrib. Zool.* 336:1–98.

Taylor, R. J. 1976. Value of clumping to prey and the evolutionary response of ambush predators. *Amer. Nat.* 110:13–29.

Thapar, V. 1986. *Tiger: Portrait of a Predator.* London: Collins Sons.

Van Orsdol, K. G. 1982. Ranges and food habits of lions in Rwenzori National Park, Uganda. *Symp. Zool. Soc. London* 49:325–340.

Van Orsdol, K. G. 1984. Foraging behaviour and hunting success of lions in Queen Elizabeth National Park, Uganda. *African J. Ecol.* 22:79–99.

Van Orsdol, K. G., Hanby, J. P., and Bygott, J. D. 1985. Ecological correlates of lion social organization (*Panthera leo*). *J. Zool.* 206:97–112.

Van Valkenburgh, B. 1985. Locomoter diversity within past and present guilds of large predatory mammals. *Paleobiol.* 11:406–428.

Waser, P. M. 1980. Small nocturnal carnivores: Ecological studies in the Serengeti. *African J. Ecol.* 18:167–185.

Waser, P. M., and Jones, W. T. 1983. Natal philopatry among solitary mammals. *Quart. Rev. Biol.* 58:355–390.

Werner, E. E., and Hall, D. J., 1974. Optimal foraging and the size selection of prey by the Bluegill Sunfish (*Lepomis macrochirus*). *Ecology* 55:1216–1232.

Wrangham, R. W. 1982. Mutualism, kinship, and social evolution. In: King's College Sociobiology Group, eds. *Current Problems in Sociobiology,* pp. 269–290. Cambridge: Cambridge Univ. Press.

Wrangham, R. W., and Rubenstein, D. I. 1986. Social evolution in birds and mammals. In: D. I. Rubenstein & R. W. Wrangham, eds. *Ecological Aspects of Social Evolution,* pp. 452–470. Princeton, N.J.: Princeton Univ. Press.

CHAPTER **11**

The Advantages and Disadvantages of Small Size to Weasels, *Mustela* Species

CAROLYN M. KING

The weasels (*Mustela* spp.) are a group of small mustelid carnivores that originated in the late Pliocene and are now distributed throughout the Holarctic region. *Mustela erminea,* the stoat or ermine, is circumboreal north of about 40°N. *M. nivalis* is sympatric with *erminea* over most of the same area. It includes two distinct subspecies, the common weasel of western Europe and Britain (*M. n. vulgaris* Erxleben 1777), and the least weasel of northern Scandinavia, USSR, and North America (*M. n. nivalis* Linnaeus 1766), which are different in appearance and range (Stolt 1979) but interbreed in captivity (F. Frank, pers. comm.). A third species, *M. frenata,* the long-tailed weasel, is confined to America, from about 50°N to about 15°S.

All have the characteristic "weasel" look (Figure 11.1): small (all under 600 mm total length) with long, thin bodies, short legs, flattish triangular heads, bright black eyes, and long whiskers. They form a size-graded set of two (rarely three) sympatric carnivores, in which *frenata* (or, in the absence of *frenata, erminea*) is always the largest, and *nivalis* is always the smallest; where all three are sympatric, *erminea* is the middle-sized one. All have rather short fur, brown above and white or cream below, turning totally white in winter in the northern races. All are, to different degrees, specialist predators of small mammals, willing to take birds, insects, lizards, or invertebrates if hungry, but preferring to concentrate on whatever varieties of rodents and lagomorphs are provided by the local fauna; and they hunt these prey very effectively, with seemingly boundless energy.

The way of life of the weasels is extremely successful; as a group, they are without doubt the most abundant mammalian carnivores in the world. But their small size and specialized hunting strategies give them high efficiency as predators at the price of substantial inefficiency in physiology and uncertainty in reproductive success. These strategies and their penalties, and the ecological consequences that follow, are the subject of this chapter.

302

Figure 11.1.
A typical weasel,
Mustela erminea.
(Drawing by
L. Douglas.)

The Evolutionary Origin of Small Size in Weasels

The evolutionary history of many groups of mammals shows a tendency toward gradual increase in size, by no means universal but common enough to have been labeled "Cope's Rule" (Stanley 1973; Peters 1983:192). Evolutionary trends toward decreased size are much less common, perhaps because at any one time "all the smaller niches will be occupied, [so] the only way to conquer new worlds is to make larger niches. It is only through . . . some peculiar change of conditions that the smaller niches will be vacated [or created], and there might be a selection pressure for a reversion to smaller size" (Bonner 1965:190). I have suggested that the weasels are an example of this process (King 1983a, 1984a). In the Miocene period, when the subfamily Mustelinae originated, the ancestors of the weasels were forest-dwelling hunters, probably somewhat similar to martens. Several marten-like carnivores are known from the Miocene (see Martin, this volume), and by the early Pliocene there were at least three separate lines of true *Martes* already established, as well as some other forms intermediate between *Martes* and *Mustela* (Anderson 1970:122). Throughout the Pliocene period the northern climate was cooling toward the Pleistocene; the forests were being replaced by grassland; and the evolution of the voles was presenting a new niche for predators small enough to make best use of an abundant but unreliable resource.

Such circumstances would favor a decrease in size and a shift along the *r*–K spectrum in the direction of a more *r*-selected or opportunistic life-history strategy (King and Moors 1979a). In the late Pliocene the tundra and the earliest lemmings appeared (Kowalski 1980), and also the first of the modern weasels, *M. plioerminea* and *M. pliocaenica* in Eurasia (Kurtén 1968) and *M. rexroadensis* in North America (Kurtén and Anderson 1980). The small size of these predators was, I suggest, originally an adaptation for hunting voles on Pliocene grasslands; but as climatic cooling continued and conditions became more and more severe, small size also became a useful preadaptation for hunting lemmings under snow. The timing of these events implies that the characteristic that later became one of the most critical advantages of small size to the northern weasels, the ability to use the nests of rodents and the insulating snow blanket to escape the infinite heat sink of the clear night sky (Pruitt 1978), started as a side benefit of the more obvious advantage of the ability to pursue voles through their runways in matted grass.

Factors Influencing Body Size in Contemporary Weasels

The consequences of small size to mammals in general have often been reviewed (e.g., Bourlière 1975; Panteleev 1981; Clutton-Brock and Harvey 1983; Peters 1983; Schmidt-Nielsen 1984; Gittleman 1985). When weasels are considered alone, only a certain subset of factors need be listed (Figure 11.2):

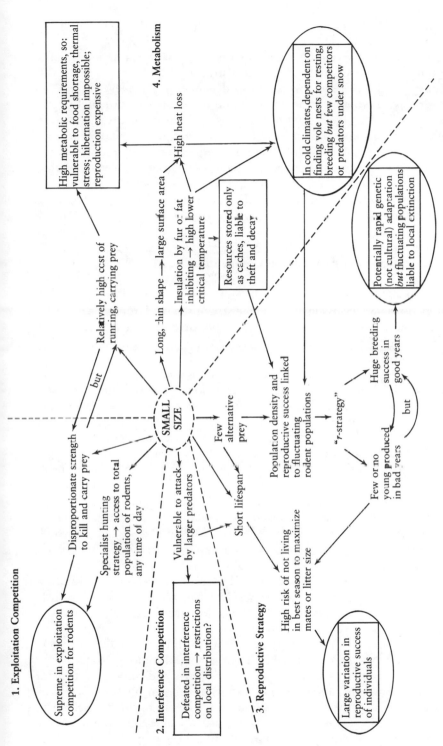

Figure 11.2. The advantages (*circled*) and disadvantages (*boxed*) of small size to the weasels as a group. There are four main subject areas, linked together at various levels. The advantage in exploitation competition (*area 1*) is balanced by the disadvantage in interference competition (*area 2*); in reproductive strategy (*area 3*) there is a net balance at both individual and population levels; only in metabolism (*area 4*) are the advantages outweighed by the disadvantages. It seems likely that ecological energetics is especially significant in determining body size in weasels, and/or the advantage in foraging efficiency is important enough to compensate.

The content within the figure:

1. Exploitation Competition

High metabolic requirements, so; vulnerable to food shortage, thermal stress; hibernation impossible; reproduction expensive

4. Metabolism

Relatively high cost of running, carrying prey

Disproportionate strength to kill and carry prey

Specialist hunting strategy → access to total population of rodents, any time of day

Supreme in exploitation competition for rodents

but

Long, thin shape → large surface area

High heat loss

Insulation by fur or fat inhibiting → high lower critical temperature

Resources stored only as caches, liable to theft and decay

In cold climates, dependent on finding vole nests for resting, breeding *but* few competitors or predators under snow

SMALL SIZE

Few alternative prey

Short lifespan

Population density and reproductive success linked to fluctuating rodent populations

"r-strategy"

Huge breeding success in good years

but

Few or no young produced in bad years

Potentially rapid genetic (not cultural) adaptation *but* fluctuating populations liable to local extinction

Vulnerable to attack by larger predators

Defeated in interference competition → restrictions on local distribution?

2. Interference Competition

High risk of not living in best season to maximize mates or litter size

3. Reproductive Strategy

Large variation in reproductive success of individuals

hunting strategy and exploitation competition, predation and interference competition, metabolism, and reproductive strategy.

The interplay of profit and loss can be understood on three levels. First and most obviously, there are the advantages and disadvantages of small size to weasels as a group compared with other mammalian carnivores. However, although all weasels may be considered small by comparison with other carnivores, not all are equally small; there are degrees of smallness within the group, caused by substantial interspecific and sexual variation. Hence, we may look at the same list of advantages and disadvantages of *relative smallness* as applied, second, to the species of weasels separately, and third, to the sexes within each species.

Hunting Strategy and Exploitation Competition

A predator has first to find a suitable prey and then to kill and process it without incurring a net loss in energy. The size of the predator strongly influences both the energy equations and the risks involved, in different ways according to the size of the prey.

Small rodents are relatively easy to kill but often hard to find. Weasels observed in enclosures can easily overpower mice or voles exposed in the open, although even in such artificial conditions they do not always catch every one (Erlinge et al. 1974; Nams 1981). Rodents familiar with the enclosure can often detect a hunting weasel first (Jamison 1975); failing that, they may escape by freezing or by rapid flight along known routes, deliberately dodging through dense cover and around obstacles to throw off the pursuer (Metzgar 1967; King 1985). In the wild, where rodents are free to hide in or escape to much larger areas of more complex cover than can be provided in any enclosure, the work of hunting must greatly increase, especially during the periodic population declines of voles, or in habitats where small rodents are generally scarce. For example, each male *nivalis* living in a deciduous woodland observed in England must have had to search through miles of tunnels on its large home range (7–15 ha) in order to find, every day, enough of the 21–39 rodents hidden on each hectare (King 1980a). Smaller ranges would not have been viable, and, as it was, the resident weasels were often seriously undernourished. When the density of rodents dropped lower still, all weasels disappeared from the area (Hayward 1983).

Weasels therefore run a greater risk of failing to find a prey than of being injured while attempting to capture it. Their optimum hunting strategy is to maximize their ability to search through the burrows and runway systems of rodents and lagomorphs, and through all kinds of cover into which these prey could escape. Because the senses of weasels are adapted to function either in full sunlight or underground (Gewalt 1959), they are able to hunt at any time of day and to reach a far higher proportion of the prey population (including nestlings) than is available to larger predators. The smallest weasels (*M. n. nivalis*) can go wherever a vole can go, even into its own nest; in cold climates they are very efficient at hunting under snow (Formosov 1946) and regularly take over the nests of their most recent prey (McLean et al. 1974; Madison et al. 1984). Larger weasels (*erminea* and *frenata*) can still follow voles into thick cover (log piles, tumbled rocks) that would exclude a fox or cat, and can enter the burrows of watervoles (*Arvicola* spp.) and rabbits. They will take a variety of mammals up to sciurid/lagomorph size, as well as birds; but changes in the numbers of small rodents still influence their density and population dynamics (Edson 1933; Erlinge 1983; King 1983b, 1983c).

Unfortunately, although weasels are the world experts at finding rodents, their ultimate dependence on rodents may, in some temperate habitats, put weasels at a long-term disadvantage in exploitation competition with larger predators. A varied fauna of generalist predators with other prey available can sometimes exert a very strong collective pressure on small rodents, enough to hold their populations low all the year round and reduce the breeding success of female weasels (Erlinge 1983). Weasels come into their own in the far north, where they can continue to survive on a sparse population of voles or lemmings long after other predators have turned to hunt other prey or moved elsewhere (Fitzgerald 1981).

Hunters small enough to enter rodent runways must still not be too small to execute a kill; weasels therefore make up extra size in length rather than girth, and use it by wrapping their long bodies around a catch, which helps to contain its struggles (as described by, e.g., Heidt 1972). They do not have to sacrifice muscular strength for size, since the force that an individual muscle can exert is the same, per unit of cross-sectional area, in mammals of any size (Schmidt-Nielsen 1984:163). On the contrary, weasels appear to be relatively stronger than larger predators; no lion can run at speed carrying a carcass of half its own weight. The difference is a simple result of scaling. With decreasing size, the mass of an animal decreases in proportion to the third power of its length (L^3), but the cross-sectional area of its muscles (which determines the force they can exert) decreases only as the square of their length (L^2). Hence, the force exerted by muscles, relative to mass, increases in proportion to the decrease in body size. This apparently disproportionate strength of weasels is one of the mechanical advantages of small size, and there are others. For example, although the incremental cost of moving one unit of body weight over one unit of vertical distance is independent of size (about 1.36 ml O_2 per kg per vertical meter: Schmidt-Nielsen 1984:175), the increase in metabolic rate attributable to the vertical component, relative to the resting rate, is much smaller in lighter animals. It makes little difference to a weasel whether it is running straight, up, or down; climbing trees or steep mountainsides takes hardly any extra effort.

The economics of hunting also have some particular disadvantages for a small predator. The energy cost of running is relatively high in small animals, because they have to take many more steps to move one unit of body mass over one unit of distance, each step requiring work in proportion to mass. Hence, the foraging range of weasels is limited, even though they depend entirely on fresh meat, a scarce food resource, and therefore need relatively large home ranges (Gittleman and Harvey 1982). For the same reason, when local food supplies fail, long-distance migration is not a feasible option for weasels. Another problem is that the energy cost of carrying prey, to a safe place or to the young, increases in direct proportion to the added load: for example, if the load is 50% of the body mass, oxygen consumption increases by 50% (Schmidt-Nielsen 1984:176). Weasels routinely carry prey at least that heavy; even the smallest of them have the strength, but the cost is high. Finally, the prey resources in any local community are nested; whatever is food for the smaller predators is also food for the larger, and weasels cannot defend the stocks of live prey on their home ranges against larger competitors. Sometimes they can be seriously affected by this competition (Erlinge 1983). Oksanan et al. (1985) suggest that the weasels' habit of making caches of dead prey in inaccessible places (Rubina 1960; Parovshchikov 1963) is not a consequence of "surplus killing" but a positive strategy for the smaller members of a predator guild competing with larger ones for access to unpredictable resources.

Predators and Interference Competition

Weasels are small enough to be vulnerable to attack by almost any other predator. Most species of raptors and larger carnivores have been recorded as killing weasels occasionally, although the victim is not always eaten; Macdonald (1977) reported that a red fox (*Vulpes vulpes*) appeared to find weasel carcasses distasteful. Weasels caught too far from cover have little defense except extreme pugnacity. In a face-to-face encounter with a larger carnivore, a determined weasel with its sharp, explosive bark, bared teeth, fearless attitude, and powerful anal scent glands can sometimes effectively deter an attack; but a swooping raptor is a more dangerous enemy. If the raptor's talons do not pierce the weasel's body at once, it might twist around in the raptor's grasp and attack the bird's throat, perhaps forcing it to release its hold (Seton 1929; Burnham 1970). A better defense is the black tail tip of the two larger weasel species, which Powell (1982) has shown to be a classic predator-deflection spot. A stooping hawk may be confused into grasping at the end of the weasel's thin tail and missing its body. The smallest species, *nivalis,* has no spot because, Powell suggests, it has too short a tail. This does not explain why it does not have a longer tail, which would presumably allow it to have a spot too. Powell suggests that *nivalis* has a short tail for reasons of heat conservation, and I would add that it may be less vulnerable to raptors because it spends relatively more time under snow, thick cover, or underground.

The vulnerability of weasels to predation may, perhaps, restrict their distribution (e.g., they avoid open spaces: Musgrove 1951), and it gives them a distinct disadvantage in interference competition with the many larger predators that also feed on small mammals; it does not, however, necessarily mean that the body size, population density, or dynamics of weasels is controlled by interaction with or predation by larger predators. Powell's (1973) model was based on a number of assumptions (e.g., that weasel populations are not limited by food) that now appear unlikely (Delattre 1983; Erlinge 1983; King 1983b), so it is not acceptable itself even though it led directly to Powell's (1982) elegant experiments. Likewise, Ralls and Harvey (1985) have thrown doubt on the validity of character displacement as a factor controlling body size in North American weasels, as once advocated by McNab (1971).

Metabolism

Three features of the weasel's hunting strategy have important consequences for their physiology: (1) their long, thin shape exposes a relatively large surface area to the air (Brown and Lasiewski 1972); (2) their unique niche depends on their ability to search through narrow runways and dense cover to find hidden rodents not available to other predators, and this active technique is very expensive in energy (see above); and (3) their ability to move through confined spaces would be compromised by heavy body insulation, so their fur must be relatively short (Freuchen and Salomonsen 1959; Casey and Casey 1979) and any subcutaneous fat confined to dips in the body outline (Appendix 11.1).

Weasels therefore suffer a very high rate of loss of body heat. Their lower critical temperatures are so high that they almost never reach a state of thermoneutrality (Casey and Casey 1979). The resting metabolism of a weasel depends on ambient temperature; in cold climates, throughout most of the weasels' range, the energy cost of thermoregulation for an inactive weasel may be up to three (Sandell 1985) to six times BMR (Chappell 1980). Arctic weasels can avoid this huge expenditure only by resting in a borrowed nest (Casey and Casey 1979; Chappell 1980), especially if they improve its insulating properties by lining the inside with rodent fur (McLean et al. 1974).

Weasels cannot, of course, stay in the shelter of their nests indefinitely, but opinions are divided as to whether low temperatures restrict their movements when they are actively hunting. Casey and Casey (1979:162) point out that, if the arctic weasels have to generate up to six times BMR merely to maintain body temperature, they "might have little capacity in reserve for energy generation during activity," especially as "heat loss should be even greater in active animals due to forced convection." If this is so, we would expect the arctic weasels to keep to the shelter of the snow cover even when they are active. Tracking studies in USSR report that weasels do not venture above the snow when the air temperature falls below a certain minimum, measured at −13°C in western Siberia by Kraft (1966). Formosov (1946:79) found that "common voles which ran onto the snow to escape an ermine, which had dug into their nest, froze in a distance of 3—4 meters from their snow hole" when the temperature was −12° to −15°C.

A contrary view is expressed by Sandell (1985). He estimates that weasels do not need to generate more than three times BMR even when the ambient temperature is −30°C, and that, since 75% of the energy expended during activity is released as heat, a running animal does not need to spend energy on thermoregulation. Sandell concludes from this that active weasels are, in practice, almost independent of ambient temperature; in Sweden he has snow-tracked *erminea* after nights with temperatures down to −25°C, and at the same time seen tracks of voles running for tens of meters above the snow (M. Sandell, pers. comm.). Casey and Casey (1979) and Sandell (1985) worked in habitats differing in important conditions (e.g., permafrost, light regime), and calculated the energy requirements of weasels from different models, so it is not clear whether this disagreement is real or technical. Much depends on the estimated rate of loss of body heat during activity at low ambient temperatures. Nevertheless, it is still clear that the northern weasels are absolutely dependent on insulated rodent nests when they are resting or breeding; and presumably they might often choose to hunt under the snow when it is less cold there than on the surface.

The physiological consequences of small size for weasels in general are practically all seriously disadvantageous, and the northern weasels survive at what appears to be the limit of their metabolic capacity (Casey and Casey 1979). Their constant need for shelter and frequent meals is a handicap in years when rodent populations are low and winter nests few; they can store resources only as caches, liable to theft and decay, not as fat; they are vulnerable to temporary food shortages, but hibernation, torpor (at least in adults), and migration are all impossible; and the additional energy required for reproduction may be hard to find except in years when rodents are abundant. These are all serious problems for small homeotherms living in a cold climate. Weasels do have the incidental advantage that the subnivean habitat is relatively free of competitors and predators, but those hazards are much less serious than the constant danger of chilling and starvation.

Reproductive Strategy

All mammals live for roughly the same length of physiological time, that is, about 200—250 million breaths and about 800—1200 million heartbeats (Peters 1983:122; Schmidt-Nielsen 1984:146). The pulse of a weasel runs at about 400—500 beats per minute (measured by Tumanov and Levin 1974 on male and female *nivalis* and *erminea*), which gives it a physiological lifespan of about 3—6 years. In captivity both those species can live that long, but in the wild, very few indeed; the average age at death of *nivalis* is <1 year, and of *erminea* <1.5 years, and the maximum age attained (infrequently) 3+ and 8 years, respectively (King 1980b; Erlinge 1983; Debrot 1984; Grue and King 1984).

Lifespan is therefore scaled to size, so that small animals "must rush to complete their life histories in the face of an early expiration date" (Calder 1983:217). This introduces a new, acute problem for weasels, as regards timing. For all temperate and northern mammals the resources vital for reproductive success are seasonal, linked to a fixed annual cycle of plant growth; but for weasels the ordinary annual cycle is overlain by multi-annual fluctuations in the abundance of their key resource. The northern weasels are supreme specialists in the exploitation of unstable populations of small rodents, and for that they must be small; but small size automatically speeds up their physiology and reduces their lifespan to the extent that the periods of most favorable conditions for breeding (the peaks in rodent numbers every three or four years) are longer apart than the average weasel's expectation of life at birth.

Weasels in general are, relative to the larger carnivores, typical "*r*-strategists" (King and Moors 1979a) and this is a high risk–high reward policy, with advantages and disadvantages at both the individual and the population levels. In a cold or stressful environment individuals may not always be able to provide enough energy both for normal metabolism and for reproduction at the same time. Yet all weasels must seize every possible opportunity to breed, and some must manage to produce a few young even in poor years; but their full potential productivity can be achieved only when food is not limiting (i.e., at rodent peaks). Small rodents are the ideal prey for breeding weasels throughout their range, and the reproductive output of weasel populations is strongly correlated with the distribution and density of rodents even where other prey are available (Erlinge 1981; King 1983b, 1983c).

Individuals living at a time of peak rodent numbers are likely to achieve high reproductive success. King (1981, 1983c) documented the changes in fecundity and productivity of the *erminea* living in New Zealand beech forests through two population irruptions of rodents (feral *Mus musculus* and, to a lesser extent, *Rattus rattus*). The unusual breeding success of the post-seedfall years was certainly measurable in terms of increased numbers of young reared by the adult females, and the equivalent statistic (not measured) would presumably be increased numbers of matings for males. But both males and females are short-lived on average and run a high risk of missing the best season. When the density of weasels is low, males have a lesser chance of finding several mates, or even one; when the density of rodents is low, females have a lesser chance of producing their full potential number of young, or even any at all (Tapper 1979; King 1983a, 1983b). Over the general population, then, there is always a large variation in the breeding success of individuals. For the adults, the rewards are great in the years when the chances of success are high, but the losses disastrous in all other years. For the young, growth is rapid and hunting instinctive (i.e., they do not need the extended apprenticeship required to learn the complex hunting skills of larger carnivores: Bekoff et al. 1984), so they can leave the family and disperse with minimum delay; but very few will survive the following winter.

At the population level, the rapid production of young, high population turnover, and huge rate of increase in good years are advantages in that weasels have great potential for rapid genetic adaptation; but these are also disadvantages, because weasel populations fluctuate a great deal and are very liable to local (though not to total) extinction (King and Moors 1979a, 1979b). The short contact between the generations also means that weasels are confined to a somewhat conservative way of life, in contrast to the elaborate and flexible social patterns developed by some of the larger carnivores (Bourlière 1975).

Body Size in Relation to Species and Sex

The general advantages and disadvantages of small size in the weasels as a group (Figure 11.2) also apply, to different degrees, to individual weasels of each sex and species. There are also additional considerations, not entered into Figure 11.2, that apply only to relationships between sexes or species; for example, small size is probably a disadvantage to males in competition for mates. The balance of profit and loss in relation to local resources is therefore different for each of up to six size classes of weasels living in a given place. This adjustable relationship may help to explain two important puzzles about weasels: the coexistence of similar species within the weasel set, and their strong sexual dimorphism.

Coexistence of Similar Species

The problem of the nearly universal coexistence of at least two species of weasels throughout the northern Holarctic, despite the extensive overlap in their ecological requirements, has frequently been discussed. Rosenzweig (1966) concluded that it must depend on predation of the smaller by the larger species; Powell and Zielinski (1983) invested much thought and computer time in the question and concluded that, in theory, it ought not to be possible except by continual local extinction and recolonization. King and Moors (1979b) suggested that coexistence might be permitted by an unstable balance of the different size-related advantages enjoyed by each species, as determined by the environment. For example, in Europe *nivalis* is superior to *erminea* in the exploitation of small rodents and is able to respond to vole peaks immediately by producing extra summer litters but is more vulnerable to local extinction; *erminea* is superior in interference and is able to exploit larger prey but is restricted by obligatory delayed implantation to producing only one litter a year. In a patchy environment of sufficient size, each species can find, at least locally and temporarily, conditions that suit its strategy; *nivalis* can avoid confrontations with *erminea,* and *erminea* can avoid dependence on a single prey resource.

The assumptions on which this idea was based have since been checked by field work. Pounds (1981) confirmed that *nivalis* is much more efficient in hunting small rodents, and Erlinge and Sandell (1985) confirmed that *erminea* does dominate *nivalis* in the field. That does not prove the idea correct; nevertheless, any significant change in the size distribution of prey does have differential results explicable, at least in part, as consequences of a shift in a dynamic balance of opposing forces. For example, the removal of rabbits by myxomatosis in Britain in 1953–55 put *erminea* at a disadvantage in competition both with *nivalis* and, perhaps, also with larger predators (Erlinge 1983), whereas *nivalis* greatly increased. Conversely, when both were transported to New Zealand, *nivalis* was disadvantaged by the absence there of voles, whereas *erminea* thrived (King and Moors 1979b).

Obviously, no such simple proposition could provide the whole answer. The net relative effect of close competitors of different sizes on each other involves a whole suite of spatial, temporal, and other components even more difficult to quantify than body size, prey size, and dominance (Hespenheide 1973). We can observe that these two species do coexist, at least partly because they are of different sizes, although it is clear that interspecific competition plays no *direct* role in determining the local mean body size in either species (Ralls and Harvey 1985; Harvey and Ralls 1985). They do deal in totally different ways with the same problem, that of adjusting reproductive effort to wildly variable probabilities of success (King 1981). No further conclusions are possible at present.

Sexual Dimorphism

Male and female weasels have in common the goal of minimizing the chances of leaving no young, given that their expectation of life is short and their chances of successful breeding vary greatly from year to year. However, there the similarity ends, because each sex has evolved different breeding strategies, and they cooperate as little as possible. For both sexes, body size is one of the most important factors involved, but for different reasons.

Females invariably bring up their young alone (Erlinge 1979), and in order to maximize the number of weaned young surviving at the end of each attempt, they must value foraging efficiency above all things, for two reasons. First, the energy demands of solo parenthood are enormous. During lactation the female's own food requirements increase by a massive 80–100% even in a temperate climate (East and Lockie 1964; Hayward 1983), and later, until they are fully independent, she has to supply dead prey for her young as well. Her requirements may amount to a total of 500–600% above normal (Sandell

1985). The female can accomplish this best when rodents are abundant, since rodents are the ideal resource on which to rear young weasels; they pose no risk for the mother to kill, are rapidly replaced, are not too heavy to carry (but see above on strength and loading), are not too big to be taken into a nest or cache whole, and are supplied in a convenient waterproof skin that helps to retard decay. Females therefore are always under strong pressure to stay small enough to be maximally efficient rodent hunters (Simms 1979; Pounds 1981). They can still kill larger prey, aided by the "shock effect" described by Hewson and Healing (1971), but this takes more energy and risk and so is less efficient as a means of feeding the young than concentrating on rodents if they are available. Incidentally, the argument that females stay small so as to minimize their own energy requirements for maintenance and thereby are able to channel more of their catch into their young (Moors 1980) is valid only "if smaller and larger females were equally efficient hunters, which may not be the case . . . nothing is known about the relationship between weasel size and hunting efficiency" (Ralls and Harvey 1985:162). It is worth pointing out, though, that the energy costs of running and carrying loads are inversely related to size (see above); I wonder whether females hunting prey of a given size, available at a given density, might even have to expend relatively more energy than males to search for, kill, and carry home each meal; the advantage of their smaller appetites, if it is one, might soon be canceled out. It would be interesting to see this prediction tested.

The second reason why females must maximize foraging efficiency is that small young (less than 5–7 weeks old in the case of *erminea*) are unable to maintain their body temperature when left in the nest alone. While the female is out hunting, the young huddle together, and if the nest temperature falls below 10–12°C, they will enter into a reversible cold rigor, with reduced sensitivity, cardiac and respiratory function (Segal 1975). Full homiothermy at environmental temperatures down to 0°C is achieved only when their fur is fully grown, at the age of about two months. This mechanism is an advantage to the young, which can channel all their energy into rapid growth during the periods that the female is present. But from the female's point of view it is a strong reason to minimize the time she is away from the nest during that first few weeks, since the young in torpor are not growing and are also vulnerable to interference or predation. The breeding female should therefore maximize her total hunting efficiency by starting with a time-minimizing strategy when the young are very small, and changing gradually to an energy-maximizing one when they are older. This idea is not supported by evidence, but, again, it could be tested.

Males, by contrast, are polygynous or promiscuous, and attempt to maximize the number of matings they can achieve each season. Large size is a positive advantage in competition for females, because it is a reliable indicator of fighting ability, and this is important even if damaging fights are usually avoided. Larger size is also often correlated with age and, by implication, with willingness to escalate a confrontation (Erlinge 1977; Sandell 1986). In very cold climates the advantage to a dominant male of larger size during the very brief breeding season must be (at least above a certain limit) counteracted by other penalties jeopardizing survival through the much longer nonbreeding season.

If large size is really so vital for reproductive success in males, the question arises as to why male size should be so extremely variable. One possible answer is that males may often fail to reach their full potential size (Ralls and Harvey 1985; R. A. Powell, pers. comm.). Whatever the genetic heritage of an individual male, his actual size at maturity may be set by whether or not he is well fed by his mother. Weasels in captivity commonly reach much higher weights than in the wild, especially the males (East and Lockie 1964; Hayward 1983); and local patterns of sexual dimorphism are variable and sometimes conflicting (Ralls and Harvey 1985). A simple way to test this idea would be to compare the degree of sexual dimorphism in cohorts of weasels born in years of good and bad supplies of small rodents. This requires year-class age determination of adults, which has only recently been proved reliable (Grue and King 1984), and then the analysis of a large sample of weasels collected in both good and bad years (Kopein, 1969; Powell and King, in preparation).

Geographical Variation in the Body Size of Weasels

Size in mammals is affected by many factors, some synergistic, some interacting, and some opposing; the observed phenotype is always a compromise between them. Eventually it may be possible to construct a model to explain body size in weasels, but we do not have enough data at present, either on the actual variation in their body size (especially in Eurasia) or on their ecological requirements, to do that. However, it is possible to make a few suggestions.

The Northern Hemisphere Continents

The patterns of geographical variation in the separate species of *Mustela* are complex and contradictory; for example, the smallest local races of *erminea* in Eurasia are found in the north and east, whereas the smallest in North America are in the southwest. However, in weasels in general, the continental-scale pattern is the same in both the Old and the New Worlds; the largest or only local species is relatively small in the far north (mean male condylobasal lengths in *erminea*, always the largest species in the north, not exceeding 43–46 mm right around the Pole), and relatively large in the south (*frenata* in the southern United States, and *nivalis* in Egypt, reach 50–53 mm or more) (Kratochvil 1977a, 1977b; Osborn and Helmy 1980; Ralls and Harvey 1985).

The first and most obvious explanation to check is Bergmann's Rule, which states that mammals living in very cold climates tend to be larger than related mammals of similar habits living in milder conditions (McNab 1971; Gittleman 1985; Erlinge 1987). Weasels ought to be prime examples of this generalization, because they are so extremely sensitive to thermal stress; their metabolic inefficiency is costly in the northern parts (the great majority) of their range. In fact, as in the order Carnivora in general (Gittleman 1985), Bergmann's Rule fails to explain the observed variation, either in individual species or in the weasels as a group (Ralls and Harvey 1985). Only the *erminea* in North America are substantially larger in the north (Figure 11.3C), and then only by comparison with their exceptionally small relatives farther south in that continent; they are not larger than the *erminea* living at the same latitudes in Eurasia. The *nivalis* of Europe are just as substantially larger in the south (Figure 11.3A). Within defined areas spanning a range of climates, Bergmann's Rule does not explain the observed variation among local populations of *erminea* in the USSR (Petrov 1962), New Zealand (King and Moody 1982), or Europe (Erlinge 1987).

There are two critical assumptions behind Bergmann's Rule: that latitude and temperature are correlated, and that all the species being compared are equally exposed to ambient air temperature. Ralls and Harvey (1985) found that the skull lengths of North American weasels were, indeed, about as well correlated with temperature as with latitude; but in Eurasia the simple latitude-temperature correlation is confused by low-latitude ranges of high mountains, on which some exceptionally small weasels live (Reichstein 1957; Morosova-Turova 1965; Heptner et al. 1967). A direct correlation between weasel size and local temperature shows that both sexes of both species of Eurasian weasels are smaller in colder climates—the opposite of Bergmann's prediction (Table 11.1). But this does not mean that weasels disprove the rule. The smaller, cold-climate weasels sheltering under snow are not directly exposed to ambient temperature, as their southern and lowland relatives are, so it is logically invalid to apply the rule to weasels anyway. Besides, the relationship between size and temperature cannot be crucial, since there is usually at least as great a difference between the males and the females living in one place as there is between the northern and the southern members of either sex. We need, then, to explain in other terms the general north-south variation in body size of the largest or only local weasel species.

Sandell (1985) has proposed a model predicting the optimal body size in male and female *erminea* in terms of ecological energetics and sexual selection. His basic assumption is that the total energy budget of any animal, expressed as a multiple of BMR, is limited. This seems reasonable, especially for weasels, which often have to operate their energy budgets with little to spare. Sandell assumes that, in winter, both sexes will be under pressure to minimize their daily energy expenditure, and that the best way to do that is to increase foraging efficiency, thereby reducing the total time out of the nest. Hence, for any combination of values for foraging efficiency and ambient temperature,

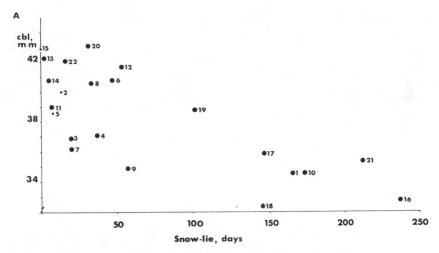

Figure 11.3. Geographical variation in skull size (condylobasal length in *A* and *B*; basal length of Hall [1951] in *C* and *D*) of male weasels with duration of snow-lie. Smaller symbols refer to island populations. *A*. Eurasian *nivalis* (N.B.: in Egypt, mean condyloincisive length of males is 50 mm: Osborn and Helmy 1980). *B*. Eurasian *erminea*. *C*. North American *erminea*. *D*. *frenata*.

Key to sources and locations·

Eurasia
1. Scandinavia (Stroganov 1962; Reichstein 1957)
2. Scotland (King 1977; King and Moody 1982)
3. Denmark (Fog 1969)
4. Northern Germany (Reichstein 1957)
5. England (King 1977; King and Moody 1982)
6. Poland (Reichstein 1957)
7. Central Germany (Reichstein 1957)
8. Czechoslavakia (Kratochvil 1977a, 1977b)
9. Southern Germany (Reichstein 1957)
10. Switzerland (Reichstein 1957)
11. Western France (Beaucournu and Grulich 1968)
12. Roumania (Barbu 1968)
13. Italy (Miller 1912)
14. Spain (Reichstein 1957)
15. Sardinia (Beaucournu and Grulich 1968)
16. Northern Siberia (Stroganov 1962)
17. Central European Russia (Morozova-Turova 1965; Heptner et al. 1967)
18. Transbaikal (Morozova-Turova 1965; Stroganov 1962)
19. Southern European Russia (Morozova-Turova 1965)
20. Turkmenia (Morozova-Turova 1965)
21. Tien Shan Mountains (Morozova-Turova 1965; Heptner et al. 1967)
22. Trans-Caucasus (Morozova-Turova 1965)
23. Islay Island (off western Scotland) (Miller 1912)
24N. Northern Ireland (Fairley 1981)
24S. Southern Ireland (Fairley 1981)
25. Terschelling Island (off Holland) (van Soest et al. 1972)
26. Holland (van Soest et al. 1972)
27. Eastern Siberia (Stroganov 1962)
28. Karaginski Island (off Kamchatka) (Stroganov 1962)
29. Western Siberia (Stroganov 1962)
30. Kamchatka (Vershinin 1972)
31. Shantar Island (Sea of Okhotsk) (Petrov 1956)

32. Altai Mountains (Stroganov 1962)
33. Northern Caucasus (Heptner et al. 1967)

North America
All the points plotted are local means taken from Hall (1951), and some are based on small samples. The data analyzed by Ralls and Harvey (1985) are much more detailed, comprehensive, and accurate but not available to be plotted against snow-lie in the same way as the Eurasian data.

34. Greenland
35. Point Barrow, Alaska
36. Southampton Island
37. Great Slave Lake
38. Southeastern Alaska
39. Admiralty Island, Alaska
40. Queen Charlotte Island
41. Vancouver Island
42. Newfoundland
43. Vancouver district
44. Olympic Peninsula
45. Cascade Mountains
46. Minnesota
47. Idaho
48. New York and Pennsylvania
49. Coastal Oregon
50. Colorado
51. Southeastern British Columbia
52. Southern Alberta
53. Maine
54. Massachusetts
55. Northwestern California
56. Southwestern California
57. San Joaquin Valley
58. Los Angeles
59. Arkansas
60. Georgia
61. Southern Texas

(*continued*)

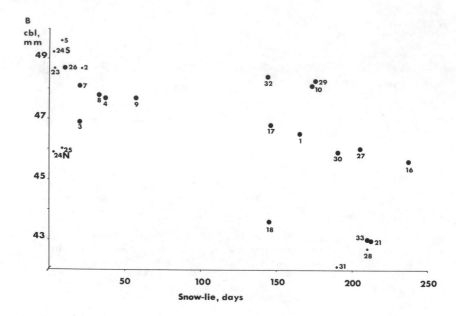

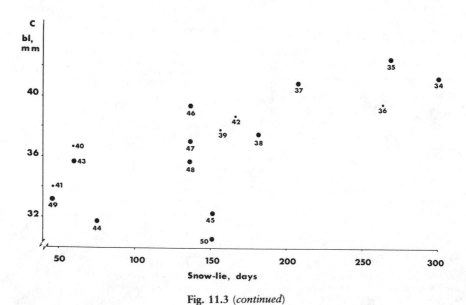

Fig. 11.3 (*continued*)

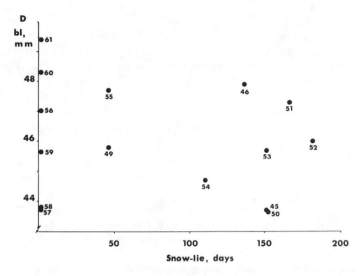

Fig. 11.3 (*continued*)

Table 11.1. Geographical variation in northern hemisphere mainland weasels (condylobasal length) with respect to latitude and climate

		Spearman rank correlation coefficients[a]		
		nivalis	*erminea*	*frenata*
Males				
Eurasia	Latitude	−0.63	n.s.	—
	Temperature[b]	0.87	0.60	—
	Snow-lie[c]	−0.77	−0.69	—
North America	Latitude	n.s.	0.67	n.s.
	Temperature	n.s.	−0.69	n.s.
Females				
Eurasia	Latitude	−0.45	n.s.	—
	Temperature	0.67	0.68	—
	Snow-lie	−0.62	−0.67	—
North America	Latitude	−0.44	0.67	0.16
	Temperature	n.s.	−0.64	n.s.

[a]Coefficients given are significant at $P < 0.05$ or better. n.s. = not significant.

[b]Mean annual dry bulb temperature recorded at one or several meteorological stations representing the areas sampled. Eurasian data collected by C. M. King (unpublished), North American by Ralls and Harvey 1985.

[c]Analyzed as the number of days per year of stable snowcover over 2.5 cm thick. No snow-lie analysis given by Ralls and Harvey; but the weasels that live in regions of heavy snow cover (*nivalis* and *erminea* in Canada and Alaska) are much smaller than those that live in the snow-free south (*frenata* in southern U.S.A.). The correlation between snow cover and size in five samples of European *erminea* reappears in fig. 1 of Erlinge (1987).

there will be an optimum body size, which in winter will be the same for both males and females. In the breeding season, however, the equations leading to success are different: intense sexual selection favors a larger size in males than in females. The actual body size observed is a compromise between the different seasonal optima.

Sandell confined his model to *erminea* living in the conditions he had studied in Sweden. I propose the following hypothesis, an extension of Sandell's model, to explain the general, continental-scale southward increase in the body size of the largest or only local weasel species.

My balance sheet of the profits and losses associated with small size (Figure 11.2), is not exhaustive and cannot assign relative values to the factors identified. Nevertheless, it confirms Sandell's assumption that ecological energetics is one of the most important considerations—more specifically, the effect of ambient temperature on metabolism and foraging efficiency. In arctic and alpine regions environmental conditions above the snow are severe (Pruitt 1978), and the ability to escape them is the condition for survival of small mammals there (Formosov 1946). When the air temperature is mild and the wind-chill factor low, weasels may emerge onto the surface, and their tracks are often seen (Teplov 1948; Nyholm 1959; Fitzgerald 1977); but they are still absolutely dependent on subnivean nests and prey. Therefore, the primary needs to avoid exposure to ambient temperature and to retain access to rodent tunnels and nests impose an upper limit to the size of arctic and alpine weasels. This overrides all other considerations, such as the reproductive advantage of larger size to males. The same restrictions apply to the small mammals hunted by the weasels. Hence, the advantages of avoiding thermal stress and of maximizing foraging efficiency reinforce each other in favoring small size in northern weasels.

By contrast, in the milder climates of lowland southern United States and southwestern Europe, the lesser need to avoid exposure to the air relaxes the restraints on sexual selection favoring larger size in males. Dawkins and Krebs (1979) point out that competition favors males that are slightly larger than the current population mode, whatever the current mode may be; hence, a general increase in the mean body size of both sexes (female size is "dragged upwards" by selection for increased male size, for reasons explained by Lande 1980) may be expected, if permitted by the net balance of energy economy as outlined by Sandell (1985). Moreover, in the south there are various larger prey such as lagomorphs and sciurids whose populations are more stable than those of small rodents, and which could more easily be caught by larger weasels. It is true that small rodent populations are relatively more stable in the south (Hansson and Henttonen 1985); but they are still capable of pronounced fluctuations over the long term (Southern and Lowe 1982), and they also serve as staple prey for a variety of generalist predators (Erlinge 1983). These considerations would make small size and extreme specialization on small prey less viable strategies in the south than in the north, and reinforce sexual selection in favoring larger size in southern weasels.

If this hypothesis is correct, parallel southward increases in the body sizes of the largest or only local weasel species and in its prey would be expected. However, since correlation does not necessarily imply causation, this is merely an observation, not an explanation. For example, it is not possible to predict in advance whether the northern weasels are small because their prey are small or because both are constrained by the same environmental hazards. As Erlinge (1987) pointed out, "The correlation between stoat body size and prey availability . . . can be interpreted either as a causal relationship or as an effect of an alternative process of adaptation."

I propose that this combination of energetics, size of available prey, and sexual selection (in unknown proportions) explains why the niche for a weasel-shaped carnivore allows only smaller individuals in severe climates, but larger ones in milder climates. In Eurasia the mean body sizes of both sexes in local populations of *nivalis* and *erminea* are inversely correlated with the mean number of days of snow cover per year (Figure 11.3A, B; Table 11.1). In North America neither *nivalis* nor *erminea* is larger in the south; the niche for a large southern weasel with relatively generalist food habits was already occupied by *frenata* (and presumably also by its predecessor and possible ancestor, *rexroadensis*) by the time the ancestors of the present *erminea* crossed the Bering bridge from Siberia in the middle Pleistocene or before (Kurtén and Anderson 1980). Of course, *frenata* can travel across snow and burrow into it (Fitzgerald 1977); its southern niche is determined more by prey diversity than by the snow cover itself (Gamble 1981, cf. Simms 1979).

The basic idea presented here was suggested by the size distribution of the living species; but there is some other evidence that small size actually is advantageous to cold-climate weasels, both in the Pleistocene and now. First, fossil *M. palerminea* (the direct ancestor of the contemporary *erminea*) from the cold phases of the middle Pleistocene are smaller than those from the warm phases (Kurtén 1960); and fossil *nivalis* from Polish caves dated to the Eemian (last) interglacial period resemble modern Polish *nivalis vulgaris*, whereas fossil *nivalis* dated to the following Weichselian glacial period are smaller, like the modern boreal *nivalis nivalis*, which no longer lives as far south as Poland (Wojcik 1974). Second, among > 4000 skulls of *erminea* collected north of Tjumen (57°N, 65°E) over the years 1959–64, there was a progressive decrease with age in mean condylobasal length in every annual cohort, in both sexes; and the smaller animals were fatter. To Kopein (1969) these were signs that the smaller individuals were better adapted, and lived longer, in that severe environment than did the larger ones.

It is important to note that this idea is a generalization and applies only to the largest weasel species in a local set (or to the only one, if only one is present). Like all other generalizations, ranging from simple verbal hypotheses to complex mathematical models, it is not entirely true but may have value as "a lie which makes you see the truth" (R. H. MacArthur, as quoted by Crowell 1986:59). It does not apply to the smaller species, except indirectly as follows. Where two species of weasels are sympatric, the smaller one is always the more strongly specialized on small rodents (King and Moors 1979b). If this strategy is less successful in the south (for reasons suggested above), it may explain why the smaller species do not extend as far south as the larger ones.

Most generalizations are difficult to test, and this one will be too, for several reasons. First, the key concept in Sandell's model, foraging efficiency, is hard to define. There is an almost endless list of variables that might contribute to it, and practically nothing is known about what determines it in weasels. Second, the hypothesis includes two other considerations, ecological energetics and sexual selection, and all three must be taken into account if we are to determine which is cause and which effect. Simple correlations dealing with only one variable at a time tend to be inadequate. For example, of the five hypotheses on geographic variation in North American weasels tested by Harvey and Ralls (1985), all except prey-size distribution were rejected. There is indeed a simple general relationship between body size of weasels and their prey (Moors 1980; Erlinge 1987), but certainly the equations that govern the dynamics of hunting must include many other characters of the prey besides body size. For example, the relationship between body size in weasels and the average size of the local voles and their tunnels is closer in females than in males (Simms 1979; Pounds 1981), which implies that males and females are differentially sensitive to some other factor(s) besides prey size. Ecological energetics and sexual selection are both powerful forces whose consequences for body size must affect males and females differently. It seems unlikely that any single-factor hypothesis will suffice to explain body size, which is inevitably a compromise between a range of possible optima.

The Northern Hemisphere Islands

Permanent populations of weasels can live only on large islands: off the coast of Britain the lower limit is approximately 60 km^2 (King and Moors 1979b). Data on body size in island weasels are sparse and statistically inadequate, and the origin and date of colonization is seldom known. However, island weasels are often at least slightly different from those of the nearest mainland, either larger or smaller, and *erminea* in Ireland has been wrongly quoted as an example of a theoretical generalization (Hutchinson 1959).

For many years the only measurements of Irish weasels available were from the northern part of the island, where *M. erminea hibernica* (Figure 11.3B, point 24N) is roughly intermediate in size between mainland British *M. e. erminea* (points 2, England, and 5, Scotland) and *M. nivalis*. These small northern animals were assumed to represent all Irish *erminea;* this assumption was linked with the absence of *nivalis* from Ireland, and the conclusion was drawn that *erminea* in Ireland must be one of the classic examples of character displacement (Hutchinson 1959; Williamson 1972:117). This explanation has been decisively quashed by Fairley's (1981) data, showing that, although *erminea* in the north of Ireland is indeed much smaller than in Britain, only 250 km away in the south it is as large or larger (point 24S). Hutchinson also took the very large *nivalis* living in the Mediterranean, in the absence of *erminea,* as the reverse example, but that is disputable too. There is a clear north-south increase in the size of *nivalis* in Europe, with (on present data) no clear step past the southern limit of *erminea;* all the Mediterranean races of *nivalis* are large, both on the mainland (Figure 11.3A, points 13, Italy, and 14, Spain) and on islands (point 15, Sardinia).

A special explanation was offered by van Soest et al. (1972) to account for the relatively small size of the *erminea* living on the island of Terschelling, off the coast of Holland (Figure 11.3B, points 25, 26). These weasels are very heavily infested with *Skrjabinglyus nasicola,* a damaging cranial parasite (91%, cf. 23% on the nearby mainland coast), which van Soest et al. suggest may stunt their growth. However, the size differences were not tested, and there is no evidence of the same effect elsewhere (King 1977; King and Moody 1982).

The real explanation for the variation in size of weasels in Ireland, Terschelling, and other islands is still unknown. My guess is that the mean body size of the weasels on any island will drift toward whatever gives them, in the local conditions, the best year-round, long-term compromise between the advantages of small size (Figure 11.2) and the upward pull of sexual selection; and the point of balance will be related, in some way we do not yet understand, both to the climate and to the size distribution of the prey. Unfortunately, we have no precise data on the size ranges and relative abundance of prey available on any of the northern hemisphere islands occupied by weasels, or even any certainty as to what characteristics of the prey fauna to measure. It is not necessarily a simple correlation between the body sizes of predator and prey, or of the diameters of weasel bodies and rodent tunnels (Simms 1979), since there is a well-documented tendency for rodents to be larger on islands (Lomolino 1985), whereas weasels seem to be either smaller or larger (Figure 11.3). If the observed size of the weasels on each island is a unique local compromise, no generalized theoretical model will explain the whole pattern unless it includes detailed information on the ecology of the weasels living in all the island habitats, how long they have been there, and where they came from. But there is one group of islands, the New Zealand archipelago, where some at least of the required information is available, and any attempt to devise a general explanation of what determine body size in weasels must take account of what is happening there.

New Zealand

Both *nivalis* and *erminea* were introduced into New Zealand over the 20 or so years after 1884 (King and Moors 1979b; King 1984b). Both were relatively large, since they came from British stock, among the largest in western Europe (Figure 11.3A, B; Kratochvil 1977a, 1979b). They found in New Zealand an environment radically different from home, with a wider range of climates (warmer in the north, colder in the high mountains) and a fauna of potential prey which was (and is) completely different in size distribution (King and Moors 1979b). There were no voles, and feral house mice (*Mus musculus*) were the only rodents under 50 g; on the other hand, there were Norway and ship rats (*Rattus norvegicus, R. rattus*) and European rabbits (*Oryctolagus cuniculus*), all in great numbers, as well as (at that time) still considerable numbers of large and unwary native birds. There were also numerous species of large (3 g) flightless native insects (Orthoptera), collectively known by their Maori name *weta*. These have been called "invertebrate mice" or "insect rodents" (Stevens 1980:255) because they to some extent held the niche occupied by small rodents elsewhere in the world.

In the 100 years since their arrival in the alpine beech forests of New Zealand, *erminea* of both sexes have become, on average, slightly larger than their British ancestors (assumed to have remained the same); in the lower-altitude mixed podocarp forests females are larger but males are not (Figure 11.4). This shift in mean size is precisely dated and consistent at all ages and even on small local scales (King and Moody 1982). The male *erminea* living in the foothills of the Southern Alps are probably the largest in the world, and near the top of the range for male *frenata*; the females are larger than any Eurasian female *erminea,* and near the middle of the range for female *frenata*. This pattern has not yet been shown to be genetic; but even simple phenetic changes, if as consistent as that, may be taken as evidence of adaptation in the broadest sense (Clutton-Brock and Harvey 1979), if only as indicating some change in the conditions of life for *erminea* in New Zealand. It seems most likely that the key factor is the size distribution and local abundance of the prey available. For example, it is no longer profitable for females to remain small so as to specialize on rodents. Perhaps this is the reason why they have become larger in all habitats. They still take what small rodents there are more often than do males (King and Moody 1982:63) and the abundant *weta;* but they also catch large prey (rats and lagomorphs) as often as do males (King and Moody 1982:68). However, there is no simple correlation between the body sizes of New Zealand *erminea* in general and their prey. If there were, we would expect to be able to predict the position of a point for New Zealand on the plot neatly relating prey size and body weight in *erminea* given by Erlinge (1987). But when this is done (from the list of prey items given by King and Moody 1982) according to the formula and prey weight loadings given by Erlinge (1987), the

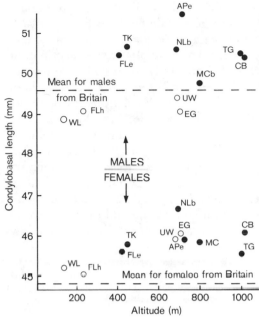

Figure 11.4. Mean condylobasal length of adult male and female *erminea* in New Zealand. ○ = mixed native podocarp and hardwood forests; ● = southern beech (*Nothofagus* spp.) forests and tussock grassland. Key to localities, further information, and statistics in King and Moody (1982). The mean condylobasal lengths of adult *erminea* from Britain (ages classified by the same criteria) are shown by the dotted lines. (Courtesy *New Zealand Journal of Zoology*.)

Key to abbreviations:

APe	Arthur's Pass (eastern side)
CB	Craigieburn
EG	Egmont
FLe	Fiordland (Eglinton Valley)
FLh	Fiordland (Hollyford Valley)
MC	Mount Cook
MCb	Mount Cook (Ball Hut Road)
NLb	Nelson Lakes (beech forest)
TG	Tangariro
TK	Takaro
UW	Urewera
WL	Westland

mean prey size of *erminea* in New Zealand works out at about the same figure as for the much smaller *erminea* in northern Sweden.

There are two reasons for this. One is that the formula requires the frequencies of occurrences of prey items eaten to be loaded according to prey body weight. The results are hugely influenced by the weight assigned to the large prey species, which depends on some assessment of the age of the individuals taken and the number of meals eaten from each carcass. The other reason is that *erminea* in New Zealand take large numbers of insects, especially *weta*, whereas the diet analyses of European *erminea* quoted by Erlinge reported only vertebrates. The high frequencies of insects distort the relative values for vertebrate prey. Yet we have shown, using a different logic and loading system (King and Moody 1982:71), that although insects were frequently found in our samples (in 41% of all guts containing food), they contributed only a small proportion (<10%) of the biomass of food eaten. New Zealand *erminea* depend for most of their sustenance (>50%) on large prey (lagomorphs, possums, and rats). If insects are omitted and the data recalculated from the frequencies of vertebrate prey alone, the prey size index works out, as one would expect, at about the same as for *erminea* from Britain.

Whether or not it is valid to omit the numerous but nutritionally unproductive insects in such comparisons depends on the foraging strategy of *erminea* in New Zealand. Nothing is known about how they hunt insects, or whether they pick them up only in passing or deliberately search them out, so this intriguing problem remains unresolved. Neither is there any simple correlation between local variations in the body size of *erminea* and the local distribution of any particularly favored large mammalian prey. For example, the size distribution of local populations shown in Figure 11.4 is positively correlated in both sexes with the distribution of rabbits. The larger, beech forest–tussock grassland *erminea* tend to eat relatively more rabbits than the smaller ones living in the podocarp-hardwood forests ($r_3 = 0.57$, $P < 0.05$ in males; $r_3 = 0.22$, $P > 0.05$ in females). But we cannot thereby conclude that the male alpine New Zealand *erminea* have become larger than their British ancestors in order to exploit rabbits more efficiently, because they generally eat rabbits less often than do British *erminea* (King and Moody 1982). Overall, they take as many large mammalian prey as do their relatives in Britain, by making up the total with possums and rats; but these are more common on the podocarp-hardwood forests. New Zealand would be a fertile field for further study on the ecology and economics of hunting by weasels.

Conclusions

The natural history of the weasels is very largely the story of how these adaptable little carnivores profit from the advantages, and cope with the disadvantages, of their diminutive stature. Almost every aspect of their lives is controlled by their small size, but, in the four most important ones (Figure 11.2), the list of the advantages and disadvantages seems to be more or less balanced in all respects except that of physiology. In competitive relationships the weasels' supremacy in exploitation, compared with the larger carnivores and raptors, is countered by their vulnerability to interference. In reproductive strategy the gains and losses that follow from being small roughly cancel each other out at both individual and population levels. Only with respect to metabolism are the pluses far outweighed by the minuses, the more so the colder the climate. Of course, none of the factors identified in Figure 11.2 can be quantified; they are most unlikely to be equal, of if they were, the way of life of a weasel would be impossible, especially in the Arctic. As Bergmann recognized long ago, for most mammals small size is most demanding and dangerous in cold climates. Paradoxically, the weasels have turned this argument on its head; for them it is their small size that provides a passport to survival and an unassailable advantage over larger predators in the exploitation of small rodents in the Arctic.

Weasels, like most carnivores, are not easy to study; they are scarcer, more intelligent, and more wide-ranging than mice and voles, and the literature on weasels is only just beginning to get past the straight descriptive stage that students of small rodents left behind years ago. Anyone who has pitted wits against weasels in the field knows, all too well, that it can be difficult to obtain large quantities of statistically respectable data from them. Weasel populations are also very unstable, and many conscientious field workers who deserved better have been disappointed when the weasels they expected to study declined to cooperate. On the other hand, weasels are certainly more abundant than larger carnivores, potentially able to supply large samples; and with the development in recent years of many new field techniques, they are beginning to provide some good opportunities to test theoretical ideas (Sandell 1985; Sandell, this volume). This review suggests a considerable list of ideas that could be tested and questions that might be answered by future field observers and modelers. The great difficulties will be in deciding how to identify and define the critical parameters, especially foraging efficiency and size distribution of *available* prey, and how to determine their effects on body size without falling into circular arguments. For example, the difficulty of interpreting the body size of *erminea* in New Zealand in terms of the most acceptable current hypothesis (Erlinge 1987) underlines (1) the need to understand the comparison between the sizes of predator and prey in terms of foraging strategy rather than simple morphometrics, and (2) the possibility that other factors besides prey size may be important. Erlinge (1987:37) concluded that "since alternative explanations for size variation in the small mustelids are not supported by existing data, evidence suggests that size variation in the stoat [*erminea*] is caused primarily by regional differences in the size frequency distribution of their available prey." Perhaps it is time to cease relying on existing data, to collect new information that could test alternative or additional ideas, and to design some critical experiments capable of distinguishing between or ranking competing explanations.

How far does the above discussion apply to other small carnivores, such as *Martes* species (the martens and fishers) and the viverrids? Some of it certainly does. All carnivores have to balance the energy equations of hunting, though not all take the risks that weasels do; metabolic restraints presumably apply in some form to all mammals. On the other hand, only weasels have the *combination* of (1) such an extremely stressful, energy-intensive way of life (the net result of inefficient heat conservation, active hunting technique, and wide distribution in cold climates); (2) such great uncertainty in reproductive success (the net result of a short average lifespan and dependence on fluctuating prey resources for maximum productivity); (3) such vulnerability to interference from so many other predators. *Martes* species also live in cold climates but are well protected against heat loss by their thick fur. The reproductive success of some martens is also strongly affected by the population fluctuations of voles (Bayevsky 1956; Weckwerth and Hawley 1962), but established adults probably have more chance of living to the next breeding season than adult weasels do. Viverrids (civets, genets, and mongooses), native to the tropics and subtropics of Africa and Asia, tend to be omnivorous; all are much larger and longer lived than weasels. Weasels alone have perfected the danger-

ous art of being the smallest carnivore of all, in the most challenging environment in the world.

Acknowledgments

I thank B. M. Fitzgerald, J. L. Gittleman, K. Ralls, and M. Sandell for many
helpful comments; and the editor of the *New Zealand Journal of Zoology* for
permission to reproduce Figure 11.4.

Appendix 11.1: Distribution of Adipose Tissue in Weasels

In the course of dissecting many hundreds of weasels (*M. nivalis* in Britain
and *M. erminea* in New Zealand: King 1977; King and Moody 1982:100), I
observed that adipose tissue was stored only in a few discrete sites, and that fat
was laid down in these in a certain order and withdrawn in the reverse order. If
there was any fat present at all, it was first deposited along the spine and
around the kidneys. The abdominal mesenteries were used next, although
deposition there was moderate until the later stages. With increasing fatness,
sites outside the body cavity were used, filling in the dips in the body outline
from the posterior forward; beginning under the tail, around the testes and in
the loins, and then in the angles of the hind legs, in the brachial pit and the
angles of the forelegs. As deposition progressed, the first sites were still being
used: a weasel that had reached the stage of using the front limb sites had a
relatively greater amount in the body cavity. This pattern ensures that even a
grossly fat weasel has the same streamlined profile characteristic of the species
and does not become too rotund to enter rodent tunnels and nests.

References

Anderson, E. 1970. Quaternary evolution of the genus *Martes* (Carnivora, Mustelidae).
    Acta Zool. Fennica 130:1–132.
Barbu, P. 1968. [Systematics and ecology of the weasel *Mustela nivalis* L., from forests
    in the districts of Ilfor and Prahova, Roumania]. *Travaux du Muséum d'histoire
    naturelle "Grigore Antipa"* (Bucharest) 8:991–1002 (in French).
Bayevsky, Y. B. 1956. [Changes in fertility of the Barguzin sable]. *Byulleten' Moskov-
    skogo Obshchestva Ispytatelei Prirody-otdel Biologii* 61(6):15–25 (in Russian:
    translation available from Department of Internal Affairs, Wellington, N.Z.).
Beaucournu, J. C., and Grulich, I. 1968. [On the weasels of Corsica]. *Mammalia*
    32:341–371 (in French).
Bekoff, M., Daniels, T. J., and Gittleman, J. L. 1984. Life history patterns and the
    comparative social ecology of carnivores. *Ann. Rev. Ecol. Syst.* 15:191–232.
Bonner, J. T. 1965. *Size and Cycle.* Princeton, N.J.: Princeton Univ. Press.

Bourlière, F. 1975. Mammals, small and large: The ecological implications of size. In: F. B. Golley, K. Petrusevicz & L. Ryszkowski, eds. *Small Mammals, Their Productivity and Population Dynamics,* pp. 1–8. Cambridge: Cambridge Univ. Press.

Brown, J. H., and Lasiewski, R. C. 1972. Metabolism of weasels: The cost of being long and thin. *Ecology* 53:939–943.

Burnham, P. M. 1970. Kestrel attempting to prey on weasels. *British Birds* 63:338.

Calder, W. A. III. 1983. Ecological scaling: Mammals and birds. *Ann. Rev. Ecol. Syst.* 14:213–230.

Casey, T. M., and Casey, K. K. 1979. Thermoregulation of arctic weasels. *Physiol. zool.* 52:153–164.

Chappell, M. A. 1980. Thermal energetics and thermoregulatory costs of small arctic mammals. *J. Mamm.* 61:278–291.

Clutton-Brock, T. H., and Harvey, P. H. 1979. Comparison and adaptation. *Proc. Roy. Soc. Lond.* B 205:547–565.

Clutton-Brock, T. H., and Harvey, P. H. 1983. The functional significance of variation in body size among mammals. In: J.F. Eisenberg & D. G. Kleiman eds. *Advances in the Study of Mammalian Behaviour,* 632–663. Special Publications no. 7. Lawrence, Kans.: American Society of Mammalogists.

Crowell, K. L. 1986. A comparison of relict vs. equilibrium models for insular mammals of the Gulf of Maine. *Biol. J. Linn. Soc.* 28:37–64.

Dawkins, R., and Krebs, J. R. 1979. Arms races between and within species. *Proc. Roy. Soc. Lond.* B 205:489–511.

Debrot, S. 1984. [The structure and dynamics of a stoat (*Mustela erminea*) population]. *Rev. Ecol. Terre Vie* 39:77–88 (in French).

Delattre, P. 1983. Density of weasel (*Mustela nivalis* L.) and stoat (*Mustela erminea* L.) in relation to water vole abundance. *Acta Zool. Fennica* 174:221–222.

East, K., and Lockie, J. D. 1964. Observations on a family of weasels (*Mustela nivalis*) bred in captivity. *Proc. Zool. Soc. Lond.* 143:359–363.

Edson, J. M. 1933. A visitation of weasels. *Murrelet* 14:76–77.

Erlinge, S. 1977. Agonistic behaviour and dominance in stoats (*Mustela erminea* L.). *Z. Tierpsychol.* 44:375–388.

Erlinge, S. 1979. Adaptive significance of sexual dimorphism in weasels. *Oikos* 33:233–245.

Erlinge, S. 1981. Food preference, optimal diet and reproductive output in stoats *Mustela erminea* in Sweden. *Oikos* 36:303–315.

Erlinge, S. 1983. Demography and dynamics of a stoat *Mustela erminea* population in a diverse community of vertebrates. *J. Anim. Ecol.* 52:705–726.

Erlinge, S. 1987. Why do European stoats *Mustela erminea* not follow Bergmann's rule? *Holarctic Ecology* 10:33–39.

Erlinge, S., Jonsson, B., and Willstedt, H. 1974. [Hunting behavior and the choice of prey of captive weasels]. *Fauna och Flora* (Stockholm) 69:95–101 (in Swedish).

Erlinge, S., and Sandell, M. 1985. Coexistence of stoat *Mustela erminea* and weasel *Mustela nivalis:* Social dominance, scent communication and reciprocal distribution. In: M. Sandell. Ecology and behaviour of the stoat *Mustela erminea* and a theory on delayed implantation. Ph.D. thesis, Univ. Lund, Sweden. 115 pp.

Fairley, J. S. 1981. A north-south cline in the size of the Irish stoat. *Proc. Roy. Irish Acad.* B 81:5–10.

Fitzgerald, B. M. 1977. Weasel predation on a cyclic population of the montane vole (*Microtus montanus*) in California. *J. Anim. Ecol.* 46:367–397.

Fitzgerald, B. M. 1981. Predatory birds and mammals. In: L. C. Bliss, J. B. Cragg, D. W. Heal, & J. J. Moore, eds. *Tundra Ecosystems: A Comparative Analysis,* pp. 485–508. Cambridge: Cambridge Univ. Press.

Fog, M. 1969. Studies on the weasel (*Mustela nivalis*) and the stoat (*Mustela erminea*) in Denmark. *Danish Rev. Game Biol.* 6:1–14.

Formosov, A. N. 1946. *Snow Cover as an Integral Factor in the Environment and Its Importance in the Ecology of Mammals and Birds,* translated from the Russian by W. Prychodko & W. O. Pruitt. Edmonton: Boreal Institute for Northern Studies, Univ. Alberta, Occasional Publication no. 1 (1973).

Freuchen, P., and Salomonsen, F. 1959. *The Arctic Year.* London: Putnam Press.

Gamble, R. L. 1981. Distribution in Manitoba of *Mustela frenata longicauda* Bonaparte, the longtailed weasel, and the interrelation of distribution and habitat selection in Manitoba, Saskatchewan and Alberta. *Canadian J. Zool.* 59:1036–1039.

Gewalt, W. 1959. [Visual discrimination in mustelids]. *Zool. Beitr.* (Berlin) 5:117–175 (in German).

Gittleman, J. L. 1985. Carnivore body size: Ecological and taxonomic correlates. *Oecologia* 67:540–554.

Gittleman, J. L., and Harvey, P. H. 1982. Carnivore home range size, metabolic needs and ecology. *Behav. Ecol. Sociobiol.* 10:57–63.

Grue, H., and King, C. M. 1984. Evaluation of age criteria in New Zealand stoats (*Mustela erminea*) of known age. *New Zealand J. Zool.* 11:437–443.

Hall, E. R. 1951. *American Weasels.* Lawrence: Univ. Kansas Publication no. 4 (Museum of Natural History).

Hansson, L., and Henttonen, H. 1985. Gradients in density variations of small rodents: The importance of latitude and snow cover. *Oecologia* 67:394–402.

Harvey, P. H., and Ralls, K. 1985. Homage to the null weasel. In: P. J. Greenwood, P. H. Harvey & M. Slatkin, eds. *Evolution: Essays in Honour of John Maynard Smith,* pp. 155–171. Cambridge: Cambridge Univ. Press.

Hayward, G. F. 1983. The Bioenergetics of the Weasel *Mustela nivalis* L. D.Phil. thesis, Univ. Oxford, U.K.

Heidt, G. A. 1972. Anatomical and behavioural aspects of killing and feeding by the least weasel, *Mustela nivalis* L. *Proc. Ark. Acad. Sci.* 26:53–54.

Heptner, V. G., Naumov, N. P., Yurgenson, P. B., Sludskii, A. A., Chirkova, A. F., and Bannikov, A. G. 1967. [Mammals of the Soviet Union, vol. 2]. Chapters on *Mustela nivalis* and *M. erminea,* pp. 636–686, translated from the Russian by British Library, Boston Spa, Yorkshire, LS23 7BQ, UK, translation # RTS 6458.

Hespenheide, H. A. 1973. Ecological inferences from morphological data. *Ann. Rev. Ecol. Syst.* 4:213–229.

Hewson, R., and Healing, T. D. 1971. The stoat, *Mustela erminea,* and its prey. *J. Zool. (Lond.)* 164:239–244.

Hutchinson, G. E. 1959. Homage to Santa Rosalia, or, Why are there so many kinds of animals? *Amer. Nat.* 93:145–159.

Jamison, V. C. 1975. Relative susceptibility of resident and transient *Peromyscus* subjected to weasel predation. M.A. thesis, Univ. Montana.

King, C. M. 1977. The effects of the nematode parasite *Skrjabingylus nasicola* on British weasels (*Mustela nivalis*). *J. Zool. (Lond.)* 182:225–249.

King, C. M. 1980a. The weasel *Mustela nivalis* and its prey in an English woodland. *J. Anim. Ecol.* 49:127–159.

King, C. M. 1980b. Population biology of the weasel *Mustela nivalis* on British game estates. *Hol. Ecol.* 3:160–168.

King, C. M. 1981. The reproductive tactics of the stoat, *Mustela erminea,* in New Zealand forests. In: J. A. Chapman & D. Pursley, eds. *Proceedings of the Worldwide Furbearer Conference,* pp. 443–468. Frostburg, Md.: Worldwide Furbearer Conference, Inc.

King, C. M. 1983a. The life history strategies of *Mustela nivalis* and *M. erminea*. *Acta Zool. Fennica* 174:183–184.

King, C. M. 1983b. Factors regulating mustelid populations. *Acta Zool. Fennica* 174:217–220.

King, C. M. 1983c. The relationships between beech (*Nothofagus* sp.) seedfall and populations of mice (*Mus musculus*), and the demographic and dietary responses of stoats (*Mustela erminea*), in three New Zealand forests. *J. Anim. Ecol.* 52:141–166.

King, C. M. 1984a. The origin and adaptive advantages of delayed implantation in *Mustela erminea*. *Oikos* 42:126–128.

King, C. M. 1984b. *Immigrant Killers: Introduced Predators and the Conservation of Birds in New Zealand*. Auckland: Oxford Univ. Press.

King, C. M. 1985. Interactions between woodland rodents and their predators. In: J. R. Flowerdew, J. Gurnell, & J. H. W. Gipps, eds. *The Ecology of Woodland Rodents: Bank Voles and Woodmice*, pp. 219–247. Symposia of the Zoological Society of London no. 55. Oxford: Clarendon Press.

King, C. M., and Moody, J. E. 1982. The biology of the stoat (*Mustela erminea*) in the National Parks of New Zealand. *New Zealand J. Zool.* 9:49–144.

King, C. M., and Moors, P. J. 1979a. The life history tactics of mustelids, and their significance for predator control and conservation in New Zealand. *New Zealand J. Zool.* 6:619–622.

King, C. M., and Moors, P. J. 1979b. On co-existence, foraging strategy and the biogeography of weasels and stoats (*Mustela nivalis* and *M. erminea*) in Britain. *Oecologia* 39:129–150.

Kopein, K. I. 1969. The relationship between age and individual variation in the ermine. *Akademija Nauk SSSR, Uraljskij Filial* 71:106–112. Translated in *Biology of Mustelids: Some Soviet Reserach*, vol. 2:132–138 (New Zealand Department of Scientific and Industrial Research, Bulletin no. 227, ed. by C. M. King, 1980).

Kowalski, K. 1980. Origin of mammals of the arctic tundra. *Folia Quaternaria* 51:3–16.

Kraft, V. A. 1966. Influence of temperature on the activity of the ermine in winter. *Zoologicheskii Zhurnal* 45:148–150. Translated in *Biology of mustelids: Some Soviet Research*, pp. 104–107 (British Library, Boston Spa, Yorkshire, LS23 7BQ, U.K., ed. by C. M. King, 1975).

Kratochvil, J. 1977a. Sexual dimorphism and status of *Mustela nivalis* in Central Europe (Mammalia, Mustelidae). *Acta Sci. Nat. Brno* 11:1–42.

Kratochvil, J. 1977b. Studies on *Mustela erminea* (Mustelidae, Mammalia) I. Variability of metric and mass traits. *Folia Zool.* 26:291–304.

Kurtén, B. 1960. Chronology and faunal evolution of the earlier European glaciations. *Commentat. Biol. Helsingfors* 21:1–52.

Kurtén, B. 1968. *Pleistocene Mammals of Europe*. London: Wiedenfeld & Nicholson.

Kurtén, B., and Anderson, E. 1980. *Pleistocene Mammals of North America*. New York: Columbia Univ. Press.

Lande, R. 1980. Sexual dimorphism, sexual selection and adaptation in polygenic characters. *Evolution* 34:292–305.

Lomolino, M. V. 1985. Body size of mammals on islands: The island rule re-examined. *Amer. Nat.* 125:310–316.

Macdonald, D. W. 1977. On food preference in the red fox. *Mamm. Rev.* 7:7–23.

McLean, S. F. Jr., Fitzgerald, B. M., and Pitelka, F. A. 1974. Population cycles in Arctic lemmings: Winter reproduction and predation by weasels. *Arctic & Alpine Res.* 6:1–12.

McNab, B. K. 1971. On the ecological significance of Bergmann's Rule. *Ecology* 52:845–854.

Madison, D., Fitzgerald, R., and McShea, W. 1984. Dynamics of social nesting in overwintering meadow voles: Possible consequences for population cycling. *Behav. Ecol. Sociobiol.* 15:9–17.

Metzgar, L. H. 1967. An experimental comparison of screech owl predation on resident and transient white-footed mice (*Peromyscus leucopus*). *J. Mamm.* 48:387–391.

Miller, G. S. 1912. *Catalogue of the Mammals of Western Europe*. London: British Museum (Natural History).

Moors, P. J. 1980. Sexual dimorphism in the body size of mustelids (Carnivora): The roles of food habits and breeding systems. *Oikos* 34:147–158.

Morosova-Turova, L. G. 1965. Geographical variation in weasels in the Soviet Union. In: D. P. Dement'ev et al., eds. [Game and Fur Animals: Their biology and commercial exploitation]. Moscow, Translated in *Biology of Mustelids: Some Soviet Research*, pp. 7–38 (British Library, Boston Spa, Yorkshire, LS23 7BQ, U.K., ed. by C. M. King, 1975).

Musgrove, B. F. 1951. Weasel foraging patterns in the Robinson Lake area, Idaho. *Murrelet* 32:8–11.

Nams, V. 1981. Prey selection mechanisms of the ermine (*Mustela erminea*). In: J. A. Chapman & D. Pursley, eds. *Proceedings of the Worldwide Furbearer Conference*, pp. 861–882. Frostburg, Md.: Worldwide Furbearer Conference, Inc.

Nyholm, E. S. 1959. Stoats and weasels and their winter habitat. *Soumen Riista* 13:106–116. Translated in *Biology of Mustelids: Some Soviet Research*, pp. 118–131 (British Library, Boston Spa, Yorkshire, LS23 7BQ, U.K. ed. by C. M. King, 1975).

Oksanen, T., Oksanen, L., and Fretwell, S. D. 1985. Surplus killing in the hunting strategy of small predators. *Amer. Nat.* 126:328–346.

Osborn, D. J., and Helmy, I. 1980. The contemporary land mammals of Egypt (including Sinai): *Mustela nivalis*. *Fieldiana: Zoology* n.s. 5:406–409.

Panteleev, P. A. 1981. Evolutionary strategy of changes in body size in mammals. *Sov. J. Ecol.* 12:315–321.

Parovshchikov, V. Y. 1963. A contribution to the ecology of *Mustela nivalis* Linnaeus, 1766, of the Arkhangel'sk North. *Vestnik Ceskoslovenske Spolecnosti zoologicke* 27:335–344, Translated in *Biology of Mustelids: Some Soviet Research*, pp. 84–97 (British Library, Boston Spa, Yorkshire, LS23 7BQ, U.K., ed. by C. M. King, 1975).

Peters, R. H. 1983. *The Ecological Implications of Body Size*. Cambridge: Cambridge Univ. Press.

Petrov, O. V. 1956. Sexual dimorphism in the skull of *Mustela erminea* L. *Vestnik Leningrad Univ.* 15:41–56. Translated in *Biology of Mustelids: Some Soviet Research*, pp. 55–78 (British Library, Boston Spa, Yorkshire, LS23 7BQ, U.K., ed. by C. M. King, 1975).

Petrov, O. V. 1962. The validity of Bergmann's Rule as applied to intraspecific variation in the ermine. *Vestnik Leningrad Univ., Biol. Ser.* 9;144–148. Translated in *Biology of Mustelids: Some Soviet Research*, pp. 30–38 (British Library, Boston Spa, Yorkshire, LS23 7BQ, U.K., ed. by C. M. King, 1975).

Pounds, C. J. 1981. Niche overlap in sympatric populations of stoats (*Mustela erminea*) and weasels (*Mustela nivalis*) in northeast Scotland. Ph.D. thesis, Univ. Aberdeen. 326 pp.

Powell, R. A. 1973. A model for raptor predation on weasels. *J. Mamm.* 54:259–263.

Powell, R. A. 1982. Evolution of black-tipped tails in weasels: Predator confusion. *Amer. Nat.* 119:126–131.

Powell, R. A., and Zielinski, W. J. 1983. Competition and coexistence in mustelid communities. *Acta Zool. Fennica* 174:223–227.

Pruitt, W. O., Jr. 1978. *Boreal Ecology.* Studies in Biology no. 91. London: Edward Arnold.

Ralls, K., and Harvey, P. H. 1985. Geographic variation in size and sexual dimorphism of North American weasels. *Biol. J. Linn. Soc.* 25:119–167.

Reichstein, H. 1957. [Skull variation in European weasels (*Mustela nivalis* L.) and stoats (*Mustela erminea* L.) in relation to distribution and sex]. *Z. Säugetierk.* 22:151–182 (in German).

Rosenzweig, M. L. 1966. Community structure in sympatric carnivora. *J. Mamm.* 47:602–612.

Rubina, M. A. 1960. Some features of weasel (*Mustela nivalis* L.) ecology based on observations in the Moscow region. *Byulletin' Moskovskogo Obshchestva Ispytalelei Prirody-Otdel Biologii* 65:27–33. (British Library, Boston Spa, Yorkshire, LS23 7BQ, U.K., translation no. RTS 2292).

Sandell, M. 1985. Ecological energetics and optimum body size in male and female stoats *Mustela erminea:* Predictions and test. In: Ecology and behaviour of the stoat *Mustela erminea* and a theory on delayed implantation. Ph.D. thesis, Univ. Lund, Sweden. 115 pp.

Sandell, M. 1986. Movement patterns of male stoats *Mustela erminea* during the mating season: Differences in relation to social status. *Oikos* 47:63–70.

Schmidt-Nielsen, K. 1984. *Scaling: Why is Animal Size So Important?* Cambridge: Cambridge Univ. Press.

Segal, A. N. 1975. Postnatal growth, metabolism and thermoregulation in the stoat. *Sov. J. Ecol.* 6:28–32.

Seton, E. T. 1929. *Lives of Game Animals.* New York: Doubleday, Doran.

Simms, D. A. 1979. North American weasels: Resource utilisation and distribution. *Canadian J. Zool.* 57:504–520.

Southern, H. N., and Lowe, V. P. W. 1982. Predation by tawny owls (*Strix aluco*) on bank voles (*Clethrionomys glareolus*) and wood mice (*Apodemus sylvaticus*). *J. Zool. (Lond.)* 198:83–102.

Stanley, S. M. 1973. An explanation for Cope's rule. *Evolution* 27:1–26.

Stevens, G. 1980. *New Zealand Adrift. The Theory of Continental Drift in a New Zealand Setting.* Wellington: A. H. & A. W. Reed.

Stolt, B.-O. 1979. Colour patterns and size variation of the weasel, *Mustela nivalis*, in Sweden. *Zoon* 7:55–62.

Stroganov, S. U. 1962. *Carnivorous Mammals of Siberia.* Translated by Israel Program for Scientific Translations, Jerusalem (1969).

Tapper, S. C. 1979. The effect of fluctuating vole numbers (*Microtus agrestis*) on a population of weasels (*Mustela nivalis*) on farmland. *J. Anim. Ecol.* 48:603–617.

Teplov, V. P. 1948. The problem of sex ratio in ermine. *Zoologicheski Zhurnal* 27:567–570. Translated in *Biology of Mustelids: Some Soviet Research*, pp. 98–103 (British Library, Boston Spa,Yorkshire, LS23 7BQ, U.K. ed. by C. M. King, 1975).

Tumanov, I. L., and Levin, V. G. 1974. Age and seasonal changes in some physiological characters in the weasel (*Mustela nivalis* L.) and ermine (*Mustela erminea* L.). *Vestnik Zoologii* 2:25–30. Translated in *Biology of Mustelids: Some Soviet Research*, vol. 2:192–196 (New Zealand Department of Scientific and Industrial Research Bulletin 227, ed. by C. M. King, 1980).

van Soest, R. W. M., van der Land, J., and van Bree, P. J. H. 1972. *Skrabingylus nasicola* (Nematoda) in skulls of *Mustela erminea* and *Mustela nivalis* (Mammalia) from the Netherlands. *Beaufortia* 20:85–97.

Vershinin, A. A. 1972. The biology and trapping of the ermine in Kamchatka. *Byulletin' Moskovskogo Obshchestva Ispytatelei Prirody-Otdel Biologii* 77:16–26.

Translated in *Biology of Mustelids: Some Soviet Research*, 2:11–23. New Zealand Department of Scientific and Industrial Research, Bulletin 227, ed. by C. M. King 1980).

Weckwerth, R. P., and Hawley, V. D. 1962. Marten food habits and population fluctuations in Montana. *J. Wildl. Mgmt.* 26:55–74.

Williamson, M. 1972. *The Analysis of Biological Populations*. Special Topics in Biology. London: Edward Arnold.

Wojcik, M. 1974. Remains of Mustelidae (Carnivora, Mammalia) from the late Pleistocene deposits of Polish caves. *Acta Zoologica Cracoviensia* 19:75–90.

CHAPTER 12

Basal Rate of Metabolism, Body Size, and Food Habits in the Order Carnivora

BRIAN K. MCNAB

Organisms expend energy for a variety of tasks, including body maintenance, movement, resource acquisition, courtship, reproduction, and growth.
Energy expenditure is greatest in species that have high costs of maintenance
(e.g., endotherms), high activity levels (due either to extended periods of activity or to the use of expensive forms of locomotion, such as flight and
elaborate courtship rituals), expensive means or extended periods of resource
acquisition, high rates of reproduction, high postnatal growth rates, and extended periods of parental care.

The complex pattern of energy expenditure of mammals and birds, because
of their endothermy and extensive activity, makes comparison of one species
with another difficult: two species may have the same total expenditure but
differ radically in the apportioning of energy; two species may apportion
energy similarly but have markedly different total expenditures; or two species
may differ both in total expenditures and energy apportionment. Some factors,
such as body size, appear to affect all elements in an expenditure, whereas
others, such as food habits, differentially affect specific elements of an expenditure, as well as the total expenditure. Therefore, a simplification usually is
required to compare the expenditures of two or more species, because all
elements of a field expenditure are known only for a few species, and then only
under specific environmental conditions. To compare species, one must use an
expenditure that is equivalent in all species. Such an expenditure in endotherms is the basal rate of metabolism. The basal rate is a measure of the
minimal cost of maintenance at a normal body temperature during the usual
period of rest, when an adult is post absorptive. Variation in total field expenditures seems to correlate with variation in basal rate (McNab 1980).

This chapter explores the variation in the basal rate of metabolism of mammals belonging to the order Carnivora. Members of this order show great
variation in body size (here, body mass) and in food habits (including verte-
brate-, invertebrate-, leaf-, and fruit-eating specialists, as well as those having a
mixed diet or scavanging habits). As a result, they are expected to show appre-

335

ciable differentiation in the level of energy expenditure, including basal and field rates of metabolism.

One difficulty arising from using a food habit as the name for an order having a diverse set of food habits is that many "carnivores" are not carnivorous, especially if invertebrativory is excluded from carnivory. Here, the somewhat awkward term "carnivoran" is used to evade the implication for food habits of the term "carnivore."

The Basal Rate of Metabolism in Carnivorans

Data available on the basal rate of metabolism for 43 species of carnivorans belonging to nine families are summarized in Table 12.1. Most of these data were derived from the literature, although data for 14 species were obtained recently in my laboratory and will be published in detail elsewhere. A few other species have had rates of metabolism reported (e.g., raccoon dog, *Nyctereutes procyonoides* [Korhonen and Harri 1984]; African hunting dog, *Lycaon pictus* [Taylor et al. 1971]; cheetah, *Acinonyx jubatus* [Taylor and Roundtree 1973]), but in each case question exists whether standard conditions were met.

Most carnivoran families are underrepresented in Table 12.1, both in terms of species number and ecological diversity, a condition that is most notable in the families Canidae, Ursidae, Herpestidae, and Viverridae; moreover, the Ailuropodidae (assuming that the giant panda is not an ursid) is not represented at all. Yet, on the whole, enough of the diversity in size and food habits found in the order is represented to permit this preliminary examination to be made of the influence of these factors on energy expenditure. Other potentially important factors, including climate, activity level, and familial affiliation, are also examined. In some cases, measurements of basal rate in a species have been made by several observers, most notably in the fennec fox (*Fennicus zerda*), arctic fox (*Alopex lagopus*), and coyote (*Canis latrans*). Only one value is given in Table 12.1 for each species; that value was chosen by the criteria that the individuals had an adult body mass, thermoneutrality was clearly defined, the measurements were made during the inactive period, and the individuals were postabsorptive. Beyond these "objective" criteria, the lowest rate was chosen under the assumption that higher basal rates might have been contaminated by activity. Nevertheless, some of the variation reported, especially for the coyote and arctic fox, may have a biological basis and significance (see Discussion).

Interspecific variability in basal rate of metabolism is measured either in terms of the standard deviation of the mean of several species or, in a regression of basal rate on body mass, by r^2 and by the standard error of estimate, $Sy \cdot x$. With respect to regressions, emphasis here is placed on $Sy \cdot x$ because 1 −

r^2 is a poor measure of the dispersion of data around a curve (Smith 1980, 1984; McNab 1988).

The Influence of Body Mass on Carnivoran Basal Rates

Body mass is generally acknowledged to be the most important factor setting the level of basal rate in mammals (McNab 1980, 1986b; Muller 1985) and is often considered to be the only important factor (Kleiber 1932, 1961; Scholander et al. 1950). Recently, several reevaluations of the relation between basal rate and body mass in mammals have been made, most notably by Hayssen and Lacy (1985), Elgar and Harvey (1987), and McNab (1988). Among the 43 species assembled in Table 12.1, variation in body mass "accounts for" 67% ($=100r^2$) of the variation in mass-specific basal rate. The fitted relationship

$$\dot{V}_{O_2}/m = 4.05m^{-0.288}$$

for all 43 species at carnivoran masses is slightly lower than the Kleiber curve

$$\dot{V}_{O_2}/m = 3.42m^{-0.25}$$

and 17% higher than the mean mammalian curve described by McNab

$$\dot{V}_{O_2}/m = 3.45m^{-0.287}.$$

Nevertheless, appreciable residual variation exists around the fitted curve: $Sy \cdot x = 0.388$, which means that the central 68% of the observations fall between 41 and 244% of the value expected from mass, a factorial difference of 6.0:1!

In the manner of Hayssen and Lacy (1985) and Elgar and Harvey (1987), the residual variation can be described relative to taxonomic affiliation. For example, measured canids tend to have intermediate basal rates (i.e., two of six species are greater than the mean carnivoran curve, two are similar to the mean curve, and two are below: 2-2-2, assuming that an expected value falls between 91 and 109% of a value expected from the mean carnivoran curve). Ursids (0-2-1) and herpestids (1-1-1) are also intermediate. Procyonids (0-0-3), red panda (0-0-1), viverrids (0-1-5), and hyaenids (0-0-2), however, are low. In contrast, mustelids (9-2-1) and felids (6-0-1) are high.

The difficulty with accepting this "analysis" as definitive is that little understanding issues from the statement that canids have intermediate basal rates and felids have high basal rates, correct though it may generally be. Such a statement has been viewed as a taxonomic character (see, for example, McKenna's argument [1975] that the low body temperature and "poor" temperature regulation of xenarthrans is a retention of primitive features), which ignores the adaptive nature of energy expenditure (and by implication the

Table 12.1. Basal rates of metabolism in fissiped carnivorans

Species	Food habit[a]	Mass (g)	Basal rate: $cm^3 \cdot g^{-1} \cdot h^{-1}$	Basal rate: %[b]	Reference
Canidae					
Fennec (*Fennecus zerda*)	I	1106	0.36	67	Noll-Banholzer 1969
Kit fox (*Vulpes macrotis*)	V	1868	0.50	109	Golightly and Ohmart 1983
Arctic fox (*Alopex lagopus*)	V	3600	0.38	100	Casey et al. 1979
Red fox (*Vulpes vulpes*)	V	5010	0.50	143	Irving et al. 1955
Crab-eating fox (*Cerdocyon thous*)	M−	5444	0.28	82	Hennemann et al. 1983
Coyote (*Canis latrans*)	M+	1000	0.27	96	Golightly and Ohmart 1983
Ursidae					
Brown bear (*Ursus arctos*)	M+	136000	0.128	96	Watts 1988
Black bear (*U. americanus*)	M−	143000	0.081	61	Watts 1988
Polar bear (*U. maritimus*)	V	204000	0.119	100	Watts 1988
Procyonidae					
Crab-eating raccoon (*Procyon cancrivorus*)	M−	1160	0.40	75	Scholander et al. 1950
Kinkajou (*Potos flavus*)	F	2400	0.32	74	Müller and Kulzer 1977
Coati (*Nasua nasua*)	M−	4000	0.25	68	Chevillard-Hugot et al. 1980
Ailuridae					
Red panda (*Ailurus fulgens*)	L	5740	0.158	47	McNab 1988b
Mustelidae					
Least weasel (*Mustela nivalis*)	V	77	2.29	197	Casey and Casey 1979
Ermine (*M. erminea*)	V	210	1.48	170	Iversen 1972
Long-tailed weasel (*M. frenata*)	V	297	0.95	122	Brown and Lasiewski 1972
Spotted skunk (*Spilogale putorius*)	M+	624	0.47	75	Kilgore, pers. comm.
American mink (*Mustela vison*)	V	660	0.74	119	Farrell and Wood 1968
European pine marten (*Martes martes*)	V	920	0.80	140	Iversen 1972
American pine marten (*M. americana*)	V	1038	0.66	120	Worthen and Kilgore 1981
North American badger (*Taxidea taxus*)	V	9000	0.30	103	Harlow 1981
European otter (*Lutra lutra*)	V	10000	0.45	161	Iversen 1972
Eurasian badger (*Meles meles*)	M+	11050	0.27	96	Iversen 1972

Table 12.1. (*Continued*)

Species	Food habit[a]	Mass (g)	Basal rate cm³ · g⁻¹ · h⁻¹	%[b]	Reference
Wolverine (*Gulo gulo*)	V	12700	0.46	170	Iversen 1972
Sea otter (*Enhydra lutris*)	I	40000	0.64	337	Iversen and Krog 1973
Herpestidae					
Slender mongoose (*Herpestes sanguineus*)	V	500	0.76	112	Kamau et al. 1979
Small Indian mongoose (*H. auropunctatus*)	V	611	0.66	103	Ebisu and Whit tow 1976
Meerkat (*Suricata suricatta*)	I	850	0.37	64	Müller and Lo- jewzski 1986
Viverridae					
Large-spotted genet (*Genetta tigrina*)	M+	1732	0.44	94	Hennemann and Konecny 1980
Small-tooth palm civet (*Arctogalidia trivirgata*)	M−	2013	0.30	67	pers. observ.
Fanaloka (*Fossa fossa*)	M+	2260	0.40	91	pers. observ.
Common palm civet (*Paradoxurus hermaphroditus*)	M−	3410	0.21	54	pers. observ.
African palm civet (*Nandinia binotata*)	F	4270	0.27	75	pers. observ.
Binturong (*Arctictis binturong*)	F	14280	0.156	61	pers. observ.
Hyaenidae					
Aardwolf (*Proteles cristatus*)	I	7710	0.25	81	McNab 1984
Striped hyena (*Hyaena hyaena*)	M+	35120	0.171	86	pers. observ.
Felidae					
Margay (*Felis wiedii*)	V	3616	0.28	74	pers. observ.
Bobcat (*Lynx rufus*)	V	7880	0.47	152	pers. observ.
Ocelot (*Felis pardalis*)	V	10416	0.31	111	pers. observ.
Mountain lion (*F. concolor*)	V	41150	0.25	132	pers. observ.
Jaguar (*Panthera onca*)	V	68900	0.184	113	pers. observ.
African lion (*Panthera leo*)	V	98000	0.176	119	pers. observ.
Tiger (*P. tigris*)	V	138200	0.177	132	pers. observ.

Note. Taxonomy follows Ewer 1973.

[a]Food habits: I:invertebrates, V:vertebrates, F:fruit, M+: mixed diets with over 25% volume intake as vertebrates, M−: mixed diets with less than 25% volume.

[b]% rate of metabolism = 100(measured rate) / $3.45m^{-0.288}$.

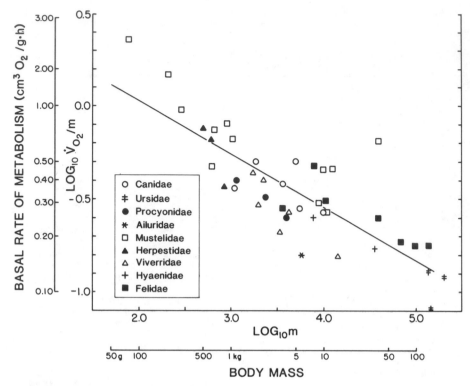

Figure 12.1. Log_{10} mass-specific basal rate of metabolism in carnivorans as a function of log_{10} body mass in relation to taxonomic affiliation. Data taken from Table 12.1. The curve is fitted by least squares for all carnivorans (see text).

flexibility of body temperature and temperature regulation). Thus, some canids in fact have high basal rates and at least one felid has a low basal rate (Figure 12.1). An emphasis on the taxonomic correlations of basal rate tends to ignore the significance of high, intermediate, or low rates of metabolism, and specifically why the correlation with mass accounts for only 67% of the variation in basal rate, that is, why $Sy \cdot x = 0.388$. The principal reason why such residual variation in energy expenditure remains is undoubtedly that basal rate is functionally associated with factors other than mass. Three such factors are food habits, activity level, and climate.

The Influence of Food Habits on Carnivoran Basal Rates

Recent evidence (McNab 1969, 1983, 1986b) has suggested that the principal correlate of the residual variation in the basal rate of eutherian mammals is food habits. That is, at masses greater than approximately 100 g, vertebrate-eaters and grazers have basal rates equal to, or greater than, the Kleiber curve, whereas invertebrate-, fruit-, and leaf-eaters have basal rates less than expected from the Kleiber relation (McNab 1986b). Species with mixed diets generally have intermediate basal rates.

This differentiation in rate of metabolism with respect to food habits has been related to various factors that restrict energy (and, generally, nutrient) availability, on the one hand, or that show little or no such restriction, on the other. The former factors are associated with, or "lead to," low rates of metabolism, the latter "permitting"—but not requiring—high rates of metabolism (McNab 1986b). Factors that appear to restrict energy (or nutrient) availability include a seasonal variation in food availability; a low energy (or nutrient) content of food, either inherently or as produced by foraging methods, as in ant/termite-eaters (McNab 1984); and the use of chemical or mechanical defenses by prey to diminish predation. These factors are more important to large consumers because of the large volume of food required by a large mass (McNab 1986b), which reemphasizes the conclusion (McNab 1971) that total rate of metabolism is the ecologically significant rate of metabolism, mass-specific rates principally referring to turnover rates (Morrison 1960; Lindstedt and Calder 1981). The use of mass-specific rates, however, facilitates a graphic analysis of residual variation.

The pattern seen between basal rate and food habits among mammals in general is found among carnivorans (Figure 12.2): carnivorans that feed principally, or exclusively, on vertebrates have basal rates that are close to, or greater than, the values expected from the fitted carnivoran curve, irrespective of whether these habits are found in canids, mustelids, herpestids, or felids. (Recent measurements by Okarma and Koteja [1987] indicated a high basal rate [141% of the carnivoran value] in the vertebrate-eating gray wolf.) The one vertebrate-eating specialist known to have a low basal rate is the margay (*Felis wiedii*). In contrast, invertebrate-eating specialists belonging to the families Canidae, Herpestidae, and Hyaenidae have low basal rates. Frugivorous viverrids and the folivorous red panda (*Ailurus fulgens*) also have low basal rates. A mixed diet that includes significant proportions (i.e., >25% of total volumetric intake on an annual basis) of vertebrates is associated with an intermediate basal rate (varying from 75 to 96% of the carnivoran curve; mean 91%) in canids, ursids, procyonids, mustelids, viverrids, and hyaenids, whereas a mixed diet that includes few vertebrates (i.e., <25% of total intake), and thus is primarily a mixture of invertebrates and fruit, is associated with a low basal rate (varying from 54 to 82%; mean 68%), as is to be expected. This latter mixture occurs in some canids, ursids, procyonids, and viverrids. A remarkable exception to the pattern seen between basal rate and food habits is found in sea otters (*Enhydra lutris*), which, in spite of combining fish and invertebrates in a diet, have the highest basal rate measured in a carnivoran relative to a mass standard.

The association of food habits with basal rate is found in bears: the polar

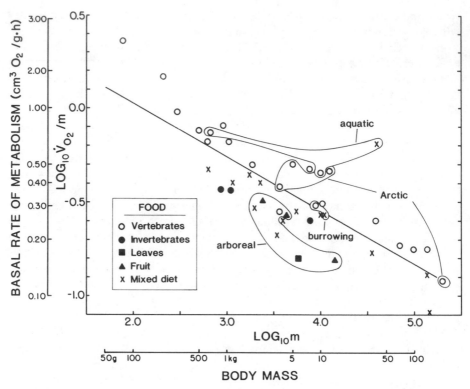

Figure 12.2. $\log_{10}$ mass-specific basal rate of metabolism in carnivorans as a function of $\log_{10}$ body mass in relation to food habits, arboreality (activity level), and climate. Data taken from Table 12.1. The curve is fitted by least squares for all carnivorans (see text).

bear (*Ursus maritimus*) has a high basal rate and nearly exclusive vertebrate-eating habits; the brown bear (*U. arctos*) has an intermediate rate and a mixed diet that includes many vertebrates; and the black bear (*U. americanus*) has the lowest basal rate and an omnivorous diet. The heavy fur coat found in the tropical sloth bear (*Melursus ursinus*) may reflect a low basal rate in association with a marked seasonal commitment to termite eating (Laurie and Seidensticker 1977).

As noted, the two exceptions to the general correlation of basal rate with food habits were the sea otter, which has an unexpectedly high basal rate, and the margay, which has an unusually low rate. Iversen and Krog (1973) measured the otter's basal rate at 3.37 times the value expected from the mean carnivoran curve (Figure 12.1), and Morrison et al. (1974) measured its basal rate at 2.99 times mass expectations, so little doubt exists that this species has a very high basal rate. One possible explanation for the sea otter's high basal rate relates to the kinds of foods it eats. Much of the diet consists of fish and macroinvertebrates (Kenyon 1969) like abalone, sea urchins, mussels, and clams. Most large terrestrial invertebrate-eaters eat ants and termites but as a consequence ingest large quantities of sand, soil, and carton (McNab 1984), which reduces the density of available energy (and nutrients) in food. The marine invertebrates eaten by otters are large and are consumed after being shelled, so here invertebrate eating, mechanically and energetically, may be much more like vertebrate eating than it is to feeding on soil invertebrates. Other marine mammals feeding on macroinvertebrates, such as the walrus (*Odobenus rosmarus*), also appear to have a basal rate equal at least to Kleiber's value (Iversen and Krog 1973).

The margay, which has an unusually low basal rate for a vertebrate-eater, has been shown by Konecny (unpubl. ms.) to eat significant amounts of fruit (found in 14.4% of the scats) and arthropods (33.3% of the scats). Whether they do so opportunistically, or whether they rely on fruit and arthropods, is unclear. Equally unclear is whether the low basal rate in the margay reflects the mixed diet. In the cases of the margay and the sea otter a factor other than diet may influence basal rate. For margays, this factor is activity level, and for sea otters, it is "climate."

The Influence of Activity Level on Carnivoran Basal Rates

The basal rate of metabolism in mammals is correlated with their general level of activity, as measured by the proportion of body mass that is muscle (McNab 1978, 1986b, S. D. Thompson and T. I. Grand, pers. comm.). This correlation is most clearly demonstrated by sedentary, arboreal species, which are characterized by small muscle masses. This coupling may be most prevalent in arboreal mammals because the level of predation may be reduced in forest canopies.

The coupling of activity level and basal rate is found in the order Carnivora (Figure 12.2). Thus, the category "fruit- and leaf-eating specialists" is actually "arboreal, fruit- and leaf-eaters" because the only carnivorans presently constituting that category belong to the genera *Potos*, *Ailurus*, *Nandinia*, and *Arctictis*. Of the six carnivorans that constitute the mixed-diet group that eats few vertebrates, namely, *Procyon*, *Ursus*, *Arctogalidia*, *Paradoxurus*, *Nasua*, and *Cerdocyon*, two of the lowest basal rates, relative to the carnivoran curve, are found in *Arctogalidia* and *Paradoxurus*, the two arboreal species (Table 12.1).

The most interesting example of the influence of activity level on basal rate in carnivorans is found in the margay, an arboreal predator. In all other cases, arboreal habits are found in species that have food habits associated with low basal rates (e.g., folivory and frugivory), a condition that does not easily permit the determination of the factor responsible for the low rates. In the case of the margay, a low basal rate occurs *in spite of* the predominantly vertebrate-eating habits of this cat. Its low basal rate appears to reflect this cat's highly arboreal habits and its presumptively low muscle mass. Another cat, the clouded leopard (*Neofelis nebulosa*), would be expected from this view also to have a low basal rate, given its tropical, arboreal habits.

The Influence of Climate on Carnivoran Basal Rates

In 1950 Scholander et al. maintained that climate had no effect on the level of basal rate of metabolism in terrestrial mammals. Since 1963, however, various studies (McNab and Morrison 1963; McNab 1966; Hulbert and Dawson 1974; Shkolnik and Schmidt-Nielsen 1976) concluded that basal rate indeed varies with climate, broadly defined: with many exceptions, cold-climate species have been found to have high basal rates, whereas warm-climate and desert-dwelling species have low basal rates. Burrowing mammals weighing more than 100 g also have low basal rates, irrespective of food habits, which has usually been interpreted to be a response to high burrow temperatures (McNab 1966, 1979; Contreras 1986) and thus might be called a climatic factor. On the other hand, marine mammals have generally been said to have high basal rates (Scholander et al. 1950; Kanwisher and Sundnes 1965), apparently in response to the cold waters in which these species live.

Several problems exist with these interpretations and measurements. First, the association of basal rate with climate is difficult to separate from the association with food habits. Thus, desert dwellers that are characterized by low basal rates often use foods that may be (mast crops eaten by heteromyid rodents) or are (invertebrates and fruits eaten by the fennec fox) associated with low basal rates. Are these low rates linked with climate or with food habits, or with both? Second, Lavigne et al. (1986) have argued that most of the early measurements of rate of metabolism in seals and dolphins were not taken under standard conditions and thus were artificially high. They have summarized recently collected data to show that cold-water pinnipeds have basal rates that are approximately those expected from the Kleiber relation.

When the data assembled in Table 12.1 are examined for an influence of climate, independent of food habits, relatively little is seen (Figure 12.2). Arctic terrestrial canids and mustelids have basal rates that are similar to those expected from mass, although some aquatic mustelids, most notably the sea otter, have high basal rates. So another explanation for the high basal rate of the sea otter might be the otter's occurrence in cold water coupled with a small mass (for a marine mammal). The basal rates found in bears are correlated with climate, as well as with food habits: the highest rate is found in the Arctic polar bear, an intermediate rate in the northerly distributed brown bear, and the lowest rate in the southerly distributed black bear. Finally, the influence of burrowing in the case of the Eurasian badger (*Meles meles*) and the North American badger (*Taxidea taxus*) is shown in their basal rates.

Even though the correlations of basal rate with climate among carnivorans are compatible with observations on other mammals, they are difficult to interpret. First of all, measurements of coyotes indicate a lower basal rate in central than in Arctic North America (Shield 1972; Golightly and Ohmart 1983). This difference can be interpreted as a response to climate, but it may also be correlated with a change in food habits, if the food intake in an Arctic setting demonstrates a greater dependence on vertebrates. Second, if the sea otter's very high basal rate reflects a climatic adaptation in association with a small mass, why does the significantly smaller European otter (*Lutra lutra*) and the even smaller (semi-aquatic) American mink (*Mustela vison*) have relatively smaller increases in basal rate? (Could these differences reflect the degree to which these species are committed to an aquatic life?) Even the correlation of a low basal rate with burrowing habits is compromised in the case of the Eurasian badger by its extensive use of invertebrates as food (Kruuk and Parish 1981).

Quantitative Analysis

Analysis thus indicates that body mass, food habits, activity level, and (possibly) climate influence the level of basal rate of metabolism in carnivorans. Here an attempt is made to ascertain the magnitude of these contributions. A complication is that one factor (body mass) is quantitative, whereas the others are qualitative. A second difficulty is that many of these factors are associated with each other: for example, most arboreal carnivorans either have mixed diets or are specialists on fruit, and these habits are generally found in species with an intermediate mass and a tropical distribution. Therefore, any analysis of the comparative importance of these factors is subject to factor interaction.

The extent of factor interaction is demonstrated by changes in the apparent contribution of factors to basal rate with changes in the order in which the factors are considered. For example, mass is examined first or last, and activity and climate are examined before and after food habits (Figure 12.3). If no change in apparent contribution occurs with factor order, no factor interaction exists. In each analysis, the effect of a factor is determined by dividing a particular set of species into a number of subsamples based on that factor. Then a pooled, weighted statistic is calculated from subsample statistics to compare with the pooled statistic existing before sample subdivision. The change in the statistic associated with the subdivision is ascribed to the factor that was introduced. Another factor then can be introduced, with the appropriate subdivisions and calculations repeated. Note that as categories are sequentially subdivided, a mathematic limit is reached, so that no category of one or two species can be used.

Two statistics are used. One is r^2, which indicates the proportion of variation in y that is accounted for by variation in x, when y is regressed on x. To examine the influence of various factors on r^2, one must regress basal rate on

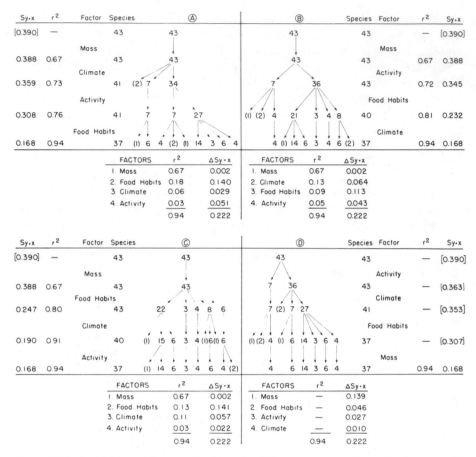

Figure 12.3. Partitioning of r^2 and residual variation in basal rate of metabolism by the sequential division of 43 carnivorans into various groups based on their food habits, climate, and activity level, and on the influence of body mass (see text). In the absence of a regression, variation in basal rate is expressed in terms of standard deviation, which is enclosed in brackets to differentiate it from the standard error of the estimate.

body mass, so mass must be the first factor examined, which precludes variation in factor order to include mass. Such analyses are found in Figures 12.3A, B, and C. Another statistic, standard error of estimate ($Sy·x$), or its equivalent in samples without a regression, standard deviation, measures the variation around the regression or mean, respectively. As noted, $Sy·x$ is a better measure of residual variation than $1 - r^2$ (Smith 1980, 1984; McNab 1988a), especially because r^2 reflects the slope of the regression (and therefore the units of y) and because r^2 is not a linear estimate of the correlation of y with x.

Analysis I (Figure 12.3A)

In this analysis, body mass, climate, activity level, and food habits were examined in that order. The division based on climate permitted the 43 species to be segregated into three groups (two burrowers [North American badger and Eurasian badger], seven cold-acclimated species [three aquatic/semi-aquatic and four Arctic], and 34 remaining species). This division produces a small increase in r^2 and a small reduction in $Sy \cdot x$. With the addition of activity level, the category of 34 species is divided into seven arboreal species and 27 terrestrial species, a division that leads to another small increase in r^2, but a marked reduction in $Sy \cdot x$. Finally, each of the three remaining categories is subdivided by food habits; as a result, r^2 shows a marked increase (to 0.94) and $Sy \cdot x$ shows a large decrease (to 0.168). From the viewpoint of the ability to account for the variation in basal rate, mass is quantitatively the most important factor by far, followed by food habits, climate, and activity level. The introduction of food habits produced the greatest reduction in residual variation, followed by activity level, climate, and mass.

Analysis II (Figure 12.3B)

The second analysis examines in order mass, activity, food habits, and climate. After mass, climate makes the greatest contribution to r^2, followed by food habits and activity. In this analysis food habits makes the greatest contribution to a reduction in residual variation, followed by climate, activity, and mass.

Analysis III (Figure 12.3B)

Here mass, food habits, climate, and activity are examined. After mass, food habits, climate, and activity level contribute to r^2 in that order, but in terms of $Sy \cdot x$ the most important factor is food habits, followed by climate, activity level, and mass.

Analysis IV (Figure 12.3D)

Unlike the other analyses, this one extracts body mass last, so that r^2 cannot be calculated as other factors are added. The reduction in $Sy \cdot x$ is most affected by the addition of mass, followed in magnitude by food habits, activity, and climate.

These analyses indicate a high degree of interaction among the factors. With respect to $Sy·x$, body mass is more important than food habits when food habits are extracted before mass, but food habits are more important when they are extracted after mass. Generally, the relative importance of a factor seems to be diminished if it is the first factor incorporated within an analysis. This interaction issues from the correlation of food habits with body mass (see also Gittleman 1985): invertebrativory and frugivory occur principally at small to intermediate masses. And low activity levels occur mainly in arboreal species that eat leaves, fruits, or a mixed diet, but also in the arboreal, verte-brate-eating margay.

Analyses I, II, and III were so constructed that the influence of each factor other than mass on r^2 and $Sy·x$ was examined first, second, and third, thereby diminishing the effect of position. As a consequence, average magnitudes of the influence of the factors can be calculated without regard to their position in the analysis. Body mass ($r^2 = 0.67$), food habits (0.13), climate (0.09), and activity level (0.04) collectively account for 94% of the variation in basal rate of metabolism in members of the order Carnivora.

An analysis of $Sy·x$ in I, II, and III, however, indicates that food habits make the greatest contribution to a reduction in $Sy·x$ of basal rate (mean = 0.131), followed by climate (0.050), activity (0.039), and mass (0.002). The apparent contradiction between the analysis of r^2 and $Sy·x$ is because these statistics measure different aspects of the relation of basal rate to the four parameters. Thus, at the end of each analysis the four factors can account for 94% of the variation in mass-specific basal rate, but the residual variation is great enough that $Sy·x = 0.168$, which means that 68% of the values should fall between 70 and 147% of the fitted curve—a 2.1-fold variation in basal rate. Clearly, body mass and food habits are the two most important factors setting the basal rate of carnivorans, with climate and activity level having more limited effects.

This view contradicts the contention (Hayssen and Lacy 1985; Elgar and Harvey 1987) that food habits, and presumably activity levels and climatic distribution, are so intertwined with phylogeny that they cannot be separated in their influence on rate of metabolism. What is truly intertwined are the effects of body mass and of food habits on rate of metabolism. The use of ecological mavericks within carnivoran families suggests not only that food habits are an important factor associated with the level of energy expenditure but throws doubt, as well, on the concept that phylogeny has a predominant influence on the level of energy expenditure independent of body size and food habits (for an exception, see McNab 1986a). For example, most mustelids have high basal rates (Figure 12.1), apparently in association with a strictly vertebrativorous diet, yet the spotted skunk (*Spilogale putorius*) has a low basal rate. It has a mixed diet, eating many insects and some fruits in summer and small vertebrates in winter. Equally, the arboreal margay has a low basal rate in spite of its predominantly vertebrate-eating habits and membership in the Felidae. After all, most close relatives of an organism are ecologically similar to that organism, which, given that some correspondence exists between physiology and ecology, implies that close relatives are also likely to be physiologically similar.

Discussion

The available data on carnivorans suggest that basal rate of metabolism is correlated with body mass, food habits, climate, and activity level, although these factors are not completely independent of each other. Little residual variation in basal rate remains to be potentially ascribed to phylogeny, at least within the order Carnivora. That is, palm civets tend to have low rates of metabolism *because* they feed principally on invertebrates, fruit, or a mixture of these food items, *not* because they are viverrids, except in the sense that palm-civets generally tend to have these habits and thus this level of energy expenditure. The depression in rate of metabolism, however, may be exaggerated in palm-civets by arboreal habits, sluggish behavior, and small muscle masses. In contrast, felids generally have high rates of metabolism, not because they are felids (whatever that might mean), but in relation to a strictly vertebrate-eating habit. Even if all felids had high basal rates, that does not mean that a high basal rate is a phylogenetic character, but simply that cats are uniformally committed to vertebrate-eating habits. So when a carnivoran evolutionarily strays from a vertebrate-eating diet, its rate of metabolism is almost surely modified to reflect its new diet. One such radical dietary shift is found in the arboreal, folivorous red panda; its basal rate is similar to those found in the xenarthran three-toed sloth (*Bradypus varigatus*) and two-toed sloth (*Choloepus hoffmanni*), which are also arboreal and folivorous (McNab 1978).

The correlation of basal rate with food habits, climate, and activity level may have consequences for carnivorans other than as a direct impact on energy budgets. McNab (1980) argued that several aspects of reproduction in eutherians are associated with basal rate (and thus level of energy expenditure) independent of the influence of body mass: as basal rate increases, gestation period decreases, postnatal growth rate increases, and fecundity increases. In fact, Gittleman and Oftedal (1987) showed that folivorous and frugivorous carnivorans, such as the red panda and the binturong (*Arctictis binturong*) both of which have low basal rates, have low growth rates. Consequently, mammals with high basal rates tend to have a large r_{max}, the maximal population growth constant. Of course, the influence of mass and basal rate on reproduction may be limited in many ways. For example, at large masses total fecundity is so reduced that a high basal may be associated only with a reduction in generation time, or the correlation between growth rate and rate of metabolism may be modified by a shift from precociality to altriciality. A broad-scaled exploration of the relationship between reproduction and energetics in mammals has recently appeared as a volume in the Symposia of the Zoological Society of London (Loudon and Racey 1987; see also Gittleman and Thompson 1988; Oftedal and Gittleman, this volume).

Unfortunately, the data available on the parameters of reproduction and rate of metabolism are too limited at present for an extensive analysis of this relationship in carnivorans, but one aspect of the potential connection is worthy of speculative exploration. Although most independent measurements of the rate of metabolism in one species differ only by 5% or so, some species belonging to the family Canidae have been reported to have much greater differences in basal rate. Such differences have been reported in the coyote (Shield 1972; Golightly and Ohmart 1983) and the arctic fox (Scholander et al. 1950; Casey et al. 1979; Korhonen et al. 1983); whether these differences reflect technical difficulties in measurement or individual and population differences within these species is unclear.

If these differences are indeed correct, they may have significant consequences for reproduction. For example, Braestrup (1941) maintained that the arctic fox is locally differentiated into two "types," what he called "lemming" foxes and "coast" foxes. Lemming foxes usually live inland in Arctic North America, in the Canadian Arctic archipelago, northern and eastern Greenland, and Eurasia. They feed heavily on lemmings, approximately 95% of the individuals are white, and they have a high reproductive output (individual litters commonly are six to eight, regularly as large as 15, and occasionally as great as 20 to 24!); these foxes have highly cyclic population fluctuations that track lemming populations. Coast foxes, however, live within a few kilometers of the coast and on islands that have no lemmings, including western and southern Greenland and Iceland. They feed mainly on sea birds and marine life along the coast and on birds and hares in the interior, have a high frequency of the blue color-phase (up to 100% on small islands), and have a lower reproductive output (litters average six, with the maximum rarely exceeding ten); coast foxes have reduced population amplitudes, which may in part be cued to population fluctuations in ptarmigan and hare. In some cases (northeast Greenland), the foxes living along bird cliffs have a high percentage of the blue color variant, whereas the nearby inland populations are principally white, suggesting that "coast" and "lemming" populations may be sharply defined in terms of local topography. Coast foxes living on St. Paul's Island in the Pribilofs do not feed on an indigenous lemming, even during years when seabird populations are low. The Arctic fox, thus, may be behaviorally and physiologically differentiated into two forms; one reflection of this differentiation may be that lemming foxes have higher rates of metabolism than do coast foxes, a difference that may be related to their differential fecundity. Rate of energy expenditure may, therefore, be correlated (at least on the population level) with coat color: those populations (and individuals?) with blue coats may have lower basal rates than those with white coats. The published diversity in measured basal rate has not been compared with population character-istics or, even, to coat color. Other canids (jackals, *Canis* spp.) appear to have fecundities that vary with geography (P. Moehlman, pers. comm.), which raises the question as to whether a geographic pattern occurs in rate of metab-olism and whether such patterns reflect the genetic basis of adaptation.

Conclusions

Rate of energy expenditure is generally acknowledged to be an important parameter of existence because all biologically important activities require energy and because energy availability may occasionally be limited. The basal rate of metabolism, which is used here as an index of the level of energy expenditure, varies in members of the order Carnivora with a series of factors, including (in order of diminishing influence) body mass, food habits, climate, and activity level. This variation potentially has consequences for a species' reproductive output. The analysis proposed here clearly demonstrates that various aspects of the biology of carnivorans, including body size, food habits, activity level, reproduction, and physiology, are all interconnected. The signifi-cance of these facets in the life of a mammal is taken out of context when they are studied in isolation. The most radical consequences for the life history of an organism stem from a marked change in its body size or food habits. A change in diet has repeatedly occurred in the order Carnivora, producing much of its diversity in anatomy, physiology, and behavior.

Acknowledgments

I thank many people for the loan of mammals that were used for the mea-surement of rate of metabolism: Gene and Rusti Schuler (of the Wild Animal Retirement Village, Waldo, Florida), Miles Roberts and Chris Wemmer (of the U.S. National Zoo), and Jack Brown (Santa Fe Community College teaching zoo, Gainesville, Florida), as well as the staff of the Apple Valley (Minneapolis) Zoo and the Roosevelt Park Zoo, Minot, North Dakota. Delbert Kilgore permitted me to quote his measurements on the spotted skunk, and Paul D. Watts (Churchill, Manitoba) thoughtfully permitted me to use his data on bears. John Gittleman (University of Tennessee) kindly invited this contribu-tion and made helpful suggestions to improve it as did Robert T. Golightly (Humboldt State University).

References

Braestrup, F. W. 1941. A study on the Arctic fox in Greenland. *Meddelelser om Grøn-land* 131:1–101.

Brown, J. H., and Lasiewski, R. C. 1972. Metabolism of weasels: the price of being long and thin. *Ecology* 53:939–943.

Casey, T. M., and Casey, K. K. 1979. Thermoregulation of Arctic weasels. *Physiol. Zool.* 52:153–164.

Casey, T. M., Withers, P. C., and Casey, K. K. 1979. Metabolic and respiratory responses of Arctic mammals to ambient temperature during the summer. *Comp. Biochem. Physiol.* 64A:331–341.

Chevillard-Hugot, M.-C., Muller, E. F., and Kulzer, E. 1980. Oxygen consumption, body temperature, and heart rate in the coati (*Nasua nasua*). *Comp. Biochem. Physiol.* 65A:305–309.

Contreras, L. C. 1986. Bioenergetics and distribution of fossorial *Spalacopus cyanus* (Rodentia): Thermal stress, or cost of burrowing? *Physiol. Zool.* 59:20–28.

Ebisu, R. J., and Whittow, G. C. 1976. Temperature regulation in the small Indian mongoose (*Herpestes auropunctatus*). *Comp. Biochem. Physiol.* 54A:309–313.

Elgar, M. A., and Harvey, P. H. 1987. Basal metabolic rates in mammals: Allometry, phylogeny and ecology. *Funct. Ecol.* 1:25–36.

Ewer, R. F. 1973. *The Carnivores.* Ithaca, N.Y.: Cornell Univ. Press.

Farrell, D. J., and Wood. A. J. 1968. The nutrition of the female mink (*Mustela vison*). I. The metabolic rate of the mink. *Canadian J. Zool.* 46:41–45.

Gittleman, J. L. 1985. Carnivore body size: Ecological and taxonomic correlates. *Oecologia* 67:540–554.

Gittleman, J. L., and Oftedal, O. T. 1987. Comparative growth and lactation energetics in carnivores. *Symp. Zool. Soc. London* 57:41–77.

Gittleman, J. L., and Thompson, S. D. 1988. Energy allocation in mammalian reproduction. *Amer. Zool.* 28:863–875.

Golightly, R. T., and Ohmart, R. D. 1983. Metabolism and body temperature of two desert canids: Coyotes and kit foxes. *J. Mamm.* 64:624–635.

Harlow, H. J. 1981. Metabolic adaptations to prolonged food deprivation by the American badger *Taxidea taxus*. *Physiol. Zool.* 54:276–284.

Hayssen, V., and Lacy, R. C. 1985. Basal metabolic rates in mammals: Taxonomic differences in the allometry of BMR and body mass. *Comp. Biochem. Physiol.* 81A:741–754.

Hennemann, W. W.,III, and Konecny, M. J. 1980. Oxygen consumption in large spotted genets, *Genetta tigrina*. *J. Mamm.* 61:747–750.

Hennemann, W. W., III, Thompson, S. D., and Konecny, M. J. 1983. Metabolism of crab-eating foxes, *Cerdocyon thous*: Ecological influences on the energetics of canids. *Physiol. Zool.* 56:319–324.

Hulbert, A. J., and Dawson, T. J. 1974. Thermoregulation in perameloid marsupials from different environments. *Comp. Biochem. Physiol.* 47A:591–616.

Irving, L., Krog, H., and Monson, M. 1955. The metabolism of some Alaskan mammals in winter and summer. *Physiol. Zool.* 28:173–185.

Iversen, J. A. 1972. Basal energy metabolism of mustelids. *J. Comp. Physiol.* 81:341–344.

Iversen, J. A., and Krog, J. 1973. Heat production and body surface area in seals and sea otters. *Norwegian J. Zool.* 21:51–54.

Kamau, J. M. Z., Johansen, K., and Maloiy, G. M. O. 1979. Thermoregulation and standard metabolism of the slender mongoose (*Herpestes sanguineus*). *Physiol. Zool.* 52:594–602.

Kanwisher, J., and Sundnes, G. 1965. Physiology of a small cetacean. *Hvalrådets Skrifter* 48:45–53.

Kenyon K. W. 1969. The sea otter in the eastern Pacific Ocean. *N. Amer. Fauna* 68:1–352.

Kleiber, M. 1932. Body size and metabolism. *Hilgardia* 6:315–353.

Kleiber, M. 1961. *The Fire of Life*. New York: J. Wiley.

Konecny, M. 1988. Movement patterns and food habits of four sympatric carnivore species in Belize, Central America. Unpublished ms. Available from M. Konecny, University of Florida, Gainesville.

Korhonen, H., and Harri, M. 1984. Seasonal changes in thermoregulation of the raccoon dog (*Nyctereutes procyonoides* Gray 1834). *Comp. Biochem. Physiol.* 77A:213–219.

Korhonen, H., Harri, M., and Asikainen, J. 1983. Thermoregulation of polecat and raccoon dog: A comparative study with stoat, mink and blue fox. *Comp. Biochem. Physiol.* 74A:225–230.

Kruuk, H. and Parish, T. 1981. Feeding specialization of the European badger *Meles meles* in Scotland. *J. Anim. Ecol.* 50:773–788.

Laurie, A., and Seidensticker, J. 1977. Behavioural ecology of the sloth bear (*Melursus ursinus*). *J. Zool. (Lond.)* 182:187–204.

Lavigne, D. M., Innes, S., Worthy, G. A. J., Kovacs, K. M., Schmitz, O. J., and Hickie, J. P. 1986. Metabolic rates of seals and whales. *Canadian J. Zool.* 64:279–284.

Lindstedt, L. S., and Calder, W. A. 1981. Body size, physiological time, and longevity of homeothermic animals. *Quart. Rev. Biol.* 56:1–16.

Loudon, A. S. I., and Racey, P. 1987. *Reproductive Energetics in Mammals*. Symposium no. 57. London: Zoological Society of London.

McKenna, M. C. 1975. Toward a phylogenetic classification of the Mammalia. In: W. P. Luckett & F. S. Szalay, eds. *Phylogeny of the Primates, A Multidisciplinary Approach*, pp. 21–46. New York: Plenum Press.

McNab, B. K. 1966. The metabolism of fossorial rodents: A study of convergence. *Ecology* 47:712–733.

McNab, B. K. 1969. The economics of temperature regulation in Neotropical bats. *Comp. Biochem. Physiol.* 31:227–268.

McNab, B. K. 1971. On the ecological significance of Bergmann's rule. *Ecology* 52:845–854.

McNab, B. K. 1978. Energetics of arboreal folivores: Physiological problems and ecological consequences of feeding on an ubiquitous food supply. In: G. G. Montgomery, ed. *The Ecology of Arboreal Folivores*, pp. 153–162. Washington, D.C.: Smithsonian Institution Press.

McNab, B. K. 1979. The influence of body size on the energetics and distribution of fossorial and burrowing rodents. *Ecology* 60:1010–1021.

McNab, B. K. 1980. Food habits, energetics, and the population biology of mammals. *Amer. Nat.* 116:106–124.

McNab, B. K. 1983. Ecological and behavioral consequences of adaptation to various food resources. In: J. F. Eisenberg & D. G. Kleiman, eds. *Advances in the Study of Mammalian Behavior*, pp. 664–697. Special Publication no. 7. Lawrence, Kans.: American Society of Mammalogists.

McNab, B. K. 1984. Physiological convergence amongst ant-eating and termite-eating mammals. *J. Zool. (Lond.)* 203:485–510.

McNab, B. K. 1986a. Food habits, energetics, and the reproduction of marsupials. *J. Zool. (Lond.)* 208:595–614.

McNab, B. K. 1986b. The influence of food habits on the energetics of eutherian mammals. *Ecol. Monog.* 56:1–19.

McNab, B. K. 1988a. Complications in scaling basal rate of metabolism in mammals. *Quart. Rev. Biol.* 63:25–54.

McNab, B. K. 1988b. Energy conservation in a tree-kangaroo (*Dendrolagus matschiei*) and the red panda (*Ailurus fulgens*): *Physiol. Zool.* 61:280–292.

McNab, B. K., and Morrison, P. R. 1963. Body temperature and metabolism in subspecies of *Peromyscus* from arid and mesic environments. *Ecol. Monogr.* 33:63–82.

354   *Brian K. McNab*

Morrison, P. R. 1960. Some interrelations between weight and hibernation function. *Bull. Mus. Comp. Zool.* 124:75–91.
Morrison, P. R., Rosenmann, M., and Estes, J. A. 1974. Metabolism and thermoregulation in the sea otter. *Physiol. Zool.* 47:218–229.
Müller, E. F. 1985. Basal metabolic rates in primates—The possible role of phylogenetic and ecological factors. *Comp. Biochem. Physiol.* 81A:707–711.
Müller, E. F., and Kulzer, E. 1977. Body temperature and oxygen uptake in the kinkajou (*Potos flavus*, Schreber), a nocturnal tropical carnivore. *Arch. Internat. Physiol. Biochem.* 86:153–163.
Müller, E. F., and Lojewski, U. 1986. Thermoregulation in the meerkat (*Suricata suricatta* Schreber, 1776). *Comp. Biochem. Physiol.* 83A:217–224.
Noll-Banholzer, U. 1969. Body temperature, oxygen consumption, evaporative water loss and heart rate in fennec. *Comp. Biochem. Physiol.* 62A:585–592.
Okarma, H., and Koteja, P. 1987. Basal metabolic rate in the gray wolf in Poland. *J. Wild. Mgt.* 51:800–801.
Scholander, P., Hock, R., Walters, V., and Irving, L. 1950. Adaptation to cold in Arctic and tropical mammals and birds in relation to body temperature, insulation, and basal metabolic rate. *Biol. Bull.* 99:259–271.
Shield, J. 1972. Acclimation and energy metabolism of the dingo, *Canis dingo*, and the coyote, *Canis latrans*. *J. Zool. (Lond.)* 168:483–501.
Shkolnik, A., and Schmidt-Nielsen, K. 1976. Temperature regulation in hedgehogs from temperate and desert environments. *Physiol. Zool.* 49:56–64.
Smith, R. J. 1980. Rethinking allometry. *J. Theor. Biol.* 87:97–111.
Smith, R. J. 1984. Allometric scaling in comparative biology: Problems of concept and method. *Amer. J. Physiol.* 246:R152–R160.
Taylor, C. R., and Roundtree, V. J. 1973. Temperature regulation and heat balance in running cheetahs: A strategy for sprinters? *Amer. J. Physiol.* 224:848–851.
Taylor, C. R., Schmidt-Nielsen, K., Dmi'el, R., and Fedak, M. A. 1971. Effect of hyperthermia on heat balance during running in the African hunting dog. *Amer. J. Physiol.* 20:823–827.
Watts, P. D. 1988. Whole body thermal conductance of denning ursids. *J. Thermal Biol.* In press.
Worthen, G. L., and Kilgore, D. L., Jr. 1981. Metabolic rate of pine marten in relation to air temperature. *J. Mamm.* 62:624–628.

CHAPTER 13

Patterns of Energy Output During Reproduction in Carnivores

OLAV T. OFTEDAL AND
JOHN L. GITTLEMAN

Reproduction is energetically expensive (Harvey 1986; Loudon and Racey 1987; Gittleman and Thompson 1988). The pregnant female requires energy and nutrients for the synthesis of fetal, placental, uterine, and mammary tissues. Lactation involves an even greater drain of nutrients and energy. During reproduction energy expenditure may also rise as a consequence of increases in metabolic rate and activity level (Thompson and Nicoll 1986). To support the energetic costs of late pregnancy and lactation, maternal food intake must increase and/or the energy accumulated prior to reproduction or during early pregnancy must be mobilized (Loveridge 1986; Gittleman and Thompson 1988). Thus female reproduction involves substantial commitment of nutritional resources: for example, the energy required by a female ungulate to rear a single offspring from conception to weaning is similar to maintenance energy needs for approximately 100–150 days (Oftedal 1985).

Reproductive patterns undoubtedly influence the magnitude of maternal expense. Litter size is particularly important (Millar 1979; Mattingley and McClure 1982). In ungulates, the production of twins increases the energy requirements of the mother by about 25% over the requirements of a mother with a singleton (Oftedal 1985). Domestic cats (*Felis domesticus*) with five kittens consume twice as much energy as do cats with two kittens during the period of intensive lactation (Loveridge 1986). Interspecific comparisons indicate that the milk energy outputs of females with large litters are often two to three times higher than those of mothers with single young, at least at peak lactation (Oftedal 1984b). Litter mass and litter growth rate are valuable indicators of the magnitude of nutrient transfer from mother to young (Gittleman and Oftedal 1987).

The carnivores offer an excellent opportunity to examine variation in energy output during reproduction in that they exhibit such a wide range in maternal weight (0.06–320 kg), birth weight (3–1650 g), litter size (1–8.8 young), postnatal growth rate (1.5–161 g/d), and duration of the lactation period (30–730 days) (Gittleman 1986a; Gittleman and Oftedal 1987). Differing patterns

355

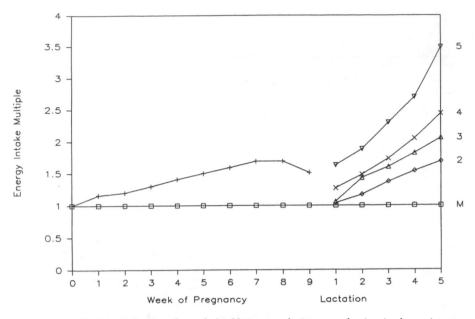

Figure 13.1. Changes in intake of metabolizable energy during reproduction in domestic cats. Energy intake is expressed relative to energy intake for maintenance of constant body weight in nonreproductive cats (indicated by horizontal line at a value of 1.0). An increase in energy intake above maintenance is represented by an increase in the energy intake multiple (= energy intake at given stage of reproduction / energy intake at maintenance) above a value of 1.0. During lactation, energy intakes are illustrated separately for litters of two ($\Diamond$), three ($\triangle$), four (X), and five ($\triangledown$) kittens. Data are taken from Loveridge (1986) and represent 10 cats during pregnancy and 15 cats for each litter size during lactation.

of reproductive output must place very different behavioral and nutritional demands on female carnivores. In some species conspecific helpers and mates assist in food acquisition, defense of the denning area, and/or transport of young from one site to another (see Gittleman 1985; Moehlman, this volume), but the burden of prenatal and postnatal nutrient transfer is largely borne by the mother.

Quantitative comparison of the reproductive strategies of mammals requires measurement of the energetic costs of reproduction, including the energy and mass in reproductive products (e.g., fetal and birth weights, milk yield), the daily metabolic rates of reproductive females, and the intake and utilization of dietary energy (Oftedal 1985; Kunz and Nagy 1987; Gittleman and Thompson 1988). Unfortunately, the little direct work on reproductive energetics in carnivores has been largely restricted to domestic species. In a recent study of domestic cats Loveridge (1986) demonstrated that maternal energy intake increases 1.5–1.7-fold during pregnancy, and that the magnitude of increase during lactation varies greatly as a function of litter size (Figure 13.1). At peak lactation (ca. 4–5 weeks) maternal energy intake reached 2.5–3.0

times maintenance levels when large litters were reared. Energy intake did not directly reflect reproductive expenditure, however, since the cats accumulated energy in body stores during pregnancy and depleted body stores during lactation, with the magnitude of gain and loss related to litter size (Loveridge 1986). Yet even for domestic species data on reproductive energy expenditure are incomplete. For example, there are no reliable studies of milk yield in the domestic cat, and the few studies of milk composition have produced contradictory results (Gittleman and Oftedal 1987). Milk energy output has been directly measured for only four carnivore species, two of which (mink, *Mustela vison*; and dog, *Canis familiaris*) are domesticated.

Given this situation, it is not possible to specify the energetic costs of reproduction in carnivores. Nonetheless, one may develop estimates of energy output during reproduction using a factorial method (see Oftedal 1985). Data on birth weight, litter size, and chemical composition of newborn carnivores are used to estimate the energy transferred from maternal to fetal tissues during pregnancy. Data on milk yield, milk composition, and postnatal growth are used to estimate the energy transferred from mother to young during lactation. We are particularly interested in relating the magnitude of these estimated energy transfers to maternal metabolic size and to carnivore reproductive strategies. We also compare carnivore patterns of energy transfer to patterns in another more completely studied group, the ungulates (orders Artiodactyla, Perissodactyla, and Proboscidea).

The data reported in this chapter are taken largely from recent reviews of life history patterns, parental care, and postnatal growth in carnivores (Gittleman 1986a, 1986b; Gittleman and Oftedal 1987). Readers are referred to these review papers for the original sources of the data. It is our hope that future research will evaluate some of the assumptions made in this paper, and that the approximations developed herein will be replaced by direct studies of the energetics of carnivore reproduction.

Reproductive Output during Pregnancy

Pregnancy involves the deposition of energy and nutrients both in fetal growth and in the development of associated supportive tissues such as the uterus, placenta, and mammary gland. In ungulates the fetus at birth represents about 82% of the total energy accumulation of the gravid uterus (i.e., fetuses, placentas, uterus) (Robbins and Moen 1975; Agricultural Research Council 1980), or about 68% of the combined energy accumulation in the gravid uterus and mammary glands, since energy in the mammary glands is equivalent to 25% of the energy in the gravid uterus (Rattray et al. 1974). Since the energy in the term fetus is such a large proportion of maternal energy deposition and is easily measured following parturition, neonatal energy content is a valuable indicator of maternal reproductive effort during pregnancy.

Table 13.1. Body composition of domestic carnivores at birth

Species	N	Water (%)	Dry matter (%)	Fat (%)	Protein (%)	Ash (%)	Gross energy (kcal/g)
Dog[a]	2	80.1	19.9	1.37	14.5	2.27	0.99
Cat[a]	2	79.9	20.1	1.69	14.0	2.57	0.97
Mink[b]	6	82.0	18.0	2.14	13.1	1.84	0.97

[a]Data from Thomas 1911. See Mundt et al. 1981 for additional sources of data on neonatal dogs.
[b]Data from Oftedal 1981.

Energy Content of Neonatal Carnivores

Although data on the chemical composition of newborn carnivores are available only for a few domestic species, the values are quite uniform (Table 13.1). During fetal and early postnatal development, most mammals undergo a progressive decline in body water as a percentage of body weight; thus, species born at an earlier stage of development tend to be higher in water content as well as lower in fat content (Adolph and Heggeness 1971). High water and low fat contents result in relatively low energy content. The high water (80–82%), low fat (1.4–2.1%), and low energy (0.97–0.99 kcal/g) contents of domestic carnivores indicate the relatively immature state of these species at birth. By contrast, precocial young of domestic ungulates contain 20–40% more energy per unit weight (1.2–1.4 kcal/g) (Oftedal 1985) than do domestic carnivores.

The domestic carnivore data are probably representative for carnivores born at a similar developmental stage (e.g., canids, small felids, most mustelids). For calculation of interspecies trends, we have opted to use the average caloric content (0.98 kcal/g; Table 13.1) in estimating the gross energy content (kcal) of newborn carnivores. This calculation may underestimate energy content in species that are born with their eyes open, such as the sea otter (*Enhydra lutris*), malagasy civet (*Fossa fossa*), spotted hyena (*Crocuta crocuta*), and some *Panthera* species (Ewer 1973; Hemmer 1976), as neonates of these species may be at a more advanced stage of physiological development and may therefore be lower in water and higher in fat and energy, but no data are available. Conversely, the small neonates of the American black bear (*Ursus americanus*) contain about 84% water at birth (Oftedal 1988). In comparison with other carnivores, bears and pandas are considered especially altricial (Schaller et al. 1985; Gittleman 1988) and would be expected to be lower in fat and energy content at birth.

Litter Weight and Energy Content

The litter weight at birth has been calculated for 56 carnivore species using litter size and birth weight data compiled by Gittleman (1986b) (Table 13.2).

Table 13.2. Litter weight and energy content at birth

Species	Female weight (kg)	Litter size	Birth weight (g)	Litter weight (%)	Litter[a] energy per MBS (kcal/kg$^{0.75}$)	Daily[b] energy per MBS (kcal · kg$^{-0.75}$ · d^{-1})
Canidae						
Gray wolf	31.1	5.5	425.0	7.5	174	2.8
(*Canis lupus*)						
Coyote	9.7	6.2	225.0	14.4	249	4.0
(*C. latrans*)						
African hunting dog	22.2	8.8	365.0	14.5	308	4.4
(*Lycaon pictus*)						
Dhole	13.8	4.3	275.0	8.6	162	2.6
(*Cuon alpinus*)						
Arctic fox	2.9	7.1	66.0	16.2	207	3.9
(*Alopex lagopus*)						
Red fox	3.9	4.8	105.0	12.9	178	3.3
(*Vulpes vulpes*)						
Fennec	1.5	2.8	34.8	6.5	70	1.3
(*Fennecus zerda*)						
Gray fox	3.3	3.8	107.5	12.4	164	2.6
(*Urocyon cinereoargenteus*)						
Colpeo fox	6.7	5.0	168.0	12.5	198	3.4
(*Dusicyon culpaeus*)						
Crab-eating fox	6.0	3.1	140.0	7.2	111	2.0
(*Cerdocyon thous*)						
Maned wolf	23.0	1.8	360.0	2.8	60	1.0
(*Chrysocyon brachyurus*)						
Ursidae						
Brown bear	298.5	2.0	1000.0	0.67	27	0.4
(*Ursus arctos*)						
Black bear	97.0	2.5	285.0	0.73	23	0.3
(*U. americanus*)						
Polar bear	320.0	1.9	641.6	0.38	16	—
(*Thalarctos maritimus*)						
Ailuropodidae						
Giant panda	96.8	1.5	104.8	0.13	4	0.1
(*Ailuropoda melanoleuca*)						
Procyonidae						
Ring-tailed cat	0.87	3.0	28.0	9.7	91	1.8
(*Bassariscus astutus*)						
Kinkajou	2.00	1.5	170.5	12.8	149	1.4
(*Potos flavus*)						
Coati	5.00	4.0	140.0	11.2	164	2.2
(*Nasua narica*)						
Raccoon	6.70	3.8	105.9	6.0	95	1.5
(*Procyon lotor*)						
Mustelidae						
Weasel	0.080	5.8	3.00	21.8	113	2.7
(*Mustela nivalis*)						
Least weasel	0.058	4.8	1.42	11.8	57	1.5
(*M. rixosa*)						
Long-tailed weasel	0.230	6.0	3.10	8.1	55	2.3
(*M. frenata*)						

(*continued*)

Table 13.2. (Continued)

Species	Female weight (kg)	Litter size	Birth weight (g)	Litter weight (%)	Litter[a] energy per MBS (kcal/kg$^{0.75}$)	Daily[b] energy per MBS (kcal · kg$^{-0.75}$ · d^{-1})
American pine marten (*Martes americana*)	0.770	2.6	28.00	9.5	87	3.3
Fisher (*M. pennanti*)	2.250	2.7	28.00	3.4	40	0.5
Sable (*M. zibellina*)	1.030	3.0	32.50	9.5	93	3.3
Wolverine (*Gulo gulo*)	10.350	2.8	99.20	2.7	47	4.7
Zorilla (*Ictonyx striatus*)	0.630	2.3	15.00	5.5	48	1.3
White-naped weasel (*Poecilogale albinucha*)	0.250	2.0	4.00	3.2	22	0.7
European badger (*Meles meles*)	10.900	3.0	103.50	2.9	51	1.2
Striped skunk (*Mephitis mephitis*)	2.000	6.0	33.00	9.9	115	1.8
Spotted skunk (*Spilogale putorius*)	0.430	4.3	15.90	15.9	126	4.2
European otter (*Lutra lutra*)	7.100	2.5	285.00	10.0	161	2.4
Viverridae						
African palm civet (*Nandinia binotata*)	3.20	1.8	56.0	13.2	41	0.6
Common palm civet (*Paradoxurus hermaphroditis*)	2.70	3.3	95.5	11.7	147	—
Binturong (*Arctictis binturong*)	13.00	3.0	319.0	7.4	137	1.5
Fanaloka (*Fossa fossa*)	1.60	1.0	82.5	5.2	57	0.7
Falanouc (*Eupleres goudoti*)	2.10	1.5	150.0	10.7	126	—
Ring-tailed mongoose (*Galidia elegans*)	0.81	1.0	47.5	5.9	55	0.7
Banded mongoose (*Mungos mungo*)	1.23	3.8	20.0	6.2	64	1.1
Meerkat (*Suricata suricatta*)	0.72	4.0	30.5	16.9	153	2.0
Hyaenidae						
Brown hyena (*Hyaena brunnea*)	43.9	2.3	693	3.6	92	—
Spotted hyena (*Crocuta crocuta*)	55.3	2.0	1500	5.4	145	1.3
Felidae						
European wild cat (*Felis silvestris*)	4.33	3.3	137.0	10.4	148	2.2
Jungle cat (*F. chaus*)	6.65	2.9	135.5	5.9	93	1.4
Serval (*Leptailurus serval*)	10.40	2.4	143.5	3.3	58	0.8

Table 13.2. (*Continued*)

Species	Female weight (kg)	Litter size	Birth weight (g)	Litter weight (%)	Litter[a] energy per MBS (kcal/kg^{0.75})	Daily[b] energy per MBS (kcal · kg^{-0.75} · d^{-1})
Bengal cat (*Prionailurus bengalensis*)	3.30	2.5	83.0	6.3	83	1.2
Fishing cat (*Prionailurus viverrinus*)	6.30	2.5	92.5	3.7	57	0.6
Mountain lion (*Puma concolor*)	39.60	2.5	400.0	2.5	62	0.9
Ocelot (*Leopardus pardalis*)	10.75	2.5	250.0	5.8	103	1.4
Geoffroy's cat (*L. geoffroyi*)	2.20	2.0	65.0	5.9	71	1.0
African lion (*Panthera leo*)	135.50	2.6	1650.0	3.2	106	1.0
Tiger (*P. tigris*)	131.00	2.5	1255.0	2.4	79	0.8
Leopard (*P. pardus*)	39.30	2.6	549.3	3.6	89	0.9
Jaguar (*P. onca*)	77.60	2.5	816.6	2.6	77	0.7
Snow leopard (*P. uncia*)	32.50	2.8	442.6	3.8	89	0.9
Cheetah (*Acinonyx jubatus*)	60.00	3.8	287.5	1.8	50	0.6

Note. Data from Gittleman 1986b, in which references are cited. Values for *Mustela nivalis* and *M. rixosa* modified according to Heidt et al. 1968 and Heidt 1970. Species eliminated because data were inconsistent: *Lynx lynx, L. rufus, Enhydra lutris, Mustela erminea*. MBS = metabolic body size.

[a]Calculated from litter weight assuming an energy content of 0.98 kcal/g.

[b]Litter energy per metabolic size divided by gestation length. Periods of delayed implantation excluded in calculation (see text). Taxonomy and order of species follows Ewer 1973, except that the giant panda is considered to belong to Ailuropodidae.

These estimates may include some error since birth weight, litter size, and maternal weight data often derive from different litters and populations, and data are from both wild and captive animals. Among carnivores, litter weight is positively correlated with maternal body weight ($r = 0.84$); the scaling exponent of litter weight to maternal body weight is 0.86 (Gittleman 1986a). If litter weight is expressed as a percentage of maternal weight, the range is very large (0.1–22%). Litter weights of most carnivores are within a somewhat smaller range (3–16%), which is comparable to the range of ungulate litter weights (3–15%) (Oftedal 1985). Most canids have rather heavy litters (7–16%); most felid litters are relatively light (2–6%); and ailuropodid and ursid litters are very light (0.1–0.7%) (Table 13.2).

Litter energy values calculated from litter weights are expressed relative to maternal metabolic size in Table 13.2 so that data on energy deposition during pregnancy may be related to maternal energy requirements. We have used maternal weight to the power 0.75 as an estimate of maternal metabolic size (Kleiber 1965) to allow comparison with other mammalian groups, although McNab (this volume) reports that oxygen consumption of carnivores is proportional to the power 0.73. It is unlikely that these two exponents are significantly different. Estimated litter energy content per maternal metabolic size ranges from 4 kcal/kg$^{0.75}$ for the giant panda (*Ailuropoda melanoleuca*), to over 300 kcal/kg$^{0.75}$ for the African hunting dog (*Lycaon pictus*). As with litter weight, familial differences emerge: most canids are high (>110 kcal/kg$^{0.75}$), most mustelids and felids are lower (40–110 kcal/kg$^{0.75}$), and ailuropodids and ursids are very low (<30 kcal/kg$^{0.75}$) (Table 13.2). Procyonids and hyaenids tend to have litters of intermediate to high estimated energy content (90–160 kcal/kg$^{0.75}$), whereas the data for viverrids are variable (40–150 kcal/kg$^{0.75}$). These data represent first approximations since they derive from data from a variety of original sources and are based on the assumption that all carnivore neonates have the same energy-to-mass ratio.

Energy Deposition in Relation to Maternal Maintenance

The deposition of energy in the litter represents a small proportion of the maternal daily energy budget. Daily requirements for energy are usually expressed in terms of metabolizable energy (ME), which is defined as ingested energy minus energy lost in excreta, namely, in feces, urine, and methane, although methane production is negligible in carnivores (Maynard et al. 1979). In a carnivore ME is roughly equivalent to the energy available to the animal for metabolic purposes, but a portion of ME is inevitably lost as heat. Domestic dogs, cats, minks, and foxes require about 100–150 kcal ME per kg$^{0.75}$ to support maintenance needs (National Research Council 1982, 1985, 1986). Maintenance requirements for wild canids, felids, and mustelids may be somewhat higher because of their increased activity and thermoregulatory demands (MacDonald et al. 1984). Species with reduced resting metabolic rates such as the red panda (*Ailurus fulgens*) (see McNab, this volume) would be expected to have lower maintenance needs. Nonetheless, the energy contained in the litter at birth represents at most about two or three times the daily ME requirements for maintenance, and in the majority of species is probably no greater than daily maintenance needs.

The average daily rate of energy deposition in the litter may be calculated for the entire gestation period to compare to daily maintenance needs. For this calculation, periods of delayed implantation (see Mead, this volume), if of known duration, have been excluded from the duration of gestation (Gittleman 1986a) since energy is not required for embryonic or fetal development at this stage. The daily energy increment is clearly very small, $0.1-4.4$ kcal/kg$^{0.75}$ per day (Table 13.2). These average values reflect the relatively low nutritional demands of early and middle pregnancy but underestimate deposition rates in late gestation. Fetal growth follows a sigmoid curve, with most growth occurring in the last 20% of gestation (see Oftedal 1985).

The Efficiency of Energy Deposition

The energetic requirement for producing a litter during pregnancy is much greater than simply the energy deposited in fetuses, however. In the case pregnant cattle (*Bos taurus*) and sheep (*Ovis aries*) only about 13% of the increase in ME requirement above the maintenance level is deposited in the gravid uterus and 11% in the fetus itself (Agricultural Research Council 1980); the remainder is incorporated in mammary tissues or is used for metabolic purposes (including the respiration of fetal tissues). These estimates are based on animals that are neither augmenting nor depleting energy stores. If carnivores exhibit similar efficiency of energy deposition in fetal tissues, the total ME required to produce a litter can be estimated as litter energy at birth divided by 0.11, or in other words, approximately $36-2800$ kcal/kg$^{0.75}$, depending on species. This range is equivalent to maintenance ME requirements for about $0.3-23$ days. By contrast, estimated ME requirements for pregnancy in ungulates range from approximately 20 days' maintenance in small species (4 kg) to approximately 40 days in species of a size (400 kg) comparable to the largest carnivores (Oftedal 1985). The apparent discrepancy between carnivores and ungulates stems in part from the more altricial state (and hence lower energy density) of the neonatal carnivore (see Martin 1984), and in part from differences in litter mass relative to maternal metabolic size (Eisenberg 1981; Gittleman 1986b).

Energy Output during Lactation

Postnatal Litter Growth

Postnatal growth rate and lactation performance are inevitably linked by the dependence of suckling young on milk as the primary source of nutrients for synthesis of body mass (Oftedal 1981). Since net accumulation of mass is possible only if nutrient transfer from the mother exceeds the maintenance requirements of the young, it is evident that postnatal growth rate is an indirect measure of the magnitude of maternal energy output during lactation (Gittleman and Oftedal 1987). Data on the growth rates of carnivores during the period from birth to weaning are provided in Table 13.3. Litter growth rate is calculated from average growth rate (of individual offspring) and average litter

Table 13.3. Postnatal growth rate in relation to maternal metabolic size

Species	Female body weight (kg)	Individual growth rate (g/d)	Litter growth weight (g/d)	Litter growth per MBS (gk·g$^{0.75}$·d^{-1})
Canidae				
Gray wolf	31.1	161.0	885.8	67.3
(*Canis lupus*)				
Coyote	9.7	31.0	192.2	35.0
(*C. latrans*)				
Dhole	13.8	69.3	298.0	41.6
(*Cuon alpinus*)				
Red fox	3.9	17.5	84.0	30.3
(*Vulpes vulpes*)				
Gray fox	3.3	12.8	48.6	19.8
(*Urocyon cinereoargenteus*)				
Maned wolf	23.0	42.0	75.6	7.2
(*Chrysocyon brachyurus*)				
Ursidae				
Black bear	97.0	77.0	192.5	6.2
(*Ursus americanus*)				
Polar bear	320.0	120.2	228.4	3.0
(*Thalarctos maritimus*)				
Ailuropodidae				
Giant panda	96.8	78.8	78.8	2.6
(*Ailuropoda melanoleuca*)				
Ailuridae				
Red panda	4.9	6.0	11.1	3.4
(*Ailurus fulgens*)				
Procyonidae				
Ring-tailed cat	0.9	6.9	12.9	14.0
(*Bassariscus astutus*)				
Coati	5.0	12.0	48.0	14.4
(*Nasua narica*)				
Raccoon	6.7	24.3	92.3	22.2
(*Procyon lotor*)				
Mustelidae				
Ermine	0.3	2.5	15.0	37.0
(*Mustela erminea*)				
Weasel	0.1	1.5	10.4	58.5
(*M. nivalis*)				
Fisher	2.3	15.0	40.5	21.7
(*Martes pennanti*)				
Wolverine	10.4	79.3	158.6	27.4
(*Gulo gulo*)				
Tayra	4.4	21.7	76.0	25.0
(*Eira barbara*)				
Zorilla	0.6	5.1	11.6	17.0
(*Ictonyx striatus*)				
White-naped weasel	0.3	1.8	3.6	8.9
(*Poecilogale albinucha*)				
European badger	10.9	24.7	74.1	12.4
(*Meles meles*)				
Striped skunk	2.0	7.9	47.4	28.2
(*Mephitis mephitis*)				
Spotted skunk	0.4	4.8	20.4	40.6
(*Spilogale putorius*)				

Table 13.3. (*Continued*)

Species	Female body weight (kg)	Individual growth rate (g/d)	Litter growth weight (g/d)	Litter growth per MBS (gk·g$^{0.75}$·d^{-1})
River otter (*Lutra canadensis*)	7.8	26.7	26.7	5.7
Viverridae				
Binturong (*Arctictis binturong*)	13.0	31.8	95.3	13.9
Slender mongoose (*Herpestes sanguineus*)	0.4	2.7	6.8	13.5
Marsh mongoose (*Atilax paludinosus*)	3.3	14.3	43.0	17.6
Meerkat (*Suricata suricatta*)	0.7	6.2	24.9	32.5
Hyaenidae				
Striped hyena (*Hyaena hyaena*)	26.6	63.5	158.7	13.6
Spotted hyena (*Crocuta crocuta*)	55.3	96.0	192.0	9.5
Felidae				
European wild cat (*Felis silvestris*)	4.3	12.0	39.6	13.3
Jungle cat (*F. chaus*)	6.7	21.5	64.5	15.5
Sand cat (*F. margarita*)	2.2	12.0	33.0	18.3
Bengal cat (*Prionailurus bengalensis*)	3.3	12.2	30.5	12.5
Fishing cat (*P. viverrinus*)	6.3	11.0	27.5	22.6
African golden cat (*Profelis aurata*)	6.2	27.4	54.7	13.9
Caracal (*Caracal caracal*)	9.7	25.0	75.0	13.6
Mountain lion (*Puma concolor*)	39.6	32.0	80.0	5.1
Lynx (*Lynx lynx*)	17.8	35.7	82.1	9.5
African lion (*Panthera leo*)	135.5	122.0	317.2	8.0
Tiger (*P. tigris*)	131.0	86.0	215.0	5.6
Leopard (*P. pardus*)	39.3	33.0	85.8	5.5
Jaguar (*P. onca*)	77.6	48.0	120.0	4.6
Snow leopard (*P. uncia*)	32.5	48.0	134.4	9.9
Clouded leopard (*Neofelis nebulosa*)	17.0	23.0	46.0	5.5
Cheetah (*Acinonyx jubatus*)	60.0	50.0	190.0	8.8

Note. Adapted from Gittleman and Oftedal 1987, in which original references are cited. MBS = metabolic body size.

size, but as these two measurements were often made on different litters, the calculated litter growth rates are approximate. Litter growth rate increases with maternal size, but the scaling factor (0.50) is lower than might be expected from studies on other mammals (Gittleman and Oftedal 1987). Many species with heavy body weights (ursids, the giant panda, large felids) exhibit disproportionately slow growth, causing a reduction in slope of the line of best fit.

If litter growth is expressed relative to maternal metabolic size, most canids and some mustelids exhibit high rates of litter growth (20–40 $g/kg^{0.75}$); the hyaenids, small felids, and some viverrids have intermediate litter growth rates (10–20 $g/kg^{0.75}$); and the large felids, ursids, and pandas have slow-growing litters (3–10 $g/kg^{0.75}$ (Table 13.3). Litter growth rate appears to be influenced by maternal diet: herbivorous species, including folivores such as the red panda and the giant panda and frugivores such as the binturong (*Arctictis binturong*) (Gittleman 1986b), tend to grow less rapidly than more omnivorous/ carnivorous species (Gittleman and Oftedal 1987). Mothers of species that feed on foods of low digestible energy content such as bamboo (Dierenfeld et al. 1982) may be limited in their ability to acquire and process sufficient energy to support high milk output and rapid postnatal litter growth. A similar explanation has been put forth by McNab (this volume) to explain the low resting metabolic rates of the red panda and the giant panda, but a relationship between maternal metabolic rate during lactation and milk energy output has not been established. The form of parental care may also influence litter growth rate: species with biparental or communal caring systems (e.g., canids) have higher relative growth rates than do maternal species (Gittleman and Oftedal 1987). The cooperation of mates and helpers in acquiring prey may increase the amount of energy that can be channeled to offspring.

Milk Energy Content

Differences in postnatal growth undoubtedly reflect differences in nutrient intakes of the young as determined by milk composition and yield. Milk composition has been reported for 31 species of carnivores, but most of these species are represented by only one or a few samples that have been obtained opportunistically and may not be typical of the species studied (Gittleman and Oftedal 1987). More complete series of samples covering much or most of the lactation period have been analyzed for only eleven carnivore species (Gittleman and Oftedal 1987). Carnivore milks exhibit substantial variation in energy content, much of which is related to lactation stage. An increase in dry matter and energy content in late lactation is apparently typical of a number of carnivores, such as arctic fox (*Alopex lagopus*), domestic dog, American mink, and several bear species (Gittleman and Oftedal 1987). Given the effects of lactation stage on milk composition, interspecific comparisons should be re-

Table 13.4. The energy content in carnivore milks at midlactation

Species	n	Stage (days)	Gross energy (kcal/g)
Domestic dog (*Canis familiaris*)	25	7–37	1.5
Gray wolf (*C. lupus*)	1	28	1.4
Arctic fox (*Alopex lagopus*)	?	5–30?	1.9
Red fox (*Vulpes vulpes*)	3	28–35	1.1
Raccoon dog (*Nyctereutes procyonoides*)	22	7–59	1.0
Malayan sun bear (*Helarctos malayanus*)	1	90	1.6
Brown bear (*Ursus arctos*)	9	60–95	2.2
Black bear (*U. americanus*)	8	60	2.8
Raccoon (*Procyon lotor*)	1	38	0.9
American mink (*Mustela vison*)	20	10–27	1.2
Striped skunk (*Mephitis mephitis*)	15	20–48	2.0
Sea otter (*Enhydra lutris*)	1	'mid'	2.7
Domestic cat (*Felis domesticus*[a])	45?	6–38	0.9–1.7

Note. Adapted from review by Gittleman and Oftedal 1987, in which sources are cited. Data converted to kcal using the conversion 1 kcal = 4.184 kJ. Species are arranged taxonomically according to Ewer 1973.

[a]Because of discrepancies among published data on domestic cats, it is not possible to provide a single estimate of energy content, so a range of mean values is given

stricted to data from a comparable stage of lactation (Oftedal 1984a). Midlactation may be defined as the period about peak lactation during which milk is relatively constant in composition. At midlactation most carnivores produce milks containing 1–2 kcal/g, with somewhat higher values reported for bears and sea otters (Table 13.4). Carnivore milks are of moderate to high energy density in comparison with those of other terrestrial mammals (Oftedal 1984a). The higher milk energy content of some carnivore species—for instance, the arctic fox and the sea otter—is thought to reflect an adaptation to the presumed elevation of neonatal metabolic requirements in arctic and/or aquatic environments (see Oftedal et al. 1987a for parallel discussion with regard to otariid seals). In hibernating bears milks may be high in fat and energy as a means of utilizing maternal body fat while conserving maternal lean body mass (Ramsay and Dunbrack 1986; Gittleman and Oftedal 1987).

Milk and Energy Yield

In nondomestic species the milk intake by suckling young can be determined by hydrogen isotope methodology (Oftedal 1984b). Milk and energy intakes have been measured for neonates of four carnivore species at about peak lactation (Table 13.5). This period represents the time of maximal nutrient drain on the mother and may indicate maternal capacity. Across a broad size range of mammals, milk intake (g/d) is proportional to body weight, although

Table 13.5. Milk intakes of the young and milk yields in four carnivore species at the time of peak lactation

Species	Litters (n)	Young (n)	Time of peak (days)	Intake by young[a]			Output by mother[b]		
				Milk intake (g/d)	Energy intake (kcal/d)	Energy per MBS $(kcal \cdot kg^{-0.83} \cdot d^{-1})$	Milk yield (g/d)	Energy output (kcal/d)	Energy per MBS $(kcal/kg^{0.75})$
Black bear (Ursus americanus)	3	8	75	168.0	470.0	236	620	1720	58
Domestic dog (Canis familiaris)	5	25	26	175.0	257.0	221	1050	1540	231
Striped skunk (Mephitis mephitis)	5	19	31	26.9	53.0	240	151	297	167
American mink (Mustela vison)	6	23	20	24.5	28.9	210	119	140	145

Note. Adapted from Gittleman and Oftedal 1987, in which original sources are cited. Data converted to kcal using the conversion 1 kcal = 4.184 kJ.
[a]Energy intake expressed relative to metabolic body size (MBS) of suckling young, i.e., $W^{0.83}$.
[b]Energy output expressed relative to metabolic body size of mother, i.e., $W^{0.75}$.

there is considerable variation among species (Oftedal 1981); among the four carnivores daily milk intake ranges from 7 to 27% of the weight of the young.

Energy intakes of the young can be calculated from milk intake and the energy density of milk at midlactation. Among terrestrial mammals, milk energy intake scales to body weight of the young to the power 0.83, an exponent that is significantly different from 0.75 (Oftedal 1981, 1984b). If the energy intakes of the four carnivore species are expressed relative to weight $(kg)^{0.83}$, all values fall within 7% of the estimated mammalian norm, 225 $kcal \cdot kg^{-0.83} \cdot d^{-1}$. As many potential errors can occur in estimating both milk composition and milk intake (Oftedal 1984b), such close correspondence is encouraging and suggests that milk energy intake can be predicted with a reasonable level of accuracy from body weight of the young.

Multiplying the mean energy intake of suckling young by the litter size provides an estimate of the maternal milk energy output (Table 13.5). Milk energy output at peak lactation tends to be proportional to both maternal metabolic size (weight$^{0.75}$) and litter size (Oftedal 1984a). Large-littered species such as domestic pigs (*Sus scrofa*) and laboratory rats (*Rattus norvegicus*) have very high daily energy outputs (ca. 240 $kcal \cdot kg^{-0.75} \cdot d^{-1}$.) as compared with ungulates with single young (ca. 65–90 $kcal \cdot kg^{-0.75} \cdot d^{-1}$) (Oftedal 1984b, 1985). A domestic dog with five young approaches the energy output of large-littered pigs and rats, but mink and striped skunks (*Mephitis mephitis*) have intermediate energy outputs (Table 13.5). The American black bear has a very low output of milk energy (less than an ungulate with a single offspring) even though it produces very energy dense milk (Tables 13.4, 13.5).

Predicted Milk Energy Output

As the young develop, their daily energy requirements increase (Oftedal 1981); at the point where maternal energy output in milk is no longer sufficient to match the rising requirements, the young are obliged to seek additional foods, either solid foods or, in some communal species, milk from other mothers. It is likely that peak lactation occurs at about the time of initial ingestion of solid foods in most carnivores. Since the milk energy intakes of suckling young are closely related to metabolic mass ($W^{0.83}$) at the lactation peak, peak milk energy output (EO, kcal) of the mother can be predicted from litter metabolic mass (LMM, equal to litter size times mean metabolic mass of the young): EO = 227 × LMM. Predicted EO values range from 65 to 300 $kcal \cdot kg^{-0.75} \cdot d^{-1}$ among 29 species for which appropriate weight data are available (Table 13.6).

The predicted values are simply a first approximation, given potential errors in estimating weights, litter size, and time of first solids intake from a variety of published sources. Most carnivores fall within a moderately high range of predicted energy output (150–220 $kcal \cdot kg^{-0.75} \cdot d^{-1}$). Most carnivores ex-

Table 13.6. Predicted milk energy outputs from litter mass at first intake of solid foods

Species	Maternal weight (kg)	Time of first solids (days)	Weight of young (kg)	Litter size (n)	Litter metabolic mass (kg$^{0.83}$)	Predicted peak milk energy output	
						Daily (kcal)	Per MW$^{0.75}$ (kcal · kg$^{-0.75}$ · d^{-1})
Canidae							
Domestic dog (*Canis familiaris*)	12.7	28	1.27	6.0	7.32	1650	245
Coyote (*C. latrans*)	9.7	28	0.87	6.0	5.35	1200	219
Dhole (*Cuon alpinus*)	13.8	30	2.35	4.3	8.73	1970	275
Red fox (*Vulpes vulpes*)	3.9	30	0.63	3.9	2.66	598	215
Gray fox (*Urocyon cinereoargenteus*)	3.3	30	0.49	3.8	2.11	474	194
Ursidae							
Black bear (*Ursus americanus*)	81.6	120	4.5	3.0	10.4	2350	87
Ailuropodidae							
Giant panda (*Ailuropoda melanoleuca*)	96.8	180	15	1.0	9.47	2130	69
Ailuridae							
Red panda (*Ailurus fulgens*)	4.9	90	0.66	1.9	1.31	295	90
Procyonidae							
Ring-tailed cat (*Bassariscus astutus*)	0.87	36	0.18	3.0	0.72	163	181
Coati (*Nasua narica*)	5.00	49	0.73	4.0	3.07	691	207
Raccoon (*Procyon lotor*)	6.70	63	1.00	3.8	3.80	855	205
Mustelidae							
Ermine (*Mustela erminea*)	0.22	21	0.019	8.0	0.30	67	209

Species							
Weasel (M. nivalis)	0.06	18	0.014	4.7	0.14	31	255
American mink (M. vison)	0.96	23	0.110	4.8	0.75	169	174
Zorilla (Ictonyx striatus)	0.63	39	0.170	2.0	0.46	103	146
White-naped weasel (Poecilogale albinucha)	0.25	46	0.050	2.4	0.20	45	127
Striped skunk (Mephitis mephitis)	2.16	35	0.220	5.7	1.61	363	201
Spotted skunk (Spilogale putorius)	0.52	37	0.160	3.0	0.66	147	240
Viverridae							
Binturong (Arctictis binturong)	13.00	56	2.10	3.0	5.55	1250	182
Meerkat (Suricata suricatta)	0.72	27	0.20	4.0	1.04	235	300
Felidae							
Domestic cat (Felis domesticus)	3.6	32	0.41	3.9	1.86	419	160
European wild cat (F. silvestris)	4.3	32	0.52	3.3	1.92	432	145
Bengal cat (Prionailurus bengalensis)	3.3	44	0.62	2.5	1.68	378	154
Lynx (Lynx lynx)	17.8	50	1.86	2.3	3.84	864	100
African lion (Panthera leo)	136.0	56	8.48	2.6	15.30	3450	87
Tiger (P. tigris)	131.0	56	6.07	2.5	11.20	2510	65
Leopard (P. pardus)	39.3	42	1.94	2.6	4.50	1010	65
Jaguar (P. onca)	77.6	70	4.18	2.5	8.19	1840	71
Cheetah (Acinonyx jubatus)	60.0	33	1.94	3.8	6.58	1480	69

Note. Adapted from Gittleman and Oftedal 1987, in which original sources of data are cited. Data converted to kcal using the conversion 1 kcal = 4.184 kJ.

hibiting a tendency for communal rearing, such as gray wolf (*Canis lupus*), red fox (*Vulpes vulpes*), dhole (*Cuon alpinus*), and meerkat (*Suricata suricatta*), appear to achieve high milk energy outputs (220–300 kcal · kg$^{-0.75}$ · d^{-1}) although the African lion (*Panthera leo*) does not. By contrast, the American black bear, red panda, giant panda, cheetah (*Acinonyx jubatus*), and large cats of the genus *Panthera* have low predicted milk energy outputs (65–90 kcal · kg$^{-0.75}$ · d^{-1}), comparable to those of ungulates with single young. These low energy outputs may reflect limitations imposed by hibernation (black bear), may be a consequence of being restricted to a diet low in digestible energy (red panda, giant panda) or may reflect a limitation of prey capture rate on maternal milk-producing ability in very large predators (Gittleman and Oftedal 1987).

Milk Energy Output in Relation to Maternal Maintenance

Milk energy output at peak lactation represents a considerable addition to maternal energy needs. For most carnivores listed in Table 13.6, the estimated milk energy outputs *per day* are equivalent to maintenance ME requirements for about one or two days (100–250 kcal/kg$^{0.75}$). If the assumption is made that the net efficiency of milk energy production is about 70% (as in domestic ungulates; see Agricultural Research Council 1980), lactating carnivores need to increase ME intakes by approximately 150–300% to meet the energy requirements of peak milk production. In the case of domestic cats measured ME intakes increase by approximately 100–200% during intensive lactation when litters of 4–5 kittens are reared (Table 13.1), but these cats also lose about 20% of body weight, so it appears that this level of intake is not sufficient to cover the energetic demands of intensive lactation.

The total or aggregate amount of milk energy transferred from mother to young is not known for any carnivore, but it is clearly much greater than the energy deposited in fetuses during pregnancy. In all carnivores examined, the estimated energy output in milk during one day at peak lactation is similar to or greater than the estimated energy content of the entire litter at birth (Tables 13.2, 13.6).

A first estimate of the aggregate amount of milk energy transferred over lactation can be calculated from average daily milk yield, average milk energy content, and lactation length (Oftedal 1985). Unfortunately, the shape of the lactation curve has not been described for any carnivore. Among ungulates, the average daily yield over lactation is typically approximately 60–80% of the peak daily yield (Oftedal 1985). If we assume an average daily energy yield that is 70% of peak yield and a lactation length of 60 days (see Gittleman 1986b for a tabulation of weaning ages of carnivores), a carnivore with a peak energy yield of 200 kcal · kg$^{-0.75}$ · d^{-1} would have a calculated total energy output of 8400 kcal/kg$^{0.75}$. Given a net efficiency of milk energy secretion of 70%, the calculated ME requirement for lactation would be 12,000 kcal/kg$^{0.75}$

or the equivalent of about 100 days of maintenance ME. Though hypothetical, this value is similar to estimates of the ME required for lactation by ungulates (ca. 90–120 days of maintenance ME) (Oftedal 1985).

Summary: Patterns of Reproductive Output in Carnivores

The order Carnivora is large and diverse, exhibiting a wide array of reproductive patterns (Ewer 1973; Eisenberg 1981; Gittleman 1986a, 1986b; Gittleman and Oftedal 1987). It is not surprising that patterns of energy output during reproduction should differ along phylogenetic lines and between species. For example, in relation to maternal metabolic size, canids tend to invest more energy in their litters than do felids both during gestation (Table 13.2) and at peak lactation (Table 13.6), but within each family there is substantial variation. Some of the variation may stem from error of estimation since the numerical values developed herein are based on both extrapolation and predictive relationships that require further validation. Yet it is not so much the existence of differences but their interpretation in relation to causal pathways or adaptive significance that is of particular interest. In this context we will briefly discuss allometric, phylogenetic, and life history correlates of the estimated energy outputs of carnivores during pregnancy and lactation.

Physiological, behavioral, and life history parameters are all influenced by body size (Calder 1984); allometric correlations of life histories are so ubiquitous as to be considered routine by some workers (Harvey 1986). Among carnivores, correlations of 0.84 or higher have been observed between maternal body weight and each of the following variables: birth weight, litter weight, postnatal growth rate, and length of lactation (Gittleman 1986b; Gittleman and Oftedal 1987). Peak milk energy output also bears an allometric relationship to maternal weight in mammals (Linzell 1972; Hanwell and Peaker 1977; Oftedal 1984a, 1984b), and milk energy intake of the young is related to neonatal weight (Oftedal 1981, 1984b). Although such allometric effects must be accounted for in functional studies of interspecific differences, it is not always obvious what scaling factor is most appropriate. For example, in most mammalian studies, growth rate scales to adult body weight raised to a power of about 0.75 (range 0.69–0.83), but in carnivores the scaling factor is only 0.58 (Gittleman and Oftedal 1987). We have opted herein to express energetic values in relation to maternal metabolic size ($W^{0.75}$) since interspecific comparisons of metabolic rate and maintenance requirements are usually expressed in this fashion (Kleiber 1975; Hudson and Christopherson 1985; National Research Council 1985, 1986).

Traditionally, mammalian life histories and reproductive patterns have been viewed as extremely malleable, to adjust to environmental conditions (Lack 1954; Sadleir 1969). If so, interspecific variation should be great and phylogenetic effects weak or nonexistent. Yet in our analysis we have repeatedly

Table 13.7. Estimated daily energy outputs during pregnancy and peak lactation, by carnivore family

	Daily energy deposition during pregnancy (kcal · kg$^{-0.75}$ · d^{-1})	Postnatal growth rate (g · kg$^{-0.75}$ · d^{-1})	Milk energy output at peak lactation (kcal · kg$^{-0.75}$ · d^{-1})
Canidae	2.85 ± 0.35(11)	34 ± 8.37(11)	230 ± 13.9(5)
Ursidae	0.35 ± 0.05(2)	4.6 ± 1.60(2)	87(1)
Ailuridae and Ailuropodidae	0.10(1)	3.0 ± 0.40(2)	80 ± 10.5(2)
Procyonidae	1.73 ± 0.18(4)	16.9 ± 2.7(3)	198 ± 8.4(3)
Mustelidae	1.94 ± 0.27(13)	25.7 ± 4.7(11)	193 ± 17.8(7)
Viverridae	1.10 ± 0.23(6)	19.4 ± 4.5(4)	—
Hyaenidae	1.3(1)	11.6 ± 2.1(2)	241 ± 59.0(2)
Felidae	1.03 ± 0.113(14)	10.8 ± 1.31(16)	102 ± 13.4(9)

Note. Data expressed relative to maternal metabolic size (kg$^{0.75}$). Data from Ailuridae and Ailuropodidae combined for purposes of presentation. Values in parentheses refer to number of species for which data are available (see Tables 13.2, 13.3, 13.6).

observed familial differences in parameters of reproductive energy output, after maternal size is factored out. Estimated energy transfer rates from mother to young during pregnancy and peak lactation, as well as an indicator of mass transfer during lactation (postnatal growth rate), are summarized by carnivore family in Table 13.7. Both the mean estimate and the standard error of the mean have been calculated; further statistical analysis is not warranted given that these estimates are ratios that probably do not follow a normal distribution, and most represent extrapolated or predicted values. Each of the three estimates is somewhat different: one represents the *average* rate of energy deposition in the young during prenatal growth; the second represents the *average* rate of mass deposition during postnatal growth, and the third represents the *peak* rate of energy output in milk. The numbers of species represented are rather small for some large families (e.g., Viverridae and Procyonidae). Nonetheless, certain consistent familial characteristics emerge: (1) the canids appear to have high transfer rates by all three measures, that is, both during pregnancy and lactation; (2) the procyonids and mustelids tend to have moderately high transfer rates by the three measures during both pregnancy and lactation; (3) the felids appear to have the lowest transfer rates of the major families during both pregnancy and lactation; (4) the ursids and pandas (Ailuridae, Ailuropodidae) appear to have particularly low transfer rates in both pregnancy and lactation.

Despite these familial patterns, estimated values for individual species vary substantially. The extent to which this represents true interspecific variation versus variation attributable to error in our species estimates is difficult to assess without further study. Is litter energy content of fennec foxes (*Fennecus zerda*) and maned wolves (*Chrysocyon brachyurus*) really only ⅓ that of most canids, once corrected for maternal metabolic size (Table 13.2)? Is the energy transfer rate during lactation actually lower for striped polecats (*Ictonyx striatus*) and white-naped weasels (*Poecilogale albinucha*) than for most other mustelids, as both postnatal growth rates (Table 13.3) and predicted milk energy outputs (Table 13.6) indicate? Where consistent patterns emerge among species groups, the trends are more convincing. For example, small felids exhibit higher rates of energy transfer during lactation than do large felids, whether assessed by growth rate (Table 13.3) or by predicted milk energy output (Table 13.6). Small felids also have higher resting metabolic rates than do large felids (McNab, this volume).

After controlling for allometric and phylogenetic effects, one can relate some features of reproductive output in carnivores to maternal diet and mode of parental care. Herbivorous species (bears, pandas) have slow individual and litter growth rates and perhaps low milk energy yields. Much plant matter (especially structural material such as leaves and stems) is low in digestible energy, especially for carnivores with simple digestive systems (Dierenfeld et al. 1982; Gittleman 1988). McNab (this volume) has demonstrated a relationship between dietary habit and basal metabolic rate, as well as between basal metabolic rate and various life histories. As McNab argues, low metabolic rates may be adaptive to diets low in energy content and/or high in toxic compounds. Consequently, low reproductive effort (e.g., low milk energy output, low postnatal growth rate) may parallel low metabolic rates for species with nutritionally poor diets. Contrary to this physiological explanation, though, is that low growth rates and low predicted milk energy outputs of large felids are not associated with particularly low metabolic rates. These comparative results should encourage further work on the physiology of felids.

Although the comparative data are sparse, we have also detected that species with biparental (e.g., coyote, *Canis latrans*; brown hyena, *Hyaena brunnea*) or communal care (e.g., gray wolf, meerkat) have higher milk energy outputs and higher litter growth rates than do species with maternal care. There appear to be several reasons to explain why species with paternal care or conspecific helpers have different reproductive outputs (see Gittleman 1985; Gittleman and Oftedal 1987; Moehlman, this volume). First, mothers that have helpers to bring them food may be able to produce more milk for growing young (e.g., African hunting dog: Malcolm and Marten 1982; black-backed jackal, *Canis mesomelas*: Moehlman, this volume); direct feeding of young by helpers may also contribute to high growth rates in late lactation (most Canidae: Kleiman and Brady 1978). Second, communal care may also provide protection for the young. Increased protection may allow mothers and/or helpers to spend more time and effort in providing nutrients (milk and food) rather than in vigilance, or may allow the young to devote their ingested nutrients to growth rather than self-defense. Thus, the mechanisms whereby communal behaviors increase nutrient transfer to the young and consequently increase growth rates and overall energy output may be quite indirect.

In summary, comparative data on carnivores suggest that reproductive output is tied to allometric and phylogenetic (familial) constraints and, after removing these constraints, to dietary and parental characteristics. Further work on dietary efficiency, metabolic rate, nutrient deposition, and milk output, in conjunction with behavioral studies, is needed to evaluate the assumptions and predictions presented in this chapter, and to determine the errors in generalizing these assumptions and predictions to most carnivores.

References

Adolph, E. F., and Heggeness, F. W. 1971. Age changes in body water and fat in fetal and infant animals. *Growth* 35:55–63.

Agricultural Research Council. 1980. *The Nutrient Requirements of Ruminant Livestock*. Farnham Royal, Eng.: Commonwealth Agricultural Bureaux.

Calder, W. A. III. 1984. *Size, Function, and Life History*. Cambridge: Harvard Univ. Press.

Dierenfeld, E. S., Hintz, H. F., Robertson, J. B., Van Soest, P. J., and Oftedal, O. T. 1982. Utilization of bamboo by the giant panda. *J. Nutr.* 112:636–641.

Eisenberg, J. F. 1981. *The Mammalian Radiations*. Chicago: Univ. Chicago Press.

Ewer, R. F. 1973. *The Carnivores*. Ithaca, N.Y.: Cornell Univ. Press.

Gittleman, J. L. 1985. Functions of communal care in mammals. In: P. J. Greenwood, P. H. Harvey & M. Slatkin, eds. *Evolution: Essays in honor of John Maynard Smith*, pp. 187–205. Cambridge: Cambridge Univ. Press.

Gittleman, J. L. 1986a. Carnivore brain size, behavioral ecology, and phylogeny. *J. Mamm.* 67:23–36.

Gittleman, J. L. 1986b. Carnivore life history patterns: Allometric, ecological and phylogenetic associations. *Amer. Nat.* 127:744–771.

Gittleman, J. L. 1988. The behavioral energetics of lactation in a herbivorous carnivore, the red panda (*Ailurus fulgens*). *Ethology.* In press.

Gittleman, J. L., and Oftedal, O. T. 1987. Comparative growth and lactation energetics in carnivores. *Symp. Zool. Soc. London* 57:41–77.

Gittleman, J. L., and Thompson, S. D. 1988. Energy allocation in mammalian reproduction. *Amer. Zool.* 28:863–875.

Harvey, P. H. 1986. Energetic costs of reproduction. *Nature* 321:648–649.

Hanwell, A., and Peaker, M. 1977. Physiological effects of lactation on the mother. *Symp. Zool. Soc. London.* 41:297–312.

Heidt, G. A. 1970. The least weasel *Mustela nivalis* Linnaeus. Developmental biology in comparison with other North American *Mustela*. *Publications of the Museum, Michigan State Univ.*, Biological Series no. 4:227–282.

Heidt, G. A., Petersen, M. K., and Kirkland, G. L. 1968. Mating behavior and development of weasels (*Mustela nivalis*) in captivity. *J. Mamm.* 49:413–419.

Hemmer, H. 1976. Gestation period and postnatal development in felids. *World's Cats* 3:143–164.

Hudson R. J., and Christopherson, R. J. 1985. Maintenance metabolism. In: R. J. Hudson & R. G. White, eds. *The Bioenergetics of Wild Herbivores*, pp. 121–142. Boca Raton, Fla.: CRC Press.

Kleiber, M. 1975. *The Fire of Life: An Introduction to Animal Energetics* (rev. ed.) New York: John Wiley.

Kleiman, D. G., and Brady, C. A. 1978. Coyote behavior in the context of recent canid research: Problems and perspectives. In: M. Bekoff, ed. *Coyotes*, pp. 163–188. New York: Academic Press.

Kunz, T. H., and Nagy, K. N. 1987. Methods of energy budget analysis. In: T. H. Kunz, ed. *Ecological and Behavioral Methods for the Study of Bats*. Washington, D.C.: Smithsonian Institution Press.

Lack, D. 1954. *The Natural Regulation of Animal Numbers*. Oxford: Clarendon Press.

Linzell, J. L. 1972. Milk yield, energy loss in milk, and mammary gland weight in different species. *Dairy Sci. Abstr.* 34:351–360.

Loudon, A., and Racey, P. A., eds. 1987. *The Reproductive Energetics of Mammals*. Symposia of the Zoological Society of London no. 57. Oxford: Oxford Univ. Press.

Loveridge, G. G. 1986. Bodyweight changes and energy intake of cats during gestation and lactation. *Anim. Technol.* 37:7–15.

MacDonald, M. L., Rogers, Q. R., and Morris, J. G. 1984. Nutrition of the domestic cat, a mammalian carnivore. *Ann. Rev. Nutr.* 4:521–562.

Malcolm, J. R., and Marten, K. 1982. Natural selection and the communal rearing of pups in African wild dogs (*Lycaon pictus*). *Behav. Ecol. Sociobiol.* 10:1–13.

Martin, R. D. 1984. Scaling effects and adaptive strategies. *Symp. Zool. Soc. Lond.* 51:87–117.

Mattingley, D. K., and McClure, P. A. 1982. Energetics of reproduction in large-littered cotton rats (*Sigmodon hispidus*). *Ecology* 63:183–195.

Maynard, L. A., Loosli, J. K., Hintz, H. F., and Warner, R. G. 1979. *Animal Nutrition* (7th ed.) New York: McGraw-Hill.

Millar, J. S. 1979. Energetics of lactation in *Peromyscus maniculatus*. *Canadian J. Zool.* 57:1015–1019.

Mundt, von, H.-C., Thomee, A., and Meyer, H. 1981. Zur Energie- und Eiweissversorgung von Säugwelpen über die Muttermilch. *Kleintier Praxis* 26:353–360.

National Research Council. 1982. *Nutrient Requirements of Mink and Foxes*. Washington, D.C.: National Academy of Sciences.

National Research Council. 1985. *Nutrient Requirements of Dogs*. Washington, D.C.: National Academy of Sciences.

National Research Council. 1986. *Nutrient Requirements of Cats*. Washington, D.C.: National Academy of Sciences.

Oftedal, O. T. 1981. Milk, protein and energy intakes of suckling mammalian young: A comparative study. Ph.D. dissert., Cornell Univ., Ithaca, N.Y.

Oftedal, O. T. 1984a. Body size and reproductive strategy as correlates of milk energy yield in lactating mammals. *Acta Zoologica Fenn.* 171:183–186.

Oftedal, O. T. 1984b. Milk composition, milk yield and energy output at peak lactation: A comparative review. *Symp. Zool. Soc. Lond.* 51:33–85.

Oftedal, O. T. 1985. Pregnancy and lactation. In: R. J. Hudson & R. G. White, eds. *The Bioenergetics of Wild Herbivores*, pp. 215–238. Boca Raton, Fla: CRC Press.

Oftedal, O. T. 1988. Body composition of neonatal black bears. Unpublished data.

Oftedal, O. T., Boness, D. J., and Tedman, R. A. 1987a. The behavior, physiology and anatomy of lactation in the Pinnipedia. *Current Mamm.* 1:175–245

Oftedal, O. T., Iverson, S. J., and Bonness, D. J. 1987b. Milk and energy intake of suckling California sea lion (*Zalophus californicus*) pups in relation to sex, growth, and predicted maintenance requirements. *Physiol. Zool.* 60:560–575.

Ramsay, M. A., and Dunbrack, R. L. 1986. Physiological constraints on life history phenomena: The example of small bear cubs at birth. *Amer. Nat.* 127:735–743.

Rattray, P. V., Garrett, W. N., East, N. E., and Hinman, N. 1974. Growth, development and composition of the ovine conceptus and mammary gland during pregnancy. *J. Anim. Sci.* 38:613-626

Robbins, C. T., and Moen, A. N. 1975. Uterine composition and growth in pregnant white-tailed deer. *J. Wildl. Mgmt.* 39:684–691.

Sadleir, R. M. F. S. 1969. *The Ecology of Reproduction in Wild and Domestic Mammals*. London: Methuen.

Schaller, G. B., Jinchu, H., Wenshi, P., and Jing, Z. 1985. *The Giant Pandas of Wolong.* Chicago: Univ. Chicago Press

Thomas, K. 1911. Über die Zusammensetzung von Hund und Katze während der ersten Verdoppelungsperioden des Geburtgewuchtes. *Archiv für Anatomie und Physiologie, Physiologischer Abteilung* 1:9–38.

Thompson, S.D., and Nicoll, M. E. 1986. Basal metabolic rate and energetics of reproduction in therian mammals. *Nature* 321:690–693.

PART III

EVOLUTION

INTRODUCTION

Simpson began *Tempo and Mode in Evolution* (1944) with the following: "The basic problems of evolution are so broad that they cannot hopefully be attacked from the point of view of a single scientific discipline. Synthesis has become both necessary and more difficult as evolutionary studies have become more diffuse and more specialized" (p. xv). Following the wave of studies singularly devoted to natural selection theory as it applies to adaptation and the resulting criticism surrounding such an approach (Clutton-Brock and Harvey 1979; Gould and Lewontin 1979; Mayr 1983), the field is returning to Simpson's vision of using theory and methodologies from different fields, a healthy development. Each chapter in this part falls under the heading of a particular trait or methodology yet adopts a pluralistic approach to carnivore evolution by considering not only adaptive evolution but also modes of constraint (allometry, genetic drift) and historical process.

The first three chapters consider different forms of morphological or physiological constraints and the ways these forms might affect the ability of a carnivore to enter an unused niche, utilize new or varied food resources, or adjust reproductive parameters to meet inclement environmental conditions. Taylor analyzes the five primary locomotor patterns in carnivores and the anatomical characters associated with them. Of particular importance is the observation that, even though carnivores have distinct anatomical constraints on locomotor movements, most species maintain the flexibility to traverse many habitat types and procure various food resources in spite of these constraints. In a similar vein, Van Valkenburgh reviews and quantitatively tests hypotheses related to variation in dental morphology. Carnivore dentition, including canine shape, premolar size and shape, carnassial blade length, and postcarnassial molar size, not only relates to diet and body size but also to guild structure and number of species represented in a community. Coupled with historical evidence, this type of analysis gives support to the idea that broad range evolutionary explanations may be inferred from what often appears to be rather scant dental characters. Focusing on physiological constraints, Mead reviews the extensive comparative and experimental literature on delayed implantation. Although this chapter seems more specialized than the preceding chapter—because delayed implantation is observed only in a small percentage of all carnivores—Mead clearly shows the full complexity of carnivore reproduction by focusing on this one aspect; hormonal, anatomical, ecological, and evolutionary factors are all integrated in the control of delayed implantation.

The final three chapters in the volume are, in many ways, the most controversial. The evolutionary divergence of carnivores and pinnipeds, the rate of evolutionary change, the relative merits of different kinds of information (e.g., molecular versus fossil data) in piecing together evolutionary scenarios or the phylogenetic relationships of extant carnivore families, all are topics that have been the focus of intense discussion and disagreement. Although many of the conclusions in these chapters do not agree on phylogenetic branching sequences or classifications, arguments from the respective camps are clearly presented so that further research and debate may develop on solid ground. In summarizing the data on molecular and biochemical evolution, Wayne, Benveniste, Janczewski, and O'Brien argue that the phylogeny of carnivores was defined early in its evolutionary history and that the larger, more specialized carnivores (canids, felids, ursids) have higher speciation and extinction rates. Further, the authors show that the DNA hybridization data indicate a monophyletic origin of the pinnipeds, in contrast to some morphological work (see Wozencraft). Following a thorough review of the long history of carnivore classification, Wozencraft uses 100 skull, postcranial, and soft anatomy characters to reassess carnivore phylogeny. Some of the controversial highlights in this analysis are that (1) the herpestids and viverrids are placed in separate families, (2) the red and giant pandas are included in the Ursidae, and (3) the close relationships of the Otariidae with Ursidae and the Phocidae with Mustelidae are recognized. Finally, Martin elegantly describes the ancestral, fossil carnivore groups and those biogeographic factors that gave rise to extant groups, thus selecting for particular "ecomorphs," or morphological types, suited for specific adaptive zones. With this background, Martin presents a detailed taxonomic review of fossil carnivores, complete with line drawings depicting certain species, and argues that a combination of faunal interchanges, efficient foraging strategies, and an influx of available prey led to the successful radiation of the carnivores.

As evidenced by the chapters in this part, we are entering an exciting and prolific period for evolutionary studies of carnivores. Taxonomic, phylogenetic, and morphological questions will undoubtedly continue to flourish. We hope, though, that some of the ideas presented here will provoke more investigations into the mechanisms of carnivore evolution, especially variables related to genetic and fitness effects.

JOHN L. GITTLEMAN

References

Clutton-Brock, T. H., and Harvey, P. H. 1979. Comparison and adaptation. *Proc. Royal Soc. London* B205:547–565.

Gould, S. J., and Lewontin, R. C. 1979. The spandrels of San Marco and the Panglossian paradigm: A critique of the adaptationist programme. *Proc. Royal Soc. London* B205:581–598.

Mayr, E. 1982. How to carry out the adaptationist program? *Amer. Nat.* 121:324–334.

Simpson, G. G. 1944. *Tempo and Mode in Evolution*. New York: Columbia Univ. Press.

CHAPTER 14

Locomotor Adaptations
by Carnivores

MARK E. TAYLOR

Carnivores exhibit a wide range of locomotor behaviors. However, an animal's morphology limits its range of movements and therefore provides a constraint to certain locomotor activities. For instance, the body proportions and morphology of the sea otter (*Enhydra lutris*) make it an excellent swimmer, whereas it has difficulty moving on land. Likewise, the arboreal specializations of the ringtail (*Bassariscus astutus*), which allow it to perform complex acrobatic movements, restrict its abilities for other forms of locomotion such as running or digging. Some carnivores improve the effect of their locomotor skills with behavioral modifications; for example, the group hunting behavior of the African lion (*Panthera leo*) and the gray wolf (*Canis lupus*) allows them to catch prey that they would not be able to catch as individuals (Mech 1970; Schaller 1972). Therefore, from an evolutionary viewpoint, one must recognize that behavioral adaptations may be as important as morphological adaptations. However, although the behavior of many carnivores is poorly known, it is possible to infer a great deal from their morphological adaptations and to use this information in a predictive way to understand their role in particular ecosystems.

Ambulatory Carnivores

Most carnivores incorporate the walk into their repertoire of gaits. The walk is a symmetrical gait in which each foot is on the ground more than half the time (Hildebrand 1982). A few species are primarily ambulatory and rarely use a faster gait. These ambulatory carnivores have, for the most part, given up an active predatory existence and have become specialized herbivores (e.g., giant panda, *Ailuropoda melanoleuca*) or omnivores (e.g., raccoon, *Procyon lotor*; skunk, *Mephitis mephitis*).

382

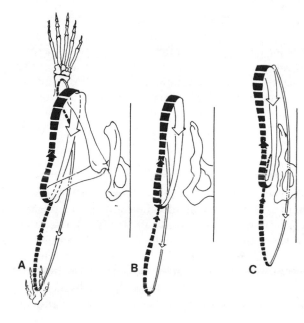

Figure 14.1. The excursions of the knee (*wide line*) and foot (*narrow line*) relative to the pelvis as seen from the dorsal aspect in a walking raccoon (*Procyon lotor*) (A), a fox (B), and a cat (C). The darkened half of each line represents the nonpropulsive (swing) phase; the light half, the propulsive phase. (From Jenkins and Camazine 1977, courtesy The Zoological Society of London.)

Locomotion

Many authors have studied walk patterns in carnivores, from Muybridge (1887) in the latter part of the last century to more recent authors (Hildebrand 1961, 1976, 1982, 1985b; M. E. Taylor 1970, 1971; Jenkins 1971; Jenkins and Camazine 1977; Dagg 1979; Goslow and Van de Graaf 1982; Hurst et al. 1982). Although the walk has been studied extensively in the dog and cat, they are cursors, and for basic walk patterns species such as the raccoon or opossum (*Didelphis virginiana*) are more typical (Jenkins 1971). Raccoons abduct their limbs more than do dogs and cats, and do not move them in a parasagittal plane (Figure 14.1). Also, the resting stance of raccoons is far more variable, whereas the dog and cat, which are obligatory digitigrade species, do not show much tendency to vary femoral posture while stationary (Jenkins and Camazine 1977).

In a study of seven large carnivores (polar bear, *Ursus maritimus;* black bear, *U. americanus;* tiger, *Panthera tigris;* lion, *P. leo;* cougar, *Felis concolor;* cheetah, *Acinonyx jubatus;* and spotted hyena, *Crocuta crocuta*) Dagg (1979) found that they all had similar walking patterns, with lateral supporting legs used to a large extent, and diagonal supporting legs rarely used. The legs of these larger carnivores are placed well beneath the body, so that the animal can readily balance on two lateral legs while swinging the other two forward. Large carnivores have relatively unstable walking patterns in which the center of gravity is not between diagonal supporting legs, whether they live in open or forested habitats.

Many carnivores walk while foraging, though more cursorial species readily break into a slow trot. Some of the smaller carnivores do not usually move faster than a walk; for example, a skunk could not move beyond a fast walk on a treadmill (Goslow and Van de Graaf 1982), though faster gaits have been recorded.

Morphology

In a study on the effects of the clavicle on locomotion Jenkins (1974) found that the weight-bearing shoulder of the ambulatory raccoon moves linearly and obliquely at about a 20° angle to the sagittal plane. He also showed that the clavicle acts as a strut under compression during walking, preventing the shoulder from pressing in on the thoracic cage. The orientation of the linear movement of the scapula varies in different species, depending on thoracic conformation and the presence or absence of a clavicle. In cursorial species the ribs form a nearly sagittal thoracic wall, so the scapulae are nearly perpendicular to the horizontal and there is no clavicle; in the raccoon the thorax widens abruptly and flares the scapulae. The shape of the thorax and orientation of the humeral joints are probably as important in determining movements of the shoulder as the presence or absence of a clavicle.

In ambulatory carnivores such as the raccoon the shape and position of the hip joint is such that the femur projects laterally rather than being in the sagittal plane (Figure 14.1). The articular cartilage within the acetabulum is also positioned medially, which permits greater abduction of the hind limb. The position of the fovea capitis on the head of the femur also indicates the orientation of the hind limb. In felids and canids the fovea are approximately centered in the acetabular fossae when the femurs are sagitally oriented. In raccoons and other ambulatory carnivores fovea position is approximately 15° from the meridian and 25–45° from the equator, a feature consistent with the relatively large and variable abduction employed in both posture and gait (Jenkins and Camazine 1977). The wide range of excursion movements made possible by the hip joint in species like raccoons and skunks allows them to walk over uneven terrain. Canid and felid patterns of hip movement in which there is relatively little abduction is recognized as characteristic of cursorial mammals (Gregory 1912; Howell 1944; Jenkins 1971). Likewise, movements in the ankle of the skunk are far more variable than in the cat, and the primary ankle extensors, m. soleus, mm. medial and lateral gastrocnemius, and m. plantaris, are capable of producing greater forces through a wider range of ankle angles than in more cursorial species. However, the cat is capable of storing more elastic energy in its tendons, indicative of specialization toward a

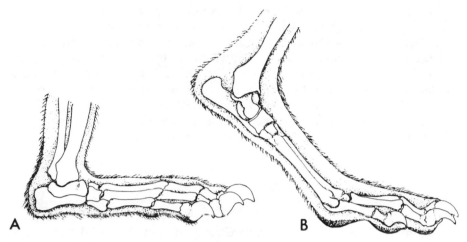

Figure 14.2. Medial view of hind feet of (A) red panda (*Ailurus fulgens*), plantigrade, and (B) fox (*Vulpes vulpes*), digitigrade.

jumping and sprinting way of life (Goslow and Van de Graaff 1982; Wingerson 1983).

The feet of ambulatory carnivores such as bears, raccoons, and pandas are typically plantigrade, and weight is transmitted to the ground through the tarsals or carpals, metapodials, and phalanges (Figures 14.2A, 14.6A). These characters are indicated superficially by the presence of thenar and hypothenar pads as well as interdigital and digital pads (Brown and Yalden 1973). Whether a foot is plantigrade or not is also indicated by the metapodial/propodial ratio, as M. E. Taylor (1971, 1974, 1976) showed in the case of African viverrids. Davis (1964) indicated that the giant panda, brown bear (*Ursus arctos*), and black bear also have low humero-metacarpal indices indicative of a plantigrade manus.

Cursorial Carnivores

Cursorial carnivores forage over large areas, move to new sources of food when immediate supplies fail, and take advantage of seasonal variability of climate and food (Gittleman 1985; Hildebrand 1985b).

There are three types of cursors: (1) those that are capable of prolonged trotting but that do not normally move very fast (e.g. hyenas and many canids), (2) those that run fast and depend upon both speed and stamina in overhauling ungulate prey (e.g., African hunting dog, *Lycaon pictus;* and gray wolf), (3) those that are sprinters, capable of very rapid acceleration, but that maintain a high speed for only short distances (e.g., cheetah).

Carnivore structure is the result of a compromise between the ability to catch prey and the ability to kill it. Felids use retractile claws to pull down prey, but because of this the structure of the foot cannot be perfectly digitigrade. The exception is the cheetah, which has lost the retractability of its claws and knocks over its prey (Gonyea 1978). Canids depend on their teeth for biting prey and then bringing it down; consequently, their feet do not perform the same dual function as those of felids. Genets (*Genetta* spp.) have retractile claws but use a neck bite for killing vertebrate prey. Knowledge of such behaviors is important to the investigation of morphological constraints on locomotion.

Locomotion

Cursorial carnivores use the trot, pace, run, half-bound, and gallop to move quickly or to cover large distances. The gait used depends on the species and is related to the animal's size and morphology. Energetic considerations make gait transitions profitable, and animals change gaits at speeds proportional to the square roots of their leg lengths (Alexander 1984). Smith and Savage (1956) pointed out that for large mammals, lifting the center of gravity up and down is an extravagant use of energy, whereas swinging the legs back and forth is more economical. These factors are important in understanding limb shape and stance.

In a trot, alternate sets of fore and hind feet on opposite sides of the body swing in unison, so that the body is suspended between alternating diagonal limbs (Figure 14.3*A*) (Hildebrand 1961; M. E. Taylor 1970; Brown and Yalden 1973; Dagg 1976). Many of the mustelids with long arched backs and relatively short limbs do not use this gait but go from a walk to the bound (Dagg 1973; Gambaryan 1974; Williams 1983).

In the pace the fore and hind feet on the same side of the body swing more or less together, which avoids interference between fore and hind feet (Brown and Yalden 1973; Hildebrand 1982). This gait is found in larger canids, felids, and ursids (Dagg 1973).

Running gaits include periods of suspension when all feet are off the ground and the limbs swing as diagonal pairs, with usually two floating phases in any stride (Figure 14.3*B*). The distance the body moves forward while it is unsupported increases the stride length. Some species such as the bear may run, but it has a very short floating phase (Gambaryan 1974).

The half-bound (Figures 14.3*C*, 14.4), a form of galloping, is characterized by one forefoot's touching the ground first, followed by the other, whereas the hind feet touch down together (M. E. Taylor 1970). Many of the smaller carnivores, particularly the mustelids, use this method of progression (Tarasoff et al. 1972; Gambaryan 1974; Williams 1983). The relative duration of the floating phase is important for cursorial carnivores in attaining high speeds.

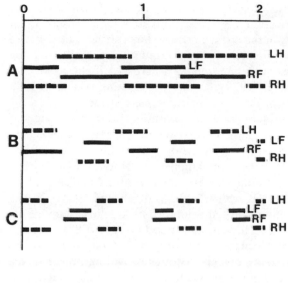

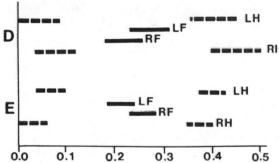

Figure 14.3. Gait patterns showing footfall sequences of (A) white-tailed mongoose (*Ichneumia albicauda*) walking; (B) banded mongoose (*Mungos mungo*) running; (C) banded mongoose galloping using the half-bound; (D) cheetah (*Acinonyx jubatus*) using the rotary gallop with left lead; (E) cheetah using the rotary gallop with right lead. The period that each foot is on the ground is shown by the length of the respective line. The letters L, R, H, and F mean left, right, hindfoot, and forefoot, respectively. Time is in seconds. (A–C, after M. E. Taylor 1970; D–E modified from Hildebrand 1961, all courtesy *Journal of Mammalogy*.)

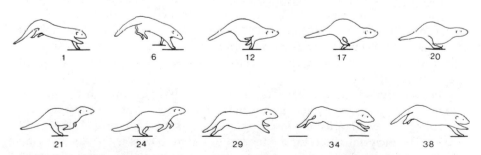

Figure 14.4. Banded mongoose (*Mungos mungo*) using the half-bound, a form of galloping; the film speed was 64 frames per s and the frame numbers are indicated. (From M. E. Taylor 1970, courtesy *Journal of Mammalogy*.)

The cheetah is able to gallop as fast as it does by means of two such long phases, where the hind step constitutes 9%, the front step 16%, crossed flight 24%, and extended flight 51% of the distance covered in a stride. The distance covered in such a stride is approximately 7 m. The hind step of a galloping dog constitutes 10% of the stride, the front step 17%, crossed flight 18%, and extended flight 55% (Gambaryan 1974), so that the floating phases in both animals constitute about 75% of the time for a complete stride.

For most felids the ability to accelerate to maximum speed in the shortest time seems to be essential for hunting success. When the cheetah moves quickly, the half-bound is abandoned and a rotary gallop is used (Hildebrand 1959). This form of gallop, which allows the animal maneuverability, is distinguished from the half-bound in that the hind feet are not placed on the ground together, but one follows the other (Figure 14.3D, E) (Hildebrand 1985b). When rapid acceleration and not maximum speed is important to felids, many of the cursorial adaptations discussed by Howell (1944) and Gray (1968) to achieve a high maximum steady-state velocity are inappropriate (Gonyea 1978). To accelerate a mass to a high velocity quickly requires providing power to the limbs and having muscle origins farther out on lever arms. If high velocity can be achieved slowly, then muscle insertions can be located closer to the fulcra. Since felids do not maintain sustained chases but need to accelerate rapidly, the adaptations required are different from those of cursors such as wolves and African hunting dogs.

Morphology

Cursors have longer legs than do ambulatory species, and their length is associated with the distal elements, for example, the high propodial-epipodial and propodial-metapodial indices (Howell 1944; Hildebrand 1954, 1985b; M. E. Taylor 1974; Gonyea 1978; Van Valkenburgh 1985).

The scapula of cursors is relatively flat and rectangular, and contributes to a distinct increase in stride length (Smith and Savage 1956; Ewer 1973; M. E. Taylor 1974; Hildebrand 1982). In cursorial carnivores the clavicle is often reduced to a vestige (Hildebrand 1982) because its presence would inhibit maximum fore limb acceleration and would serve to deflect the shoulder laterally during the propulsive stroke, with consequent loss of forward thrust (Jenkins 1974).

The humerus of cursors appears to be relatively light, though the greater tuberosity is generally larger than in noncursors and protrudes proximally, providing a large area for the insertion of the m. supraspinatus (M. E. Taylor 1974). The bicipital notch for the tendon of the m. biceps is not distinct, and the bicipital tuberosity on the radius is located proximally, indicating weak flexor ability of the forearm (M. E. Taylor 1974). Small differences in position of the insertion of muscles have a major effect on the out forces that can be

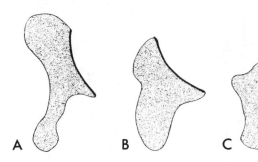

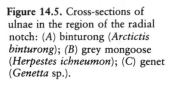

Figure 14.5. Cross-sections of ulnae in the region of the radial notch: (*A*) binturong (*Arctictis binturong*); (*B*) grey mongoose (*Herpestes ichneumon*); (*C*) genet (*Genetta* sp.).

produced. In cursors the insertions are nearer the joint, increasing the amplitude of arc of movement but decreasing its force (Hildebrand 1982). Movement between the humerus and radius and ulna is limited to the anteroposterior plane by the shape of the trochlea and capitulum (M. E. Taylor 1974) and by the shape of the olecranon fossa (Gonyea 1978). In the cheetah and in several felids inhabiting grasslands, the olecranal inclination was found to be low and to be associated with the pendulumlike movement of the lower limb (Gonyea 1978). Here the adaptive trend is toward restriction of elbow movement to flexion and extension in a sagittal plane. Elbow stability is maximized through the congruency of a deep trochlea with the corresponding surfaces of the radius and ulna (Jenkins 1973). The radial head lies in a deep radial notch of the ulna, which increases stability while running, although in canids and hyaenids the radial notch is positioned more anteriorly (Gonyea 1978). In all felid species examined by Gonyea (1978) the radial notch was found to face laterally, but among the Viverridae there is considerable variation, the cursorial forms facing anteriorly (Figure 14.5*B*) and the arboreal forms facing laterally (Figure 14.5*A, C*) (M. E. Taylor 1974). There is a tendency in cursors for the radius to become more anteriorly placed relative to the ulna as it becomes the major load-bearing bone (Hildebrand 1982).

Metapodials are closely adjoined in at least their proximal half and are roughly cylindrical in section (Figure 14.6*B*) (M. E. Taylor 1974). Movement of the carpus is slight and predominantly in the anteroposterior plane, with a large pisiform directed ventrally and associated with flexion of the manus rather than with ulna deviation.

The pelvis is characterized by wide flared ilia, between which is wedged the sacrum. Femoral condyles are set well back from the long axis of the femur, and a deep patellar groove with high marginal ridges is present (M. E. Taylor 1976). Radiographs show a marked concentration of trabecullae in the anterior region of the tibia, reflecting the predominance of tensile forces associated with increased development of crural extensors (M. E. Taylor 1976). The tibial crest is generally well developed. With increasing cursorial ability there is a reduction in the importance of the fibula, and in the cheetah it has become fused to the tibia in midshaft (Hildebrand 1982). Bellies of muscles of the crus

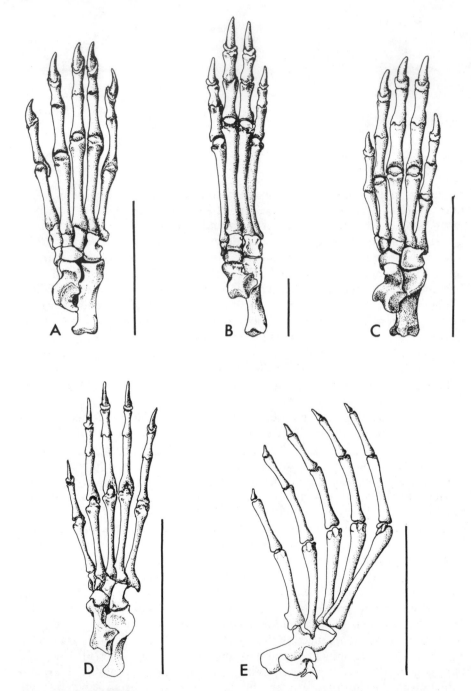

Figure 14.6. Hind feet of (*A*) raccoon (*Procyon lotor*); (*B*) cheetah (*Acinonyx jubatus*); (*C*) badger (*Taxidea taxus*); (*D*) kinkajou (*Potos flavus*); (*E*) sea otter (*Enhydra lutris*). (*E* modified from Howard 1973b). Scale bar = 5 cm.

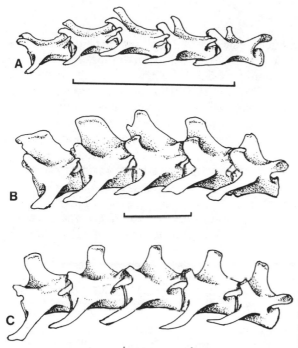

Figure 14.7. Lumbar vertebrae of three cursorial carnivores: (*A*) marten (*Martes americana*); (*B*) gray wolf (*Canis lupus*); (*C*) cheetah (*Acinonyx jubatus*). Scale bar = 5 cm.

are located proximally on the relatively long tibia, to reduce moments of inertia of the leg, and the long-muscle tendons are associated with storage of elastic energy used in the propulsive phase of locomotion (Goslow and Van de Graaff 1982). Adductor and abductor muscles are reduced or oriented such that they have a flexion extension function (Hildebrand 1982). There is also a trend toward reduction of the antebrachial and manual flexor musculature, with consequent diminution of the medial epicondyle of the humerus (Jenkins 1973). The shape of articular surfaces of metatarsal-phalangeal joints limit medio-lateral movement of the phalanges, and the claws are short and non-retractile (Figure 14.6*B*) (M. E. Taylor 1976).

Species that have rigid backbones, such as hyenas, ursids, canids (Figure 14.7*B*), and large mustelids trot or gallop and do not use the bound (Dagg 1973; M. E. Taylor 1976). Those species that use the bound or half-bound flex and extend the back to help increase speed (Figure 14.7*A, C*). The vertebral columns consequently are more flexible and do not have the heavy dorsal spines associated with gallopers (Gambaryan 1974; Hildebrand 1982). The muscles of the back and abdominal wall in bounding carnivores constitute a substantial fraction of the mass of the body and are obviously important in forceful extension or flexion of the back (Alexander and Jayes 1981), contributing considerably to the speed of these cursors (Hildebrand 1982).

Fossorial Carnivores

Hildebrand (1982) uses the term "scratch diggers" for carnivores that use their feet to scratch and move soil rather than their teeth or muzzles. Therefore their fossorial adaptations are mainly associated with the feet.

Fossorial carnivores can be divided into those that excavate often extensive burrow systems without using preexisting holes (e.g., European badger, *Meles meles;* American badger, *Taxidea taxus;* and honey badger, *Mellivora capensis*) and those that do not initiate their own burrow systems but modify preexisting ones (stink badger, *Mydaus javanensis;* Burmese ferret badger, *Melogale personata;* Chinese ferret badger, *M. moschata;* suricate, *Suricata suricatta;* banded, dwarf, and yellow mongooses, *Mungos mungo, Helogale parvula,* and *Cynictis pencillatta,* respectively).

Other carnivores, such as foxes, may dig holes, but they are not generally considered to be fossorial since they spend most of their time above ground.

Locomotion

The locomotion of fossorial carnivores consists of digging activities and methods of progressing above ground (surficial locomotion). Few authors have studied digging by fossorial carnivores (Lampe 1976; Quaife 1978).

Digging by the American badger is composed of two major activities, soil cutting and soil shifting (Quaife 1978). Soil cutting is used to break hard soil, usually at the surface, and involves putting as much force as possible at the tips of the claws. The forefeet are used for digging and may be used alternately. When the soil has been sufficiently loosened, it is shifted back. After a hole is started, the badger may rest its head on the opposite side of the hole to support its forequarters, while using both forefeet to excavate (Perry 1939; Lampe 1976; Quaife 1978). The hind feet are less specialized and are usually used only in helping to move the soil backward (Lampe 1976; Quaife 1978; Long and Killingley 1983). The depth and extent of burrows depends largely on soil conditions. Digging by carnivores such as the suricate has not been studied, though they are capable of excavating extensive burrow systems (Lynch 1980).

The size and shape of the animal affects its above-ground locomotion. Neal (1948) describes the European badger as running with the hind feet overlapping the forefeet slightly. Lampe (1976) describes the American badger as using a slow walk while searching for food. There are no reports of its using fast gaits such as the run or the gallop. Details on the movements of other badgers are lacking. Surficial locomotion of the suricate or banded mongoose is either a walk, a run, or a half-bound (Figures 14.3C, 14.4) (M. E. Taylor 1970; Kingdon 1977).

Morphology

Mammals that spend much of their lives underground exhibit a variety of adaptations for fossorial life. These may include a short tail, small or vestigial external ears, small eyes, and a longish snout (Long and Killingley 1983). The digging apparatus in these carnivores is restricted to the limbs, largely the fore limbs, and the head, teeth, and nose do not play a significant role. Fossorial carnivores produce large out forces at their claws and are heavily muscled, with the insertions of the muscles being as far out on any limb segment as practical. The muscles are particularly bulky, and movements are forceful but not rapid (Hildebrand 1985a).

Since large out forces must be generated by the limbs and body, the scapula and associated shoulder musculature are well developed. The major muscles of the power stroke of the forearm at the shoulder are the mm. pectoralis, latissimus dorsi, serratus anterior, and teres major (Quaife 1978). The modifying effect of these muscles on the shape of the mammalian scapula has been noted by Wolffson (1950), and it is not unusual to see quite a variation in the form of the scapula (M. E. Taylor 1971). There are gradations in the rugosity of the blade in different badgers, and powerful diggers such as the American badger have a large posterior flange for the origin of the m. teres major (Smith and Savage 1956; Quaife 1978; Hildebrand 1982), whereas semi-fossorial species such as the banded mongoose have a more generalized scapula (M. E. Taylor 1971).

The humeri of badgers are relatively massive and characterized by large heads, short diaphyses, and heavy flanges for the insertions of the mm. pectoralis, deltoid, latissimus dorsi, and teres major. They also have a large distal end with large medial epicondyles and smaller lateral epicondyles (Quaife 1978). Great muscular forces and leverages are reflected in the internal buttressing of the bone; in the American badger there is substantial trabecular bone in the diaphysis compared with a carnivore like the dog (Quaife 1978). Large flanged humeral epicondyles provide for the origins of well-developed forearm flexors and extensors. A deep trochlea provides added stability, reducing the possibility of lateral displacement of the forearm.

Out forces produced by the m. triceps group are related to the relative length of the olecranon to the ulnar shaft; the longer the olecranon, the greater the out forces that can be transmitted to the claws (Smith and Savage 1956; Quaife 1978; Van Valkenburgh 1985). The ratio of in-lever to out-lever is about 1 : 6 in nonfossorial carnivores like the dog but drops to 1 : 3.5 in badgers (Quaife 1978; Hildebrand 1982). The cross-sectional shape of the ulna in the American badger is comparable to that of an I-beam, giving considerable strength for a small amount of material, provided that the forces associated with digging are limited to one plane. The radius is short and robust, with a wide distal end articulating with the scapholunar.

The forefoot of the European badger is digitigrade, with about 10° of hyperextension of the carpus (Yalden 1970). The American badger can hyperextend its wrist to approximately 50°, indicating that it is more plantigrade than the European badger (Quaife 1978). The carpus of the European badger is relatively broad and allows only about 20° of ulnar deviation. There is a large pisiform associated with flexion and ulnar deviation, and the metacarpals and phalanges are short, so that large out forces from the forearm are directed to the claws. The American badger usually places the forefeet on the substratum with the brachium slightly medially rotated; the weight is therefore placed more on the lateral aspect of the carpo-metacarpal portion of the manus. This positioning provides protection for the long claws as the feet are moved forward (Quaife 1978).

The terminal phalanges of the semi-fossorial banded and dwarf mongooses are much longer than those of the generalized viverrid (M. E. Taylor 1974, 1976). The falculae grow at a much greater rate in fossorial species than in any other species, and zoo animals that do not have the opportunity to dig soon have very overgrown claws (Ewer 1973). Associated with the lengthening of the claws for digging is the development of the forearm flexor musculature, and M. E. Taylor (1974) comments on the relative width of the medial epicondyle in the banded mongoose, which provides a large area for the origins of the mm. flexor carpi ulnaris, flexor carpi radialis, and flexor profundus digitorum. This is comparable to the adaptation in the badgers but is not so extreme. The trochlea of the banded mongoose is also relatively deep, to provide stability at the elbow. However, the limb proportions of these semi-fossorial species resemble those of the more generalized viverrids (M. E. Taylor 1976).

Arboreal Carnivores

Locomotion

Arboreal carnivores use three primary strategies for obtaining their food. (1) The fast predators move swiftly in trees or on the ground in pursuit of their prey (marten, *Martes americana;* ringtail, *Bassariscus astutus;* and the more arboreal genets such as the servaline genet, *Genetta servalina*). Hildebrand (1982) refers to these carnivores as the jumping and leaping category of climbers. (2) Some species use stealth to approach their prey. Hildebrand (1982) refers to them as the grasping group of arboreal species (e.g., palm civet, *Nandinia binotata;* binturong, *Arctictis binturong;* many of the cats; and the raccoon). (3) Others are slow moving and largely vegetarian, consuming fruits and shoots, like the pandas, bears, and kinkajou (*Potos flavus*). For rapid movements through trees, small size is required (Fleagle and Mittermeier 1980) and reflexes must be fast. However, the fossa (*Cryptoprocta ferox*), which is large (10–20 kg) and normally moves slowly in trees, has been observed to climb a 20-m tree in 4–5 s (Albignac 1970). Stealthy carnivores require greater strength in their limbs to maintain control as they stalk prey, and adaptations in these species involve greater out forces on the fore and hind feet. Finally, herbivorous carnivores do not leap in the canopy, nor do they search for vertebrate prey. They may stay in individual trees for days on end while they feed on particular fruits. Locomotion of arboreal carnivores has been little studied, although information on a number of species is available (Hurrell 1968; M. E. Taylor 1970; Sokolov and Sokolov 1971; Trapp 1972; Albignac 1970; Ewer 1973; Gonyea 1978; Nowak and Paradiso 1983; Jenkins and McClearn 1984; Roberts and Gittleman 1984; Gittleman 1985; Van Valkenburgh 1985).

ARBOREAL WALKING

When the substrate on which an animal moves is horizontal, or nearly so, it can walk without slipping backward. However, as the angle of the substrate increases, the amount of friction required to prevent the animal from slipping must increase (Figure 14.8) (Cartmill 1985). There comes a point when regular walking is impossible (i.e., when downward forces exceed friction), and then the animal must use a grasping form of locomotion to hang on. Some carnivores like palm civets and red pandas (*Ailurus fulgens*) (Figure 14.10C–F) use claws to cling to the bark (Hurrell 1968), and different principles are involved (Cartmill 1985).

The main distinction between arboreal and terrestrial walking is the amount of contact surface area between the feet and substrate. Friction is important, but other criteria, such as diameter of the support and whether the animal can get its feet opposing one another, are also important (Hildebrand 1982; Cartmill 1985). Larger animals ensure that their feet make adequate contact with the substrate by supinating both fore and hind feet. The feet are also partially supinated when protracted so that any tendency to slip can be corrected rapidly (M. E. Taylor 1970). In horizontal walking the body is kept close to the branch, but when the support is relatively vertical, the front part of the body is usually kept close to it with the hind part farther away (Cartmill 1985). The tail is generally used as a balancing organ (e.g., the red panda and the martens), but in two species (the binturong and kinkajou) it is prehensile.

VERTICAL LOOPING

This method of locomotion, described by M. E. Taylor (1970) for the palm civet, is a form of controlled progression up and down vertical supports in which the animal moves head first (Figure 14.9). It holds on with the forefeet; the hind feet are brought forward, the back flexing. Then the hind feet grasp the support, the forelegs are removed, and the body and legs are extended, the

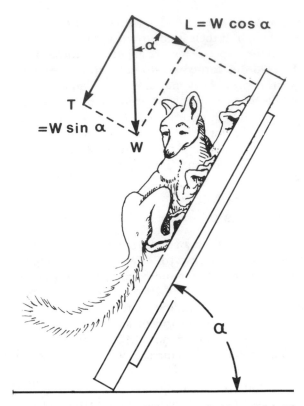

Figure 14.8. The relationship between slope and friction in climbing animals. The animal's weight (W) can be analyzed into forces normal to its support (load L) and tangential to its support (T) The animal will slide down the incline when the tangential force exceeds the force of static friction— that is when $W \sin \alpha > \mu_s W \cos \alpha$ (where μ = coefficient of static friction). (Drawing by M. Cartmill from Cartmill 1979, courtesy *American Journal of Physical Anthropology*, Alan R. Liss, Inc., publisher.)

forefeet then gripping at the new location. The speed of this locomotion varies. Trapp (1972) reports the use of vertical looping by the ringtail, and Wemmer and Watling (1986) observed its use by the Sulawesi palm civet (*Macrogalidia musschenbroekii*). Climbers in the jumping and leaping category are not able to produce the control to perform this locomotion, and M. E. Taylor (1970) reports the genet climbing upward, but having to jump down.

JUMPING

Jumping is an integral part of arboreal locomotion, though not for bears and the giant panda. Most jumping involves the rapid extension of the hind legs (e.g., *Genetta* spp.) (M. E. Taylor 1970). Ringtails move fast and are capable of making a series of jumps, ricocheting off intermediate structures (Trapp 1972). Both Hurrell (1968) and Sokolov and Sokolov (1971) describe the

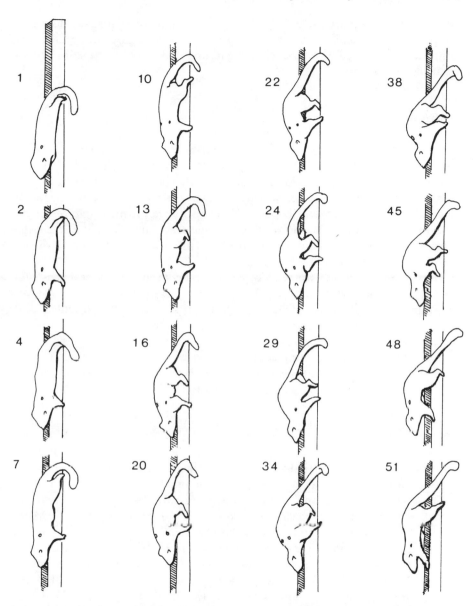

Figure 14.9. Palm civet (*Nandinia binotata*) climbing down a vertical support, using a vertical looping mode of progression. Each stride represents about 50 cm. The tail is foreshortened because the end is bent toward the camera. The film speed was 64 fps and the frame numbers are indicated. (Modified from M. E. Taylor 1970.)

jumping ability of martens, giving jump distances of up to 10 m; such distances are probably unusual, 5 m being a more reasonable figure. In species that jump, the tail is used as a balancing organ and is believed to function some-what like a parachute.

Morphology

Most arboreal species have relatively slim bones that are not rugose with heavy tuberosities (Sokolov and Sokolov 1971; M. E. Taylor 1974, 1976; Leach and Dagg 1976; Leach and deKleer 1978). The fore limb may be in part stabilized by a small clavicle, which is present in felids and some arboreal viverrids. The scapula typically has an obvious flange for the origin of the m. teres major, which is an important flexor of the brachium. There is also a large acromion that overhangs the glenoid fossa and provides for the origin of the m. acromiodeltoid, an important abductor of the limb.

There is a large groove for the biceps tendon between the humeral head and greater tuberosity, which indicates strong flexor ability of the arm. However, the humerus is relatively cylindrical with a relatively flat trochlea and capitulum, allowing the ulna to rock on the humerus in supination (M. E. Taylor 1974). The olecranon is not particularly long and is angled forward so that the limb cannot be completely extended as in cursorial species (M. E. Taylor 1971). However, the lateral epicondyle is well developed, as in fossorial species, so that there is a large area for the origin of the flexors of the manus (M. E. Taylor 1971, 1974; Gilbert 1973). The degree of supination in the forearm is regulated by the shape of the radial notch, and in arboreal species it is laterally oriented (Figure 14.5A, C) (M. E. Taylor 1974; Gonyea 1978). For controlled climbing, the bicipital tuberosity of the radius in aboreal species is both prominent and situated more distally than in ambulatory or cursorial species, allowing powerful flexion of the fore limb.

The feet of jumping and leaping species (Figure 14.10A, B) are not so modified as in scansorial species, whose feet are adapted for gripping (M. E. Taylor 1971, 1974, 1976; Cartmill 1985; Van Valkenburgh 1985, 1986). In scansorial species considerable movement occurs in the wrist, and there is a large radial sesamoid and pisiform associated with radial and ulnar deviation (M. E. Taylor 1971, 1974). The metacarpals of such species are relatively short, but the proximal phalanges are long (M. E. Taylor 1971, 1974, 1976; Van Valkenburgh 1985, 1986). The first metacarpal is distinctly waisted, and its proximal articular surface is such that it can be adducted to a large degree. Foot pads are large and fleshy as in the palm civet (Figure 14.10C, D), or completely furred as in the red panda (Figure 14.10E, F), allowing a surface for grasping branches of various sizes (Cartmill 1974, 1985; M. E. Taylor 1974, 1976). The claws of many species are retractile and are of use in climbing as well as in catching prey.

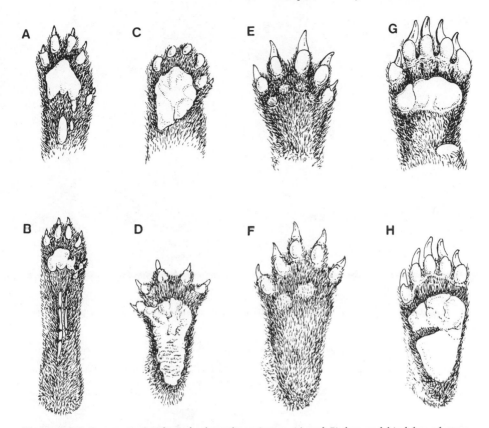

Figure 14.10. Representative feet of arboreal carnivores. (*A* and *B*) fore and hind feet of genet (*Genetta tigrina*); (*C* and *D*) fore and hind feet of palm civet (*Nandinia binotata*); (*E* and *F*) fore and hind feet of red panda (*Ailurus fulgens*); (*G* and *H*) fore and hind feet of black bear (*Ursus americanus*). (*A–D* redrawn from M. E. Taylor 1971; *G* and *H* modified from Davis 1964.)

The adaptations of the hind limb are, for the most part, less obvious than in the fore limb (Sokolov and Sokolov 1971; M. E. Taylor 1976). However, to be an effective slow climber, an animal must be able to hold on to a branch with the hind feet alone. Consequently, it must be able to adduct the femur effectively and transmit forces through the feet to the branch. There must be sufficient mobility in the joints to allow supination of the foot so that appropriate forces can be applied to a variety of branch diameters. An alternative strategy is to reverse the foot so that it is pointing backward and the animal is able to hang from its feet when descending vertically head first. Jenkins and McClearn (1984) and Trapp (1972) have shown how the ringtail, kinkajou, and margay (*Felis wiedii*) accomplish this by rotating the foot in the region of the talocrural joint (Figure 14.11). It is then possible for the animal to use its extensors as flexors for descent and exert an appropriate amount of control.

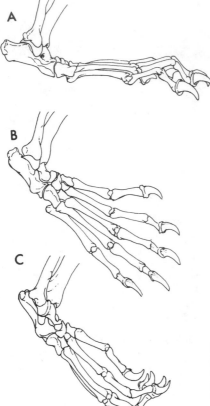

Figure 14.11. Hind foot reversal in the kinkajou (*Potos flavus*). A right foot is shown in lateral view in postures typical of (*A*) a stance on a horizontal surface, (*B*) foot inversion, and (*C*) foot reversal. (From Jenkins and McClearn 1984, courtesy *Journal of Morphology*.)

Aquatic Carnivores

Carnivores show a range of aquatic adaptations, from species like mink (*Mustela vison*), which have relatively few structural adaptations yet spend much of their time in the water, to the sea otter, which rarely comes on land and has many adaptations for an aquatic existence. Many carnivores swim yet do not treat water as a three-dimensional habitat. The otters represent the most well-known aquatic species, but there are others such as the otter civet (*Cynogale bennetti*), polar bear, mink, and the marsh mongoose (*Atilax paludinosus*) that are also accomplished swimmers.

Locomotion

Swimming may occur at or beneath the surface of the water, and the methods are different. All aquatic carnivores spend some time on land, and their terrestrial locomotion is affected by the degree of their aquatic adaptations.

SWIMMING

Dagg and Windsor (1972) examined swimming in a number of non-specialized swimmers and found that they swam using terrestrial gait sequences. They floated either with their backs out of the water (mink and the domestic cat) or with their backs submerged and only their heads tilted up (polecat, *Mustela putorius;* and skunk). The amount of submergence depends in part on fat content and on fur condition, both of which may vary with the time of the year.

The mink uses a power-and-recovery method of swimming underwater, involving alternate use of all four limbs with either diagonally opposite legs or ipsilateral legs in the power stroke. Neither the two fore limbs nor the two hind are used in a simultaneous power stroke; each limb is moved through an arc of approximately 130°. However, when swimming at the surface, mink use the forefeet, occasionally using a hind foot for turning or diving (Dunstone 1979). The marsh mongoose is a capable swimmer and diver, often remaining submerged for at least 15 s, and it normally swims with only its head and a small part of its back exposed (M. E. Taylor 1970; Nowak and Paradiso 1983). Little is known of the swimming ability of the otter civet. In the case of the polar bear, unilateral paddling of the fore limbs has been reported, with the hind limbs trailing passively behind (DeMaster and Stirling 1981).

The otters, which are for the most part proficient swimmers, include a number of species, of which the most studied are the Canadian river otter (*Lutra canadensis*), the European otter (*L. lutra*), and the sea otter (Harris 1968; Tarasoff et al. 1972; Kenyon 1975). The sea otter spends most of its time in water, rarely coming out on land, and is distinctly awkward in terrestrial locomotion. River otters are good swimmers but can also travel long distances over land. The swimming of the Canadian river otter involves the use of forefeet, hind feet, and tail or any combination of these (Tarasoff 1972). This otter has two methods of aquatic locomotion, a thrust-recovery movement of the limbs, and a carangiform movement of the tail in a vertical direction (Tarasoff et al. 1972). The giant otter (*Pteronura brasiliensis*) uses its feet for slow swimming but for swimming rapidly depends on undulations of the tail, using its feet for steering (Nowak and Paradiso 1983). The Canadian river otter (Figure 14.12A) moves the hind feet in a 120° arc, moving the leg parallel to the body and maximally expanding the foot with its plantar surface pushing the water caudad. The foot does not rise above the level of the back or break the surface (Tarasoff et al. 1972). In the recovery stroke the otter moves the foot forward in a flexed position. The fore limbs may also be involved and make movements like those of the hind, though no obvious pattern of use emerges. During rapid swimming this otter uses a carangiform mode of locomotion in which most of the movement originates from the lumbar region. It keeps the hind feet parallel with the long axis of its body, the plantar surfaces facing upward and acting together to help produce the forward thrust. It keeps the anterior part of the body, with the forelimbs held close to the chest,

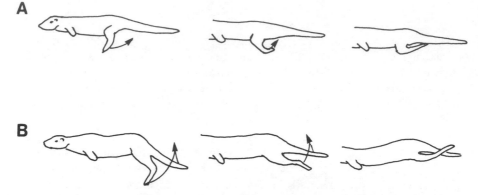

Figure 14.12. Sequence of body, limb, and tail movements of (*A*) the otter (*Lutra canadensis*) and (*B*) sea otter (*Enhydra lutris*) during aquatic locomotion. (From Tarasoff et al. 1972, courtesy *Canadian Journal of Zoology*.)

relatively rigid (Tarasoff et al. 1972). During this movement the sacrum moves through an arc of 90°, the frequency of the strokes varies, and there may be gliding in between power strokes.

The sea otter uses three forms of aquatic locomotion, (1) craniocaudal sweeps of the pelvic limbs, often involving bending of the lumbar, sacral, and caudal regions (Figure 14.12*B*); (2) vertical thrust-and-recovery movements of the pelvic limbs while it is on its back at the surface of the water; and (3) horizontal sweeps of the dorsoventrally flattened tail while at the surface of the water, with twisting of the distal parts (Tarasoff et al. 1972).

Terrestrial Locomotion

Terrestrial locomotion of aquatic carnivores involves walking, running, and bounding. Those species that have few anatomical adaptations for swimming have a relatively typical terrestrial walk, either using a diagonal sequence (Figure 14.13) or a pace. The polar bear uses a terrestrial walk (Dagg 1979) and has been studied at various speeds (Hurst et al. 1982), including running up to 25 mph (Peterson 1966). The mink holds the head low and the back more or less horizontal and uses a diagonal sequence walk in both the slow and fast walks (Dunstone 1979; Williams 1983). River otters have an arched back during terrestrial locomotion, and the sea otter has a strongly arched back, but because of its large hind feet (Figure 14.6*E*) and relatively short hind legs, a normal walk is barely possible (Murie 1940; Kirkpatrick et al. 1955; Kenyon 1975). Unlike the sea otter, the river otter is capable of a run in which the gait sequence is similar to a walk, but faster. Except for the polar bear, aquatic carnivores use a half-bound for rapid locomotion (Tarasoff et al. 1972; Williams 1983), although in the case of the sea otter only the lighter juveniles are able to use this form of locomotion (Kirkpatrick et al. 1955; Kenyon, 1975). Williams (1983) found that there was a distinct break in the energetic

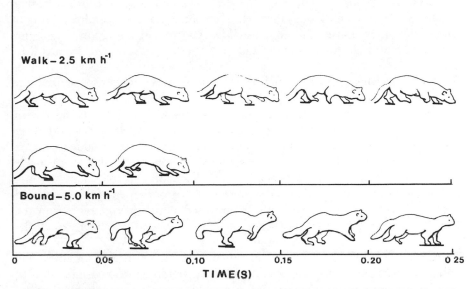

Figure 14.13. Walk and half-bound gaits of the mink (*Mustela vison*). One walking stride at 2.5 km h^{-1} took 0.35 s to complete. The bound cycle at 5.0 km h^{-1} was completed in 0.25 s. Note the use of spinal flexion during the bound, which is absent when walking. (From Williams 1983, courtesy *Journal of Experimental Biology*.)

cost of locomotion between a fast walk at 3.9 km/h and the half-bound at 5.0 km/h (Figure 14.13). Incremental transport costs were 36% lower while bounding than for walking at comparable speeds. Mink normally went from a fast walk directly into the bound, with no intermediate gait such as the pace or trot. The river otter, when moving rapidly across country with snow on the ground, may incorporate sliding into its run or bound, lifting its feet off to the side and sliding on its belly (Severinghaus and Tanck 1947; Liers 1951; Peterson 1966; Harris 1968).

Morphology

Two important criteria are associated with swimming: producing the forward thrust by means of the body and/or limbs, and minimizing turbulence and drag while moving through a dense fluid medium. The morphological features of aquatic carnivores reflect these criteria.

The amount of energy expended in swimming is related to whether the animal must swim to stay afloat. Some animals have a low specific gravity due to the presence of fat or air trapped in the fur and therefore do not have to swim to stay afloat (Dagg and Windsor 1972). Flotation in the sea is easier than in fresh water.

For species such as mink there are virtually no special locomotor adaptations (Howell 1930), though Dunstone (1979) comments on the fur's retaining a layer of air and aiding buoyancy. He also notes that the cylindrical body is similar to other nonspecialized mustelids. The marsh mongoose shows no obvious adaptations for swimming and the fur becomes completely wet (M. E. Taylor 1970). The polar bear, which is also a noted swimmer, has large oarlike forepaws (DeMaster and Stirling 1981) but is otherwise unspecialized for swimming.

Both the sea otter and the river otter have body shapes that comply with hydrodynamic criteria for aquatic propulsion (Lighthill 1969; Tarasoff et al. 1972), having a fusiform shape and hind limbs modified to produce a lunate border for the carangiform mode of propulsion. The tail in the river otter is slightly dorsoventrally flattened near its base (Tarasoff et al. 1972) and is more important in producing forward thrust than is that of the sea otter. Associated with this are chevron bones in the river otter, which are present from the fourth caudal vertebra posteriorly and are associated with a greater vascularization of the large tail; they are not present in the sea otter (W. P. Taylor 1914). The cervical vertebrae of aquatic species are relatively shorter (Sokolov and Sokolov 1970), and Bisaillon et al. (1976) provide percentage means with standard deviations of the cervical segments compared with the thoracolumbar segments. For the sea otter the cervical segment is 18.9% ± 1.26; for the river otter, 27.5% ± 1.98; for the mink, 31.0% ± 1.32; and for the dog, 40.6% ± 2.14. The shortness of the cervical region is associated with streamlining of the body and the development of the thoracolumbar and caudal regions for propulsion.

The fore limbs in otters are generally reduced in length and surface area (Ondrias 1960, 1961; Tarasoff et al. 1972) and are highly mobile on the chest wall, since there is no clavicle (Howard 1973a). In the sea otter the digits of the forefoot are not individualized; and if objects are to be grasped, they must be held between the hand and distal forearm. The transverse creases of the forefoot do not allow much flexion of the palm (Howard 1973a). The antebrachium and forefoot are short, particularly when compared with the hind foot (Figure 14.6E) (Howard 1973b).

The pelvis of the sea otter is comparatively heavy and more nearly parallel to the vertebral column than that of the river otter (Tarasoff 1972), and the ilia are flared anteriorly. There is no ligamentum teres in the sea otter, which allows greater movement of the femur (W. P. Taylor 1914). There is a shortening of the propodial and epipodial elements and an elongation of the metapodials (Smith and Savage 1956; Ondrias 1960). With this change in proportions, the muscle insertions on the crus have moved distally; the more aquatic species produce the thrust with the foot, rather than moving the whole limb. The proportion of the hind limb protruding from the body contours is reduced in otters. Both the metatarsals and phalanges are elongated, and a generous web of skin exists between the digits of the hind feet, so that the foot becomes twice as wide when the digits are spread. The length of the individual digits of the river otter are IV = III > V = II > I; the sea otter is quite different, with the fifth digit being longest, the digit lengths being V > IV > III > II > I (Tarasoff et al. 1972).

Conclusions and Discussion

Although only a few of the many carnivores have been studied with regard to both their locomotion and morphology, five distinct locomotor categories are identified within the order.

Ambulatory carnivores are characterized by nonspecialized limb structure in which the limbs do not move in a sagittal plane. Walking is the preferred gait, and most species have plantigrade feet. Gaits used often involve lateral sequences such as the pace, and the limbs are placed beneath the body. Various morphological features such as the shape of the thorax and hip permit abduction of the limbs.

Cursorial carnivores, perhaps the most typical of the order, are specialized for different kinds of pursuit. There are species capable of prolonged trotting, usually at moderate speeds, those that are capable of fast pursuit over long distances, and those that are capable of short bursts of speed. The locomotor gaits used also depend upon the animal's size, smaller species using the bound or half-bound for rapid locomotion, and larger species using the trot and gallop. Cursorial carnivores have a digitigrade stance and are characterized by long limbs, in which the metapodials are long relative to the propodials. Morphological features restrict movement to the anteroposterior plane and reduce inertia by having muscles located proximally. The insertion of limb muscles close to the joints assists with rapid limb movement.

Fossorial carnivores are scratch diggers, their digging largely accomplished by means of their forefeet. Morphological specializations for digging enhance their ability to dig burrows but limit their ability to move on the surface. Badgers, which are specialized for digging, do not use rapid gaits, whereas some of the semi-fossorial mongooses use a bounding gait to move rapidly above ground. The morphological features of fossorial species involve the maximization of forces to the claws, which is attained by the shortening of the limbs, particularly the distal elements and the insertion of muscles relatively distally on the limb bones. Movements are restricted by deep articular surfaces to prevent dislocation, and long bones have extensive trabecullae in their diaphyses, reflecting the strong forces produced by flexor musculature on the diaphyses.

Arboreal carnivores use three different strategies for obtaining food. Jumping and leaping species pursue their prey rapidly, moving from the ground into trees and utilizing much of the canopy. Grasping species move much slower and in a controlled fashion, using stealth to catch their prey. The third category consists of large and mainly herbivorous species, which climb slowly. Jumping and leaping species are relatively small; they have plantigrade forefeet and digitigrade hind feet and use their tail for balancing during leaps. Grasping species have plantigrade fore and hind feet with relatively fleshy pads, and their tails may be used for balancing; in two species they are prehensile.

From the mink to the sea otter, aquatic carnivores show a range of morphological adaptations. The extent of aquatic specialization affects terrestrial locomotion, and specialized swimmers like the sea otter are ungainly on land. The feet are provided with webs in most species, and the sea otter departs from the basic mammalian digit proportions in having the fifth digit of the hind foot the longest and the first digit the shortest. The more aquatic the carnivore, the less the limb protrudes from the body contour.

Information concerning the behavior of carnivores has greatly increased with the use of cinephotography, cineradiography, infrared telescopes, radiotelemetry, and the like, and the anatomy of many of the commoner species is well known. However, there are many species about which little is known and others that are difficult to study in their natural habitat. Many questions concerning the morphology and behavior of carnivores remain to be answered.

Acknowledgments

I thank all the many people who have contributed to this chapter. In particular I thank C. S. Churcher, J. L. Gittleman, and A. P. Russell, who have read various drafts of the manuscript and added much to the content. Any errors are, however, mine, and the particular biases reflect my judgment rather than that of the reviewers. I also thank the staff of the Mammalogy Department at the Royal Ontario Museum for their support, in particular Sophie Poray-Swinarski for drawing many of the figures. And last, I thank my wife, Ingrid Taylor, for her encouragement, criticism, and support.

References

Albignac, R. 1970. Notes ethologiques sur quelques carnivores malagaches: le *Cryptoprocta ferox* (Bennett). *Terre Vie* 24:395–402.
Alexander, R. McN. 1984. Walking and running. *Amer. Sci.* 72:348–354.
Alexander, R. McN., and Jayes, A. S. 1981. Estimates of the bending moments exerted by the lumbar and abdominal muscles of some mammals. *J. Zool. (Lond.)* 194:291–303.
Bisaillon, A., Piérard, J., and Larivière, N. 1976. Le segment cervical des carnivores (Mammalia: Carnivora) adaptés à la vie aquatique. *Canadian J. Zool.* 54:431–436.
Brown, J. C. and Yalden, D. W. 1973. The description of mammals—2: Limbs and locomotion of terrestrial mammals. *Mamm. Rev.* 3:107–134.
Cartmill, M. 1974. Pads and claws in arboreal locomotion. In F. A. Jenkins, Jr., ed. *Primate Locomotion*, pp. 45–83. New York: Academic Press.

Cartmill, M. 1979. The volar skin of primates: Its frictional characteristics and their functional significance. *Amer. J. Phys. Anthropol.* 50:497–510.

Cartmill, M. 1985. Climbing. In: M. Hildebrand, D. M. Bramble, K. F. Liem & D. B. Wake, eds. *Functional Vertebrate Morphology,* pp. 73–88. Cambridge: Harvard Univ. Press.

Dagg, A. I. 1973. Gaits in mammals. *Mamm. Rev.* 3:135–154.

Dagg, A. I. 1976. *Running, Walking and Jumping: The Science of Locomotion.* London: Taylor and Francis.

Dagg, A. I. 1979. The walk of the large quadrupedal mammals. *Candian J. Zool.* 57:1157–1163.

Dagg, A. I., and Windsor, D. E. 1972. Swimming in northern terrestrial mammals. *Canadian J. Zool.* 50:117–130.

Davis, D. D. 1964. The giant panda: A morphological study of evolutionary mechanisms. *Field. Zool. Mem.* 31:1–339.

DeMaster, D. P., and Stirling, I. 1981. *Ursus maritimus.* Mammalian Species no. 145. Lawrence, Kans.: American Society of Mammalogists.

Dunstone, N. 1979. Swimming and diving behavior of the mink. *Carnivore* 2:56–61.

Ewer, R. F. 1973. *The Carnivores.* Ithaca, N.Y.: Cornell Univ. Press.

Fleagle, J. G., and Mittermeier, R. A. 1980. Locomotor behavior, body size and comparative ecology of seven Surinam monkeys. *Amer. J. Phys. Anthrop.* 52:301–314.

Gambaryan, P. P. 1974. *How Mammals Run: Anatomical Adaptations.* New York: John Wiley & Sons.

Gilbert, B. M. 1973. *Mammalian Osteo-Archaeology,* Columbia: Missouri Archaeological Society.

Gittleman. J. L. 1985. Carnivore body size: Ecological and taxonomic correlates. *Oecologica* 67:540–554.

Gonyea, W. J. 1978. Functional implications of felid forelimb anatomy. *Acta Anatomica* 102:111–121.

Goslow, G. E., and Van de Graaff, K. 1982. Hindlimb joint angle changes and action of the primary ankle extensor muscles during posture and locomotion in the striped skunk (*Mephitis mephitis*). *J. Zool. (Lond.)* 1982:405–419.

Gray, J. 1968. *Animal Locomotion.* New York: W. W. Norton.

Gregory, W. K. 1912. Notes on the principles of quadrupedal locomotion and on the mechanism of the limbs in hoofed animals. *Ann. New York Acad. Sci.* 22:267–294.

Harris, C. J. 1968. *Otters: A Study of the Recent Lutrinae.* London: Wiedenfeld and Nicolson.

Hildebrand, M. 1954. Comparative morphology of the body skeleton in recent Canidae. *Univ. California Publ. Zool.* 52:399–470.

Hildebrand, M. 1959. Motions of the running cheetah and horse. *J. Mamm.* 40:481–495.

Hildebrand, M. 1961. Further studies on locomotion of the cheetah. *J. Mamm.* 42:84–91.

Hildebrand, M. 1976. Analysis of tetrapod gaits: General considerations and symmetrical gaits. In: R. M. Herman, S. Grillner, P. S. G. Stein, & D. G. Stuart, eds. *Neural Control of Locomotion,* pp. 203–236. New York: Plenum.

Hildebrand, M. 1982. *Analysis of Vertebrate Structure.* New York: John Wiley & Sons.

Hildebrand, M. 1985a. Digging of quadrupeds. In: M. Hildebrand, D. M. Bramble, K. F. Liem & D. B. Wake, eds. *Functional Vertebrate Morphology,* pp. 89–109. Cambridge: Harvard Univ. Press.

Hildebrand, M. 1985b. Walking and running. In: M. Hildebrand, D. M. Bramble, K. F. Liem & D. B. Wake, eds. *Functional Vertebrate Morphology,* pp. 38–57. Cambridge: Harvard Univ. Press.

408   *Mark E. Taylor*

Howard, L. D. 1973a. Muscular anatomy of the forelimb of the sea otter (*Enhydra lutris*). *Proc. California Acad. Sci.* 39:411–500.
Howard, L. D. 1973b. Muscular anatomy of the hind limb of the otter *Enhydra lutris*. *Proc. California Acad. Sci.* 40:335–416.
Howell, A. B. 1930. *Aquatic Mammals: Their Adaptations to Life in the Water.* Springfield, Ill.: Charles C Thomas.
Howell, A. B. 1944. *Speed in Animals.* New York: Haffner.
Hurrell, H. G. 1968. *Pine martens.* Forestry Commission: Forest Record No. 64. London: Her Majesty's Stationery Office. 23 pp.
Hurst, R. J., Leonard, M. L., Beckerton, P., and Oritsland, N. A. 1982. Polar bear locomotion: Body temperature and energetic cost. *Canadian J. Zool.* 60:222–228.
Jenkins, F. A. 1971. Limb posture and locomotion in the Virginia opossum (*Didelphis marsupialis*) and in other non-cursorial mammals. *J. Zool. (Lond.)* 165:303–315.
Jenkins, F. A. 1973. The functional anatomy and evolution of the mammalian humero-ulnar articulation. *Amer. J. Anat.* 137:281–298.
Jenkins, F. A. 1974. The movement of the shoulder in claviculate and aclaviculate mammals. *J. Morphol.* 144:71–84.
Jenkins, F. A., and Camazine, S. M. 1977. Hip structure and locomotion in ambulatory and cursorial carnivores. *J. Zool. (Lond.)* 181:351–370.
Jenkins, F. A., and McClearn, D. 1984. Mechanisms of hind foot reversal in climbing mammals. *J. Morphol.* 182:197–219.
Kenyon, K. W. 1975. *The Sea Otter in the Eastern Pacific Ocean.* New York: Dover.
Kingdon, J. 1977. *East African Mammals: An Atlas of Evolution in Africa,* vol. 3A: *Carnivores.* New York: Academic Press.
Kirkpatrick, C. M., Stullken, D. E., and Jones, R. D. 1955. Notes on captive sea otters. *Arctic* 8:46–59.
Lampe, R. S. 1976. Aspects of the predatory strategy of the North American badger, *Taxidea taxus.* Ph.D. dissert., Univ. Minnesota Minneapolis. 103 pp.
Leach, D., and Dagg, A. I. 1976. The morphology of the femur in marten and fisher. *Canadian J. Zool.* 54:559–565.
Leach, D., and deKleer, V. S. 1978. The descriptive and comparative postcranial osteology of marten and fisher: The axial skeleton. *Canadian J. Zool.* 56:1180–1191.
Liers, E. E. 1951. Notes on the river otter (*Lutra canadensis*). *J. Mamm.* 32:1–9.
Lighthill, M. J. 1969. Hydrodynamics of aquatic animal propulsion. *Ann. Rev. Fluid Mech.* 1:413–446.
Long, C. A., and Killingley, C. A. 1983. *The Badgers of the World.* Springfield, Ill.: Charles C Thomas.
Lynch, C. D. 1980. Ecology of the suricate, *Suricata suricata,* and yellow mongoose, *Cynictis penicillata,* with special reference to their reproduction. *Memoirs of the National Museum Bloemfontein.* 14:1–145.
Mech, L. D. 1970. *The Wolf.* Garden City, N.Y.: Natural History Press.
Murie, O. J. 1940. Notes on the sea otter. *J. Mamm.* 21:119–131.
Muybridge, E. 1887 (1957 rpt.). *Animals in Motion.* New York: Dover. 74 pp. and 183 plates.
Neal, E. 1948. *The Badger.* London: Collins.
Nowak, R. M., and Paradiso, J. L. 1983. *Walker's Mammals of the World,* 4th ed. Baltimore: Johns Hopkins Univ. Press.
Ondrias, J. C. 1960. Secondary sexual variation and body skeletal proportions in European Mustelidae. *Arkiv för Zoologi.* 12:577–583.
Ondrias, J. C. 1961. Comparative osteological investigations on the front limbs of European Mustelidae. *Arkiv för Zoologi.* 13:311–320.
Perry, M. L. 1939. Notes on a captive badger. *Murrelet* 20:49–53.

Peterson, R. L. 1966. *The Mammals of Eastern Canada.* Toronto: Oxford Univ. Press.

Quaife, L. R. 1978. The form and function of the North American badger (*Taxidea taxus*) in relation to its fossorial way of life. M.Sc. thesis, Univ. of Calgary. 197 pp.

Roberts, M. S., and Gittleman, J. L. 1984. Mammalian Species no. 222. *Ailurus fulgens.* Lawrence, Kans.: American Society of Mammalogists.

Schaller, G. B. 1972. *The Serengeti Lion: A Study of Predator-Prey Relationships.* Chicago: Univ. Chicago Press.

Severinghaus, C. W., and Tanck, J. E. 1947. Speed and gait of an otter. *J. Mamm.* 29:71.

Smith, J. M., and Savage, R. J. G. 1956. Some locomotory adaptations in mammals. *J. Linn. Soc. (Zoology).* 42:603–622.

Sokolov, A. S., and Sokolov, I. I. 1970. Some special features of the locomotory organs of the river and sea otters associated with their mode of life. *Byulleten Moskovskogo obshchestva ispytatelei prirody, Otdel biologicheskii* 75:5–17.

Sokolov, I. I., and Sokolov, A. S. 1971. Some features of the locomotory organs of *Martes martes* L. associated with its mode of life. *Byulleten Moskovskogo obshchestva ispytatelei prirody, Otdel biologicheskii* 76:40–51.

Tarasoff, F. J. 1972. Comparative aspects of the hindlimbs of the river otter, sea otter, and harp seal. In: R. J. Harrison, ed. *Functional Anatomy of Mammals,* pp. 333–359. New York: Academic Press.

Tarasoff, F. J., Bisaillon, A., Pierard, J., and Whitt, A. P. 1972. Locomotory patterns and external morphology of the river otter, sea otter, and harp seal (Mammalia). *Canadian J. Zool.* 50:915–927.

Taylor, M. E. 1970. Locomotion in some East African viverrids. *J. Mamm.* 51:42–51.

Taylor, M. E. 1971. The comparative anatomy of the limbs of East African Viverridae (Carnivora) and its relationship with locomotion. Ph.D. dissert., Univ. of Toronto. 233 pp.

Taylor, M. E. 1974. The functional anatomy of the forelimb of some African Viverridae (Carnivora). *J. Morphol.* 143:307–336.

Taylor, M. E. 1976. The functional anatomy of the hindlimb of some African Viverridae (Carnivora). *J. Morphol.* 148:227–254.

Taylor, W. P. 1914. The problem of aquatic adaptation in the Carnivora, as illustrated in the osteology and evolution of the sea otter. *Univ. California Publ. Geol.* 7:465–495.

Trapp, G. R. 1972. Some anatomical and behavioral adaptations of ringtails, *Bassariscus astutus. J. Mamm.* 53:549–557.

Van Valkenburgh, B. 1985. Locomotory diversity within past and present guilds of large predatory mammals. *Palaeobiology* 11:406–428.

Van Valkenburgh, B. 1986. Skeletal indicators of locomotory behavior in living and extinct carnivores. *J. Vert. Palaeobiology* 11:406–428.

Wemmer, C., and Watling, D. 1986. Ecology and status of the Sulawesi Palm Civet *Macrogalidia musschenbroekii* Schlegel. *Biol. Conserv.* 35:1–17.

Williams, T. M. 1983. Locomotion in the north American mink, a semi-aquatic mammal. II. The effect of an elongate body on running energetics and gait patterns. *J. Exp. Biol.* 105:283–295.

Wingerson, L. 1983. The lion, the spring and the pendulum. *New Scientist* 97:237–239.

Wolffson, D. M. 1950. Scapula shape and muscle function, with special reference to the vertebral border. *Amer. J. Phys. Anthropol.* 8:331–338.

Yalden, D. W. 1970. The functional morphology of the carpal bones in carnivores. *Acta Anat.* 77:481–500.

CHAPTER 15

Carnivore Dental Adaptations and Diet:
A Study of Trophic Diversity within Guilds

BLAIRE VAN VALKENBURGH

The order Carnivora includes a remarkable array of feeding types and dental morphologies, ranging from pure meat eaters with large cutting carnassial teeth to frugivores with broad crushing teeth. These very different dental forms have evolved from less specialized forms over the course of the Cenozoic, largely as a result of different functional regions of the tooth row being emphasized (Butler 1946; Savage 1977). More so than many other groups (e.g., artiodactyls, rodents), the carnivorans have retained a versatile dentition, with different teeth adapted for cutting meat, crushing bone, and grinding insects and fruits (Figure 15.1). This versatility has led to the evolution of divergent dental patterns and diets within the order, presumably largely as a result of competition for food.

The inherent versatility of the carnivoran tooth row is apparent within a single dog skull (Figure 15.2). There are four functionally distinct areas. The anterior teeth, consisting of the canines and incisors, are used for display, defense, killing prey, and dismembering carcasses (A, Figure 15.2). Directly distal to these are the premolars (B, Figure 15.2), which function as piercers in some species (e.g., canids) and as crushers in others (e.g., hyenas, Figure 15.1). These are followed by the primary cutting tools, the carnassials, composed of the upper fourth premolar (P^4) and the lower first molar (M_1) (C, Figure 15.2). The M_1 is often a two function tooth, where the anterior half, the trigonid, acts as a blade and the posterior half, the talonid, acts as a grinding basin. The remainder of the tooth row, the postcarnassial molars (M^{1-2}, M_{2-3}), are devoted to grinding (D, Figure 15.2).

Bones, meat, fruit, and insects differ in texture and hardness and thus are more efficiently fractured by teeth with different designs. For example, meat can be considered a soft food that is much more readily comminuted by a bladed than a pointed tooth. By contrast, bone is hard and brittle and broken more easily by conical teeth. And plant material, although highly variable in texture, can be fractured into small particles with a mortar-and-pestle tooth design (see Lucas 1979 for detailed discussion of food texture and tooth

410

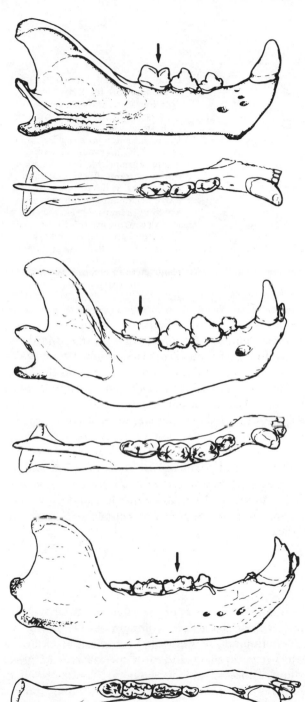

Figure 15.1. Dental diversity among carnivores. Right mandibles of a meat specialist, the puma (*top*); a meat/bone eater, the spotted hyena (*middle*); and an omnivore, the brown bear (*bottom*) shown in both lateral and occlusal views. All are drawn to the same anteroposterior length. In each, the carnassial tooth is indicated by the arrow. Compare the development of grinding molars behind the carnassial in the three, as well as the relative size of the premolars.

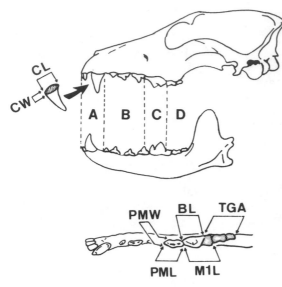

Figure 15.2. The functional regions of the carnivore tooth row and dental measurements. A, canines and incisors; B, premolars, C, carnassials; D, postcarnassial molars. CL, maximum anteroposterior length of canine, CW, maximum mediolateral width of canine. BL, blade length of carnassial tooth. PMW, maximum mediolateral width of premolar. PML, maximum anteroposterior length of premolar. M1L, maximum anteroposterior length of carnassial. TGA, total grinding area of lower molars *(shaded area)*, was measured from color slides by a polar planimeter; all other measurements were taken with dial calipers.

shape). Given these form-function correlations, the dietary habits of animals can often be inferred from their dentition (cf. Cope 1879, 1889; Matthew 1901, 1909; Butler 1946, Crusafont-Pairo and Truyols-Santonja 1956, 1957, 1966; Van Valen 1969). Teeth preserve well in the fossil record and thus can be used to reconstruct the likely diets of extinct species and to study dietary separation among possible competitors in ancient communities.

In this chapter measurements of tooth shape and size are shown to predict aspects of diet and predatory behavior for a sample of 47 extant Carnivora. With the addition of body weight as an indicator of prey size, the measurements are then applied to an ecological problem, dietary separation within predator guilds in several modern communities with contrasting environments: the savannah-woodland mosaic of East Africa, the lowland rainforest of Malaysia, and the cool temperate montane forest of Yellowstone National Park in western North America.

Materials and Methods

The study sample consists of all the members of the three guilds, as well as species whose geographic ranges are outside these communities (Table 15.1). Included are omnivores (e.g., coatimundi, *Nasua nasua*; American black bear; *Ursus americanus*), bone eaters (e.g., hyenas), and meat specialists (e.g., gray wolf, *Canis lupus*; African lion, *Panthera leo*). Excluded are species smaller than 7 kg, as well as molluscivores (e.g., sea otter, *Enhydra lutris*), insectivores

Table 15.1. The sources for behavioral information, body weights, and values of the dental indices of the species used for guild analyses

Species	Behavioral references	Log body weight	CS	RPS	PMD	RBL	RGA
Meat group							
Bobcat	1–5	1.00	79.7	1.92	.45	1.00	—
(*Lynx rufus*)							
Canadian lynx	5,6	1.04	77.0	1.91	.45	1.00	—
(*L. canadensis*)							
Jagourundi	10	0.87	69.0	1.96	.52	1.00	—
(*Felis yagourundi*)							
Golden cat	7,8	1.03	79.7	2.19	.51	1.00	—
(*F. aurata*)							
Temminck's cat	7,8	1.18	77.0	2.05	.43	1.00	—
(*F. temmincki*)							
Fishing cat	7,8	0.95	72.0	2.03	.48	1.00	—
(*F. viverrina*)							
Serval	9,15	1.14	74.5	2.09	.46	1.00	—
(*F. serval*)							
Clouded leopard	8,11	1.30	78.0	2.43	.49	1.00	—
(*Neofelis nebulosa*)							
Caracal	9,12,13	1.22	69.0	2.00	.49	1.00	—
(*Caracal caracal*)							
Snow leopard	11,14	1.72	82.9	2.08	.46	1.00	—
(*Uncia uncia*)							
Puma	16,17	1.77	82.1	1.81	.45	1.00	—
(*Puma concolor*)							
Cheetah	9,18–20,25	1.76	78.4	2.00	.45	1.00	—
(*Acinonyx jubatus*)							
Jaguar	10,21,22	1.72	82.9	2.08	.46	1.00	—
(*Panthera onca*)							
Leopard	9,19,23,25	1.65	72.8	2.44	.49	1.00	—
(*P. pardus*)							
African lion	9,19,24–26	2.21	72.7	2.33	.51	1.00	—
(*P. leo*)							
Tiger	7,27	2.21	76.7	2.30	.49	1.00	—
(*P. tigris*)							
Gray wolf	29–33	1.65	53.6	2.43	.58	0.72	0.66
(*Canis lupus*)							
Dhole	7,27	1.23	58.8	2.01	.48	0.74	0.66
(*Cuon alpinus*)							
African hunting dog	9,20,25,34	1.34	64.2	2.53	.49	0.72	0.57
(*Lycaon pictus*)							
Bush dog	17,35	0.95	72.4	1.99	.49	0.72	0.55
(*Speothos venaticus*)							
Meatbone group							
Spotted hyena	9,24,36,41	1.72	71.5	3.60	.68	0.92	0.12
(*Crocuta crocuta*)							
Striped hyena	9,38,39	1.51	70.8	3.54	.58	0.79	0.30
(*Hyaena hyaena*)							
Brown hyena	37,40	1.61	71.0	4.25	.70	0.84	0.31
(*H. brunnea*)							
Meat/nonvertebrate group							
Large spotted civet	8	0.93	73.0	2.40	.49	0.66	0.84
(*Viverra megaspila*)							

(continued)

Table 15.1. (*Continued*)

Species	Behavioral references	Log body weight	Dental indices				
			CS	RPS	PMD	RBL	RGA
Large Indian civet (*V. zibetha*)	7,8	0.93	73.1	2.42	.48	0.62	0.92
African civet (*Civettictis civetta*)	9,42	1.03	79.7	2.64	.60	0.50	1.16
Chilean fox (*Dusicyon culpaeus*)	35,43	1.10	65.1	2.20	.45	0.68	0.79
Crab-eating fox (*Cerdocyon thous*)	35,44	0.85	59.4	2.00	.47	0.59	0.99
Maned wolf (*Chrysocyon brachyurus*)	17,35	1.36	63.7	2.35	.50	0.57	1.08
Golden jackal (*Canis aureus*)	9,45	0.85	56.2	2.22	.46	0.64	0.90
Side-striped jackal (*C. adustus*)	9	0.85	59.2	2.25	.49	0.66	0.93
Black-backed jackal (*C. mesomelas*)	9,45–47	0.85	60.5	2.24	.47	0.66	0.75
Coyote (*C. latrans*)	48–52	1.06	60.5	2.14	.44	0.66	0.76
Red fox (*Vulpes vulpes*)	53–57	0.86	73.1	2.23	.44	0.67	0.81
Ratel (*Mellivora capensis*)	9,57	1.00	76.5	2.74	.69	0.67	0.40
Wolverine (*Gulo gulo*)	58–62,82	1.38	77.4	2.66	.61	0.68	0.63
N. American badger (*Taxidea taxus*)	58,63–65	0.93	73.1	2.24	.51	0.57	0.76
European badger (*Meles meles*)	17,66	1.11	81.3	1.50	.52	0.40	1.49
Coatimundi (*Nasua nasua*)	10,17,67	0.95	57.7	1.96	.64	0.66	1.46
Polar bear (*Ursus maritimus*)	68,69	2.57	71.3	0.94	.55	0.52	1.83
Nonvertebrate/meat group							
Binturong (*Arctictis binturong*)	7,8	1.02	60.8	2.38	.68	0.63	1.19
Raccoon (*Procyon lotor*)	70–73	0.98	72.0	2.14	.67	0.58	1.32
Raccoon dog (*Nyctereutes procyonoides*)	57,74,75	0.85	66.0	1.79	.48	0.62	0.99
Asiatic black bear (*Selenarctos thibetanus*)	7,76	2.00	63.2	1.37	.56	0.52	1.96
Spectacled bear (*Tremarctos ornatus*)	17,77	2.13	63.3	1.01	.64	0.52	2.06
American black bear (*Ursus americanus*)	58,78,79	2.18	66.2	1.06	.56	0.48	2.17
Brown bear (*U. arctos*)	58,80–82	2.42	72.4	1.25	.59	0.48	2.23

Note. Diet categories and dental indices are described in the text. CS, canine shape; RPS, relative premolar size, PMD, premolar shape; RBL, relative blade length; and RGA, relative grinding area. Behavioral references are listed in abbreviated form at the end of the chapter.

(e.g., aardwolf, *Proteles cristatus*; and sloth bear, *Melursus ursinus*), and frugivores (e.g., kinkajou, *Potos flavus*). The study is thus focused on terrestrial
species, jackal size or larger, that eat meat with some regularity and are likely
to compete when sympatric.

Each species was assigned to a dietary category (described below) on the
basis of published scat studies and behavioral observations (Table 15.1). Fifteen measurements were taken on two skulls of wild-caught adults, one male
and one female, of each species. The measurements include skull length, face
length, jaw depth, crown height, anteroposterior and mediolateral diameter of
upper canine, length and width of the largest lower premolar and first two
molars, and grinding area and cutting blade length of the lower molars (Figure
15.2). From the larger sample of 15 measurements, five morphometric ratios
are derived that reflect aspects of prey-killing behavior and diet (the relative
proportions of meat, bone, and fruit or insects). The correspondence between
behavior and morphology as described by these ratios is explored with bivariate plots and statistics.

The measured skulls are housed in the United States National Museum in
Washington, D.C., and the American Museum of Natural History in New
York.

Dietary Categories

The four dietary categories are defined according to the volume, or frequency of occurrence, of meat, bone, and nonvertebrate (e.g., plant material and
insects) foods, as well as observations on feeding and hunting behavior.
1. Meat: greater than 70% meat
2. Meat/bone: greater than 70% meat with the addition of large bones
3. Meat/nonvertebrate: 50–70% meat, with fruit and/or insects making up the
   balance
4. Nonvertebrate/meat: less than 50% meat, with fruit and/or insects predomi-
   nating

These categories are of necessity broad. Many species exhibit considerable
seasonal or geographical shifts in food choice which make more precise
classification difficult and unrealistic (cf. Kay et al. 1978).

Morphometric Ratios

In all cases except one (premolar size, PMS), the ratios consist of one dental
measure over another, rather than over body weight (Table 15.1, Figure 15.2).
Thus, they are estimates of tooth shape rather than size.

Upper Canine Shape (CS)

The cross-sectional shape of the upper canine tooth was estimated as the ratio of its mediolateral width to its anteroposterior length at the dentine-enamel junction (CW/CL, Figure 15.2).

Premolar Shape and Size (PMD, RPS, respectively)

The shape of the largest lower premolar (the fourth in all sampled species except the hyenas, where the third is largest) was measured as the ratio of maximum mediolateral width to maximum anteroposterior length(PMW/PML, Figure 15.2). To gauge relative premolar size (RPS), I divided the maximum width of the largest lower premolar (PMW, Figure 15.2) by the cube root of body weight.

Relative Blade Length (RBL)

The relative proportion of the first lower molar devoted to slicing as opposed to grinding is estimated by the ratio of the anteroposterior length of the trigonid measured along the buccal margin divided by maximum M_1 length (BL/M1L, Figure 15.2).

Relative Grinding Area (RGA)

The relative proportion of the molar area devoted to grinding as opposed to slicing is estimated by dividing the square root of the total grinding area of the molars (TGA, Figure 15.2) by the total blade length of the carnassial (BL, Figure 15.2). The entire occlusal area of the lower second and third (if present) molars, as well as that of the talonid of the lower M_1, was measured with a polar planimeter. The area estimates were made from color transparencies of the lower molars, taken with the occlusal surface parallel to the plane of focus of the camera. This estimate of grinding area differs from that of Kay (1975, 1977) and Kay et al. (1978), who measured individual wear facet areas, but is suitable for carnivorans because they tend to wear the entire occlusal surface as a flat plane.

The Guilds

The species composition of each guild is listed in Table 15.2. Relevant climatic, floral and faunal characteristics of each community can be found in Van Valkenburgh (1985). The guilds include all the nonaquatic species within the community which capture and consume prey. Predators smaller than jackals (7 kg) are excluded because the evidence for strong competitive interactions

Table 15.2. Predator guild composition

Serengeti
  African lion (*Panthera leo*)
  Leopard (*P. pardus*)
  Cheetah (*Acinonyx jubatus*)
  Caracal (*Caracal caracal*)
  Serval (*Felis serval*)
  African hunting dog (*Lycaon pictus*)
  Blackbacked jackal (*Canis mesomelas*)
  Golden jackal (*C. aureus*)
  Sidestriped jackal (*C. adustus*)
  Spotted hyena (*Crocuta crocuta*)
  Striped hyena (*Hyaena hyaena*)
  Ratel (*Mellivora capensis*)
  African civet (*Civettictis civetta*)
Malaysia
  Tiger (*Panthera tigris*)
  Leopard (*P. pardus*)
  Clouded leopard (*Neofelis nebulosa*)
  Temminck's cat (*Felis temmincki*)
  Fishing cat (*F. viverrina*)
  Dhole (*Cuon alpinus*)
  Binturong (*Arctictis binturong*)
  Large spotted civet (*Viverra megaspila*)
Yellowstone
  Puma (*Puma concolor*)
  Lynx (*Lynx canadensis*)
  Bobcat (*L. rufus*)
  Gray wolf (*Canis lupus*)
  Coyote (*C. latrans*)
  Red fox (*Vulpes vulpes*)
  Wolverine (*Gulo gulo*)
  Badger (*Taxidea taxus*)
  American black bear (*Ursus americanus*)
  Brown bear (*U. arctos*)

Note. For references, see Van Valkenburgh 1985.

among these animals is weak in comparison with that for larger carnivores (see Van Valkenburgh 1985).

The diversity of predator dental morphologies and of body sizes within each guild is portrayed graphically with three-dimensional graphs, as in a previous study on locomotor diversity (Van Valkenburgh 1985). Log body weight (LBW) and two morphometric indices, relative blade length (RBL) and premolar size (RPS), are used as axes of the volume. Body weights were taken from the literature.

Four three-dimensional graphs are presented, one for each of the three guilds and one containing the entire sample of 48 carnivorans. This last portrays the volume of the defined morphological space currently occupied by living Carnivora and provides a framework in which to view each guild.

Because the three-dimensional plots can display only the morphological differences among sympatric predators in three characters, the Euclidean distance between species in six-dimensional morphospace (one dimension per morphometric index) was used as a measure of species dispersion within each guild. Guilds were compared on the basis of (1) the average length of the links of a minimum spanning tree connecting all guild members, and (2) the average distance between each species and the guild centroid (determined as the mean value of each of the six characters) (for details, see Van Valkenburgh 1985).

Results and Discussion

Correspondence between Behavior and Morphology

Because three of the dietary groups are dominated by single families (meat—Felidae; meat/nonvertebrate—Canidae; nonvertebrate/meat—Ursidae), it is important to consider allometric trends within families before proceeding with a discussion of functional differences. Regressions of the log of each of the morphometric indices against log body weight showed there were no significant changes in ratio values with increasing body size for ursids, canids, and felids. In every case, slopes were either not significantly different from zero or extremely close to zero. Thus the differences discussed below among dietary groups are not a result of simple size increase within carnivore families.

CANINE SHAPE

The upper canines of the meat and meat/bone species (open and closed circles, Figure 15.3) tend to be more round in cross-section than those of the meat/nonvertebrate and nonvertebrate/meat groups (triangles and diamonds, Figure 15.3), although there is considerable overlap (CS, Table 15.3). In a separate study of canine shape and strength characteristics of large predators, it is clear that prey-killing behavior explains canine shape better than diet does (Van Valkenburgh and Ruff 1987). Canine shape reflects the stresses incurred during biting. Felids have rounder, more robust canines than do canids because the killing bite of felids is deeper and more forceful (Ewer 1973). Canids have relatively narrow canines that are used to produce more shallow, slashing wounds. Hyaenids have relatively forceful jaws and canines shaped like those of felids (CS, Table 15.1) but kill like canids (Kruuk 1972). Notably, some fossil dogs, (e.g., *Osteoborus* spp.) of suspected bone-eating habits (Matthew and Stirton 1930; Dalquest 1969) had round canines like those of modern hyenas. Thus, bone-eating habits appear to be associated with stronger canines in some canids as well as hyaenids, perhaps because crushing bones requires greater bite strength and increases the risk of canine breakage (see Van Valkenburgh and Ruff 1987 for details).

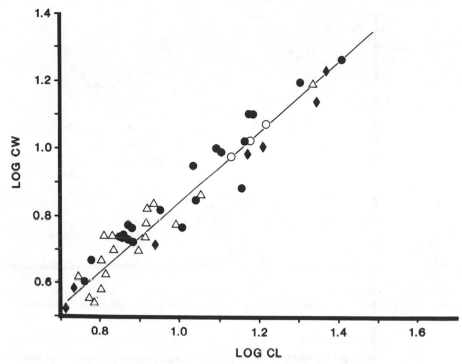

Figure 15.3. Log/log plot of mediolateral canine width (CW, Figure 15.2) against anteroposterior canine length (CL, Figure 15.2), both in millimeters. Regression line: $y = 1.012x - 0.166$. Standard error of slope $= 0.038$. Correlation coefficient (r) $^-$ 0.97. Meat group, solid circles; Meat/bone group, open circles; meat/nonvertebrate group, triangles; nonvertebrate/meat group, diamonds.

Table 15.3. Mean values $(\bar{X})$ and standard deviation (SD) of each morphometric variable for the four diet groups

Group		Log body weight	Dental indices				
			CS	PMD	RPS	RBL	RGA
Meat	$\bar{X}$	1.39	73.9[3,4]	2.14[2,4]	.48[2]	.94[3,4]	.07[3,4]
($n = 20$)	SD	.403	7.98	.216	.032	.115	.15
Meat/bone	$\bar{X}$	1.60	71.1	3.79[1,3,4]	.65[1,3]	.85[3,4]	0.0[3,4]
($n = 3$)	SD	.127	.354	.391	.062	.067	0.0
Meat/nonvertebrate	$\bar{X}$	1.09[4]	69.6[1]	2.18[2,4]	.53[2,4]	.61[1,2]	.48[1,2,4]
($n = 17$)	SD	.409	10.3	.430	.076	.080	.237
Nonvertebrate/meat	$\bar{X}$	1.65[3]	66.2[1]	1.57[1,2,3]	.60[3]	.55[1,2]	.87[1,2,3]
($n = 7$)	SD	.666	4.32	.540	.071	.064	.247

Note. Table 15.1 lists the species included in each group and the source of the behavioral data used to classify each species. The diet categories are defined in the text. Abbreviations for the morphometric indices are listed in the note to Table 15.1. A superscript indicates that the mean is significantly different at the .05 level or better (Student's t, two-tailed test) from that of another group: 1, significantly different from the meat group; 2, meat/bone; 3, meat/nonvertebrate; 4, nonvertebrate/meat.

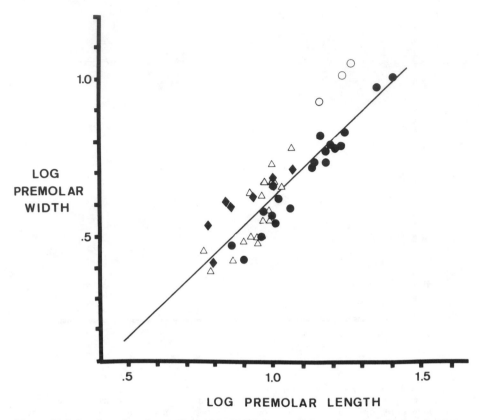

Figure 15.4. Log/log plot of premolar width (PMW, Figure 15.2) against premolar length (PML, Figure 15.2), both in millimeters. Symbols and abbreviations as in Figure 15.3. Regression line: $y = 0.90x - 1.65$. Standard error = 0.77. Correlation coefficient (r) = 0.93.

PREMOLAR SHAPE

Members of the meat/bone group and nonvertebrate/meat group have broader premolars than do members of the meat and meat/nonvertebrate groups (Figure 15.4; PMD, Table 15.3). Thus, there is a trend of increasing premolar roundness that corresponds to a shift away from a predominantly meat diet toward one that includes more nonvertebrate foods or bone.

The advantage of round as opposed to bladed teeth for crushing hard foods is clear; they can withstand much higher pressures before breaking (Lucas 1979). The advantage of using premolars rather than postcarnassial molars as crushers is less clear, since the jaw muscles can exert more force closer to the mandibular joint. However, the size (diameter) of the bone or fruit that can be manipulated between the teeth is relatively greater in the premolar region

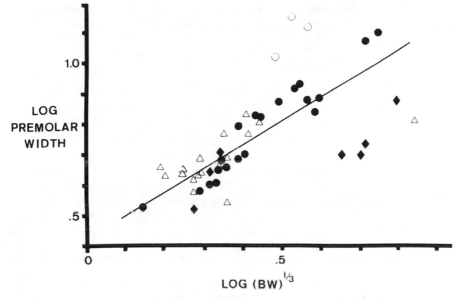

Figure 15.5. Log/log plot of premolar width (PMW, Figure 15.2), in millimeters, against the cube root of body weight, in kilograms. Symbols and abbreviations as in Figure 15.3. Regression line: $y = 0.79x + 0.41$. Standard error = 1.13. Correlation coefficient (r) = 0.84.

because gape is less restricted (Ewer 1954; Crusafont-Pairo and Truyols-Santonja 1957; Savage 1977).

PREMOLAR SIZE

Of the four diet groups, the meat/bone group exhibits by far the widest premolars relative to body weight (Figure 15.5; RPS, Table 15.3). Unlike premolar shape (PMD), the measure of premolar size easily separates the meat/bone eaters from the nonvertebrate/meat group. The differences in premolar size among the other three groups are less pronounced but suggest that species that eat little meat tend to have relatively small premolars (RPS, Table 15.3). This is largely a result of the fact that bears, which dominate the nonvertebrate/meat group, have unusually narrow premolars for their size. Even the most carnivorous, the polar bear (*Ursus maritimus*), has relatively slender premolars (RPS, Table 15.1). Why bears should have such narrow premolars is not clear.

The relatively large premolars of bone eaters are an adaptation to the heavy wear imposed by their diet. Observations on the correspondence between age and degree of dental attrition in African lions (Smuts et al. 1978) and hyenas (Kruuk 1972; Mills 1982) demonstrate much more rapid wear in hyenas.

Consequently, hyenas would lose the ability to crush bone at a much younger age if their teeth were not relatively large (cf. Kay 1975).

RELATIVE BLADE LENGTH

Meat and meat/bone eaters have longer blades relative to total carnassial length than do members of the meat/nonvertebrate group, which in general have longer blades than do members of the nonvertebrate/meat group (Figure 15.6; RBL, Table 15.3). The value for the meat group is biased by the predominance of felids within the group, all of which have a blade length–M_1 length ratio of 1 (RBS, Table 15.1) and thus fall on a straight line in Figure 15.5. Nevertheless, if cats are removed from the sample, the differences in mean RBL value between the meat group and the omnivore groups remain significant (P < 0.01, Student's T, 2-tailed). Thus, the measure of blade length stands as a good indicator of meat content of the diet. The positive correlation between relative trigonid blade length and meat eating reflects the blade's primary

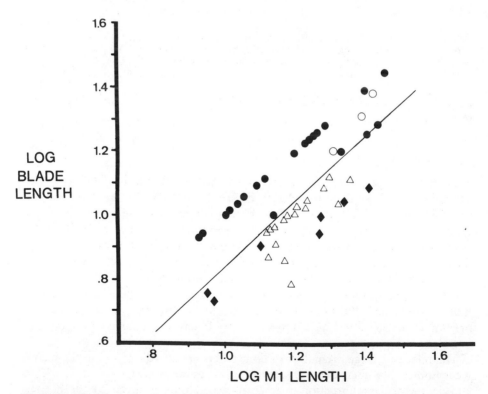

Figure 15.6. Log/log plot of blade length (BL, Figure 15.2) against first lower molar length (M1L, Figure 15.2), both in millimeters. Symbols and abbreviations as in Figure 15.3. Regression line: $y = 1.05x - 2.21$. Standard error = 1.12. Correlation coefficient (r) = 0.87.

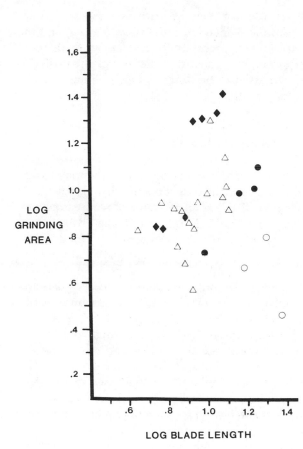

Figure 15.7. Log/log plot of the square root of total grinding area (TGA, Figure 15.2) against blade length (BL, Figure 15.2) in millimeters. Symbols and abbreviations as in Figure 15.3. Regression line is not shown because the slope approximated zero and the correlation coefficient is less than 0.10.

function as a meat slicer (Cope 1879, 1889; Matthew 1909; Butler 1946; Crusafont-Pairo and Truyols-Santonja 1956, 1957; Van Valen 1969; Savage 1977).

RELATIVE GRINDING AREA

In general, the species with the most fruit or insects in their diet have the largest grinding area relative to their cutting blade length (Figure 15.7; RGA, Table 15.3). Species with somewhat more meat in their diet have less grinding area, and the meat- and meat/bone-eating species have the least of all. In fact, most of the meat-eating species are missing from Figure 15.6 because they are felids, all of which have no grinding area on their carnassial and are without postcarnassial molars (RGA, Table 15.1). The loss of the M_2 in felids reflects their strictly carnivorous diet. The canids in the meat group eat slightly more nonvertebrate foods and have retained a limited amount of grinding area

(RGA, Table 15.1). Hyenas have also lost their postcarnassial molars but are not as carnivorous as felids; both the striped hyena (*Hyaena hyaena*) and the brown hyena (*H. brunnea*) are known to eat fruits and insects with some regularity (Table 15.1). Ewer (1954) suggested that the loss of the posterior molars in hyenas allowed the carnassial to be closer to the jaw joint, thereby increasing bite force at the cutting tooth.

Dental and Trophic Diversity within the Sample

The results of the bivariate analysis suggest that differences among extant predators in diet content are best indicated by two of the five dental indices: (1) premolar size relative to body weight (RPS) as an indicator of bone eating, and (2) the proportion of the carnassial blade devoted to slicing (RBL) as an indicator of the relative importance of meat versus nonvertebrate foods in the diet. In Figure 15.8, these two variables and body weight (LBW) are used to define a morphological volume in which all 47 carnivore species are plotted. Body weight is included because of its demonstrated importance as a determinant of life history traits, foraging radius, metabolic requirements, interspecific dominance, and prey size (Rosenzweig 1966, 1968; McNab 1971; Schaller 1972; Eaton 1979; Eisenberg 1981; Lamprecht 1981; Gittleman and Harvey 1982; Gittleman 1985, 1986). Although social behavior and phylogeny can complicate some of these correlations (cf. Eaton 1979; Lamprecht 1981; Gittleman 1985), it is generally true that similarities in body weight between sympatric carnivore species imply broad similarities in their ecology.

Each of the four diet groups is characterized by a limited range of premolar size and blade length values, and thus each occupies a separate region of the volume floor (Fig. 15.8). Position on the blade-length axis (RBL) reflects the relative proportions of meat and nonvertebrate foods in the diet. The strictest meat species, the cats, have high RBL values and cluster along the volume's left edge (RBL > 1, Figure 15.8). More omnivorous members of the meat group, such as the gray wolf and African hunting dog (*Lycaon pictus*) (Cl, Lp; Figure 15.8), have retained a talonid on their lower molar, have lower blade length values than the felids, and are located near the center of the volume (RBL near 0). Adjacent to them, with still shorter cutting blades, are the members of the meat/nonvertebrate group (triangles). Members of the remaining group have the relatively shortest cutting blades (low, negative RBL values) and form a clump near the right edge of the volume (diamonds, Fig. 15.8).

The premolar axis (RPS) distinguishes meat/bone specialists from all others and reveals species within other groups which tend toward bone eating. Meat-bone eaters exhibit a combination of relatively wide premolars (high RPS) and long cutting blade (high RBL), and thus are positioned in the left, rear corner (open circles, Figure 15.8).

In general, the species in the center of the volume are more diverse tax-

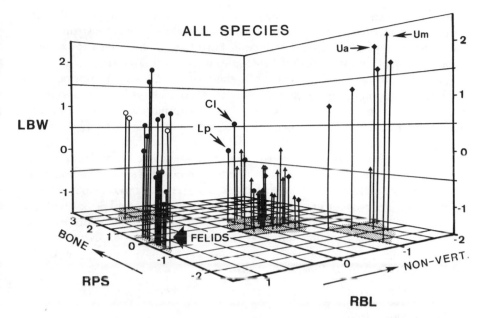

Figure 15.8. Body size and dental morphospace. All species listed in Table 15.1 are plotted. As in Figures 15.3–7, species are represented by symbols that indicate their dietary classification. The axes are: log body weight (LBW); premolar width/cube root body weight (RPS); and lower M1 blade length/total M1 length (RBL). Units are standardized normal deviates of values shown in Table 15.1. The shaded area indicates the range of RPS and RBL values observed in the sample of 47 Carnivora. Abbreviations are as follows: Lp, *Lycaon pictus;* Cl, *Canis lupus;* Ua, *Ursus arctos;* and Um, *Ursus maritimus.*

onomically and more generalized in their diet than those near the edges. Mustelids, canids, viverrids, and procyonids all occur within the central clump, whereas each outside cluster is dominated by species from a single family. The ursids dominate the nonvertebrate/meat group, the hyenas form the meat/bone group, and the felids split from the other meat species to form their own clump along the left edge (Table 15.1, Figure 15.8).

For canids, felids, and hyaenids, the two measures of premolar size and blade length appear to work as both taxonomic and functional indices; and their small range of premolar size and blade length values might be assumed to indicate a low diversity of feeding types within each. This is generally true, but in at least one case dental morphology appears to be more influenced by ancestry than function. The polar bear (*Ursus maritimus*), is quite predacious and yet its dentition deviates only slightly from its more omnivorous ancestor, the brown bear (*Ursus arctos*) (Um vs. Ua, Figure 15.8) (Kurtén 1964). Apparently, the functional demands of consuming prey are adequately met with teeth that seem more appropriate for grinding fibrous foods than slicing meat. However, the polar bear rarely coexists with other, more dentally suited and perhaps more efficient meat eaters such as wolves, and the bear might switch to other foods such as fruits in the presence of competition. It appears that the influence of ancestry rather than function on dental shape can be strong if the functional requirements are not stringent. Fortunately, the polar bear appears to be the exception rather than the rule, and the morphological indices remain as good indicators of diet (as defined by the four categories) for most species.

The distribution of taxa within the volume as four clumps results in unfilled regions of the volume. Obviously, there are combinations of premolar size, blade length, and body weight that are not exhibited by any of the 47 species of extant Carnivora. For example, there is no species of large size (LBW > 1) in the meat/nonvertebrate group. Similarly, there are no species with premolars of average or narrow width (RPS near 0) that have retained a small grinding basin on the carnassial (RBL = 0–1); they have wider premolars, no basin at all, or a larger basin. This suggests the existence of adaptive valleys and peaks, with the valleys being defined by those unoccupied areas of the volume. Work in progress on fossil predator guilds indicates that portions of the unoccupied areas were once filled, and thus that the peaks and valleys have shifted over evolutionary time (cf. Van Valkenburgh 1988).

Guild Comparisons

The morphological volume with all 47 species provides a framework in which to view each of the guild volumes. If competition for food is important within the large predator guild, we would not expect all guild members to be within the same diet category, and thus in the same part of the volume. Furthermore, we might expect that environments that produce a greater abundance and diversity of prey should have more meat and meat/bone specialists. In such environments the degree of morphologic similarity among predators in the measured features is predicted to be relatively great.

As expected, all three Recent guilds include carnivorans from several diet categories, but they differ in species richness and the diet categories represented. The Serengeti has the most species, with 13 carnivorans, drawn from each of the four diet categories except nonvertebrate/meat (Figure 15.9C, Table 15.4). Yellowstone has ten predators, representing each diet group except meat/bone (Figure 15.9A, Table 15.4). Malaysia has the smallest number of species, eight, and like Yellowstone is without meat/bone specialists (Figure 15.9B, Table 15.4). In both tropical guilds, Malaysia and Serengeti, the meat and meat/bone species make up more than half of the guild, whereas the omnivores dominate the Yellowstone (Table 15.4).

The average morphological distance between predators is greater in the Yellowstone than the Serengeti or Malaysia (Table 15.5). However, if the bears (which are marginal members of the predator guild) are excluded from the Yellowstone, the dispersion among the remaining species is similar. Statistical significance of the differences in dispersion can be tested for the DFC but

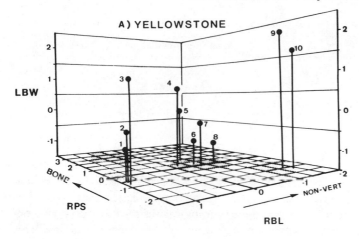

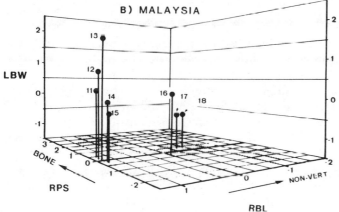

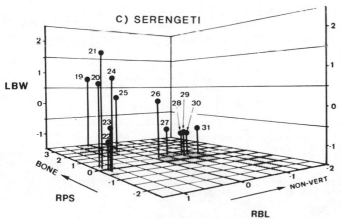

Figure 15.9. Yellowstone (*A*), Malaysia (*B*), and Serengeti (*C*) guilds. Symbols and axes as in Figure 15.8. The shaded area indicates the range of RPS and RBL values observed in the sample of 47 Carnivora (Figure 15.8). Species are as follows (scientific names are given in Table 15.1): 1, bobcat; 2, Canadian lynx; 3, puma; 4, gray wolf; 5, wolverine; 6, red fox; 7, coyote; 8, North American badger; 9, brown bear; 10, American black bear; 11, clouded leopard; 12, leopard; 13, tiger; 14, Temminck's cat; 15, fishing cat; 16, dhole; 17, large spotted civet; 18, binturong; 19, spotted hyena; 20, leopard; 21, African hunting dog; 27, ratel; 28, side-striped jackal; 29, black-backed jackal; 30, golden jackal; 31, African civet.

Table 15.4. Number of species within diet categories in each guild

Guild	Meat	Meat/bone	Meat/nonvertebrate	Nonvertebrate/meat
Serengeti	6	2	5	0
Yellowstone	4	0	4	2
Hemphillian	4	1	3	0
Orellan	6	0	3	0

not the MST measure because of a lack of independence of link lengths in the latter. A comparison of DFC values finds no significant differences between guilds, with or without bears included (P > 0.20, Mann-Whitney, 2-tailed) (Zar 1984). Given the greater number of predators in the Serengeti, it is perhaps surprising that the average morphological distance between two Serengeti predators is not less than in the other, less species-rich guilds. This contrasts with the results of a previous study of locomotor morphologies within the same guilds which showed a greater similarity among the African predators (Van Valkenburgh 1985). It suggests that the degree of morphologic resemblance (in body weight and the dental indices) between sympatric species has been limited similarly in all three communities. The limits may be determined largely by competition for food, a process likely to occur in all three communities, despite their different histories and environments.

The greater species richness of the Serengeti guild is probably due to the much greater richness and abundance of terrestrial herbivore prey in the savannah as compared with either the Malaysian rainforest or Yellowstone temperate forest. The Serengeti fauna includes 24 species of terrestrial herbivores (Schaller 1972); Malaysia has 11 (Medway 1969, 1971), and Yellowstone has six (Weaver 1978). In east Africa the biomass of herbivores is estimated to be 7418 kg/km^2 (Sinclair 1977); in Malaysia the same estimate is 492 (Eisenberg and Seidensticker 1976), and for Yellowstone it is 390 (Weaver 1978). The increased availability of low-stature vegetation in the Serengeti supports a greater diversity of herbivores, which in turn has encouraged greater diversification within the predator guild, particularly among the meat specialists.

Table 15.5. Morphological distance characteristics of each guild

Guild	Species (n)	MST[a]	SD[b]	DFC[c]	SD[b]
Serengeti	13	1.49	.605	2.12	.540
Malaysia	8	1.52	.733	1.68	.675
Yellowstone, with bears	10	1.67	1.06	2.33	.784
without bears	8	1.48	.769	1.78	.119

[a]MST = mean link length of the minimum spanning tree.
[b]SD = standard deviation of the mean for each guild.
[c]DFC = mean distance from the guild centroid.

For example, in east Africa there are two specialized bone-crushers, the hyenas, whereas in Malaysia there is none and Yellowstone has but two taxa with a tendency toward bone eating, the gray wolf and wolverine (*Gulo gulo*). Bone eaters are likely to be favored in environments where the probability of finding a carcass is high. The Serengeti is ideal; herbivore biomass is high and circling vultures mark the presence of carrion (Houston 1979). In the rainforest carcasses are less common, and decompose more rapidly, and soaring scavengers are obscured by tree cover. Although winter kills of ungulates may be fairly common in Yellowstone, biomass levels are probably still too low to make scavenging a daily possibility.

The greater diversity of meat eaters in Malaysia as opposed to Yellowstone also reflects increased availability of prey. Although the terrestrial herbivore biomass levels are similar in the two environments, rainforests have much higher densities and richness of arboreal prey (cf. Eisenberg and Thorington 1973; Eisenberg and McKay 1974; Eisenberg et al. 1979). There are no biomass data available for Malaysian forests, but a comparison of Malaysian and North American forest mammal diversity reveals a four-fold difference in the number of taxa (17 in Louisiana against 77 in Malaysia) (Emmons et al. 1983).

The large number of omnivores in Yellowstone is surprising, given that plant foods of all kinds are extremely scarce for several months each year, but can be explained by two behavioral adaptations: hibernation and food switching. Three of the six omnivores spend all or most of the winter in dens: the North American badger (*Taxidea taxus*), American black and brown bears (Craighead and Mitchell 1982; Lindzey 1982; Pelton 1982). The remaining three, the red fox (*Vulpes vulpes*), coyote (*Canis latrans*), and wolverine, switch from a summer diet that includes a considerable amount of fruits and insects to a winter diet of meat and carrion (diet references in Table 15.1).

Omnivores make up only 25% of the Malaysian guild despite the relative wealth of fruits in the rainforest compared with the savannah or temperate forest. The addition of the Malaysian sun bear, *Helarctos malayanus* (which was considered too frugivorous to include in the predator guild), would bring the total number of omnivores in Malaysia to three, as opposed to five in Yellowstone and six in east Africa. This still seems a significant difference, and it may be that primates, birds, and rodents that eat plant and insect foods exclusively are replacing generalized omnivores in Malaysia. Similarly, the absence of non-meat/vertebrate-eating species such as bears in the Serengeti might be due to competitive exclusion by suids and large primates, such as the baboons (*Papio* spp.), which forage on the ground. There was a bear species in Africa five and a half million years ago, *Agriotherium africanus*, and its disappearance has been attributed to concommitant declines in food availability and annual rainfall (Hendey 1980).

The comparisons of the three guilds suggest that their differences can be explained largely by present-day abundances and diversity of prey. However, differences in prey diversity among the three guilds seem less easily explained by environmental factors. Whereas the low diversity of terrestrial herbivores in the rainforest as opposed to the savannah undoubtedly reflects the former's lack of low-stature vegetation, the same cannot be said of Yellowstone. The montane community includes grasslands, open woodlands, and areas of secondary growth forest, and yet it has only six terrestrial herbivore taxa (Meagher 1973; Weaver 1978).

It appears that this low ungulate diversity is due more to extinction without replacement than some persistent environmental limitation. Ten to fifteen thousand years ago, the plains surrounding Yellowstone contained almost as many ungulate species as exist now in the Serengeti (Hibbard et al. 1965; Kurtén and Anderson 1980). The contemporaneous guild of predators included an additional four meat (sabertooth cats, *Smilodon fatalis, Homotherium serum*; lion *Panthera leo-atrox*; and cheetah-like cat, *Miracinonyx trumani*) and one bone-meat (dire wolf, *Canis dirus*) species, bringing the total guild diversity to fifteen. It may be that the decline to present levels of large mammal diversity accompanied a decline in environmental conditions such that large numbers of herbivores could not persist. However, the enormous size of bison herds in the nineteenth century suggests that grassland productivity was high (Roe 1951; McDonald 1981, 1984). More likely, processes that would have increased diversity, such as speciation and immigration, were stymied by human interference (habitat destruction and restriction, as well as hunting) (cf. McDonald 1984).

Both the tropical faunas appear to have suffered less from Pleistocene events. The rainforest community of Southeast Asia persisted throughout the Pleistocene, and the large mammals that vanished seem to have been grazing species that would have existed outside the rainforest community (Verstappen 1975). In east Africa the number of large predators declined about two million years ago, but several medium-sized species appeared to fill out the guild, and the overall change in species richness was minor (Savage 1978; Klein 1984; Walker 1984). Thus their herbivore biomass and diversity levels appear to more accurately reflect current environmental conditions than do those of Yellowstone.

This study of three recent guilds of large predators is a first step toward understanding the evolution and function of dental differences among sympatric carnivores. At present, the analysis is being expanded to include fossil as well as modern guilds. Future morphometric analyses will use multivariate approaches to assess the relative contribution of each dental feature to dietary separation. Some teeth, such as canines, are likely to reflect selection for functions other than feeding behavior, such as display. Indices are also being developed to distinguish more subtle differences in diet among omnivores through the use of microwear features as visualized in the scanning electron microscope (e.g., Teaford and Walker 1984).

Summary and Conclusions

The relative importance of bone, meat, and nonvertebrate foods in the diets of 47 carnivorans is predicted from measurements of canine shape, premolar size and shape, carnassial blade length, and postcarnassial molar size. Following this, body weight and the five dental indices are used to examine the diversity of dietary types represented within three Recent guilds of large predators. Results show that guilds differ primarily in the number of member species, and the relative representation of the four dietary types: meat, meat/bone, meat nonvertebrate, and nonvertebrate/meat.

Among the sampled guilds, species richness of predators is highest in the community that exhibits the highest levels of terrestrial herbivore biomass and diversity, the Serengeti. Moreover, the two tropical communities, Serengeti and Malaysia, support a greater diversity of meat specialists (e.g., felids) and prey species than does the temperate Yellowstone community. The occurrence of meat/bone-eating species, such as hyenas, within a guild appears to depend on the predictability of discovering carcasses. Open, dry, or cool habitats with scavenging birds improve carcass predictability and should favor meat/bone specialists.

Historical evidence suggests that the apparently depauperate Yellowstone community is the result of extinction without replacement.

Acknowledgments

I thank L. Perkins for assistance in measuring specimens; R. T. Bakker for suggestions on dental measurements; and R. T. Bakker, A. R. Biknevicius, J. Gittleman, J. B. C. Jackson, R. F. Kay, S. M. Stanley, and R. K. Wayne for critical review at various stages. This work was supported in part by The Johns Hopkins University and the American Association of University Women.

References

Butler, P. M. 1946. The evolution of carnassial dentitions in the Mammalia. *Proc. Zool. Soc. Lond.* 116:198–220.
Cope, E. D. 1879. The origin of the specialized teeth of the Carnivora. *Amer. Nat.* 13:171–173.
Cope, E. D. 1889. The mechanical causes of the development of the hard parts of the Mammalia. *J. Morph.* 3:232–236.
Craighead, J. J., and Mitchell, J. A. 1982. Grizzly bear (*Ursus arctos*). In: J. A. Chapman & G. A. Feldhammer, eds. *Wild Mammals of North America*, pp. 653–663. Baltimore: Johns Hopkins Univ. Press.
Crusafont-Pairo, M., and Truyols-Santonja, J. 1956. A biometric study of the evolution of fissiped carnivores. *Evolution* 10:314–332.

Crusafont-Pairo, M., and Truyols-Santonja, J. 1957. Estudios masterometricos en la evolucion de los fissipedos. *Boletin del Institute de Geologica y Minera España* 68:83–224.

Crusafont-Pairo, M., and Truyols-Santonja, J. 1966. Masterometry and evolution, again. *Evolution* 20:204–210.

Dalquest, W. W. 1969. Pliocene carnivores of the Coffee Ranch (type Hemphill) local fauna. *Bull. Texas Memorial Mus.* 15:1–44.

Eaton, R. L. 1979. Interference competition among carnivores: A model for the evolution of social behavior. *Carnivore* 2:9–16.

Eisenberg, J. F. 1981. *The Mammalian Radiations.* Chicago: Univ. Chicago Press.

Eisenberg, J. F., and McKay, G. M. 1974. Comparison of ungulate adaptations in the New World and Old World tropical rainforests with special reference to Ceylon and the rainforests of Central America. In: V. Geist & F. Walther, eds. *The Behavior of Ungulates and Its Relation to Management,* pp. 585–602. Morges, Switzerland: IUCN Publications, new series 24.

Eisenberg, J. F., O'Connell, M. A., and August, P. V. 1979. Density, productivity and distribution of mammals in two Venezuelan habitats. In: J. F. Eisenberg, ed. *Vertebrate ecology in the Northern Neotropics,* pp. 187–207. Washington, D.C.: Smithsonian Institution Press.

Eisenberg, J. F., and Seidensticker, J. 1976. Ungulates in Southern Asia: A consideration of biomass estimates for selected habitats. *Bio. Conserv.* 10:293–308.

Eisenberg, J. F., and Thorington, R. 1973. A preliminary analysis of a neotropical mammal fauna. *Biotropica* 5:150–161.

Emmons, L. H., Gautier-Hion, A., and Dubost, G. 1983. Community structure of the frugivorous-folivorous mammals of Gabon. *J. Zool.* 199:209–222.

Ewer, R. F. 1954. Some adaptive feature in the dentition of hyaenas. *Ann. Mag. Nat. Hist.* 7:188–94.

Ewer, R. F. 1973. *The Carnivores.* Ithaca, N.Y.: Cornell Univ. Press.

Gittleman, J. L. 1985. Carnivore body size: Ecological and taxonomic correlates. *Oecologia* 67:540–554.

Gittleman, J. L. 1986. Carnivore life history patterns: Allometric, phylogenetic and ecological associations. *Amer. Nat.* 127:744–771.

Gittleman, J. L., and Harvey, P. H. 1982. Carnivore home-range size, metabolic needs and ecology. *Behav. Ecol. Sociobiol.* 10:57–63.

Hendey, Q. B. 1980. *Agriotherium* (Mammalia, Ursidae) from Langebaanweg, South Africa, and relationships of the genus. *Ann. South African Mus.* 81:1–109.

Hibbard, C. W., Ray, C. E., Savage, D. W., Taylor, D. W., and Guilday, J. E. 1965. Quaternary mammals of North America. In: H. E. Wright & D. G. Frey, eds. *The Quaternary of the United States,* pp. 509–525. Princeton, N.J.: Princeton Univ. Press.

Houston, D. C. 1979. The adaptations of scavengers. In: A. R. E. Sinclair & M. Norton-Griffiths, eds. *Serengeti: Dynamics of an Ecosystem,* pp. 263–286. Chicago: Univ. Chicago Press.

Kay, R. F. 1975. The functional adaptations of primate molar teeth. *Amer. J. Phys. Anthropol.* 43:195–216.

Kay, R. F. 1977. Molar structure and diet in extant Cercopithecidae. In: K. Joysey & P. Butler, eds. *Function and Evolution of Teeth,* pp. 309–339. London: Academic Press.

Kay, R. F., Sussman, R. W., and Tattersall, J. 1978. Dietary and dental variations in the genus *Lemur,* with comments concerning dietary-dental correlations among Malagasy primates. *Amer. J. Phys. Anthropol.* 49:119–128.

Klein, R. G. 1984. Mammalian extinctions and Stone Age people in Africa. In: P. S.

Martin & L. G. Klein, eds. *Quaternary Extinctions*, pp. 553–573. Tucson: Univ. Arizona Press.

Kruuk, H. 1972. *The Spotted Hyena*. Chicago: Univ. Chicago Press.

Kurtén, B. 1964. The evolution of the polar bear, *Ursus maritimus* Phipps. *Acta Zool. Fennica* 108:1–26.

Kurtén, B., and Anderson, E. 1980. *Pleistocene Mammals of North America*. New York: Columbia Univ. Press.

Lamprecht, J. 1981. The function of social hunting in larger terrestrial carnivores. *Mamm. Rev.* 11:169–179.

Lindzey, F. G. 1982. Badger (*Taxidea taxus*). In: J. A. Chapman & G. A. Feldhammer, eds. *Wild Mammals of North America*, pp. 653–663. Baltimore: Johns Hopkins Univ. Press.

Lucas, P. W. 1979. The dental-dietary adaptations of mammals. *Neues Jahrbuch für Geologie und Paläontologie Monatschafte* 8:486–512.

McDonald, J. N. 1981. *North American Bison: Their Classification and Evolution*. Berkeley: Univ. California Press.

McDonald, J. N. 1984. The reordered North American selection regime and late Quaternary megafaunal extinctions. In: P. S. Martin & L. G. Klein, eds. *Quaternary extinctions*, pp. 404–439. Tucson: Univ. Arizona Press.

McNab, B. K. 1971. On the ecological significance of Bergmann's rule. *Ecology* 52:845 851.

Matthew, W. D. 1901. Tertiary mammals of northeastern Colorado. *Mem. Amer. Mus. Nat. Hist.* 1:353–447.

Matthew, W. D. 1909. The Carnivora and Insectivora of the Bridger Basin, middle Eocene. *Mem. Amer. Mus. Nat. Hist.* 9:291–567.

Matthew, W. D., and Stirton, R. A. 1930. Osteology and affinites of *Borophagus*. *Univ. California Publ. Geol. Sci.* 19:171–217.

Meagher, M. M. 1973. *The Bison of Yellowstone National Park*. Scientific Monograph Series, no. 14. Washington, D.C.: U.S. National Park Service.

Medway, G. G. 1969. *The Wild Mammals of Malaya*. London: Oxford Univ. Press.

Medway, G. G. 1971. Importance of Taman Negara in the conservation of mammals. *Malay Nature J.* 24:212–214.

Mills, M. G. L. 1982. Notes on age determination, growth and measurements of brown hyaenas, *Hyaena brunnea*, from the Kalahari Gemsbok National Park. *Koedoe* 25:55–61.

Pelton, M. R. 1982. Black bear (*Urus americanus*). In: J. A. Chapman & G. A. Feldhammer, eds. *Wild Mammals of North America*, pp. 653–663. Baltimore: Johns Hopkins Univ. Press.

Roe, R. C. 1951. *The North American Buffalo*. Toronto: Univ. Toronto Press.

Rosenzweig, M. L. 1966. Community structure in sympatric carnivora. *J. Mamm.* 47:602–612.

Rosenzweig, M. L. 1968. The strategy of body size in mammalian carnivores. *Amer. Midland Nat.* 80:299–315.

Savage, R. J. G. 1977. Evolution in carnivorous mammals. *Paleontology* 20:237–271.

Savage, R. J. G. 1978. Carnivora. In: V. J. Maglio & H. B. S. Cooke, eds. *Evolution of African Mammals*, pp. 249–267. Cambridge, Mass: Harvard Univ. Press.

Schaller, G. B. 1972. *The Serengeti Lion*. Chicago: Univ. Chicago Press.

Sinclair, A. R. E. 1977. *The African Buffalo*. Chicago: Univ. Chicago Press.

Smuts, G. L., Anderson, J. L., and Austin, J. C. 1978. Age determination of the African lion (*Panthera leo*). *J. Zool.* 185:115–146.

Teaford, M., and Walker, A. 1984. Quantitative differences in dental microwear between primate species with different diets and a comment on the presumed diet of *Sivapithecus*. *Amer. J. Phys. Anthropol.* 64:191–200.

Van Valen, L. 1969. Patterns of dental growth and adaptation in mammalian carnivores. *Evolution* 23:96–117.

Van Valkenburgh, B. 1985. Locomotor diversity within past and present quilds of large predatory mammals. *Paleobiology* 11:406–428.

Van Valkenburgh, B. 1988. Tropic diversity in past and present guilds of large predatory mammals. *Paleobiology* 14:155–173.

Van Valkenburgh, B., and Ruff, C. B. 1987. Canine tooth strength and killing behaviour in large carnivores. *J. Zool.* 212:1–19.

Verstappen, H. T. 1975. On palaeo climates and landform development in Malesia. In: G. Bartstra & W. A. Casparie, eds. *Modern Quaternary Research in Southeast Asia*, pp. 3–36. Rotterdam: A. A. Balkema.

Walker, A. 1984. Extinction in hominid evolution. In: M. H. Nitecki, ed. *Extinction*, pp. 119–152. Chicago: Univ. Chicago Press.

Weaver, J. 1978. *The Wolves of Yellowstone*. Natural Resources Report no. 14. Washington, D.C.: U.S. Department of the Interior, National Park Service.

Zar, J. H. 1984. *Biostatistical Analysis* (2nd. ed). Englewood Cliffs, N.J.: Prentice-Hall.

Behavioral References For Table 15.1

1. Marston, M. A. 1942. *J. Wildlf. Mgmt.* 6:328–337.
2. Matson, J. R. 1948. *J. Mamm.* 29:69–70
3. Hamilton, W. J., and Hunter, R. P. 1939. *J. Wildlf. Mgmt.* 3:99–103.
4. Miller, G. J., and Carron, R. 1976. *I. V. C. Occ. Paper* 4.
5. Jones, J. H., and Smith, N. S. 1979. *J. Wildlf. Mgmt.* 43:666–672.
6. Nellis, C. H., and Keith, L. B. 1968. *J. Wildlf. Mgmt.* 32:718–722.
7. Saunders, J. K. 1963. *J. Wildlf. Mgmt.* 27:384–390.
8. Prater, S. H. 1965. Bombay Nat. Hist. Soc; Bombay, India
9. Lekagul, B., and McNeeley, J. A. 1977. Bangkok Assoc. Conserv. Wildlf.
10. Leopold, A. S. 1959. Univ. Calif. Press; Berkeley.
11. Walker, E. P. 1964. Johns Hopkins Univ. Press; Baltimore.
12. Smithers, R. H. N. 1978. *South African J. Wildlf. Res.* 8:29–37.
13. Skinner, J. D. 1979. *J. Zool.* 189:523–525.
14. Blomqvist, L. 1978. *Internatl. Pedigree Bk. Snow Leopards* 1:6–21.
15. Smithers, R. H. N. 1978. *Fauna and Flora* 33.
16. Hornocker, M. 1970. *Wildlf. Monogr.* 21:1–39.
17. Nowak, R. M., and Paradiso, J. L. 1983. Johns Hopkins Univ. Press; Baltimore.
18. Eaton, R. L. 1970. *J. Wildlf. Mgmt.* 34:56–67.
19. Schaller, G. B. 1972. Univ. Chicago Press; Chicago.
20. Frame, G., and Frame, L. 1981. E. P. Dutton; New York.
21. Schaller, G., and Vasconcelos, J. M. 1978. *Z. Säugetierk.* 43:296.
22. Mondolfi, E., and Hoogesteijn, R. 1984. *Proc. Internatl Cat Symp.*; Texas.
23. Pienaar, U. de V. 1969. *Koedoe* 12:108–176.
24. Smuts, G. L. 1979. *South African Tydskr. Natuurnav.* 9:19–25.
25. Kruuk, H., and Turner, M. 1967. *Mammalia* 31:1–27.
26. Eloff, F. C. 1973. *J. South African Wildlf Mgmt. Assoc.* 3:59–63.
27. Johnsingh, A. J. T. 1981. Ph.D. dissert., Madura Univ.; India.
28. Seidensticker, J. 1976. *Biotropica* 8:225–234.

29. Mech, L. D. 1966. *Parks of the U.S., Fauna Ser.* 7:1–210.
30. Mech, L. D. 1970. Natural History Books; Chicago.
31. Carbyn, L. N. 1975. Ph.D. dissert., Univ. Toronto, Ontario.
32. Cowan, I. M. 1947. *Canadian J. Res.* 25(D):139–174.
33. Voight, D. R. et al. 1976. *J. Wildlf. Mgmt.* 40:663–668.
34. Estes, R. D., and Goddard, J. 1967. *J. Wildlf. Mgmt.* 31:52–69.
35. Langguth, A. 1975. Pp. 192–206. In: Fox, M. W., ed. Van Nostrand Reinhold; New York.
36. Kruuk, H. 1972. Univ. Chicago Press; Chicago.
37. Mills, M. G. L. 1978. *Z. Tierpsychol.* 48:113–141.
38. Macdonald, D. W. 1978. *Israel J. Zool.* 27:189–198.
39. Kruuk, H. 1976. *East African Wildlf. J.* 14:91–111.
40. Owens, M. J., and Owens, D. D. 1978. *East African Wildlf. J.* 16:113–135.
41. Bearder, S. K. 1977. *East African Wildlf. J.* 15;263–280.
42. Guy, P. 1977. *South African J. Wildlf. Res.* 7:87–88.
43. Jaksic, F. et al. 1980. *J. Mamm.* 61:254–260.
44. Bisbal, F., and Ojasti, J. 1980. *Acta Biol. Venez.* 10:469–496.
45. Lamprecht, J. 1978. *Z. Säugetierk.* 43:210–223.
46. Hall-Martin, A. J., and Botha, B. P. 1980. *Koedoe* 23:157–162.
47. Bothma, J. D. 1971. *Zool. Africana* 6:195–203.
48. Murie, A. 1944. U.S. Dept. Interior, Fauna Ser. 4.
49. Ogle, T. F. 1971. *Northwest Sci.* 45: 213–218.
50. Gipson, P. S. 1974. *J. Wildlf. Mgmt.* 38:848–853.
51. MacCracken, J. G., and Hansen, R. M. 1982. *Great Basin Nat.* 42:45–49.
52. Kleiman, D. G., and Brady, C. A. 1978. Pp. 163–188. In: Bekoff, M., ed. Academic Press; New York.
53. Scott, J. G. 1943. *Ecol. Monogr.* 13:428–479.
54. Goszczynski, J. 1976. *Acta Theriol.* 19:1–18.
55. Frank, L. G. 1979. *J. Zool.* 183:526–532.
56. Errington, P. L. 1937. *Ecology* 18:53–61.
57. Novikov, G. A. 1962. Israel Prog. Scientific Pub.; Jerusalem.
58. Chapman, J. A., and Feldhammer, G. A. 1983. Johns Hopkins Univ. Press; Baltimore.
59. Krott, P. 1960. *Monogr. Wildsäuget.* 13:1–159.
60. Haglund, B. 1966. *Viltrevy* 4:81–283.
61. Hornocker, M. G., and Hash, H. S. 1981. *Canadian J. Zool.* 59:1286–1301.
62. Myhre, R., and Myrberget, S. 1975. *J. Mamm.* 56:752–757.
63. Messick, J. P., and Hornocker, M. G. 1981. *Wildlf. Monogr.* 76:1–53.
64. Snead, E., and Hendrickson, G. O. 1942. *J. Mamm.* 23:380–390.
65. Errington, P. L. 1937. *J. Mamm.* 18:213–216.
66. Skoog, P. 1970. *Viltrevy* 7:1–97.
67. Kaufmann, J. H. 1962. *Univ. California Pub. Zool.* 60:95–222.
68. Stirling, I. et al. 1977. *Canadian Wildlf. Ser. Occ. Pap.* 33:1–64.
69. Jonkel, C. et al. 1976. *Canadian Wildlf. Ser. Pap.* 26:1–42.
70. Hamilton, W. J. 1936. *Ohio J. Sci.* 36:131–140
71. Baker, R. et al. 1945. *J. Wildlf. Mgmt.* 9:45–56.
72. Wood, J. C. 1954. *J. Mamm.* 35:406–415.
73. Harman, D. M., and Stains, H. J. 1979. *Amer. Mus. Novit.* 2679:1–24.
74. Ikeda, H. et al. 1979. *Japanese J. Ecol.* 29:35–48.
75. Viro, P., and Mikkola, H. 1981. *Z. Säugetierk.* 46:20–26.
76. Schaller, G. 1969. *J. Bombay Nat. Hist. Soc.* 66:156–159.

77. Peyton, B. 1980. *J. Mamm.* 61:639–652.
78. Bennett, J. et al. 1943. *J. Mamm.* 24:25–31.
79. Franzmann, A. W. et al. 1980. *J. Wildlf. Mgmt.* 44:764–768.
80. Pearson, A. M. 1975. *Canadian Wildlf. Serv. Rep.* 34:1–86.
81. Murie, A. 1981. *Scientific Monogr. Ser.* 14:1–251.
82. Haglund, B. 1974. *Le Naturaliste Canadien* 101:457–466.

CHAPTER 16

The Physiology and Evolution of
Delayed Implantation in Carnivores

RODNEY A. MEAD

The duration of pregnancy in many, but not all, species of carnivores is much longer than one would predict on the basis of body size. The prolonged gestation is due in part to an arrest in embryonic development that can last from a few days to ten months, depending on the species. This form of embryonic diapause, referred to as obligate delay of implantation, occurs in seven of the 12 families of living carnivores (Mustelidae, Ursidae, Ailuropodidae, Ailuridae, Phocidae, Otaridae, and Odobenidae). Occurrence of delayed implantation within a given family, subfamily, or genus is not uniform (Table 16.1).

It has been suggested that delayed implantation occurs in all genera of bears and pandas (Table 16.2); however, unimplanted blastocysts have been observed only in the American black bear (*Ursus americanus*) (Wimsatt 1963), brown bear (*U. arctos*) (Craighead et al. 1969; Dittrich and Kronberger, 1963), Himalayan black bear (*U. thibetanus*) (Dittrich and Kronberger 1963), and sloth bear (*Melursus ursinus*) (Puschman et al. 1977). Embryonic diapause is also suspected to occur in the polar bear (*U. maritimus*), Malayan sun bear (*Helarctos malayanus*), and both the giant and red pandas (*Ailuropoda melanoleuca* and *Ailurus fulgens*), as gestation periods of these species are somewhat prolonged. In the case of the sun bear gestation ranges from 95 days or less (Dathe 1970) to 230 days (McCusker 1974). Assuming that all reports are correct, delayed implantation may not always occur in this species. Alternatively, the bears in question have been misidentified or this genus may consist of two closely related species, only one of which exhibits delayed implantation.

Delayed implantation occurs in nearly all species of pinnipeds whose reproductive cycles have been investigated (Table 16.3), but remains to be demonstrated in such species as the leopard seal (*Hydrurga leptonyx*), Ross seal (*Ommatophoca rossi*), northern elephant seal (*Mirounga angustirostris*), and New Zealand seal lion (*Neophoca hookeri*). Embryonic diapause in many pinnipeds begins while the female is lactating; however, renewed embryonic

437

Table 16.1. Reproductive characteristics of female Mustelidae

Species	Distribution	Breeding season	Gestation	Delay[a]	Litter size	Parturition	References
Lutrinae							
Aonyx capensis (African clawless otter)	Africa	varies	63 days	N.D.	2–5	varies	Ansell 1960; Rosevear 1974; Kingdon 1977
A. cinerea (Oriental small-clawed otter)	Asia	varies	60–64 days	N.D.	1–6	varies	Leslie 1970; Timmis 1971; Duplaix-Hall 1975
Enhydra lutris (sea otter)	Pacific Ocean, N. Amer., USSR	varies	6–7 months	L.D.	1–2	varies	Novikov 1956; Sinha et al. 1966; Brosseau et al. 1975
Lutra canadensis (American river otter)	N. Amer.	Mar.–Apr.	245–365 days	L.D.	2–4	Mar.–Apr.	Hamilton and Eadie 1964; Duplaix-Hall 1975
L. lutra (European otter)	Europe, Asia	Feb.–Apr.	60–62 days	N.D.	2–5	Apr.–May	Novikov 1956; Duplaix-Hall 1975
L. maculicollis (spotted-necked otter)	Africa	July	60 days	N.D.	2–3	Sept.	Procter 1963
L. perspicillata (Indian smooth-coated otter)	Asia	Aug., varies	60–62 days	N.D.	—	Oct., varies	Desai 1974; Duplaix-Hall 1975
Pteronura brasiliensis (flat-tailed or giant otter)	S. Amer.	July–Aug.	65–70 days	N.D.	1–5	Aug.–Oct.	Harris 1968; Trebbau 1972; Autuori & Deutsch 1977
Melinae							
Arctonyx collaris (hog badger)	Asia	Apr.–Sept.	5–9.5 months	L.D.(?)	2–4	Feb.	Parker 1979
Meles meles (European badger)	Europe, Asia	Feb.–Mar.	345–365 days	L.D.	2–6	Feb.	Neal and Harrison 1958; Harrison 1963; Canivenc and Bonnin 1981
Taxidea taxus (American badger)	N. Amer.	July–Aug.	7–8.5 months	L.D.	2–3	Mar.–Apr.	Wright 1966
Mellivorinae							
Mellivora capensis (honey badger or ratel)	Africa, Asia	Not known	6 months(?)	suspected	2	Not known	Rosevear 1974
Mephitinae							
Conepatus mesoleucus (hog-nosed skunk)	N. Amer., S. Amer.	Feb.	2 months	N.D.	2–4	Apr.–May	Patton 1974

Species	Region	Mating season	Gestation	Delayed implantation	Litter size	Birth season	References
Mephitis mephitis (striped skunk)	N. Amer.	Feb.–Apr.	59–77 days	S.D.	1–10	May–June	Wade-Smith & Richmond 1975, 1978; Wade-Smith et al. 1980
Spilogale gracilis[b] (western spotted skunk)	N. Amer.	Sept.–Oct.	210–260 days	L.D.	1–6	Apr.–June	Mead 1968b; Greensides & Mead 1973; Mead 1981
S. putorius (eastern spotted skunk)	N. Amer.	Mar.–July	45–55 days	N.D.	5	May–Aug.	Mead 1968a
S. pygmaea (pygmy spotted skunk)	Mexico	Mar.–June	48 days	N.D.	2–6	May–Aug.	Teska et al. 1981
Mustelinae							
Eira barbara (tayra)	Central & S. Amer.	Apr.	63–67 days	N.D.	3	Jun.–July	Encke 1968; Poglayen-Neuwall 1978
Gulo gulo (wolverine)	USSR, N. Amer.	May–July	8–9 months	L.D.	3–4	Mar.	Wright & Rausch 1955; Rausch & Pearson 1972
Ictonyx striatus (zorilla or striped pole-cat)	Africa	Aug.–Nov.	35–44 days	N.D.	1–3	Sept.–Dec.	Ball 1978; Rowe-Rowe 1978
Martes americana (American marten)	N. Amer.	July–Aug.	259–276 days	L.D.	2–5	Mar.–Apr.	Pearson & Enders 1944; Wright 1963
M. flavigula (yellow-throated marten)	Asia	Oct.–Nov.	172–190 days	L.D.	1–5	Mar.–Apr.	Roberts 1977; Andriuskevicius (pers. comm.) 1982
M. foina (stone or beech marten)	Europe	July	236–274 days	L.D.	1–8	Mar.–Apr.	Novikov 1956; Stubbe, 1968; Stubbe et al. 1981; Canivenc et al. 1981; Madsen and Rasmussen 1985
M. martes (European pine marten)	Europe	July	230–270 days	L.D.	3–8	Mar.–Apr.	Stubbe 1968; Canivenc et al. 1969; Canivenc 1970
M. pennanti (fisher)	N. Amer.	Mar.–Apr.	327–358 days	L.D.	3–4	Mar.–Apr.	Enders & Pearson 1943; Eadi & Hamilton 1958; Wright & Coulter 1967
M. zibellina (sable)	Asia	June–July	253–297 days	L.D.	1–5	Apr.–May	Novikov 1956; Bernatskii et al. 1976
Mustela altaica (mountain weasel)	Asia	Feb.–Mar.	40 days	N.D.	7–8	Apr.–May	Novikov 1956; Roberts 1977; Tumanov 1977
M. erminea (short-tailed weasel or stoat)	N. Amer., Europe	May–July	10–11 months	L.D.	4–13	Apr.–May	Watzka 1940; Deanesly 1943; Lavrov 1944; Wright 1963

(*continued*)

Table 16.1. (*Continued*)

Species	Distribution	Breeding season	Gestation	Delay[a]	Litter size	Parturition	References
M. frenata (long-tailed weasel)	N. Amer., S. Amer.	July	9 months	L.D.	6–9	Apr.–May	Wright 1942; Wright 1963
M. nivalis (least weasel)	N. Amer., Europe	Spring, summer	34–37 days	N.D.	1–7	Spring, summer, fall	Deanesly 1944; Hartman 1964; Heidt 1970
M. eversmanni (steppe polecat)	Europe, Asia	Mar.–Apr.	36–41 days	N.D.	3–17	Apr.–May	Ognev 1931; Schmidt 1932; Novikov 1956; Stroganov 1962
M. nigripes (black-footed ferret)	N. Amer.	Mar.–Apr.	42–45 days	N.D.	2–6	May	Hillman & Carpenter 1983; T. Thorne, pers. comm.
M. putorius (ferret)	Europe, Asia	Mar.–July	40–42 days	N.D.	2–12	May–Aug.	Robinson 1918; Hammond & Walton 1934
M. lutreola (European mink)	Europe	April	40–43 days	N.D.	2–7	May	Moshonkin 1981, 1983
M. sibirica (Kolinsky mink)	Asia	Feb.–Apr.(?)	34 days	N.D.	2–10	Apr.–May(?)	Novikov 1956; Tumanov 1977
M. vison (mink)	N. Amer., Europe, Asia	Mar.–Apr.	40–75 days	S.D.	1–17	Apr.–May	Hansson 1947; Enders 1952
Poecilictis libyca (N. African striped weasel)	Africa	Feb.–May	37 days	N.D.	1–3	Mar.–June	Petter 1959; Rosevear 1974
Poecilogale albinucha (African striped weasel)	Africa	Aug.–Mar.	31–33 days	N.D.	1–3	Sept.–Apr.	Rowe-Rowe 1978
Vormela peregusna (marbled polecat)	Europe, Asia	Apr.–June	8–11 months	L.D.	1–8	Jan.–Apr.	Mendelssohn et al. 1988

[a]N.D. = no delayed implantation; S.D. = short period of delayed implantation; L.D. = long period of delayed implantation.
[b]The eastern and western populations of spotted skunks have not officially been recognized as separate species. However, for the purpose of clarity of this paper, the western population is referred to as *S. gracilis*.

Table 16.2. Reproductive characteristics of female Ursidae, Ailuropodidae, and Ailuridae

Species	Distribution	Breeding season	Gestation	Delay	Parturition	Sources
Ailuridae						
Ailurus fulgens (red panda)	Burma, Nepal, China	Jan.–Mar.	112–158 days	Suspected	May–Aug.	Roberts & Gittleman 1984
Ailuropodidae						
Ailuropoda melanoleuca (giant panda)	China	Apr.–May	132–148 days	Suspected	Sept.	Peking Zoo 1974
Ursidae						
Helarctos malayanus (Malayan sun bear)	SW. China	varies	95–106 days, 174–230 days	Suspected	varies	Dathe 1963, 1970; McCusker 1974
Melursus ursinus (sloth bear)	India	May–July	6–7 months	Duration unknown	June–Jan.	Jacobi 1975; Laurie & Seidensticker 1977; Puschman et al. 1977
Tremarctos ornatus (spectacled bear)	S. Amer.	May–Aug.	5.5–8.5 months	Suspected	Jan.–Mar.	Dathe 1967; Bloxam 1977
Ursus americanus (American black bear)	N. Amer.	Jun.–July	7–8 months	5 months	Jan.–Feb.	Hamlett 1935; Wimsatt 1963; Erickson et al. 1964; Daniel 1974
U. arctos (brown bear)	N. Amer.	May–July	7–8 months	5 months	Jan.–Feb.	Dittrich & Kronberger 1963; Craighead et al. 1969
U. maritimus (polar bear)	Arctic	Feb.–May	228–303 days	Suspected	Nov.	Dittrich 1961; Volf 1963
U. thibetanus (Himalayan black bear)	Asia	Jun.–July	7–8 months	Duration unknown	Jan.–Feb.	Dittrich & Kronberger 1963

Table 16.3. Reproductive characteristics of female pinnipeds known to exhibit an obligate delay of implantation

Species	Distribution	Breeding season	Gestation (*months*)	Delay (*months*)	Parturition	References
Phocidae						
Cystophora cristata (hooded seal)	Arctic, N. Atlantic	Feb.–Mar.	11.5–12	3–5	Feb.–Mar.	Bertram 1940; Harrison 1969
Erignathus barbatus (bearded seal)	Arctic, Japan, Canada	May–Jun.	11.5–12	2–3	May–June	McLaren 1958; Harrison & Kooyman 1968; Harrison 1969
Halichoerus grypus (grey seal)	U.S.S.R., Iceland, Britain, Norway	Sept.–Oct.	11.5–12	3.5	Sep.–Oct.	Blackhouse & Hewer 1956; Harrison 1963; Hewer & Blackhouse 1968; Bonner 1972
Leptonychotes weddelli (Weddell seal)	Antarctic	Oct.–Dec.	10	2	Aug.–Nov.	Mansfield 1958; Harrison 1969; Kooyman 1981
Lobodon carcinophagus (crabeater seal)	Antarctic	Nov.–Dec.	9–10	2	Sep.–Nov.	Bertram 1940; Harrison 1969; Øritsland 1970
Mirounga leonina (southern elephant seal)	Subantarctic Islands	Sept.–Nov.	11–11.5	3–4	Oct.–Nov.	Harrison et al. 1952; Laws 1956; Ling & Bryden 1981
Phoca groenlandica (harp seal)	N. Pacific, SW. China	Jan.–Apr.	11.5–12	3	Jan.–Apr.	Harrison 1963; Harrison & Kooyman 1968; Ronald & Healey 1981

P. hispida (ringed seal)	Arctic, Finland, Canada	Mar.–Aug.	12	3–4	Mar.–May	Harrison & Kooyman 1968; Harrison 1969; Frost & Lowry 1981
P. largha (large or spotted seal)	Arctic, N. Atlantic	Jun.–July	10–11	2–3	Apr.–May	Harrison 1969; Bigg 1981
P. vitulina (harbor seal)	N. Pacific, N. Atlantic	July	10–11	2–3	May–July	Fisher 1954; Harrison 1963, 1969; Bonner 1972; Bigg & Fisher 1974
Odobenidae						
Odobenus rosmarus (walrus)	N. Pacific, N. Atlantic	Jan.–Feb.	15	4–5	Apr.–June	Fay 1981, 1982
Otariidae						
Arctocephalus pusillus (Cape fur seal)	SW. Africa, SE. Australia	Nov.–Dec.	12	3–4	Nov.–Dec.	Harrison 1969
A. australis (South American fur seal)	S. Amer.	Nov.–Dec.	12	4	Nov.–Dec.	Rand 1955
Callorhinus ursinus (Northern fur seal)	N. Pacific	Jun.–July	12	4	Jun.–July	Craig 1964; Baker et al. 1970; Daniel 1981
Eumetopias jubatus (Steller's sea lion)	N. Pacific	May–July	11–11.5	3.5	May–June	Harrison 1969; Schusterman 1981
Otaria flavescens (South American sea lion)	S. Amer.	Dec.–Jan.	12	3–4	Dec.–Jan.	Hamilton 1939; Daniel 1981
Zalophus californianus (California sea lion)	N. Pacific	Jun.–July	11–11.5	3–3.5	May–July	Odell 1981

development and implantation are not initiated when lactation ceases (see Boshier 1981 for diagram illustrating this point). Another interesting feature of pinniped reproduction is that the postimplantation period lasts seven to eight and a half months in nearly all species (Laws 1956) except in the walrus (*Odobenus rosmarus*) (Fay 1981).

Unfortunately, no experimental work has been conducted on the mechanisms that control embryonic diapause in bears, and only a few such studies have been conducted with the fur seal (*Callorhinus ursinus*) (Daniel 1981). Consequently we must rely upon extensive studies with three species of mustelids—mink (*Mustela vison*), European badger (*Meles meles*), and western spotted skunk (*Spilogale gracilis*)—for insights into the mechanisms that control this fascinating process.

The Physiology of Delayed Implantation

Changes in the Embyro

During diapause, embryonic development becomes arrested at the blastocyst stage. Each blastocyst is surrounded by a relatively tough acellular membrane, the zona pellucida, that is retained in all carnivores until a few hours before implantation. Delayed implanting blastocysts consist of a single layer of squamous trophoblast cells that forms a hollow sphere around a spherical knot of cells known as the inner cell mass (Figure 16.1). Although embryonic development does not progress beyond the blastocyst during diapause, blastocysts of mustelids such as the short-tailed weasel (*Mustela erminea*), sable (*Martes zibellina*), mink, European badger, American badger (*Taxidea taxus*), and western spotted skunk gradually increase in diameter (Deanesly 1943; Neal and Harrison 1958; Baevsky 1963; Wright 1966; Canivenc and Bonnin 1981; Mead and Rourke 1985). Blastocyst diameter also increases during diapause in nonmustelid carnivores such as the black bear (Wimsatt 1963) and several seals (Daniel 1971; Boshier 1981). Total cell numbers also increase throughout the period of delayed implantation in most species; however, this increase in cell number is restricted to the trophoblast in the western spotted skunk and perhaps American badger (Wright 1966; Mead 1968a). Trophoblast cell number has also been reported to increase slowly in delayed implanting blastocysts of the European badger (*Meles meles*) (Harrison 1963) and seals (Rand 1955; Laws 1956; Smith 1966; Daniel 1971). On the other hand, there is no evidence of increased trophoblast cell number during delayed implantation in blastocysts of the mink, sable or short-tailed weasel (Baevsky 1963; Shelden 1972). Several lines of evidence indicate that blastocysts of carnivores are metabolically active during diapause. Oxygen consumption of delayed implanting blastocysts of the black bear, northern fur seal (*Callorhinus ursinus*), and mink was as high or higher than that of "activated" rabbit blastocysts

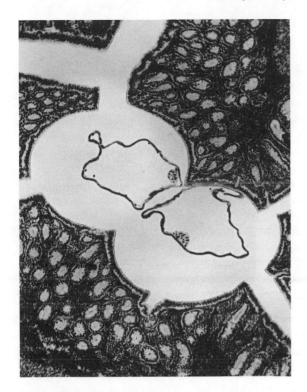

Figure 16.1. Two unimplanted blastocysts in the uterus of the western spotted skunk. Note that both embroys consist of a single layer of trophoblast cells and a spherical inner cell mass. The zona pellucida (*outermost covering*) surrounds each blastocyst until implantation occurs. (From Mead 1968b, courtesy of *Journal of Mammalogy.*)

(Gulyas and Daniel 1967). DNA synthesis must occur in trophoblast cells of those species in which an increase in cell number has been reported. RNA synthesis occurs in delayed implanting blastocysts of the mink (Gulyas and Daniel 1969) and spotted skunk (Mead and Rourke 1985) but not the northern fur seal. Protein synthesis, likewise, occurs in blastocysts of the spotted skunk, mink, and northern fur seal during delayed implantation (Gulyas and Daniel 1969; Rourke and Mead 1982). Moreover, detailed ultrastructural studies of delayed implanting skunk blastocysts also suggest that reduced metabolic activity occurs during diapause (Enders et al. 1986).

A few days before implantation one can detect numerous cytological and metabolic changes within the blastocysts, which are now undergoing renewed development. These include a rapid increase in cell numbers and polyribosomes in both the trophoblast and inner cell mass (Mead 1968b; Enders et al. 1986). The inner cell mass loses its spherical appearance and forms the embryonic disc, and a new germ layer (endoderm) differentiates from the inner cell mass. There is also a dramatic increase in endocytotic activity in the trophoblast cells (Enders et al. 1986). Such cytological changes in skunk blastocysts are extremely well correlated with the marked increase in RNA and protein synthesis observed in activated preimplantation embryos (Rourke and Mead 1982; Mead and Rourke 1985).

Effects of Photoperiod

Changes in day length synchronize implantation within the population. In some species, such as the European badger, decreasing day length is the trigger (Canivenc and Bonnin 1981), whereas increasing day length is required in most species, for example the mink (Hansson 1947), American marten (*Martes americana*) (Pearson and Enders 1944), sable (Belyaev et al. 1951), and western spotted skunk (Mead 1971). Light is not, however, absolutely essential for induction of renewed embryonic development or implantation, as both occur in blind animals, albeit out of synchrony with those exposed to a natural photoperiod (Kirk 1962; Mead 1971; May and Mead 1986). Thus there appears to be an endogenous rhythm that can trigger implantation, but it lacks the precision required to synchronize this reproductive event with seasonal changes in the environment so that the young can be born at an appropriate time of year. Several investigators have suggested that changes in day length play little or no role in timing implantation in pinnipeds and have also implied that this process is controlled by some endogenous rhythm (Harrison 1963; Daniel 1981; Spotte 1981).

The Pineal

Although the pineal gland is not required for implantation (Mead 1972), it plays a major role in mediating the effects of light on synchronizing the time of blastocyst implantation in the population of spotted skunks. Changes in day length are perceived by the eyes. The neural input from the retina passes along the optic nerves and reaches the suprachiasmatic nuclei (SCN) within the hypothalamus via the retinohypothalmic tract (May et al. 1985). A multisynaptic neural pathway passes from the SCN to the pineal via the superior cervical ganglia (SCG). Increased neuronal activity within postganglionic fibers of the SCG occurs during periods of darkness and results in increased enzyme activity and melatonin levels in the pineal gland of rats and presumably other mammals (Nishino et al. 1976). Consequently, melatonin is secreted in greater concentrations at night and for a longer duration during the long nights of winter in such species as the ferret (*Mustela putorius*) (Baum et al. 1986) and mink (Ravault et al. 1986). Injections or Silastic pellets containing melatonin result in a further delay of blastocyst implantation in the western spotted skunk (May and Mead 1986). The site and mechanism of action of melatonin remains unknown. Administration of melatonin to mink does cause a decrease in plasma prolactin levels (Martinet et al. 1983; Rose et al. 1985). Other experiments have conclusively demonstrated that treatments which suppress prolactin secretion in mink result in decreased luteal function and prevent or further delay blastocyst implantation (Papke et al. 1980; Martinet et al. 1981), whereas manipulative treatments that increase plasma prolactin levels shorten the duration of the preimplantation period in mink (Murphy 1983). Other lines of indirect evidence, to be discussed later, suggest that melatonin may be acting via mechanisms other than the suppression of prolactin secretion to prevent implantation in the European badger.

The Pituitary

Numerous expreriments have indicated that the pituitary and ovaries must be present and functional for implantation in mustelids. Although similar studies have not been carried out in other carnivores with delayed implantation, there is every reason to believe that these endocrine organs are essential for regulation of virtually all aspects of pregnancy. It has been postulated that insufficient secretion of gonadotropic hormones from the pituitary results in delayed implantation in mustelids. This hypothesis was initially supported by histochemical studies that indicated that the gonadotrophs of the anterior pituitary of the mink (Baevskii 1964) and European badger (Herlant and Canivenc 1960) were less abundant or less active during diapause, when compared to those of postimplantation specimens. Hypophysectomy prior to the expected time of implantation prevented nidation and abolished the preimplantation increase in plasma progesterone in the western spotted skunk (Mead 1975) and mink (Murphy and Moger 1977). Such studies confirmed that the pituitary was absolutely essential for normal ovarian function during pregnancy. On the other hand, removal of the uterus containing unimplanted blastocysts had no significant effect on pre- and postimplantation progesterone secretion in the spotted skunk (Mead and Swannack 1978) or postimplantation luteal function in mink (Canivenc et al. 1966), thus indicating that the embryos and placentae were not essential sources of gonadotropic hormones. Other evidence that supports the hypothesis that delayed implantation may result from insufficient gonadotropin secretion comes from the measurement of plasma levels of gonadotropins in the western spotted skunk (Foresman and Mead 1974), European badger (Canivenc and Bonnin 1981), and mink (Martinet et al. 1981). Such measurements suggest that luteinizing hormone (LH) and prolactin increase in parallel with the rise in blood levels of progesterone. Administration of bromoergocryptine, a dopamine agonist that inhibits prolactin secretion, to mink during the preimplantation period inhibited the rise in serum prolactin and progesterone and prevented blastocyst implantation, whereas the administration of exogenous prolactin caused a premature rise in serum progesterone and hastened implantation (Papke et al. 1980; Martinet et al. 1981). Moreover, prolactin can maintain progesterone secretion in pregnant hypophysectomized mink, thus suggesting that it is the primary gonadotropin responsible for stimulating increased ovarian activity at the time of implantation (Murphy et al. 1981). Similar observations have now been made in the spotted skunk (Berria et al. 1988). However, prolactin is not known to play such a central role in the European badger (Canivenc and Bonnin 1981:36; Canivenc and Laffargue 1956). The gonadotropic hormones responsible for stimulating increased ovarian activity before implantation remain to be elucidated in this species. All of these studies, however, are consistent with the hypothesis that delayed implantation results from insufficient pituitary activity even though different pituitary hormones may be responsible in different species.

The Ovaries

Ovariectomy of most mammals during early pregnancy results in termination of pregnancy. Carnivores and mustelids in particular fit this pattern. Ovariectomy of the European badger (Canivenc and Laffargue 1958; Neal and Harrison 1958), long-tailed weasel (*Mustela frenata*) (Wright 1963), short-tailed weasel (Shelden 1972), mink (Møller 1974; Murphy et al. 1982), and western spotted skunk (Mead 1981; Mead et al. 1981) during the preimplantation period results in decreased progesterone secretion and eventual death of the blastocysts. Such experiments dramatically emphasize that the ovaries must be present for long-term blastocyst survival and implantation. The corpus luteum is the only constituent of the ovary that consistently undergoes pronounced morphological and physiological changes that temporally coincide with the cessation and subsequent renewed embryonic development and implantation in mustelids (Mead 1981, 1986), black bear (Wimsatt 1963; Erickson et al. 1964; Foresman and Daniel 1983), and pinnipeds (Boshier 1981; Daniel 1981). Plasma progesterone levels increase before renewed embryonic development in all species of carnivores so far examined that exhibit a delay of implantation. This coincides with a marked increase in diameter of the corpora lutea and cytological changes in the luteal cells. However, all attempts to induce implantation in intact or ovariectomized mustelids (Mead 1981; Mead et al. 1981) or northern fur seal (Daniel 1981) by administering progesterone, other synthetic progestins such as medroxyprogesterone acetate (Murphy et al. 1982), or a combination of estrogens and progestins (Cochrane and Shackelford 1962; Shelden 1973) have consistently failed to induce implantation. However, experiments with mink and ferrets (*Mustela putorius*) clearly indicate that the corpus luteum is the only ovarian compartment needed to induce implantation in these species (Foresman and Mead 1978; Murphy et al. 1983; Mead 1986). If the corpora lutea are indeed responsible for somehow inducing renewed embryonic development and blastocyst implantation, what are they secreting that promotes these activities? In vitro studies regarding steroid metabolizing potential of corpora lutea of the western spotted skunk (Ravindra et al. 1984) and domestic ferret (Kintner and Mead 1983) indicate that progesterone is the predominant steroid produced but corpora lutea of both species also have the enzymatic capability to aromatize androgens to estrogens. However, plasma estrogen levels appear to decline during the peri-implantation period in the spotted skunk (Ravindra and Mead, 1984) and European badger (Mondain-Monval et al. 1980). Since neither steroid alone or in combination can induce implantation, it has been hypothesized that the corpora lutea may secrete a nonsteroidal compound which acts in concert with progesterone to induce implantation. Mead et al. (1988) have recently reported that extracts from ferret (*Mustela putorius*) corpora lutea will induce implantation in ovariectomized progesterone treated ferrets. Preliminary findings are consistent with the hypothesis that the luteal implantation inducing factor is a protein.

The Uterus

How ovarian hormones induce renewed embryonic development and implantation is another unanswered question. The most plausible hypothesis that is consistent with the bulk of existing data is that the ovarian hormones act on the uterus, rather than on the embryos, to modify the uterine environment in such a manner that renewed embryonic development can occur. That the uterine environment changes before activation of the embryos and that this response is under hormonal control is not in question. Numerous studies indicate that the uterus undergoes striking cytological changes during the peri-implantation period (Enders and Given 1977; Schlafke et al. 1981). Likewise, the quantity of various constituents of uterine fluid appear to undergo dramatic changes. However, such studies should be interpreted with caution as changes in uterine fluid volume are unknown for any carnivore. Consequently, changes in uterine fluid constituent concentrations are unknown. Moreover, flushing the uterus usually results in some damage to the luminal epithelium, thereby resulting in contamination of the uterine fluid sample with nonsecretory products. In spite of such criticisms, the total protein content of uterine fluid from the ferret (Daniel 1970), northern fur seal (Daniel 1971), and western spotted skunk (Fazleabas et al. 1984) have been reported to increase dramatically in parallel with increased diameter of the blastocysts. However, there is no conclusive evidence that these proteins are essential for renewed embryonic development.

Origin of the uterine luminal contents remains unknown in most species. Most of the proteins and probably other constituents of uterine fluid are of serum origin, as evidenced by similar electrophoretic mobilities of serum and uterine fluid proteins. A small percentage of the proteins are believed, however, to be synthesized and secreted by the uterus. For example, the uterus of the western spotted skunk incorporates radio-labeled amino acids into two proteins (mw > 200,000 and mw 43,000) throughout most of the preimplantation period. As the time of implantation approached, there was a quadrupling of the amount of radioactivity incorporated into these two proteins, and a third radio-labeled protein (mw 24,000) was detectable for the first time (Mead et al. 1979). Synthesis and/or secretion of all three proteins was stimulated in ovariectomized skunks by progesterone but was inhibited by estradiol-17β. All three of these proteins are known to be of uterine origin as uterine explants were capable of incorporating radio-labeled amino acids into these same three proteins in vitro (R. Mead, pers. observ.). However, the biological actions of these proteins remains unknown. A potent inhibitor of plasminogen activator was also present in uterine flushings of the western spotted skunk during all stages of pregnancy tested (Fazleabas et al. 1984). Since the total amount of inhibitory activity increased after implantation occurred, it was deemed unlikely that this protease inhibitor was in any way responsible for the delay of implantation. A more plausible explanation of its function would be that it serves to neutralize any plasminogen activator in uterine fluid, thereby indirectly protecting the uterus and its contents from the destructive action of plasmin.

There are at least three possible ways in which changes in the uterine environment might initiate resumption of embryonic development. (1) An influx of nutrients and/or ions into the uterine fluid, regardless of their origin, could conceivably trigger renewed embryonic development. (2) The rapidly changing hormonal milieu that is temporally correlated with renewed embryonic development may stimulate synthesis of uterine-specific factors that enhance embryonic development. (3) Changes in ovarian hormones might inhibit synthesis of a uterine fluid constituent that had been inhibiting or retarding embryonic development. Each hypothesis has its proponents; however, each is supported by very limited amounts of data obtained from studies with noncarnivores that exhibit embryonic diapause (mice, rats, roe deer). Other studies involving rodents and pigs suggest that the developing embryo may also play an active role in triggering implantation through its secretion of catechol estrogens and prostaglandins (Pakrasi and Dey 1982, 1983; Kantor et al. 1985; Mondschein et al. 1985). These substances in turn are believed to modify the uterine environment in the immediate vicinity of the embryo, thereby facilitating implantation. Unfortunately, our knowledge regarding implantation in carnivores has not progressed to an equivalent stage, and there are no data to support or refute any of these ideas in any carnivore.

The Evolution of Delayed Implantation

Numerous investigators have speculated on the ecological significance and/ or selective pressures that might have favored development of delayed implantation. Most believe embryonic diapause to be a reproductive specialization derieved from the more primitive condition consisting of a short gestation period of constant duration.

One can identify five basic hypotheses regarding the ecological significance of delayed implantation. The first of these states that delayed implantation is of selective advantage in that it permits the young to be born as early as possible in the spring, thereby permitting them somewhat longer to develop physically and behaviorally before facing the rigors of winter (Fries 1880; Prell 1927, 1930; Harrison and Kooyman 1968; Canivenc as quoted in Enders 1981; King, 1984). Prell (1927) and King (1984) have suggested that a cooling environment such as would be encountered during periods of glaciation served as the ecological factor that produced selection pressure for early parturition and rapid maturation of the young.

Murr (1929) is the only investigator to suggest that delayed implantation evolved as a direct result of a reduction in body temperature. Although some carnivores that exhibit embryonic diapause become quite lethargic and experience reduced body temperature during the winter (e.g., the American black bear), this reduction in core temperature is not the cause of delayed implantation, as mating and initiation of embryonic diapause occurs in the summer (June).

Other investigators have suggested that selection pressures favored a later mating season. However, without a delay of implantation the young would be born in late fall or winter, which would necessitate the invention of embryonic diapause to postpone parturition until a more favorable time of year. Each proponent of this hypothesis has given this idea a different slant. For example, Lack (1954) suggested that delayed implantation allowed species to avoid breeding during inclement seasons. Wright (1963) suggested that the late breeding season would be advantageous to species like the short-tailed weasel in that adult males might assist the female in feeding her litter and in exchange would inseminate the female and her sexually precocial offspring. Sandell (1985) has proposed that obligate delay of implantation has evolved as a mechanism that permits females to select the best possible mates by creating the most competition between males. According to this hypothesis, the best time for such mate competition would be when food was most abundant so that each male could spend the maximum amount of time competing for estrous females. But why wouldn't strenuous mate competition occur regardless of time of mating? One could also argue that if mating occurred at a much less favorable time of year, only the most fit males would be able to locate and successfully mate with several females during the breeding season.

Renfree (1978) has suggested that the primary selective advantage of embryonic diapause that is common to all species is that it ensures synchrony of one or more reproductive processes. Harrison and Kooyman (1968) proposed that delayed implantation enabled parturition and mating to be synchronized in pinnipeds. Stenson (1985) suggested that delayed implantation permitted American river otters (*Lutra canadensis*) to synchronize the time of birth without a similar synchronization in mating. This would allow mating to occur over a relatively long period of time. Alternatively, Stenson proposed that delayed implantation might have temporally synchronized parturition and breeding. Stenson believes this might be of advantage in that lactating females would be much more restricted to the vicinity of their dens, making it easier for males to locate estrous females at this time. This hypothesis would not fit species, like the western spotted skunk, that do not mate until lactation has ceased.

Another hypothesis proposes that delayed implantation facilitates the prolonged separation of the sexes after the breakup of the harems in pinnipeds (Harrison and Kooyman 1968). If the male were to remain with the female, he would be competing for the same food resource. However, a prolonged pregnancy and a restricted breeding season brings the male and female into direct competition for a finite resource for only brief periods each year (Short as quoted in Enders 1981).

Another novel hypothesis suggests that delayed implantation evolved as a mechanism for small voracious predators, such as long and short-tailed weasels, to limit their population size (Heidt 1970). If all species of weasels were capable of producing two or more litters per year, as is the case with the diminutive least weasel (*Mustela nivalis*), the prey base might not support the population.

Many of these hypotheses suffer from the fact that they attempt to identify a single selection pressure or single ecological advantage that would be of common benefit to all species. This approach does not appear reasonable or warranted, for the following reason. An obligate or seasonal delay of implantation is known to occur in one or more species belonging to six orders of mammals (Marsupialia, Edentata, Chiroptera, Artiodactyla, Insectivora, and Carnivora). Although this trait is broadly distributed among mammals, close scrutiny of its occurrence within each order reveals a spotty distribution (see Table 16.1). This strongly suggests that delayed implantation has evolved independently several different times in most genera exhibiting this trait. If this is indeed the case, there is no compelling reason to suspect that it evolved in response to the same selection pressure each time it appeared. One would also predict that delayed implantation would have conveyed some specific adaptive advantage to at least some species in each order of mammals; otherwise, one would not expect to encounter the trait in such diverse groups. However, the selective advantage derived from this reproductive trait need not be the same for each species.

If one accepts the argument that delayed implantation has evolved independently several different times in the past, it should also stand to reason that it must have been a relatively easy evolutionary step to accomplish. One of the main requirements would be that blastocysts could survive for long periods without undergoing continuous development. Blastocysts of the domestic ferret, a species that does not exhibit delayed implantation, can withstand short periods of experimentally induced diapause (Foresman and Mead 1978), suggesting that the ability of the blastocyst to withstand diapause may be relatively common in carnivores even though not all species employ this strategy.

The selection pressures that brought about the evolution of delayed implantation were undoubtedly varied, and the selective advantage or advantages conferred by this trait probably differed for each species or genus. Unfortunately, we lack sufficient knowledge of the ecological conditions that prevailed at the time the trait evolved and thus can only guess as to the selection pressures that were operating and the selective advantages that might have been derived from this specialization. The spotty distribution of delayed implantation and the fact that one cannot readily identify any distinct ecological advantage of this feature in today's environment (Hamlett 1935) also suggests that once this trait evolved it was not so simple to abandon. Such a trait would be retained provided it was not selected against. Consequently, functional values of this reproductive specialization might today be secondary or fortuitous rather than a circumstance of the original evolutionary design.

The genus *Spilogale* can be used as an example to postulate how delayed implantation might have evolved, but the model may not apply to other species. Van Gelder's (1959) distribution map for the genus *Spilogale* shows the greatest diversity of forms occurring in Central America. Moreover, Van Gelder considers the pygmy spotted skunk (*S. pygmaea*) to be more primitive than other existing members of the genus, in that it more closely resembles fossil forms. This further suggests that the genus first evolved in Central America or southern Mexico and subsequently migrated north. The pygmy spotted skunk has a short gestation of approximately 48 days, with no known period of delayed implantation (Teska et al. 1981). Pregnant or lactating females have been captured from May through August, thus suggesting that this species produces two litters per year (Teska et al. 1981). If we accept Van Gelder's assertion that this species is primitive, then delayed implantation, which is known to occur only in the more northernly distributed western spotted skunk (*S. gracilis*), must be considered an advanced or specialized condition. Both the fossil and recent distribution records (reviewed by Van Gelder 1959) suggest that members of this genus have been extending their range northward after recession of the glaciers, at which time some spotted skunks invaded regions east of the continental divide whereas others inhabited regions west of the divide. The mountains have served as an isolating barrier, as the eastern population has a short gestation period with no known period of delayed implantation (Mead 1968a), 64 chromosomes (Hsu and Mead 1969), and in some southern parts of its range it may produce two litters per year (Gates 1937; Van Gelder 1959). The western population has 60 chromosomes (Hsu and Mead 1969), a distinctly different shaped baculum (Mead 1967), and a very long gestation period accompanied by delayed implantation (Mead 1968b). The latter may have evolved in the following manner.

Spotted skunks evolved in Central America and Mexico, and all originally had an extended breeding season and often produced two litters per year. This assumption is supported by field data accompanying museum specimens collected from this region (Van Gelder 1959). As spotted skunks gradually extended their range northward, they encountered greater seasonal variability in day length. This caused onset of the spring breeding season to be somewhat delayed in the more northern latitudes, which in turn caused delay of the onset of a second estrus after parturition. Females that bred during the second estrus were now pregnant at a time of year when the duration of darkness each day was rapidly increasing. This resulted in the secretion of melatonin from the pineal for a longer time each day, which in turn altered hypothalamic-pituitary function. Seasonal changes in hypothalamic-pituitary response to feedback from ovarian hormones, as is known to occur in ewes (Legan and Karsch 1979; Goodman et al. 1982), could result in reduced ovarian function and inadequate preparation of the uterus. Consequently, embryonic development became retarded but did not result in death of the blastocysts because of their innate ability to undergo discontinuous development. The blastocysts continued their diapause until the following spring, at which time increasing day length resulted in the secretion of melatonin for a shorter time each day. This resulted in altered hypothalamic-pituitary response to the feedback of ovarian hormones and increased secretion of prolactin. This stimulated increased luteal function, which in turn promoted resumption of embryonic development, and implantation. Lactation further delayed onset of estrus, which now occurred in late summer or early fall and thus effectively limited females to a single litter per year.

Delayed implantation may have been of some ecological advantage in that young skunks could be born approximately 15–30 days earlier than those of females breeding in the spring. Such a differential in parturition dates occurs between eastern and western spotted skunks inhabiting the more northern limits of their range today. Since these small carnivores rapidly reach adult size (Crabb 1944), an additional 30 days would permit them to develop better survival skills and thus increase their chance of surviving a long, harsh winter. Note that this scenario does not require a specific selection pressure for its occurrence. However, it does require a neutral or positive selection pressure for its retention. The hypothesis does not adequately explain why females inhabiting regions east of the continental divide failed to develop delayed implantation. Perhaps they did but the trait was selected against for reasons that are not now apparent. On the other hand, it is possible that only some populations had the ability to produce two litters per year, and this trait was not possessed by the initial invaders of the east side of the continental divide. Consequently, females of this population would not have faced the dilemma of having to delay the birth of a second litter resulting from a postpartum mating.

Summary

Delayed implantation is a reproductive specialization that has evolved independently within several families and genera of carnivores. More data regard-

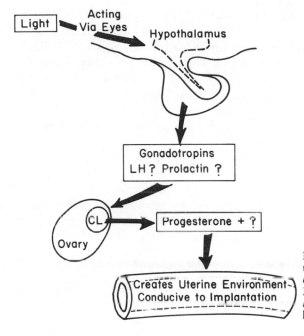

Figure 16.2. The major steps in the physiological regulation of delayed implantation in mustelids. Many but not all events depicted are presumed to be applicable to bears and pinnipeds.

ing the phylogenetic relationships of all carnivores and the presence or absence of this reproductive trait in other species is needed before we can more adequately comment on its evolutionary history. Although delayed implantation was undoubtedly of some selective advantage in the past, it is currently of questionable ecological significance for many species. For example, of what value can it be to sea otters (*Enhydra lutris*), which give birth during all months of the year (Sinha et al. 1966; Estes, this volume)?

Figure 16.2 summarizes the current theory regarding the physiological control of delayed implantation in mustelids. Light acting via the eyes induces a change in both hypothalamic and pituitary function. The altered secretion of pituitary hormones stimulates increased ovarian activity, particularly within the corpus luteum. The latter increases in size and secretes increased quantities of progesterone and some additional proteinaceous factor that induces renewed embryonic development and implantation. This presumably occurs by altering the uterine environment, rendering it more conducive to embryonic growth and development.

Although this theory is consistent with most of the existing data, it has several weaknesses. The site of action of melatonin remains unknown. Consequently, we do not understand how light alters hypothalamic-pituitary function. In other species, the pulsatile pattern of episodic hypothalamic hormone secretion changes during the transition from anestrus to estrus. Perhaps a similar change occurs during the transition from diapause to implantation in carnivores, yet this remains to be documented. More studies are needed to determine the chemical identity of the luteal protein factor that acts in conjunction with progesterone to induce implantation. Do the ovarian hormones induce renewed embryonic development and implantation by acting directly on the embryos, the uterus, or both? Are the physiological mechanisms that control delayed implantation the same in all carnivores? Answers to some of these intriguing questions may never be answered as populations of many of these magnificant animals are threatened due to habitat destruction, environmental toxins, and competition with humans.

Acknowledgments

Portions of this work were supported by a grant from the National Institutes of Child Health and Human Development (HD 06556).

References

Ansell, W. F. H. 1960. *Mammals of Northern Rhodesia.* Lusaka: Government Printer.

Autuori, M. P., and Deutsch, L. A. 1977. Contribution to the knowledge of the giant Brazilian otter *Pteronura braziliensis. Zoologischer Garten* 47:1–8.

Baevskii, Y. B. 1964. Changes in the anterior lobe of the hypophysis, corpora lutea of pregnancy, and the thyroid in the mink (*Mustela vison*) associated with implantation of embryos. *Doklady Akademii Nauk SSSR* 157:1493–1495.

Baevsky, U. B. 1963. The effect of embryonic diapause on the nuclei and mitotic activity of mink and rat blastocysts. In: A. C. Enders, ed. *Delayed Implantation*, pp. 141–153. Chicago: Univ. Chicago Press.

Baker, R. C., Wilke, F., and Baltzo, C. H. 1970. The northern fur seal. *U.S. Fish and Wildlife Service Circular* no. 336:1–19.

Ball, M. P. 1978. Reproduction in captive born zorillas at the National Zoological Park, Washington. *Internat. Zoo Yearb.* 18:140–143.

Baum, M. J., Lynch, H. J., Gallagher, C. A., and Deng, M. H. 1986. Plasma and pineal melatonin levels in female ferrets housed under long or short photoperiods. *Biol. Reprod.* 34:96–100.

Belyaev, D. K., Pereldik, N. S., and Portnova, N. T. 1951. Experimental reduction of the period of embryonal development in sables (*Martes zibellina* L.). *J. General Biol.* 12:260–265.

Bernatskii, V. G., Snytko, E. G., and Nosova, H. G. 1976. Natural and induced ovulation in the sable (*Martes zibellina* L). Translated by Consultants Bureau, N.Y. from *Doklady Akademii Nauk SSSR* 230:1238–1239.

Berria, M., Joseph, M. M., and Mead, R. A. 1988. Role of prolactin and luteinizing hormone in regulating timing of implantation in the spotted skunk. *Biol. Reprod.* 38: Suppl. 1, Abst. 407.

Bertram, G. C. L. 1940. *The Biology of the Weddell and Crabeater Seals, with a Study of the Comparative Behavior of the Pinnipedia.* Scientific reports of the Graham Land Expedition, 1934–37. London: British Museum of Natural History.

Bigg, M. A. 1981. Harbour seal—*Phoca vitulina* and *P. largha.* In: S. H. Ridgway & R.

J. Harrison, eds. *Handbook of Marine Mammals*, pp. 1–27. New York: Academic Press.

Bigg, M. A., and Fisher, H. D. 1974. The reproductive cycle of the female harbour seal off south-eastern Vancouver Island. In: R. J. Harrison, ed. *Functional Anatomy of Marine Mammals*, pp. 1–27. New York: Academic Press.

Blackhouse, K. M., and Hewer, H. R. 1956. Delayed implantation in the gray seal *Halichoerus grypus* (Fab). *Nature* 178:550.

Bloxam, Q. 1977. Breeding the spectacled bear *Tremarctos ornatus* at Jersey Zoo. *Internat. Zoo Yearb.* 17:158–161.

Bonner, W. N. 1972. The grey seal and the common seal in European waters. *Oceanography and Marine Biology: An Annual Review* 10:461–507.

Boshier, D. P. 1981. Structural changes in the corpus luteum and endometrium of seals before implantation. *J. Reprod. Fert. Supplement.* 29:143–149.

Brosseau, C., Johnson, M. L., Johnson, A. M., and Kenyon, K. W. 1975. Breeding the sea otter *Enhydra lutris* at Tacoma Aquarium. *Internat. Zoo Yearb.* 15:144–147.

Canivenc, R. 1970. Photopériodisme chez quelques mammifères a nidation différée. In: J. Benoit & I. Assenmacher, eds. *La photorégulation de la reproduction chez les oiseaux et les mammifères*, pp. 453–466. Paris: Centre National de la Recherche Scientifique.

Canivenc, R., and Bonnin, M. 1981. Environmental control of delayed implantation in the European badger (*Meles meles*). *J. Reprod. Fert. Supplement.* 29:25–33.

Canivenc, R., Bonnin-Laffargue, M., and Lajus, M. 1966. L'utérus gravide a-t-il une fonction lutéotrope chez le vison (*Mustela vison*)? *Comptes rendus seances de la Société de Biologie* 160:2285–2287.

Canivenc, R., Bonnin-Laffargue, M., and Lajus-Boue, M. 1969. Induction de nouvelles générations lutéales pendant la progestation chez la martre européene (*Martes martes* L.). *Comptes rendus Académie Sciences* (Paris) ser. D 269:1437–1440.

Canivenc, R., Laffargue, M. 1956. Présence de blastocystes libres intra-utérins au cours de la lactation chez le blaireau européen (*Meles meles*). *Comptes rendus séances de la Société de Biologie* 150:1193–1196.

Canivenc, R., and Laffargue, M. 1958. Action de differents équilibres hormonaux sur la phase de vie libre de l'oeuf fécondé chez le blaireau européen (*Meles meles* L.). *Comptes rendus séances de la Société de Biologie* 152:58–61.

Canivenc, R., Mauget, C., Bonnin, M., and Aitken, J. 1981. Delayed implantation in the beech marten *Martes foina*. *J. Zool.* 193:325–332.

Cochrane, R. L., and Shackelford, R. M. 1962. Effects of exogenous oestrogen alone and in combination with progesterone on pregnancy in the intact mink. *J. Endocrinol.* 25:101–106.

Crabb, W. D. 1944. Growth, development and seasonal weights of spotted skunks. *J. Mamm.* 25:213–221.

Craig, A. M. 1964. Histology of reproduction and the estrus cycle in the female fur seal (*Callorhinus ursinus*). *J. Fish Res. Board Canada* 21:773–811.

Craighead, J. J., Hornocker, M. G., and Craighead, F. C. 1969. Reproductive biology of young female grizzly bears. *J. Reprod. Fert. Supplement.* 6:447–475.

Daniel, J. C., Jr. 1970. Coincidence of embryonic growth and uterine protein in the ferret. *J. Embryol. Exp. Morphol.* 24:305–312.

Daniel, J. C., Jr. 1971. Growth of the preimplantation embryo of the northern fur seal and its correlation with changes in uterine protein. *Devel. Biol.* 26:316–322.

Daniel, J. C., Jr. 1974. Conditions associated with embryonic diapause during reproduction in the black bear. In: *Second Eastern Workshop on Black Bear Management and Research*, Progress Report—A contribution from Federal Aid to Wildlife Restoration, pp. 103–111. [Gatlinburg, Tenn: Great Smokey Mountain National Park.]

Daniel, J. C., Jr. 1981. Delayed implantation in the northern fur seal (*Callorhinus ursinus*) and other pinnipeds. *J. Reprod. Fert. Supplement.* 29:35–50.

Dathe, H. 1963. Beitrag zur Fortplanzungsbiologie des Malaien Bären, *Helarctos m. malayanus* (Raffl.) *Zeitschrift für Säugetierkunde* 28:155–162.

Dathe, H. 1967. Bemerkungen sur Aufzucht von Brillenbären, *Tremarctos ornatus* (Cuv.), im Tierpark Berlin. *Zoologischer Garten* 34:105–133.

Dathe, H. 1970. A second generation birth of captive sun bears. *Internat. Zoo Yearb.* 10:79.

Deanesly, R. 1943. Delayed implantation in the stoat (*Mustela mustela*). *Nature* 151:365–366.

Deanesly, R. 1944. The reproductive cycle of the female weasel (*Mustela nivalis*). *Proc. Zool. Soc. London* 114:339–349.

Desai, J. H. 1974. Observations on the breeding habits of the Indian smooth otter *Lutrogale perspicillata*. *Internat. Zoo Yearb.* 14:123–124.

Dittrich, L. 1961. Zur Werfzeit des Eisbären (*Ursus maritimus*). *Säugetier. Mitt.* 9:12–15.

Dittrich, L., and Kronberger, H. 1963. Biologisch-anatomische Untersuchungen über die Fortpflanzungsbiologie des Braunbären (*Ursus arctos L.*) und anderen Ursiden in Gefangenschaft. *Zeitschrift für Säugetierkunde* 28:129–155.

Duplaix-Hall, N. 1975. River otters in captivity: A review. In: R. D. Martin, ed. *Breeding Endangered Species in Captivity*, pp. 315–327. New York: Academic Press.

Eadie, W. R., and Hamilton, W. J. 1958. Reproduction in the fisher in New York. *New York Fish and Game J.* 5:77–83.

Encke, W. 1968. A note on breeding and rearing of tayras *Eira barbara* at Krefeld Zoo. *Internat. Zoo Yearb.* 8:132.

Enders, A. C. 1981. Embryonic diapause—Perspectives. *J. Reprod. Fert.*, Suppl. 29:229–241.

Enders, A. C., and Given, R. L. 1977. The endometrium of delayed and early implantation. In: R. M. Wynn, ed. *Biology of the Uterus*, pp. 203–243. New York: Plenum Press.

Enders, A. C., Schlafke, S., Hubbard, N. E., and Mead, R. A. 1986. Morphologic changes in the blastocyst of the western spotted skunk during activation from delayed implantation. *Biol. Reprod.* 34:423–437.

Enders, R. K. 1952. Reproduction in the mink. *Proc. Amer. Phil. Soc.* 96:691–755.

Enders, R. K., and Pearson, O. P. 1943. The blastocyst of the fisher. *Anat. Rec.* 85:285–287.

Erickson, A. W., Nellor, J., and Petrides, G. A. 1964. *The Black Bear in Michigan*. Michigan State University Agricultural Experiment Station Research Bulletin no. 4. [East Lansing: Michigan State Univ.]

Fay, F. H. 1981. Walrus—*Odobenus rosmarus* (Linnaeus, 1758). In: S. H. Ridgway & R. J. Harrison, eds. *Handbook of Marine Mammals*, 1:1–23. New York: Academic Press.

Fay, F. H. 1982. Ecology and biology of the Pacific walrus *Odobenus rosmarus divergens*, Illiger. *North American Fauna* no. 74:188–197. Washington, D.C.: U.S. Fish and Wildlife Service.

Fazleabas, A. T., Mead, R. A., Rourke, A. W., and Roberts, R. M. 1984. Presence of an inhibitor of plasminogen activator in uterine fluid of the western spotted skunk during delayed implantation. *Biol. Reprod.* 30:311–322.

Fisher, H. D. 1954. Delayed implantation in the harbour seal, *Phoca vitulina* L. *Nature* 173:879–880.

Foresman, K. R., and Daniel, J. C., Jr. 1983. Plasma progesterone concentrations in pregnant and non-pregnant black bears (*Ursus americanus*). *J. Reprod. Fert.* 68:235–239.

Foresman, K. R., and Mead, R. A. 1974. Pattern of luteinizing hormone secretion during implantation in the spotted skunk (*Spilogale putorius latifrons*). *Biol. Reprod.* 11:475–480.

Foresman, K. R., and Mead, R. A. 1978. Luteal control of nidation in the ferret (*Mustela putorius*). *Biol. Reprod.* 18:490–496.

Fries, S. 1880. Über die Fortpflanzung von *Meles taxus*. *Zoologischer Anzeiger* 3:486–492.

Frost, K. J., and Lowry, L. F. 1981. Ringed, baikal, and caspian seals—*Phoca hispida*, *Phoca sibirica* and *Phoca caspica*. In: S. H. Ridgway & R. J. Harrison, eds. *Handbook of Marine Mammals*, pp. 29–53. New York: Academic Press.

Gates, W. H. 1937. Spotted skunks and bobcat. *J. Mamm.* 18:240.

Goodman, R. L., Bittman, E. L., Foster, D. L., and Karsch, F. J. 1982. Alterations in the control of luteinizing hormone pulse frequency underlie the seasonal variation in estradiol negative feedback in the ewe. *Biol. Reprod.* 27:580–589.

Greensides, R. D., and Mead, R. A. 1973. Ovulation in the spotted skunk (*Spilogale putorius latifrons*). *Biol. Reprod.* 8:576–584.

Gulyas, B. J., and Daniel, J. C., Jr. 1967. Oxygen consumption in diapausing blastocysts. *J. Cell. Physiol.* 70:33–36.

Gulyas, B. J., and Daniel, J. C., Jr. 1969. Incorporation of labeled nucleic acid and protein precursors by diapausing and nondiapausing blastocysts. *Biol. Reprod.* 1:11–20.

Hamilton, J. E. 1939. The leopard seal (*Hydrurga leptonyx* De Blainville). *Discovery Reports* 18:239–264.

Hamilton, W. J., and Eadie, W. R. 1964. Reproduction in the otter, *Lutra canadensis*. *J. Mamm.* 45:242–252.

Hamlett, G. W. D. 1935. Delayed implantation and discontinuous development in the mammals. *Quart. Rev. Biol.* 10:432–447.

Hammond, J., and Walton, A. 1934. Notes on ovulation and fertilization in the ferret. *J. Exp. Biol.* 11:307–319.

Hansson, A. 1947. The physiology of reproduction in mink (*Mustela vison*, Schreb.) with special reference to delayed implantation. *Acta Zool.* 28:1–136.

Harris, C. J. 1968. *Otters—A Study of the Recent Lutrinae*. London: Weidenfeld and Nicolson.

Harrison, R. J. 1963. A comparison of factors involved in delayed implantation in badgers and seals in Great Britain. In: A. C. Enders, ed. *Delayed Implantation*, pp. 99–114. Chicago: Univ. Chicago Press.

Harrison, R. J. 1969. Reproduction and reproductive organs. In: H. T. Anderson, ed. *Biology of Marine Mammals*, pp. 253–348. New York: Academic Press.

Harrison, R. J., and Kooyman, G. L. 1968. General physiology of the Pinnipedia. In: R. J. Harrison, R. C. Hubbard, R. S. Peterson, & C. E. Rice, eds. *The Behavior and Physiology of Pinnipeds*, pp. 212–296. New York: Appleton-Century-Crofts.

Harrison, R. J., Matthews, L. H., and Roberts, J. M. 1952. Reproduction in some Pinnipedia. *Trans. Zool. Soc. London* 27:437–540.

Hartman, L. 1964. The behaviour and breeding of captive weasels (*Mustela nivalis*). *New Zealand J. Sci.* 7:147–156.

Heidt, G. A. 1970. The least weasel *Mustela nivalis* Linnaeus. Developmental biology in comparison with other North American *Mustela*. *Publications of the Museum Michigan State Univ., Biological Series* 4:227–282.

Herlant, M., and Canivenc, R. 1960. Les modifications hypophysaires chez la femelle du Blaireau (*Meles meles* L.) au cours du cycle annuel. *Comptes rendus Hebdomadaires des séances Académie des Sciences* (Paris) 250:606–608.

Hewer, H. R., and Blackhouse, K. M. 1968. Embryology and foetal growth of the grey seal, *Halichoerus grypus*. *J. Zool.* 155:507–533.

Hillman, C. N., and Carpenter, J. W. 1983. Breeding biology and behavior of captive black-footed ferrets. *Internat. Zoo Yearb.* 23:186–191.

Hsu, T. C., and Mead, R. A. 1969. Mechanisms of chromosomal changes in mammalian speciation. In: K. Benirschke, ed. *Comparative Mammalian Cytogenetics*, pp. 9–17. New York: Springer-Verlag.

Jacobi, E. F. 1975. Breeding sloth bears in Amsterdam Zoo. In: R. D. Martin, ed. *Breeding Endangered Species in Captivity*, pp. 351–356. New York: Academic Press.

Kantor, B. S., Dey, S. K., and Johnson, D. C. 1985. Catechol oestrogen induced initiation of implantation in the delayed implanting rat. *Acta Endocrinol.* 109:418–422.

King, C. M. 1984. The origin and adaptive advantages of delayed implantation in *Mustela erminea. Oikos* 42:126–128.

Kingdon, J. 1977. *East African Mammals*. New York: Academic Press.

Kintner, P. J., and Mead, R. A. 1983. Steroid metabolism in the corpus luteum of the ferret. *Biol. Reprod.* 29:1121–1127.

Kirk, R. J. 1962. The effect of darkness on the mink reproductive cycle. *Amer. Fur Breeder* 35:20–21.

Kooyman, G. L. 1981. Weddell seal—*Leptonychotes weddelli*. In: S. H. Ridgway & R. J. Harrison, eds. *Handbook of Marine Mammals*, pp. 275–296. New York: Academic Press.

Lack, D. 1954. *The Natural Regulation of Animal Numbers*. London: Oxford Univ. Press.

Laurie, A., and Seidensticker, J. 1977. Behavioural ecology of the sloth bear (*Melursus ursinus*). *J. Zool.* 182:187–204.

Lavrov, N. P. 1944. Biology of ermine reproduction (*Mustela erminea* L.). *People's Commissariat for Procurement, U.S.S.R. Transactions of the Central Laboratory of Biology of Game Animals* 6:124–149.

Laws, R. M. 1956. The elephant seal (Mirounga leonina Linn.). III. The physiology of reproduction. *Falkland Islands Dependencies Survey Scientific Reports* 15:1–66.

Legan, S. J., and Karsch, F. J. 1979. Neuroendocrine regulation of the estrous cycle and seasonal breeding in the ewe. *Biol. Reprod.* 20:74–85.

Leslie, G. 1970. Observations on the oriental short-clawed otter *Aonyx cinerea* at Aberdeen Zoo. *Internat. Zoo Yearb.* 10:79–81.

Ling, J. K., and Bryden, M. M. 1981. Southern elephant seal—*Mirounga leonina*. In: S. H. Ridgway & R. J. Harrison, eds. *Handbook of Marine Mammals*, pp. 297–327. New York: Academic Press.

McCusker, J. S. 1974. Breeding Malayan sun bears at Forth Worth Zoo. *Internat. Zoo Yearb.* 15:118–119.

McLaren, I. A. 1958. Some aspects of growth and reproduction of the bearded seal, *Erignathus barbatus* (Erxleben). *J. Fish. Res. Board Canada* 15:219–227.

Madsen, A. B., and Rasmussen, A. M. 1985. Reproduction in the stone marten *Martes foina* in Denmark. *Natura Jutlandica Natural History Museum*, 8000 Aarhus C. Denmark 21:145–148.

Mansfield, A. W. 1958. The breeding and reproductive cycle of the Weddell seal (*Leptonychotes weddelli*, Lesson). *Falkland Islands Dependencies Survey, Scientific Reports* 18:1–41.

Martinet, L., Allain, D., and Meunier, M. 1983. Regulation in pregnant mink (*Mustela vison*) of plasma progesterone and prolactin concentrations and regulation of onset of the spring moult by daylight ration and melatonin injections. *Canadian J. Zool.* 61:1959–1963.

Martinet, L., Allais, C., and Allain, D. 1981. The role of prolactin and LH in luteal function and blastocyst growth in mink (*Mustela vison*). *J. Reprod. Fert. Supplement* 29:119–130.

May, R., DeSantis, M., and Mead, R. A. 1985. The suprachiasmatic nuclei and retinohypothalamic tract in the western spotted skunk. *Brain Res.* 339:378–381.

May, R., and Mead, R. A. 1986. Evidence for pineal involvement in timing implantation in the western spotted skunk. *J. Pineal Res.* 3:1–8.

Mead, R. A. 1967. Age determination in the spotted skunk. *J. Mamm.* 48:606–616.

Mead, R. A. 1968a. Reproduction in eastern forms of the spotted skunk (genus *Spilogale*). *J. Zool. (Lond.)* 156:119–136.

Mead, R. A. 1968b. Reproduction in western forms of the spotted skunk (genus *Spilogale*). *J. Mamm.* 49:373–390.

Mead, R. A. 1971. Effects of light and blinding upon delayed implantation in the spotted skunk. *Biol. Reprod.* 5:214–220.

Mead, R. A. 1972. Pineal gland: Its role in controlling delayed implantation in the spotted skunk. *J. Reprod. Fert.* 30:147–150.

Mead, R. A. 1975. Effects of hypophysectomy on blastocyst survival, progesterone secretion and nidation in the spotted skunk. *Biol. Reprod.* 12:526–533.

Mead, R. A. 1981. Delayed implantation in the Mustelidae with special emphasis on the spotted skunk. *J. Reprod. Fert. Supplement* 29:11–24.

Mead, R. A. 1986. Role of the corpus luteum in controlling implantation in mustelid carnivores. *Ann. New York Acad. Sci.* 476:25–35.

Mead, R. A., Concannon, P. W., and McRae, M. 1981. Effect of progestins on implantation in the western spotted skunk. *Biol. Reprod.* 25:128–133.

Mead, R. A., Joseph, M. M., Neirinckx, S. 1988. Partial characterization of a luteal factor that induces implantation in the ferret. *Biol. Reprod.* 38:798–803.

Mead, R. A., and Rourke, A. W. 1985. Accumulation of RNA in blastocysts during embryonic diapause and the periimplantation period in the western spotted skunk. *J. Exp. Zool.* 235:65–70.

Mead, R. A., Rourke, A. W., and Swannack, A. 1979. Uterine protein synthesis during delayed implantation in the western spotted skunk and its regulation by hormones. *Biol. Reprod.* 21:39–46.

Mead, R. A., and Swannack, A. 1978. Effects of hysterectomy on luteal function in the western spotted skunk (*Spilogale putorius latifrons*). *Biol. Reprod.* 18:379–383.

Mendelssohn, H., Ben-David, M., and Hellwing, S. 1988. Reproduction and growth of the marbled polecat (*Vormela peregusna syriaca*) in Israel. *J. Reprod. Fert. Abstract Series No.* 1:20.

Møller, O. M. 1974. Plasma progesterone before and after ovariectomy in unmated and pregnant mink, *Mustela vison. J. Reprod. Fert.* 37:367–372.

Mondain-Monval, M., Bonnin, M., Canivenc, R., and Scholler, R. 1980. Plasma estrogen levels during delayed implantation in the European badger (*Meles meles* L.). *Gen. Comp. Endocrinol.* 41:143–149.

Mondschein, J. S., Hersey, R. M., Dey, S. K., Davis, D. L., and Weisz, J. 1985. Catechol estrogen formation by pig blastocysts during the preimplantation period: Biochemical characterization of estrogen-$2/4$-hydroxylase and correlation with aromatase activity. *Endocrinology* 117:2339–2346.

Moshonkin, N. N. 1981. Potential polyestricity of the mink (*Lutreola lutreola*). *Zoologicheskii Zhurnal.* 60:1731–1734.

Moshonkin, N. N. 1983. The reproductive cycle in females of the European mink (*Lutreola lutreola*). *Zoologicheskii Zhurnal.* 62:3–15.

Murphy, B. D. 1983. Precocious induction of luteal activation and termination of delayed implantation in mink with the dopamine antagonist pimozide. *Biol. Reprod.* 29:658–662.

Murphy, B. D., Concannon, P. W., and Travis, H. F. 1982. Effects of medroxyprogesterone acatate on gestation in mink. *J. Reprod. Fert.* 66:491–497.

Murphy, B. D., Concannon, P. W., Travis, H. F., and Hansel, W. 1981. Prolactin: The hypophyseal factor that terminates embryonic diapause in mink. *Biol. Reprod.* 25:487–491.

Murphy, B. D., Mead, R. A., and McKibbin, P. E. 1983. Luteal contribution to the termination of preimplantation delay in mink. *Biol. Reprod.* 28:497–503.

Murphy, B. D., and Moger, W. H. 1977. Progestins of mink gestation: The effects of hypophysectomy. *Endocrine Research Communications* 4:45–60.

Murr, E. 1929. Zur Erklärung der verlängerten Tragdauer bei Säugertieren. *Zoologischer Anzeiger* 85:113–129.

Neal, E. G., and Harrison, R. J. 1958. Reproduction in the European badger (*Meles meles* L.). *Trans. Zool. Soc. London* 29:67–131.

Nishino, H., Koizumi, K., and McBrooks, C. 1976. The role of suprachiasmatic nuclei of the hypothalamus in the production of circadian rhythm. *Brain Res.* 112:45–59.

Novikov, G. A. 1956. *Carnivorous Mammals of the Fauna of the USSR.* Zoological Institute of the Academy of Science of the USSR, no. 62. Jerusalem: Israel Program for Scientific Translations, 1962.

Odell, D. K. 1981. California sea lion—*Zalophus californianus.* In: S. H. Ridgway & R. J. Harrison, eds. *Handbook of Marine Mammals*, pp. 67–97. New York: Academic Press.

Ognev, S. I. 1931. *Mammals of Eastern Europe and Northern Asia*, vol. 2: *Carnivora (Fissipedia).* Jerusalem: Israel Program for Scientific Translations, 1962.

Øritsland, T. 1970. Sealing and seal research in the Southwest Atlantic pack ice, September–October 1964. In: M. W. Holdgate, ed. *Antarctic Ecology*, pp. 367–376. London: Academic Press.

Pakrasi, P. L., and Dey, S. K. 1982. Blastocyst is the source of prostaglandins in the implantation site in the rabbit. *Prostaglandins* 24:73–77.

Pakrasi, P. L., and Dey, S. K. 1983. Catechol estrogens stimulate synthesis of prostaglandins in the preimplantation rabbit blastocyst and endometrium. *Biol. Reprod.* 29:347–354.

Papke, R. L., Concannon, P. W., Travis, H. F., and Hansel, W. 1980. Control of luteal function and implantation in the mink by prolactin. *J. Anim. Sci.* 50:1102–1107.

Parker, C. 1979. Birth, care and development of chinese hog badgers (*Arctonyx collaris albogularis*) at Metro Toronto Zoo. *Internat. Zoo Yearb.* 19:182–185.

Patton, T. S. 1974. Ecological and behavioral relationships of the skunks of Tans-Pecos Texas. PhD. dissert., Texas A&M Univ., College Station, Tex. 199 pp.

Pearson, O. P., and Enders, R. K. 1944. Duration of pregnancy in certain mustelids. *J. Exp. Zool.* 95:21–35.

Peking Zoo. 1974. On the breeding of the giant panda and the development of its cubs. *Acta zool. sinica* 20:151–154.

Petter, F. 1959. Reproduction en captivité du zorilla du Sahara *Poecilictis libyca.* *Mammalia* 23:378–380.

Poglayen-Neuwall, I. 1978. Breeding, rearing and notes on the behavior of tayras (*Eira barbara*). *Internat. Zoo Yearb.* 18:134–140.

Prell, H. 1927. Uber doppelte Brunstzeit und verlängerte Tragzeit bei den einheimischen Arten der Mardergattung *Martes* pinel. *Zoologischer Anzeiger* 74:122–128.

Prell, H. 1930. Die verlängerte Tragzeit der einheimischen *Martes* arten: Ein Erklärungsversuch. *Zoologischer Anzeiger* 88:17–31.

Procter, J. 1963. A contribution to the natural history of the spotted-necked otter (*Lutra maculicollis* Lichtenstein) in Tanganyika. *East African Wildl. J.* 1:93–102.

Puschman, W., Schuppel, K. F., and Kronberger, H. 1977. Detection of blastocyst in uterine lumen of Indian bear *Melursus u. ursinus.* In: R. Ippen & H. D. Schrader, eds. *Sickness in Zoos*, pp. 389–391. Berlin, East Germany: Akad. Verlag.

Rand, R. W. 1955. Reproduction in the female Cape fur seal, *Arctocephalus pusillus* (Schreber). *Proc. Zool. Soc. London* 124:717–740.

Rausch, R. A., and Pearson, A. M. 1972. Notes on the wolverine in Alaska and the Yukon Territory. *J. Wildl. Mgmt.* 36:249–268.

Ravault, J. P., Martinet, L., Bonnefond, C., Claustrat, B., and Brun, J. 1986. Diurnal variations of plasma melatonin concentrations in pregnant or pseudopregnant mink (*Mustela vison*) maintained under different photoperiods. *J. Pineal Res.* 3:365–374.

Ravindra, R., Bhatia, K., and Mead, R. A. 1984. Steroid metabolism in corpora lutea of the western spotted skunk (*Spilogale putorius latifrons*). *J. Reprod. Fert.* 72:495–502.

Ravindra, R., and Mead, R. A. 1984. Plasma estrogen levels during pregnancy in the western spotted skunk. *Biol. Reprod.* 30:1153–1159.

Renfree, M. B. 1978. Embryonic diapause in mammals—A developmental strategy. In: M. C. Clutter, ed., *Dormancy and Developmental Arrest*, pp. 1–46. New York: Academic Press.

Roberts, M. S., and Gittleman, J. L. 1984. *Ailurus fulgens*. Mammalin Species no. 222. Lawrence, Kans.: American Society of Mammalogists.

Roberts, T. J. 1977. *The Mammals of Pakistan*. London: Ernst Benn.

Robinson, A. 1918. The formation, rupture and closure of ovarian follicles in ferrets and ferret-polecat hybrids and some associated phenomena. *Trans. Roy. Soc. Edinburgh* 52:303–362.

Ronald, K., and Healey, J. P. 1981. Harp seal—*Phoca groenlandica*. In: S. H. Ridgway & R. J. Harrison, eds. *Handbook of Marine Mammals*, pp. 55–87. New York: Academic Press.

Rose, J., Stormshak, F., Oldfield, J., and Adair, J. 1985. The effects of photoperiod and melatonin on serum prolactin levels of mink during the autumn molt. *J. Pineal Res.* 2:13–19.

Rosevear, D. R. 1974. *The Carnivores of West Africa*. British Museum of Natural History, no. 723. London: British Museum of Natural History.

Rourke, A. W., and Mead, R. A. 1982. Blastocyst protein synthesis during obligate delay of implantation and embryo activation in the western spotted skunk. *J. Exp. Zool.* 221:87–92.

Rowe-Rowe, D. T. 1978. Reproduction and postnatal development of South African mustelines (Carnivore: Mustelidae). *Zool. Africana* 13:103–114.

Sandell, M. 1985. Ecology and behaviour of the stoat, *Mustela erminea*, and a theory on delayed implantation. Ph.D. dissert., Univ. Lund, Sweden. 155pp.

Schlafke, S., Enders, A. C., and Given, R. L. 1981. Cytology of the endometrium of delayed and early implantation with special reference to mice and mustelids. *J. Reprod. Fert. Supplement* 29:135–141.

Schmidt, F. 1932. Der Steppeniltis (*Putorius eversmanni* Less.). *Deutsche Pelztierzüchter* 7:453–458.

Schusterman, R. J. 1981. Steller sea lion – *Eumetopias jubatus*. In: S. H. Ridgway & R. J. Harrison, eds. *Handbook of Marine Mammals*, pp. 119–141. New York: Academic Press.

Shelden, R. M. 1972. The fate of the short-tailed weasel, *Mustela erminea*, blastocysts following ovariectomy during diapause. *J. Reprod. Fert.* 31:347–352.

Shelden, R. M. 1973. Failure of ovarian steroids to influence blastocysts of weasels (*Mustela erminea*) ovariectomized during delayed implantation. *Endocrinology* 92:638–641.

Sinha, A. A., Conaway, C. H., and Kenyon, K. W. 1966. Reproduction in the female sea otter. *J. Wildl. Mgmt.* 30:121–130.

Smith, M. S. R. 1966. *Studies on the Weddell Seal (Leptonychotes weddelli* Lesson) *in*

McMurdo Sound Antarctica. Ph.D. dissert., Univ. Canterbury, Christchurch, New Zealand.

Spotte, S. 1981. Photoperiod and reproduction in captive female northern fur seals. *Mamm. Rev.* 11:31–35.

Stenson, G. B. 1985. The reproductive cycle of river otters. Ph.D. dissert., Univ. British Columbia, Vancouver.

Stroganov, S. U. 1962. *Carnivorous Mammals of Siberia.* pp. 351–359. Jerusalem: Israeli Program Scientific Translations, 1969.

Stubbe, M. 1968. Zur Populationbiologie der *Martes*—Arten. *Beiträge zur Jagd- und Wildforschung* 104:195–203.

Teska, W. R., Rybak, E. N., and Baker, R. H. 1981. Reproduction and development of the pygmy spotted skunk (*Spilogale pygmaea*). *Amer. Midland Nat.* 105:390–392.

Timmis, W. H. 1971. Observations on breeding the Oriental short-clawed otter *Amblonyx cinerea* at Chester Zoo. *Internat. Zoo Yearb.* 11:109–111.

Trebbau, P. 1972. Notes on the Brazilian giant otter *Pteronura brasiliensis. Zoologischer Garten* 41:152–156.

Tumanov, I. L. 1977. On potential polyestrus in some species of the Mustelidae. *Zoologicheskii Zhurnal* 56:619–625.

Van Gelder, R. G. 1959. A taxonomic revision of the spotted skunks (Genus *Spilogale*). *Bull. Amer. Mus. Nat. Hist.* 117:233–392.

Volf, J. 1963. Bemerkungen zur Fortpflanzungsbiologie der Eisenbären, *Thalarctos maritimus* (Phipps) in Gefangenschaft. *Z. Säugetierk.* 28:163–166.

Wade-Smith, J., and Richmond, M. E. 1975. Care, management, and biology of captive striped skunks (*Mephitis mephitis*). *Lab. Anim. Sci.* 25:575–584.

Wade-Smith, J., and Richmond, M. E. 1978. Reproduction in captive striped skunks (*Mephitis mephitis*). *Amer. Midland Nat.* 100:452–455.

Wade-Smith, J., Richmond, M. E., Mead, R. A., and Taylor, H. 1980. Hormonal and gestational evidence for delayed implantation in the striped skunk, *Mephitis mephitis. General and Comparative Endocrinol.* 42:509–515.

Watzka, M. 1940. Mikroskopisch-anatomische Untersuchungen über die Ranzzeit ünd Tragdauer des Hermelins (*Putorius ermineus*). *Zeitschrift für mikroskopisch-anatomische Forschung* 48:359–374.

Wimsatt, W. A. 1963. Delayed implantation in the Ursidae, with particular reference to the black bear (*Ursus americanus* Pallas). In: A. C. Enders, ed. *Delayed Implantation*, pp. 49–76. Chicago: Univ. Chicago Press.

Wright, P. L. 1942. Delayed implantation in the long-tailed weasel (*Mustela frenata*), the short-tailed weasel (*Mustela cicognani*), and the marten (*Martes americana*). *Anat. Rec.* 83:341–353.

Wright, P. L. 1963. Variations in reproductive cycles in North American mustelids. In: A. C. Enders, ed. *Delayed Implantation*, pp. 77–97. Chicago: Univ. Chicago Press.

Wright, P. L. 1966. Observations on the reproductive cycle of the American badger (*Taxidea taxus*). In: I. W. Rowlands, ed. *Comparative Biology of Reproduction in Mammals*, pp. 27–45. New York: Academic Press.

Wright, P. L., and Coulter, M. W. 1967. Reproduction and growth in Maine fishers. *J. Wildl. Mgmt.* 31:70–87.

Wright, P. L., and Rausch, R. 1955. Reproduction in the wolverine, *Gulo gulo. J. Mamm.* 36:346–355.

CHAPTER 17

Molecular and Biochemical Evolution of the Carnivora

ROBERT K. WAYNE, RAOUL E. BENVENISTE,
DIANNE N. JANCZEWSKI, AND STEPHEN J. O'BRIEN

The fissiped carnivores include eight distinct families that are traditionally grouped into two superfamilies: the Canoidea (or Arctoidea) and the Feloidea (or Aeluroidea). The Canoidea include the bear, dog, raccoon, and weasel families; and the Feloidea include the cat, hyena, mongoose, and civet families. Both groups are extremely heterogeneous with respect to the morphology and life history of their constituents. They include taxa that are entirely carnivorous, insectivorous, and omnivorous and that have cursorial, arboreal, fossorial, and aquatic habits. Such wide-ranging adaptations have led to several instances of parallel and convergent evolution of morphologic traits which have confounded the efforts of taxonomists to relate certain taxa.

Over the last few years a variety of molecular techniques has been applied to determine the evolutionary relationships within and among several carnivore families (Sarich 1969a, 1969b, 1973; Collier and O'Brien 1985; O'Brien et al. 1985, 1987; Goldman et al. 1987; Wayne and O'Brien 1987; Wayne et al. 1987a, 1987b). In this chapter we review the relationships of three carnivore families derived from these studies: the Canidae (dogs), the Ursidae (bears), and the Felidae (cats). We also present a phenogram of the Carnivora, including carnivore species from each of the eight families plus species from two pinniped families, the Otariidae (sea lions) and the Phocidae (earless seals).

The trees we present were derived from evolutionary distance estimates obtained from several molecular techniques, including (1) DNA hybridization, (2) protein electrophoresis, (3) measurement of albumin immunological distance (AID), and (4) high-resolution G-banding of karyotypes. Evolutionary trees were constructed using published phenetic algorithms designed to analyze distance matrices (Fitch and Margoliash 1967; Sneath and Sokal 1973; Dayhoff 1976; Fitch 1981). In this chapter we present the deduced phylogenies, an assessment of the various aspects of confidence and ambiguity for each topology, and an interpretive review of the implications of the molecular results in the context of morphologic and fossil data.

465

Molecular Procedures

In 1962 Zuckerkandl and Pauling suggested that mutations in genomic DNA accumulated in a stochastic but steady fashion that was roughly related to elapsed time. The genetic difference between individuals from different species would therefore be proportional to the amount of time that had passed since they last shared a common ancestor. By assuming that the "molecular clock hypothesis" is valid (Wilson et al. 1977; Nei 1978; Thorpe 1982), one can relate phenetic trees derived from molecular data to absolute time by calibration with a fossil date. For instance, the time of separation of Old and New World primates is approximately 30–50 millions years before present (M.Y.B.P.) (Radinsky 1978); thus, an absolute time scale can be placed on trees of the primate order based on this date, and the rate of molecular evolution can be calculated and used in trees of other groups. However, it is generally preferable to use a calibration date based on species from the group of interest since the rate of gene evolution of different taxonomic groups may vary (Benveniste et al. 1977; Brownell 1983; Britten 1986). The reader is referred to Wilson et al. (1977) and Thorpe (1982) for a technical discussion of the molecular clock hypothesis and to Gribbin and Cherfas (1982) for an excellent description of the contributions of molecular techniques to our understanding of human evolution.

One karyological and three molecular procedures have been utilized in the study of carnivores. We would encourage the use of several procedures because confirmation of evolutionary relationships with multiple, independent methods tends to reveal incorrect deductions and thereby minimizes error in phylogenetic inference (Gribbin and Cherfas 1982; O'Brien et al. 1985; Ayala 1986). For example, all of the methods described here were used to assess the relationships of the giant panda, and a consistent phylogeny was derived (O'Brien et al. 1985).

The first procedure we employ is DNA hybridization (Kohne et al. 1972; Benveniste 1976, 1985). This method involves the hybridization of radioactively labeled cellular DNA of one species to the cellular DNA of other species and measures the stability of DNA hybrids that are formed. Two measurements can be derived from these experiments: first, the percentage of hybridization between species A and B; and second, the difference between a melting profile of heterologous DNA hybrids and that of homologous DNAs. The latter measurement, termed ΔTm, is directly proportional to the extent of base pair mismatching. The ΔTm (or ΔTmR, which is ΔTm corrected for the normalized final percentage of hybridization) values are compiled in a table that is used to construct phenetic trees. DNA hybridization data are particularly powerful for species that diverged 10–60 M.Y.B.P. but less sensitive for comparisons of recently diverged taxa (Sibley and Ahlquist 1983; Benveniste 1985).

The second method involves the estimation of genetic distance (D) (Nei

1972, 1978). In this procedure a series of soluble proteins or isozymes are separated electrophoretically and compared. The extent of genetic difference between two species is then estimated based on differences in allele frequencies at each genetic locus (for a summary of genetic distance data, see Avise and Aquadro 1981). The distance value, D, provides an estimate of the average number of gene differences per locus between individuals from two populations. Under the constraints of certain assumptions relating to the electrophoretic detection of mutations and the relative rates of nucleotide substitution (Nei 1972, 1978), the genetic distance estimates increase proportionately with the amount of time elapsed since the populations shared a common ancestor. Most of our genetic distance estimates have been derived from a group of about 50 genetic loci; however, in a few cases larger data sets have been developed using two-dimensional (2D) gel electrophoresis, which usually resolves over 300 protein gene products (Goldman et al. 1988).

The range of evolutionary time resolved by isozyme genetic distance varies depending on the gene-enzyme systems employed, because different proteins evolve at different rates. Retrospectively, our own results suggest that D values correlate well with divergence times of 0.1–10 M.Y.B.P. Outside these limits, the relationship between D and evolutionary time may not always be linear. Thus, when a tree is calibrated with a single or several fossil dates, the precise timing of the other divergence nodes may not define a straight line through the origin (see O'Brien et al. 1985).

The third method that we employed for use in carnivore systematics is the measurement of albumin immunological distance (AID). This procedure measures the immunological distance between species based upon amino acid substitutions in homologous proteins. Substitutions are detected by the displacement of titration curves in a microcomplement fixation assay. Briefly, several rabbits are immunized with purified serum albumin from species A. The rabbit antiserum is then pooled and titered against albumin from species A. In an evolutionary distance determination, albumin from species B is preincubated with titered antiserum against species A. The adsorbed antiserum is then retested against the homologous species A albumin. The remaining antibodies bind to antigen A and fix complement in an amount quantitatively related to the amount of amino acid sequence difference between the two species. When several antisera against different species are prepared, a matrix of immunological distances can be constructed and used for estimating relationships.

Like DNA hybridization, AID is useful for measurements between species that are rather distant (5–50 million years). This is because albumin evolves rather slowly (compared, for example, with transferrin) such that 1 AID unit $\cong$ 0.6 million years (Maxson and Wilson 1975; Sarich 1973; Thorpe 1982). Divergence times of 1–2 M.Y.B.P. represent the lower limit that can be resolved by this technique (Collier and O'Brien 1985).

Karyological Measurements

The analysis of comparative chromosome morphology has revealed a great deal about the evolution of the Carnivora. In our laboratory, skin biopsies from a large number of carnivore species have been used to establish primary tissue culture lines, and these have been employed to prepare high-resolution G-banded karyotypes. The alignment of banding patterns between chromosomes of several different species has allowed the construction of minimum-distance phylogenetic trees based upon the principles of maximum parsimony. The efficacy of this procedure has been demonstrated since chromosomes of different species found to be homologous by G-banding also share homologous linkage (syntenic) groups (Nash and O'Brien 1982; O'Brien and Nash 1982).

With certain interesting exceptions, the carnivore order has a largely conservative karyotype (Wurster-Hill and Gray 1973, 1975; Wurster-Hill and Centerwall 1982; Dutrillaux and Couturier 1983; Couturier et al. 1986). The Felidae is the prototype family. The domestic cat has 19 chromosomes, 16 of which are invariant in all 37 species of felids. Furthermore, of these 16 chromosomes, 15 are present in several other carnivore families (Procyonidae, Mustelidae, Viverridae, Hyaenidae). In two families, Canidae and Ursidae, there has been a dramatic reorganization of the primitive carnivore karyotype (O'Brien et al. 1985; Nash and O'Brien 1987; Wayne et al. 1987a, 1987b). The conclusions of these studies and those of Wurster-Hill and Gray (1973, 1975) and Wurster-Hill and Centerwall (1982) have been incorporated into our discussion of carnivore evolution.

Relationships of the Felidae

Because of a special fascination that zoologists and naturalists have for cultural, aesthetic, and scientific aspects of the cat family, an extensive literature has accumulated regarding its taxonomy. Although there is a good consensus with respect to species identification of most of the 37 extant felid species, there is little agreement among the various published classification schemes, which range from a division of the Felidae into 19 distinct genera (Ewer 1973) to as few as two genera, *Felis* and *Acinonyx* (Romer 1968) (see Nowak and Paradiso 1983).

In an attempt to resolve such taxonomic and evolutionary ambiguities, we have used two molecular approaches: albumin immunological distance (Collier and O'Brien 1985) and isozyme genetic distance (O'Brien et al. 1987). The AID tree defines three major lineages in the evolution of the Felidae (Figure 17.1). These include the ocelot lineage made up of the small South American cats, the domestic cat lineage including the small Mediterranean cats, and the Panthera lineage consisting of a heterogeneous array of large and small cats

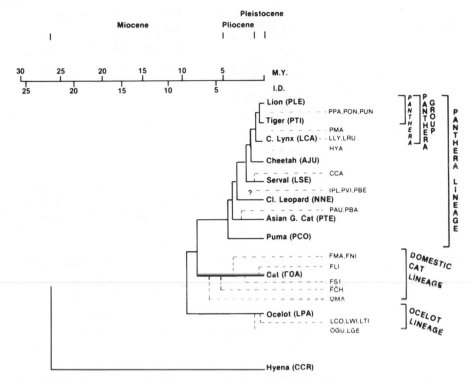

Figure 17.1. UPGMA tree of ten index cat species (*bold print*) based on average reciprocal micro-complement fixation measurements (Sneath and Sokal 1973; Collier and O'Brien 1985). The additional species were added to the tree on the basis of one-way immunological distance (see Collier and O'Brien 1985). Basis of time scale equivalence: 1 immunological distance unit (I.D.) = 0.6 million years (Collier and O'Brien 1985). The geologic time scales uses the convention of Hsu et al. (1984). Codes: AJU, *Acinonyx jubatus;* FCA, *Felis catus;* LPA, *Leopardus pardalis;* LSE, *Leptailurus serval;* LCA, *Lynx canadensis;* NNE, *Neofelis nebulosa;* PLE, *Panthera leo;* PTI, *Panthera tigris;* PTE, *Profelis temmincki;* PCO, *Puma concolor;* CCA, *Caracal caracal;* FCH, *Felis chaus;* FLI, *Felis libyca;* FMA, *Felis margarita;* FNI, *Felis nigripes;* FSI, *Felis silvestris;* HYA, *Herpailurus yagouaroundi;* IPL, *Ictailurus planiceps;* LWE, *Leopardus wiedi;* LTI, *Leopardus tigrina;* LGE, *Leopardus geoffroyi;* LCO, *Lynchailurus colocolo;* LLY, *Lynx lynx;* LRU, *Lynx rufus;* OGU, *Oncifelis guigna;* OMA, *Otocolobus manul;* PON, *Panthera onca;* PPA, *Panthera pardus;* PUN, *Panthera uncia;* PMA, *Pardofelis marmorata;* PBA, *Pardofelis badia;* PBE, *Prionailurus bengalensis;* PVI, *Prionailurus viverrinus;* PAU, *Profelis aurata;* FBI, *Felis bieti;* LPA, *Lynx pardina;* MIR, *Mayailurus iriomotensis;* OJA, *Oreailurus jacobita;* PRU, *Prionailurus rubiginosa;* CCR, *Crocuta crocuta.* Common names are given in the text.

(see discussion below). The common ancestor of extant felids is fairly recent and, according to Figure 17.1, begins approximately 10 m.y.b.p.

Two of the felid groups are also delineated by DNA hybridization data (Table 17.1). Hybridization of radio-labeled domestic cat DNA to DNA of other felids is greatest to species in the domestic cat lineage (Table 17.1, Figure 17.1). Similarly, radio-labeled African lion DNA hybridizes to a greater extent with DNA from species in the pantherid group. Each of these index species

Table 17.1. The thermal stability (ΔTmR) of hybrids formed between the unique sequence DNA of domestic cat or African lion and the DNA of other cat species

Species	Code	Radioactive DNA probe	
		Domestic cat	African lion
Domestic cat (*Felis catus*)	(FCA)	0.0	2.6
European wild cat (*F. silvestris*)	(FSI)	0.3	2.9
Sand cat (*F. margarita*)	(FMA)	1.0	2.9
Jungle cat (*F. chaus*)	(FCH)	1.1	NT
African wild cat (*F. libyca*)	(FLI)	1.1	NT
Black-footed cat (*F. nigripes*)	(FNI)	1.4	2.3
Caracal (*Caracal caracal*)	(CCA)	2.2	2.7
Leopard cat (*Prionailurus bengalensis*)	(PBE)	2.2	2.5
Margay (*Leopardus weidi*)	(LWE)	2.5	2.6
African lion (*Panthera leo*)	(PLE)	2.8	0.0
Tiger (*P. tigris*)	(PTI)	3.2	1.2
Leopard (*P. pardus*)	(PPA)	2.8	1.1
Snow leopard (*P. uncia*)	(PUN)	3.2	1.2

Source. Modified from O'Brien et al. 1987.

Note. Codes for the individual species, listed in the caption of Figure 17.1, can be used to relate data in this table to those in the figure. NT = not tested.

shows similar ΔTmRs to species outside their respective groups, which is to be expected because species outside the two groups should all be equidistant to the index species.

The Ocelot Lineage

The smaller South American cats form a coherent, closely related group that, according to Figure 17.1, recently diverged within the last 2–3 million years. Paradoxically, their divergence from the other cats is a rather ancient one, occurring approximately 10 M.Y.B.P. Six extant species, all endemic to South America, are included in this lineage. They have been divided into as many as three and as few as one genus (Hemmer 1978; Nowak and Paradiso 1983). This group does not include three cats whose present-day range extends into South America: the jaguarundi (*Herpailurus yagouaroundi*); the puma (*Puma concolor*); and the jaguar (*Panthera onca*). Samples of the Andean mountain cat (*Oreailurus jacobita*) were not available so its position is uncertain.

This grouping of cats is supported by morphologic and karyological studies. Glass and Martin (1978) used a multivariate analysis of cranial measurements to assess morphologic similarity of several large and small cat species. They conclude that there is a close association of margay (*Leopardus weidi*) (LWI), ocelot (*L. pardalis*) (LPA), and little spotted cat (*L. tigrina*) (LTI), as suggested by Figure 17.1. Herrington (1983) also found that cats of the ocelot lineage form a consistent group in a cladistic analysis of skeletal morphology and karyology. However, the results of her phenetic analysis show a variable grouping of these cats which is perhaps due to convergent evolution with other small cats. Comparative karyology offers strong evidence for grouping of the small South American cats. The cats in the ocelot lineage have a diploid number of 36, in contrast with all other felid species, which have a diploid number of 38. The ocelot lineage species all share a unique metacentric chromosome, C3, resulting from the fusion of two acrocentric F-group chromosomes present in cats with 38 chromosomes (Wurster-Hill and Centerwall 1982).

Two important characteristics of the ocelot lineage are its ancient, divergence from other cat lineages and the recent divergence of the extant species. The early divergence of the ocelot lineage is suggested by the early appearance of small "*Felis*" species in North America 13–15 m.y.b.p. (Savage and Russell 1983; Werdelin 1985). More recently, during the Pliocene and Pleistocene (0.5–5 m.y.b.p.), there were a number of small North American cats that shared similarities with the present-day South American species (Kurtén and Anderson 1980; Werdelin 1985). We can conclude from the fossil data and Figure 17.1 that the Plio-Pliestocene North American species that was ancestral to the modern South American taxa left no extant North American or Old World descendents.

The recent radiation of South American small cats coincides with the appearance of the Panamanian land bridge 2–3 m.y.b.p. Before this time, South America was devoid of terrestrial, placental carnivores (Patterson and Pascual 1972). Therefore, the rapid divergence of the small South American cats was perhaps promoted by reduced competition from the native South American fauna. Moreover, isolation and divergence may have been fostered by ice age–induced faunal fragmentation during the Pleistocene (Gingerich 1984). Fagan and Wiley (1978) suggest that the small South American cats diverged rapidly in morphology as a consequence of neotenic retention of juvenile characters.

The Domestic Cat Lineage

Species of the genus *Felis* (Ewer 1973) are small cats that were generally derived from ancestors that inhabited the Mediterranean basin (Kurtén 1968). The AID results place these species in a cluster that is likely to be monophyletic, which we refer to as the domestic cat lineage. Unlike the ocelot lineage, this group apparently diverged over a wider time span than did the South American cats (Figure 17.1). According to the AID phylogeny the Pallas cat (*Otocolobus manul*) (OMA), separated first in the evolution of this group, followed by the jungle cat (*Felis chaus*) (FCH), and finally the other five species in *Felis* (Nowak and Paradiso 1983).

This AID branching order is supported by the morphologic and behavioral observations of Hemmer (1976, 1978), who suggests that the jungle cat is a primitive offshoot of the genus *Felis* and that the Pallas cat is a separate but related genus. Other morphologic work by Herrington (1983) further supports the notion that the genus *Felis* is a monophyletic group. The six species of *Felis* all have an identical karyotype (2N = 38) that is distinct from that of all other felid species and differs by a single chromosome, E4, from that of the Pallas cat (Wurster-Hill and Centerwall 1982).

An important molecular confirmation of the monophyletic aspect of the genus *Felis* is based on the distribution of endogenous retroviral families present in the chromosomal DNA of several felid species. Retroviruses are RNA-containing viruses originally discovered in association with leukemia and sarcoma in several vertebrate species (Weiss et al. 1982). In many mammalian species multiple copies of retroviral genomes are present in their normal DNA (Benveniste and Todaro 1974a). The domestic cat (*Felis catus*) has at least two retroviral families, designated FeLV (about 8–10 copies) and RD-114 (about 20 copies), present in the normal DNA of all individuals (Benveniste and Todaro 1974a; Benveniste et al. 1975; Reeves and O'Brien 1984).

The evolutionary history of these felid endogenous retroviruses has been well studied (Benveniste and Todaro 1974a, 1974b; Benveniste et al. 1975; reviewed by Benveniste 1985). These investigators hybridized RD-114–specific probes to cellular DNA from a number of different mammals, including over 20 felid species. Although cellular DNA from all the tested cats showed greater than 90% homology with domestic cat DNA, only five species, all members of the genus *Felis*—domestic cat, jungle cat, sand cat (*Felis margarita*), black-footed cat (*F. nigripes*), and African wild cat (*F. lybica*)—contained DNA sequences homologous to the RD-114 genome. Furthermore, nucleic acid sequences related to RD-114 were found in the cellular DNA of Old World monkeys, the greatest degree of cross-hybridization existing between RD-114 and baboon cellular DNA. These results indicate that RD-114 retroviral genes were originally of primate origin and had been introduced into the germ line of the ancestor of modern *Felis* species at a point before the divergence of species in the genus *Felis,* but after the divergence of Pallas cat and other felid lineages (see Figure 17.1).

At approximately the same time as this event, the second endogenous retroviral family, FeLV, was also introduced into an ancestor of the modern *Felis* species. This virus is distantly related to an Old World rodent retrovirus and also occurs only in species of *Felis*. The distribution of the two endogenous retroviral families combined with the AID results and confirmatory morphological and ethological data all converge on the monophyletic history of the genus *Felis,* as suggested by Figure 17.1.

The earliest Old World *Felis* species described so far, *Felis lunensis,* appears in the early Villafranchian (3 M.Y.B.P.) (Kurtén 1965). This would also tend to confirm the divergence time suggested by Figure 17.1. However, the first Pallas cat (OMA) appeared more recently in the mid-Pleistocene (1 M.Y.B.P.), a million years later than suggested by the AID phylogeny. European wild cat (*F. silvestris*) fossils are slightly older, the first forms appear in the lower mid-Pleistocene. Fossils of the remaining species are not well known (Kurtén

1965, 1968). In summary, the domestic cat lineage is characterized by the appearance of small, morphologically similar cats throughout its evolutionary history. The evolution of these small cats is centered in the Mediterranean basin and Asia, but some species have successfully invaded areas as diverse as the southern tip of Africa (the black-footed cat, FNI) and hostile environments such as the frozen steppes (the Pallas cat, OMA) and desert terrain (the sand cat, FMA).

The Panthera Lineage

The remaining members of the Felidae form a heterogeneous group that has diverged at various times over the last 5–7 million years (Figure 17.1). The most recent radiation led to the five species of roaring cats, genus *Panthera*, 1–2 M.Y.B.P. Closely aligned with this group is the *Lynx* genus and the marbled cat (*Pardofelis marmorata*) (PMA). The placement of *Lynx* close to *Panthera* species was not entirely expected on morphological grounds, although *Panthera, Lynx,* and marbled cat species all share an identical karyotype that is distinct from the other felid genera (Wurster-Hill and Centerwall 1982).

The AID tree does not resolve some relationships within the Panthera lineage (Collier and O'Brien 1985). These unresolved branch points include species as morphologically diverse as the puma (PCO), golden cat (*Profelis aurata*) (PAU), cheetah (*Acinonyx jubatus*) (AJU), serval (*Leptailurus serval*) (LSE), caracal (*Caracal caracal*) (CCA), and the South American jaguarundi (HYA). Although the data do not conclusively resolve the relationship among these species, they do place them all, including the cheetah (AJU), squarely on the pantherid lineage.

An independent evolutionary assessment of the pantherine lineage is provided by data derived from protein electrophoresis (Figure 17.2) (O'Brien et al. 1987). These results increase the resolution to a limited extent. The lineage leading to species of the genus *Panthera* first becomes genetically distinct 2–3 M.Y.B.P. This divergence is preceded by that of the snow leopard (*Panthera uncia*) at approximately 3 M.Y.B.P. and the clouded leopard (*Neofelis nebulosa*) at 5–6 M.Y.B.P. Within *Panthera*, the jaguar and leopard (*Panthera pardus*) may have shared a common ancestor more recently than their time of divergence from the African lion (*P. leo*) or the tiger (*P. tigris*) (Figure 17.2). Finally, the three subspecies of tigers had very small distance values, which indicated a very recent divergence (0.1–0.2 M.Y.B.P.).

The other four species that were tested (golden cat, cheetah, caracal, and puma) split off as a group approximately 8 M.Y.B.P. Apparently these four species are not as closely related to each other as are the other cats of the genus *Panthera*.

Morphologic and genetic data support the close association of the large cats and some aspects of the trees in Figures 17.1 and 17.2. The lion, tiger, leopard,

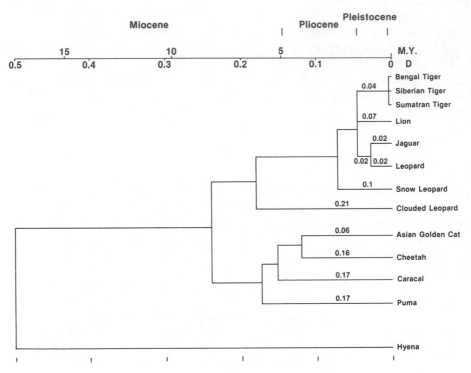

Figure 17.2. UPGMA tree of the large cats based on isozyme genetic distance data (Sneath and Sokal 1973; O'Brien et al. 1987). Basis of time scale equivalence: 0.1 genetic distance unit = 3.5 million years (O'Brien et al. 1987). The geologic time scale uses the convention of Hsu et al. (1984). Scientific names are given in the text.

and jaguar have an incompletely ossified hyoid that allows them to roar and thus unites the group (Nowak and Paradiso 1983). Herrington (1983) finds the snow leopard (*P. uncia*), the only nonroaring member of the genus, to be the most divergent member of the *Panthera* since it retains many characters found in the puma and the cheetah but has a hyoid with structural similarities to the pantherids. Hemmer (1976, 1978) recognized the distant similarities of the cheetah and the pantherids based on morphological data. The placement of the diminutive jaguarundi near the large cats and the lynx was also suggested by the morphological results of Werderlin (1981).

The earliest fossil record of a *Panthera* species is approximately 2 M.Y.B.P., which coincides with the divergence date seen in Figures 17.1 and 17.2 (Kurtén 1968; Savage 1978; Kurtén and Anderson 1980; Savage and Russell 1983). Hemmer (1976) argues that the radiation of the *Panthera* species occurred in two phases; first, a jaguar-like form spread over Africa, Europe, Asia, and North America in the early Pleistocene; and second, the differentiation of the lion, tiger, leopard, and jaguar. The tree in Figure 17.2 lends support to this hypothesis, but there is a slight suggestion that the leopard and jaguar split more recently and are more closely related to each other than to the other large cats. Historically, their ranges do not overlap; the leopard is Old World and the jaguar is New World (Kurtén 1968; Kurtén and Anderson 1980). The earliest South American record of the jaguar is from the mid-Pleistocene of Bolivia (0.5 M.Y.B.P.) (Hemmer 1976). However, early Pleistocene jaguar fossils, at approximately 1.5 M.Y.B.P., are known from North America (Hemmer 1976; Kurtén and Anderson 1980).

The African lion appeared in Europe and Africa about 0.5 M.Y.B.P. (Kurtén 1968; Savage 1978). Its relationship to the tiger is difficult to ascertain because of the morphologic similarity of these taxa (Neff 1982). Other species of *Panthera* have more incomplete fossil records. The snow leopard is known from the late Pleistocene of the USSR. The clouded leopard has a fossil record that extends into the early Pleistocene of Java, although Figures 17.1 and 17.2 suggest this lineage has been distinct since the late Miocene (Hemmer 1976).

Lynx is a relatively well-studied genus first appearing in the mid-Pliocene (3–4 M.Y.B.P.) (Kurtén 1968; Werdelin 1981). The first lynx, *L. issiodorensis*, shares features with *Felis* species and does not resemble cats of the Panthera group in which it is placed in Figure 17.1 (Werdelin 1981). However, *Lynx* and the jaguarundi (HJA) do show some morphologic similarities, as suggested by their nearness in Figure 17.1 (Werdelin 1981).

The caracal and the serval made their first appearance in the early Pleistocene (Savage 1978), which is consistent with the proposed relationships. The puma and the cheetah have been associated in a common lineage, as suggested by Figure 17.2 (Martin et al. 1977; Adams 1979). The first appearance of the genus *Acinonyx* is approximately 3 M.Y.B.P., but the cheetah lineage may extend into Hempillian times, 4–5 M.Y.B.P. (Adams 1979; Kurtén and Anderson 1980). This older date is in agreement with that suggested by Figure 17.2.

Summary

The branching order and affinities of the Felidae suggested by the trees in Figures 17.1 and 17.2 are in general agreement with morphologic and genetic studies and with the fossil record. The Felidae can be divided into three distinct groups. The most ancient divergence is represented by the ocelot lineage. The extant species of this lineage have all appeared very recently, soon after the opening of the Panamanian land bridge into South America. The second lineage includes the small Mediterranean cats. The extant species of this lineage have diverged at various times from their ancestral stock throughout the Pliocene and Pleistocene. The third group, the Panthera lineage, includes a heterogeneous array of species that evolved over an 8-million-year period. The species of the modern great cat genus *Panthera* first appeared very recently in the Pleistocene. The caracal, Asian golden cat, cheetah and puma form an older monophyletic group that diverged from species leading to cats in the genus *Panthera* approximately 5–8 M.Y.B.P. The position of the smaller cats in the Panthera lineage remains the focus of future genetic study.

Relationships of the Canidae

The molecular evolution of the Canidae was reconstructed based on isozyme genetic distances among 18 species from 13 genera (Wayne and O'Brien 1987). We present a consensus phenogram of distance-Wagner and UPGMA (unweighted pair-group method arithmetic averages) trees in Figure 17.3.

Canids appear to be separated into six distinct lineages, three of which contain only one of the studied species. The principal groupings are: the wolf-like canids, including the gray wolf (*Canis lupus*), other species in the genus

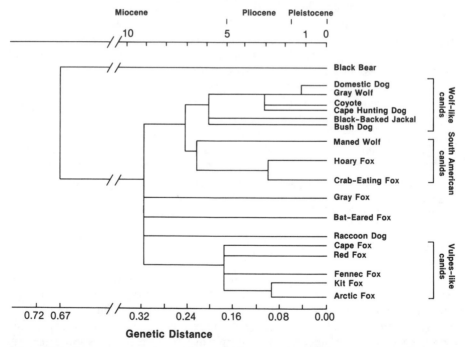

Figure 17.3. Consensus tree of the dog family based on isozyme genetic distance (Wayne and O'Brien 1987). This consensus is based on the topologies of UPGMA and distance-Wagner trees calculated by the BIOSYS-1 program of Swofford and Selander. The "strict method" algorithm (Rohlf 1982) contained in the PAUP program, version 2.4, by D. L. Swofford was used to generate the consensus tree. Time scale equivalence was determined by assuming that *Canis davisii* was the common ancestor of the wolflike canids and the South American canids (Berta 1984). Since this extinct species appeared approximately 7 M.Y.B.P., 0.1 genetic distance unit ≅ 2.5 million years. Scientific names are given in the text.

Table 17.2. The termal stability (ΔTmR) of hybrids formed between the unique sequence DNA of domestic dog, red fox, or gray fox and the DNA of other canid species

| | Radioactive DNA probe | | | | | |
| | Domestic dog | | Red fox | | Gray fox | |
	ΔTmR	SD	ΔTmR	SD	ΔTmR	SD
Domestic dog (*Canis familiaris*)	0.0	—	2.7^2	0.3	1.8^3	0.3
Gray wolf (*C. lupus*)	0.0^2	0.7	NT	—	NT	—
Black-backed jackal (*C. mesomelas*)	0.6^4	0.5	2.1^3	0.5	2.0^4	0.4
Bush dog (*Speothos venaticus*)	1.8^3	0.4	1.9^1	—	3.3^2	0.5
Maned wolf (*Chrysocyon brachyurus*)	2.0^3	0.9	2.3^1	—	3.3^3	0.9
Crab-eating fox (*Cerdocyon thous*)	1.5^3	0.4	1.2^2	0.2	2.3^3	0.2
Hoary fox (*Dusicyon vetulus*)	1.1^2	0.2	2.0^2	0.2	2.2^2	1.0
Gray fox (*Urocyon cinereoargenteus*)	2.3^3	0.3	1.8^3	0.6	0.0	—
Bat-eared fox (*Otocyon megalotis*)	1.9^3	0.2	2.5^2	0.2	1.5^3	0.3
Raccoon dog (*Nyctereutes procyonoides*)	1.7^2	0.4	1.3^2	0.3	1.8^2	0.0
Arctic fox (*Alopex lagopus*)	2.5^5	0.5	1.0^2	0.4	1.6^2	0.2
Fennec (*Fennecus zerda*)	2.3^2	0.2	1.0^1	—	NT	—
Red fox (*Vulpes vulpes*)	1.4^3	0.4	0.0	—	2.2^2	0.4

Note. Superscripts on ΔTmR values indicate number of replicate experiments that the average ΔTmR is based on. NT = not tested. Dash indicates that a value cannot be calculated.

Canis, and the African wild dog (*Lyacon pictus*); the South American canids including the maned wolf (*Chrysocyon brachyurus*), the crab-eating fox (*Cerdocyon thous*), and the hoary fox (*Dusicyon vetulus*); and the *Vulpes*-like canids including the Cape fox (*Vulpes chama*), the red fox (*V. vulpes*), the fennec (*Fennecus zerda*), the kit fox (*V. macrotis*), and the arctic fox (*Alopex lagopus*). The three monotypic lineages are: the gray fox (*Urocyon cinereoargenteus*); the bat-eared fox (*Otocyon megalotis*); and the raccoon dog (*Nyctereutes procyonoides*).

Because the canid radiation was so recent (9–10 M.Y.B.P.), the results of DNA hybridization between canid genera yielded limited information (Table 17.2). However, two of the groups delineated by the genetic distance analyses are also defined by DNA hybridization: the wolflike canids (the domestic dog, *Canis familiaris;* gray wolf; and black-backed jackal, *C. mesomelas*) and the *Vulpes*-like canids (the red fox, arctic fox, and fennec) (Table 17.2). However, because the standard error is large for most of the hybridization measurements, they cannot be used to resolve close relationships within the Canidae. A similar limitation of this procedure was also observed in the recent radiation of the extant felids (Table 17.1).

Wolflike Canids

The wolflike canids form a single cluster that, like the domestic cat lineage, includes species that have diverged at various times throughout the history of the group. The gray wolf is most closely allied with the domestic dog, followed by the coyote (*C. latrans*) and the African wild dog. The black-backed jackal and the bush dog (*Speothos venaticus*) are most distantly associated with the other wolflike canids.

Morphologic studies support the close kinship of the gray wolf and domestic dog, and their significant distance from the coyote (Lawrence and Bossert 1967; Clutton-Brock et al. 1976; Nowak 1979; Olsen 1985; Wayne 1986a, 1986b). However, the distant relationship of these taxa to their congener, the black-backed jackal, has not been recognized by systematists. Simpson (1945) placed the African wild dog in its own subfamily, Simocyoninae, along with the dhole (*Cuon alpinus*) and the bush dog. However, many of the characters that unite these taxa, such as the modified carnassial blade, represent dietary or locomotor adaptations and have developed independently in other extant and extinct canids (Berta 1979, 1984).

The branching order and time scale depicted for the wolflike canids (Figure 17.3) are consistent with fossil and biogeographic evidence. The domestic dog is an extremely recent derivative of the gray wolf (<15,000 years ago) (Olsen 1985). The extant coyote is an endemic North American canid whose evolution can be traced into the Pliocene (approximately 2 M.Y.B.P.) (Giles 1960; Kurtén 1974; Nowak 1979; Kurtén and Anderson 1980). The African wild dog and the black-backed jackal are both endemic African canids first appearing approximately 1 M.Y.B.P. and 2 M.Y.B.P., respectively (Savage 1978). Fossils from species in the genus *Lycaon* are known from the Pleistocene of Europe and provide a link between the modern African wild dog and its potential European wolflike ancestors (Kurtén 1968).

South American Canids

This group includes taxa that differ significantly in their relationship to one another. The small South American foxes, including the hoary fox and the crab-eating fox are the most closely related. The genus *Dusicyon* also includes five other fox species not available for analysis, but these are likely to be closely associated with each other and with the crab-eating fox as well (Langguth 1969). The maned wolf is distantly related to *Dusicyon* and *Cerdocyon* species, but together these taxa form a single cluster. The South American bush dog seems to be distantly allied to this group and is consistently clustered with the wolflike canids using several phenetic algorithms. Because of the unexpected placement of the bush dog in the wolflike canid group, this assignment should be considered as tentative until confirmed by other approaches.

The morphology of the South American canids has been studied by a number of authors (Langguth 1969, 1975; Clutton-Brock et al. 1976; Berta 1984), who unfortunately present conflicting results. Langguth views the maned wolf and the bush dog as distinct monotypic genera. He does not ally the bush dog with the wolflike canids. The morphologic differences he finds between the hoary fox and the crab-eating fox suggest that they should be placed in separate genera. Berta (1981, 1984) analyzes the morphology of South American canids using a cladistic approach. Her results disagree with Figure 17.3 in showing a close relationship among the bush dog, the crab-eating fox, and the Asiatic raccoon dog. These taxa are more distantly allied to species in the *Dusicyon* group. She allies the maned wolf with *Canis* species. Clutton-Brock et al. (1976) find a high degree of morphologic similarity between the hoary fox and the crab-eating fox in measurements of skulls and limb bones, and suggest placing them in a single genus, as suggested by Figure 17.3. They place the maned wolf and the bush dog in separate monotypic genera. Karyological data suggest that the maned wolf is closely associated with the wolflike canids because it has a similar karyotype (Wayne et al. 1987a). The bush dog, the crab-eating fox, and the hoary fox all have similar karyotypes that differ slightly from the maned wolf and wolflike canids (Wayne et al. 1987a). In summary, the morphologic and karyologic studies seem to suggest close association between the *Dusicyon* group and the crab-eating fox, but the positions of the maned wolf and the bush dog are unresolved.

Fossil material is scant for most of the South American canids but does offer support for some aspects of the genetic distance tree. The bush dog appears relatively recently in the mid-Pleistocene of South America and may be allied to a much larger wolflike New World genus, *Protocyon*, which appears in the late Pliocene of South America, approximately 2 M.Y.B.P. (Kurtén and Anderson 1980). The bush dog and *Protocyon* species share several morphologic features, including the presence of a modified carnassial tooth (Kurtén and Anderson 1980). The fossil record of the maned wolf is poor and extends just to the mid-Pleistocene (Berta 1981). The crab-eating fox and the hoary fox also first appear in South America in the mid-Pleistocene approximately 1 M.Y.B.P. Berta (1984) suggests that the common ancestor of these genera existed around 3–6 M.Y.B.P., which agrees with the divergence date suggested by Figure 17.3. The ancient divergence of the wolflike canids and the South American canids is supported by the presence of a good structural ancestor for both groups in North America at approximately 7 M.Y.B.P., *Canis davisi* (Savage and Russell 1983; Berta 1984). In summary, the radiation of the South American foxes (the *Dusicyon* group and crab-eating fox) is a relatively recent event that occurred approximately 2–3 M.Y.B.P. This radiation coincided with the opening of the Panamanian land bridge and may have been fostered by the absence of placental terrestrial predators in South America—a radiation analogous to that of the small South American cats. The maned wolf and the bush dog are ancient derivatives of the ancestral stock that led, respectively, to the South American foxes and the modern wolflike canids.

Vulpes-like Canids

The *Vulpes*-like canids form a single cluster that includes the cape fox, the red fox, the fennec, the kit fox, and the arctic fox. The latter two taxa have diverged more recently, whereas the remaining taxa represent more ancient divergences (Figure 17.3).

Clutton-Brock et al. (1976) find some morphologic similarities between the fennec and species in the genus *Vulpes* that support the grouping of these taxa as seen in Figure 17.3. They find the arctic fox to show some similarity to *Vulpes* species in cranial and dental characters but consider the arctic fox sufficiently distinct as to warrant its own genus. Contrary to Figure 17.3, a close affinity of the arctic fox and the kit fox is not suggested by their morphological analyses. However, the karyological evidence suggests that these two taxa are closely related since they share an identical karyotype not found in any other canid (Wayne et al. 1987b). A close relationship of these two taxa to the red fox is suggested by similarities in their chromosomal arm morphology (Yoshida et al. 1983). The G-banded karyotype of the fennec actually shows more resemblance to the wolflike canids, which indicates that either this karyotype is primitive for the *Vulpes*-like canids or that the fennec is not as closely associated with this group as suggested by Figure 17.3 (Wayne et al. 1987a, 1987b).

The Cape fox and the red fox have fossil records extending to the early mid-Pleistocene (approximately 1 M.Y.B.P.) (Kurtén 1968; Savage 1978; Savage and Russell 1983). Figure 17.3 suggests a more ancient divergence beginning in the early Pliocene (5 M.Y.B.P.). Similarly, the recent appearance of the fennec in the late Pleistocene fossil record does not support the tree in Figure 17.3. However, its ancient divergence from the *Vulpes* species is not contradicted by the fossil record since *Vulpes* species are not known from the African record for several million years preceding the first appearance of the fennec; thus, their common ancestor may have existed no later than the Pliocene (Savage 1978). The kit fox and the arctic fox both make their first appearance in the fossil record in the mid-Pleistocene. An origination near this time is suggested by Figure 17.3. The ancient divergence of these taxa from the red fox is again not opposed by the fossil record since red fox—like species are absent from North America for approximately 2 million years preceding the first appearance of the kit fox and arctic fox (Kurtén and Anderson 1980). The first recognized *Vulpes* species in the fossil record is mid-Miocene (9–12 M.Y.B.P.) (Savage and Russell 1983). Hence, the Miocene divergence of the *Vulpes* clade from other canids as seen in the genetic distance tree is strongly supported.

Monotypic Genera

The remaining three canid lineages each contain a single species and have been phyletically distinct since the origin of the recent Canidae. However, the gray fox lineage likely includes another species, the Channel Island fox (*Urocyon littoralis*). We are presently conducting genetic studies of this species.

Morphologic and taxonomic studies have generally agreed with these designations. Clutton-Brock et al. (1976) find the raccoon dog and bat-eared fox worthy of generic distinction. The bat-eared fox was given subfamilial rank by Simpson (1945). Huxley (1880) suggests that the bat-eared fox is an ancient member of the family. Similarly, Berta (1984) finds the raccoon dog to be a primitive member of the family Canidae but, in contrast to the genetic distance tree, she detects features that unite it with the crab-eating fox and bush dog.

The karyological evidence supports the notion of the ancient divergence of the monotypic taxa. All three have distinct karyotypes not found in any other canid (Wayne et al. 1987a, 1987b). The bat-eared fox and gray fox have karyotypes dominated by acrocentric chromosomes, most of which are G-band homologous with those of the wolflike canids (Wayne et al. 1987a).

The fossil record of the monotypic taxa supports ancient times of origination. The first *Urocyon* species appears in the late Miocene (6–9 M.Y.B.P.) (Kurtén and Anderson 1980). The monotypic genus *Nyctereutes* first appeared about 5 M.Y.B.P. in the European fossil record (Savage and Russell 1983; Berta 1984). It is not known from the North American record, casting doubt on its role in the evolution of the bush dog. The monotypic genus *Otocyon* has a sparse fossil record extending into the late Pliocene (3 M.Y.B.P.) of Africa (Savage and Russell 1983).

Summary

The Canidae can be divided into several groups that show various degrees of genetic similarity. The wolflike canids include the gray wolf, domestic dog, coyote, African wild dog, and black-backed jackal. Surprisingly, the bush dog may also be associated with this group. The black-backed jackal and the bush dog separated first, approximately 6 M.Y.B.P., followed by the African wild dog and the coyote at 3 M.Y.B.P. The domestic dog and gray wolf are genetically very closely related.

With the exception of the bush dog, the South American canids are only distantly associated with the wolflike canids. Unlike the small South American cats, these canids are divided into two ancient and independent lineages, one leading to the maned wolf and the other to the small fox-like canids. The latter group includes two taxa that have diverged very recently, coincident with the opening of the Panamanian land bridge in the Pliocene.

The *Vulpes*-like canids are a distinct group that are an ancient offshoot of the Canidae. One of the four distinct branches that make up this lineage leads to both the kit fox and arctic fox. The number of distinct branches making up this group may increase as more fox species are sampled. There are three monotypic genera: *Urocyon* (the gray fox), *Nyctereutes* (the raccoon dog), and *Otocyon* (the bat-eared fox), which are not closely related to any of the sampled canid species.

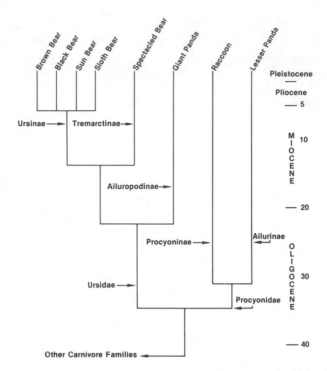

Figure 17.4. A consensus tree of the bears, giant panda, and a New and Old World procyonid based on four molecular and one karyological methodology. The phylogeny was calibrated by alignment with trees developed for the great apes using the time of divergence between humans and orangutans at 16 M.Y.B.P. (O'Brien et al 1985). Scientific names are given in the text.

Relationships of the Ursidae

The family Ursidae consists of eight distinct species, which have been organized into as many as seven genera (Ewer 1973; Stains 1984). Ursids are large, stocky animals with a New and Old World distribution, excluding Australia and Africa. Because of the low number of species and the extended time during which the bears have evolved, this family provides an excellent group for comparing trees based on different genetic approaches. For this reason, we have used each of the molecular and karyological procedures outlined in the introduction and have derived a consensus tree of the bear family and of several procyonids, including the raccoon (*Procyon lotor*) and lesser or red panda (*Ailurus fulgens*) (Sarich 1973; O'Brien et al. 1985; Goldman et al. 1987; Nash and O'Brien 1987) (Figure 17.4). The consensus tree indicates that between 30 and 40 M.Y.B.P. the progenitor of modern ursids and procyonids split into two lineages. Within 10 million years of that event the procyonid group split into the Old World procyonids represented today by the red panda, and New World procyonids (for example, raccoons, coatis, olingos, kinkajous). Approximately 18–25 M.Y.B.P. the ancestor of the giant panda (*Ailuropoda melanoleuca*) diverged from the ursid line. The next divergence is between the ursine bears and the spectacled bear (*Tremarctos ornata*), which occurred between 12–15 M.Y.B.P. The lineages leading to the remaining species—the brown bear (*Ursus arctos*), the black bear (*U. americanus*), the sun bear (*U. malayanus*), and the sloth bear (*U. ursinus*)—first became distinct 5–7 M.Y.B.P.

The chromosomal evolution of this family has been particularly informative (O'Brien et al. 1985; Nash and O'Brien 1987). All six members of the Ursinae subfamily have a nearly identical, largely acrocentric, karyotype (2N = 74). Most of these acrocentric chromosomes are homologous to the chromosome arms of the giant panda's 42 mostly metacentric chromosomes (Nash and O'Brien 1987). The spectacled bear also has a low-numbered, largely metacentric karyotype (2N = 52), which consists of chromosomes derived from fusions of chromosome arms found in the Ursinae karyotype. Interestingly, the Ursinae chromosome fusions seen in the giant panda are all combinations of chromosomes that are different from the fusion combinations seen in the spectacled bear. Thus, we have hypothesized that the primitive ursid karyotype was a high-numbered acrocentric karyotype like that seen in the Ursine bears, and that the spectacled bear and giant panda have reorganized karyotypes that were derived independently from the primitive ursid karyotype. The procyonids have a metacentric karyotype containing many primitive chromosomes found in the Felidae and other carnivore families (Wurster-Hill and Centerwall 1982).

Previous taxonomic studies generally agree with the divisions of the Ursidae as outlined by the tree in Figure 17.4. The six bears of the genus *Ursus* are usually considered together in the subfamily Ursinae (Hall 1981). Some authors further subdivide them into several genera: *Thalarctos* (polar bear), *Ursus* (brown and black bear), *Melursus* (sloth bear), *Helarctos* (sun bear), and *Selenarctos* (Asiatic black bear) (Ewer 1973; Stains 1984). Generic distinction of these taxa may be warranted on morphologic grounds, but for the four ursine bears used in this study generic distinction is not well supported since they are genetically close. The spectacled bear is often placed in the subfamily Tremarctinae and has been considered a very primitive bear (Kurtén 1966; Hall 1981; Nowak and Paradiso 1983). Its distant association with the Ursine bears in Figure 17.4 corroborates this designation. The controversy surrounding the affinities of the giant panda appears to have been resolved by the use of several molecular approaches that provide concordant results (Figure 17.4; see O'Brien et al. 1985). These studies confirm the finding of Davis (1964), whose extensive anatomical study suggested that the giant panda was more closely allied with ursids than with procyonids.

The fossil history of the Ursidae is in good agreement with the branching scheme presented in Figure 17.4 (Kurtén 1964, 1966, 1968, 1986; Thenius 1979; Kurtén and Anderson 1980). The fossil precursor of the giant panda, *Agriarctos,* is derived separately from the Miocene genus *Ursavus,* which is the common ancestor of all recent bears. This genus first appears approximately 20 M.Y.B.P., a date that is consistent with the molecular tree (Thenius 1979; Kurtén 1986; Van Valen 1986). The first tremarctine bear, *Plionarctos,* appears in the late Pliocene (Kurtén 1966). The divergence of this lineage from that leading to the ursine bears probably dates to the mid-to late Miocene (Kurtén 1966; Savage and Russell 1983). The ursine bears have a recent record with the appearance of the extant species in the mid-Pleistocene (Kurtén 1968; Kurtén and Anderson 1980). Their common ancestor appears earlier in the Old World during the early Pliocene, 4–5 M.Y.B.P. (Kurtén and Anderson 1980). Thus, the branch order and timing of the tree in Figure 17.4 correspond with the fossil evidence.

In summary, the recent Ursidae are a heterogeneous family made up of eight species, six of which are the likely result of a recent and contemporaneous radiation. The remaining ursids, the spectacled bear and the panda bear, represent monotypic, ancient lineages. The former is restricted to South America and branched from the main line of ursid evolution about 15 M.Y.B.P. The latter is endemic to China and is the living relict of any isolated bear lineage that extends 20 M.Y.B.P.

Relationships of the Carnivora

A molecular tree of the Carnivora based on hybridization of unique sequence DNA is presented in Figure 17.5. The first division of the Carnivora corresponds to the superfamilial division of the order into canoid and feloid carnivores. The former includes the dog, weasel, raccoon, and bear families and the latter the mongoose, civet, hyena, and cat families. Among the canoid carnivores, the most ancient division is between the canids and the remaining canoid families and occurred approximately 50 M.Y.B.P. Subsequent to this division several distinct lineages of canoid and feloid carnivores appeared simultaneously. This divergence time is estimated at 40 M.Y.B.P. based on data from the fossil record and is the basis for the time scale in Figure 17.5 (Tedford 1975; Radinsky 1977, 1982; Flynn and Galiano 1982).

The outline of this tree is in good agreement with immunological distance phylogenies derived by Sarich (1969a, 1969b, 1973) and Seal et al. (1970). The DNA distance matrix provides an independent confirmation of these pioneering studies. For a thorough discussion of the relationships of carnivore families based on morphology and the fossil record, see the chapters by Martin and by Wozencraft in this volume. The origin and composition of each of these modern families based on DNA hybridization data are described below.

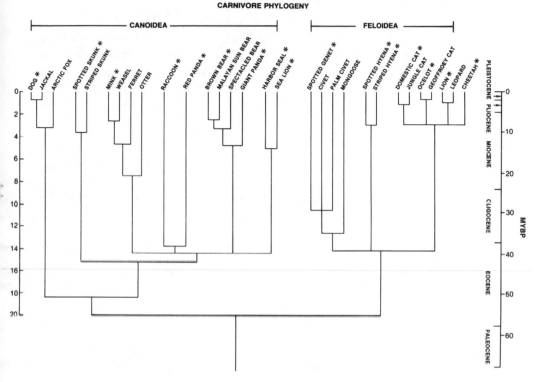

Figure 17.5. Phenetic tree based on thermal stability of DNA hybrids. This tree was derived from the Fitch-Margoliash algorithm (1967) contained in the Kitsch program of J. Felsenstein (University of Washington). The time scale was based on a fossil divergence time of approximately 40 M.Y.B.P. for all the modern carnivore families with the exception of the Canidae, which is considered a more ancient divergence (see text for discussion). 1 $\triangle$TmR = 2.8 million years. * = ^{3}H-labeled unique sequence cellular DNA from these species was used as the hybridization probe. The divergence time of felids and hyaenids (40 M.Y.B.P.) does not agree with those suggested by the AID or protein electrophoresis analysis (Figures 17.1 and 17.2). This difference is most likely due to a lack of linearity between these genetic distance measures and the time since reproductive isolation for species that have large divergence times. The species tested were domestic dog (*Canis familiaris*), black-backed jackal (*C. mesomelas*), arctic fox (*Alopex lagopus*), spotted skunk (*Spilogale putoris*), striped skunk (*Mephitis mephitis*), mink (*Mustela vision*), weasel (*M. frenata*), ferret (*M. putorius*), otter (*Lutra canadensis*), raccoon (*Procyon lotor*), red panda (*Ailurus fulgens*), brown bear (*Ursus arctos*), Malayan sun bear (*Helarctos malayanus*), spectacled bear (*Tremarctos ornatus*), giant panda (*Ailuropoda melanoleuca*), harbor seal (*Phoca vitulina*), sea lion (*Eumetopias jubatus*), spotted genet (*Genetta genetta*), oriental civet (*Viverra tangalunga*), palm civet (*Paradoxurus hermaphroditus*), mongoose (*Herpestes* spp.), spotted hyena (*Crocuta crocuta*), striped hyena (*Hyaena hyaena*), domestic cat (*Felis catus*), jungle cat (*F. chaus*), ocelot (*Leopardus pardalis*), Geoffroy's cat (*L. geoffroyi*), African lion (*Panthera leo*), leopard (*P. pardus*), cheetah (*Acinonyx jubatus*).

Canidae

The origin of this family is the most ancient of the arctoid carnivores. The extant species are thus an extremely recent offshoot of a lineage that has remained phyletically distinct for 50 million years. The fossil relatives of the Canidae include several groups of now extinct carnivores that were very diverse throughout the Cenozoic (see discussion in Martin, this volume). Among the recent canids, the DNA hybridization difference is largest between the arctic fox and the domestic dog and suggests a divergence time of approximately 10 M.Y.B.P. between these taxa. The domestic dog and black-backed jackal divergence is considerably more recent, approximately 2 M.Y.B.P. This branching order is in agreement with the tree in Figure 17.3, which shows the divergence of the arctic fox and domestic dog to be more ancient relative to that of the domestic dog and black-backed jackal.

Mustelidae

The mustelid family appears to have separated early into two lineages (Figure 17.5). One lineage contains the modern skunks, and the other contains otters, and weasels. This ancient division occurred at approximately the same time as the division among other canoid families (40 M.Y.B.P.). Within the skunk lineage the divergence of the two extant genera occurred fairly recently, about 10 M.Y.B.P., which is contemporaneous with the origin of most of the canid lineages (Figure 17.3). Within the weasel lineage the divergence of the otter (*Lutra canadensis*) is shown as beginning approximately 20–25 M.Y.B.P. The remaining mustelids used in this study are all members of the genus *Mustela* and show divergence times within the last 15 million years.

These groupings of mustelids correspond with the traditional taxonomic arrangements of the Mustelidae based on morphologic similarity. The skunks are most often placed in their own subfamily, as are both the otters and weasels (Nowak and Paradiso 1983). The earliest known skunks are found in the Miocene (Kurtén and Anderson 1980), but the 40 M.Y.B.P. divergence date in Figure 17.4 suggests that undifferentiated and genetically distinct skunk ancestors existed throughout the Oligocene. The contemporaneous divergence of the lineages leading to the recent skunks and the canids is supported by the fossil evidence because an early skunk, *Martinogale*, occurs with *Canis davisi*, the likely progenitor of the South American and wolflike canids, in the late Miocene Edson Local Fauna (Harrison 1983). The first otter is late Oligocene, approximately corresponding with the 22 M.Y.B.P. divergence time in Figure 17.4. This morphologically unusual group appears rather suddenly with a complete suite of otter-like characteristics (Savage 1957).

Procyonidae and Ursidae

The divergence of the Procyonidae from other carnivore families occurred approximately 40 M.Y.B.P. and was contemporaneous with the divergence of the other canoid carnivore families except the Canidae (Figure 17.5). As previously noted, the procyonids consist of two ancient lineages, one leading to the Old World lesser or red panda and the other leading to the New World species. An ancient divergence of the subfamilies is suggested by their placement into two separate subfamilies and their geographic separation (Hunt 1974; Nowak and Paradiso 1983). The fossil record of the Procyonidae extends into the Oligocene of the Old and New World. Representatives of the modern subfamilies are apparent by the mid-Miocene (18–20 M.Y.B.P.) (Kurtén 1968; Kurtén and Anderson 1980; Baskin 1982; Savage and Russell 1983). Figure 17.5 suggests that the divergence of the modern subfamilies preceded this date by at least 15 million years.

Within the Ursidae, the DNA hybridization data (Figure 17.5) is topologically equivalent to the consensus tree (Figure 17.4). The relative divergence time of the giant panda derived from the DNA results, however, is more recent. For this reason, we advocate a time range of 18–25 million years for the divergence time of the giant panda from the ursine lineage.

Pinnipedia

Earless seals, sea lions, and walruses have been placed in a separate order or suborder from the fissiped carnivores (Simpson 1945; Nowak and Paradiso 1983). The DNA hybridization data suggest that earless seals and sea lions are most closely related to the canoid carnivores and form a separate branch radiating at about the same time (around 40 M.Y.B.P.) as most of the other canoid families. Within this cluster, the divergence of the earless seals and sea lions occurred approximately 15 M.Y.B.P.

Considerable controversy surrounds the origins of the pinnipeds. Simpson (1945) believed the pinnipeds to be a suborder of the Carnivora closely allied to the canoid carnivores. Immunologic data supported this notion (Sarich 1969a, 1969b; Seal et al. 1970). Recent morphologic work on extant and fossil species has contradicted this conclusion by suggesting that the pinnipeds are diphyletic, with the sea lions and walruses more closely allied with the bears, and with the earless seals showing affinity with the mustelids (Hunt 1974; Tedford 1975). The DNA hybridization tree supports the immunological conclusions and would suggest that the earless seals and sea lions be placed in the same canoid family. The association of the pinnipeds in a single group is also supported by recent studies of highly repetitive DNA and protein sequences (Arnason 1986; de Jong 1986).

Viverridae and Herpestidae

The Viverridae (civets and genets) is a diverse family that includes three subfamilies. According to Figure 17.5, the subfamilies represented in this study are a result of ancient divergences (30 M.Y.B.P.). The Herpestidae (mongoose) diverged before the Viverridae lineage at approximately 35 M.Y.B.P.. The ancient divergence of these taxa is in striking contrast to the more recent radiation of extant genera in other carnivore families (for example, Canidae, Felidae). (See Wozencraft [1984, this volume] for a discussion of morphologic studies of these taxa.) The fossil record of the Viverridae is poor but indicates an origin around 40 M.Y.B.P. The appearance of the modern subfamilies is more recent, beginning in the Miocene (Gregory and Hellman 1939).

Hyaenidae

The hyaenids are represented by two of the four extant species, the spotted hyena (*Crocuta crocuta*) and striped hyena (*Hyaena hyaena*). The divergence time of these two taxa is approximately 10 M.Y.B.P. and is nearly contemporaneous with the divergence of the major cat lineages (Figures 17.1, 17.2, and 17.5). The large DNA sequence difference between the hyena and the other feloid carnivores (40 M.Y.B.P.) suggests an ancient divergence and places them equally close to the viverrids and felids.

The first fossil hyena is mid-Miocene, considerably later than the divergence time suggested by Figure 17.5 (Kurtén 1968; Savage 1978). The extant species both appear in the Pleistocene record, which is later than expected from Figure 17.5. However, *Hyaena*-like species do appear in the Pliocene (Kurtén 1968).

Felidae

DNA hybridization and fossil data indicate that the divergence of felids from other feloid carnivores occurred approximately 40 M.Y.B.P. However, the radiation of the 37 extant species into the three lineages described earlier (Figure 17.1) occurred much later, approximately 10 M.Y.B.P. This latter date agrees precisely with the date suggested by isozyme distance data and data from the fossil record discussed previously. Within the Felidae radiation, the DNA hybridization method cannot resolve intergeneric relationships as well as other methods because of the recent time period involved (see Table 17.1).

An interesting result is the apparently large distance between the cheetah and the African lion (Figure 17.5). This indicates that the cheetah may have diverged earlier from other felid lineages than suggested by Figures 17.1 and 17.2. The fossil record suggests such an interpretation because cheetah fossils are among the oldest of any species in the Panthera lineage (Adams 1979). Also, the cheetah has many unique morphologic features that set it apart from other pantherids. Thus, we suggest that the cheetah is an ancient offshoot of the Panthera lineage (Figure 17.1) as a reasonable consensus based on the combined genetic, morphologic, and paleontologic evidence.

Conclusions

The order Carnivora includes a diverse array of taxa that vary considerably in size, diet, and locomotor specializations (Ewer 1973; Van Valkenburgh 1985, 1987, this volume; Gittleman 1986a, 1986b; Martin, this volume). The evolutionary relationships of the order suggest that many lineages were defined early in its history, and the subsequent diversification within these lineages has produced numerous convergences in feeding, locomotor, and social behavior. Notable exceptions to this pattern are the otters, seals, and skunks. In the former two groups the necessity to adapt to an aquatic lifestyle may have limited their potential morphologic diversity. It is puzzling that the skunks as small, generalized carnivores have remained at low diversity throughout their history. Other low-diversity carnivore families, such as the bears or the hyenas, are large in body size and more specialized. The nearly contemporaneous appearance of the modern carnivore families at 40 M.Y.B.P. suggests that their origin may have been associated with a distinct event such as the extinction of potential predators or enhanced opportunity for genetic isolation and specialization (Radinsky 1977, 1982; see Martin, this volume).

Within each of the families there is a wide range of divergence times. Many of the smaller, more generalized carnivore families such as the procyonids, viverrids, and mustelids contain taxa with ancient divergence times. In contrast, the larger carnivores such as the cats, dogs, and bears contain lineages that have more recent originations. These families have shown numerous radiations of now extinct groups throughout their history which have apparently left no living descendents (see Martin, this volume). This suggests that species turnover is more rapid within the larger, more specialized carnivores; that is to say, speciation and extinction rates are higher. The smaller carnivore families generally have more living taxa and contain more ancient lineages. The difference between these carnivore groups is perhaps indicative of the stability of their respective ecological roles; more specialized carnivores are more likely to suffer dramatic extinctions because of their limited dietary and locomotor flexibility (Kurtén and Anderson 1980; Guilday 1984; Diamond 1984). The smaller, generalized carnivores are collectively more able to succeed in a variety of settings, and thus each lineage has more temporal stability.

References

Adams, D. B. 1979. The cheetah: Native American. *Science* 205:1155–1158.

Arnason, U. 1986. Pinniped phylogeny enlightened by molecular hybridizations using highly repetitive DNA. *Mol. Biol. Evol.* 3:356–365.

Avise, J. C., and Aquadro, C. F. 1981. A comparative summary of genetic distance in vertebrates. *Evol. Biol.* 14:114–126.

Ayala, F. J. 1986. On the virtues and pitfalls of the molecular evolutionary clock. *J. Hered.* 77:226–235.

Baskin, J. A. 1982. Tertiary Procyoninae of North America. *J. Vert. Paleo.* 2:71–93.

Benveniste, R. E. 1976. Evolution of type C viral genes: Evidence for an Asian origin of man. *Nature* 261: 101–108.

Benveniste, R. E. 1985. The contributions of retroviruses to the study of mammalian evolution. In: R. J. MacIntyre, ed. *Molecular Evolutionary Genetics*, pp. 359–417. New York: Plenum.

Benveniste, R. E., Callahan, R., Sherr, C. J., Chapman, V., and Todaro, G. J. 1977. Two distinct endogenous type C viruses isolated from the Asian rodent *Mus cervicolor*: Conservation of virogene sequences in related rodent species. *J. Virol.* 21:849–852.

Benveniste, R. E., Sherr, C. J., and Todaro, G. J. 1975. Evolution of type C viral genes: Origin of feline leukemia virus. *Science* 190:886–888.

Benveniste, R. E., and Todaro, G. J. 1974a. Multiple divergent copies of endogenous C-type virogenes in mammalian cells. *Nature* 252:170–173.

Benveniste, R. E., and Todaro, G. J. 1974b. Evolution of C-type viral genes: Inheritance of exogenously acquired viral genes. *Nature* 252:456–459.

Berta, A. 1979. Quarternary evolution and biogeography of the larger South American Canidae (Mammalia: Carnivora). Ph.D. dissert. Univ. California, Berkeley. 262 pp.

Berta, A. 1981. Evolution of large canids in South America. *Anais II Congresso Latino-Americano de Paleontologia*. Porte Alegre 2:835–845.

Berta, A. 1984. The Pleistocene bush dog. *Speothos pacivorus* (Canidae) from the Lagoa Santa caves, Brazil. *J. Mamm.* 65:549–559.

Britten, R. J. 1986. Rates of DNA sequence evolution differ between taxonomic groups. *Science* 231:1393–1398.

Brownell, E. 1983. DNA/DNA hybridization studies of muroid rodents: Symmetry and rates of molecular evolution. *Evolution* 37:1034–1051.

Clutton-Brock, J., Corbett, G. B., and Hills, M. 1976. A review of the family Canidae with a classification by numerical methods. *Bull. British Mus. Zool.* 29:119–199.

Collier, G. E., and O'Brien, S. J. 1985. A molecular phylogeny of the Felidae: Immunological distance. *Evolution* 39:473–487.

Couturier, J., Razafimahatratra, E., Dutrillaux, B., Warter, S., and Rumpler, Y. 1986. Chromosome evolution in the Malagasy Carnivora. I. R-banding studies of *Cryptoprocta ferox, Fossa fossa, Galidia elegans*, and *Mungotictis decemlineata. Cytogenet. Cell Genet.* 41:1–8.

Davis, D. 1964. The giant panda: A morphological study of evolutionary mechanisms. *Fieldiana Zoology Memoirs.* 3:1–339.

Dayhoff, M. O. 1976. Survey of new data and computer methods of analysis. In: M. O. Dayhoff, ed. *Atlas of Protein Sequence and Structure*, vol. 5, supp. 2, pp. 1–8. Washington, D.C.: National Biomedical Research Foundation.

de Jong, W. W. 1986. Protein sequence evidence for monophyly of the Carnivore families Procyonidae and Mustelidae. *Mol. Biol. Evol.* 3:276–281.

Diamond, J. M. 1984. Historic extinctions: A Rosetta Stone for understanding prehistoric extinctions. In P. S. Martin & R. G. Klein, eds. *Quaternary Extinctions*. Tucson: Univ. Arizona Press.

Dutrillaux, B., and Couturier, J. 1983. The ancestral karyotype of Carnivora: Comparison with that of Platyrrhine monkeys. *Cytogenet. Cell Genet.* 35:200–208.

Ewer, R. F. 1973. *The Carnivores*. Ithaca, N.Y.: Cornell Univ. Press.

Fagan, R. M., and Wiley, K. S. 1978. Felid paedomorphosis with special reference to *Leopardus*. *Carnivore* 1:72–81.

Fitch, W. M. 1981. A non-sequential method for constructing trees and hierarchical classifications. *J. Mol. Evol.* 18:30–37.

Fitch, W. M., and Margoliash, E. 1967. Construction of phylogenetic trees. *Science* 155:279–284.

Flynn, J. M., and Galiano, H. 1982. Phylogeny of early Tertiary Carnivora, with a description of a new species of *Protictis* from the middle Eocene of Northwestern Wyoming. *Amer. Mus. Novitates* 2632:1–16.

Giles, E. 1960. Multivariate analysis of Pleistocene and Recent coyotes (*Canis latrans*) from California. *Univ. California Publ. Geol. Sci.* 36:369–90.

Gingerich, P. G. 1984. Pleistocene extinctions in the context of origination-extinction equilibria in cenozoic mammals. In: P. S. Martin & R. G. Klein, eds. *Quaternary Extinctions*, pp. 211–222. Tucson: Univ. Arizona Press.

Gittleman, J. L. 1986a. Carnivore brain size, behavioral ecology and phylogeny. *J. Mamm.* 67:23–36.

Gittleman, J. L. 1986b. Carnivore life history patterns: Allometric, phylogenetic and ecological associations. *Amer. Nat.* 127:744–771.

Glass, G. E., and Martin, L.D. 1978. A multivariate comparison of some extant and fossil Felidae. *Carnivore* 1:80–87.

Goldman, D., O'Brien, S. J., and Giri, P. R. 1988. The molecular phylogeny of the bears as indicated by two-dimensional electrophoresis. *Evolution*. In press.

Gregory, W. K., and Hellman, M. 1939. On the evolution and major classification of the civets and allied fossil and recent Carnivora: A phylogenetic study of skull and dentition. *Proc. Amer. Philos. Soc.* 81:309–392.

Gribbin, J., and Cherfas, J. 1982. *The Monkey Puzzle: Reshaping the Evolutionary Tree*. New York: Pantheon Books.

Guilday, J. E. 1984. Pleistocene extinctions and environmental change: Case study of the Appalachians. In: P. S. Martin & R. G. Klein, eds. *Quaternary Extinctions*. Tucson: Univ. Arizona Press.

Hall, E. R. 1981. *The Mammals of North America* (2nd ed.). New York: John Wiley and Sons.

Harrison, J. A. 1983. The Carnivora of the Edson Local Fauna (Late Hemphillian), Kansas. *Smithsonian Contributions to Paleobiology* 54:26–42.

Hemmer, H. 1976. Fossil history of the living Felidae. In: R. L. Eaton, ed. *The World's Cats*, vol. 3, no. 2: *Contributions to Biology, Ecology, Behavior and Evolution*, pp. 1–14. Seattle: Carnivore Research Institute.

Hemmer, H. 1978. The evolutionary systematics of the living Felidae. Present status and current problems. *Carnivore* 1:71–79.

Herrington, S. J. 1983. Systematics of the Felidae: A quantitative analysis. M.S. thesis, Univ. Oklahoma, Norman. 136 pp.

Hsu, K. J., la Brecque, J., Percival, S. F., Wright, R. C., Gombose, A. M., Pisciotto, K., Tucker, P., Peterson, N., McKenzie, J. A., Weissert, H., Karpoff, A. M., Carman, M. F. Jr., and Schreiber, E. 1984. Numerical age of the Cenozoic biostratigraphic datum levels: Results of the South Atlantic drilling. *Geological Society of America Bull.* 95:863–876.

Hunt, R. M., Jr. 1974. The auditory bulla in Carnivora: An anatomical basis for reappraisal of carnivora evolution. *J. Morph.* 143:21–76.

Huxley, T. H. 1880. On the cranial and dental characters of the Canidae. *Proc. Zool. Soc. London* 16:238–288.

Kohne, D. E., Chiscon, S. A., and Hoyer, B. H. 1972. Evolution of primate DNA sequences. *J. Hum. Evol.* 1:627–644.

Kurtén, B. 1964. The evolution of the polar bear, *Ursus maritimus* Phipps. *Acta Zool. Fennica* 108:1–26.

Kurtén, B. 1965. On the evolution of the European wild cat, *Felis silvestris* Schreber. *Acta Zool. Fennica* 111:1–29.

Kurtén, B. 1966. Pleistocene bears of North America. I. Genus *Tremactos*, spectacled bears. *Acta Zool. Fennica* 115:1–120.

Kurtén, B. 1968. *Pleistocene Mammals of Europe.* Chicago: Aldine.

Kurtén, B. 1974. A history of coyote-like dogs in North America (Canidae, Mammalia). *Acta Zool. Fennica* 140:1–38.

Kurtén, B. 1986. Reply to "A molecular solution to the riddle of the giant panda's phylogeny". *Nature* 318:487.

Kurtén, B., and Anderson, E. 1980. *Pleistocene Mammals of North America.* New York: Columbia Univ. Press.

Langguth, A. 1969. Die südamerikanischen Canidae unter besonderer Berücksichtigung des Mähenwolfes *Chrysocyon brachyurus* Illiger. *Zeitschrift für wissenschaftliche Zoologie* 179:1–88.

Langguth, A. 1975. Ecology and evolution in the South American canids. In: M. W. Fox, ed. *The Wild Canids*, pp. 192–206. New York: Van Nostrand Reinhold.

Lawrence, B., and Bossert, W. H. 1967. Multiple character analysis of *Canis lupus*, *latrans* and *familiaris* with a discussion of the relationship of *Canis niger*. *Amer. Zool.* 7:223–232.

Martin, L. D., Gilbert, B. M., and Adams, D. B. 1977. A cheetah-like cat in the North American Pleistocene. *Science* 195:981–982.

Maxson, L. R., and Wilson, A. C. 1975. Albumin evolution and organismal evolution in tree frogs (Hylidae). *Syst. Zool.* 24:1–15.

Nash, W. G., and O'Brien, S. J. 1982. Conserved regions of homologous G-banded chromosomes between orders in mammalian evolution: Carnivores and primates. *Proc. Natl. Acad. Sci.* 79:6631–6635.

Nash, W. G., and O'Brien, S. J. 1987. A comparative chromosome banding analysis of the Ursidae and their relationship to other Carnivores. *Cytogenet. Cell Genet.*, 45:206–12.

Neff, N. A. 1982. *The Big Cats.* New York: Abrams Inc.

Nei, M. 1972. Genetic distances between populations. *Amer. Nat.* 106:283–292.

Nei, M. 1978. Estimation of average heterozygosity and genetic distance from a small number of individuals. *Genetics* 89:583–590.

Nowak, R. M. 1979. North American Quaternary *Canis*. *Mongr. Mus. Nat. Hist. Univ. Kansas* 6:154.

Nowak, R. M., and Paradiso, J. L. 1983. *Walker's Mammals of the World* (4th ed.), vol. 2. Baltimore: Johns Hopkins Univ. Press.

O'Brien, S. J., Collier, G. E., Benveniste, R. E., Nash, W. G., Newman, A. K., Simonson, J. M., Eichelberger, M. A., Seal, U. S., Bush, M., and Wildt, D. E. 1987. Setting the molecular clock in Felidae: The great cats, *Panthera*. In: R. L. Tilson, ed. *Tigers of the World*. pp. 10–27. Park Ridge, N.J.: Noyes Publications.

O'Brien, S. J., and Nash, W. G. 1982. Genetic mapping in mammals: Chromosome map of domestic cat. *Science* 216:257–265.

O'Brien, S. J., Nash, W. G., Wildt, D. E., Bush, M. E., and Benveniste, R. E. 1985. A molecular solution to the riddle of the giant panda's phylogeny. *Nature* 317:140–144.

Olsen, S. J. 1985. *Origins of the Domestic Dog: The fossil record.* Tucson: Univ. Arizona Press.

Patterson, B., and Pascual, R. 1972. The fossil mammal: Fauna of South America. In: A. Keast, F. C. Erk & B. Blass, eds. *Evolution, Mammals and Southern Continents,* pp. 247–309. Albany: State Univ. New York Press.

Radinsky, L. 1977. Brains of early Carnivores. *Paleobiology* 3:333–349.

Radinsky, L. 1978. Do albumin clocks run on time? *Science* 200:1182–1185.

Radinsky, L. 1982. Evolution of skull shape in carnivores. 3: The origin and early radiation of the modern carnivore families. *Paleobiology* 8:177–195.

Reeves, R. H., and O'Brien, S. J. 1984. Molecular genetic characterization of the RD-114 gene family of endogenous feline retroviral sequences. *J. Virol.* 52:164–171.

Rohlf, F. J. 1982. Consensus indices for comparing classifications. *Math. Biosci.* 59:131–144.

Romer, A. S. 1968. *Notes and Comments on Vertebrate Paleontology.* Chicago: Univ. Chicago Press.

Sarich, V. 1969a. Pinniped origins and the rate of evolution of carnivore albumins. *Syst. Zool.* 18:286–295.

Sarich, V. 1969b. Pinniped phylogeny. *Syst. Zool.* 18:416–422.

Sarich, V. 1973. The giant panda is a bear. *Nature* 245:218–220.

Savage, D. E., and Russell, D. E. 1983. *Mammalian Paleofaunas of the World.* London: Addison-Wesley.

Savage, R. J. G. 1957. The anatomy of *Potamotherium* an Oligocene lutrine. *Proc. Zool. Soc. London* 129:151–244.

Savage, R. J. G. 1978. Carnivora. In: V. J. Maglio & H. B. S. Cooke, eds. *Evolution of African Mammals,* pp. 249–267. Cambridge: Harvard Univ. Press.

Seal, U. S., Phillips, N. I., and Erickson, A. W. 1970. Carnivora systematics: Immunological relationships of bear albumins. *Comp. Biochem. Physiol.* 32:33–48.

Sibley, C. G., and Ahlquist, J. E. 1983. Phylogeny and classification of birds based on the date of DNA-DNA hybridization. In: R. F. Johnston, ed. *Current Ornithology,* 1:245–288. New York.: Plenum Press.

Simpson, G. G. 1945. The principles of classification and a classification of the mammals. *Bull. Amer. Mus. Nat. Hist.* 85:1–350.

Sneath, P. H. A., and Sokal, R. R. 1973. *Numerical Taxonomy.* San Francisco: W. H. Freeman.

Stains, H. J. 1984. Carnivores. S. Anderson & J. K. Jones, eds. *Orders and Families of Recent Mammals of the World,* pp. 491–522. New York: Wiley and Sons.

Tedford, R. H. 1975. Relationships of Pinnipeds to other carnivores (Mammalia). *Syst. Zool.* 25:363–374.

Thenius, E. 1979. Zur systematischen und phylogenetischen Stellung des Bambusbären: *Ailuropoda melanoleuca* David (Carnivora, Mammalia). *Z. Säugetierk.* 44:286–305.

Thorpe, J. P. 1982. The molecular clock hypothesis: Biochemical evolution, genetic differentiation and systematics. *Ann. Rev. Ecol. Syst.* 13:139–168.

Van Valen, L. M. 1986. Palaeontological and molecular views of panda phylogeny. *Nature* 319:428.

Van Valkenburgh, B. 1985. Locomotor diversity in past and present guilds of large predator mammals. *Paleobiology* 11:406–428.

Van Valkenburgh, B. 1987. Skeletal indicators of locomotor behavior in living and extinct carnivores. *J. Vert. Paleo.* 7:162–182.

Wayne, R. K. 1986a. Cranial morphology of domestic and wild canids: The influence of development on morphologic change. *Evolution* 40:243–261.

Wayne, R. K. 1986b. Limb morphology of domestic and wild canids: The influence of development on morphologic change. *J. Morphol.* 187:301–319.

Wayne, R. K., Nash, W. G., and O'Brien, S. J. 1987a. Chromosomal evolution of the Canidae: I. Species with high diploid numbers. *Cytogenet. Cell Genet.* 44:123–133.

Wayne, R. K., Nash, W. G., and O'Brien, S. J. 1987b. Chromosomal evolution of the Canidae. II. Species with low diploid numbers. *Cytogenet. Cell Genet.* 44:134–141.

Wayne, R. K., and O'Brien, S. J. 1987. Allozyme divergence within the Canidae. *Syst. Zool.* 36:339–355.

Weiss, R., Teich, N., Varmus, H., and Coffin, J. 1982. *RNA Tumor Viruses.* New York: Cold Spring Harbor Press.

Werdelin, L. 1981. The evolution of lynxes. *Ann. Zool. Fennici* 18:37–71.

Werdelin, L. 1985. Small Pleistocene felines of North America. *J. Vert. Paleo.* 5:194–210.

Wilson, A. C., Carlson, S. S., and White, T. J. 1977. Biochemical evolution. *Ann. Rev. Biochem.* 46:573–639.

Wozencraft, W. C. 1984. A phylogenetic reappraisal of the Viverridae and its relationship to other Carnivora. Ph.D. dissert., Univ. Kansas, Lawrence. 1108 pp.

Wurster-Hill, D. H. 1975. The interrelationship of chromosome banding patterns in procyonids, viverrids, and felids. *Cytogenet. Cell Genet.* 15:306–331.

Wurster-Hill, D. H., and Centerwall, W. R. 1982. The interrelationships of chromosome banding patterns in canids, mustelids, hyena, and felids. *Cytogenet. Cell Genet.* 34:178–192.

Wurster-Hill, D. H., and Gray, C. W. 1973. Giemsa banding patterns in the chromosomes of twelve species of cats (Felidae). *Cytogenet. Cell Genet.* 12:377–397.

Wurster-Hill, D. H., and Gray, C. W. 1975. The interrelationships of chromosome banding patterns in procyonids, viverrids, and felids. *Cytogenet. Cell Genet.* 15:306–331.

Yoshida, M. A., Takagi, N., and Sasaki, M. 1983. Karyotypic kinship between the blue fox (*Alopex lagopus* Linn.) and the silver fox (*Vulpes vulpes* Desm.). *Cytogenet. Cell Genet.* 35:190–194.

Zuckerkandl, E., and Pauling, L. 1962. Molecular disease, evolution, and genic heterogeneity. In: M. Kasha & B. Pullman, eds. *Horizons in Biochemistry,* pp. 189–225. New York: Academic Press.

CHAPTER 18

The Phylogeny of the
Recent Carnivora

W. CHRIS WOZENCRAFT

A Historical Perspective

Konrad Gesner (1551), in one of the first widely distributed bestiaries, grouped animals that eat meat, a procedure that Linnaeus (1758) followed and identified as the order Ferae. The grouping of mammals that were carnivorous was further refined by Geoffroy Saint-Hilaire and Cuvier (1795), Cuvier (1800, 1817), Gray (1821), and Temminck (1835–41). They inferred relationships among species and groups of species primarily on the basis of morphological similarities in dentition.

Much of the impetus for classifying carnivores resulted from curators of large collections, who publicized descriptive catalogues of their collections and who were forced to consider taxonomic arrangements and the rationale behind the decisions made in these lists. Thus the first truly comprehensive attempts to classify the Carnivora were based on the nature (i.e., skins and skulls) of the specimens in museums. Notable among these were the catalogues of Schreber (1778), Geoffroy Saint-Hilaire (1803), Desmarest (1820), Gray (1825, 1843, 1869), Temminck (1835–41), Jentink (1887, 1892) and Troussart (1898–99). These early catalogues and their implied taxonomies greatly influenced the way biologists viewed the affinities among carnivores and suggested comparative studies to clarify further these associations.

When European morphologists began comparative surveys of the rapidly expanding collections, they encountered problems of homoplasy (i.e., parallel and convergent characters), which seemed to align taxa in "unnatural" arrangements. The ability to eat meat and its morphological implications was clearly not a good criterion when used alone, as this included the bats and insectivores in one early classification (Linnaeus 1748) and the opossum (*Didelphis virginiana*) in another (Blümenbach 1791; Cuvier 1800). Cuvier (1800) proposed another set of criteria to distinguish major subgroups within the Carnivora and grouped animals into plantigrade and digitigrade assemblages. Although most morphologists today would recognize the high amount of parallelisms in this feature (these groups also included marsupials and insectivores), the basic idea can still be seen in the taxonomy of Flower (1869b), which has served as the basis for many of the more recent classifications.

H. N. Turner (1848) was intrigued by the conservative nature of the basicranium and the consistent variation shown by this area at the family level. His classification (Table 18.1) marked a shift in systematic studies of carnivores to factors other than those directly influenced by food habits. In the latter half of the nineteenth century, discussions concerning the relationships among carnivores revolved around large suites of characters, assembled from detailed comparative morphological studies of the skull, postcrania, and soft anatomy and the similiarities in the complex features they shared (Gervais 1855; Flower 1869b; Huxley 1880; Mivart 1882a, 1882b, 1885a, 1885b). Flower and Lydekker (1891) summarized the implications of these studies in their classification, a widely accepted standard until 1945 (Table 18.1).

On the American continent the focus shifted from studies among Recent carnivores to the relationships between extinct carnivores and Recent taxa. Spurred by the recognition of an entirely new suborder of extinct carnivorous mammals, the Creodonta (Cope 1875), paleontologists began to question not only the associations among families but the relationships of carnivores to other mammalian orders. Cope (1880, 1882), Wortman and Matthew (1899), and Matthew (1901, 1909) led a focus on more restricted suites of characters that could be identified in fossil taxa. This effectively shifted the emphasis back to the delineation of relationships based on dentition, as teeth were the most common fossil remains. As early as 1869 Flower warned of problems in the elucidation of higher level relationships resulting from a heavy reliance on dental characters. Nevertheless, a clear diagnostic character for defining the Carnivora, the presence of the carnassial shear on P4/m1, emerged from these studies.

On the basis of morphological similarities in the skull, Gregory and Hellman (1939) proposed a classification somewhat different from those of Turner (1848) or Flower and Lydekker (1891) (Table 18.1). Gregory and Hellman suggested that hyaenids were more closely related to viverrids and that herpestids were a distinct group. They believed mustelids were an intermediate group between canoids and feloids.

G. G. Simpson's (1945) classification (Table 18.1), a standard yardstick for most post–World War II studies, reflected the paleontological training of the Cope-Matthew tradition. His classification, like that of Gregory and Hellman (1939), included extinct taxa and therefore leaned heavily toward data discernible from the fossil record, especially teeth; it placed less emphasis on other morphological studies such as those by van Kampen (1905) and van der Klaauw (1931), as well as the many papers on external morphology by R. I. Pocock. He accepted the arguments of Turner (1848) and Mivart (1885b) for the inclusion of the pinnipeds as a subgroup of the Carnivora but was reluctant to consider them diphyletic. He considered the basic dichotomy between fel-

Table 18.1. Major phenetic classifications of the Recent Carnivora 1845–1945

Turner (1848)	Flower and Lydekker (1891)	Gregory and Hellman (1939)	Simpson (1945)
Family Felidae Subfamily Viverrina	Suborder Carnivora vera Section Aeluroidea Viverridae	Suborder Fissipedia Superfamily Feloidea Viverridae Herpestidae	Suborder Fissipedia Superfamily Feloidea Viverridae
Subfamily Hyaenina	Hyaenidae Proteleidae	Hyaenidae	Hyaenidae
Subfamily Felina	Felidae	Felidae	Felidae
Family Canidae (Subfamily Canina)	Section Cynoidea Canidae	Superfamily Canoidea Canidae	Superfamily Canoidea Canidae
Family Ursidae Subfamily Ursina Subfamily Procyonina Subfamily Ailurina[a]	Section Arctoidea Ursidae Procyonidae	Ursidae Procyonidae	Ursidae Procyonidae
Subfamily Mustelina Family Phocidae	Mustelidae Suborder Pinnipedia	Superfamily Musteloidea Mustelidae Suborder Pinnipedia Superfamily Otarioidea	Mustelidae Suborder Pinnipedia
Subfamily Arctocephalina Subfamily Trichechina	Otariidae Trichechidae	Otariidae Odobaenidae [sic] Superfamily Phocoidea	Otariidae Odobenidae
Subfamily Phocina	Phocidae	Phocidae	Phocidae

Note. Turner 1848 and Gregory and Hellman 1939 were primary studies.
[a]Included in the Procyonidae by Flower and Lydekker 1891, Gregory and Hellman 1939, and Simpson 1945.

iforms and caniforms, suggested by Cuvier (1800), Turner (1848), Flower (1869b), and Winge (1895, 1924) to represent a natural division. Simpson's (1945) scheme placed those groups with primitive dentitions (i.e., canids and some viverrids) as basal carnivores and those with advanced dental arrays (i.e., felids and ursids) at more derived levels. However, his classification of carnivores did not coincide with other morphological studies on the Carnivora; Pocock (1916b, 1919), van der Klaauw (1931), and Gregory and Hellman (1939) revealed the distinctiveness of the herpestids, and Hough (1944, 1948, 1953) pointed out the derived nature of the canids.

The four major classifications of carnivores (Table 18.1) are strikingly similar and differ little from Turner's original proposal. They are based on similar methodologies; they lean heavily on the comparative morphology of the skull and dentition; and they phenetically group taxa based on the distribution of similar features.

After Simpson's (1945) classification there were several publications that more clearly elucidated characters among carnivores. Segall (1943), Davis and Story (1943), and Story (1951) discussed in detail the relationship between the bullar region and the internal carotid system. Butler (1946) identified the primitive carnivore dentition and discussed the evolution of carnassial dentitions. Scheffer (1958) summarized much of what was known concerning pinniped biology. Davis's (1964) thorough morphological study of the giant panda (*Ailuropoda melanoleuca*) demonstrated that it was a highly specialized bear. Radinsky (1980, 1981a, 1981b, 1982, 1984) studied the functional anatomy of the skulls of the Carnivora. His analyses suggested that great zygomatic arch width, long temporal fossa, long tooth rows, small brain size, and the forward placement of the carnassial are primitive carnivore features.

During this period there were intensive investigations into evolutionary mechanisms operating at the molecular and biochemical level. Serological work by Leone and Wiens (1956) and Pauly and Wolf (1957) supported the caniform/feliform split and the inclusion of the pinnipeds within caniforms. Pioneering work by Fredga (1972) clarified genic evolution within the Herpestidae. Fay et al. (1967) reviewed the cytogenetic evidence for the evolution of the pinnipeds. In a more extensive cytogenetic study Arnason (1974) supported the monophyly theory of pinniped origin. Wurster and Benirschke (1967, 1968), Wurster-Hill and Gray (1975), and Dutrillaux and Couturier (1983) discussed the cytogenetic evolution of carnivore families and investigated karyotypic trends within the group. Their studies suggested; (1) monophyletic origin for pinnipeds; (2) separation of herpestids and viverrids; (3) close affinity between the viverrids and hyaenids; and (4) the distinctiveness of the mustelids (Figure 18.1A).

Two complementary publications, J. E. King's *Seals of the World* (1964 [1st ed.]; 1983 [2nd ed.] and R. F. Ewer's *The Carnivores* (1973), stand out as landmark references on the biology and evolution of carnivores. King and Ewer combined a survey of primary literature with their own research to

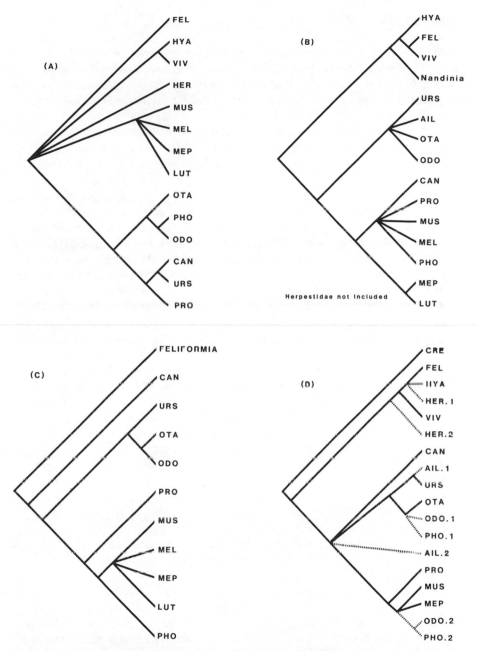

Figure 18.1. Phylogenetic trees of the Recent Carnivora: (A) modified from Wurster and Benirschke 1968 on the basis of cytogenetic data; (B) based on bullar types identified in Hunt 1974; (C) modified from Tedford 1976 on the basis of morphological data; and (D) based on papers presented at the 1986 American Society of Mammalogists symposium "Evolution of the Carnivora" (see text). Abbreviations: AIL = *Ailurus;* CAN = Canidae; CRE = Creodonta; FEL = Felidae; HER = Herpestidae; HYA = Hyaenidae; LUT = Lutrinae; MEL = Melinae; MEP = Mephitinae; MUS = Mustelinae; ODO = Odobeninae; OTA = Otariidae; PHO = Phocidae; URS = Ursidae (including *Ailuropoda*); VIV = Viverridae.

provide a comprehensive discussion of behavior, evolution, anatomy, and ecology of the wide diversity of terrestrial and aquatic carnivores. Unfortunately, the aquatic/terrestrial division of the two books tended to reinforce the notion of separate carnivore orders rather than highlight their common evolutionary heritage.

Before 1950 the distinction between primitive and derived conditions were not necessarily made; furthermore, "primitive" was usually interpreted as simple or older in the fossil record and "derived" as more complex and/or recent. If a character was found in both a fossil and Recent carnivore, the character was often assumed to be primitive, and therefore the Recent species was also considered primitive.

Darwin's original definition of evolution was "descent with modification," and although most classifications had attempted to reflect modification, few had incorporated descent or ancestry into their classification schemes. Many systematists recognized that classifications should reflect not only change but descent or phylogeny as well, and therefore phylogenetic hypotheses should be based on monophyletic groups. A more rigorous definition of "primitive" and "derived" appeared in 1950 (English translation, 1966) with the publication of Hennig's treatise on phylogenetic systematics. A method was proposed by Hennig and followed up by others to evaluate how one determines the primitive or derived nature of a feature, and a theoretical basis was given for incorporation of descent or ancestry and modification into systematics. Taxa are identified by examining phenotypes, and then a hypothesis is inferred from the distribution of shared derived attributes (synapomorphies). Previously some taxa had been grouped on comparative similarities in primitive features (pleisomorphies), which does not reflect the genealogical history of the group (Hennig 1966). The reconstruction of phylogeny should be based on synapomorphies that indicate a common evolutionary lineage (i.e., a monophyletic group). Although characters are identified based on similarity, the homology and polarity of the feature (i.e., primitive or derived) is determined by other methods (e.g., out-group comparison).

Several major phylogenetic hypotheses were advanced between 1975 and 1988, using this methodology. McKenna (1975) outlined a higher level taxonomy for the Mammalia, which was a departure from Simpsonian ideas. McKenna grouped the order Carnivora with two extinct orders, the Cimolesta and the Creodonta within the grandorder Ferae. Flynn and Galiano (1982) strongly defended the idea of a natural dichotomy in the Carnivora (Feliformia versus Caniformia), a split that they believed occurred in the early Paleocene. They placed six families in the Feliformia: Didymictidae, Viverravidae, and the Recent monophyletic group: Herpestidae, Viverridae, Felidae, and Hyaenidae. Hunt (1987) suggested placing *Nandinia binotata* separate from the Viverridae as the most primitive feloid family. Flynn and Galiano (1982) followed Tedford (1976) and others who placed the pinnipeds (Otariidae, Phocidae) within the Caniformia, which also included the Ursidae, Mustelidae, Procyonidae, Canidae and the extinct families Miacidae and Nimravidae. The position of the nimravids has been controversial, with proponents of its placement in the feliforms (Hunt 1987), caniforms (Flynn and Galiano 1982), and as the sister group to the rest of the Carnivora (Neff 1983). Novacek (1986), Novacek and Wyss (1986), and Shoshani (1986) presented phylogenetic trees of the relationships among Recent mammalian orders. Novacek restricted data to skull morphology, whereas Shoshani relied on both cranial and postcranial anatomy. Novacek and Wyss, though using a smaller number of characters, reflected morphology in both skeletal and soft anatomy. All are similar in their placement of the order Carnivora. Shoshani (1986) believed the carnivores to be an early offshoot of the monophyletic group leading to the Primates, Scandentia, Dermoptera, and Chiroptera. Novacek (1986:94), through grouping these taxa together in the Epitheria, stated that "the orders show either too few similarities or too many conflicting traits to justify a close association with one or another of these larger groups." This placement was also reflected in the six-branch polytomy presented in Novacek and Wyss (1986). These two trees suggest that the Carnivora may be one of the earliest sister groups to the clade leading to the majority of the Eutheria (but excluding many insectivores, edentates, and pholidotans).

Recent attempts to investigate relationships among carnivores have focused on the basicranium, a region originally emphasized by Turner (1848). It should not be surprising that the basicranium is crucial to our understanding of carnivoran phylogeny, as many functional aspects of the animal's biology can be inferred from this region and it is an ideal area to focus studies on higher level relationships. Several organ systems are concentrated here, such as cranial nervation, blood circulation, balance, mastication, head and neck muscle attachment, and hearing. The basicranium is an extremely complex region with functional, structural, and physiological mechanisms that affect the morphology of a small area. As Turner (1848) originally pointed out, it is relatively conservative in nature and shows little variation below the family level. The morphology and physiology of many of the organ systems can be inferred from foramina, grooves, and other bony structures, thus allowing comparison with fossil taxa. The importance of this area in understanding carnivoran relationships was emphasized in early studies by Flower (1869b), Mivart (1882a, 1882b, 1885a, 1885b), and van der Klaauw (1931). More recently, studies by Butler (1948), McDowell (1958), Bugge (1978), Presley (1979), MacPhee (1981), Neff (1983), and Wible (1986, 1987) have clarified much of our understanding of the homology and development of the basicranial region in various groups of mammals and demonstrated its importance in elucidating relationships. R. M. Hunt (1974) published a reappraisal of carnivoran evolution based on a hypothesis for the transformation of auditory bullar characters. Although Hunt did not explicitly do so, a tree can be constructed based on his bullar types (Figure 18.1B). Three years later Novacek (1977) summarized the development of the auditory bulla and placed this information in the broader perspective of eutherian evolution.

A symposium on the evolution of the Carnivora was held at the 66th annual meeting of the American Society of Mammalogists, in Madison, Wisconsin, on 16 June 1986. Five carnivore biologists presented reviews of current ideas concerning phylogenetic hypotheses based on morphological examination of Recent and extinct carnivores, ecological and behavioral studies, and bio-chemical and molecular research, as shown in the composite tree (Figure 18.1D). There were several areas of consensus among these biologists that are useful to list here: (1) extinct Creodonta represent the closest sister taxon to the Carnivora (McKenna 1975; Flynn and Galiano 1982); (2) among Recent mammalian orders, the clade leading to the Primates, Scandentia, Chiroptera, and Dermoptera is the closest branch to the Carnivora (Novacek 1986; Shoshani 1986); (3) the order Carnivora has two main lineages, the Feliformia and the Caniformia (Kretzoi 1945; Flynn and Galiano 1982); (4) the aquatic carnivore families (otariids, odobenines, and phocids) are a subgroup of the Caniformia and should be included within the Carnivora (Turner 1848; Mivart 1885b; Gregory and Hellman 1939; Tedford 1976); (5) among the terrestrial carnivores, the otariids are most closely related to the ursids (Mivart 1885b; Mitchell and Tedford 1973; Wyss 1987); (6) the Herpestidae and Viverridae are distinct monophyletic families (Pocock 1919; Gregory and Hell-man 1939; Neff 1983; Wozencraft 1984b; Hunt 1987); and (7) among the Recent fissipeds, the mustelids are most closely related to the procyonids (Ted-ford 1976; Schmidt-Kittler 1981; Neff 1983; Wozencraft 1984b).

The carnivore symposium also highlighted areas at the family level where more research is needed. The position of the phocids and odobenines in rela-tion to the terrestrial carnivores was questioned by Berta and Deméré (1986) and Wyss (1987). Pocock (1919), Gregory and Hellman (1939), Wurster and Benirschke (1967), Wozencraft (1984a, 1984b) and Hunt (1987) agreed on the separation of the mongooses (Herpestidae) from the civets (Viverridae), al-though they did not agree on whether herpestids were closer to the hyenas (Hunt 1987) or were the most primitive sister group to the other three taxa (Bugge 1978; Wozencraft 1984b). Some paleontologists (Beaumont 1964; Hemmer 1978) have suggested that the fossa (*Cryptoprocta ferox*) is related to the felids, a position not followed by others (Gray 1865; Petter 1974; Wozen-craft 1984b). Ginsburg (1982), Wozencraft (1984a, 1984b), and Decker and Wozencraft (in press) have argued that the red panda (*Ailurus fulgens*) is more closely related to the ursids than the procyonids, a position that Sarich (1976) and Todd and Pressman (1968) hinted at with biochemical based phe-nograms. Whereas O'Brien et al. (1985:3) maintained that "the lesser panda and procyonids share a common ancestor at or subsequent to ursid diver-gence." Of these research areas, the controversy that has the most impact on the arrangement of families concerns the relationships of the pinnipeds to the terrestrial carnivores.

On the Origin(s) of the "Pinnipeds" or Aquatic Carnivora

Before Turner's (1848) classification of the Carnivora, the pinnipeds had been placed in Pisces (Linnaeus 1758), in an order with elephants and sirenians (Brisson 1756), and in the order "Palmata" with the beaver (*Castor canadensis*) and platypus (*Ornithorhynchus anatinus*) (Blümenbach 1788). Murie (1870, 1872, 1874) and Allen (1880) compared the morphology of representative taxa and suggested that the walrus (*Odobenus rosmarus*) be grouped with the otariids and commented on their ursine features; however, Doran (1879) suggested that the pinnipeds were monophyletic based on a survey of mammalian ear ossicles.

Mivart (1885b) examined the morphology of the Recent groups (otariids, odobenines, and phocids) and was impressed by the similiarity of the phocids to the lutrines and the otariids to the ursids. Although he did not go as far as proposing that the groups be thus arranged, the data he presented suggested such an arrangement. Howell (1929) did not accept Mivart's arguments concerning diphyletic origins. He maintained that the number and kinds of convergences that would be necessary to maintain the diphyletic theory made it unlikely and believed that the fossil evidence for pinnipeds was earlier than that for ursids or lutrines.

The similarities between otariids and ursids, on one hand, and mustelids and phocids, on the other, was emphasized by van Kampen (1905) and van der Klaauw (1931) on the basis of characters in the basicranium. Savage (1957) provided more support for a lutrine-phocid clade with his description of an Oligocene lutrine, *Potamotherium,* that appeared intermediate in many characteristics. McLaren (1960) addressed the question directly in his paper "Are the Pinnipedia biphyletic?" and concluded that Mivart's suggestion was correct; there was a close association of the Otariidae, Odobeninae, and Ursidae, representing one monophyletic group, and the Mustelidae and the Phocidae, representing another. Mitchell and Tedford (1973) and Repenning (1976) argued strongly for the ursine origin of the otariids and also for the inclusion of the walrus in this monophyletic group. Repenning et al. (1979) concluded that the otariids and odobenines originated in the Pacific whereas the phocids probably originated in the Atlantic.

Hunt's (1974) analysis of the auditory bulla provided support for the diphyletic hypothesis (Figure 18.1*B*) with distinct bullar types being shared by respective clades. He also outlined similarities in bullar characters between the mephitine-lutrine clade and the ursids, which has further confused the situation; Tedford (1976) interpreted the mephitine-lutrine condition as secondarily derived and concisely summarized many of these ideas (Figure 18.1C), adding dental and postcranial features to support the diphyletic hypothesis.

At this point most paleo- and neomammalogists accepted a diphyletic origin of the group. However, beginning in the late 1960s and during the 1970s, cytogenetic and biochemical phenetic studies suggested an alternative monophyletic origin. Fay et al. (1967), Anbinder (1969, 1980), and Arnason (1974, 1977) presented cytogenetic evidence that argued for monophyly. Sarich (1969), Seal (1969), Seal et al. (1971), and Arnason and Widegren (1986) concluded, based on biochemical similarities, that monophyly could be supported. This has caused morphologists to take a second look at the anatomy of the respective groups.

The controversy over the origin of these groups centers on the treatment of parallel and convergent characters. Aquatic adaptations shared by pinnipeds appear to outweigh other suggested character affinities. Phenetic approaches exclude parallel or convergent possibilities, in this particular example, excluding anything but a monophyletic origin. If one wishes to discern the relationships among two or more taxa that share some derived characteristics, then at least two possibilities exist: either these features were derived from a common ancestor, or they represent parallel or convergent evolution.

As with many phylogenetic analyses, the crux of the argument depends on how one handles homoplasy. One could argue that aquatic animals would share more aquatic features with each other than with terrestrial animals; moreover, the characters that are shared with the terrestrial groups (not those that exclude them) will best indicate phylogeny. Therefore, characters that convey genealogical information must be teased apart from those that simply indicate an aquatic adaptation. Unfortunately, the sheer number and complexity of morphological changes that are necessary to change a terrestrial carnivore to a pelagic marine mammal tend to mask possible relationships. The congruence in the phenetic approaches of the biochemical and karyological data may be tracking these aquatic adaptations as well.

It is therefore appropriate to ask what specializations one would expect an aquatic mammal to have. Loss of hair, flippers, changes in respiratory physiology, fat storage, behavioral and ecological changes, and loss or reduction of various external features (sweat glands, lacrimal gland, external pinnae, and so forth) would seem obvious aquatic adaptations. Less obvious are changes in morphological features that affect hearing, vision, food habits, balance, blood circulation, and foraging. The ancestor of the pinnipeds was terrestrial; however, many terrestrial adaptations would be less than adequate for an aquatic environment. The terrestrial ear and eye may actually impede perception in an aquatic environment (Repenning 1972).

Wyss (1987), in one of the first phylogenetic approaches to the question of monodiphyly in pinnipeds, precluded diphyly by considering only one terrestrial outgroup, the Ursidae. Furthermore, his final tree (Wyss 1987, fig. 7) does not include outgroups, nor does it include characters (in his text and published by others) that are inconsistent with his tree. He used 43 morphological characters, but of these 35 are directly related to being aquatic. Moreover, many of these characters are actually parts of character complexes (e.g., the five malleus characters) responding to single functional pressures (Repenning 1972). Some characters listed are directly correlated features (e.g., epitympanic recess large—ear ossicles large); others could not be verified on U.S. National Museum (USNM) specimens (e.g., the walrus and phocids having a similarly inflated auditory bullae). No measurement data were presented to support size difference claims. Twelve characters are allometric and show a much different distribution when size is eliminated as a factor (e.g., large baculum). Finally, the polarity for some characters as assigned by Wyss would be quite atypical (e.g., diverging palate, considered by Wyss to be derived, is primitive on the basis of outgroup comparison).

Repenning (1972), Mitchell and Tedford (1973), Hunt (1974), Tedford (1976), and Muizon (1982a) outlined many derived features shared by otariids, odobenines, and ursids on one hand and mustelids and phocids on the other, a conclusion supported by this study (Figure 18.2). In this analysis monophyly could be supported only if one weighted size, tooth loss, simplification, and regeneration in various lineages. Clearly, more research needs to be done and the fossil taxa need to be included in the overall picture. The Recent walrus has traditionally been placed in a separate family because of its highly

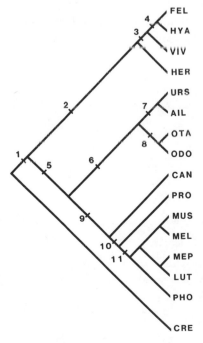

Figure 18.2. Phylogeny of the Recent Carnivora as determined from 100 morphological characters of the skull, dentition, postcranial anatomy, and soft anatomy. Abbreviations after Figure 18.1 (see Appendix 18.1 for characters).

derived nature. The phylogeny presented here supports inclusion of the walruses as a subfamily of otariids, following Mitchell and Tedford (1973).

A Phylogenetic Reappraisal of the Recent Carnivora

Our current ideas concerning the phylogeny and classification of carnivores are based on the analysis of characters traditionally used to delineate groups and relationships among taxa. I have attempted to integrate these key morphological innovations, using phylogenetic theory as outlined by Hennig (1966) and Wiley (1979, 1981). These texts should be consulted for more information concerning phylogenetic methodology. The basic thesis of this section is as follows: (1) if one compiles key morphological innovations viewed as significant in the literature (limited here to 100 characters) and (2) organizes and interprets their polarity (i.e., primitive or derived nature) using phylogenetic theory, one can then (3) present a comprehensive synthesis or summary in the form of a phylogenetic tree that will integrate morphological trends.

The phylogenetic hypothesis that follows is summarized in a classification (Table 18.2) and phylogram (Figure 18.2). The bracketed numbers in the headings refer to the numbers in the phylogram. The discussion traces the branching sequence of this tree and discusses at the appropriate nodes other data that may conflict with this hypothesis. At these points, important trends in the evolution of the group will be discussed. A more detailed analysis and discussion can be found in the original sources.

This analysis was based on features reported in the literature and representing the morphology of the skull, postcranial, and soft anatomy (see Appendix

Table 18.2. A phylogenetic classification of the families of Recent Carnivora (based on the phylogeny in Figure 18.2)

Order Carnivora Bowdich 1821
    Suborder Feliformia Kretzoi 1945
        Superfamily Feloideaoidea Fischer de Waldheim 1817
            Family Herpestidae Bonaparte 1845
            Family Viverridae Gray 1821
            Family Hyaenidae Gray 1821
            Family Felidae Fischer de Waldheim 1817
    Suborder Caniformia Kretzoi 1945
        Superfamily Ursoidea
            Family Ursidae Fischer de Waldheim 1817
            Family Otariidae Gray 1825
        Superfamily Canoidea Fischer de Waldheim
            Family Canidae Fischer de Waldheim 1817
            Family Procyonidae Gray 1825
            Family Mustelidae Fischer de Waldheim 1817
            Family Phocidae Gray 1821

Note. See appendix (Wozencraft, this volume) for taxa included in each family.

18.1). Character states, once confirmed by examination of specimens, were used in a computer algorithm (PAUP, version 2.4, developed by David Swofford, Illinois Natural History Survey) to determine the most parsimonious trees. Interpretation of qualitative characteristics were coded using outgroup comparison after methods proposed by Wiley (1981) and Watrous and Wheeler (1981). The resulting tree (Figure 18.2) had 197 steps (i.e., character transformations) and a Consistency Index of 0.569.

Order Creodonta

Cope (1875) recognized the Creodonta as a suborder of the Insectivora, whereas Wortman (1901) placed the suborder in the Carnivora. Matthew (1909) considered the creodonts to consist of five extinct families (Hyaenodontidae, Oxyaenidae, Miacidae, Mesonychidae, and Arctocyonidae). Schlosser (1887) had earlier suggested the separation of the miacids from the creodonts, but it was not until Kretzoi (1945) and Simpson (1945) that this gained wide acceptance. Van Valen (1969) suggested that the Mesonychidae and the Arctocyonidae were archaic ungulates and should be transferred to the condylarths. Currently, most paleontologists recognize the Creodonta, as thus revised as the sister group to the Carnivora.

Creodonts can be separated from the Carnivora by the location of the carnassials at M1/m2 or M2/m3, fissured ungual phalanges, and separate scaphoid and lunar bones (Denison 1938; Savage 1977). The late Cretaceous or early Paleocene *Cimolestes* or a *Cimolestes*-like paleoryctine is often suggested as the most likely common ancestor to the Creodonta and the Carnivora (Gregory and Simpson 1926; MacIntyre 1966; Van Valen 1966).

Order Carnivora [1]

The P4/m1 principal carnassial shear remains as the central character complex that unites the Carnivora despite its secondary loss in some taxa; moving anteriorly or posteriorly along the tooth row from this locus are trends in tooth reduction. The primitive dentition (I3/3, C1/1, P4/4, M3/3) has spatulate subequal incisors, large canines, and m1 paraconid < metaconid < protoconid (Flynn and Galiano 1982). Flynn and Galiano suggested that the fusion of the carpal bones in the Carnivora may not have been a unique event; independent fusion may have occurred in the caniforms and feliforms. The bulla consists of three ossified elements (rostral and caudal entotympanic, and ectotympanic) (Hunt 1974). The postcranial anatomy is primitive in its retention of a calcaneal fibular facet (Szalay and Decker 1974; Novacek 1980) and in the medial position of the lesser trochanter on the femur (the third trochanter is absent).

Suborder Feliformia, Superfamily Feloidea [2]

Kretzoi (1945) abandoned Flower and Lydekker's (1891) classification (Fissipedia and Pinnipedia) and arranged the families differently into two suborders, the Feliformia and Caniformia. Tedford (1976) placed all of Simpson's (1945) superfamily Feloidea into the Feliformia and all of the superfamily Canoidea into the Caniformia, a natural division that was followed by Flynn and Galiano (1982). Simpson (1945) rejected Flower's "Aeluroidea" as it was not based on a taxon included within the group. The Feloidea is here understood to refer to the monophyletic group consisting of the Herpestidae, Viverridae, Hyaenidae, and Felidae; the Feliformia includes extinct families not considered here.

The viverrids, herpestids, hyaenids, and felids have been recognized since Gray (1837) as a monophyletic group, except by Hough (1953), who suggested uniting the felids, canids, and the extinct nimravids into the Cynofeloidea on the basis of the presence of a septum in the bulla and a number of primitive features. Most feloids have a bulla divided by a bilaminar (ecto- and entotympanic) septum except for the African palm civet (*Nandinia binotata*), in which the entotympanic is cartilaginous, and the hyaenids, which have secondarily lost the entotympanic portion of the septum (Hunt 1974). Flynn and Galiano (1982) recognized this group on the basis of four relative characters and two qualitative characters (M3 lost, m2 hypoconulid ≥ hypoconid). Wortman and Matthew (1899) believed the carnassial notch also to be diagnostic.

Primitive feloids share the derived condition of a large, complex entotympanic, whereas the ectotympanic is primitively smaller. However, in herpestids and hyaenids the ectotympanic is secondarily enlarged, and accompanying this enlargement is a change in the orientation of bullar elements. In the Hyaenidae the ectotympanic is expanded ventrally to the entotympanic (Hunt 1974; 1987), whereas in herpestoids there is a coequal anterior–posterior division (Pocock 1919).

Herpestidae. All primary systematic studies that have investigated the relationships of the mongooses to the civets and other carnivores have either placed the mongooses in a separate family or suggested that they do not share any derived features with the viverrids (Winge 1895; van Kampen 1905; Pocock 1916b, 1919; van der Klaauw 1931; Gregory and Hellman 1939; Wurster and Benirschke 1968; Fredga 1972; Thenius 1972; Radinsky 1975; Bugge 1978; Neff 1983; Wozencraft 1984b; Hunt 1987). In the literature review and the corresponding data matrix compiled for this study, no synapomorphies were found that united the viverrids and the herpestids.

In 1916 Pocock proposed the name Mungotidae, believing the generic name *Herpestes* to be invalid based on arguments advanced by Thomas (1882).

Allen (1924) demonstrated that Thomas was in error and that the correct family name is Herpestidae (see the appendix to this volume for subfamily and species list).

Bonaparte (1845) and Gray (1865) recognized the distinctiveness of the mongooses, but Flower and Lydekker (1891) placed them at the subfamily level without giving rationale. Between 1891 and 1945 four systematic works argued strongly for the recognition of the mongooses as a separate family, among them Gregory and Hellman (1939). Simpson (1945), who otherwise relied heavily on Gregory and Hellman (1939), viewed their recognition of the Herpestidae as an "unnecessary complication" that offered no advantage over a three-family system. If Gregory and Hellman had viewed the herpestids as the sister group to the viverrids, then raising this group to the family level would be unwarranted. However, the relative placement of the mongooses on their tree and references in the text to a "civet (viverrid) hyena-cat group" sharing a common ancestor suggest that they viewed the mongooses as a more primitive outgroup to the above-mentioned monophyletic group, a conclusion that this study supports. Combining the civets and mongooses into a single family would make the taxon paraphyletic. Simpson's (and others) reluctance to recognize the mongooses may be related to the primitive "carnivore" dental pattern found in the viverrine civets and some herpestine mongooses. Matthew (1909) and Gregory and Hellman (1939) were impressed by the similarities between the dentition of extinct viverravids and the Recent viverrine civets, and they proposed taxonomic groupings based on these plesiomorphic characters.

The mongooses are characterized by the uniquely derived nature of their anal sac and the structure of the auditory bulla. The major component of the enlarged anal sac secretion is a carboxylic acid, a by-product of bacterial metabolism (Gorman et al. 1974) that parallels development in hyaenids and mustelids. The slight lateral expansion of the mastoid coupled with the more narrow paroccipital process permits the mastoid to be seen from the posterior aspect of the skull (Petter 1974). A distinguishing feature of the Herpestidae is the complex arrangement of the bullar elements, with an expanded ectotympanic and a circular external auditory meatal tube that is unique among the families of Carnivora (van der Klauuw 1931). Most mongooses have an internal cusp on the third upper premolar, variable in development and vestigial in some individuals.

Petter (1974) suggested that the Malagasy mongooses (galidiines) and civets (cryptoproctines) share a common ancestry; however, the basicranial characters used by her to unite both groups are considered primitive (van der Klauuw 1931; Hunt 1974; Novacek 1977). The uniquely derived bullar complex that the galidiines share with the herpestines makes derivation of Malagasy carnivores from a common ancestor unlikely. Fredga (1972) concluded, after an extensive karyological study of the mongooses, that the karyotype of Galidii-

nae is primitive. Galidiinae thus appear to be an early offshoot of the her-
pestine-mungotine lineage and separate from the phylogenetic history of other
Malagasy carnivores.

Derived Feloids (Felidae, Hyaenidae, Viverridae) [3]. Characters involving
bullar structure, circulatory pattern, and anterior position of basioccipital/
basisphenoid suture suggest that viverrids, felids, and hyaenids share a common ancestor not shared with the mongooses (appendix to this volume, and
Figure 18.2) *contra* Simpson (1945) and Hunt (1987). They all share a suite of
characters associated with their complex elongate bulla (Hunt 1974), with
corresponding modifications of the carotid circulatory system (Bugge 1978).
The paroccipital process is wide and protrudes beyond the ventralmost extension of the bulla in the Viverridae and Hyaenidae, whereas the development of
the mastoid can be viewed from the posterior aspect of the skull in felids (van
der Klaauw 1931). The ectotympanic is similar in all three groups, a primitively shaped C that is secondarily expanded medially in the hyaenids (van der
Klaauw 1931).

Hunt (1987) argued for a herpestid-hyaenid clade on the basis of five synapomorphies: (1) similar ontogenetic development of the bulla, (2) nonretractile claws, (3) anal pouch, (4) lack of ear bursa, and (5) a similar auditory
meatal tube. His study showed that there are some similarities in the development and morphology of the bullae in some mongooses and the hyaenids.
However, there are characteristics of the petrosal and carotid canal that also
support a felid-hyaenid clade (Bugge 1978; Wozencraft 1984b). Nonretractile
claws would be coded as primitive when outgroup comparison is used. The
anal glands, present at least in a rudimentary state in nearly all carnivores,
empty into a clearly defined large pouch only in a few (mustelids, herpestids,
Cryptoprocta ferox, and hyaenids). The morphology of the anal pouch and the
variation in number and location of anal glands in each of these groups are
dissimilar enough to call into question their homology (Flower 1869a; Davis
and Story 1949; Pocock 1916c; 1916b). The ear bursa is absent in a variety of
carnivore families (mustelids, ursids, otariids, hyaenids, some herpestids). If
one were to unite hyaenids and herpestids on the loss of the ear bursa, this
would cause its reversal in at least some herpestids. The absence of the bursa in
hyaenids "suggests affinity with the mongooses. But the arrangement of the
main cartilages of the ear is not in the least like that of the mongooses" (Pocock
1916c:333). Finally, the presence of the external auditory meatal tube is quite
different in hyaenids and herpestids. In hyaenids there is an inflation of the
ectotympanic ventral to the meatal tube. The tube itself is a ventrally continuous C-shaped surface that forms a complete cylindrical tube. In the herpestids
the tube is continuous only on the dorsal surface. On the ventral surface the
posterior and anterior growths from the arms of the ectotympanic meet along
the midline, forming an incomplete suture that closes with age (van der
Klaauw 1931).

Viverridae. The diversity of the civets has puzzled previous workers and presented problems in identifying their common ancestor. Particularly problematic genera include the Malagasy falanouc (*Eupleres goudoti*) and fossa, and the African palm civet. Gray (1865), Flower (1869), Mivart (1882a), and Pocock (1915a, 1915b, 1916a) included all of these taxa within the Viverridae. The African palm civet is the only civet that does not possess the typical "viverrid" bulla, and it lacks true internal septa (Hunt 1974, 1987); the ectotympanic is the primitive C shape, but the ectotympanic remains cartilaginous throughout the animal's adult life. The cartilage is slightly reduced from the typically more inflated nature of the viverrid bulla, but otherwise it occupies the same position and has the same elongated nature.

Hunt (1987) united the viverrids and felids (and therefore excluded the African palm civet) on nine synapomorphies, of which seven are direct reflections of an interpretation of the relative inflation and chambering of the bullae. Statements concerning relative inflation of a chamber are difficult to evaluate; however, measurements of the hypotympanic cavity height of USNM specimens of Asiatic linsangs (*Prionodon*) and the small toothed palm civet (*Arctogalidia trivirgata*) show less inflation than the African palm civet. Furthermore, we could not verify other characters used by Hunt to distinguish between viverrids and the African palm civet (e.g., pocketing of the mastoid; lack of inbending of the ectotympanic in the African palm civet). Hunt argued that two characters were primitive for the African palm civet (separating it from viverrids): the separation of the hypoglossal foramen from the posterior lacerate foramen, and the paroccipital process not being closely applied to the bullae. The wide separation of these two foramina are only found elsewhere in the Caniformia, suggesting that the condition in other feliforms is derived. The application of the paroccipital process to the bullae varies within the caniforms with both states being present, making the determination of the polarity of this character uncertain.

The fossa would make a good primitive cat on the basis solely of tooth structure, and Beaumont (1964) and Hemmer (1978) have suggested such a relationship to the Felidae. However, all of the cranial, postcranial, and soft anatomical features suggest a close affinity with the viverrine civets (see Appendix 18.1). The dental similarities are thus convergent to felids and may be related to its ecological status as a large carnivore on Madagascar.

The falanouc presents a more complex problem because of its suite of autapomorphic (i.e., unique) traits that make it difficult to identify shared derived features. The incisiform canines, extension of the lacrimal onto the rostrum, late fusion of the carpals, and M1/M2 carnassial shear are all primitive features more typical of creodonts than of the Carnivora. However, the bulla is highly modified in a manner typical of viverrids (Mivart 1882a; Pocock 1915a).

Civets receive their name from the presence of scent ("civet") glands external to the anal region. Perineal scent glands are present in all but one subfamily of viverrids (cryptoproctines), although they have been secondarily lost in some species. Another feature that distinguishes the viverrids from among all other Recent carnivores is the loss of the cruciate sulcus on the cortex of the cerebellum.

Felids and Hyaenids [4]

Felids and hyaenids share a suite of modifications that are oriented toward the greater shearing function of the carnassials; the talonid on m1 is reduced and the metaconid is lost. The parastyle on m1 is elongated and rotated more in line with the axis of the mandible. P4 is elongated and the protocone is reduced. Some crushing teeth (M1/m2) are reduced or lost. In the basicranium the internal carotid is reduced and the alisphenoid canal is absent (Davis and Story 1943; Bugge 1978). The hallux is reduced or absent in both families (Mivart 1882a). Furthermore, Pauly and Wolfe (1957) pointed out serological resemblances between felids and hyaenids.

Felidae. The felid character that stands out most dramatically is the hardest one to define. The general impression upon viewing a felid skull is that of a short rostrum and a rounded or dome-shaped dorsal profile with forward-pointing orbits. This is least developed in the large cats and most noticeable in the smaller *Felis* species and the cheetah (*Acinonyx jubatus*). The anterior palatine foramen is located in the primitive position, along or directly adjacent to the palatine-maxilla suture. Because of the reversals necessary to maintain this condition and that of the other feloids as derived, I suggest that this character is reversed in felids. The bullar type is a reflection of the degeneration of the internal carotid and the lack of an external constriction between the ectotympanic and entotympanic portions (Hunt 1974). The internal carotid is developed only in the embryonic and early postnatal stages and is completely lost in adults (Davis and Story 1943; Bugge 1978). Reduction and loss in dentition have been taken to extremes, coupled with further development of the carnassial shear.

Hyaenidae. It has been suggested by some authors (Gregory and Hellman 1939; Thenius and Hofer 1960; Romer 1966; Hunt 1987) that some or all of the Hyaenidae were derived as an offshoot of the early viverrid or herpestid lineage. Flower (1869b) first suggested the relationship to the herpestids because of the inflated nature of the ectotympanic bulla. Although it is true that both groups have inflated ectotympanic bullae, the orientation of the respective elements in the two groups are quite dissimilar. *Proteles, Hyaena,* and *Crocuta* species do not differ in their internal carotid circulation, a condition similar to that found in the felids (Bugge 1978); and Hunt (1974, 1987) united the three genera based on the nature of the bullar structure. The presence or absence of the carotid groove is a matter of individual variation; it is usually absent, although a small vestigial groove can sometimes be detected (Pocock 1916a).

SUBORDER CANIFORMIA [5]

Simpson's (1945) Canoidea is placed here within the suborder Caniformia following Tedford (1976) and Flynn and Galiano (1982). Two superfamilies (Ursoidea, Canoidea) are recognized among the Recent caniforms, following the conventions of Linnaean hierarchy as presented by Wiley (1979). The Ursoidea are the basal group, characterized by the most primitive bullar construction and basicranial arterial circulation (Hunt 1974; Novacek 1977; Bugge 1978). They have secondarily modified their dentition and postcranial skeleton.

The position of the Canidae within this group has been controversial; it has historically been placed in its own superfamily (=Cynoidea Flower 1869b), as the basal group for the ursoids (=Canoidea Simpson 1945), and as a sister group to the felids (=Cynofeloidea Hough 1953). The canids retain the most primitive dentition among the caniforms (Butler 1946), with little tooth reduction or loss. They have some of the general primitive features pointed out by Radinsky (1982) such as a long rostrum, widely displaced zygomata, and the retention of m3. Their bulla are quite atypical of the caniforms, with an elongated entotympanic portion reminiscent of many feliforms and an entotympanic septum (see discussion in canid section).

This analysis of characters showed two equally parsimonious trees for the caniformes. The first tree placed the canids as the first branch (in a position similar to that in Figure 18.1C). With the canids in this position, parallel development of several basicranial characters are required. The second branching pattern (Figure 18.2) minimizes reversals in bullar and carotid circulatory characters.

Superfamily Ursoidea (Ursidae, Otariidae) [6]

Simpson (1945) rejected Flower's Cynoidea because it was not based on a taxon included within group. He considered Flower's "Arctoidea" and Fischer de Waldheim's (1817) "Ursoidea" objectionable "because no one considers the bears as nuclear or typical in the group as it is now constituted" (p. 222). The first objection is valid, but the second and third are not; the next available name is Ursoidea.

The European fossil, *Cephalogale,* has been considered either an early canid or ursid. It lacks the derived crushing dentition typical of true bears but rather has the more primitive dentition typical of the canids. Schlosser (1899, 1902) and Dehm (1950) have argued on the basis of dental evidence the origin of the

Ursidae from *Cephalogale*. Ginsburg (1966) added basicranial characters and suggested that *Cephalogale* belongs in the Hemicyoninae, a subfamily of the Ursidae (see also Martin, this volume).

The ursoid lineage is characterized by ten synapomorphies; among these are the lack of contact between the jugal and lacrimal, the shape of the lacrimal, the formation of a hypomastoid fossa, and the morphology of M2. The red panda is the only recent taxon with a primitively long tail and posteriorly diverging palate, possibly suggesting that it may have branched off before the ursid-otariid lineage. However, there are extinct taxa in the ursid-otariid lineage with these primitive features.

Ursidae (including Ailurus) [7]. Disagreements concerning the phylogenetic relationships of the red and giant panda to other carnivores have generated a great deal of controversy since the discovery of the "pandas" in the early 1800s. Mivart (1885b), Gregory (1936), and Simpson (1945) suggested placing both species with the Procyonidae on the basis of phenetic similarities in dentition. Some workers have suggested uniting the red and giant panda in their own family (Pocock 1921; Thenius and Hofer 1960). However, Davis (1964), Chorn and Hoffmann (1978), Thenius (1979), Hendey (1980) and O'Brien et al. (1985) place the giant panda with the true bears on the basis of a variety of derived morphological features. This leaves the relationship of the red panda to other taxa unresolved.

The red panda shares nine synapomorphies with ursines but have further modifications not typical of the true bears; of these, four are dental characters with parallel development in the procyonids and mustelids and three are reversals to the primitive condition. Although Ginsburg (1982) supported the *Ailurus*-ursid clade based on fossil evidence and dental arguments, the strongest support lies with the basicranium (Hunt 1974; Bugge 1978). Furthermore, the adaptive type represented by the red panda falls within the range of adaptive types known from the ursid fossil record.

Ursinae. The bears share a variety of features that concern the orbit, dentition, and the auditory bulla. The lacrimal is reduced to the point of being a vestigial rim of bone around the naso-lacrimal foramen. The paroccipital-mastoid ridge encloses the hypomastoid fossa and the stylomastoid foramen. The derived uninflated bulla is relatively flat and quite dwarfed by the mastoid, squamosal, and basioccipital processes (Flower 1869b; Hunt 1974). All bears share an emphasis on the molars and a marked reduction of the anterior premolars, which are completely lost with age in many bears. P4 is reduced and has lost most of its shearing function. The M2 of bears takes on a unique oblong shape that is more elongated along the lingual edge, making the tooth the longest in the upper palate. In the mandibular dentition the second molar is larger than the first and more robust; the first still retains some of the shearing nature of the trigonid; however, both lower molars are elongate, and the second is clearly adapted for crushing.

Ailurinae. The branching pattern (Figure 18.2) suggests placing the red panda within the Ursidae. There has been a hesitancy to do this because of the superficial resemblance of the red panda to the North American raccoon (*Procyon lotor*). The raccoon is probably one of the more derived procyonids (Baskin 1982; Decker and Wozencraft, in press), and inclusion of the red panda in this family would make the procyonids paraphyletic (Schmidt-Kittler 1981). The Ailurinae is here understood to include only the monotypic genus *Ailurus* and represents the first out group to the Ursine bears. Many of the differences between the red panda and the bears is related to their feeding ecology (Mayr 1986).

Otariidae (including Odobenus*) [8].* The otariids, a well-defined taxon, have been recognized as a sister group to the ursids since Flower (1869b) and Mivart (1885b). The walrus has been separated by some and included by others at the family level (see appendix to this volume). The main arguments for separation have revolved around the highly derived nature of the Recent walrus. The main arguments for inclusion of the walrus are phylogenetic (Tedford 1976) or based on the similarities between fossil walruses and otariids (Mitchell and Tedford 1973). There appears to be little disagreement among morphologists that it represents the sister group to the remaining otariids (Ling 1978; King 1983), a conclusion confirmed by this analysis; it is kept here in the more conservative placement.

Cytogenetic evidence supports previous morphological studies and highlight the walrus's uniqueness. However, Fay et al. (1967) suggested an intermediate position between the otariids and the phocids. Wyss (1987) argued that walruses have a mosaic of a primitive "ursid" type bulla with some derived aquatic modifications; he proposed a phocid-*Odobenus* clade. However, most of his characters are allometric when the morphological features are scaled to body size; they do not unite the phocids with the walrus and exclude the otariids. Wyss pointed out that the phocids and the walrus have abdominal testes, a derived feature for the Carnivora. The large number of reversals that would be necessary to support a phocid-walrus clade suggest that this feature is a parallel development. A non–size-related feature used by Wyss to support the walrus-phocid clade concerns the length of the suture between the rostral branch of the premaxilla and the nasals. In most phocids the nasals are so retracted that this contact is small, if there is any common suture at all. In many otariids there appears to be a somewhat longer common suture. In the walrus there is an even longer suture (not a reduced contact as suggested by Wyss), but the situation is not homologous to either the phocid or otariid condition. The premaxilla shares a long suture along the ventral edge of the nasal so that it is not visible from the dorsal surface.

The extensive comparisons of Recent and fossil otariids and odobenines made by Mitchell and Tedford (1973) and Repenning and Tedford (1977) outline many synapomorphies that suggest a walrus-otariid clade, a conclusion supported by this analysis. The Recent walrus appears to be highly autapomorphic when compared with fossil taxa; these are in need of revision, and ongoing analysis may clarify the position of these taxa.

Superfamily Canoidea (Canidae, Procyonidae, Mustelidae, Phocidae) [9]

The superfamily Canoidea, as here defined, excludes the ursids (including *Ailurus*) and otariids (including *Odobenus*) *contra* Simpson (1945) and Tedford (1976) (Figure 18.1C). The division of the Recent Caniformia into two clades follows conventions of Linnaean hierarchy (2 and 3), as presented by Wiley (1979).

This group is united on the derived nature of the basicranial region (see Appendix 18.1). The entotympanic ossification centers have further expanded and form an inflated single chambered bulla (Hunt 1974). The meatal tube has retained its primitive C shape with little expansion either medially or laterally (van der Klaauw 1931). The external and internal carotid artery systems are connected by two anastomoses, starting a trend to emphasize the external system as the main blood supply to the brain (Bugge 1978).

Canidae. In general, the canids have many of the plesiomorphies for general skull morphology identified by Radinsky (1980, 1981a) and the primitive dental pattern, with a bicuspid talonid on m1 (Butler 1946). However, exceptions exist such as the African hunting dog (*Lycaon pictus*), which has lost the entoconid and metaconid on m1, creating a long shearing surface from the protoconid to the posterior hypoconid. The paroccipital process is large and is connected to the caudal surface of the bulla by a ridge of bone.

The construction of the bulla has caused some controversy over the systematic position of the canids. Flower (1869b) suggested that the low entotympanic septum in the bullar chamber was an intermediary state between the undivided bulla of procyonids and mustelids and the fully divided bulla of the feloids; Hough (1953) believed the structure to be totally homologous to the posterior half of the bilaminar septum in felids. However, the exact ontogenic development and homology of the septum is unclear, with some mustelids also having entotympanic pseudoseptae. Regardless of how septal homology is interpreted, other features shared with the canoids point to the independent derivation of the low entotympanic septum. The internal carotid artery is retained as a major artery running from the posterior to the medial lacerate foramen. The arterial shunt is not developed as with the felids or hyaenids, nor is it completely excluded from the cranial cavity as in the procyonids (Bugge 1978).

"*Parvorder Mustelida*" (sensu *Tedford 1976*) (*procyonids, mustelids, and phocids*) [10]. This group was originally identified by Swainson (1835) as a monophyletic group. Gregory and Hellman (1939) applied the superfamily name to include only mustelids, believing that the Mustelidae need be distinguished from the rest of the "arctoids" because of overall similarity to herpestids. Tedford (1976) recognized the parvorder Mustelida and included the Mustelidae and the Phocidae in the Musteloidea Swainson (1835). Schmidt-Kittler (1981) referred to the procyonids and the mustelids as the Musteloidea and did not discuss the phocids. His comparative anatomical analysis showed derived features of the petrosal sinuses and the roof of the meatus. The asymmetrical branching sequence (Figure 18.2) does not necessarily require a taxonomic label (Wiley 1979).

The suite of changes at this point suggest an early radiation from other lineages. The alisphenoid canal is lost and the bullar type is changed by the movement of the posterior carotid foramen anteriorly along the medial wall of the entotympanic. The internal carotid artery also becomes totally enclosed by entotympanic bone. The suprameatal fossa develops in the posttympanic process of the squamosal but is secondarily lost in some taxa.

Procyonidae. The suprameatal fossa, a distinctive feature of the procyonids, is also variously formed in the mustelids and phocids (see comments, Musteloidea section). Schmidt-Kittler's (1981) pioneering studies suggested a polarity for this feature and traced its development among Recent and fossil specimens. This fossa is also found in some ursids and viverrids but is never expanded posteriorly to the extent found in the Procyonidae. The bulla consists of a single chamber and the mastoid protrudes laterally. The procyonids have modified P4 by the development of a hypocone on the posterior lingual cingulum, and the upper molars have become much more bunodont and quadrate, corresponding to a frugivorous/omnivorous diet. In the cranial circulation pattern, his group has lost the X anastomosis (Bugge 1978) that is present in the canids and mustelids. *Bassariscus* and *Bassaricyon* represent the most primitive genera (Baskin 1982; Decker and Wozencraft, in press).

"*Musteloidea*" (sensu *Tedford 1976*) (*mustelids, phocids*) [11]. The loss of the carnassial notch at this point excludes some fossils (e.g., *Potamotherium, Plesictis, Plesiogale,* and *Leptarctus*), which have been considered early mustelids by some authors. It is possible that the loss of the notch occurred more than once (see Mustelidae section). The second upper molar is greatly reduced or absent. Schmidt-Kittler (1981) suggested that the suprameatal fossa, a characteristic of the Mustelida, has moved in mustelids to a more dorsal position. Muizon (1982a) believed that he had identified the fossa in phocids; however, this homology has been questioned by Wyss (1987). Tedford (1976) pointed out that the loss of the postscapular fossa also distinguishes this group.

Mustelidae. The Mustelidae, the most diversified of all the families of Carnivora, may be paraphyletic (Muizon 1982b). Its members are united by the loss of the carnassial notch (Wortman 1901) and the enlargement of the anal sac, from which the family derives its name. The constriction of M1 has been used as a character in many keys, but the constriction is absent in some taxa or is only slight. The mustelines and melines share an inflated bullae in which the hypotympanic sinus is inflated posterior to the promontorium (Tedford 1976). The honey badger (*Mellivora capensis*) is here included with mustelines, as it shares with true weasels the loss of the m1 metaconid and the anterior directed external auditory meatus (Pocock 1921). The Melinae are united by a broad posterior cingulum on P4 and the loss of the suprameatal fossa (Qui Zhanxiang and Schmidt-Kittler 1982). Additionally, these two subfamilies share the canoid type of inflated bulla and a dentition without much modification.

Radinsky (1973) suggested that mephitines are the sister group to melines, but the number and kinds of synapomorphies that the otters and skunks share suggest otherwise. The mephitine-lutrine bulla, like the ursid-otariid bulla, consists basically of a large ectotympanic plate with little inflation, with the caudal and medial entotympanic portions forming the connecting walls to the basioccipital/petrosal complex (Hunt 1974). The otters and skunks differ from the ursoid bulla in the development of the position and pathway of the internal carotid (Bugge 1978).

Savage (1957) described the anatomy of *Potamotherium*, which he identified as an Oligocene lutrine. McLaren (1960) later allied *Potamotherium* with the phocids and lutrines. He believed that "the late Oligocene otter *Potamotherium* has many anatomical foreshadowings of the Phocidae" (p. 26). *Potamotherium* has a carnassial notch, a feature lacking in the Recent mustelids and phocids. It is possible that the loss of the carnassial notch occurred more than once among carnivores. Among extant groups, the otariids, mustelids, some procyonids, and phocids all have lost this notch.

Muizon (1982b) suggested an asymmetrical branching pattern with the following order: melines, mustelines, mephitines, lutrines, and the phocids last; this would require a new taxonomy to reflect the paraphyletic nature of the previously recognized family. The mustelids as depicted here have two major groups (Figure 18.2) and represent the only Recent carnivore family with two major bullar types. As the homophyly of the traditionally recognized group remains intact in this hypothesis, I have retained the more conservative arrangement.

Phocidae. The relationship of phocids to other terrestrial carnivores is difficult to discern because of the great number of autapomorphies. They are the most aquatically adapted family within the Carnivora (King 1983) and show an array of adaptations in cranial, dental, postcranial, and soft anatomical structures. When derived features unique to the phocids are teased out, one is left with a mosaic of mainly canoid and some musteloid features. Functional morphological studies have shown a unique adaptation to underwater hearing unlike other pinnipeds (Repenning 1972), and their ecology and behavior reflect this (King 1983). Their relationship to the mustelids is unclear. Muizon (1982a) suggested that they be united with the lutrines, whereas Hunt (1974) showed derived features shared with the mustelines (but not the lutrines). The retraction of the nasals on the rostrum and the enlargement of the maxilla in the orbit are seen in most taxa (Wyss 1987).

Summary

The phylogeny of the Carnivora that is now most familiar is Turner's (1848), which was further refined by later workers and widely accepted in its recent form, Simpson's (1945) classification. It was based primarily on phenetic groupings; Hennig (1966), Wiley (1981), and others have shown that phenetic methods do not portray evolutionary descent or common ancestry.

The phylogeny of the Carnivora that is presented here traces the genealogy of the respective lineages based on the acquisition of key morphological innovations. These innovations are functionally related to the ability of carnivores to locate (modifications to sight and sound reception) and catch prey (changes in limb morphology), to utilize food resources (changes in dentition), and to invade new environments (aquatic). The phylogeny suggested here may be a new arrangement, but is basically a synthesis of ideas well established in the literature interpreted within a phylogenetic framework.

Although the ideas here are based on previous publications, this phylogenetic hypothesis differs from the Turner-Simpson classification in five important ways: it (1) recognizes the Herpestidae; (2) clarifies the correct superfamily name, Feloidea for the Recent Feliformia; (3) includes the red and greater panda in the Ursidae; (4) includes the walrus in the Otariidae; and (5) recognizes the close relationship of the Otariidae to the Ursidae and the Phocidae to the Mustelidae. Moreover, the branching sequence of the phylogeny suggests that the ursoids and the herpestids are the most primitive members of their respective suborders.

There appears to be large areas of common agreement among comparative morphologists as to the homology and development of various structures used to indicate a common evolutionary heritage. The tree presented here (Figure 18.2) to a great extent agrees with previously suggested relationships (Figure 18.1). The basic assumption throughout this paper is that classifications should be a reflection of our best phylogenetic hypothesis concerning the relationships of the included taxa. Many of the biochemical and cytogenetic studies done to date on the Carnivora have been phenetic in design and may be good reflections of overall similarity, but one needs to be cautious in the interpretation of their results. Phenetic studies do not take into account descent and sidestep the issue of convergence and parallelism; similarity may not necessarily indicate common ancestry, and they deal with characters at different levels of universality for which homology is uncertain (Wiley 1981). However, as the analysis of the pinnipeds in this chapter illustrates, phylogenetists are far from agreement as to how homoplasy should be handled. Synapomorphies can be misinterpreted because of the complex relationships and strong adaptive forces that are working on a complex of interrelated characters. The bottom line in any phylogenetic hypothesis does not lie strictly with a numerical count of characters or consistency ratios, but rather with the careful analysis of the synapomorphies and interpretation of their biological significance.

Acknowledgments

I extend my sincerest thanks to A. Berta, A. R. Biknevicius, D. M. Decker, R. S. Hoffmann, T. Holmes, K. Koopman, L. D. Martin, J. Mead, R. H. Tedford, R. W. Thorington, L. A. Wozencraft, A. Wyss, and C. J. Young, all of whom provided helpful comments and useful suggestions on revisions of the manuscript. This work was supported by the University of Kansas, Museum of Natural History, and the National Museum of Natural History, Smithsonian Institution.

Appendix 18.1. Phylogenetic Characters

The following list of characters and character-state codes were used as the basis for the hypothesis presented (Figure 18.2). Characters were obtained from the literature and then verified. Their inclusion in this type of analysis may suggest an arrangement that the original paper did not. Where available, earlier published discussions are referenced. These references either directly or indirectly provide the basis for the assigned polarities. Simple characters that can be verified by observation (e.g., the loss of a tooth) are not referenced. The taxonomic unit (OTU) and their abbreviations are given in Figure 18.1. OTUs with the apomorphic condition are listed after that character state; OTUs not listed for a character are presumed to be plesiomorphic.

Rostrum Palatal Region

1. palatine/maxilla, anterior palatine foramina (Pocock 1921): 0 = located at palatine/maxilla suture; 1 = well anterior in the maxilla (HYA VIV HER MEP LUT). [The anterior palatine foramina appears primitively and in most ontogenetic sequences at the palatine/maxilla suture. In some feloids it migrates forward before birth. There is a deep groove running anteriorly from this foramen in some pinnipeds. This groove gradually closes over creating a false tube; usually, there is a well-defined suture and not the complete closure found in herpestoids. It is individu-

ally variable in most taxa, and the primitive condition is found in all three pinniped families.]

2. palatine, relative size (excluding mesopterygoid) (Pocock 1921): 0 = midline length less than midline length of maxilla; 1 = midline length greater than midline length of maxilla (PHO). [This character is reflective of the contribution of the maxilla to the palate.]

3. palatine, relative size (including mesopterygoid) (Pocock 1921): 0 = midline length less than midline length of maxilla; 1 = midline length greater than midline length of maxilla (HER URS AIL PRO MUS MEL MEP LUT PHO). [Closely related to character no. 2, this reflects the development of the posterior extension of the palate over the pterygoids.]

4. palate, posterior width (Wyss 1987): 0 = significantly wider than width measured at canines; 1 = nearly equal to width at canines (URS OTA ODO). [Wyss reversed this polarity.]

5. maxilla, infraorbital canal (Novacek 1986): 0 = anterior opening anterior to nasolacrimal foramen; 1 = anterior opening ventral or posterior to nasolacrimal opening (FEL OTA ODO AIL PRO MUS MEL MEP LUT PHO). [Probably better coded as "long rostrum." All taxa with long rostra have the primitive anteriorly placed foramen.]

6. premaxilla, rostral process (Mivart 1885a; Gromova et al. 1968; Novacek 1986): 0 = broad contact with nasal; 1 = narrow contact with nasal (LUT PHO).

7. nasals, posteriormost edge (King 1983): 0 = V shape or convergent; 1 = W shape or divergent (OTA ODO). [The condition in odobenids is different than the condition in otariids. In juvenile odobenids there is a slight divergent shape; in most adults this becomes a straight line, perpendicular to the skull midline.]

Orbital Region

8. lacrimal (Gregory 1920): 0 = present, with orbital flange; 1 = vestigial, restricted to area around foramen (URS AIL).

9. lacrimal, fusion to maxilla (King 1971): 0 = remains distinct throughout the adult life; 1 = fuses early to the maxilla (ODO? MUS MEL MEP LUT? PHO). [The acceptance of LUT with the derived state depends upon the subjective measure of "early."]

10. lacrimal, nasolacrimal foramen (Gregory 1920): 0 = present; 1 = absent (PHO). [In some otariids the foramen is vestigial and often lost in old adults.]

11. inferior oblique muscle fossa (Davis 1964): 0 = widely separate from nasolacrimal foramen; 1 = closely adjacent to nasolacrimal foramen (URS AIL).

12. jugal (King 1983): 0 = reaches lacrimal; 1 = reduced, does not reach

lacrimal (URS OTA ODO AIL MUS MEL MEP LUT PHO) [In *Ailurus* and the ursids the lack of contact between the jugal and lacrimal is caused by the reduction of the lacrimal and the reduction of the jugal.]

13. squamosal, dorsal process (King 1983): 0 = absent; 1 = present (PHO). [King described this as "jugal-squamosal join interlocking," and it refers to the anteriormost portion of the squamosal on the zygomatic arch with a dorsal process.]

14. palatine, orbital wing (Muller 1934): 0 = reaches lacrimal, broad contact; 1 = reaches lacrimal, narrow contact (URS AIL LUT); 2 = does not reach lacrimal (OTA ODO PHO).

15. frontal, supraorbital process (Novacek 1986); 0 = small process; 1 = long process (FEL HYA VIV HER OTA). [The odobenids have a frontal process that is closely appressed to the lacrimal flange. It may be interpreted as being homologous to the process; the rearrangement of the rostral region has caused a loss of the normal space separating these processes.]

16. alisphenoid, canal (Turner 1848; Mivart 1885a; Pocock 1916b; Novacek 1986): 0 = present; 1 = absent (FEL HYA PRO MUS MEL MEP LUT PHO).

17. orbitosphenoid, optic foramen (Repenning and Tedford 1977): 0 = separate rostral borders; 1 = common rostral border (i.e., interorbital septum) (OTA).

18. sphenopalatine foramen (Muller 1934): 0 = enclosed in palatine; 1 = enlarged, eclispses orbitosphenoid (OTA ODO). [This is a function of two variables, the enlargement of the orbital vacuity that includes the sphenopalatine foramen and the elongation of the orbitosphenoid anteriorly from the optic foramen.]

Basicranial Region

19. squamosal, postglenoid foramen (Flower 1869b; Flower and Lydekker 1891; Hough 1953): 0 = present, anterior to bullae; 1 = vestigial/absent (FEL HYA VIV HER OTA ODO PHO).

20. squamosal, suprameatal fossa (Schmidt-Kittler 1981; Muizon 1982b): 0 = vestigial/absent; 1 = present, dorsal to external auditory meatus (URS AIL); 2 = present, dorsal and posteriodorsal to external auditory meatus (PRO); 3 = ventrally closed, fossa reduced (MUS MEL MEP LUT PHO). [The presence of the suprameatal fossa in phocids has been questioned by Wyss (1987).]

21. basioccipital, hypoglossal foramen (Turner 1848; Mivart 1885a): 0 = separate from posterior lacerate foramen; 1 = closely adjacent or confluent with posterior lacerate foramen (FEL HYA VIV HER).

22. basioccipital, posterior lacerate foramen (Turner 1848; Mivart 1885a): 0 = small; 1 = large (URS OTA ODO AIL CAN LUT PHO).
23. petrosal/basioccipital suture (van der Klaauw 1931): 0 = petrosal closely adjacent/attached to basioccipital; 1 = petrosal widely separated from basioccipital (FEL HYA VIV HER URS OTA ODO AIL). [Not usually visible from the ventral side of the skull.]
24. auditory bulla; carotid canal, posterior opening (van der Klaauw 1931; Hunt 1974): 0 = adjacent to posterior lacerate foramen; 1 = considerably anterior to posterior lacerate foramen (HYA VIV HER PRO MUS MEL MEP LUT PHO); 2 = absent (carotid canal absent) (FEL).
25. auditory bulla, carotid canal, posterior opening (Mitchell and Tedford 1973): 0 = vertical (i.e., perpendicular to basisphenoid) or absent; 1 = horizontal (MUS MEL).
26. auditory bulla, carotid canal, anterior opening (Mitchell and Tedford 1973): 0 = vertical or absent; 1 = horizontal (PHO). [Visible only from bullar chamber.]
27. auditory bullae, hypotympanic cavity (=bullar chamber) (Flower 1869b; van der Klaauw 1931): 0 = inflated; 1 = not inflated (OTA MEP AIL LUT).
28. auditory bullae, entotympanic ossification (van der Klaauw 1931; Hunt 1974; Flynn and Galiano 1982): 0 = unossified; 1 = ossified (FEL HYA VIV HER URS OTA ODO AIL CAN PRO MUS MEL MEP LUT PHO). [*Nandinia* has an unossified entotympanic bullae.]
29. ectotympanic (van der Klaauw 1931; Hunt 1974; Tedford 1976): 0 = inflated; 1 = not inflated (AIL URS OTA ODO).
30. ectotympanic, external auditory meatal tube (Mivart 1885a; van der Klaauw 1931): 0 = absent; 1 = present (HYA HER URS OTA ODO AIL PRO MUS MEL MEP LUT PHO). [The formation of the external auditory tube in the herpestids is different and may not be homologous to that found in other taxa.]
31. ectotympanic, septum (Flower 1869b; van der Klaauw 1931; Hunt 1974): 0 = absent; 1 = present (FEL HYA VIV HER).
32. entotympanic, septum (van der Klaauw 1931; Hunt 1974): 0 = absent; 1 = present (FEL VIV HER CAN?). [See text for discussion.]
33. entotympanic, caudal (van der Klaauw 1931; Hunt 1974): 0 = not inflated or absent; 1 = developed, inflated along anterior/posterior axis (CAN PRO MUS MEL PHO); 2 = greatly inflated, clearly separated from ectotympanic part of bullae (FEL HYA VIV HER).
34. caudal entotympanic, medial portion (van der Klaauw 1931): 0 = not inflated/absent; 1 = greatly inflated (FEL VIV CAN PRO MUS MEL PHO).
35. tympanohyal (Mitchell and Tedford 1973): 0 = closely associated with stylomastoid foramen; 1 = separated from stylomastoid foramen (OTA AIL URS ODO MEL LUT PHO).

36. tympanohyal (Mitchell and Tedford 1973): 0 = posterior to styl-
    omastoid foramen; 1 = anterior to stylomastoid foramen (PHO).
37. paroccipital process (Turner 1848; Flower 1869b; van der Klaauw
    1931): 0 = protrudes posteriorly; 1 = cupped around posterior edge of
    entotympanic (FEL HYA VIV HER). [See text for discussion of CAN]
38. paroccipital process (Turner 1848; Flower 1869b; van der Klaauw
    1931): 0 = short/vestigial/absent; 1 = long (HYA VIV URS AIL CAN).
39. petrosal, hypomastoid fossa (Flower 1869b; Mivart 1885a): 0 = absent;
    1 = present (URS OTA ODO AIL). [Laterally defined by the pre-
    tromastoid ridge that connects the paroccipital process with the
    mastoid.]
40. petrosal, inferior petrosal sinus (Davis and Story 1943; Davis 1964;
    Hunt 1974): 0 = small; 1 = large (OTA ODO AIL); 2 = very large
    (URS).
41. petrosal, internal acoustical meatus (Repenning 1972; Wyss 1987): 0 =
    present; 1 = absent (PHO).
42. petrosal, fenestra cochleae (round window) (Repenning 1972): 0 =
    opens into middle ear; 1 = opens externally (PHO).
43. petrosal, fenestra cochleae (round window) (Repenning 1972): 0 = ap-
    proximately equal in size to oval window, cochlear fossula not devel-
    oped; 1 = considerably greater in size than oval window, cochlear fos-
    sula well developed (OTA ODO PHO).
44. petrosal, whorl of cochlea (Repenning 1972): 0 = posteriolateral orien-
    tation; 1 = transverse to skull (PHO).
45. petrosal, mastoid, lateral process (Flower 1869b; Mivart 1885a): 0 =
    absent; 1 = present (URS OTA ODO AIL CAN PRO MUS MEL MEP
    LUT PHO). [The determination of the lateral extent of the process is
    somewhat subjective.]
46. petrosal, mastoid, epitympanic sinus (van der Klaauw 1931; Tedford
    1976; Schmidt-Kittler 1981): 0 = absent; 1 = present, separated from
    hypotympanic sinus (MEP).
47. petrosal, pit for insertion of tensor tympani (Repenning 1972): 0 =
    present, deep pocket; 1 = shallow fossa or absent (OTA ODO MEP LUT
    PHO).
48. malleus, muscular process (Doran 1879; Wyss 1987): 0 = absent; 1 =
    present, small (VIV HER URS AIL MEL MEP LUT); 2 = present, large
    (FEL HYA CAN PRO MUS).
49. malleus, processus gracilis and anterior lamina (Doran 1879; Wyss
    1987): 0 = small/vestigial; 1 = well developed (FEL HYA VIV URS AIL
    CAN PRO MUS).
50. major a2 arterial shunt (Bugge 1978): 0 = small; 1 = large, intracranial
    rete (FEL HYA).
51. major a4 arterial shunt (Story 1951; Bugge 1978): 0 = absent; 1 =

present (VIV HER CAN PRO MUS MEL); 2 = present, intracranial rete (FEL HYA).

52. major anastomosis X (Bugge 1978): 0 = absent; 1 = present (CAN PRO MUS MEL MEP LUT).
53. major anastomosis Y (Bugge 1978): 0 = absent; 1 = present (HER).
54. Course of the internal carotid (Hough 1953; Hunt 1974; Bugge 1978; Presley 1979; Wible 1986, 1987): 0 = between petrosal and entotympanic; 1 = in sulcus between basioccipital and entotympanic (HER PRO MUS MEL MEP LUT PHO); 2 = enclosed in entotympanic groove (VIV HYA); 3 = exterior to entotympanic (FEL).
55. cruciate sulcus (Radinsky 1975): 0 = present; 1 = absent (VIV). [Radinsky (1975) suggested that this feature developed independently several times within the Carnivora.]

Dentition

56. carnassial shear (Cope 1880; Matthew 1909): 0 = M1/m2 and/or M2/m3; 1 = P4/m1 (FEL HYA VIV HER CAN PRO MUS MEL MEP LUT PHO AIL ODO URS OTA).
57. I1: 0 = present; 1 = absent (ODO).
58. I1, I2, transverse grooves (King 1983): 0 = absent; 1 = present (OTA URS? AIL?).
59. i1 (Cobb 1933; King 1983): 0 = present; 1 = absent (OTA ODO PHO).
60. i2: 0 = present; 1 = absent (ODO).
61. i3: 0 = present; 1 = absent (ODO).
62. C1: 0 = present, large; 1 = present, very large (ODO).
63. Buccal cingulum, upper molars: 0 = small/not developed; 1 = greatly enlarged, with buccal cusps (AIL).
64. P1: 0 = present; 1 = absent (FEL ODO AIL MEP LUT).
65. P3, lingual cusp: 0 = absent; 1 = present (HER AIL PRO).
66. P4, parastyle (Flynn and Galiano 1982): 0 = vestigial/absent; 1 = present (FEL HYA VIV HER).
67. P4, metastyle blade (Flynn and Galiano 1982): 0 = V or slit-shaped notch; 1 = without notch (OTA ODO MUS MEL MEP LUT PHO). [The loss of the carnassial notch in the pinnipeds is caused by the simplification of the entire tooth, a process not necessarily homologous to the loss of the notch in mustelids.]
68. P4, protocone (Flynn and Galiano 1982): 0 = medial or posterior to paracone; 1 = anterior to paracone (FEL HYA VIV HER URS OTA ODO AIL CAN PRO MUS MEL MEP LUT PHO).
69. P4 hypocone: 0 = absent; 1 = present (AIL PRO). [Ursids have a posterolingual cusp; it is not clear if this is homologous to the hypocone of others.]

70. P4, talon: 0 = vestigial/absent; 1 = wide basin (MEP LUT).
71. p1: 0 = present; 1 = absent (FEL HYA ODO MEP LUT).
72. p2: 0 = present; 1 = absent (FEL ODO).
73. m1, talonid: 0 = large, or tooth simplified; 1 = reduced/vestigial (FEL HYA).
74. M1, hypocone: 0 = absent; 1 = present (URS AIL CAN PRO).
75. M1, constriction: 0 = absent; 1 = present (MUS MEL MEP LUT).
76. M1, size (Flynn and Galiano 1982): 0 = large; 1 = vestigial/absent (FEL HYA).
77. M2: 0 = present; 1 = absent (FEL HYA ODO MUS MEL MEP LUT PHO).
78. M2, hypocone: 0 = absent; 1 = present (URS AIL).
79. M3: 0 = present; 1 = absent (FEL HYA VIV HER URS OTA ODO AIL CAN PRO MUS MEL MEP LUT PHO).
80. m2: 0 = present; 1 = absent (FEL HYA OTA ODO PHO).
81. m3: 0 = present; 1 = absent (FEL HYA VIV HER OTA ODO AIL PRO MUS MEL MEP LUT PHO).
82. M2/m2 size (Flynn and Galiano 1982): 0 = smaller than M1/m1; 1 = larger than M1/m1 (URS).

Postcranial

83. baculum (Mivart 1885a; Flower and Lydekker 1891): 0 = small and simple/absent; 1 = long, stylized (URS OTA ODO AIL CAN MUS MEL MEP LUT PHO); 2 = anteriorly bilobed (PRO).
84. scapula, postscapular fossa (Davis 1964; Tedford 1976): 0 = absent; 1 = present (URS AIL PRO).
85. scapula, teres major process (Matthew 1909; Tedford 1976; Flynn and Galiano 1982): 0 = small or absent; 1 = large (URS OTA ODO AIL PRO MUS MEL MEP LUT PHO).
86. tail: 0 = long; 1= vestigial (URS OTA ODO PHO).
87. scaphoid and lunar (Cope 1880; Flynn and Galiano 1982): 0 = unfused; 1 = fused (FEL HYA VIV HER URS OTA ODO AIL CAN PRO MUS MEL MEP LUT PHO). [Flynn and Galiano (1982) suggested that these elements fused independently in the caniform and feliform lineages.]
88. hallux (Geoffroy Saint-Hilaire and Cuvier 1795; Turner 1848): 0 = present; 1 = reduced/absent (FEL HYA CAN).
89. calcanea, fibular facet (Matthew 1909; Flynn and Galiano 1982): 0 = present; 1 = absent (URS OTA ODO AIL CAN PRO MUS MEL MEP LUT PHO).
90. femur, third trochanter (Gromova et al. 1968): 0 = present; 1 = absent (FEL HYA VIV HER URS OTA ODO AIL CAN PRO MUS MEL MEP

LUT PHO). [All Recent Carnivora lack the third trochanter; however, it is present in some miacids.]

External/ Soft Anatomy

91. Cowper's (bulbourethral) gland (Turner 1848; Mivart 1885a): 0 = absent/small; 1 = present, well developed (FEL HYA VIV HER).
92. prostate (Turner 1848; Mivart 1885a): 0 = small/vestigial; 1 = large, bilobed ampulla (FEL HYA VIV HER).
93. kidneys (Turner 1848; Mivart 1885a): 0 — simple; 1 — conglomerate/renculate (URS OTA ODO LUT PHO).
94. external pinnae (Turner 1848; Mivart 1885a; Pocock 1914, 1915a, 1915b, 1916c, 1917, 1921): 0 = present, large; 1 = present, small (OTA LUT); 2 = absent (ODO PHO).
95. testes (King 1983): 0 = scrotal; 1 = abdominal (ODO PHO).
96. anal glands (Mivart 1885a; Pocock 1916c, 1916b, 1921): 0 = simple; 1 = enlarged with enlarged anal sac (HYA HER MUS MEL MEP LUT).
97. perineal scent glands (Turner 1848; Mivart 1882; Pocock 1915a, 1915b): 0 = absent; 1 = present (VIV).
98. astragalus, process (King 1983): 0 = absent; 1 = present (PHO).
99. psoas major muscle, distal insertion (Muizon 1982b): 0 = femur, second trochanter; 1 = ilium (PHO).
100. phalanges, terminal claws (Cope 1880): 0 = normal, smooth; 1 = fissured (CRE).

References

Allen, J. A. 1880. *History of North American pinnipeds: A monograph of the walruses, seal-lions, sea-bears and seals of North America.* U.S. Geological and Geographical Survey of the Territories; Miscellaneous Publication 12. Washington, D.C.: Government Printing Office.

Allen, J. A. 1924. Carnivora collected by the American Museum Congo Expedition. *Bull. Amer. Mus. Nat. Hist.* 47:73–281.

Anbinder, E. M. 1969. [On the question of systematics and phylogeny of pinnipeds, family Phocidae. In: N. N. Vorontsov, ed. *Mammals: Evolution, Karyology, Systematics, Faunistics,* pp. 23–25. Novosibirsk: Acad. Sci. U.S.S.R., Siberian Branch.]

Anbinder, E. M. 1980. [*Karyology and Evolution of the Pinnipeds.* Moscow: Nauka.]

Arnason, U. 1974. Comparative chromosome studies in Pinnepedia. *Hereditas* 76:179–226.

Arnason, U. 1977. The relationship between the four principal pinniped karyotypes. *Hereditas* 87:227–242.

Arnason, U., and Widegren, B. 1986. Pinniped phylogeny enlightened by molecular hybridizations using highly repetitive DNA. *Mol. Biol. Evol.* 3:356–365.

Baskin, J. A. 1982. Tertiary Procyoninae (Mammalia: Carnivora) of North America. *J. Vert. Paleontol.* 2:71–93.

Beaumont, G. de. 1964. Remarques sur la classification des Felidae. *Eclogae Geologicae Helvetiae* 57:837–845.

Berta, A., and Deméré, T. A. 1986. *Callorhinus gilmorei* n. sp., (Carnivora: Otariidae) from the San Diego formation (Blancan) and its implications for otariid phylogeny. *Trans. San Diego Soc. Nat. Hist.* 21(7):111–126.

Blümenbach, J. F. 1788. *Handbuch der Naturgeschichte.* Göttingen: J. C. Dieterich.

Blümenbach, J. F. 1791. *Handbuch der Naturgeschichte* (4th ed.). Göttingen: J. C. Dieterich.

Bonaparte, C. L. J. L. 1845. *Catalogo methodico dei mammiferi Europei.* Milan: L. di Giacomo Pirola.

Bowdich, T. E. 1821. *An analysis of the natural classification of Mammalia.* Paris: J. Smith.

Brisson, M. J. 1756. *Le regnum animale in classes IX distributum, sive synopsis methodica sistens generalem animalium distributionem in classes IX, & duraum primarum classium, quadrupedum scilicet & cetaceorum, particularem divisionem in ordines, sectiones, genera & species.* Paris: C. J. B. Bauche.

Bugge, J. 1978. The cephalic arterial system in carnivores, with special reference to the systematic classification. *Acta Anat.* 101: 45–61.

Butler, P. M. 1946. The evolution of carnassial dentitions in the Mammalia. *Proc. Zool. Soc. London* 116:198–220.

Butler, P. M. 1948. On the evolution of the skull and teeth in the Erinaceidae, with special reference to fossil material in the British Museum. *Proc. Zool. Soc. London* 118:446–500.

Chorn, J., and Hoffmann, R. S. 1978. *Ailuropoda melanoleuca. Mamm. Species* 110:1–6.

Cope, E. D. 1875. On the supposed Carnivora of the Eocene of the Rocky Mountains. *Proc. Acad. Nat. Sci. Philadelphia* 27:444–448.

Cope, E. D. 1880. On the genera of the Creodonta. *Proc. Amer. Phil. Soc.* 19:76–82.

Cope, E. D. 1882. On the systematic relations of the Carnivora Fissipedia. *Proc. Amer. Philos. Soc.* 20:471–475.

Cuvier, G. L. C. F. D. 1800. *Leçons d'anatomie comparée de G. Cuvier: Recueillies et publiées sous ses yeux par C. Duméril,* vol. 1. Paris: Baudouin.

Cuvier, G. L. C. F. D. 1817. *Le règne animal distribué d'après son organisation, pour servir de base à l'histoire naturelle des animaux et d'introduction à l'anatomie comparée. Les mammifères,* vol. 1. Paris: Deterville.

Davis, D. D. 1964. The giant panda: A morphological study of evolutionary mechanisms. *Fieldiana Zool. Mem.* 3:1–339.

Davis, D. D., and Story, H. E. 1943. The carotid circulation in the domestic cat. *Field Mus. Nat. Hist. Zool. Ser.* 28:1–47.

Davis, D. D., and Story, E. H. 1949. The female external genitalia of the spotted hyena. *Fieldiana Zool.* 31:277–283.

Decker, D. M., and Wozencraft, W. C. In press. A phylogenetic analysis of Recent procyonid genera with comments on the relationship of *Ailurus. J. Mamm.*

Dehm, R. 1950. Die Raubtiere aus dem Mittel-Miocän (Burdigalium) von Wintershof-West bei Eichstätt in Bayern. *Bayerische Akademie der Wissenschaften. Abhandlungen* (Munich) 58:1–141.

Denison, R. W. 1938. The broad-skulled Pseudocreodi. *Ann. New York Acad. Sci.* 38:163–256.

Desmarest, A. G. 1820. *Mammalogie, ou description des espèces de mammifères,* vol. 1: *Première partie, contenant les ordres des bimanes, des quadrumunes et des carnassiers.* Encyclopédie Méthodique. Paris: V. Agasse.

Doran, A. H. G. 1879 (for 1878). The mammalian ossicula auditûs. *Trans. Linn. Soc. London, Zool.*, 2nd ser. 1:371–497.

Dutrillaux, B., and Couturier, J. 1983. The ancestral karyotype of Carnivora: Comparison with that of platyrrhine monkeys. *Cytogenet. Cell. Genet.* 35:200–208.

Ewer, R. F. 1973. *The Carnivores*. Ithaca, N.Y.: Cornell Univ. Press.

Fay, F. H., Rausch, V. R., and Feltz, E. T. 1967. Cytogenetic comparison of some pinnipeds (Mammalia: Eutheria). *Canadian J. Zool.* 45:773–778.

Fischer de Waldheim, G. 1817. Adversaria Zoologica. *Mémoire Société Impériale Naturelle* (Moscow) 5:368–428.

Flower, W. H. 1869a. On the anatomy of the Proteles, *Proteles cristatus* (Sparrman). *Proc. Zool. Soc. London* 1869:474–496.

Flower, W. H. 1869b. On the value of the characters of the base of the cranium in the classification of the Order Carnivora, and on the systematic position of *Bassaris* and other disputed forms. *Proc. Zool. Soc. London* 1869:4–37.

Flower, W. H., and Lyddekker, R. 1891. *An Introduction to the Study of Mammals Living and Extinct*. London: Adam & Charles Black.

Flynn, J. J., and Galiano, H. 1982. Phylogeny of early Tertiary Carnivora, with a description of a new species of *Protictis* from the middle Eocene of northwestern Wyoming. *Amer. Mus. Novitates* 2725:1–64.

Fredga, K. 1972. Comparative chromosome studies in mongooses (Carnivora, Viverridae) 1. Idiograms of 12 species and karyotype evolution in Herpestinae. *Hereditas* 71:1–74.

Geoffroy Saint-Hilaire, E. 1803. *Catalogue des mammifères du Muséum National d'Histoire Naturelle*. Paris: Muséum National d'Histoire Naturelle.

Geoffroy Saint Hilaire, E., and Cuvier, F. G. 1795. Mémoire sur une nouvelle division des mammifères, et les principes qui doivent servir de base dans cette sorte de travail, lu à la Societe d'Histoire naturelle, le premier floreal de l'an troisieme. *Mag. Encyclopédique* 2:164–187.

Gervais, F. L. P. 1855. *Histoire naturelle des Mammifères avec l'indication de leurs moeurs, et de leurs rapports avec les arts, le commerce et l'agriculture*. Paris: L. Curmer.

Gesner, K. 1551. *Historiae animalium*, vol. 1: *De quadrupedibus viviparis*. Zurich: C. Froschovervm.

Ginsburg, L. 1966. Les amphicyons des phosphorites du Quercy. *Ann. Paleontol.* 52:23–64.

Ginsburg, L. 1982. Sur la position systématique du petit panda, *Ailurus fulgens* (Carnivora, Mammalia). *Geobios (Lyon), Mémoire Spécial* 6:259–277.

Gorman, M. L., Nedwell, D. B., and Smith, R. M. 1974. An analysis of the contents of the anal scent pockets of *Herpestes auropunctatus* (Carnivora: Viverridae). *J. Zool. (Lond.)* 172:389–399.

Gray, J. E. 1821. On the natural arrangement of vertebrose animals. *Med. Reposit.* 15:296–310.

Gray, J. E. 1825. Outline of an attempt at the disposition of the Mammalia into tribes and families with a list of the genera apparently appertaining to each tribe. *Ann. Philos.* new ser. 10:337–344.

Gray, J. E. 1837 (for 1836). Characters of some new species of Mammalia in the society's collection, with remarks upon the dentition of the Carnivora, and upon the value of the characters used by M. Cuvier to separate the plantigrade from the digitigrade Carnivora. *Proc. Comm. Sci. & Corres. Zool. Soc. London* 1836:87–88.

Gray, J. E. 1843. *List of the specimens of Mammalia in the collection of the British Museum*. London: British Museum (Natural History) Publication.

Gray, J. E. 1865 (for 1864). A revision of the genera and species of viverrine animals (Viverridae) founded on the collection in the British Museum. *Proc. Zool. Soc. London* 1864:502–579.

Gray, J. E. 1869. *Catalogue of carnivorous, pachydermatous and edentate mammals in the British Museum.* London: British Museum (Natural History) Publication.

Gregory, W. K. 1920. Studies of the comparative myology and osteology, no. IV: A review of the evolution of the lacrimal in vertebrates with special reference to that of mammals. *Bull. Amer. Mus. Nat. Hist.* 42:95–263.

Gregory, W. K. 1936. On the phylogenetic relationships of the giant panda (*Ailuropoda*) to other carnivores. *Amer. Mus. Novitates* 878:1–29.

Gregory, W. K., and Hellman, M. 1939. On the evolution and major classification of the civets (Viverridae) and allied fossil and recent Carnivora: Phylogenetic study of the skull and dentition. *Proc. Amer. Philos. Soc.* 81:309–392.

Gregory, W. K., and Simpson, G. G. 1926. Cretaceous mammal skulls from Mongolia. *Amer. Mus. Novitates* 225:1–20.

Gromova, V. I., Dubrovo, I. A., and Yanovskaya, N. M. 1968. Order Carnivora. In: V. I. Gromova & Y. A. Orlov, eds. *Fundamentals of Paleontology* 13:241–310. Washington, D.C.: U.S. Department of Commerce, Clearinghouse for Federal Scientific and Technical Information, no. TT 67–51241.

Hemmer, H. 1978. The evolutionary systematics of living Felidae: Present status and current problems. *Carnivore* 1:71–79.

Hendey, Q. B. 1980. *Agriotherium* (Mammalia: Ursidae) from Langebaanweg, South Africa, and relationships of the genus. *Ann. South African Mus.* 81:1–109.

Hennig, W. 1950. *Grundzüge einer Theorie der phylogenetischen Systematik.* Berlin: Deutscher Zentralverlag.

Hennig, W. 1966. *Phylogenetic Systematics.* Urbana: Univ. Illinois Press.

Hough, J. R. 1944. The auditory region in some Miocene carnivores. *J. Paleontol.* 18:470–479.

Hough, J. R. 1948. The auditory region in some members of the Procyonidae, Canidae, and Ursidae: Its significance in the phylogeny of the Carnivora. *Bull. Amer. Mus. Nat. Hist.* 92:67–118.

Hough, J. R. 1953. Auditory region in North American fossil Felidae: its significance in phylogeny. *U.S. Geol. Surv. Professional Paper* 243g:95–115.

Howell, A. B. 1929 (for 1928). Contribution to the comparative anatomy of the eared and earless seals (genera *Zalophus* and *Phoca*). *Proc. U.S. Nat. Mus.* 73:1–142.

Hunt, R. M. 1974. The auditory bulla in Carnivora: An anatomical basis for reappraisal of carnivore evolution. *J. Morphol.* 143:21–76.

Hunt, R. M. 1987. Evolution of the Aeluroid Carnivora: Significance of auditory structure in the nimravid cat *Dinictis*. *Amer. Mus. Novitates* 2886:1–74.

Huxley, T. H. 1880. On the cranial and dental characters of the Canidae. *Proc. Zool. Soc. London* 1880:238–288.

Jentink, F. A. 1887. Catalogue ostéologique des mammifères. *Muséum d'Histoire Naturelle des Pays-Bas* 9:1–360.

Jentink, F. A. 1892. Catalogue systématique des mammifères (Singes, Carnivores, Ruminants, Pachyderms, Sirénes, et Cétacés). *Muséum d'Histoire Naturelle des Pays-Bas* 11:1–208.

King, J. E. 1964. *Seals of the World.* London: British Museum (Nat. Hist.).

King, J. E. 1971. The lacrimal bone in the Otariidae. *Mammalia* 35:465–470.

King, J. E. 1983. *Seals of the World* (2nd ed.). Ithaca, N.Y.: Cornell Univ. Press.

Kretzoi, M. 1945. Bemerkungen über das Raubtiersystem. *Annales Historico Naturales Musei Nationalis Hungarici* (Budapest) 38:59–83.

Leone, C. A. and Wiens, A. L. 1956. Comparative serology of carnivores. *J. Mamm.* 37:11–23.

Ling, J. K. 1978. Pelage characteristics and systematic relationships in the Pinnipedia. *Mammalia* 42:305–313.

Linneaus, C. 1748. *Systema Naturae, sistens Regna tria Naturae, in classes et ordines genera et species redacta, tabulisque aeneis illustrata* (6th ed.). Stockholm.

Linneaus, C. 1758. *Systema Naturae per regna tri naturae, secundum classis, ordines, genera, species cum characteribus, differentiis, synonymis locis* (10th ed.). Stockholm.

McDowell, S. B. 1958. The greater Antillean insectivores. *Bull. Amer. Mus. Nat. Hist.* 115:113–214.

MacIntyre, G. T. 1966. The Miacidae (Mammalia, Carnivora). Part I. The systematics of *Ictidopappus* and *Proticitis. Bull. Amer. Mus. Nat. Hist.* 131:119–209.

McKenna, M. C. 1975. Toward a phylogenetic classification of the Mammalia. In: W. P. Luckett & F. S. Szalay, eds. *Phylogeny of the Primates: A Multidisciplinary Approach,* pp. 21–46. New York: Plenum Press.

McLaren, I. A. 1960. Are the Pinnipedia biphyletic? *Syst. Zool.* 9:18–28.

MacPhee, R. D. E. 1981. Auditory regions of Primates and Eutherian Insectivores: Morphology, ontogeny, and character analysis. *Contrib. primatol.* 18:1–282.

Matthew, W. D. 1901. Additional observations on the Creodonta. *Bull. Amer. Mus. Nat. Hist.* 14:1–38.

Matthew, W. D. 1909. The Carnivora and Insectivora of the Bridger basin, Middle Eocene. *Mem. Amer. Mus. Nat. Hist.* 9:291–586.

Mayr, E. 1986. Uncertainty in science: Is the giant panda a bear or raccoon? *Nature* 323:769–771.

Mitchell, E., and Tedford, R. H. 1973. The Enaliarctinae, a new group of extinct aquatic Carnivora and a consideration of the origin of the Otariidae. *Bull. Amer. Mus. Nat. Hist.* 151:201–284.

Mivart, St. G. 1882a. A classification and distribution of the Aeluroidea. *Proc. Zool. Soc. London* 1882:135–208.

Mivart, St. G. 1882b. Notes on some points in the anatomy of the Aeluroidea. *Proc. Zool. Soc. London* 1882:459–520.

Mivart, St. G. 1885a. On the anatomy, classification, and distribution of the Arctoidea. *Proc. Zool. Soc. London* 1885:340–404.

Mivart, St. G. 1885b. Notes on the Pinnipedia. *Proc. Zool. Soc. London* 1885:484–500.

Muizon, C. de. 1982a. Phocid phylogeny and dispersal. *Ann. South African Mus.* 89:175–213.

Muizon, C. de. 1982b. Les relations phylogénétiques des Lutrinae (Mustelidae, Mammalia). *Geobios (Lyon), Mémoire Spécial,* 6:259–277.

Muller, J. 1934. The orbitotemporal region in the skull of the Mammalia. *Archives Neerlandaises Zoologie* 1:118–259.

Murie, J. 1870. Researches upon the anatomy of the Pinnipedia. Part 1. On the walrus (*Trichechus rosmarus* Linn.). *Trans. Zool. Soc. London* 7:411–464.

Murie, J. 1872. Researches upon the anatomy of the Pinnipedia. Part 2. Descriptive anatomy of the sea-lion (*Otaria jubata*). *Trans. Zool. Soc. London* 7:527–596.

Murie, J. 1874. Researches upon the anatomy of the Pinnipedia. Part 3. Descriptive anatomy of the sea-lion (*Otaria jubata*). *Trans. Zool. Soc. London* 8:501–582.

Neff, N. A. 1983. The basicranial anatomy of the Nimravidae (Mammalia: Carnivora): Character analyses and phylogenetic inferences. Ph.D. dissert. City University of New York, New York.

Novacek, M. J. 1977. Aspects of the problem of variation, origin and evolution of the Eutherian auditory bulla. *Mamm. Rev.* 7:131–149.

Novacek, M. J. 1980. Cranioskeletal features in tupaiids and selected Eutheria as phylogenetic evidence. In: W. P. Luckett, ed. *Comparative Biology and Evolutionary Relationships of Tree Shrews*, pp. 35–93. New York: Plenum Press.

Novacek, M. J. 1986. The skull of Leptictid insectivorans and the higher-level classification of Eutherian mammals. *Bull. Amer. Mus. Nat. Hist.* 183:1–111.

Novacek, M. J., and Wyss, A. R. 1986. Higher-level relationships of the Recent Eutherian orders: Morphological evidence. *Cladistics* 2(3):257–287.

O'Brien, S. J., Nash, W. G., Wildt, D. E., Bush, M. E., and Benveniste, R. E. 1985. A molecular solution to the riddle of the giant panda's phylogeny. *Nature* 317:140–144.

Pauly, L. K., and Wolfe, H. R. 1957. Serological relationships among members of the order Carnivora. *Zoologica* 42:159–166.

Petter, G. 1974. Rapports phylétiques des viverrides (Carnivores Fissipèdes). Les formes de Madagascar. *Mammalia* 38:605–636.

Pocock, R. I. 1914. On the feet and other external features of the Canidae and Ursidae. *Proc. Zool. Soc. London* 1914:913–941.

Pocock, R. I. 1915a. On some of the external characters of the genus *Linsang* with notes upon the genera *Poians* and *Eupleres. Ann. & Mag. Nat. Hist.* ser. 8, 16:341–351.

Pocock, R. I. 1915b. On the feet and glands and other external characters of the Paradoxurinae genera *Paradoxurus, Arctictis, Arctogalidia,* and *Nandinia. Proc. Zool. Soc. London* 1915:387–412.

Pocock, R. I. 1916a. The alisphenoid canal in civets and hyaenas. *Proc. Zool. Soc. London* 1916:442–445.

Pocock, R. I. 1916b. On the external characters of the mongooses (Mungotidae). *Proc. Zool. Soc. London* 1916:349–374.

Pocock, R. I. 1916c. On some of the external structural characters of the striped hyaena (*Hyaena hyaena*) and related genera and species. *Ann. & Mag. Nat. Hist.,* ser 8, 17:330–343.

Pocock, R. I. 1917. On the external characters of the Felidae. *Ann. & Mag. Nat. Hist.* ser. 8, 19:343–352.

Pocock, R. I. 1919. The classification of the mongooses (Mungotidae). *Ann. & Mag. Nat. Hist.* ser. 9, 3:515–524.

Pocock, R. I. 1921. The auditory bulla and other cranial characters in the Mustelidae. *Proc. Zool. Soc. London* 1921:473–486.

Presley, R. 1979. The primitive course of the internal carotid artery in mammals. *Acta Anat.* 103:238–244.

Qui Zhanxiang and Schmidt-Kittler, N. 1982. On the phylogeny and zoogeography of the Leptarctines. (Carnivora, Mammalia). *Palaeontol. Z.* 56:131–145.

Radinsky, L. B. 1973. Are stink badgers skunks? Implications of neuroanatomy for mustelid phylogeny. *J. Mamm.* 54:585–593.

Radinsky, L. B. 1975. Viverrid neuroanatomy: Phylogenetic and behavioral implications. *J. Mamm.* 56:130–150.

Radinsky, L. B. 1980. Analysis of carnivore skull morphology. *Amer. Zool.* 20:784.

Radinsky, L. B. 1981a. Evolution of skull shape in carnivores. I. Representative modern carnivores. *Biol. J. Linn. Soc.* 15:369–388.

Radinsky, L. B. 1981b. Evolution of skull shape in carnivores. 2. Additional modern Carnivores. *Biol. J. Linn. Soc.* 16:337–355.

Radinsky, L. B. 1982. Evolution of skull shape in carnivores. 3. The origin and early radiation of the modern carnivore families. *Paleobiology* 8:177–195.

Radinsky, L. B. 1984. Basicranial axis length v. skull length in analysis of carnivore skull shape. *Biol. J. Linn. Soc.* 22:31–41.

Repenning, C. A. 1972. Underwater hearing in seals: Functional morphology. In: R. J. Harrison, ed. *Functional Anatomy of Marine Mammals*, 1:307–331. New York: Academic Press.

Repenning, C. A. 1976. Adaptive evolution of seal lions and walruses. *Syst. Zool.* 25:375–390.

Repenning, C. A., Ray, C. E., and Grigorescu, D. 1979. Pinniped biogeography. In: J. Gray & A. J. Boucot, eds. *Historical Biogeography, Plate Tectonics, and the Changing Environment*, pp. 357–369. Corvallis: Oregon State Univ. Press.

Repenning, C. A., and Tedford, R. H. 1977. Otarioid seals of the Neogene. *U.S. Geol. Surv. Professional Paper* 992:1–93.

Romer, A. S. 1966. *Vertebrate Paleontology* (3rd ed.). Chicago: Univ. Chicago Press.

Sarich, V. M. 1969. Pinniped origins and the rate of evolution of carnivore albumins. *Syst. Zool.* 18:286–295.

Sarich, V. M. 1976. Transferrin. *Trans. Zool. Soc. London* 33:165–171.

Savage, R. J. G. 1957. Early miocene mammal faunas of the Telhyan region. In: C. G. Adams & D. V. Ager, eds. *Aspects of Telhyan Biogeography*, pp. 247–282. Systematics Association Publication no. 7. London: The Systematics Association.

Savage, R. J. G. 1977. Evolution in carnivorous mammals. *Paleontology* 20:237–271.

Scheffer, V. B. 1958. *Seals, Sea Lions, and Walruses: A Review of the Pinnipedia*. Stanford, Calif.: Stanford Univ. Press.

Schlosser, M. 1887 (for 1886). Über das Verhältnis der Cope'schen Creodonta zu den übrigen Fleischfressern. *Morphologie Jahrbuch* 12:287–298.

Schlosser, M. 1899. Creodonten und Carnivoren des europäischen Tertiärs. *Beiträge zur Paläontologie und Geologie* 7:225–368.

Schlosser, M. 1902. Beiträge zur Kenntniss der Säugethierreste aus den sudent schen Bohnerzen. *Geologische und Palaontologische Abhandlungen* (Jena) 5:1–144.

Schmidt-Kittler, N. 1981. Zur Stammesgeschichte der marderverwandten Raubtiergruppen (Musteloidea, Carnivora). *Eclogae Geologicae Helvetiae* 74:753–801.

Schreber, J. C. D. 1778. *Die Säugthiere in Abbildungen nach der Natur mit Beschreibungen* 3:289–590. Erlangen: Walter.

Segall, W. 1943. The auditory region of Arctoid carnivores. *Field Mus. Nat. Hist. Zool. Ser.* 29:33–59.

Seal, U. S. 1969. Carnivora systematics: A study of hemoglobins. *Comp. Biochem. Physiol.* 31:799–811.

Seal, U. S., Erickson, A. W., Sniff, D. B., and Hofman, R. J. 1971. Biochemical, population genetic, phylogenetic and cytological studies of Antarctic seal species. In: G. Deacon, ed. *Proceedings of a Symposium on Antarctic ice and water masses. Tokyo, Japan, 19 September 1970*, pp. 77–95. Cambridge: Scientific Committee on Antarctic Research.

Shoshani, J. 1986. Mammalian phylogeny: Comparison of morphological and molecular results. *Mol. Biol. Evol.* 3:222–242.

Simpson, G. G. 1945. The principles of classification and a classification of mammals. *Bull. Amer. Mus. Nat. Hist.* 85:1–350.

Story, H. E. 1951. The carotid arteries in the Procyonidae. *Fieldiana Zool.* 32:477–557.

Swainson, W. 1835. On the natural history and classification of quadrupeds. In: *The Cabinet Cyclopaedia*, vol. 121. London: Longman Rees, Orme, Brown, Green and Longman.

Szalay, F. S., and Decker, R. L. 1974. Origins, evolution, and function of the tarsus in the late Cretaceous Eutheria and Paleocene primates. In: F. A. Jenkins, ed. *Primate Locomotion,* pp. 223–259. New York: Academic Press.

Tedford, R. H. 1976. Relationship of Pinnipeds to other carnivores (Mammalia). *Syst. Zool.* 25:363–374.

Temminck, C. J. 1835–41. *Monographies de Mammalogie,* vol. 2. Leiden: C. C. Van der Hoek.

Thenius, E. 1972. *Grundzüge der Verbreitungsgeschichte der Säugetiere.* Stuttgart: Gustav Fischer.

Thenius, E. 1979. Zur systematischen und phylogenetischen Stellung des Bambusbären: *Ailuropoda melanoleuca* David (Carnivora, Mammalia). *Z. Säugetierk.* 44:286–305.

Thenius, E., and Hofer, H. 1960. *Stammesgeschichte der Säugetiere.* Berlin: Springer Verlag.

Thomas, O. 1882. On African mongooses. *Proc. Zool. Soc. London.* 1882:59–93.

Todd, N. B., and Pressman, S. R. 1968. The karyotype of the lesser panda (*Ailurus fulgens*) and general remarks on the phylogeny and affinities of the panda. *Carn. Genet. Newsl.* 5:105–108.

Troussart, E. L. 1898–99. *Catalogus mammalium tam Viventium quam fossilium.* Berlin: R. Friedlander & Sons.

Turner, H. N. 1848. Observations relating to some of the foramina at the base of the skull in Mammalia, and on the classification of the Order Carnivora. *Proc. Zool. Soc. London* 1848:63–88.

van der Klaauw, C. J. 1931. The auditory bulla in some fossil mammals. *Bull. Amer. Mus. Nat. Hist.* 62:1–341.

van Kampen, P. N. 1905. Die Tympanalgegend des Säugetierschädel. *Morphogie Jahrbuch* 34:321–722.

van Valen, L. 1966. Deltatheridia, a new order of mammals. *Bull. Amer. Mus. Nat. Hist.* 8:638–665.

van Valen, L. 1969. The multiple origins of the placental carnivores. *Evol.* 23:118–130.

Watrous, L. E., and Wheeler, Q. D. 1981. The out-group comparison method of character analysis. *Syst. Zool.* 30:1–11.

Wible, J. R. 1986. Transformations in the extracranial course of the internal carotid artery in mammalian phylogeny. *J. Vert. Paleontol.* 6:313–325.

Wible, J. R. 1987. The eutherian stapedial artery: Character analysis and implications for superordinal relationships. *Zool. J. Linnean Society* 91:107–135.

Wiley, E. O. 1979. An annotated Linnaean hierarchy, with comments on natural taxa and competing systems. *Syst. Zool.* 28:309–337.

Wiley, E. O. 1981. *Phylogenetics: The Theory and Practice of Phylogenetic Systematics.* New York: John Wiley & Sons.

Winge, H. 1895. *Jordfundne og nulevende Rovdyr (Carnivora) fra Lagoa Santa, Minas Geraes, Brasilien Copenhagen* (Museo Lundii) 2(4):1–103.

Winge, H. 1924. *Pattedyr Slaegter* (Copenhagen Universitet. Zoologische Museum. Copenhagen. H. Hagerups Forlag) 2:1–313.

Wortman, J. L. 1901. Studies of Eocene Mammalia in the Marsh collection, Peabody Museum. Part 1. Carnivora. *Amer. J. Sci.* ser. 4, 11:333–348.

Wortman, J. L., and Matthew, W. D. 1899. The ancestry of certain members of the Canidae, the Viverridae and Procyonidae. *Bull. Amer. Mus. Nat. Hist.* 12:109–138.

Wozencraft, W. C. 1984a. Evolution of the Carnivora: The basicranial evidence. (Abstract.) American Society of Mammalogists 64th annual meeting (Abstracts); p. 8.

Wozencraft, W. C. 1984b. A phylogenetic reappraisal of the Viverridae and its relationship to other Carnivora. PhD. dissert. Univ. Kansas, Lawrence.

Wurster, D. H., and Benirschke, K. 1967. Chromosome numbers in thirty species of carnivores. *Mamm. Chrom. Newsl.* 8:195–196.

Wurster, D. H., and Benirschke, K. 1968. Comparative cytogenetic studies in the Order Carnivora. *Chromosoma* 24:336–382.

Wurster-Hill, D. H., and Gray, C. W. 1975. The interrelationships of chromosome banding patterns in Procyonids, Viverrids and Felids. *Cytogenet. Cell. Genet.* 15:306–331.

Wyss, A. R. 1987. The walrus auditory region and the monophyly of pinnipeds. *Amer. Mus. Novitates* 2871:1–31.

CHAPTER 19

Fossil History of the
Terrestrial Carnivora

LARRY D. MARTIN

Carnivores, because of their position on the ecological pyramid, are considerably rarer than their prey. They are also often intelligent and solitary animals, so that their chances of dying in a fossilizing environment are not very good. Exceptions occur when herbivores are trapped in a situation that can also entrap carnivores. The La Brea Tar Pits in California, the most famous example of such a baited trap, have produced one of the highest numbers of fossil carnivorans found in any single locality. Cave sites are also important concentrating localities, as many carnivores use caves and fissures as lairs. Mixnitz Cave in Austria may have contained the remains of nearly 50,000 cave bears (Kurtén 1976a). Outside of such sites fossil carnivorans are rare, and many taxa are known only from dentitions. In spite of these problems, the fossil record of carnivorans is good and provides answers to questions concerning the evolution of adaptation and diversity in carnivores.

Flower's (1869) basic separation of the modern carnivorans into the Arctoidea, Cynoidea, and Aeluroidea forms the basis for most carnivore classifications. Paleontologists have also recognized an extinct group of primitive carnivorans, the Miacoidea (Simpson 1945). Flynn and Galiano (1982) and Wozencraft (this volume) provide good summaries of the history of carnivore classification, including fossils.

Origin of the Carnivora

Throughout the Mesozoic (64–200 million years ago), mammals were small, only rarely reaching the size of a domestic cat. Dominated by the dinosaurian radiation, it seems likely that early mammals were nocturnal and arboreal. At the end of the Cretaceous (64 million years ago), the dinosaurs along with much of the other fauna became extinct, but some mammals persisted. Most of these surviving mammals were either insectivorous or omnivorous, much like the modern opposum.

The absence of specialized terrestrial carnivores created an ecological situation that presently occurs on islands and that allowed the evolution of large, flightless ground-nesting birds (Martin 1983). The early Tertiary avian radiation included an extinct order (Gastornithiformes) of large birds with very big heads, such as species of *Gastornis* and *Diatryma*. These birds may have been direct competitors with the newly evolving mammalian carnivores. They lacked the characteristic tearing beak of modern carnivorous birds, but at that time almost all their available prey were so small that they could have been swallowed whole.

The late Cretaceous genus, *Cimolestes*, is considered a basal carnivore group. The largest species of this genus might have reached the size of a ferret. According to Lillegraven (1969:81), *Cimolestes* "was obviously specialized for a carnivorous way of life" and the "carnassial function was distributed evenly throughout the molar series." In the later Creodonta the carnassial (shearing) function of the dentition was concentrated at M1/m2 (Oxyaenidae) or M2/m3 (Hyaenodontidae). The Carnivora restrict this function to P4/m1.

The basic differentiation of mammalian teeth is best understood in terms of the lever system of the jaws, which follows a simple nutcracker arrangement, with the greatest force occuring near the fulcrum. Crushing teeth tend to occupy that position and are thus usually molars. Shearing activity may be shared with the premolars. In the Carnivora the last upper premolar (P4) has its shearing blade emphasized. The shear occludes against the anterolabial surface of m1. In some carnivores (cats) this shear is enhanced at the expense of grinding capability, and the posterior molars are lost or reduced as the shearing teeth move closer to the fulcrum of the jaw. In other forms (e.g., ursids) the crushing function is emphasized, and the posterior molars may become larger than the carnassials.

Mesozoic carnivorous mammals were almost entirely insectivorous, and the earliest mammals that show specializations for biting off portions of flesh (i.e., possessing carnassials) are from the early Paleocene. During the late Paleocene and early Eocene mammals inhabited worldwide tropical forests. Dermopterans (flying lemurs), crocodilians, and large tortoises occurred on Ellesmere Island (Dawson et al. 1976) above 80° north latitude during the late Paleocene and early Eocene, a distribution that indicates a lack of effective climatic zonation at that time.

The earliest known Carnivora are small, arboreal forms belonging to the Viverravidae, that had already lost M3/m3 and were very similar to the modern Viverridae. In fact, Gregory and Hellman (1939) thought that the Viverridae could be traced directly into this group. Some modern workers consider the viverravids to be an extinct sidebranch (Gingerich and Winkler 1985), although they are commonly included in the feliform clade near the viverrids on the basis of their molar reduction (Flynn and Galiano 1982).

A second group of viverrid-like Carnivora, the Miacidae, appear in the early Eocene. The Miacidae differ from the Viverravidae in the retention of M3/m3 and in having a less sectorial dentition. Both families have been included as subfamilies within a "Miacidae" that would then contain the roots of both the feliform and caniform radiations. The postcranial skeleton is similar in both groups, and they tend to be small (the largest being about the size of a red fox, *Vulpes vulpes*) and arboreally adapted.

The earliest significant nonarboreal habitats in the Tertiary may have been restricted (as they are in the modern rainforest) to areas of periodic disturbance, usually along the margins of lakes and streams. Large semi-aquatic and stream-marginal herbivores developed in these regions. The predators in this habitat were an extinct order of carnivorous mammals, the Creodonta, containing two families, the Oxyaenidae and the Hyaenodontidae. The oxyaenids were cat- or viverrid-like, whereas the hyaenodontids show early adaptations toward pursuit (as opposed to ambush or accidental discovery modes that must have characterized prey capture in the early Tertiary). When compared with modern mammalian carnivores (Radinsky 1977), all of these "archaic carnivores" were relatively small-brained with large olfactory lobes. In two genera (*Apataelurus* and *Machaeroides*) the Creodonta produced saber-toothed predators, a morphological system for killing large prey (see Martin 1980) that became prevalent in later faunas.

Biogeography

Throughout the Tertiary the tropics progressively contracted toward the equator, and the modern latitudinal zonation formed. New latitudinal climatic zones appeared first at high latitudes and then migrated to lower ones, following the contracting tropics (Martin 1983). The communities adapted to these new climatic zones evolved first at high latitudes and then extended southward. We have little fossil evidence from these high latitudinal regions, which probably explains many of the gaps between groups in the fossil record.

During the early Tertiary there were two main connections between North America and Eurasia. One of these ran from North America northeast to Europe (McKenna 1975) and maintained a high faunal similarity between Europe and North America until about the middle Eocene. During the time when this connection was last available, tropical conditions persisted at high latitudes, permitting the existence of a Holarctic tropical fauna. This connection ended some time in the Eocene, leaving the western connection with Asia (Beringia) as the fundamental route for faunal interchange.

The Beringian route lies at a high latitude, and as the tropics contracted, climatic filtering effects began to develop. Animals that originated at low latitudes might have had difficulty crossing the new habitats evolving in the region of Beringia. Such radiations might remain endemic or at some later time evolve a lineage able to make the high-latitude passage. On the other hand, groups that originated in the Holarctic could eventually follow the contracting tropics southward and appear simultaneously without antecedents in Europe and North America. Such an appearance would require a high-latitude distribution across Beringia. The absence of antecedents would indicate that the high-latitude region was actually the place of origin.

The southern continents maintained substantial isolation during much of the Tertiary. During the early Oligocene, Africa contained creodonts, but there is no evidence of true Carnivora. In the early and middle Miocene, amphicyonids, viverrids, herpestids, felids, nimravids, and a few surviving creodonts occurred in Africa. During the late Miocene and Pliocene, ursids, canids, mustelids, and additional felids reached Africa, and the modern African carnivore community evolved.

South America remained isolated until the late Miocene, when procyonids first appeared there. These are sometimes ascribed to waif dispersal, but ground sloths from South America appeared in North America at about the same time, and some sort of land connection would seem to be required for this two-way dispersal. An extensive endemic small felid radiation in South America may have been based in *"Felis" stouti* from the North American late Miocene (Schultz and Martin 1972; Glass and Martin 1978). Canids appeared in the South American early Pleistocene with a dhole-like fox, *Protocyon* (Berta 1981). *Canis* species arrived sometime later, as did a species of *Smilodon*, *Felis* (*Puma*) *concolor*, the jaguar (*Panthera onca*), and species of the ursid genera *Arctodus* and *Tremarctos*. Extinction of the endemic marsupial carnivore genera, *Thylacosmilus* and *Borhyaena*, as well as many of the immigrant placental carnivores, left South America with large carnivores, the jaguar, mountain lion, and Andean bear, (*Panthera, Puma* and *Tremarctos*) that had gone from North America to South America in the Pleistocene.

Although we often emphasize the taxonomic composition of floras, animals may be more influenced by the structure of plant communities (Martin et al. 1985). One of the most important structural changes during the Tertiary was the development of open spaces beyond the stream margins. In middle latitudes this seems to have begun in the late Eocene, with modern open community structures appearing in the Oligocene. At this time we find the earliest examples of typical open country organisms, including composite flowers, grasses, fossorial rodents, and grazing ungulates with high-crowned teeth.

The early open areas may have been small with comparatively large densities of trees, as the carnivores in these open spaces usually show arboreal adaptation (Van Valkenburgh 1985). In the modern fauna cats are adapted to the marginal region between trees and open areas. The Oligocene was probably the time of the greatest dominance of catlike forms. Many of these were sabertoothed "cats" belonging to the archaic family Nimravidae. Their relationship to true cats (Felidae) is hotly debated (Martin 1980), but, at best, most of their catlike features are independent acquisitions. Unequivocal cats do not appear until the early Miocene and were derived from small, arboreal viverrid-like carnivores probably in northern Asia, although the transitional forms are still unknown.

True canids first appeared in the Oligocene with the small, arboreally adapted genus *Hesperocyon*, but most pursuit predator niches were held by hyaenodontid creodonts and by a canid sister group, the Amphicyonidae. Canids first appeared in North America in the Oligocene and are restricted to that continent until the late Tertiary. Amphicyonids filled the canid adaptive zone in Eurasia, and hyaenodontids have late survivors in Africa and India.

The earliest mustelid-like forms also appeared in the Oligocene, and their evolution seems to correspond to the earliest radiations of terrestrial and fossorial rodents. Interestingly, most early carnivores with the characteristic short face and elongated cranium do not seem to be mustelids, and true mustelids appeared in the earliest Miocene at about the same time as the earliest cats. One mustelid ecomorph, *Palaeogale lagophaga*, is known from abundant material in both Europe and North America (Simpson 1946). Flynn and Galiano (1982) considered the genus *Palaeogale* to be feloid but based their argument almost entirely on dental characters.

In North America the end of the Oligocene was characterized in some regions by an arid-climate biota, and the Miocene shows early evidence of widespread tropical savanna. The Miocene fauna contains major radiations of mustelids and large pursuit predators (primarily canids in North America). Canids achieved their modern distribution largely in the late Miocene/Pliocene. Felids diversified during the late Miocene, and the last nimravid became extinct at the end of the Kimballian (late Miocene).

Arctic conditions were the last to appear, and the Ice Age may only represent the point where the tropical contraction permitted arctic climates to reach middle latitudes. These last climatic zones may have created larger areas of treeless spaces than otherwise had been present and part of the initiative for the evolution of social hunters and large-brained pursuit predators. In the early forested communities olfaction took precedence over sight, and interactions among groups of hunting predators may have been difficult to maintain. Group hunting also may be in part a reaction to herding behavior, and the formation of herds is also more practical in the open habitats that dominated the middle-latitude Plio-Pleistocene.

Iterative Evolution

White (1984) recently emphasized the recognition of morphological similarities due to similar selective pressures. These similarities result in recognizable morphological "types" for specific adaptive zones. The similar adaptive types are called ecomorphs. Ecomorphs generally do not evolve in proximity to each other but initially must be separated either geographically or temporally.

Examples of geographic ecomorphs include the many Australian marsupials that are similar to placental morphotypes (Wilson 1962).

Temporal isolation is provided by extinction. There appears to have been a regular pattern of extinctions followed by the reevolution of ecomorphs throughout the Tertiary (Martin 1985). These extinction cycles created an array of chronologically separated but very similar carnivores and have greatly confused carnivoran systematics. Detailed analysis shows that many of these animals are distant phylogenetically from the forms that they most resemble.

Although it seems reasonable to expect similarity of structure to be greater in proportion to the nearness of relationship, natural selection is a powerful force that has created detailed similarities even against very different genetic backgrounds. This suggests that the number of viable solutions to carnivoran adaptations is actually small and the probability of convergent adaptations is relatively high. It appears that in the fossil Carnivora the probability of homoplasy is three to five times greater than that for synapomorphy for many features. This would preclude simple numerical procedures in the evaluation of most proposed phylogenies.

For carnivorans I recognize civet-like, catlike, mustelid-like, and doglike ecomorphs, and there are subdivisions within these categories.

Civet-like Ecomorph

Arboreal adaptations are probably original for the Carnivora, and the modern civets may most closely approximate the habitus of the earliest carnivorans. There is a subsection of this lifestyle that deserves special mention, the development of omnivorous forms that include fruit in their diet. Civets, procyonids, and some canids, including species of *Oxetocyon, Cynarctus,* and *Urocyon,* show adaptations in this direction.

Catlike Ecomorph

The catlike ecomorph is primarily adapted to open areas in the proximity of trees, and all cats have some capacity to climb trees. Such activities are limited in the largest cats simply because of size (Taylor, this volume). Cats commonly stalk or ambush prey and make the final capture with a sudden pounce. Killing is done with a bite to the back of the neck, a killing behavior that is widely distributed and probably primitive within the Carnivora (Eisenberg and Leyhausen 1972; Leyhausen 1979). It is supplemented by a learned behavior, the throat bite (Ewer 1973). Elaboration of the throat bite to take larger prey has resulted in a special group of feliform carnivores—the so-called saber-toothed "cats." All modern cats have upper and lower canines with round cross-sections (Figure 19.1) so that their canines are roughly cone-shaped

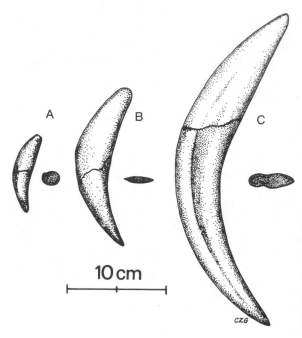

10 cm

Figure 19.1. Labial and cross-sectional views of upper canines: A, conical-toothed cat, African lion (*Panthera leo*); B, scimitar-toothed cat (*Machairodus coloradensis*); C, dirk-toothed cat (*Barbourofelis fricki*). (Modified from Martin 1980, courtesy Nebraska Academy of Sciences, Inc.)

(termed "conical-toothed" cats: Martin 1980). The cross-section of the upper canine is flattened to give a bladelike structure in saber-toothed cats, and the lower canine is reduced until it functionally becomes one of the incisors. In conical-toothed cats the killing bite is provided by the carnassials, as in most other carnivores; the canines serve to maintain a grip on the prey. In saber-toothed cats the upper canines are the primary killing organ.

The earliest saber-toothed mammalian carnivores are two genera of creodonts, *Apataelurus* and *Machaeroides*. Both genera share many saber-tooth attributes (Martin 1984), including bladelike upper canines and reduced lower canines, reduction of anterior premolars, development of a dependent flange on the lower jaw, and reduction of the coronoid process. Both creodont genera were extinct before the appearance of the earliest saber-toothed Carnivora. These early saber-toothed feliforms belong to an extinct family, the Nimravidae, first proposed by Cope (1880).

Conical-toothed Cats

Conical-toothed cats stalk their prey until they are able to close with it quickly. Although they may be capable of great speed, they tire quickly and do not pursue their prey as do canids. They are remarkably homogeneous in osteological characters. All have posteriorly recurved conical canines. The upper and lower canines are about equal in size, and the upper canines fit behind the lowers. The incisors are small, spatulate (usually with three cusps), and arranged in a straight line. The anterior premolars are usually large and multiple-rooted. The lower carnassial (m1) has a deep, narrow carnassial notch. The upper carnassial nearly always has a large protocone (except in the case of the cheetah, *Acinonyx jubatus*, and species of *Miracinonyx*). The face may be short, as in the cheetah, or long, as in the African lion (*Panthera leo*). The occiput is inclined posteriorly to the upper tooth row. The paramastoids and the paraoccipitals are small and separate. The upper carnassial is always ventral to the glenoid fossa. The limbs are variable in length ranging from the relatively short robust limbs of the jaguar to the highly elongated limbs of the cheetah. Except for the cheetah, which has secondarily lost claw retraction, all conical-toothed cats have retractile and hooded claws. They are all digitigrade, but a few, like the ocelot, (*Felis pardalis*) have remarkable powers of flexion of their feet (see Taylor, this volume).

The oldest known conical-toothed cat is an early Miocene nimravid, *Dinaelurus crassus*. *D. crassus* has the auditory bulla incompletely ossified and a large carotid foramen (inferior petrosal sinus?), as in other nimravids, but is the only nimravid known to have unserrated conical canines. It also has small incisors arranged in a straight line, as in other conical-toothed cats. The earliest true felid seems to be the middle Miocene genus *Pseudaelurus*.

Saber-toothed Cats

Saber-toothed cats all fall into one of two groups, depending on the shape of their upper canines (Figure 19.1). In one group (scimitar-toothed cats) are forms with relatively short, broad canines that usually bear coarse crenulations. The other group (dirk-toothed cats) has long, narrow upper canines with fine crenulations or with none at all. Kurtén (1963) proposed the terms "scimitar-toothed" and "dirk-toothed" for his Homotherini and Smilodontini, respectively, but I have expanded Kurtén's usage to include in each group all cats that have the characteristic canine morphology regardless of phylogenetic relationship (Schultz et al. 1970; Martin 1980). Neither type of canine is really closely comparable to a real scimitar or dirk, as the cutting edge is on the inside of the curvature. The canines of saber-toothed cats do closely resemble the Arabian curved dagger or jambiya.

In the fossil record, conical-and saber-toothed cats are commonly found together. The largest or "top predator" is usually a dirk-toothed feliform, but scimitar-toothed cats are also often lion-sized. Until the Pleistocene, conical-toothed cats ranged from the size of a domestic cat to that of a leopard. The lion- and tiger-sized feliforms are all saber-toothed in earlier faunas. Lions and tigers appear some time around the late Pliocene and early Pleistocene.

The skeletons of scimitar-toothed and dirk-toothed predators are quite different. Scimitar-toothed cats are comparatively long legged and in some forms (*Homotherium sainzelli*, Figure 19.2) may have cheetah-like skeletal proportions. Like other pursuit predators, they tend to have comparatively large brains, and some species of *Homotherium* share with the cheetah an enlargement of the optic center in the brain (Radinsky 1975).

Dirk-toothed cats are short-limbed (Figure 19.3) and are more bear- or badger-like in their skeletal proportions. They tend to develop bearlike, plantigrade hind feet and also may have developed a similar vertical posture when surveying the countryside. The fore limbs are powerfully developed, and the killing bite may have been coupled with the immobilization of the prey by the front limbs.

Anatomically, dirk-toothed cats must have been ambush predators and were most likely solitary hunters. Their brains were comparatively small and the olfactory lobes well developed. Radinsky (1975) noted the small brain of *Smilodon floridanus* but missed its significance, as he did not discriminate between scimitar- and dirk-toothed cats.

It seems likely that the saber-toothed cats would attack either the throat or the stomach because these would be the only two areas where vital organs could be reached without much risk to the sabers. The stomach would be an obvious target (Akersten 1985), as it is a large, soft area, richly supplied with blood vessels. A long tearing wound would probably result in death. However, there are good reasons for supposing that it was not the primary area for attack. It is a more easily defended region than the throat, as both the head and hind feet can be used in its defense. A wound in this area does not usually bring a quick death, and unless shock or some other type of immobilization accompanies the attack, inflicting such a wound may involve pursuit or a long struggle. A more conclusive argument against the stomach bite for dirk-toothed cats is the relatively low curvature of the stomach of large ungulates that would prevent a dirk-toothed cat from getting the stomach into its mouth. Even in the most optimistic case (see Akersten 1985), the only damage would be the cutting of strips of skin and superficial tissue. Scimitar-toothed cats may have employed some sort of stomach attack, but I believe that they primarily also used the felid throat bite. In this case, the saber-tooth does not represent a radically new killing method but only an adaptation to prey with larger necks.

Semi-fossorial Mustelid-like Carnivores

Many modern mustelids seek small vertebrates underground either by digging them out (badger-like forms) or by entering the burrows (weasel-like forms). In the late Oligocene and early Miocene similar forms evolved, in other groups.

During the late Oligocene the expansion of open spaces within the savannas reached a point where there was a large radiation of burrowing mammals,

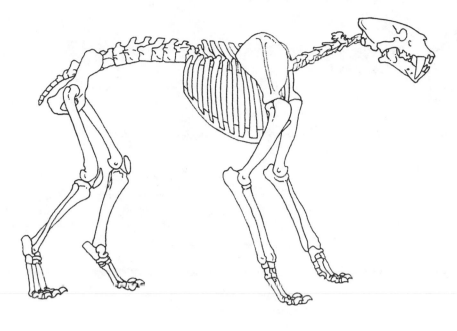

Figure 19.2. Skeletal and life restorations of the scimitar-toothed cat, *Homotherium sainzelli*, from the Pliocene of Europe.

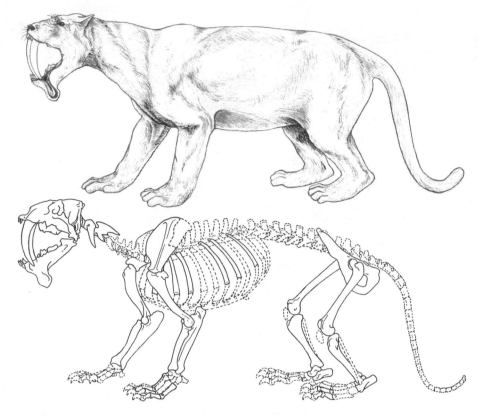

Figure 19.3. Life and skeletal restorations of *Barbourofelis fricki*. From the late Miocene of North America. (From Martin 1980, courtesy Nebraska Academy of Sciences, Inc.)

especially rodents. In North America this included archaic gophers (Entoptychinae) and fossorial beavers (Palaeocastorinae), as well as a radiation of semi-fossorial carnivores. The earliest of these are species of *Palaeogale* (Figure 19.4B), a weasel-like animal with a short face and long cranium. It appears in the Oligocene of both Eurasia and North America, and thus would seem to have a northern origin. It was treated as a mustelid by Matthew (1902) and Simpson (1946), but more recently Flynn and Galiano (1982) allied it with the Feliformia. It has feloid carnassials with a deep carnassial notch on P4 and a reduced talonid on M1. The auditory bullae are completely ossified and highly inflated, but they do not appear to have septa bullae.

The early radiation of mustelid-like carnivores was dominated by species of genera with primitive arctoid ear regions (*Mustelictis, Amphictis*) and putative procyonids (*Plesictis, Broiliana, Stromeriella, Zodiolestes*). Schmidt-Kittler (1981) provided a good discussion of the middle-ear characters that define these groups. The early Miocene genera *Promartes, Oligobunis*, and *Aeluro-*

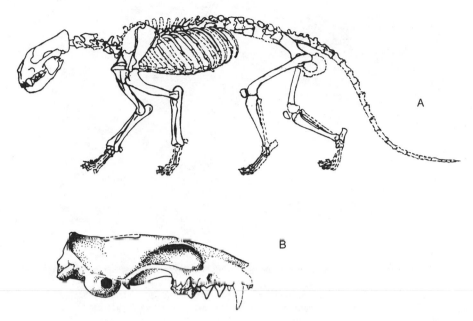

Figure 19.4. *A,* Skeletal restoration of the giant, early Miocene mustelid, *Aelurocyon brevifaces;* original about 39 cm high at the shoulder. (Modified from Riggs 1945, *Fieldiana: Geology,* vol. 9, no. 3, courtesy Field Museum of Natural History, Chicago, Ill.) *B, Palaeogale* sp. skull. (Modified from Matthew 1902.)

cyon seem to have their affinities with this promustelid radiation rather than with the modern Mustelidae.

Aelurocyon brevifaces (Figure 19.4A) from the late Arikareean is about the size of a mountain lion and is thus one of the largest known "mustelids." It was similar to the wolverine in many aspects of its body build and in the bone-crushing strength of its jaws and dentition.

Many of these early musteliform carnivores show digging adaptations. One species, *Zodiolestes daimonelixensis,* was found (Riggs 1945) curled up in the burrow (Figure 19.5) of the Miocene beaver *Palaeocastor* (see Martin and Bennett 1977 for a discussion of the fossil burrows of these beavers). *Palaeocastor* burrows (Figure 19.5) occur in concentrations much like those created by prairie dogs (*Cynomys ludovicianus*). *Zodiolestes daimonelixensis* was very likely a predator of the burrowing beavers, much as the black-footed ferret (*Mustela nigripes*) is on the prairie dog.

Doglike Ecomorphs

The doglike ecomorphs include pursuit-oriented predators. They share elongated muzzles, long cursorially adapted bodies with long tails, and digitigrade

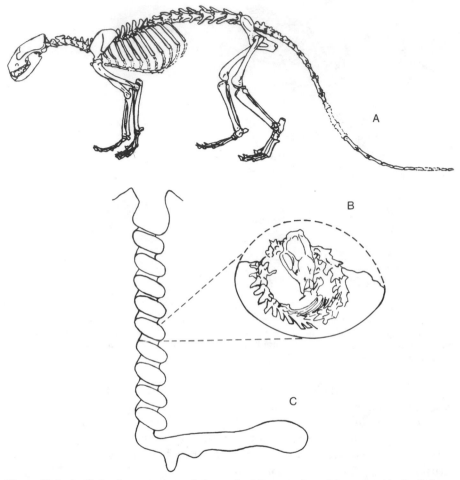

Figure 19.5. *A,* Skeletal restoration of the early Miocene, fossorial procyonid, *Zodiolestes daimonelixensis* (original about 25 cm high at the shoulder). *B,* Coiled skeleton as it was found in the burrow of the extinct fossorial beaver, *Palaeocastor fossor.* *C,* Diagram of a *Palaeocastor fossor* burrow, the trace fossil *Daimonelix* sp. *Daimonelix* burrows reach depths of 3 m. (*A, B* modified from Riggs 1945, *Fieldiana: Geology,* vol. 9, no. 3, courtesy Field Museum of Natural History, Chicago, Ill.; *C* from Martin and Bennett 1977, courtesy *Palaeogeography, Palaeoclimatology & Palaeoecology.*)

feet. The later members develop large brains and in some cases social, "pack" behavior. The earliest clear examples are the hyaenodont creodonts (Figure 19.6C). *Hyaenodon* species show up simultaneously in the early Oligocene deposits of North America and Eurasia. This is the typical signature of a group that has originated at high latitudes. The early Oligocene in North America also provides the first evidence of composite flowers and of grasses, suggesting that the hyaenodonts may have originated in the same community as plant pioneers of open spaces.

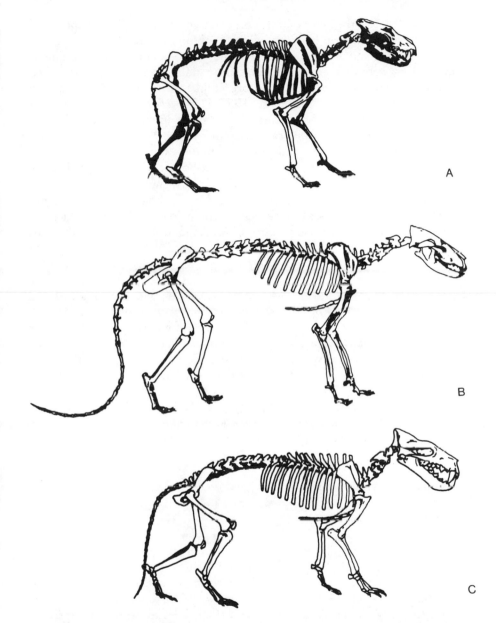

Figure 19.6. Large caniform pursuit predators. *A, Canis dirus,* an extinct true dog from the late Pleistocene of North and South America. *B, Daphoendon* sp., an extinct amphicyonid from the early Miocene of North America. *C, Hyaenodon horridus,* a creodont from the Oligocene of North America. (*B,* modified from Peterson 1910.)

In North America the hyaenodonts come in two sizes. The largest species of *Hyaenodon* is *Hyaenodon horridus*, which is about the size of a large wolf. It has a long muzzle and digitigrade feet, but the brain is small and the cranium is dominated by a huge sagittal crest and potentially bone-crushing jaws. *H. crucians* is a smaller form that was more catlike in its postcranial anatomy (Mellett 1977). According to Mellett (1977:119), "*Hyaenodon* was unquestionably the most highly evolved cursorial predator in the Oligocene."

More modern pursuit carnivora show up in the Oligocene with amphicyonids and hesperocyonine canids. One adaptation sometimes found in pursuit predators is the crushing of marrow-filled bones. Bone crushers include some hyaenodontids, amphicyonids, canids, and hyaenids.

Hyenas are carrion feeders with remarkable cranial adaptations that permit them to utilize dead carcasses more efficiently than do other scavengers. Recent studies show that they are also powerful hunters, and it seems likely that this will also apply to some of the extinct forms that resemble them anatomically.

In the highly specialized *Crocuta crocuta* the face is shortened so that the premolars become reduced in both the upper and lower jaws except for the carnassial in the upper and the p4 in the lower jaw, which is enlarged and reclined posteriorly. The third incisor in the upper jaw tends to be enlarged, and the lower jaw is deep and massive. Often the forehead is elevated above the muzzle, and there is always a very prominent sagittal crest. These same adaptations can be seen in species of the hesperocyonine canid genus *Enhydrocyon* and in the borophagine genus *Borophagus*, which was first considered a hyena.

There is also a fairly well-defined group of large pursuit predators with heavy jaws, including the genera *Hyaenodon*, *Daphoenodon*, *Amphicyon*, *Aelurodon* and the dire wolf, *Canis dirus* (Figure 19.6).

Taxonomic Review

Feliformia

VIVERRAVIDAE

The viverravidae include the oldest known Carnivora (early Paleocene), and they already show the loss of $M^3/_3$ characteristic of the *Feliformia*. They tend to enlarge the parastyle on the upper carnassial, but this is a more variable feature. Recently, Gingerich and Winkler (1985) have illustrated a viverravid basicranial region from the early Eocene of Wyoming. The ear region they illustrate is very primitive and much like that found in nimravids. Gregory and Hellman (1939) thought that viverrids might be directly derived from viverravids, but this now seems unlikely. Viverravids become extinct in the middle Eocene.

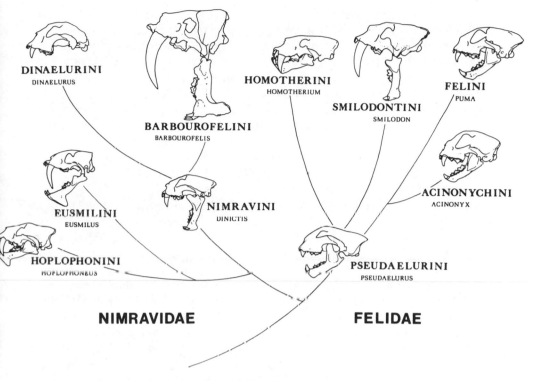

Figure 19.7. Suggested phylogeny of the Feloidea. (From Martin 1980, courtesy Nebraska Academy of Sciences, Inc.)

NIMRAVIDAE

The earliest known nimravids (Figure 19.7) are from the late Eocene or early Oligocene of Eurasia and the early Oligocene of North America. They appeared suddenly with most of their basic adaptations (retractile claws, sectorial carnassials, reduction of the posterior molars, and saber-toothed upper canines) well developed. In North America they are already divided into scimitar-toothed forms of the genus *Dinictis*, with small paramastoid processes, small dependent flanges on the ramus, and moderately long limbs, and dirk-toothed cats belonging to the genera *Hoplophoneus* and *"Eusmilus"* which have large paramastoid processes, large dependent flanges on the ramus, and short limbs. This high degree of specialization and diversification strongly suggests a longer evolutionary history, probably in northern Asia.

Matthew (1910) argued that the Felidae were derived from the ancient saber-toothed cats through the genera *Dinictis* and *Nimravus*. Both of these genera are scimitar-toothed nimravids. Matthew's interpretation has been attacked by numerous authors (De Beaumont 1964; Hunt 1974a) who feel that the nimravid saber-toothed cats differ sufficiently from the felids to suggest that the derived characters shared by the two groups could be accounted for by parallelism. I would agree that the common ancestor of the nimravids and felids must have been more viverravid than cat, and that two families are warranted. I am nevertheless skeptical of schemes that remove the Nimravidae too far from the Felidae because of shared derived characters that seem to unite them (highly sectorial carnassials, retractile and hooded claws, loss of M2–3, and the conformation of the deciduous dentition). The ear and basicranial region found in nimravids is different from that found in felids, but it does not seem to differ greatly from that described for viverravids (Gingerich and Winkler 1986) and may be primitive.

During the Oligocene the dirk-toothed Hoplophoneinae in North America included forms ranging in size from a domestic cat (*Eusmilus cerebralis*) to a jaguar (*Hoplophoneus occidentalis*). They occurred with the generally puma-sized *Dinictis* species. The genus *Dinictis* belongs to the Nimravinae, which is represented in Eurasia by the genus *Nimravus*. By the late Oligocene *Eusmilus* and *Nimravus* species occurred in both Eurasia and North America, but *Hoplophoneus* and *Dinictis* species are restricted to North America. Some species of *Dinictis* gave rise to the genus *Pogonodon*, a leopard-sized scimitar-toothed cat during the Whitneyan, and the genus *Hoplophoneus* became extinct.

One branch of the Nimravinae gave rise to the conical-toothed cat *Dinaelurus crassus* during the Arikareean. *Dinaelurus crassus* has the short face, domed skull, wide zygoma, and enlarged nares found in the cheetah and *Miracinonyx trumani* and seems to be an independent evolution of a cheetah-like cat.

In North America all feliform predators became extinct at the end of the Harrisonian, and feliform predators were absent during the early Hemingfordian.

The Felidae (Figure 19.7) developed in Eurasia during this time, as well as an unusual sidebranch of the Nimravidae, the Barbourofelini. The Barbourofelini entered Africa during the Miocene and America during the late Miocene (Clarendonian).

HERPESTIDAE

The mongooses (Herpestidae) show up in the upper Oligocene of France with excellent material of *Herpestides antiquus* (De Beaumont 1969). In Africa and Asia they filled many of the weasel and ferret niches. Herpestids and viverrids failed to get into North America, which suggests that their center of radiation was at relatively low latitudes, thus making it difficult for them to traverse northern Asia and the Bering land bridge. Mustelids, on the other hand, occur nearly worldwide and probably had a northern center of radiation.

VIVERRIDAE

The modern viverrids are often considered the most primitive living carnivores in their basic morphology, and Gregory and Hellman regarded them as little-modified extensions of some early Tertiary stock. It seems unlikely, however, that this stock would include any of the known early Tertiary Viverravidae (Gingerich and Winkler 1985).

The earliest known viverrids are from the Oligocene of Europe, but it seems likely that they have a long, undiscovered history in Africa and Asia.

HYAENIDAE

The hyenas are derived by most authors from the Viverridae sometime in the Miocene (Kurtén 1968). However, recent work on viverrid relationships (Wozencraft 1984) would seem to indicate a special relationship between cats (Felidae) and the Hyaenidae. If such a relationship should prove to be correct, then the age of the origin of the hyaenid line would have to go back to that of the origin of the cat stem. If we accept the argument that the Felidae arose from certain advanced viverrids (like *Proailurus lemanensis*) in the early or middle Miocene (Tedford 1978), we might expect to find early hyaenids at this time also.

Soon after their inception the hyaenids became doglike, with elongated muzzles, high sagittal crests, and long limbs with digitigrade feet. In the Miocene of Eurasia there are small foxlike forms belonging to the genus *Ictitherium*, and by the Pliocene there were large genera of hunting hyenas like *Chasmaporthetes* in Africa, Eurasia, and North America (Galiano and Frailey 1977; B. Kurtén, pers. comm.). The Villafranchian of Eurasia also contains abundant remains of *Hyaena* species. Later deposits in Europe are dominated by the cave hyena, (*Crocuta crocuta*), and some late Pleistocene caves in England are famous for remains of this species (Kurtén 1968).

FELIDAE

The Felidae (Figure 19.7) have the molar reduction and carnassial specialization found in Nimravids as well as bilaminar septum bullae, enlargment of the orbits (Radinsky 1975), and a cruciate sulcus on the brain. The genus *Pseudaelurus* is the basal stock, and the group quickly diversified into a series of conical- and scimitar-toothed felids ranging in size from a domestic cat ("*Felis*" *stouti*) to the size of a small jaguar (*Pseudaelurus pedionomus*). Many of the larger forms are scimitar-toothed cats, although their canines may lack serrations. However, the North American Hemphillian form *Nimravides catacopis* has coarse posterior serrations on the upper canines. *N. catacopis* is very long-limbed and about the size of an African lion. In Eurasia there was a similar genus of scimitar-toothed felid, *Dinofelis*, which totally lacked serrations on the canines. The range of *Dinofelis* spp. extends into the Villafranchian of Europe, Asia, and Africa, and the Blancan of North America (Kurtén 1973a). Another genus of scimitar-toothed cats is represented by *Machairodus*, which occurs in late Miocene deposits in Europe, Asia, and Africa. It is a large, slender-limbed cat (Figure 19.8) with anterior and posterior serrations on both the upper and lower canines. In North America it has been confused with *N. catacopis* (Martin and Schultz 1975).

The Barbourofelini became extinct throughout the world in the late Miocene. They were replaced by a new dirk-toothed felid lineage, the Smilodentini, in the Villafranchian of Eurasia and Africa and the Blancan of North America (Schultz and Martin 1970; Berta and Galiano 1983).

In North America we have a well-documented lineage leading from *Megantereon hesperus* through ?*Smilodon gracilis* and *S. fatalis* to *S. floridanus*. In this lineage we see the development of most of the progressive dirk-toothed trends, but the dependent flange on the ramus is reduced rather than enlarged, and the occipital region becomes more inclined at the same time (Martin 1984).

Smilodon species entered South America during the middle Pleistocene at about the same time that jaguars are also likely to have entered. Berta (1985) reviewed the North and South American species of *Smilodon* and concluded that all of the putative species from the middle Pleistocene through the Recent are synonyms of the South American form *S. populator*. I am presently unconvinced that this is true because of differences in details of the skull and postcranial skeleton.

Conical-toothed cats became firmly established in the Villafranchian and the Blancan, with *Felis* (*Puma*) and *F.* (*Lynx*) species apparently originating in North America and *Panthera* species in Eurasia. The oldest known *Lynx* (for a contrary view see Werdelin 1985) is *L. rexroadensis* from the late Miocene of Florida (MacFadden and Galiano 1981). A similar form must have entered Europe in the Villafranchian, where we find *L. issidorensis*. Felis (*Puma*) species remained restricted to North and South America, and in the Irvingtonian *Panthera* species entered North America in the form of *P. atrox* and *P. onca augusta* (Kurtén 1973b). Lions (*P. leo*, *P. atrox*, and *P. spelaea*) occurred in Eurasia, Africa, and North and South America. The earliest cheetahs are found in the Villafranchian of Eurasia, but North America had cheetah-like cats (Martin et al. 1977), with *Miracinonyx studeri* in the Blancan and *M. trumani* in the Late Pleistocene. Adams (1979) thought that *Miracinonyx* was really a cheetah, but it may be a parallel ecomorph. Leopards (*Panthera pardus*) also have a long history in Africa, first appearing in Villafranchian deposits. All of

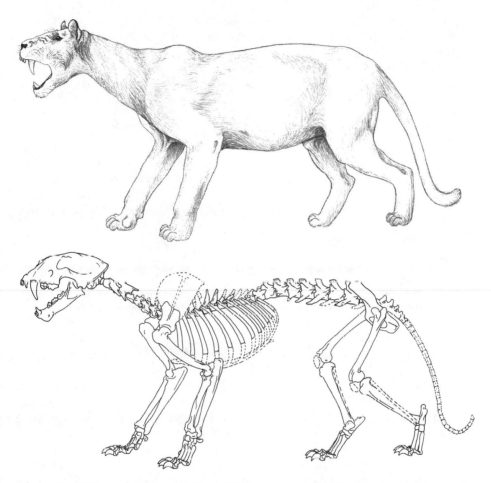

Figure 19.8. Life and skeletal restorations of *Machairodus coloradensis,* from the late Miocene of North America. (From Martin 1980, courtesy Nebraska Academy of Sciences, Inc.)

the saber-toothed cats became extinct at the end of the Pleistocene, but the conical-toothed cats in general suffered only restrictions of their ranges.

Caniformia

MIACIDAE

The removal of the Viverravidae (Wortman and Matthew 1899; Flynn and Galiano 1982) leaves a small collection of Eocene genera (*Uintacyon, Miacis,*

Vulpavus, etc.) that are small viverrid-like caniforms. It is not clear that these genera form a natural group or if they will eventually be dispersed to a variety of taxa. Morphologically they are all very similar and were probably not much different in adaptations from the viverravids.

MUSTELIDAE

All recent mustelids can be characterized by the absence of a distinct carnassial notch on P4. Carnivores of this type show up in the early Miocene of Eurasia and North America. One of the earliest is an undescribed species of *Miomustela*, a weasel-like form from the Hemingfordian of Montana and Wyoming. It is more likely a weasel ecomorph than the actual progenitor of the genus *Mustela*. Early true mustelids include the genera *Plesiogale* and *Paragale* from the Aquitanian of France.

The genus *Mustela* shows up in the Pliocene of Eurasia and North America. The oldest known forms are similar to the modern representatives in size and general anatomy. In North America the black-footed ferret developed an interesting relationship with an endemic North American social rodent, the prairie dog (*Cynomys* spp.). The oldest known black-footed ferret from the middle Pleistocene (Illinoian) of Nebraska already shows this association (Anderson 1973).

The Burdigalian in Europe and the Hemingfordian in North America was an important period of mustelid differentiation (Ginsburg 1961). Otters began to radiate at this time, and there were several large forms that may be related to the honey badgers. There were also marten-like forms and the Leptarctinae, with their short badger-like skulls and double sagittal crests. American badgers (Taxidinae) appeared as *Pliotaxidea* in the late Miocene (Hemphillian). *Pliotaxidea* already shows fossorial postcranial adaptations (Wagner 1976), and there is so little subsequent change in badgers that the Pliocene (Blancan) forms in North America are maintained in the living species, the North American badger (*Taxidea taxus*) (Martin 1984). The extinct wolverine genus *Plesiogulo* had a Holarctic distribution at this time (Harrison 1981), but true wolverines (*Gulo gulo*) do not seem to have gotten into North America until the beginning of the Pleistocene (early Irvingtonian).

Skunks seem to have a relatively short fossil record in North America with late Miocene genera *Pliogale* and *Martinogale* (Harrison 1983) and the modern genera *Mephitis*, *Conepatus*, and *Spilogale* present by the Pliocene (Blancan).

The Galictinae had a significant South American radiation producing two genera, *Galictis* and *Eira*. In the Pliocene (Blancan) of North America we find the genera *Trigonictis* and *Sminthosinus*, and in Eurasia *Enhydrictis* and *Pannonictis*. Ray et al. (1981) suggested that *Trochictis* species may stand close to the ancestry of the group. The last known occurrence of the subfamily in North America is the early Irvingtonian Cumberland Cave Local Fauna.

PROCYONIDAE

After a brief foray into musteliform habitats in the early Miocene, the pro-cyonids restricted their radiation to forms that were mostly arboreal, om-nivorous, and North American. This radiation (Procyoninae) was recently reviewed by Baskin (1982). Unfortunately, the forested habitats of pro-cyonines have not produced many fossils. The real center of their radiation may have been Central America (Baskin 1982), and much of the fossil record is from Florida and Texas. This is especially true in the late Miocene. During the early and middle Miocene, *Bassariscus* and the primitive form *Edaphocyon* occur in northern Nebraska.

Arctonasua is a late Miocene taxon with affinities with *Cyonasua* of the late Miocene of South America (Baskin 1982). *Cyonasua* is usually considered a waif immigrant, as its appearance seems to predate the land connection be-tween North and South America. However, the appearance of ground sloths in numerous North American late Miocene sites suggests a two-way dispersal in the late Miocene, and a better connection than is generally appreciated.

Baskin also describes two additional genera: *Lichnocyon*, thought to be close to *Bassaricyon*, and *Paranasua*. *Nasua* and *Procyon* are considered "sis-ter" genera, and both are close to the genus *Paranasua*. The oldest record of the genus *Procyon* is late Miocene (Hemphillian), and the genus is well known from the Pliocene (Blancan) of Kansas.

During the Oligocene the procyonid adaptive zone was largely occupied by a genera of small arctoids that show affinities with the ursids. In Europe these genera included *Cephalogale* and in North America *Parictis*, *Camplocynodon*, and *Drassonax*.

The red panda (*Ailurus fulgens*) may fall into this general category of primi-tive procyonid-like arctoids, and Ginsburg (1982) has argued that the affini-ties of the red panda is with the ursids. The genus *Sivanasua* from the late Miocene of Europe is considered the earliest Ailurine (Roberts and Gittleman 1984). The genus *Parailurus* had a Holarctic distribution, occurring in the Pliocene of Europe and North America (Tedford and Gustafson 1977). This distribution, coupled with the present distribution of the red panda, supports a central Asiatic place of origin for the Ailurinae.

URSIDAE

During the Oligocene there are a number of carnivores that increased the crushing capacity of the dentition and had an arctoid basicranial pattern. In North America these genera include *Parictis*, *Campylocynodon*, and *Dras-sonax*. In Europe there were amphicynodontines and *Cephalogale*. *Cephalogale* has derived features that show that it is an ursid (De Beaumont 1965). Mitchell and Tedford (1973) include it with the genera *Hemicyon* and *Dinocyon* in the Hemicyoninae. *Cephalogale* is a raccoon-sized carnivore, but the species of *Hemicyon* may be the size of a small brown bear, with a more doglike body. The center of the hemicyonine radiation was in Eurasia, and hemicyonids first appeared in North America in the middle Miocene (Hemingfordian). In general appearance they are more like some of the large borophagine dogs than like a modern bear.

The Miocene genus *Ursavus* makes a fairly good intermediary between the genus *Cephalogale* and the later ursines. The general pattern seems to involve a radiation in Asia and subsequent immigration to North America. During the late Miocene two gigantic forms appear, the genera *Agriotherium* and *Indarctos*, with some species of both reaching the size of the large northern subspecies of the brown bear (*Ursus arctos*). *Agriotherium* species reached Africa and lasted into the late Pliocene. *Indarctos* species may be related to the giant panda (*Ailuropoda melanoleuca*) whose fossil record is restricted to the Pleistocene. Species of both *Agriotherium* and *Indarctos* occurred in North America during the late Miocene.

The genus *Ursus* is known from a species from the late Pliocene of Europe, *Ursus minimus*. A slightly younger form, *U. etruscus*, probably gave rise to the brown bears as well as the cave bear (Kurtén 1968).

The cave bear *(U. spelaeus)* ranged throughout Europe during the Ice Age. It hibernated in caves and often died during hibernation, so that an unusually high percentage of the total cave bear mortality has been preserved (Kurtén 1968). Cave bears were hunted by humans, but they must have been fearsome prey, as their size approached that of the Kodiak bears of Alaska.

The Tremarctine bears are represented by a single living genus, *Tremarctos* (Kurtén, 1966), with no fossil record existing outside of the western hemisphere. Although their first appearance is in the Blancan (late Pliocene) Hagerman and Lisco local faunas (B. Kurtén and L. D. Martin, unpubl. data), they must have originated in northern Asia. The Hagerman Local Fauna contains a spectacled bear, *Tremarctos* sp., and the Lisco Local Fauna a short-faced bear, *Arctodus* sp.

The giant short-faced bear *(Arctodus simus)* seems to have exhibited enormous sexual dimorphism (Kurtén 1967). The males may have been the largest known terrestrial carnivore. Large individuals may have stood 6 feet at the shoulder and have weighed more than a ton. They would have dwarfed the modern giant brown bears and the polar bears (*Ursus maritimus*).

AMPHICYONIDAE

In Eurasia the early doglike forms belong to a separate family, the Amphicyonidae, that is now extinct but was once diverse. At the time that the Canidae were developing in North America, the Amphicyonidae were radiating in Eurasia. The amphicyonids are sometimes referred to as "bear-dogs," and their anatomy tends to support such a description. During most of the Oligocene and Miocene they were the dominant doglike carnivores in Eurasia, although they shared that role with the hyaenodonts. They have in the past been classified with the canids (Simpson 1945; Romer 1966), but their basicranial structure is more arctoid and they have at least one structure (ursid loop) that may ally them directly with the bears (Hunt 1977).

During the Oligocene and in particular the early Oligocene an Amphicyonid-Hyaenodontid radiation occurred in both Eurasia (Ginsburg 1961) and North America. In Eurasia the Amphicyonids include the following genera: *Cynodictis, Cynelos, Harpagophagus, Pseudamphicyon, Pseudocyonopsis, Sarcocyon, Brachycyon, Haplocyon, Amphicyon, Amphicyanis,* and *Symplectocyon* (Savage and Russell 1983). Even if we take into account the possibility of synonomy of names, this is a tremendous variety of amphicyonids, and it shows how completely the caniform adaptive zone was occupied by these animals in Europe. Some catlike forms were also produced, including the genera *Agnotherium, Thaumastocyon,* and *Tomocyon* (Kurtén 1976b). In North America at the same time there was an endemic radiation of the Daphoeninae. Daphoenines lack fully ossified auditory bullae and were all more or less coyote-sized. They include one catlike genus, *Daphoenictis* (Hunt 1974b). The other genera are *Daphoenus* and *Daphoenocyon.* The smaller size of the amphicyonid radiation in North America might be in part due to a competing radiation of nimravid cats. The North American Daphoeninae became extinct during the late Oligocene, and at this time amphicyonines began to show up in North America as immigrants from Asia. One of the most interesting of these Eurasian immigrants is the giant bear-dog *Daphoenodon superbus* (Figure 19.6), which occurred in the early Miocene (Harrisonian) of Nebraska. *D. superbus* dug large subterranean dens that have been found and excavated along with skeletons of the bear-dog itself (Hunt et al. 1983). *D. superbus* was a giant, powerfully jawed pursuit predator that resembled the later canids *Aelurodon* and *Canis (Aenocyon) dirus* (Figure 19.6). During the middle Miocene (Hemingfordian) it was replaced by *Amphicyon.* The Hemingfordian in North America was a time when amphicyonids and mustelids were especially important, and the carnivore fauna was arctoid rather than canoid dominated, as were the later faunas.

CANIDAE

The earliest canid belongs to the genus *Hesperocyon* from the early Oligocene of North America. This was a small foxlike animal with digitigrade feet and a relatively short muzzle. It was probably a good climber but was better suited for running than were any of the Eocene miacids. It was restricted to North America, as are all canids until the late Miocene. *Hesperocyon* species had an inflated, ossified auditory bullae and the distinct horizontal bullar septum characteristic of canids. The upper carnassial, as in most other canids and arctoids, lacks the enlargement of the parastyle characteristic of advanced aeluroids (Flynn and Galiano 1982). During the late Oligocene (Arikareean)

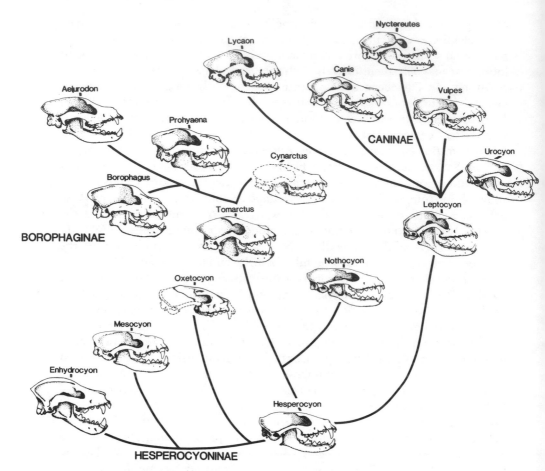

Figure 19.9. Suggested phylogeny (mostly based on Tedford 1978) of the Canidae showing the three major radiations, including three independent origins of the bone-crushing ecomorph genera; *Enhydrocyon, borophagus, Lycaon,* and of the frugivorous ecomorph genera: *Oxetocyon, Cynarctus,* and *Urocyon.*

there was a radiation of these early canids, which may be grouped together as the Hesperocyoninae (Tedford 1978). This radiation (Figure 19.9) included two small frugivorous canids (*Oxetocyon* and *Phlaocyon*), bone-crushing, hyena-like forms (*Sunkahetanka* and *Enhydrocyon*), and a coyote-sized form (*Mesocyon*). The large bone-crushing, hunting "dog" was an amphicyonid (*Daphoenodon*).

At the end of the Arikareean all of the hesperocyonine dogs became extinct except for the branch including a fox-sized genus *Nothocyon*, which gave rise to a new radiation in the Hemingfordian (Tedford 1978). This radiation (Figure 19.9) seems to have been based in the primitive canid genus *Tomarctus*. The typical *Tomarctus* was a short-faced, heavy-jawed canid with the beginnings of parastyle enlargement on the upper carnassial. During the Hemingfordian it was contemporary with a variety of amphicyonid immigrant genera from Eurasia (*Cynelos, Ysengrinia, Temnocyon,* and *Amphicyon*), and these forms filled much of the canid adaptive zone. *Amphicyon* may have ecologically replaced the earlier amphicyonid genus *Daphoenodon*, which was similar in size and overall morphology. The frugivorous niche was filled by a relative of *Tomarctus, Cynarctus,* and the bone-crushing niche may have been filled by a genus of giant mustelid, *Megalictis*. At the end of the Hemingfordian most of the amphicyonids and the large mustelids became extinct, and there was a renewed radiation of *Tomarctus*-related dogs (Borophaginae). This includes the development of hyena-like bone crushers (*Osteoborus* and *Borophagus*), a large bone-crushing hunting dog (*Aelurodon*), and another borophagine frugivorous dog (*Carpocyon*).

Modern canids (Caninae) began to radiate in the late Miocene, and one of the earliest forms is about the size of a small coyote (*Canis davisi*). It is presumably a dog of this type that crossed Beringia and entered Eurasia sometime in the late Miocene. It now seems that this immigrant from North America underwent a major radiation (Figure 19.9), probably in Asia, that resulted in the typical canids that we have in the modern fauna, including wolves, jackals, and hunting dogs. During the Villafranchian in Europe we find several kinds of wolves and a hunting dog (*Cuon*). The fox *Vulpes* and the raccoon dog (*Nyctereutes megamastoides*) also occur in these deposits. Contemporary Blancan deposits in North America show the presence of only an ancestral coyote (*Canis lepophagus*) and an ancestral gray fox (*Urocyon*).

We are thus confronted with a situation where the cradle of canid evolution (North America) seems to have played only a very minor role in the last radiation of the Canidae, and the center of that radiation must have been in Asia soon after the first canids arrived there (Nowak 1979). Wolves, hunting dogs, and the red fox all seem to be immigrants from Asia to North America at the beginning of the Pleistocene (Irvingtonian). This also seems to be about the time that canids first arrived in South America.

The oldest canid from South America is from the early Pleistocene Uquian mammal age (Berta 1981). This is a dhole-like form (*Protocyon* sp.). Somewhat later deposits contain three additional genera (*Canis, Chrysocyon,* and *Theriodictis*). Berta (1981) considers *Protocyon* and *Theriodictis* species to be allied with the indigenous South American fox complex of *Lycalopex, Dusicyon,* and *Pseudalopex* genera. The wolves present seem to be either *Canis dirus* or forms closely related to that species. At the end of the Pleistocene *Protocyon, Theriodictis,* and *Canis* species became extinct in South America, leaving only the maned wolf (*Chrysocyon brachyurus*) and the indigenous foxes.

In the North American Pleistocene there are three distinct wolf lineages, *Canis lupus, C. rufus,* and *C. dirus*. *C. dirus* is the canine version of the bone-crushing hunting dog whose role had been occupied by the borophagine genus,

Aleurodon, in the late Miocene. The coyote (*C. latrans*), and the foxes, *Vulpes* and *Urocyon* species, were also present. *Urocyon* species are somewhat similar to earlier frugivorous canids, belonging to the genus *Cynarctus*. There are also rare remains of hunting dogs, including one from the early Irvingtonian of Rock Creek Texas ("*Protocyon" texanus*), the Illinoian of the Fairbanks area of Alaska (*Cuon* sp.), and the late Pleistocene of San Josecito Cave, Nuevo Leon, Mexico (*Cuon* sp.) (Nowak and Kortlucke, pers. comm.). Tedford (1978:6) briefly suggested that jackals may also have occurred in North America.

Summary and Conclusions

Savage (1977) reported that there are 218 extinct and 98 living genera of Carnivora. This means that the modern fauna may contain nearly one-third of the total known diversity. Only a few carnivoran genera extend back before the Oligocene, and nearly all of these 218 genera are Oligocene or younger. Savage (1977) also counted 45 genera of Creodonta (all extinct) and 29 extinct out of 31 genera of carnivorous marsupials. The Creodonta are mostly Paleocene/Eocene in age, but the marsupials are primarily late Tertiary. It is clear that there has been an overall increase in world carnivore diversity since the early Tertiary.

The cause of the total increase in diversity is probably complex, but two factors may be responsible for a large part of it. One factor is occupation of Africa by carnivorans in the early Miocene and South America by the late Miocene. Faunal interchange with these two great southern continents must have resulted in an increase in diversity for the Carnivora. However, much of that increase was offset by extinction of endemic creodonts and marsupial carnivores. Most of the total increase in diversity may have been a result of establishment of climatic zonation and the appearance of new habitat opportunities.

Digging modifications, along with adaptations for pursuit, became common during the early Miocene. This occurred at the time that grasslands became abundant and represents the change from tropical forests to tropical savannas in middle latitudes (Webb 1977). During the Plio-Pleistocene the tropical savannas became more open, and eventually treeless steppes evolved. In these open savannas or parklands the advantages of herd behavior increased. The development of herd structures by herbivores must be one of the primary reasons for the development of social hunting by carnivores. The modern social hunting taxa all developed during the Plio-Pleistocene, and there is no convincing evidence for social or pack carnivores in earlier times.

Radinsky (1973) described an enlargement of the prorean gyrus on the frontal lobe of the modern canid brain associated with social pack hunting. He claimed that the enlargement inhibited the fight-or-flight response of solitary animals. This enlargement begins to be evident in endocranial casts of Miocene genus *Leptocyon* and is much enlarged in all modern dogs except the foxes. We may thus have a marker indicating the beginning of a behavioral pattern in the late Miocene that characterizes the Caninae during the Plio-Pleistocene.

Arctic adaptations seem to be very young, and the oldest tundra-adapted faunas from Alaska and Canada are probably less than two million years old. Very little of the arctic fauna is restricted to the Arctic, and the more specialized forms like the Tundra muskox, and the collared lemming did not reach middle latitudes until the late Pleistocene. It seems likely that the arctic fox and the polar bear are the youngest distinct carnivoran forms that have been recognized (Kurtén, 1964).

The other great pattern in carnivoran evolution is the cycle of extinction and reevolution of adaptive types. Almost all of the Recent Carnivora are the result of radiations that have taken place during the last seven million years. This is certainly true for the modern Felinae and Caninae. It is probably also true for ursids and mustelids. Earlier radiations almost without exception have resulted in extinct lineages.

In some ways, the prevalence of extinction is understandable. A wide variety of factors may contribute to the demise of a lineage. The truly startling aspect of the fossil record is the reappearance in great detail of adaptive types after their extinction. For instance, the earliest coyote-like predators are not the progenitors of modern coyotes and are not even canids, but belong to the extinct Amphicyonidae. The earliest musteliform carnivores are primitive arctoids as well as a possible Feliform. Modern cats are not closely related to the Oligocene nimravids, all of which are now extinct, and most of the early feline radiation was not ancestral to modern forms.

There must be adaptive zones for carnivores that disappear and then reappear. When they disappear there is extinction, and when they reappear there is reevolution. Since the beginning of the Oligocene we have had five repetitions of this pattern. Dirk-toothed cats have developed some four separate times (Figure 19.10) and large bone-crushing pursuit predators (Figures 19.6, 19.9) at least five times. Medium-sized bone-crushers have developed several times, including the genera, *Enhydrocyon* and *Borophagus* (Figure 19.9). Procyonid-like omnivore genera *Parictis*, *Oxetocyon*, *Phlaocyon*, *Cynarctus*, *Carpocyon*, and the procyonid, *Procyon*, are additional examples of the multiple evolution of ecomorphs. Many more examples can be enumerated, and it is clear that this has been a basic pattern in carnivore evolution.

The similarities between ecomorphs generated by repetitive extinctions and evolutions are extensive and profound, as there appears to be only a small number of solutions to each adaptive problem. They also support the primacy of selection over random processes and complicate the higher taxonomy of carnivorous mammals. Some relationships can be traced with fossils, and in a few cases classical transformation series are present. Unfortunately, the connecting links are often missing, because many early radiations took place at

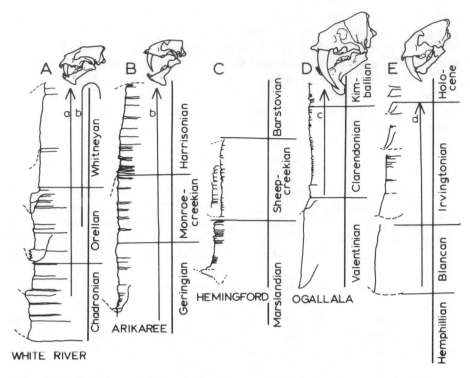

Figure 19.10. Durations of dirk-toothed cat lineages in North America: *A, Hoplophoneus; B, Eusmilus; D, Barbourofelis; E, Smilodon*. This adaptation has evolved and become extinct at least four separate times in the last 38 million years. (From Martin 1985, courtesy Nebraska Academy of Sciences, Inc.)

high latitudes and the critical fossils lie unrecovered in northern Asia and North America.

Acknowledgments

I have benefited from conversations with R. H. Tedford, C. C. Black, R. M. Hunt, L. Tanner, R. S. Hoffmann, W. C. Wozencraft, B. Kurten, E. Anderson, and C. B. Schultz, and I am grateful to the following for making specimens of carnivores available to me: C. B. Schultz and M. Voorhies, University of Nebraska State Museum; M. Green and P. Bjork, South Dakota School of Mines and Technology; J. Ostrom and M. A. Turner, Yale Peabody Museum; and R. H. Tedford, American Museum of Natural History. The illustrations are by J. B. Martin, D. Adams, M. Brooks, and C. Gregoire. The paper was critically read by R. W. Wilson, G. Ostrander, and L. Witmer.

References

Adams, D. B. 1979. The cheetah: Native American. *Science* 205:1155–1158.

Akersten, W. A. 1985. Canine function in *Smilodon* (Mammalia; Felidae; Machairo-dontinae). *Contrib. Sci., Nat. Hist. Mus. Los Angeles County.* 356:1–22.

Anderson, E. 1973. Ferret from the Pleistocene of central Alaska. *J. Mamm.* 54(3):778–779.

Baskin, J. A. 1982. Tertiary Procyoninae (Mammalia: Carnivora) of North America. *J. Vert. Paleontol.* 2(1):71–93.

Berta, A. 1981. Evolution of large canids in South America. *Anais II Congresso Latino-Americano de Paleontologia, Porto Alegre* 2:835–845.

Berta, A. 1985. The status of *Smilodon* in North and South America. *Contrib. Sci., Nat. Hist. Mus. Los Angeles County.* 370:1–15.

Berta, A., and Galiano, H. 1983. *Megantereon hesperus* from the Late Hemphillian of Florida with remarks on the phylogenetic relationships of machairodonts (Mammalia, Felidae, Machairodontinae). *J. Paleontol.* 57:892–899.

Cope, E. D. 1880. On the extinct cats of North America. *Amer. Nat.* 14:833–858.

Dawson, M. R., West, R. M., Langston, W. Jr., and Hutchison, J. H. 1976. Paleogene terrestrial vertebrates: Northernmost occurrence, Ellesmere Island, Canada. *Science* 192:781–782.

De Beaumont, G. 1964. Remarques sur la classification des Felidae. *Eclogue Geologicae Helvetiae* 57(2):837–845.

De Beaumont, G. 1965. Note sur la région auditive de quelques carnivores. *Arch. Sci., Genève* 21:213–224.

De Beaumont, G. 1969. Observations sur les Herpestinae (Viverridae, Carnivora) de l'Oligocène supérieur avec quelques remarques sur des Hyaenidae du Néogène. Première partie. *Arch. Sci., Genève* 20:179–107.

Eisenberg, J., and Leyhausen, P. 1972. The phylogenesis of killing behavior in mammals. *Z. Tierpsychol.* 30:59–93.

Ewer, R. F. 1973. *The Carnivores.* Ithaca, N.Y.: Cornell Univ. Press.

Flower, W. H. 1869. On the value of the characters of the base of the cranium in the classification of the Order Carnivora and the systematic position of *Bassaris* and other disputed forms. *Proc. Zool. Soc. London* 1869:4–37.

Flynn, J. M., and Galiano, H. 1982. Phylogeny of early Tertiary Carnivora, with a description of a new species of *Protictis* from the middle Eocene of Northwestern Wyoming. *Amer. Mus. Novitates* 2725:1–64.

Galiano, H., and Frailey, D. 1977. *Chasmaporthetes kani*, new species from China, with remarks on phylogenetic relationships of genera within the Hyaenidae (Mammalia: Carnivora). *Amer. Mus. Novitates* 2632:1–16.

Gingerich, P. D., and Winkler, D. A. 1985. Systematics of Paleocene Viverravidae (Mammalia, Carnivora) in the Bighorn Basin and Clark's Fork Basin, Wyoming. *Contrib. Mus. Paleontol. Univ. Michigan* 27:87–128.

Ginsburg, L. 1961. La faune des carnivores Miocènes de Sansan (Gers) *Mémoires du Musée National Histoire Naturelle* 9:1–190.

Ginsburg, L. 1966. Les amphicyons des phosphorites du Quercy. *Ann. Paleontol.* 52:23–64.

Ginsburg, L. 1982. Sur la position systematique du petit panda, *Ailurus fulgens* (Carnivora, Mammalia). *Geobios, mémoire spécial* 6:259–277.

Glass, G. E., and Martin, L. D. 1978. A multivariate comparison of some extant and fossil Felidae, Carnivora. *Carnivore* 1:80–88.

Gregory, W. K., and Hellman, M. 1939. On the evolution and major classification of civets (Viverridae) and allied fossil and recent Carnivora: A phylogenetic study of the skull and dentition. *Proc. Amer. Philos. Soc.* 81:309–392.

Harrison, J. A. 1981. A review of the extinct wolverine, *Plesiogulo* (Carnivora : Mustelidae), from North America. *Smith. Contrib. Paleobiol.* 46:1-27.

Harrison, J. A. 1983. The Carnivora of the Edson Local Fauna (Late Hemphillian), Kansas. *Smithsonian Contrib. Paleobiol.* 54:1–42.

Hunt, R. M. 1974a. The auditory bulla in Carnivora: An anatomical basis for reappraisal of carnivore evolution. *J. Morphol.* 143:21–76.

Hunt, R. M. 1974b. *Daphoenictis*, a cat-like carnivore (Mammalia, Amphicyonidae) from the Oligocene of North America. *J. Paleontol.* 48:1030-1047.

Hunt, R. M. 1977. Basicranial anatomy of *Cynelos* Jourdan (Mammalia: Carnivora), an Aquitanian amphicyonid from the Allier Basin, France. *J. Paleontol.* 54:826–843.

Hunt, R. M., Xue, X. X., and Kaufman, J. 1983. Miocene burrows of extinct beardogs: Indication of early denning behavior of large mammalian carnivores. *Science* 221:364–366.

Kurtén, B. 1963. Notes on some Pleistocene mammal migrations from the Palearctic to the Nearctic. *Eiszeitalter und Gegenwart* 14:96–103.

Kurtén, B. 1964. The evolution of the polar bear, *Ursus maritimus* Phipps. *Acta Zool. Fennica* 108:1–26.

Kurtén, B. 1966. Pleistocene bears of North America, I: Genus *Tremarctos*, spectacled bears. *Acta Zool. Fennica* 115:1–120.

Kurtén, B. 1967. Pleistocene bears of North America, II: Genus *Arctodus*, short-faced bears. *Acta Zool. Fennica* 117:1–60.

Kurtén, B. 1968. *Pleistocene Mammals of Europe*. London: Weidenfeld and Nicolson.

Kurtén, B. 1973a. The genus *Dinofelis* (Carnivora, Mammalia) in the Blancan of North America. *Pearce-Sellards Series Texas Memorial Museum* 19:1–7.

Kurtén, B. 1973b. Pleistocene jaguars in North America. *Comment. Biol. Soc. Sci. Fennica* 62:1–23.

Kurtén, B. 1976a. *The Cave Bear Story, Life and Death of a Vanished Animal.* New York: Columbia Univ.

Kurtén, B. 1976b. Fossil carnivora from the Late Tertiary of Bled Douarah and Cherichra, Tunisia. *Notes du Service Géologique. Tunisie* 42:177-214.

Kurtén, B., and Martin, L. D. 1988. The first appearance of the Tremarctine bears. Unpublished data.

Leyhausen, P. 1979. *Cat Behavior*. New York:Garland Press.

Lillegraven, J. A. 1969. Latest Cretaceous mammals of upper part of Edmonton Formation of Alberta, Canada, and review of marsupial-placental dichotomy in mammalian evolution. *Univ. Kansas Paleontol. Contrib.* 50:1–122.

MacFadden, B. J., and Galiano, H. 1981. Late Hemphillian cat (Mammalia : Felidae) from the Bone Valley Formation of central Florida. *J. Paleontol.* 55:218–226.

McKenna, M. C. 1975. Fossil mammals and early Eocene North Atlantic land continuity. *Ann. Missouri Bot. Garden* 62(2):335–353.

Martin, L. D. 1980. Functional morphology and the evolution of cats. *Trans. Nebraska Acad. Sci.* 7:141–154.

Martin, L. D. 1983. The origin and early radiation of birds. In: A. H. Brush & G. A. Clark, Jr., eds. *Perspectives in Ornithology*. Cambridge: Cambridge Univ. Press.

Martin, L. D. 1984. Phyletic trends and evolutionary rates. *Carnegie Mus. Nat. Hist. Special Publications* 8:526–538.

Martin, L. D. 1985. Tertiary extinction cycles and the Pliocene-Pleistocene boundary. *Institute for Tertiary-Quaternary Studies (TER-QUA) Symposium Series.* 1:33–40.

Martin, L. D., and Bennett, D. K. 1977. The burrows of the Miocene beaver *Palaeocas-*

tor western Nebraska, U.S.A. *Palaeogeography, Palaeoclimatology, Palaeoecology* 22:173–193.

Martin, L. D., Gilbert, B. M., and Adams, D. B. 1977. A cheetah-like cat in the North American Pleistocene. *Science* 165:981–982.

Martin, L. D., Rogers, R. A., and Neuner, A. M. 1985. The effect of the end of the Pleistocene on man in North America. In: J. I. Mead & D. J. Meltzer, eds., *Environments and Extinctions: Man in Late Glacial North America* pp. 15–30. Orono, Maine: Center for the Study of Early Man.

Martin, L. D., and Schultz, C. B. 1975. Cenozoic mammals from the Central Great Plains. Part 5, Scimitar-toothed cats, *Machairodus* and *Nimravides*, from the Pliocene of Kansas and Nebraska. *Bull. Univ. Nebraska State Mus.* 10:55–63.

Matthew, W. D. 1902. On the skull of *Bunaelurus*, a musteline of the White River Oligocene. *Bull. Amer. Mus. Nat. Hist.* 16:137–140.

Matthew, W. D. 1910. The phylogeny of the Felidae. *Bull. Amer. Mus. Nat. Hist.* 28:289–316.

Mellett, J. 1977. Paleobiology of North American *Hyaenodon* (Mammalia, Creodonta). *Contrib. Vert. Evol.* 1:–134.

Mitchell, E. D., and Tedford, R. H. 1973. The Enaliarctinae, a new group of extinct aquatic Carnivora and a consideration of the origin of the Otariidae. *Bull. Amer. Mus. Nat. Hist.* 151:201–284.

Nowak, R. M. 1979. North American Quaternary *Canis* Monogr., Mus. Nat. Hist., Univ. Kansas 6:1–154.

Peterson, O. A. 1910. Description of new carnivores from the Miocene of Western Nebraska. *Mem. Carnegie Mus.* 4(5):205–278.

Radinsky, L. 1973. Evolution of the canid brain. *Brain, Behavior, Evolution* 7:169–202.

Radinsky, L. 1975. Evolution of the felid brain. *Brain, Behavior, Evolution* 11:214–254.

Radinsky, L. 1977. Brains of early carnivores. *Paleobiology* 3:333–349.

Ray, C. E., Anderson, E., and Webb, S. D. 1981. The Blancan carnivore *Trigonictis* (Mammalia: Mustelidae) in the Eastern United States. *Brimleyana* 5:1–36

Riggs, E. S. 1945. Some early Miocene Carnivores. *Geol. Series Field Mus. Nat. Hist.* 9:69–114

Roberts, M. S., and Gittleman, J. L. 1984. *Ailurus fulgens. Mamm. Species* 222:1–8

Romer, A. S. 1966. *Vertebrate Paleontology* (3rd ed.) Chicago: Univ. Chicago Press.

Savage, D. E., and Russell, D. E. 1983. *Mammalian Paleofaunas of the World* London: Addison-Wesley.

Savage, R. J. G. 1977. Evolution in carnivorous mammals. *Palaeontology* 20:237–271

Schmidt-Kittler, N. 1981. Zur Stammesgeschichte der marderverwandten Raubtiergruppen (Musteloidea, Carnivora). *Eclogae Geologicae Helvetiae* 74:753–801

Schultz, C. B., and Martin, L. D. 1970. Machairodont cats from the Early Pleistocene Broadwater and Lisco local faunas. *Bull. Univ. Nebraska State Mus.* 9:1–36.

Schultz, C. B., and Martin, L. D. 1972. Two lynx-like cats from the Pliocene and Pleistocene. *Bull. Univ. Nebraska State Mus.* 9:197–203

Schultz, C. B., Martin, L. D., and Schultz, M. R. 1985. A Pleistocene Jaguar from north-central Nebraska. *Trans. Nebraska Acad. Sci.* 13:93–98.

Schultz, C. B., Schultz, M. R., and Martin, L. D. 1970. A new tribe of saber-toothed cats (Barbourofelini) from the Pliocene of North America. *Bull. Univ. Nebraska State Mus.* 9:1–31

Simpson, G. G. 1945. The principles of classification and a classification of mammals. *Bull. Amer. Mus. Nat. Hist.* 85:1–350.

Simpson, G. G. 1946. *Palaeogale* and allied early mustelids. *Amer. Mus. Novitates* 1320:1–14.

Tedford, R. H. 1978. History of dogs and cats: A view from the fossil record. In: *Nutrition and Management of Dogs and Cats*, pp. 1–10. St. Louis: Ralston Purina Co.

Tedford, R. H., and Gustafson, E. P. 1977. First North American record of the extinct *Parailurus*. *Nature* 265:621–23.

Van Valkenburgh, B. 1985. Locomotor diversity in past and present guilds of large predator mammals. *Paleobiology* 11:406–428.

Wagner, H. 1976. A New Species of *Pliotaxidea* (Mustelidae : Carnivora) from California. *J. Paleontol.* 50:107–127.

Webb, S. D. 1977. A history of savanna vertebrates in the New World. I. North America. *Ann. Rev. Ecol. Syst.* 8:355–380.

Werdelin, L. 1985. Small Pleistocene felines of North America. *J. Vert. Paleontol.* 5:194–210.

White, J. A. 1984. Late Cenozoic Leporidae (Mammalia, Lagomorpha) from the Anza-Borrego Desert, Southern California. *Carnegie Mus. Nat. Hist. Special Publications* 9:41–57.

Wilson, R. W. 1962. The significance of the geological succession of organic beings: 1859–1959. *Univ. Kansas Sci. Bull.* 42:157–178.

Wortman, J. L., and Matthew, W. D. 1899. The ancestry of certain members of the Canidae, the Viverridae, and Procyonidae. *Bull. Amer. Mus. Nat. Hist.* 12:109–139.

Wozencraft, W. C. 1984. A phylogenetic reappraisal of the Viverridae and its relationship to other Carnivora. Ph.D. dissert., Univ. Kansas, Lawrence. 1108 pages.

Classification of
the Recent Carnivora

W. Chris Wozencraft

The following list of species is taken from Honacki et al. (1982). I have compared it with the classifications of Ewer (1973) and Corbet and Hill (1986); these references should be consulted for distributions of Recent taxa and for further taxonomic information. Although Honacki et al. (1982) is used as a basis for discussion, this does not imply agreement with their taxonomy. Their classification was assembled from primary literature through 1980 by several authors who did not necessarily concur on the final taxonomic arrangement. The scheme proposed by Honacki et al. is used here as a yardstick with which additional information is compared at the end of each family list.

Ewer (1973) considered the Otariidae, Phocidae, and the genus *Odobenus* to constitute a separate order, the Pinnipedia, although she stated that it was "customary to classify them as a suborder within the Carnivora" (p. 6). The inclusion of these families within the Carnivora is well established (Tedford 1976; Wozencraft, this volume). Common English names are consistent with Ewer (1973), King (1983), MacDonald (1984), and Corbet and Hill (1986). Families are listed in a phylogenetic order consistent with the hypothesis present in Wozencraft (this volume). Species are presented in alphabetical order, grouped by subfamily; where subfamilial classification is uncertain, the taxa are included at the end of the list. In some families (Canidae, Procyonidae, Ursidae, Phocidae) either the branching sequence or lack of consensus in the recent literature does not warrant subfamilial classifications at this time. Species names preceded by an asterisk (*) are discussed in the remarks for each family.

Order Carnivora Bowdich 1821

Family Herpestidae Bonaparte 1845

Subfamily Galiidinae Gray 1864

Galidia elegans I. Geoffroy 1837, ring-tailed mongoose.

Galidictis fasciata (Gmelin 1788), broad-striped mongoose. Includes *G. ornata* and *G. striata* listed separately by Ewer (1973).
Galidictis grandidieri Wozencraft 1986, broad-striped mongoose. Described since Honacki et al. (1982).
Mungotictis decemlineata (A. Grandidier 1867), narrow-striped mongoose. Includes *Mungoictis [sic] substriata* listed separately by Ewer (1973).
*Salanoia concolor (I. Geoffroy 1837), Salano or brown mongoose. Includes *S. olivacea* listed separately by Ewer (1973). *S. unicolor* (listed by Ewer) is a junior synonym.

Subfamily Herpestinae Bonaparte 1845

Atilax paludinosus (G. Cuvier 1829), marsh mongoose.
Bdeogale crassicauda Peters 1850, bushy-tailed mongoose.
*Bdeogale jacksoni (Thomas 1894), Jackson's mongoose.
*Bdeogale nigripes Pucheran 1855, black-legged mongoose.
*Herpestes auropunctatus (Hodgson 1836), small Indian mongoose. Includes *H. palustris*.
*Herpestes brachyurus Gray 1837, short-tailed mongoose.
*Herpestes edwardsi (E. Geoffroy 1818), Indian gray mongoose.
*Herpestes fuscus Waterhouse 1838, Indian brown mongoose.
*Herpestes hosei Jentink 1903, Hose's mongoose.
Herpestes ichneumon (Linnaeus 1758), Egyptian mongoose.
*Herpestes javanicus (E. Geoffroy 1818), Javan mongoose.
Herpestes naso De Winton 1901, long-nosed mongoose.
*Herpestes pulverulentus (Wagner 1839), Cape gray mongoose.
*Herpestes sanguineus (Rüppell, 1835), slender mongoose.
*Herpestes semitorquatus Gray 1846, collared mongoose.
*Herpestes smithii Gray 1837, ruddy mongoose.
*Herpestes urva (Hodgson 1836), crab-eating mongoose.
*Herpestes vitticollis Bennett 1835, stripe-necked mongoose.
Ichneumia albicauda (G. Cuvier 1829), white-tailed mongoose.
Rhynchogale melleri (Gray 1865), Meller's mongoose.

Subfamily Mungotinae Gray 1864

Crossarchus alexandri Thomas 1910, Congo kusimanse.
Crossarchus ansorgei Thomas 1910, Angolan mongoose or kusimanse.
*Crossarchus obscurus F. Cuvier 1825, common kusimanse. Includes *C. platycephalus* listed separately by Goldman (1984).
Cynictis penicillata (G. Cuvier 1829), yellow mongoose.
*Dologale dybowskii (Pousargues 1893), Pousargues' mongoose.
*Helogale hirtula Thomas 1904, dwarf mongoose.
*Helogale parvula (Sundevall 1846), dwarf mongoose.

Liberiictis kuhni Hayman 1958, Liberian mongoose.
Mungos gambianus (Ogilby 1835), Gambian mongoose.
Mungos mungo (Gmelin 1788), banded mongoose.
Paracynictis selousi (De Winton 1896), Selous' mongoose or gray meerkat.
Suricata suricatta (Erxleben 1777), suricate or meerkat.

The stability in mongoose classification between Ewer (1973) and Honacki et al. (1982) is more a reflection of the lack of systematic studies than agreement with Ewer's classification. Although Ewer considered the mongooses part of the Viverridae, Honacki et al. (1982) and Corbet and Hill (1986) list them separately. The division between the Malagasy mongooses (Galidiinae) and the remaining two subfamilies of herpestids (Herpestinae, Mungotinae) was proposed by Gray (1865) and supported by Pocock (1919) and Petter (1974). There are five species of Malagasy mongooses, one of which, *G. grandidieri* (emendation of *G. grandidiensis,* see Wozencraft 1987) was described recently (Wozencraft 1986).

The Mungotinae is the most diverse mongoose subfamily and differs from the Herpestinae in having a derived dentition and auditory bullae. Relationships among the mungotines are poorly understood. The nature of the character differences between *Dologale* and *Helogale* species suggest that they are congeneric. The range of variation in skull measurements of the widely distributed *Helogale parvula* is inclusive of conditions found within *H. hirtula,* whose distribution is completely included within that of *parvula* (Kingdon 1977); more work needs to be done in sympatric areas. *Crossarchus obscurus* has been divided into two species by Goldman (1984), represented by two allopatric populations separated by the Dahomey Gap, a phenomenon documented for several other allopatric mammals (Booth 1954, 1958; Robbins 1978). Goldman assigned *Crossarchus* populations east of the Dahomey Gap to *C. platycephalus,* despite a 75–80% overlap with *C. obscurus* in all measured variables. Furthermore, this gap in the west African high forest zone has been attributed to human habitat modifications (Carleton and Robbins 1985). Therefore, these taxa of *Crossarchus* are recognized here as subspecies.

The Herpestinae has two monophyletic species clusters, the *Herpestes-Atilax* clade and the *Ichneumia-Bdeogale-Rhynchogale* clade. The genus *Herpestes* as classified by Honacki et al. (1982) contains 14 species. The slender (*sanguineus*) and Cape gray (*pulverulentus*) mongooses have been placed in a separate genus, *Galerella,* on the basis primarily of the absence of a lower first premolar in the adults of most specimens and the greater inflation of the auditory bullae (Rosevear 1974). Similarities between other African and Asiatic mongooses would make *Herpestes* a paraphyletic taxon with many striking convergences if *Galerella* were recognized; therefore, these species are presently maintained within *Herpestes.* Watson and Dippenaar (1987) recognize a third South African slender mongoose, *H. nigratus,* on the basis of a morphometric study of skull measurements. The Asiatic mongooses are represented by two monophyletic groups, the subgenus Urva (*urva, semitorquatus, vitticollis, brachyurus, hosei, fuscus*) and the subgenus Herpestes (*javanicus, auropunctatus, edwardsi, smithii*). Bechthold (1939), in a phenetic study of the Asiatic mongooses, considered *javanicus-auropunctatus, vitticollis-semitorquatus,* and *brachyurus-hosei-fuscus* as conspecific groups. Honacki et al. (1982) listed three species of *Bdeogale,* although Kingdon (1977) considered *B. nigripes* and *B. jacksoni* to be conspecific. Characters previously used to separate these taxa vary with the age of the animal and habitat (Rosevear 1974), and they should be considered conspecific.

Family Viverridae Gray 1821

INCERTAE SEDIS

*Nandinia binotata (Reinhardt 1830), African palm civet or two-spotted palm civet.

SUBFAMILY CRYPTOPROCTINAE GRAY 1864

*Cryptoprocta ferox Bennett 1833, fossa.
Eupleres goudoti Doyere 1835, falanouc. Includes *E. major* listed separately by Ewer (1973).
Fossa fossana P. L. S. Müller 1776, fanaloka or Malagasy civet.

SUBFAMILY VIVERRINAE GRAY 1864

*Civettictis civetta (Schreber 1777), African civet. Corbet and Hill (1986) included this species in *Viverra.*
Genetta abyssinica (Rüppell 1836), Abyssinian genet.
Genetta angolensis Bocage 1882, Angolan genet.
*Genetta felina (Thunberg 1811), South African small-spotted genet. Ewer (1973) and Corbet and Hill (1986) considered this a junior synonym of *G. genetta.*
*Genetta genetta (Linnaeus 1758), European genet or small-spotted genet.
Genetta johnstoni Pocock 1908, Johnston's genet.
*Genetta maculata (Gray 1830), large-spotted genet. Includes *G. rubiginosa* listed separately by Corbet and Hill (1986), and *G. pardina* and *G. tigrina* listed separately by Ewer (1973).
*Genetta servalina Pucheran 1855, servaline genet. Includes *G. cristata* listed separately by Corbet and Hill (1986).
Genetta thierryi Matschie 1902, Thierry's genet. *G. thierryi* is the senior synonym of *G. villiersi* (Dekeyser 1949) listed by Ewer (1973).
*Genetta tigrina (Schreber 1776), Cape genet or large-spotted genet.
Genetta victoriae Thomas 1901, giant forest genet.
Osbornictis piscivora J. A. Allen 1919, aquatic genet or Congo water civet.

Poiana richardsoni (Thomson 1842), African linsang or oyan.
Prionodon linsang (Hardwicke 1821), banded linsang.
Prionodon pardicolor Hodgson 1842, spotted linsang.
* Viverra megaspila Blyth 1862, large-spotted civet.
Viverra tangalunga Gray 1832, tangalung civet or Malay civet.
Viverra zibetha Linnaeus 1758, large Indian civet.
Viverricula indica (Desmarest 1817), small Indian civet.

SUBFAMILY HEMIGALINAE GRAY 1864

Chrotogale owstoni Thomas 1912, Owston's banded civet.
Cynogale bennettii Gray 1837, otter civet. Includes *C. Lowei.*
Hemigalus derbyanus (Gray 1837), banded palm civet.
*Hemigalus hosei (Thomas 1892), Hose's palm civet.

SUBFAMILY PARADOXURINAE GRAY 1864

Arctictis binturong (Raffles 1821), binturong.
Arctogalidia trivirgata (Gray 1832), small-toothed palm civet or three-striped palm civet.
Macrogalidia musschenbroekii (Schlegel 1879), celebes palm civet or brown palm civet.
Paguma larvata (H. Smith 1827), masked palm civet.
Paradoxurus hermaphroditus (Pallas 1777), common palm civet.
Paradoxurus jerdoni Blanford 1885, Jerdon's palm civet.
Paradoxurus zeylonensis (Pallas 1778), golden palm civet or Ceylon palm civet.

The genus *Cryptoprocta* was considered by Beaumont (1964) and Hemmer (1978) as an early offshoot of the felid stem on the basis primarily of dentition. However, features of the skull and the postcranial and soft anatomy clearly align the Malagasy fossa with the civets (Petter 1974; Köhncke and Leonhardt 1986; Laborde 1986; Wozencraft, this volume). The genera *Fossa* and *Eupleres* were placed in the Hemigalinae by Pocock (1915c) because of dental similarities, but derived cranial characters align these taxa with *Cryptoprocta* (Petter 1974; Wozencraft 1984).

The genus *Nandinia* was originally placed by Gray (1865) and Pocock (1915b) in the Paradoxurinae on the basis of similarities in the scent gland and the superficial resemblance of the feet. Pocock (1929) later reversed this position and assigned the African palm civet to the Viverrinae. Hunt (1987) argued that *Nandinia* should be separate from other families of feloids on the basis of the primitive nature of the basicranial region. Although the analysis presented by Wozencraft (this volume) places this genus within the Viverridae, its subfamilial status is uncertain.

Three major revisions of the genets (Schlawe 1980, 1981; Crawford-Cabral

1981a, 1981b; Wozencraft 1984) have occurred since Ewer's plea for work on this taxon. These studies have agreed on the existence of at least nine genet species. Schlawe tentatively suggested, and Crawford-Cabral and Wozencraft concurred, that *G. felina* is conspecific with *G. genetta*. Rosevear (1974) listed *G. cristata* Hayman 1940 as a species distinct from the *servalina* group. Crawford-Cabral (1981a) recognizes *cristata* whereas Schlawe, in Honacki et al. (1982), and Wozencraft (1984) included this taxon as a subspecies of *servalina*. Schlawe and Wozencraft recognized only one polymorphic species of large-spotted genet (*G. maculata*), whereas Crawford-Cabral separated West African (*G. pardina*) from central and east African forms (*G. rubiginosa*). Allopatric speciation problems are difficult to resolve; however, elsewhere Crawford-Cabral (1981b), Schlawe (1981), and Wozencraft (1984) agree that allopatric populations of *G. genetta* in western and eastern Africa are conspecific. As with *Crossarchus obscurus,* these taxa will be recognized here as subspecies to maintain a consistency in the rationale used in taxonomic decisions. All agreed that *G. tigrina* is restricted to the Cape population. The Zambia-Zimbabwe-Mozambique region appears especially troublesome in discerning the relationships among the large spotted forms (*tigrina, angolensis, maculata*).

The large-spotted civet, *Viverra megaspila* (*sensu* Honacki et al. 1982), has two disjunct populations, one in southern India and another in the southern regions of Indochina and the Malay penninsula. My examination of specimens at the British Museum (Natural History) confirms Pocock's (1933) placement of the south Indian form as a separate species (*V. civettina*) (Wozencraft 1984). The African civet has been listed in the genera *Viverra* and *Civettictis*. The superficial external resemblances between these taxa have led some authors to group both species in *Viverra*. Pocock (1915a) showed differences in foot pad, scent gland, and skull characters, and proposed *Civettictis*. The character differences between these taxa have not been found to occur within any other genus of viverrid and therefore support the separate generic distinction.

Family Felidae Fischer de Waldheim 1817

SUBFAMILY PANTHERINAE POCOCK 1917

*Felis marmorata Martin 1837, marbled cat. Ewer (1973) listed in the genus *Pardofelis*.

*Lynx canadensis Kerr 1792, North American lynx. Ewer (1973) and Corbet and Hill (1986) suggested this species as conspecific with *L. lynx*.

*Lynx caracal (Schreber 1776), caracal. Ewer (1973) placed this species in the genus *Caracal*, whereas Corbet and Hill (1986) included it in *Felis*.

*Lynx lynx (Linnaeus 1758), Eurasian lynx.

Lynx pardinus (Temminck 1824), Spanish lynx. Corbet and Hill (1986) considered this species conspecific with *L. lynx*.

Lynx rufus (Schreber 1776), bobcat.
Neofelis nebulosa (Griffith 1821), clouded leopard.
Panthera leo (Linnaeus 1758), lion.
Panthera onca (Linnaeus 1758), jaguar.
Panthera pardus (Linnaeus 1758), leopard or panther.
Panthera tigris (Linnaeus 1758), tiger.
Panthera uncia (Schreber 1775), snow leopard or ounce.

SUBFAMILY FELINAE FISCHER DE WALDHEIM 1817

Felis aurata Temminck 1827, African golden cat. Ewer (1973) listed in the genus *Profelis*.
Felis badia Gray 1874, bay cat or Bornean red cat. Ewer (1973) listed in the genus *Pardofelis*.
Felis bengalensis Kerr 1792, leopard or Bengal cat. Ewer (1973) listed in the genus *Prionailurus*.
Felis bieti Milne-Edwards 1892, Chinese desert cat.
Felis chaus Guldenstaedt 1776, jungle cat.
Felis colocolo Molina 1810, pampas cat. Ewer (1973) listed in the genus *Lynchailurus*.
Felis concolor Linnaeus 1771, puma, cougar, or mountain lion. Ewer (1973) listed in the genus *Puma*.
Felis geoffroyi d'Orbigny and Gervais 1844, Geoffroy's cat. Ewer (1973) listed in the genus *Leopardus*.
Felis guigna Molina 1782, kodkod. Ewer (1973) listed in the genus *Oncifelis*.
Felis iriomotensis Imaizumi 1967, Iriomote cat. Ewer (1973) listed in the genus *Mayailurus*.
Felis jacobita Cornalia 1865, Andean mountain cat. Ewer (1973) listed in the genus *Oreailurus*.
Felis manul Pallas 1776, Pallas' cat. Ewer (1973) listed in the genus *Otocolobus*.
Felis margarita Loche 1858, sand cat.
Felis nigripes Burchell 1824, black-footed cat.
Felis pardalis Linnaeus 1758, ocelot. Ewer (1973) listed in the genus *Leopardus*.
Felis planiceps Vigors and Horsfield 1827, flat-headed cat. Ewer (1973) listed in the genus *Ictailurus*.
Felis rubiginosa I. Geoffroy 1831, rusty-spotted cat. Ewer (1973) listed in the genus *Prionailurus*.
Felis serval Schreber 1776, serval. Ewer (1973) listed in the genus *Leptailurus*.
Felis silvestris Schreber 1777, European wild cat. Includes *F. libyca* listed separately by Ewer (1973; see Ragni and Randi 1986).

Felis temmincki Vigors and Horsfield 1827, Asiatic golden cat. Ewer (1973) listed in the genus *Profelis*.

Felis tigrina Schreber 1775, oncilla, tiger ocelot, or little spotted cat. Ewer (1973) listed in the genus *Leopardus*.

Felis viverrina Bennett 1833, fishing cat. Ewer (1973) listed in the genus *Prionailurus*.

Felis wiedii Schinz 1821, Margay or tree ocelot. Ewer (1973) listed in the genus *Leopardus*.

*Felis yagouaroundi E. Geoffroy 1803, jaguarundi. Ewer (1973) listed in the genus *Herpailurus*.

INCERTAE SEDIS

*Acinonyx jubatus (Schreber 1776), cheetah.

Severtzow (1858) divided the Felidae into five genera and 27 subgenera. Pocock (1917) later organized the felids into three monophyletic groups: the Pantherinae (*leo, tigris, pardus, onca, uncia, marmorata*), the Acinonychinae (*A. jubatus*), and the Felinae (14 genera). Since then, the systematics of cats has been the most studied and yet least agreed-upon among all families of carnivores. Most disagreement centers on the number and relationships among genera. This confusion appears to be related to allometric variation and morphological convergences. Cats have the greatest range in size among Recent Carnivora (Gittleman 1985), yet they show the least variation in morphology and karyotype. This is compounded by convergences in character transformations, a common feature of cat evolution (Martin 1980, 1984, this volume). The pantherine line is the best known among felids (Hemmer 1974, 1978) and probably shares a common ancestor with *Lynx* (Kral and Zima 1980; Kratochvil 1982; Collier and O'Brien 1985; Herrington 1986). It includes the large cats and the marbled cat (*marmorata*) (Werdelin 1983). The inclusion of *marmorata* warrants its generic recognition as *Pardofelis*.

The cheetah is usually placed as one of the earliest divergences from the main stem of felid evolution (Hemmer 1978; Kral and Zima 1980; Neff 1983). Kratochvil (1982) and Collier and O'Brien (1985) place the cheetah as the sister group to the *Lynx-Panthera* branch. Herrington (1986) suggested that cheetahs may share a common ancestor with Pallas' cat (*F. manul*), the exact position of which has been an enigma to most cat systematists. Hemmer (1978) and Werdelin (1983) suggested that Pallas' cat be grouped with the lynxes, but Kral and Zima (1980) and Collier and O'Brien (1985) placed Pallas' cat distinctly separate from the lynxes and cheetah and closest to the *F. margarita* group.

The caracal (*caracal*) has traditionally been placed with the lynxes based on superficial resemblances; however, Werdelin (1981) revealed that there is no phylogenetic evidence to support such a relationship; Hemmer (1978) placed the caracal with the *Felis chaus* group. Collier and O'Brien (1985) include *caracal* in the *Panthera* lineage as a sister branch to the serval. The generic designation of the remaining lynxes (*lynx, canadensis, rufus, pardina*) has also been questioned. Van Gelder (1977), on the basis of hybridization data, argued that these taxa should be included in *Felis*, a position accepted by Tumlison (1987). However, Kratochvil (1975), Werdelin (1981), and Herrington (1986) recognize the monophyletic nature of this group in *Lynx*. Werdelin (1981) suggested that *lynx* and *pardina* were distinct species in the genus *Lynx* and supported Kurtén and Rausch (1959), who considered *canadensis* and *lynx* conspecific. Tumlison (1987) considered all of these taxa conspecific.

There is some debate concerning the correct generic name for the large cats (*Panthera* versus *Leo*). Although *Panthera* Oken 1816 is the senior synonym, Oken's *Lehrbuch der Naturgeschichte* has been rejected by the International Commission on Zoological Nomenclature (ICZN Opinion 417, 1956). Corbet et al. (1974) placed before the commission an application to place *Panthera* Oken on the official list of available names. *Panthera* is here used instead of *Leo* in accordance with the provisions of Article 80 of the ICZN code.

There are three species groups represented in South America. Hemmer (1978), Kral and Zima (1980), and Herrington (1986) unite the American puma and the jaguarundi as a single clade closely related to the *Felis-Prionailurus* group. Collier and O'Brien (1985) included the puma and the jaguarundi within their pantherine group, separate from the Felinae. Groves (in Honacki et al. 1982) recognizes *Leopardus* to include *pardalis* and *wiedii*. The remaining small cats can be placed in the subgenus Oncifelis (*colocolo, guigna, jacobita, geoffroyi*) (Collier and O'Brien 1985). Four Asiatic small cats represent a single radiation and are included in the subgenus Prionailurus (*viverrina, iriomotensis, planiceps, rubiginosa*) (Pocock, 1917; Hemmer 1978).

Family Hyaenidae Gray 1821

Subfamily Hyaeninae Gray 1821

Crocuta crocuta (Erxleben 1777), spotted hyena.
Hyaena brunnea Thunberg 1820, brown hyena.
Hyaena hyaena (Linnaeus 1758), striped hyena.

Subfamily Proteiinae Geoffroy Saint Hilaire 1851

Proteles cristatus (Sparrman 1783), aardwolf. Honacki et al. (1982) listed
   Proteles in a separate family.
There is little disagreement that the aardwolf represents the first sister group to the *Hyaena-Crocuta* clade. The recognition of a separate monotypic family was based on the degree of the differences that distinguish the aardwolf from hyenas. The branching sequence of this group suggests including the aardwolf in the Hyaenidae (Wozencraft, this volume).

Family Ursidae Fischer de Waldheim 1817

Ailuropoda melanoleuca (David 1869), bamboo bear or giant panda. Ewer
(1973) placed this species in the Procyonidae, whereas Corbet and Hill
(1986) placed it in the Ailuropodidae. Mayr (1986) suggested that the
common name "panda" is misleading.

Helarctos malayanus (Raffles 1821), Malayan sun bear.

*Melursus ursinus (Shaw 1791), sloth bear.

Tremarctos ornatus (F. Cuvier 1825), spectacled bear.

*Ursus americanus Pallas 1780, American black bear.

Ursus arctos Linnaeus 1758, brown bear.

Ursus maritimus Phipps 1774, polar bear. Ewer (1973) and Corbet and Hill
(1986) place in the genus *Thalarctos*.

Ursus thibetanus G. Cuvier 1823, Asiatic black bear. Ewer (1973) and
Corbet and Hill (1986) place in the genus *Selenarctos*.

Much of the confusion in bear taxonomy has revolved around the inclusion
or exclusion of *Ailuropoda melanoleuca*. Davis's (1964) monograph on the
anatomy of the bamboo bear (giant panda) stands as one of the most thorough
of any species of mammal. He remarked, "Every morphological feature examined indicates that the giant panda is little more than a highly specialized bear"
(p. 322). Nevertheless, Ewer (1973) and Corbet and Hill (1986) did not accept
Davis's placement of *A. melanoleuca* with the bears (see Mayr 1986 for discussion). Despite the overwhelming evidence for the ursid relationship of the
bamboo bear, some English-speaking zoologists have questioned this position
because of the belief that the bear must be related to the red panda, which has
the general outward appearance of a North American raccoon (Mayr 1986).
Mayr (1986) presents convincing arguments that much of the North American
bias is related to the common names, referring to both species as "pandas,"
which implies a relationship, and to the similarity in the feeding behavior. The
placement of *A. melanoleuca* within the Ursidae follows Honacki et al. (1982)
and the phylogenetic hypothesis of Davis (1964), Chorn and Hoffmann
(1978), and Wozencraft (this volume).

Hall (1981) and Thenius (1979) placed the American black bear in *Ursus*.
The polar bear also has been suggested as a member of this polymorphic genus
(Gromov and Baranova 1981).

INCERTAE SEDIS

*Ailurus fulgens F. Cuvier 1825, red panda, lesser panda. Ewer (1973)
placed this species in the Procyonidae, whereas Corbet and Hill (1986)
placed it in the Ailuropodidae.

The genus *Ailurus* because of its superficial resemblance to *Procyon* (i.e.,
face mask, ringed tail, forefeet), has been allied with the procyonids in spite of

the conspicuous lack of derived procyonid features. Ginsburg (1982) suggested that *Ailurus* is the sister group to the Ursidae and Otariidae on the basis of skull and dental characters. Biochemical and genetic similarities suggest either an intermediate position between the procyonids and the ursids (Wurster and Benirschke 1968; Sarich 1976; O'Brien et al. 1985) or a close relationship with the bamboo bear (Tagle et al. 1986). Some phenetic studies have supported the inclusion within the procyonids (Gregory 1936), and phylogenetic studies have suggested that they represent a sister group to the ursids (Ginsburg 1982; Wozencraft 1984, this volume). Characters traditionally used to indicate family-level relationships among the Carnivora align the red panda with the bears (Hunt 1974; Wozencraft, this volume).

Family Otariidae Gray 1825

Subfamily Arctocephalinae Gray 1837

Arctocephalus australis (Zimmermann 1783), South American fur seal.
Arctocephalus forsteri (Lesson 1828), New Zealand fur seal.
Arctocephalus galapagoensis Heller 1904, Galapagos fur seal.
Arctocephalus gazella (Peters 1875), Antarctic fur seal.
Arctocephalus philippii (Peters 1866), Juan Fernandez fur seal.
Arctocephalus pusillus (Schreber 1776), South African fur seal or Afro-Australian fur seal.
Arctocephalus townsendi Merriam 1897, Guadalupe fur seal.
Arctocephalus tropicalis (Gray 1872), subantarctic fur seal.
Callorhinus ursinus (Linnaeus 1758), northern fur seal.

Subfamily Odobeninae (Gray 1869)

*Odobenus rosmarus (Linnaeus 1758), walrus. Placed in the family Odobenidae by Corbet and Hill (1986).

Subfamily Otariinae Gray 1825

Eumetopias jubatus (Schreber 1776), Steller's sea lion.
Neophoca cinerea (Peron 1816), Australian sea lion.
Otaria byronia (Blainville 1820), southern sea lion or South American sea lion.
Phocarctos hookeri (Gray 1844), Hooker's sea lion. Placed in the genus *Neophoca* by Corbet and Hill (1980).
Zalophus californianus (Lesson 1828), California sea lion.
   Scheffer's (1958) phylogeny was one of the first widely used hypothesis to explain evolutionary relationships within the otariids. He recognized three monophyletic groups: (1) *Odobenus;* (2) *Arctocephalus, Callorhinus;* and (3)

Eumetopias, Neophoca, Phocarctos, Otaria, and *Zalophus.* This grouping was defended by Ling (1978), King (1983), and Berta and Deméré (1986). Repenning (1976) presented arguments against any subfamilial grouping, among these a hybrid cross between *Arctocephalus* (Arctocephalinae) and *Zalophus* (Otariinae). There has been general agreement that the walrus represents a sister group to the Otariidae (Allen 1880; Tedford 1976; Ling 1978; King 1983), but Wyss (1987) has recently challenged this view. Wyss suggested that the walruses may be more closely related to the phocids and cited as evidence shared derived characters of the ear region (but see discussion in Wozencraft, this volume). Mitchell and Tedford (1973:279) pointed out that "walruses represent only one of a number of different [otaroid] adaptive types" and therefore should be included within the Otariidae. If known fossil walruses prove to be more closely related to the extinct Desmatophocidae, as proposed by Repenning and Tedford (1977), or to the Phocidae, as proposed by Wyss (1987), then family-level recognition of walruses may be warranted.

Family Canidae Gray 1821

Alopex lagopus (Linnaeus 1758), arctic fox.
Canis adustus Sundevall 1846, side-striped jackal.
Canis aureus Linnacus 1758, golden jackal.
Canis latrans Say 1823, coyote.
Canis lupus Linnaeus 1758, gray wolf.
Canis mesomelas Schreber 1778, black-backed jackal.
*Canis rufus Audubon and Bachman 1851, red wolf. Ewer (1973) and
     Clutton-Brock et al. (1976) questioned the validity of this species because
     of the existence of natural hybrids.
Canis simensis Rüppell 1835, simenian jackal.
Chrysocyon brachyurus (Illiger 1815), maned wolf.
Cuon alpinus (Pallas 1811), dhole or red dog.
Dusicyon australis (Kerr 1792), Falkland Island wolf (extinct).
Dusicyon culpaeus (Molina 1782), colpeo fox. Includes *D. culpaeolus* (part)
     and *D. inca* (part) listed separately by Ewer (1973).
Dusicyon griseus (Gray 1837), Argentine gray fox. Includes *D. fulvipes*
     listed separately by Ewer (1973).
Dusicyon gymnocercus (G. Fischer 1814), Azara's fox or pampas fox.
     Includes *D. culpaeolus* (part) and *D. Inca* (part) listed separately by Ewer
     (1973).
*Dusicyon microtis (Sclater 1883), small-eared zorro. Ewer (1973) listed in
     the genus *Atelocynus.*
Dusicyon sechurae Thomas 1900, sechuran fox.
*Dusicyon thous (Linnaeus 1766), crab-eating fox or common zorro. Ewer
     (1973) listed in the genus *Cerdocyon.*

Dusicyon vetulus Lund 1842, hoary fox.

Lycaon pictus (Temminck 1820), African hunting dog.

*Nyctereutes procyonoides (Gray 1834), raccoon dog.

*Otocyon megalotis (Desmarest 1821), bat-eared fox.

Speothos venaticus (Lund 1842), bush dog.

Urocyon cinereoargenteus (Schreber 1775), gray fox. Included in *Vulpes* by Corbet and Hill (1986).

Urocyon littoralis (Baird 1858), island gray fox. Included in *Vulpes* by Corbet and Hill (1986).

Vulpes bengalensis (Shaw 1800), Bengal fox.

Vulpes cana Blanford 1877, Blanford's fox.

Vulpes chama (A. Smith 1833), Cape fox.

Vulpes corsac (Linnaeus 1768), corsac fox.

Vulpes ferrilata Hodgson 1842, Tibetan sand fox.

*Vulpes macrotis Merriam 1888, kit fox. Ewer (1973) considered this species a junior synonym of *V. velox.*

Vulpes pallida (Cretzschmar 1826), pale fox or sand fox.

Vulpes rueppelli (Schinz 1825), sand fox or Rüppell's fox.

*Vulpes velox (Say 1823), swift fox.

Vulpes vulpes (Linnaeus 1758), red fox.

Vulpes zerda (Zimmermann 1780), fennec. Ewer (1973) listed in the genus *Fennecus.*

The species-level taxonomy of the canids has changed little since Ewer's (1973) list. Most of the taxonomic discussion has centered on generic and subfamilial groups. Van Gelder (1978), relying on hybridization data, proposed grouping the majority of foxes and South American canids in *Canis.* Stains (1975) placed all canids in the subfamily Caninae except three genera (*Cuon, Lycaon,* and *Speothos*) which were grouped in the subfamily Simocyoninae. Later, Stains (1984) united all canids except the genus *Otocyon* in the Caninae (placing *Otocyon* in the Otocyoninae). Langguth (1969, 1975) retained many of the monotypic genera and recognized three distinct groupings: (1) *Canis, Lycaon, Cuon;* (2) *Vulpes, Fennecus, Alopex, Otocyon,* and *Urocyon;* and (3) the South American canids. R. S. Hoffmann (unpubl. data) on the basis of a cladistic analysis, suggested including *Vulpes* in the Caninae, and placing *Speothos* as the first offshoot of the main canid tree. Clutton-Brock et al. (1976) concluded that subfamilial separations were not supported. Their phenetic study highlighted the common occurrence of convergences in canids, as in the felids. Although *Urocyon* can be differentiated on qualitative characters, they suggested its inclusion in *Vulpes.* Berta (1987) considered *Otocyon, Urocyon,* and *Vulpes* to represent a monophyletic group. Its generic designation should be maintained based on characters that it does not share with any *Vulpes.* Because of conflicting hypotheses concerning relationships among genera, subfamilial groupings are not recognized here.

In a recent cladistic study of South American canids, Berta (1984, 1987) proposed, on the basis of two masticatory characters, that the crab-eating fox and the raccoon dog may share a common ancestor. This analysis also suggested that the bush dog and the small-eared zorro are a monophyletic group sharing a common ancestor with the raccoon dog–crab-eating fox clade. This would suggest that *Dusicyon* (*sensu* Honacki et al. 1982) is a paraphyletic taxon and would warrant the recognition of the small-eared zorro in *Atelocynus*.

The red wolf of the southern United States was not recognized as a separate species by Ewer (1973) because of data on hybridization between red wolves and coyotes (Paradiso 1968) and the lack of clear discrimination from phenetic approaches (Lawrence and Bossert 1967). However, subsequent discriminant analysis suggest a definable species boundary (Elder and Hayden 1977; Nowak 1979) and recognizable red wolf–coyote crosses. Furthermore, hybridization is associated with habitat disruption by humans (Paradiso and Nowak 1972). Nowak (1979) lists qualitative characteristics to distinguish among North American species of *Canis*. Ewer (1973) considered the kit and swift fox conspecific. Studies of sympatric populations show gene pool integrity despite occasional hybrids (Packard and Bowers 1970; Rohwer and Kilgore 1973; Thornton and Creel 1975).

Family Procyonidae Gray 1825

**Bassaricyon alleni* Thomas 1880, olingo. Ewer (1973) includes this species as a junior synonym of *B. gabbii*.
**Bassaricyon beddardi* Pocock 1921, olingo. Ewer (1973) includes this species as a junior synonym of *B. gabbii*.
Bassaricyon gabbii. J. A. Allen 1876, bushy-tailed olingo or common olingo.
**Bassaricyon lasius* Harris 1932, Harris' olingo. Ewer (1973) includes this species as a junior synonym of *B. gabbii*.
**Bassaricyon pauli* Enders 1936, chiriqui olingo. Ewer (1973) includes this species as a junior synonym of *B. gabbii*.
Bassariscus astutus (Lichtenstein 1830), ring-tailed cat or cacomistle.
Bassariscus sumichrasti (Saussure 1860), Central American cacomistle. Ewer (1973) lists in the genus *Jentinkia*.
**Nasua nasua* (Linnaeus 1766), ring-tailed coatimundi. Includes *N. narica* listed separately by Ewer (1973).
Nasua nelsoni Merriam 1901, Cozumel Island coatimundi.
Nasuella olivacea (Gray 1865), lesser coatimundi or mountain coatimundi.
Potos flavus (Schreber 1774), kinkajou.
**Procyon cancrivorus* (F. Cuvier 1798), crab-eating raccoon.
**Procyon gloveralleni* Nelson and Goldman 1930, Barbados raccoon.

Corbet and Hill (1986) include this species as a junior synonym of *P. lotor.*

*Procyon insularis Merriam 1898, Tres Marias raccoon. Corbet and Hill (1986) include this species as a junior synonym of *P. lotor.*

Procyon lotor (Linnaeus 1758), raccoon.

*Procyon maynardi Bangs 1898, Bahama raccoon. Corbet and Hill (1986) include this species as a junior synonym of *P. lotor.*

*Procyon minor Miller 1911, Guadeloupe raccoon. Corbet and Hill (1986) include this species as a junior synonym of *P. lotor.*

*Procyon pygmaeus Merriam 1901, Cozumel Island raccoon. Corbet and Hill (1986) include this species as a junior synonym of *P. lotor.*

Bassariscus is considered a primitive procyonid group and *Nasua-Nasuella-Procyon* as a more derived monophyletic group (Segall 1943; Schmidt-Kittler 1981; Baskin 1982). The relationship of the genera *Bassaricyon* and *Potos* to these groups is unclear. Pocock (1921) considered *Potos* as the first outgroup to the remaining procyonids, and Segall (1943) suggested that *Potos* should be regarded as a marginal member of the musteloid stock.

There are five taxa of *Bassaricyon* recognized by Honacki et al. (1982); Poglayen-Neuwall and Poglayen-Neuwall (1965) and Ewer (1973) suggested that these forms were conspecific. Examination of specimens at the U.S. National Museum and the University of Kansas, Museum of Natural History, indicated that pelage characteristics used in the original type descriptions did not sufficiently discriminate morphotypes and therefore supported the recognition of only one species.

The genus *Procyon* is also in need of revision; *P. cancrivorous* occurs sympatrically with *P. lotor* and can be distinguished on qualitative characters. All other *Procyon* species recognized by Honacki et al. (1982) are Caribbean insular forms and are regarded as human introductions by Morgan et al. (1980) and Morgan and Woods (1986). They should be considered conspecific with *P. lotor* (Corbet and Hill 1986).

Decker and Wozencraft (in press) list a suite of morphological features to distinguish the North American Coati, *Nasua narica*, from the South American form, *N. nasua* and support their separate species recognition.

Family Mustelidae Fischer de Waldheim 1817

SUBFAMILY LUTRINAE BAIRD 1857

Aonyx capensis (Schinz 1821), Cape clawless otter.

Aonyx cinerea (Illiger 1815), Oriental small-clawed otter. Ewer (1973) listed in the genus *Amblonyx.*

Aonyx congica Lonnberg 1910, Congo otter or Zaire clawless otter.

Includes *A. microdon* and *A. philippsi* listed separately by Ewer (1973).

Enhydra lutris (Linnaeus 1758), sea otter.
Lutra canadensis (Schreber 1776), river otter.
Lutra felina (Molina 1782), sea cat or marine otter.
Lutra longicaudis (Olfers 1818), includes *L. annectens, L. enudris, L. incarum,* and *L. platensis,* listed separately by Ewer (1973).
Lutra lutra (Linnaeus 1758), European otter.
Lutra maculicollis Lichtenstein 1835, spotted-necked otter.
Lutra perspicillata I. Geoffroy 1826, smooth-coated otter.
Lutra provocax Thomas 1908, southern river otter.
Lutra sumatrana (Gray 1865), Sumatran otter or hairy-nosed otter.
Pteronura brasiliensis (Gmelin 1788), giant river otter.

Subfamily Melinae Burmeister 1850.

Arctonyx collaris F. Cuvier 1825, hog badger.
Meles meles (Linnaeus 1758), Eurasian badger.
Melogale everetti (Thomas 1895), Everett's ferret badger or Bornean ferret badger.
Melogale moschata (Gray 1831), Chinese ferret badger.
Melogale personata I. Geoffroy 1831, Burmese ferret badger. Includes *orientalis* listed separately by Ewer (1973).
Mydaus javanensis (Desmarest 1820), sunda stink-badger.
Mydaus marchei (Huet 1887), Philippines badger. Ewer (1973) listed in the genus *Suillotaxus.*

Subfamily Mephitinae Gill 1872

Conepatus chinga (Molina 1782), hog-nosed skunk.
Conepatus humboldtii Gray 1837, Patagonian hog-nosed skunk.
Conepatus leuconotus (Lichtenstein 1832), eastern hog-nosed skunk.
Conepatus mesoleucus (Lichtenstein 1832), western hog-nosed skunk.
Conepatus semistriatus (Boddaert 1784), striped hog-nosed skunk.
Mephitis macroura Lichtenstein 1832, hooded skunk.
Mephitis mephitis (Schreber 1776), striped skunk.
Spilogale putorius (Linnaeus 1758), spotted skunk.
Spilogale pygmaea Thomas 1898, pygmy spotted skunk. Ewer (1973) considered this species a junior synonym of *S. putorius.*

Subfamily Mustelinae Fischer de Waldheim 1817

Eira barbara (Linnaeus 1758), tayra. Ewer (1973) listed in the genus *Tayra.*
Galictis cuja (Molina 1782), little grison. Ewer (1973) listed in the genus *Grison.*

Galictis vittata (Schreber 1776), South American grison. Ewer (1973) listed in the genus *Grison*.

Gulo gulo (Linnaeus 1758), wolverine.

Ictonyx striatus (Perry 1810), zorilla or striped polecat.

Lyncodon patagonicus (Blainville 1842), Patagonian weasel.

Martes americana (Turton 1806), American pine marten.

Martes flavigula (Boddaert 1785), yellow-throated marten. Includes *M. gwatkinsi* listed separately by Ewer (1973).

Martes foina (Erxleben 1777), stone marten or beech marten.

Martes martes (Linnaeus 1758), European pine marten.

Martes melampus (Wagner 1841), Japanese marten.

Martes pennanti (Erxleben 1777), fisher.

Martes zibellina (Linnaeus 1758), sable.

Mellivora capensis (Schreber 1776), ratel or honey badger.

Mustela africana Desmarest 1818, Amazon or South American weasel. Ewer (1973) listed in the genus *Grammogale*.

Mustela altaica Pallas 1811, alpine weasel or mountain weasel.

Mustela erminea Linnaeus 1758, stoat or ermine.

Mustela eversmanni Lesson 1827, steppe polecat.

Mustela felipei Izor and de la Torre 1978, Columbian weasel.

Mustela frenata Lichtenstein 1831, long-tailed weasel.

Mustela kathiah Hodgson 1835, yellow-bellied weasel.

Mustela lutreola (Linnaeus 1761), European mink.

Mustela lutreolina Robinson and Thomas 1917, Javan weasel.

Mustela nigripes (Audubon and Bachman 1851), black-footed ferret.

Mustela nivalis Linnaeus 1766, least weasel. Includes *M. rixosa* listed separately by Ewer (1973).

Mustela nudipes Desmarest 1822, Malaysian weasel.

Mustela putorius Linnaeus 1758, European ferret or polecat.

Mustela sibirica Pallas 1773, Siberian weasel.

Mustela strigidorsa Gray 1853, black-striped weasel.

Mustela vison Schreber 1777, American mink.

Poecilictis libyca (Hemprich and Ehrenberg 1833), Saharan striped weasel.

Poecilogale albinucha (Gray 1864), white-naped weasel.

Vormela peregusna (Guldenstaedt 1770), marbled polecat.

INCERTAE SEDIS

Taxidea taxus (Schreber 1778), North American badger.

The Recent families of Carnivora can be characterized by a basic auditory-basicranial morphological complex that generally holds for all members in the order (Wozencraft, this volume). Of the Recent families, the mustelids show two major radiations, the Lutrine-Mephitinae and the Mustelinae-Melinae, which probably represents a very early split in their phylogeny (Hunt 1974). Although the mustelids do not show the range in body size the felids do, the proportional differences are greater (Gittleman 1985), and they clearly represent the most diversified of the families of Carnivora.

The Recent Lutrinae is divided into three monophyletic groups. Most have agreed that the sea otter is the first branch on the lutrine stem and represents an early divergence (Harris 1968; van Zyll de Jong 1972; Muizon 1982b). Of the remaining 12 species, van Zyll de Jong (1972) recognized six genera based on a phenetic analysis. He restricted *Lutra* to include only *sumatrana, maculicollis,* and *lutra.* Four species (*longicaudis, provocax, canadensis,* and *felina*), previously placed in *Lutra* by Harris (1968) and Duplaix and Roest (in Honacki et al. 1982), he placed in the genus *Lontra.* The remaining taxa (*Aonyx, Pteronura, Lutrogale=Lutra perspicillata*) he grouped into a single clade. Holmes and Hoffmann (1986) suggested that *maculicollis* shares a common ancestor with *perspicillata* and should be placed in *Hydrictis.*

In 1987 van Zyll de Jong published a phylogenetic analysis of the otters based on 12 morphological features. This analysis suggested a different arrangement than his previous study (1972), although he was hesitant to propose changes in taxonomy. Four monophyletic groups were proposed: (1) *P. brasiliensis, L. perspicillata, L. maculicollis, L. lutra,* and *L. sumatrana;* (2) *L. canadensis, L. felina, L. longicaudis,* and *L. provocax;* (3) *A. congicus* and *A. capensis;* and (4) *A. cinereus* and *E. lutris.*

The Melinae is an Old World radiation with possibly only one taxon (*Taxidea*) present in the Nearctic. *Taxidea* has been grouped with the heterogenous Melinae, initially because of the resemblance in the face mask between the Eurasian badger (*Meles*) and the North American badger. However, *Taxidea* has an auditory epitympanic sinus in the squamosal/mastoid region. This complex feature in the mustelids is present only in mephitines and the North American badger, although the complex is probably not homologous in the two groups (R. H. Tedford, pers. comm.). Nevertheless, the North American badger can be characterized by many basicranial features not present in Old World badgers, making the subfamilial position uncertain. Petter (1971) placed *Meles* and *Arctonyx* as sister groups. Long (1978) considered *Suillotaxus* (=*Mydaus marchei*) to be a subgenus of *Mydaus.* The relationship of the ferret badgers to the rest of the Melinae is unclear, but they probably represent a primitive offshoot from the main lineage (Everts 1968). Radinsky (1973) suggested that there might be a relationship between the stink badgers and the mephitines; however, he believed the neurological characters shared by the two groups were primitive.

The Mephitinae includes three genera (*Conepatus, Mephitis, Spilogale*) and represent a single radiation in the Nearctic. Ewer (1973) believed *S. pygmaea* to be conspecific with *putorius.* Van Gelder (1959) presented convincing arguments to warrant recognition of two species. Mead (1968) suggested separating *S. putorius* into eastern (subspecies *putorius, ambarvalis,* and *interrupta*) and western (subspecies *gracilis, leucoparia, latifrons, phenax*) species, western populations being characterized by delayed implantation. Hall (1981) considered the North American hog-nosed skunks (*C. mesoleucus* and *leuconotus*) to be conspecific.

Long (1978) placed *Mellivora* in the Mustelinae, and Holmes (1985) suggested that the ratel belonged to an African radiation that included *Vormela, Ictonyx, Poecilictis,* and *Poecilogale*. Holmes, in a cladistic analysis, divided the genus *Mustela* into four monophyletic groups: (1) *vison;* (2) *putorius, eversmanni, nigripes;* (3) *sibirica, lutreolina, lutreola, nudipes, strigidorsa;* and (4) *kathiah, nivalis, erminea, altaica, frenata, africana,* and *felipei*. Izorc and de la Torre (1978) reopened the question of whether *africana* and *felipei* should be placed in *Grammogale*. If Holmes's phylogeny is accepted, a generic designation for each of these groups may be warranted but not for *africana* and *felipei*.

The remaining mustelines (*Gulo, Martes, Eira, Galictis,* and *Lyncodon*) form a single clade. *Martes zibellina, M. martes, M. melampus,* and *M. americana* may be conspecific (Hagmeier 1961; Anderson 1970). However, Heptner and Naumov (1967) demonstrated that distributions of *martes* and *zibellina* are originally sympatric and that they are morphologically distinct.

Family Phocidae Gray 1825

Cystophora cristata (Erxleben 1777), hooded seal.
Erignathus barbatus (Erxleben 1777), bearded seal.
Halichoerus grypus (Fabricius 1791), gray seal.
Hydrurga leptonyx (Blainville 1820), leopard seal.
Leptonychotes weddelli (Lesson 1826), Weddell seal.
Lobodon carcinophagus (Hombron and Jacquinot 1842), crabeater seal.
Mirounga angustirostris (Gill 1866), northern elephant seal.
Mirounga leonina (Linnaeus 1758), southern elephant seal.
Monachus monachus (Hermann 1779), Mediterranean monk seal.
Monachus schauinslandi Matschie 1905, Hawaiian monk seal.
Monachus tropicalis (Gray 1850), West Indian monk seal or Caribbean monk seal.
Ommatophoca rossi Gray 1844, Ross seal.
Phoca caspica Gmelin 1788, Caspian seal.
Phoca fasciata Zimmermann 1783, ribbon seal.
Phoca groenlandica Erxleben 1777, harp seal.
Phoca hispida Schreber 1775, ringed seal.
Phoca largha Pallas 1811, spotted seal or larga seal.
Phoca sibirica Gmelin 1788, baikal seal.
Phoca vitulina Linnaeus 1758, harbor seal or common seal.
Scheffer (1958) recognized three monophyletic groups within the phocids:

(1) *Phoca, Erignathus, Halichoerus,* (2) *Monachus, Lobodon, Ommatophoca, Hydrurga, Leptonychotes,* and (3) *Crystophoca, Mirounga.* Ling (1978) and Muizon (1982a) placed *Monachus* with *Mirounga,* and separated *Erignathus* from the *Phoca-Halichoerus* clade. Muizon further recognized three clades within the Phocinae: (1) *Erignathus,* (2) *Phoca-Halichoerus,* and (3) *Histriophoca-Cystophoca.* He divided the Monachinae into four clades: (1) *Monachus,* (2) *Mirounga,* (3) *Ommatophoca-Leptonychotes,* and (4) *Lobodon-Hydrurga.* King (1983) similarly grouped *Lobodon* and *Hydrurga* in one clade and *Leptonychotes* and *Ommatophoca* in the other. Burns and Fay (1970) stated that characters used to separate *Phoca, Pusa, Pagophilus,* and *Histriophoca* were not of sufficient magnitude to warrant generic distinction; this was supported by McDermid and Bonner (1975), whose electrophoretic findings suggested lumping the taxa into one genus.

Acknowledgments

A. Berta, A. R. Biknevicius, D. M. Decker, L. R. Heaney, and R. S. Hoffmann provided many helpful comments and useful suggestions for revision of the manuscript. This work was supported by the National Museum of Natural History, Smithsonian Institution.

References

Allen, J. A. 1880. History of North American pinnipeds: A monograph of the walruses, sea-lions, sea-bears and seals of North America. U.S. Geological and Geographical Survey of the Territories, Miscellaneous Publications 12:1–785.

Anderson, E. 1970. Quaternary evolution of the genus *Martes* (Carnivora, Mustelidae). *Acta Zool. Fennica* 130:1–132.

Baskin, J. A. 1982. Tertiary Procyoninae (Mammalia: Carnivora) of North America. *J. Vert. Paleontol.* 2:71–93.

Beaumont, G. de 1964. Remarques sur la classification des Felidae. *Eclogae Geologicae Helvetiae* 57:837–845.

Bechthold, G. 1939. Die asiatischen Formen der Gattung *Herpestes. Z. Säugetierk.* 14:113–219.

Berta, A. 1984. The Pleistocene bush dog *Speothos pacivorus* (Canidae) from the Lagoa Santa Caves, Brazil. *J. Mamm.* 65:549–559.

Berta, A. 1987. Origin, diversification, and zoogeography of the South American Canidae. In: B. D. Patterson & R. M. Timm, eds. *Studies in Neotropical Mammalogy: Essays in Honor of Philip Hershkovitz.* Fieldiana Zoology New Series.

Berta, A., and Deméré, T. A. 1986. *Callorhinus gilmorei* n. sp., (Carnivora: Otariidae) from the San Diego formation (Blancan) and its implications for otariid phylogeny. *Trans. San Diego Soc. Nat. Hist.* 21:111–126.

Booth, A. H. 1954. The Dahomey gap and the mammalian fauna of the west African forest. *Revue de zoologie et de botanique africaines* 50:305–314.

Booth, A. H. 1958. The Niger, the Volta, and the Dahomey Gap as geographic barriers. *Evolution* 12:48–62.

Burns, J. J., and Fay, F. H. 1970. Comparative morphology of the skull of the řibbon seal, *Histriophoca fasciata,* with remarks on systematics of Phocidae. *J. Zool. (London)* 161:363–394.

Carleton, M. D., and Robbins, C. B. 1985. On the status and affinities of *Hylomys planifrons* (Miller, 1900) (Rodentia: Muridae). *Proc. Biol. Soc. Washington* 98:956–1003.

Chorn, J., and Hoffmann, R. S. 1978. *Ailuropoda melanoleuca. Mamm. Species* 110:1–6.

Clutton-Brock, J., Corbet, G. B., and Hills, M. 1976. A review of the family Canidae, with a classification by numerical methods. *Bull. Brit. Mus. (Nat. Hist.), Zool.* 29:117–199.

Collier, G. E., and O'Brien, S. J. 1985. A molecular phylogeny of the Felidae: Immunological distance. *Evolution* 39:473–487.

Corbet, G. B., and Hill, J. E. 1986. *A World List of Mammalian Species.* London: British Museum (Natural History).

Corbet, G. B., Hill, J. E., Ingles, J. M., and Napier, P. H. 1974. Resubmission of *Pan* Oken 1816, and *Panthera* Oken 1816, proposed conservation under the plenary powers. *Bull. Zool. Nomencl.* 31:29–43.

Crawford-Cabral, J. 1981a. Análise de dados craniométricos no género *Genetta* G. Cuvier. (Carnivora, Viverridae). *Memorias da Junta de Investigacoes Cientificas do Ultramar.* (Lisbon) 66:1 329.

Crawford-Cabral, J. 1981b. The classification of the genets (Carnivora, Viverridae, genus *Genetta*). *Boletim de Sociedade Portuguesa de Ciencias Naturais* 20:97–114.

Davis, D. D. 1964. The giant panda: A morphological study of evolutionary mechanisms. *Fieldiana Zool. Mem.* 3:1–339.

Decker, D. M., and Wozencraft, W. C. In press. The family Procyonidae. In: S. Anderson, ed. *The Mammals of South America.* Vol. 2.

Elder, W. H., and Hayden, C. M. 1977. Use of discriminant function in taxonomic determination of canids from Missouri. *J. Mamm.* 58:17–24.

Everts, W. 1968. Beitrag zur Systematik des Sonnendachse. *Z. Säugetierk.* 33:1–19.

Ewer, R. F. 1973. *The Carnivores.* Ithaca, N.Y.: Cornell Univ. Press.

Ginsburg, L. 1982. Sur la position systématique du petit panda, *Ailurus fulgens* (Carnivora, Mammalia). *Geobios (Lyon), Mémoire Spécial* 6:259–277.

Gittleman, J. L. 1985. Carnivore body size: Ecological and taxonomic correlates. *Oecologia* 67:540–554.

Goldman, C. A. 1984. Systematic revision of the African mongoose genus *Crossarchus* (Mammalia: Viverridae). *Canadian J. Zool.* 62:1618–1630.

Gray, J. E. 1865 (for 1864). A revision of the genera and species of viverrine animals (Viverridae) founded on the collection in the British Museum. *Proc. Zool. Soc. London* 1864:502–579.

Gregory, W. K. 1936. On the phylogenetic relationships of the giant panda (*Ailuropoda*) to other carnivores. *Amer. Mus. Novitates* 878:1–29.

Gromov, I. M., and Baranova, G. I. eds. 1981. *Katalog miekopitayushchikh SSSR* [Catalog of mammals of the USSR]. Leningrad: Nauka.

Hagmeier, 1961. Variation and relationships in North American martin. *Canadian Field-Nat.* 75:122–138.

Hall, E. R. 1981. *The Mammals of North America* (2nd ed.). New York: John Wiley.

Harris, C. J. 1968. *Otters: A Study of the Recent Lutrinae.* London: Weidenfeld and Nicolson.

Hemmer, H. 1974. Untersuchungen zur Stammesgeschichte der Pantherkatzen (Pantherinae). III. Zur Artgeschichte des Löwen *Panthera (Panthera) leo* (Linnaeus 1758). *Veröffent. Zool. Staatsamml. München* 17:167–280.

Hemmer, H. 1978. The evolutionary systematics of living Felidae: Present status and current problems. *Carnivore* 1:71–79.

Heptner, U. G., and Naumov, N. P. eds. 1967. [Mammals of the Soviet Union. vol. 2, pt. 1, Sea cows and Carnivores.] Moscow: Vysshaya Shkola.

Herrington, S. J. 1986. Phylogenetic relationships of the wild cats of the world. Ph.D. dissert., Univ. Kansas, Lawrence.

Hoffmann, R. S. 1988. A cladistic analysis of the Canidae. Unpublished data.

Holmes, T. 1985. A phylogeny of the Mustelinae. (Abstract.) *Am. Soc. Mammal., 65th Ann. Meeting, Abstr.* p. 60.

Holmes, T., and Hoffmann, R. S. 1986. A phylogeny of the Lutrinae and the Mephitinae. (Abstract.) *Am. Soc. Mammal., 66th Ann. Meeting, Abstr.* p. 43.

Honacki, J. H., Kinman, K. E., and Koeppl, J. W. 1982. *Mammal Species of the World: A Taxonomic and Geographic Reference.* Lawrence, Kan.: Association of Systematics Collections.

Hunt, R. M. 1974. The auditory bulla in Carnivora: An anatomical basis for reappraisal of carnivore evolution. *J. Morphol.* 143:21–76.

Hunt, R. M. 1987. Evolution of the Aeluroid Carnivora: Significance of auditory structure in the nimravid cat *Dinictis. Amer. Mus. Novitates* 2886:1–74.

Izore, R. J., and de la Torre, L. 1978. A new species of weasel (*Mustela*) from the highlands of Colombia, with comments on the evolution and distribution of South American weasels. *J. Mamm.* 59:92–102.

King, J. E. 1983. *Seals of the World* (2nd ed.). Ithaca, N.Y.: Cornell Univ. Press.

Kingdon, J. 1977. *East African Mammals: An Atlas of Evolution in Africa,* vol. 3, Part A: *Carnivores.* New York: Academic Press.

Köhncke, M., and Leonhardt, K. 1986. *Cryptoprocta ferox. Mamm. Species* 254:1–5.

Kral, B., and Zima, J. 1980. Karyosystematika celedi Felidae. *Gazella (Prague)* 2/3:45–53.

Kratochvil, J. 1975. Os penis of central European Felidae (Mammalia). *Zoologicke Listy* 24:289–296.

Kratochvil, J. 1982. Karyotyp und System der Familie Felidae (Carnivora, Mammalia). *Folia Zoologica* 31:289–304.

Kurtén, B., and Rausch, R. 1959. Biometric comparisons between North American and European mammals. II. A comparison between the northern lynxes of Fennoscandia and Alaska. *Acta Arctica* 11:21–45.

Laborde, C. 1986. Caractères d'adaptation des membres au mode de vie arboricole chez *Cryptoprocta ferox* par comparaison avec d'autres Carnivores Viverridés. *Annales des Sciences Naturelles Zoologie* ser. 13, 8:25–39.

Langguth, A. 1969. Die südamerikanischen Canidae unter besonderer Berücksichtigung des Mähnenwolfes, *Chrysocyon brachyurus* Illiger. *Zeitschrift für wissenschaftliche Zoologie* 179:1–188.

Langguth, A. 1975. Ecology and evolution in the South American canids. In: M. W. Fox, ed. *The Wild Canids: Their Systematics, Behavioral Ecology, and Evolution,* pp. 192–206. New York: Van Nostrand Reinhold.

Lawrence, B., and Bossert, W. H. 1967. Multiple character analysis of *Canis lupus, latrans* and *familiaris,* with a discussion of the relationships of *Canis niger. Amer. Zool.* 7:223–232.

Ling, J. K. 1978. Pelage characteristics and systematic relationships in the Pinnipedia. *Mammalia* 42:305–313.

Long, C. A. 1978. A listing of recent badgers of the world, with remarks on taxonomic problems in *Mydaus,* and *Melogale. Univ. Wisconsin–Stevens Point, Mus. Nat. Hist., Report on the Fauna and Flora of Wisconsin* 14:1–6.

McDermid, E. M., and Bonner, W. N. 1975. Red cell and serum protein systems of gray seals and harbour seals. *Comp. Biochem. Physiol.* 50B:97–101.

MacDonald, D. ed. 1984. *The Encyclopedia of Mammals*. New York: Facts on File.

Martin, L. D. 1980. Functional morphology and the evolution of cats. *Trans. Nebraska Acad. Sci.* 8:141–154.

Martin, L. D. 1984. Phyletic trends and evolutionary rates. *Carnegie Mus. Nat. Hist. Special Publications* 8:526–538.

Mayr, E. 1986. Uncertainty in science: Is the giant panda a bear or raccoon? *Nature* 323:769–771.

Mead, R. A. 1968. Reproduction in western forms of the spotted skunk (genus *Spilogale*). *J. Mamm.* 49:373–390.

Mitchell, E., and Tedford, R. H. 1973. The Enaliarctinae, a new group of extinct aquatic Carnivora and a consideration of the origin of the Otariidae. *Bull. Amer. Mus. Nat. Hist.* 151:201–284.

Morgan, G. S., Ray, C. E., and Arredondo, O. 1980. A giant extinct insectivore from Cuba (Mammalia: Insectivora: Solenodontidae). *Proc. Biol. Soc. Washington* 93:597–608.

Morgan, G. S., and Woods, C. A. 1986. Extinction and the zoogeography of West Indian land mammals. *Biol. J. Linnean Soc.* 28:167–203.

Muizon, C. de. 1982a. Phocid phylogeny and dispersal. *Ann. South African Mus.* 89:175–213.

Muizon, C. de. 1982b. Les relations phylogénétiques des Lutrinae (Mustelidae, Mammalia). *Geobios (Lyon), Mémoire Spécial* 6.259–277.

Neff, N. A. 1983. *The Big Cats: The Paintings of Guy Coheleach*. New York: Abrams.

Nowak, R. M. 1979. North American Quarternary *Canis*. *Univ. Kansas, Mus. Nat. Hist. Mon.* 6:1–154.

O'Brien, S. J., Nash, W. G., Wildt, D. E., Bush, M. E., and Benveniste, R. E. 1985. A molecular solution to the riddle of the giant panda's phylogeny. *Nature* 317:140–144.

Packard, R. L., and Bowers, J. H. 1970. Distributional notes on some foxes from western Texas and eastern New Mexico. *Southwestern Nat.* 14:450–451.

Paradiso, J. L. 1968. Canids recently collected in east Texas, with comments on the taxonomy of the red wolf. *Amer. Midland Nat.* 80:529–534.

Paradiso, J. L., and Nowak, R. M. 1972. *Canis rufus*. *Mamm. Species* 22:1–4.

Petter, G. 1971. Origine, phylogénie et systématique des blaireaux. *Mammalia* 35:567–597.

Petter, G. 1974. Rapports phylétiques des viverridés (Carnivores Fissipédes). Les formes de Madagascar. *Mammalia* 38:605–636.

Pocock, R. I. 1915a. On the feet and glands and other external characters of the Viverrinae, with the description of a new genus. *Proc. Zool. Soc. London* 1915:131–149.

Pocock, R. I. 1915b. On the feet and glands and other external characters of the Paradoxurinae genera *Paradoxurus*, *Arctictis*, *Arctogalidia*, and *Nandinia*. *Proc. Zool. Soc. London* 1915:387–412.

Pocock, R. I. 1915c. On some of the external characters of the genus *Linsang* with notes upon the genera *Poiana* and *Eupleres*. *Ann. Mag. Nat. Hist.* ser. 8, 16:341–351.

Pocock, R. I. 1917. The classification of existing Felidae. *Ann. Mag. Nat. Hist.* ser. 9, 1:375–384.

Pocock, R. I. 1919. The classification of the mongooses (Mungotidae). *Ann. Mag. Nat. Hist.* ser. 9, 3:515–524.

Pocock, R. I. 1921. The external characters and classification of the Procyonidae. *Proc. Zool. Soc. London* 1921:389–422.

Pocock, R. I. 1929. Carnivora. *Encyclopaedia Britannica* (14th ed.) 4:896–900.

Pocock, R. I. 1933. The civet-cats of Asia. *J. Bombay Nat. Hist. Soc.* 36:421–449.

Poglayen-Neuwall, I., and Poglayen-Neuwall, I. 1965. Gefangenschaftsbeobachtungen an Makibären (*Bassaricyon* Allen 1876). *Z. Säugetierk.* 30:321–366.

Radinsky, L. 1973. Are stink badgers skunks? Implications of neuroanatomy for mustelid phylogeny. *J. Mamm.* 54:585–593.

Ragni, B., and Randi, E. 1986. Multivariate analysis of craniometric characters in European wild cat, domestic cat, and African wild cat (genus *Felis*). *Z. Säugetierk.* 51:243–251.

Repenning, C. A. 1976. Adaptive evolution of seal lions and walruses. *Syst. Zool.* 25:375–390.

Repenning, C. A., and Tedford, R. H. 1977. Otarioid seals of the Neogene. *U.S. Geol. Survey, Prof. Paper* 992:1–93.

Robbins, C. B. 1978. The Dahomey Gap: A reevaluation of its significance as a faunal barrier to West African high forest animals. *Bull. Carnegie Mus. Nat. Hist.* 6:168–174.

Rohwer, S. A., and Kilgore, D. L. 1973. Interbreeding in the arid-land foxes, *Vulpes velox* and *V. macrotis*. *Syst. Zool.* 22:157–165.

Rosevear, D. R. 1974. *The Carnivores of West Africa*. Publication no. 723. London: Trustees of the British Museum (Natural History).

Sarich, V. M. 1976. Transferrin. *Trans. Zool. Soc. London* 33:165–171.

Scheffer, V. B. 1958. *Seals, Sea Lions, and Walruses: A Review of the Pinnipedia*. Stanford, Calif.: Stanford Univ. Press.

Schlawe, L. 1980. Zur geographischen Verbreitung der Ginsterkatzen Gattung *Genetta* G. Cuvier, 1816. *Faunistische Abhandlungen Staatliches Museum für Tierkunde Dresden* 7:147–161.

Schlawe, L. 1981. Material, Fundort, Text- und Bildquellen als Grundlagen für eine Artenliste zur Revision der Gattung *Genetta* G. Cuvier, 1816 (Mammalia, Carnivora, Viverridae). *Zoologische Abhandlungen Staatliches Museum für Tierkunde Dresden* 37:85–182.

Schmidt-Kittler, N. 1981. Zur Stammesgeschichte der marderverwandten Raubtiergruppen (Musteloidea, Carnivora). *Eclogae Geologicae Helvetiae* 74:753–801.

Segall, W. 1943. The auditory region of the arctoid carnivores. *Field Mus. Nat. Hist., Zool. Ser.* 29:33–59.

Severtzow, M. N. 1858. Notice sur la classification multisériale des Carnivores, spécialement des Félidés, et les études de zoologie générale qui s'y rattachent. *Revue Magazine de Zoologie* series 2, 10:385–393.

Stains, H. J. 1975. Distribution and taxonomy of the Canidae. In: M. W. Fox, ed. *The Wild Canids: Their Systematics, Behavioral Ecology and Evolution*, pp. 3–26. New York: Van Nostrand Reinhold.

Stains, H. J. 1984. Carnivores. In: S. Anderson & J. K. Jones, Jr., eds. *Orders and Families of Recent Mammals of the World*, pp. 491–521. New York: John Wiley and Sons.

Tagle, D. A., Miyamoto, M. M., Goodman, M., Hofmann, O., Braunitzer, G., Göltenboth, R., and Jalanka, H. 1986. Hemoglobin of pandas: Phylogenetic relationships of carnivores as ascertained with protein sequence data. *Naturwissenschaften* 73:512–514.

Tedford, R. H. 1976. Relationship of pinnipeds to other carnivores (Mammalia). *Syst. Zool.* 25:363–374.

Thenius, E. 1979. Zur systematischen und phylogenetischen Stellung des Bambusbären: *Ailuropoda melanoleuca* David (Carnivora, Mammalia). *Z. Säugetierk.* 44:286–305.

Thornton, W. A., and Creel, G. C. 1975. The taxonomic status of kit foxes. *Texas J. Sci.* 26:127–136.

Tumlison, R. 1987. *Felis lynx*. *Mamm. Species* 269:1–8.

van Gelder, R. G. 1959. A taxonomic revision of the spotted skunks (genus *Spilogale*). *Bull. Amer. Mus. Nat. Hist.* 117:229–392.

van Gelder, R. G. 1977. Mammalian hybrids and generic limits. *Amer. Mus. Novitates,* 2635:1–25.

van Gelder, R. G. 1978. A review of canid classification. *Amer. Mus. Novitates* 2646:1–10.

van Zyll de Jong, C. G. 1972. A systematic review of the Nearctic and Neotropical river otters (genus *Lutra*, Mustelidae, Carnivora). *Royal Ontario Mus., Life Sci. Contrib.* 80:1–104.

van Zyll de Jong, C. G. 1987. A phylogenetic study of the Lutrinae (Carnivora; Mustelidae) using morphological data. *Canadian J. Zool.* 65:2536–2544.

Watson, J. P., and Dippenaar, N. J. 1987. The species limits of *Galerella sanguinea* (Ruppell, 1836), *G. pulverulenta* (Wagner, 1839) and *G. nigrata* (Thomas, 1928) in Southern Africa (Carnivora: Viverridae). *Navorsinge van die Nasionale Museum Bloemfontein* 5:356–414.

Werdelin, L. 1981. The evolution of lynxes. *Ann. Zool. Fennici* 18:37–71.

Werdelin, L. 1983. Morphological patterns in the skulls of cats. *Biol. J. Linnean Soc.* 19:375–391.

Wozencraft, W. C. 1984. A phylogenetic reappraisal of the Viverridae and its relationship to other Carnivora. Ph.D. dissert., Univ. Kansas, Lawrence.

Wozencraft, W. C. 1986. A new species of striped mongoose from Madagascar. *J. Mamm.* 67:561–571.

Wozencraft, W. C. 1987. Emendation of species name. *J. Mamm.* 68:198.

Wurster, D. H., and Benirschke, K. 1968. Comparative cytogenetic studies in the Order Carnivora. *Chromosoma* 24:336–382.

Wyss, A. R. 1987. Comments on the walrus auditory region and the monophyly of pinnipeds. *Amer. Mus. Novitates* 2871:1–31.

Species and Subject Index

Common and scientific names are indexed so as to equate one with the other. For example, "Panda, red" refers the reader to "*Ailurus fulgens*," and "*Ailurus fulgens*" is followed by "(red panda)." Complete citations are given under the species' scientific name. For species whose generic names are uncertain (e.g., *Puma* or *Felis concolor*), both names are listed, but one is referred to for complete citations. Figures and tables are designated by "f" and "t," respectively, following page numbers. A species' classification in the Appendix is identified by an italicized page number following the species' scientific name.